The Palgrave Handbook of Women and Science since 1660

Claire G. Jones · Alison E. Martin · Alexis Wolf
Editors

The Palgrave Handbook of Women and Science since 1660

palgrave
macmillan

Editors
Claire G. Jones
Faculty of Humanities
and Social Sciences
University of Liverpool
Liverpool, UK

Alison E. Martin
Johannes Gutenberg University
of Mainz
Germersheim, Germany

Alexis Wolf
University of Lancaster
Lancaster, UK

ISBN 978-3-030-78972-5   ISBN 978-3-030-78973-2 (eBook)
https://doi.org/10.1007/978-3-030-78973-2

© The Editor(s) (if applicable) and The Author(s), under exclusive license to Springer Nature Switzerland AG 2022
This work is subject to copyright. All rights are solely and exclusively licensed by the Publisher, whether the whole or part of the material is concerned, specifically the rights of translation, reprinting, reuse of illustrations, recitation, broadcasting, reproduction on microfilms or in any other physical way, and transmission or information storage and retrieval, electronic adaptation, computer software, or by similar or dissimilar methodology now known or hereafter developed.
The use of general descriptive names, registered names, trademarks, service marks, etc. in this publication does not imply, even in the absence of a specific statement, that such names are exempt from the relevant protective laws and regulations and therefore free for general use.
The publisher, the authors and the editors are safe to assume that the advice and information in this book are believed to be true and accurate at the date of publication. Neither the publisher nor the authors or the editors give a warranty, expressed or implied, with respect to the material contained herein or for any errors or omissions that may have been made. The publisher remains neutral with regard to jurisdictional claims in published maps and institutional affiliations.

Cover illustration: Allan Cash Picture Library/Alamy Stock Photo

This Palgrave Macmillan imprint is published by the registered company Springer Nature Switzerland AG
The registered company address is: Gewerbestrasse 11, 6330 Cham, Switzerland

In memory of
Brandie R. Siegfried
(1963–2021)

Acknowledgements

Alison E. Martin is grateful for the help of Lena Bunje and Felix Barthelmay in assisting with editing work on sections V and VI, and would particularly like to thank them for their close attention to detail and commitment to the project in challenging circumstances.

The editors would like to thank Elske Janssen for producing the index.

Contents

Part I Introduction

1 Women in the History of Science: Frameworks, Themes
   and Contested Perspectives   3
   Claire G. Jones, Alison E. Martin, and Alexis Wolf

Part II Strategies and Networks

2 The Natural Philosophy of Margaret Cavendish,
   Duchess of Newcastle: Nature, Self-Knowing Matter,
   and the Dialogic Universe   27
   Brandie R. Siegfried

3 Navigating Enlightenment Science: The Case of Marie
   Geneviève-Charlotte Darlus Thiroux D'Arconville
   and Gabrielle Émilie Le Tonnelier De Breteuil
   and the Republic of Letters   47
   Leigh Whaley

4 'A Valuable Gift': The Medical Life of Margaret Mason,
   Lady Mount Cashell   67
   Alexis Wolf

5 Janet Taylor (1804–1870): Mathematical Instrument
   Maker and Teacher of Navigation   85
   John S. Croucher

6 Early Female Geologists: The Importance of Professional
   and Educational Societies During the Late Nineteenth
   and Early Twentieth Centuries   101
   Cynthia V. Burek

Part III Making Women Visible: Institutions, Archives and Inclusion

7   Where Are the Women? How Archives Can Reveal Hidden Women in Science 129
    Anne Barrett

8   'A Very Worthy Lady': Women Lecturing at the Royal Geographical Society, 1913–C.1940 149
    Sarah L. Evans

9   Women at the Royal Society Soirée Before the Great War 171
    Claire G. Jones

10 Career Paths Dependent and Supported: The Role of Women's Universities in Ensuring Access to STEM Education and Research Careers in Japan 195
    Naonori Kodate and Kashiko Kodate

11 Internationalism and Women Mathematicians at the University of Göttingen 223
    Renate Tobies

Part IV Cultures of Science

12 Astronomy, Education and the Herschel Family: From Caroline to Constance 247
    Emily Winterburn

13 Domestic Astronomy in the Seventeenth and Eighteenth Centuries 269
    Gabriella Bernardi

14 Darwin and the Feminists: Nineteenth-Century Debates About Female Inferiority 289
    Amanda M. Caleb

15 Women, Gender and Computing: The Social Shaping of a Technical Field from Ada Lovelace's Algorithm to Anita Borg's 'Systers' 307
    Corinna Schlombs

16 The Cultural Context of Gendered Science: India 333
    Carol C. Mukhopadhyay

17 A Seat at the Table: Women and the Periodic System 355
    Brigitte Van Tiggelen and Annette Lykknes

Part V Science Communication

18 Mediating Knowledge: Women Translating Science   381
    Alison E. Martin

19 Queen Lovisa Ulrika of Sweden (1720–1782): *Philosophe*
    and Collector   399
    Anne E. Harbers and Andrea M. Gáldy

20 Marianne North and Scientific Illustration   423
    Philip Kerrigan

21 The Cycle of Credit and Phatic Communication in Science:
    The Case of Catherine Henley   447
    Jordynn Jack

22 Rachel Carson: Scientist, Public Educator
    and Environmentalist   465
    Ruth Watts

23 Representing Women in STEM in Science-Based Film
    and Television   483
    Amy C. Chambers

Part VI Access, Diversity and Practice

24 Catalysts, Compilers and Expositors: Rethinking Women's
    Pivotal Contributions to Nineteenth-Century 'Physical
    Sciences'   505
    Mary Orr

25 'The Question Is One of Extreme Difficulty': The
    Admission of Women to the British and Irish Medical
    Profession, C. 1850–1920   529
    Laura Kelly

26 The Work of British Women Mathematicians During
    the First World War   549
    June Barrow-Green and Tony Royle

27 More Than Pioneers—How Women Became Professional
    Engineers Before the Mid-Twentieth Century   573
    Nina Baker

28 Women and Surgery After the Great War   593
    Claire Brock

29 Technology Users vs. Technology Inventors and Why We
    Should Care   611
    Wendy M. DuBow

Index   633

Notes on Contributors

Nina Baker has had a varied career, having become a merchant navy deck officer on leaving school and later taken an Engineering Design degree in her thirties, from the University of Warwick. She then gained a Ph.D. in concrete durability from the University of Liverpool. She has lived with her family in Glasgow since 1989, working variously as a materials lecturer in further education and as a research administrator and, until 2017, as an elected city councillor. Now retired from all that, her interest in promoting STEM careers for girls has led her to become an independent researcher, mainly specializing in the history of women working in engineering.

Anne Barrett M.A. AIC (Associateship of Imperial College London) is a College Archivist & Corporate Records Manager at Imperial College London. Anne has extensive experience in scientific archives and their management, which is invaluable in satisfying the diverse enquiries of internal and external users at Imperial College London, and in her own research. Externally, she works with national and international archival, records management and standards bodies, and is Chairperson of two archive charities and Trustee of three. Her most recent publication is *Women At Imperial College Past, Present and Future* (2017 World Scientific). She has also contributed articles to *Oxford Dictionary of National Biography*. In relation to International Women's celebrations at Imperial College London, she runs an annual Women Wikithon, giving the seminar attendees the opportunity to learn how to edit Wikipedia entries, and add to the number of entries for women in STEMM.

June Barrow-Green is a Professor of History of Mathematics at the Open University, Milton Keynes, UK. She is currently Chair of the International Commission on the History of Mathematics. Her research focuses on the history of nineteenth and twentieth century mathematics, particularly in Britain. Recent studies concern the role of Cambridge mathematicians during World War One, and the use of geometric surface models in research and in teaching. She is the author of *Poincaré and the Three Body Problem* (1996), an editor of the *Princeton Companion to Mathematics* (2008), and she has recently co-authored a textbook together with Jeremy Gray and Robin Wilson, *The History of Mathematics: A Source-Based Approach* (2019). She has a special interest in the history of the gender gap in mathematics and is currently working on the representation of women in mathematics as portrayed through a variety of media.

Gabriella Bernardi is a freelance journalist and writer specializing in the popularization of science. She graduated in physics from the University of Turin, and also holds a Master's degree in Journalism and Science Communication. She was awarded the Voltolino Prize for scientific journalism by the Italian Union of Scientific Journalists (UGIS). Since 2008 she has been a member of the European Union of Science Journalists' Association (EUSJA). She is a member of the IAU (International Astronomical Union) within Division C (Education, Outreach and Heritage) and Commission C3 (History of Astronomy); she also serves on the Executive Committee Working Group 'Women in Astronomy'. She has published in several newspapers and magazines that specialize in the popularization of astronomy, and is the author of six astronomy books, three of which are in English and published by Springer.

Claire Brock is an Associate Professor in the School of Arts at the University of Leicester. She is the author of three monographs, the most recent being *British Women Surgeons and their Patients, 1850–1918* (Cambridge University Press, 2017), which was funded by a two-year Wellcome Trust Research Leave Award, held between 2012 and 2014. She is currently working on a sequel to this project, exploring women in surgery during the interwar years, and a monograph on the surgical patient in the age of specialism, 1880–1930.

Cynthia V. Burek is an Emeritus Professor of Geoconservation at the University of Chester, UK. She has recently been appointed to the UNESCO IGCP Science board for her expertise in geoheritage, Geoparks and conservation. She is an honorary member of the Quaternary Research Association. Her special interests include geodiversity underpinning biodiversity, and the history of women in science, especially in geology. She has published extensively in these areas and, as an international leader in her field, is also known as a public communicator of science.

Amanda M. Caleb is a Professor of Medical Humanities at Geisinger Commonwealth School of Medicine. She received her Ph.D. in English and M.A. in Nineteenth-Century Studies from the University of Sheffield and her B.A. in English from Davidson College; she is currently pursuing her M.P.H. from the University of Alabama at Birmingham. She has published a number of articles and book chapters on topics ranging from accounts of illness in the Victorian period, to the rhetoric of British eugenics, to dementia and the role of narrative medicine.

Amy C. Chambers is a Senior Lecturer at Manchester Metropolitan University working in the fields of science communication and screen studies. Her research examines the intersection of science and entertainment media with specific focus on women and science, and discourses surrounding science and religion on screen. Recent and forthcoming publications explore the mediation of women's scientific expertise in mass media; the representation of science, gender and race in *Star Trek Discovery*; medical horror and Alice Lowe's *Prevenge*; medical research and *The Exorcist*; and women-directed science fiction film and television.

John S. Croucher Professor John Croucher is one of Australia's most prominent statisticians and authors, being recognized both nationally and internationally, including the Prime Minister's Award for Australian University Teacher of the Year. Professor Croucher holds four Ph.D.s and an international reputation for excellence in research with over 130 research papers and thirty-two books, including *Mistress of Science—The Story of the Remarkable Janet Taylor*. A Fellow of both the Royal Society of Arts and the Australian Mathematical Society, in 2015 John was made a Member of the Order of Australia (AM) for 'significant service to mathematical science in the field of statistics, as an academic, author and mentor to professional organizations'.

Wendy M. DuBow is the Director of Evaluation and a senior research scientist at the US-based National Center for Women & Information Technology (NCWIT) and an affiliate faculty member in Women and Gender Studies at the University of Colorado. She conducts mixed-methods social science research and evaluates the impact of NCWIT's programmes and resources. Dr. DuBow collects the national statistics NCWIT shares on girls' and women's participation in computing. She is also co-author of an online self-paced course for post-secondary faculty and administrators on research-based ways to recruit and retain women in computing, including how to evaluate these efforts.

Sarah L. Evans is a Research and Collections Engagement Manager at the Royal Geographical Society (with IBG). A historian by training, she carried out her doctoral research on the Society's Collections through an AHRC-funded Collaborative Doctoral Award. Her research examined women's participation in RGS-supported expeditions from 1913–1970, both mapping out the extent of that participation and then considering a number of women and their experiences in close detail. She has a particular interest in how we tell and write the histories of geography, fieldwork and exploration—and who is left out of these. She was also previously Research Assistant on the interdisciplinary AHRC-funded Hero Project, working alongside colleagues at Aberdeen, Birmingham and Durham universities. In her current role, she works to highlight and promote research-led work on the Society's Collections, sharing this work with wider audiences.

Andrea M. Gáldy gained her Ph.D. in Art History and Archaeology at the University of Manchester with a thesis on the collections of antiquities of

Cosimo I de' Medici in Florence. She held post-doctoral fellowships of the Henry Moore Foundation and at Villa I Tatti and taught for international university programmes in Florence as well as in Germany. Andrea is a founding member of the international forum *Collecting & Display* and the main editor of the series *Collecting Histories* (CSP). Her main research interests are in the field of collecting history and provenance studies, historical collections as well as their modern translations into the Digital Humanities. In 2020, she founded the research forum *Collecting Central Europe*.

Anne E. Harbers has a Master's degree in Chemistry and an M.B.A., and spent twenty-five years working globally in biotechnology, most recently in Sydney, Australia. In 2014, she completed her M.Arts in Art History at the University of Sydney and as an Independent Scholar has presented at the international forum *Collecting & Display* in Germany and in the UK, and at the *Historians of Netherlandish Art* conferences. She also presents on decorative arts for museums and the National Trust in Australia. She has published on topics relating to art, science, material culture and collecting since 2014, and prior to that on academic scientific topics. Her most recent publication, with Dr. Andrea Gáldy, is on an eighteenth-century Dutch Ladies Society for Physical Sciences, in *Women and the Art and Science of Collecting in Eighteenth-Century Europe* (Leis & Wills, eds., Routledge, 2021). She is currently undertaking her Ph.D. through Radboud University in The Netherlands, under Prof. Dr. Volker Manuth, working on a seventeenth-century Dutch still-life & marine artist.

Jordynn Jack is a Professor of English and Comparative Literature at the University of North Carolina, Chapel Hill, where she teaches courses in science writing, women's rhetorics, rhetorical theory, composition pedagogy and health humanities. She is the author of three books, *Science on the Home Front: American Women Scientists in World War II* (University of Illinois Press, 2009); *Autism and Gender: From Refrigerator Mothers to Computer Geeks* (University of Illinois Press, 2014); and *Raveling the Brain: Toward a Transdisciplinary Neurorhetoric* (Ohio State University Press, 2019).

Claire G. Jones is a Senior Lecturer in the Faculty of Humanities and Social Sciences at the University of Liverpool, UK. Her research interests focus primarily on the intersections of science, society and culture in Britain from the late-eighteenth to the early-twentieth centuries. She has published widely in these areas, including the monograph *Femininity, Mathematics and Science, 1880-1914* (Palgrave, 2009) which was the winner of the Women's History Network Book Prize.

Laura Kelly is a Senior Lecturer in the history of health and medicine at the University of Strathclyde, Glasgow with expertise in the social history of medicine in nineteenth- and twentieth-century Ireland. Her first monograph *Irish Women in Medicine, c.1880s-1920s: Origins, Education and Careers* was published by Manchester University Press in 2012 (paperback, 2015).

Her second monograph, *Irish Medical Education and Student Culture, c.1850-1950* was published by Liverpool University Press in September 2017 (paperback, 2020). She is currently working on a project on the history of contraception in modern Ireland which is funded by a Wellcome Trust research fellowship.

Philip Kerrigan completed his Ph.D. in 2009 at the University of York, where he continues to work as the project officer for a joint University and Wellcome Trust funded research centre. His thesis explores Charles Darwin's scientific and botanical writings in the artistic and cultural contexts of his time, and he has published several articles based on this work. His interests have since broadened out into the Medical Humanities and he co-edited a book of illustrated essays on mental health with the WHO Collaborating Centre for Global Health Histories. He is particularly committed to interdisciplinary ways of working and to public engagement and most recently co-authored a guide to co-producing research with members of the public.

Kashiko Kodate (Dr. Eng, University of Tokyo) is a Professor Emerita at the Faculty of Science, Japan Women's University. She is currently CEO of Photonic System Solutions Inc. and Outside Director, Hamamatsu Photonics K.K. She was a former Programme Officer of the Japan Science and Technology Agency and Specially Appointed Professor at the University of Electro-Communications, Chōfu, Japan. She has spent over forty years in optical research, ranging from micro-optics to information photonics. She co-founded and led the Japan Inter-Society Liaison Association Committee for Promoting Equal Participation of Men and Women in Science and Engineering (2002–2003). She served as the first female vice-president of the Japan Society of Applied Physics (JSAP) in 2006 and was a member of the Science Council of Japan (2005–2012). In 2010, she founded the Outstanding Female Researcher Award and Contributions to Investing in People Award at the JSAP. She has contributed to nurturing many female scientists in Japan and was granted the 2010 Prime Minister Award in recognition of this achievement.

Naonori Kodate (Ph.D. in Political Science, London School of Economics and Political Science) is Associate Professor, School of Social Policy, Social Work and Social Justice, University College Dublin (UCD), Ireland and the founding Director of the UCD Centre for Japanese Studies. His research covers comparative public policy, and science, technology and society (STS), particularly in the use of eHealth (e.g. care robots), patient safety and gender equality in STEM. He is affiliated with Hokkaido University's Public Policy Research Centre, the Institute for Future Initiatives, University of Tokyo, la Fondation France-Japon, l'Ecole des Hautes Etudes en Sciences Sociales and the Universal Accessibility & Ageing Research Centre in Nishitokyo, Japan. His books include *Japanese Women in Science and Engineering: History and Policy Change* (co-authored with Kashiko Kodate, Routledge) and *Systems*

Thinking for Global Health (co-edited with Frédérique Vallieres and Hasheem Mannan, Oxford University Press).

Annette Lykknes is a Historian of science and Professor of chemistry education at NTNU-Norwegian University of Science and Technology in Trondheim. Her research interests include the history of twentieth-century chemistry, the history of education in chemistry, women in science, collaborative couples in the sciences and teaching practices in chemistry. She has co-authored monographs in the Norwegian language on the 100th history of chemistry at NTNU (2015) and on the periodic system (2019) and is co-editor of several collected volumes, including *Women in their Element: Selected Women's Contributions to the Periodic System* (2019) and of *The Periodic System: The (Multiple) Values of an Icon* (Special issue of Centaurus, 2019), both co-edited with Brigitte Van Tiggelen.

Alison E. Martin is a Professor of British Studies at the Johannes Gutenberg Universität-Mainz in Germany, and works at the faculty in Germersheim, which specializes in Translation Studies and Interpreting. She has published widely on translation studies, from the eighteenth century to the present day, with a particular focus on travel literature, scientific writing and gender. Her most recent monograph, *Nature Translated: Alexander von Humboldt's Works in Nineteenth-Century Britain* (Edinburgh University Press, 2018; paperback 2020), explores the role played by Humboldt's female translators—key among them Helen Maria Williams—in the transmission of scientific knowledge to a general audience in the nineteenth century.

Carol C. Mukhopadhyay is a cultural anthropologist and Professor Emerita at San José State University, California. Her anthropological expertise includes gender, the family and sexuality, as well as race and ethnicity, comparative education and methodology. She has carried out extensive fieldwork in India and the US, published widely in her specialist areas, and regularly acts as a consultant for various organizations.

Mary Orr is Buchanan Chair of French at the University of St Andrews since 2016, after holding Professorships in French at the Universities of Exeter and Southampton. An expert on intertextuality and nineteenth-century French literatures that overtly engage with new scientific understanding, her publications include her ground-breaking *Flaubert's* Tentation: *Remapping Nineteenth-Century French Histories of Religion and Science* (OUP, 2008), spearheading her recent publications on transnational figures in nineteenth-century French sciences: 'Mainstream or Tributary? The Question of "Hibernian" Fishes in Thompson's *The Natural History of Ireland* (1856)', in *Nature and the Environment*, ed. Matthew Kelly (Liverpool: Liverpool University Press, 2019), pp.159–182 and 'Collecting Women in Geology: Opening the International Case of a Scottish "Cabinétière", Eliza Gordon Cumming (c. 1798–1842)', in Cynthia Burek and Bettie Higgs, (eds.), *Celebrating 100*

Years of Female Fellowship of the Geological Society: Discovering Forgotten Histories (London: Geological Society Special Publications 506, 2020).

Tony Royle is a former Royal Air Force and commercial airline pilot who is currently a Research Associate and Associate Lecturer at the Open University, Milton Keynes, UK. He is the author of *The Flying Mathematicians of World War I* (2020) and has a particular interest in the contribution of women mathematicians and scientists to the advancement of aeronautics in Britain in the early part of the twentieth century. He is an external examiner in Aviation Studies for Middlesex University, London, and his latest research considers the contributions of British mathematicians to the development of efficient aircraft propellers during World War One.

Corinna Schlombs is an Associate Professor in the Department of History at Rochester Institute of Technology. She received her Diploma in Sociology from Bielefeld University in Germany, and her Ph.D. in the History and Sociology of Science from the University of Pennsylvania. Her research focuses on technology and capitalism in transatlantic relations. She is the author of *Productivity Machines: German Appropriations of American Technology from Mass Production to Computer Automation* (MIT Press, 2019). She has published articles and book chapters on international computing and computing and gender. Most recently, her research has been supported through a National Science Foundation Scholars Award that enabled her to focus on her current book project, which investigates transatlantic transfers of productivity culture and technology in the two decades before and after World War II.

Brandie R. Siegfried was a Professor of English at Brigham Young University in Utah, US; her expertise was in Renaissance Studies in the context of English and American literature with a special interest in early modern women writers. She published widely in this area, including editing *Margaret Cavendish, Duchess of Newcastle, Poems and Fancies with The Animal Parliament* (ACMRS Press, 2018) and co-authoring *God and Nature in the Thought of Margaret Cavendish* (Ashgate, 2014).

Renate Tobies is a historian of mathematics and sciences at the Friedrich-Schiller-University of Jena (Germany). After completing her academic degrees at the University of Leipzig, she was the managing editor of the *NTM-International Journal of History of Natural Sciences, Technology and Medicine* for twenty years. In addition, she became a Visiting Professor at the University of Kaiserslautern, held a chair of history of science and technology at the University of Stuttgart, and further visiting professorships in Braunschweig, Saarbrücken, Jena, Linz and Graz. She is an Effective Member of the Académie Internationale d'Histoire des Sciences and a Foreign Member of the Agder Academy of Sciences and Letters (Kristiansand, Norway). Her main research fields are the history of mathematics and its applications, and women in mathematics, science and technology.

Brigitte Van Tiggelen is the Director for European Operations at the Science History Institute, Philadelphia, US and member of the Centre de Recherche en Histoire des Sciences, Université catholique de Louvain, Belgium. Graduated both in physics and history, she devoted her Ph.D. to chemistry in XVIIIth century Belgium. Her research interests include Belgian history of science, history of chemistry, collaborative couples and women in science, domestic science, and philosophy of chemistry, in particular the boundaries between physics and chemistry. She has authored and edited books in French and in English including *Madame d'Arconville (1720-1805): une femme de lettres et de sciences au siècle des Lumières* (2011) co-edited with Patrice Bret; *For Better Or for Worse?: Collaborative Couples in the Sciences* (2012), co-edited with Annette Lykknes and Donald L. Opitz, and *Domesticity in the Making of Modern Science* (2016), co-edited with Donald L. Opitz and Staffan Bergwik.

Ruth Watts is an Emeritus Professor of History of Education at the University of Birmingham. Her research interests are in the history of education and gender and she has published widely on these including *Women in Science: A Social and Cultural History* (Routledge, 2007). Ruth is on the Board of Editors of History of Education, ex-President of the British History of Education Society and co-convenor of the Standing Working Group on *Gender in the International Standing Conference for the History of Education* (ISCHE). She has also been involved for many years in various networks in women's history.

Leigh Whaley taught and researched European history for over thirty years. Retired from full-time teaching, Professor Whaley continues to teach two online courses: Gender and Sexuality in Europe and Western Civilization at Acadia University, Philadelphia, US. Her past publications include articles and books on subjects as wide-ranging as women in science and medicine and medieval military surgeons to female spies during World War II. She currently works for the Department of Justice, Canada, as an historical expert.

Emily Winterburn lives in Leeds and is a primary school teacher. She has written a number of commercial books including most recently her biography of Caroline Herschel, *The Quiet Revolution of Caroline Herschel*, in 2017. Emily originally studied physics before turning to History of Science (both at Manchester) as a postgraduate. Between 1998 and the birth of her second child in 2008 she worked as curator of astronomy at the Royal Observatory in Greenwich where collections included material relating to the Herschel family. At Greenwich Emily began her thesis with Imperial College, London, completing her Ph.D. in 2011. In 2014 she won the Royal Society Essay Prize for a piece on William Herschel. Since leaving Greenwich, she has worked at the University of Leeds History of Science Museum. In 2015 took part in the 'Women Writing Science' panel at the Royal Society before beginning teacher training in 2016.

Alexis Wolf is a Research Associate in the Department of English Literature and Creative Writing at Lancaster University. She has previously researched and lectured at the University of Leeds and Birkbeck, University of London. Her research focuses on women's writing of the late eighteenth and early nineteenth centuries, including in the areas of medicine and antiquarianism. She has published her research in *European Romantic Review* (2019) and served as Guest Editor for a special issue of *19: Interdisciplinary Studies in the Long Nineteenth Century* on women's archival silences (2018). She is currently working on her first book, which examines manuscript circulation within women's transnational networks in the Romantic period.

List of Figures

Fig. 6.1	The Sedgwick Club, University of Cambridge, 1900. Photograph reproduced by kind permission of the Sedgwick Museum of Earth Sciences, University of Cambridge	120
Fig. 9.1	The Ladies' Soiree 1888. Image from *The Graphic*, 16 June 1888, reproduced by kind permission of The Royal Society of London	187
Fig. 10.1	Autopsy class at Japan Women's University (JWU). Image provided courtesy of Naruse Memorial Hall, JWU	201
Fig. 10.2	Dr. Ume Tange, 1940. Image courtesy of Naruse Memorial Hall, Japan Women's University (JWU) and the Ochanomizu University Digital Archives	204
Fig. 10.3	Three pioneering women in STEM at RIKEN. From left to right: Dr. Michiyo Tsujimura, Dr. Sechi Kato, Dr. Chika Kuroda. Images provided courtesy of Hokkaidō University Archives	205
Fig. 10.4	Five female students at the Department of Physics and Mathematics, Hokkaidō University, 1939. Front row, left to right: Yurie Yokota, Akiko Soeya. Back row, left to right: Hisako Nakayama, Yoshi Kobayashi, Yuki Uenaka. Image provided courtesy of Hokkaidō University Archives	206
Fig. 10.5	Drs. Tsujimura and Kato visited by Hokkaidō students at RIKEN in 1936. Image provided courtesy of Hokkaidō University Archives	209
Fig. 10.6	Dr. Toshiko Yuasa, 1978. Image courtesy of the Ochanomizu University Digital Archives	210
Fig. 13.1	Johannes Hevelius and wife Elisabetha observing the sky with a brass sextant. Engraving from *Machinae Coelestis: Pars Prior* (1673) by Johannes Hevelius. Library of Congress: Public Domain	271

Fig. 13.2 Caroline Herschel recording her brother William's observations on the night he discovered the planet Uranus. Engraving by Paul Fouché from *Astronomie populaire: description générale du ciels* by Camille Flammarion (Paris, 1890, p. 73). Public Domain 282

Fig. 17.1 Mendeleev's Periodic Table of 1871. The platinum metals are placed in group VIII towards the right of the table. Note that they have very similar atomic weight values (Reproduced by kind permission of Jeff O. Moran) 359

Fig. 17.2 Ida and Walter Noddack in the laboratory, Berlin 1931 (Stadtarchiv Wesel O1a, 5–14-5_02) (Reproduced by kind permission of Stadtarchiv Wesel, Wesel, Germany) 366

Fig. 17.3 Analytical work in the chemistry room of the Curie Laboratory at the Institut du Radium, Paris, July 1930. From left to right: Sonia Cotelle, Marguerite Perey, Alexis Jakimach, Tchang da Tcheng (Musée Curie, MCP 1938) (Reproduced by kind permission of the Musée Curie) 369

Fig. 17.4 An all-female team: Berta Karlik (right) and Traude Bernert (left) with their experimental setup in 1943–1944 (bpk/Liselotte Orgel-Köhne, image no. 70113898, reproduced with permission of bpk Bildagentur) 372

Fig. 19.1 Queen Lovisa Ulrika of Sweden (1720–1782). Unknown artist, Lovisa Ulrika, drottning av Sverige, Nationalmuseum: Public Domain 400

Fig. 19.2 The Cabinet of Natural History at Drottningholm Palace, Sweden. Reproduced by kind permission of Alexis Daflos (photographer) and The Royal Court, Sweden 405

Fig. 19.3 A male *Calopteryx virgo*, the species to which Linnaeus (1746) gave the names 'Lovisa' and 'Ulrica' after the Princess. Photographed by Holger Gröschl, 2003: Creative Commons license 412

Fig. 26.1 The Admiralty Air Department 1918. Image from the Annie Trout Archive held at the Hartley Library at the University of Southampton. Reproduced by kind permission of the University of Southampton Archives 553

Fig. 26.2 Handley Page V/1500 aircraft: wikimedia commons. https://commons.wikimedia.org/wiki/File:Handley_Page_V-1500_in_1918.jpg. Accessed 6 January 2021 555

Fig. 26.3 Pitot tube, *Royal Society Journal* 560

Fig. 26.4 H.W. Richmond (ed.) 1925. *Textbook of Anti-Aircraft Gunnery*, vol. 1, HMSO. London: H.M.S.O., facing p. 534 (Image owned by June Barrow-Green) 565

Fig. 29.1 Race and gender composition of CIS bachelor's degrees 2019 (Reproduced from Cierra Kelley, Lyn Swackhamer and Wendy DuBow. 2021. *National Center for Women & Information Technology* [Based on National Center for Educational Statistics: Integrated Post-secondary Education System. 2019. CIP 11-Computer and Information Sciences], by kind permission of the National Center for Women and Information Technology, University of Colorado) — 615

List of Tables

Table 6.1	Medals and grants awarded to early female geologists by the geological society	106
Table 6.2	The first female fellows of the geological society 1919–1922	108
Table 6.3	Early female geologists and membership of the British Federation of University Women (BFUW)	111
Table 6.4	Chester Society of Natural Science (CSNS)—Women's Membership to 1910	115
Table 6.5	Female attendance at the Sedgwick Club, University of Cambridge	118
Table 6.6	Papers read by women members at the Sedgwick Club, University of Cambridge	121
Table 7.1	Finding women in the university archive: primary sources and strategies	144
Table 8.1	Women's Royal Geographical Society (RGS) lectures on their RGS-supported expeditionary work, 1913–1940	153
Table 9.1	Women exhibitors at Royal Society Soirées 1872–1914	188
Table 10.1	Four types of higher education institutions for women between 1870 and 1930 (selected examples)	200
Table 10.2	Proportion of female undergraduate students in Japan by discipline and subject (MEXT 2018)	214
Table 10.3	Proportion of female academic staff members in national universities in Japan by rank (JANU 2018)	215
Table 10.4	Proportion of female staff members by rank in seven former Imperial Universities and one private and two national women-only universities (JANU 2018)	216
Table 15.1	'Women as percentage of all Bachelors' recipients in the United States, by major field group, 1966–2015. Tabulated by the National Science Foundation, National Center for Science and Engineering Statistics; data from the Department of Education, National Center for Education Statistics: Integrated Postsecondary Education Data System Completions Survey	310

Table 20.1	Marianne North and scientific illustration	441
Table 24.1	List of References in Leonard Jenyns, *Manual of British Vertebrate Animals* (1835) to the Drawings in *The Fresh-Water Fishes of Great Britain* (1828-1838) by Mrs T. Edward (Sarah) Bowdich	523
Table 26.1	The numbers of women and men sitting the Cambridge mathematical tripos examination, 1908 to 1919, with concomitant output of wranglers	551

PART I

Introduction

CHAPTER 1

Women in the History of Science: Frameworks, Themes and Contested Perspectives

Claire G. Jones, Alison E. Martin, and Alexis Wolf

Until the later twentieth century, the history of science largely neglected—or simply did not see—women and their contributions to science. Frameworks of understanding cast the scientific enterprise as intrinsically male, while historiography chose to look mostly in places where women were present historically only informally at the margins, for example élite scientific societies, universities, or aristocratic gentlemen's laboratories.[1] This masculine construction of scientific activity was reinforced by a modern understanding of science as being specific in scope and limited to certain spaces, for instance the institutional laboratory, research department or observatory, and the activities that take place there. This limited perspective emerged partly out of the specialization and fragmentation of science from the late nineteenth century onwards which narrowed the range of activities that counted as science. As a result, women in specialisms such as science writing, translation, collecting and illustration, or practising science within disciplines now classified typically as arts—the

C. G. Jones (✉)
Faculty of Humanities & Social Sciences, University of Liverpool, Liverpool, UK
e-mail: c.g.jones2@liverpool.ac.uk

A. E. Martin
Johannes Gutenberg University Mainz, Germersheim, Germany
e-mail: amarti01@uni-mainz.de

A. Wolf
University of Lancaster, Lancaster, UK
e-mail: a.wolf@lancaster.ac.uk

study of ancient civilizations and architecture for instance—were excluded. Despite this, specialization also acted to open up some opportunities for women in new scientific disciplines which were yet to establish themselves and so, at their beginnings at least, were less attractive to men.[2] In addition, women who pursued their science in the domestic sphere, and/or as part of a family collaborative effort or business, were also obscured from view; a situation exacerbated by ideals of femininity which presented barriers to women seeking to practise science in the public sphere and created tension around the amateur/professional distinction. These ideals also historically prescribed certain branches of science as more suitable for the female sex, such as botany, which for much of the eighteenth century had a special affinity with femininity. In contrast to physics, botany was a taxonomic rather than an experimental science which, to a large extent, remained in the domestic sphere and considered phenomena as they occurred in nature. Ann Shteir (1997) explains how women enjoyed more culturally sanctioned access to botany than to any other science; working in their homes and gardens, women collected plants, drew them, studied them, named them, taught their children about them, and wrote popularizing books on botany.

It may have been hoped that these kinds of associations, and ideas of appropriate branches of science for women and men, would have disappeared by the twenty-first century. However, a look at data on women in STEM across different disciplines illustrates that the gender gap, especially in certain areas of science, persists. When we consider biological sciences and some specialisms in medicine, we find that women worldwide are clustered in these disciplines and are significantly under-represented in physical sciences and engineering. In addition, women of colourearn the smallest share of STEM degrees and represent the lowest numbers of practitioners (Catalyst 2020; UNESCO 2021). When women do achieve degrees in STEM subjects, they often go on to careers in teaching or science communication rather than practising as researchers. This is connected not just to ideas of suitable scientific femininity, but also to educational opportunities and choices, the culture of practice in individual disciplines, and representations of female scientists. Adding another layer of complexity, these issues alter according to culture and place and when considered from a non-Western perspective, as illustrated by chapters in this volume.

Many traditions, however, share a dominant definition of science as an epistemic product—a body of knowledge and technology—which obscures the processes of sciences; that is the methods, practices and teamwork involved in the scientific project. This has rendered invisible the work of assistants and collaborators, especially but not exclusively women. Indeed, we are only just recovering the important work of technicians and laboratory assistants of both sexes who have been essential to the project of science; here, class intersects with gender to limit not only who enjoys access to science, but also who gets the credit.[3] This policing of science and marginalizing of women went hand-in-hand with a culture that privileged men as active investigators, and in contrast took woman as its object of scientific enquiry. Ideas of women as less rational and so unsuited for scientific work can be identified from the birth of modern science in the Enlightenment period, through to nineteenth and twentieth century science influenced by evolutionary theory. The impact of these ideas was to limit opportunity for women in the sciences and, sometimes, to prompt scientific women to view themselves as somehow unnatural or different to their sex. For example, Leigh Whaley describes the transient insecurities of two female Enlightenment natural philosophers in her case study on them in this volume. Ideologically informed theories of sexed brains and women's intellectual deficit are remarkably persistent, as neuroscientist Gina Rippon demonstrates in a text which uses science to effectively debunk these ideas (2019).

The rest of this chapter will introduce the *Handbook* and illuminate these questions by mapping scholarship on the history of women in science, and by challenging male-focused histories which have misunderstood past landscapes of science and, as a result, ignored women's contributions. Although we must be careful not to overestimate the numbers of women active in science, one startling realization from the chapters and case studies presented here is that many of the women featured were well known and respected in their own time yet have dropped out of sight, unlike many of their male peers. This raises questions of historiography, how we interpret science and the past, and sociological considerations about the gendered practices, participation, and representations of science today.

From Great Women to a More Nuanced View: Scholarship on Women in Science

Although there have been efforts to reclaim and make visible women scientists of the past, still only a few (white) key figures—'heroines' of science such as Polish researcher Marie Skłodowska-Curie and British mathematician Ada Lovelace—dominate the Western popular historical imagination. A comparison of these two women reveals some interesting insights. Skłodowska-Curie (1867–1934) won the Nobel Prize in Physics in 1903, alongside Henri Becquerel and Pierre Curie, and the Nobel Prize in Chemistry in 1911 for her discovery of radium and polonium. If Skłodowska-Curie's achievements are indisputable, this is not so clear when it comes to Lovelace (1815–1852). Although an interesting footnote in the history of mathematics, her contributions to science were limited to enhancements of Charles Babbage's plans for an analytical engine, the forerunner of the computer, and the translation of a paper to which she added original notes. The latter imagined what a computer may be capable of and can be regarded as a nascent computer programme. Lovelace was the daughter of the poet Byron and his wife Anne Isabelle Milbanke, 'The Princess of Parallelograms' (Woolley 1999, p.14). Once the couple were estranged, Milbanke ensured her daughter was tutored in mathematics as a conscious counter to prevent her developing the turbulent poetic tendencies of her father (Toole 2004; Woolley 1999). As Byron's daughter, Lovelace is a compelling figure whose prominent place in the history of science is due arguably to the romance of being linked to the great romantic poet, rather than to any clear legacy to science or computing.

Echoing the place of Lovelace in the popular imagination, for a long time the recovery of women in science followed the familiar 'great man' of science model, with biographers romanticizing the 'special', 'pioneering' individual women who succeeded in this masculine sphere.[4] Although a trope persisting today, this valorization of individual women as heroines of science can be traced back to early in the twentieth century at least. In 1913, scientist and educator Rev. John Augustus Zahm, under the name H. J. Mozens, published *Woman in Science With an Introductory Chapter on Woman's Long Struggle for Things of the Mind*—a well-received text which went through a number of editions. In the preface to this, Mozens describes how he was inspired to consider this topic while in the picturesque land of the Hellenes; here, viewing representations of Aspasia 'the virgin goddess of wisdom and art and science', he was moved to investigate the intellectual achievements of women. Mozens' overblown prose describes female scientists, including his contemporaries, in language more applicable to mythical female deities than workaday, serious scientists. Skłodowska-Curie's experimental abilities are likened to magic as 'before her deft hands and fertile brain difficulties vanished as if under the magic wand of Prospero' (Mozens 1991, p. 224). In a similar vein, Mozens quotes the gendered praise spoken by the Dean of Faculty when economic entomologist, Eleanor Ormerod, was awarded an honorary Doctor of Laws by the University of Edinburgh in April 1900. Ormerod was 'entitled to be hailed as the protectoress of agriculture and the fruits of the earth [...] a beneficent Demeter of the nineteenth century' (p. 252).[5] This approach—still common in popular history at least—implicitly suggests that only special, exceptional women can succeed in science. As a result, it presents a role model so far beyond the everyday that it may deter, rather than encourage, young women to view science as a career.

This is not to suggest, however, that biographies of individual women of science are not highly valuable. Indeed, there are several chapters in this *Handbook* which present case studies that illustrate the strategies that women adopted to contribute to science and illuminate the culture and practice of particular disciplines which presented both obstacles and opportunities for them. Instead of taking a biographical approach, other scholarship in the history of women and science aims to develop frameworks with which to understand how everyday women practitioners negotiated their access to and participation in science.[6] This project aims to understand women's place in science by the processes and traditions which have determined women's access, contribution, and reception. This includes the way that modern science is defined, disciplinary differences, gendered traditions of practice, language and education, the influence of scientific spaces, institutions and networks, gendered representations, who gets scientific credit, and the intersections of science with other activities such as business, writing, translation, teaching or art. This approach is revealing many 'everyday' rather than 'exceptional' women who contributed to science and who have been obscured from view.

The scope of this *Handbook* is from 1660 when, in the Western world and beyond, the key institutions of science and the modern experimental method were emerging and establishing themselves. Chapters build on classic feminist critiques of science which demonstrated how, from this time onwards, modern science acquired privileged status as a producer of objective knowledge while simultaneously developing an overt masculine perspective and representation which expunged women and femininity (Merchant 1976; Fox Keller 1985; Schiebinger 1991). This scholarship adopted a sociologically informed history of science which lifted the discipline's veil of objectivity to make visible the masculine constructs within; this work also initiated the development of models and frameworks of understanding that better reveal women's connections to science. Many of the chapters in this *Handbook* reflect a Western perspective but, importantly, there are also studies of non-Western experience that provide an invaluable, and often striking, contrast. Caution must be taken when assuming that women's experience is uniform, even in a European and American context. For example, while women of science struggled to gain entry to scientific institutions and higher education until late in the nineteenth century, some women in Enlightenment Italy pursued a career in science, gained degrees and even a university teaching post. Laura Bassi (1711–1778) held a chair in Experimental Physics at the University of Bologna and was also a member of Italian Institute of Sciences (Findlen 1993). The history of women and science is an extremely large and complex subject and, of course, it has not been possible to cover all aspects, disciplines, and cultures here. Historical and contemporary comparisons of the operations of gender (and race) in science across different cultures is an avenue of enquiry which requires further research, especially in the context of today's global scientific world.

Part II of this *Handbook*, 'Strategies and Networks', presents analyses of how women negotiated access to science, built their own circles of support, and found ways to participate despite the barriers confronting them due to their sex. Recent scholarship has illuminated the way that 'male mentors' with connections, position and scientific facilities helped women without these advantages to access science (Pycior et al. 2006). Certainly, men have a role in supporting scientific women in the case studies presented here. However, the 'male mentor' framework of understanding can work to deprive scientific women of agency and diminish their contributions, and this is far from the reality, as this section shows. The scientific women presented here were anything but passive and powerless and, by contrast, actively cultivated networks, and opportunities to steer their careers. Education emerges as another key theme: even élite women such as Margaret Cavendish, Duchess of Newcastle (1623–1673), felt the limitations and injustice of being locked out of formal intellectual and scientific education due to her sex. Cavendish, thanks to her privileged status, was able to circumvent this and, as Brandie

R. Siegfried demonstrates, participated and published as a natural philosopher and equal in prestigious scientific networks during a career which spanned two decades. Sensitive but non-compliant towards gender norms which proscribed women's engagement in public science, Cavendish used her access to intellectual circles to acquire knowledge and connections, including with men of the newly established Royal Society of London. In this way she engaged in the key debates in natural philosophy of the time, often communicating her scientific ideas through poetry and literature.

Like Cavendish, other female natural philosophers navigating Enlightenment science proactively developed networking strategies to facilitate scientific learning and engagement despite the constraints of their sex. The latter typically included exclusion from formal institutions of science and the facilities that came with them, as well as from more informal spaces where the exchange of ideas took place such as the coffee house or café. As Leigh Whaley, in her chapter on Émilie Du Châtelet (1706–1749) and Madame D'Arconville (1720–1805) explains, women 'were forced to be creative if they wanted to practice science'. D'Arconville, a chemist and anatomist, was largely self-taught, but she sought out male mentors who could assist her scientific training and collaborated with them. Mathematician and natural philosopher Du Châtelet forged a partnership with Voltaire, but soon surpassed him and established a 'brilliant salon' or academy at her country home. Alongside scientific research, both women translated canonical scientific texts and so exploited an acceptably 'feminine' route into science, using it to go beyond translation to include extensive commentaries, engage in scientific dialogue, and contribute their own insights. Although Du Châtelet and D'Arconville cultivated relationships with the leading men of science of their day, they did not take a subordinate position to their male 'mentors' and their agency and purposeful strategies align them far from the passive, domestic Enlightenment prescription for womanhood.

Alexis Wolf, in her chapter on Margaret Mason, Lady Mount Cashell (1772–1835), introduces us to another woman who was not shy of embodying scientific—in this case medical—authority. Mason, too, adopted strategies including the writing of medical texts and the cultivation of a well-connected male mentor to facilitate her work. Mason practised in the more acceptably 'feminine' sphere of maternity and children's health, but she was also challenging the takeover of midwifery by men as the tide of professionalism tainted women's traditional dominance in this sphere as amateurish, and even unsafe. By contrast, Mason insisted that women midwives were just as capable as the professional man; at the same time, she adopted a scientific stance and was concerned with scientific theories of childrearing and medicine developed using observation and experimentation. Mason's medical text, *Advice to Young Mothers on the Physical Education of Children* (1823), was one of a small body of texts by women that together created an informal network of women disseminating and exchanging professional, medical, literary texts in Europe.

John Croucher presents us with a case study on a woman who, unlike the natural philosophers discussed above, could not rely on privileged social birth to facilitate her access to science. Indeed, the life of Janet Taylor, nautical instrument maker and mathematician, is a 'story of class and gender'. She represents a section of non-élite women working in science and technology, at a craft or guild level, who have been largely neglected until recently. Taylor ran a business providing navigation technology, and technological support and training, to the Navy and merchant shipping, operating from premises in East London. Entrepreneurialism and the needs of industry created opportunities that this able scientific woman was clever enough to exploit. Taylor was not alone in using this strategy; women participated in a thriving scientific instrument trade in Victorian London, offering bespoke and standard products and receiving custom from institutions, laboratories, and individuals. Discounting the 'invisible' women who worked with relatives in small businesses, women were registered in their own right as makers of drawing and mathematical instruments and producers of navigation and optical equipment (Morrison-Low 1991). Taylor also cultivated networks that transcended business concerns; these included engaging in scientific correspondence with the Astronomer Royal, participating in urgent debates such as that over the problems of compasses in iron ships, and working with the Admiralty.

Networking was a key strategy for scientific women, no matter their circumstances of birth, class, or discipline. Cynthia Burek's chapter on women following academic careers in geology around 1900 carefully dissects the strategies they used to facilitate access, support, and scientific exchange. Burek's analysis illustrates the way networking was indispensable to early female geologists, at the local, national, and international level, and reveals the significant number of women working in geology at the time. As well as slowly gaining entry to traditionally male geological clubs and societies, women—importantly—also created their own parallel formal and informal support networks and worked collectively to overcome the hurdles they faced as women in their discipline. Although acknowledging the support that men acting as mentors, with their privileged access to science and scientific institutions, could offer to women, this chapter also demonstrates how early female geologists identified and responded to their situation and, despite suffering continuing discrimination, were anything but powerless or lacking in agency.

Historian Londa Schiebinger has remarked that for 300 years the only permanent female presence at the élite Royal Society of London was a skeleton preserved in the anatomy cupboard (1991, p. 26). As Part III of this *Handbook*, 'Making Women Visible: Institutions, Spaces and Places of Science', reveals, the more prestigious the institution of science, the more likely was it to be hostile to women's admission. In Britain, women were not elected to a fellowship of the Royal Society of London until 1945; the French Academy of Science admitted its first female fellow in 1979 (after turning down Nobel laureate Marie Skłodowska-Curie in 1910). In China, the nuclear physicist Zehui He (1914–2011), 'the Chinese Marie Curie', was elected to the Chinese

Academy of Sciences, established in 1949, in 1980 (Fidecaro and Sutton 2011).[7] However, the admission of women was a complex issue sometimes 'fudged'. The élite Royal Astronomical Society gave two women, astronomer Caroline Herschel and mathematician Mary Somerville, honorary fellowships in 1835, but did not admit women to ordinary fellowship until 1915 (Royal Astronomical Society 2021). Other learned societies, such as the Royal Microscopical Society, accepted women's membership fees but did not allow them to attend meetings (Jones 2016). As science continued to specialize and fragment, so new societies emerged specifically to cater for amateurs, including women. For example, the British Astronomical Association was founded in 1890 to provide an alternative to the Royal Astronomical Society and was advertised as 'open to Ladies as well as Gentlemen'. Several women were active in the Association, participating in expeditions, serving on its council, and editing its journal (Olgilvie 2000, p. 77).

As suggested above, women have always contributed to science but to find them we need to look in the right places. Anne Barrett, in her case study on female scientists in the archives, in particular university archives with a focus on Imperial College London, guides us as to where and how to locate the women. Archives allow scholars to re-evaluate the scope of women's scientific discoveries across points in time, revealing their significant contributions to models of education and the fields of science, technology and medicine, despite their frequent omission from narratives surrounding breakthroughs. Archives have increasingly become a source for the reclaiming of women's scientific histories, as evidenced in initiatives set up around the globe by universities and national organizations. These projects seek to interpret and showcase the past achievements of women in science in both specific subjects and more generally. Imperial College's archives offer an important site for recovery of women's scientific involvement in an institution historically seen as a male domain. From the mid-nineteenth century onwards, women attended lectures at the College on subjects including geology, natural history and medicine, and later trained as science teachers through constituent colleges. The matriculation of women students towards the end of the nineteenth century led to their greater involvement throughout the twentieth century, with a small but influential group of women excelling in a diverse range of subjects, including engineering, mathematics and climatology. Barrett argues that by examining the archival histories of women in science at all levels—whether they made key discoveries as researchers or supported the scientific milieu as teachers or administrators—it is possible to shape future recruitment and retention of girls and women in STEM.

Sarah L. Evans' chapter similarly explores how historical documents can reframe narratives about women's involvement in scientific institutions by highlighting the numerous lectures delivered by women at the Royal Geographical Society (RGS) in the early twentieth century. Twenty-first-century depictions of the RGS have represented women as outsiders, yet the Society's own printed *Geographic Journal* as well as its administrative records reveal a much different picture, with a particular group of women geographers frequently presenting the findings of their RGS-supported expeditionary work over a period of several decades. Women took part in global expeditions, often undertaking exploratory travel alongside their husbands or within mixed expeditions, and produced significant scientific fieldwork. Some women supported the lectures of their male expeditionary counterparts through post-lecture discussions while others took the lead in presenting their own findings. Archaeologist Gertrude Caton-Thompson, for instance, repeatedly lectured on her Egyptian expeditionary work, an important example of how women's scientific practices challenged gendered norms and discourses around geographical capability. Evans recreates the various strategies by which women Fellows of the RGS assimilated within the Society's practices by adhering to the conventions of lecturing while simultaneously challenging established hierarchies through their work and presence.

Although the first women were not elected as fellows of the Royal Society of London until 1945, Claire G. Jones' chapter reveals how women participated at the Society's annual soirées (or conversaziones), both as interested guests, and as exhibitors embodying scientific authority. The 'ladies' soirée' in particular, held each summer from 1876, was a heady mix of science and society which quickly became one of the most coveted invitations of the London season. Guests and exhibitors—roles which blurred as individuals inhabited both—donned their finest dress to be entertained by science as much as to be educated about it. Fusing science and spectacle, the 'ladies' soirée' brought the wives and daughters of scientists and fellows into contact with women exhibitors, though gender inequity frequently led to the exclusion of scientifically-active women, slights which provoked pushback from suffragettes and women's rights campaigners. Nonetheless, a significant minority of women scientists did exhibit, transgressing women's exclusion from fellowship of the Royal Society itself by visibly showcasing their collecting practices, artistic interpretations, intellectual observations, and scientific techniques. Their involvement illustrates how women, whether as active scientific agents or as invested observers, helped to cultivate the growing social interest in science that characterized the late nineteenth and early twentieth centuries. Focusing on the period up to the Great War, Jones reveals the importance of networks, collaborative marriages, differing understandings of public and private, and a broad conception of what counted as science, as implicated in recovering women whose scientific expertise was often known and highly respected in their own time.

Concern about the low proportion of women involved in scientific subjects, particularly in STEM education and professions, has risen to the forefront of public debate in recent decades, with new global and national initiatives aiming to promote gender equality in the associated fields. As Naonori Kodate and Kashiko Kodate point out in their chapter, the challenges associated with integrating women within STEM subjects are not new, and can be related to endemic problems such as gendered social expectations, choices of study subjects available to girls and women, and the dynamics of the labour market. These obstacles have historical roots as well as historical remedies, including the use of women-only universities in Japan as a means of ensuring women's unique and tailored access to a STEM education. Beginning from the late nineteenth century onwards, a number of new Japanese universities were formed to promote a national agenda of educational reform, which did not initially welcome women. While the mainstream view of women's education promoted their domestic capabilities, women's teacher-training colleges increasingly enabled women to gain a scientific grounding, with women eventually gaining access to Japan's Imperial Universities in the early twentieth century. Over the course of the century, women progressed through the university system, gaining doctoral degrees in a wide range of fields and making important advances in their disciplines. Reforms gradually granted women equal access to higher education, including through the creation of women-only universities, which today make up approximately ten per cent of Japan's total universities, providing a space for women to excel in subjects from which they have traditionally been discouraged or excluded. As Kodate and Kodate's chapter shows, institutions hold powerful potential for encouraging the growth of women's involvement in STEM, not only through observing their historical achievements, but also by clearing a path for women's scientific futures.

The impact of one institution, the 'little university of Göttingen', in providing higher mathematical education to women, and so leading the way for other institutions to do the same, is the subject of the chapter by Renate Tobies. Women taking postgraduate degrees at Göttingen in the late nineteenth and early twentieth centuries, then a renowned centre of mathematical research, initially travelled from overseas as German women were not permitted entry to university. This 'experiment' with foreign women from America, Britain and elsewhere proved a success and paved the way for German women to officially study mathematics at university-level too. Again, women were supported by male professors, most significantly Felix Klein (1845–1925), who used determination and creativity to bring talented female mathematicians to Göttingen. These women were not passive recipients of male largesse, however; they also supported each other by setting up their own mathematics club and feminine networks, and when professors themselves sent their best students back to Göttingen to undertake postgraduate research. The importance of male mentors in facilitating women's access to science has been mentioned several times already (and Klein clearly fulfilled this role); however, women were also vulnerable to the actions of powerful men who had a non-inclusionary agenda. Tobies reveals how the conservative university *Kurator* of Göttingen was a staunch opponent of women becoming students and prevented at least one suitably qualified American female mathematician from attending lectures there.

Science is a broad term that encompasses many different disciplines— mathematical sciences, technology, biology, physics, astronomy, engineering, medicine and more—each with its own traditions, aesthetics, and gender colouring; these change according to geography too. The idea of hard and soft sciences, with the latter particularly suited to women, is a hierarchy which still has meaning today and one which, arguably, reflects the gender balance within each discipline, rather than any intrinsic difficulty. Part IV of this *Handbook*, 'Cultures of Science', presents some analyses and case studies to understand the impact of these differing traditions on women, both positive and negative. The culture of science practised in a domestic setting emerges as a key theme; women who collaborated with men in the home have been typically assigned the role of assistant, regardless of the nature of their participation. This was something identified by Margaret Rossiter in 1993 and coined 'The Matthew/Matilda effect' (Rossiter 1993). As she explains, this typically works to ensure it is the male collaborator who receives recognition and credit, and who is mostly recovered by the history of science. For example, in 1903 Marie Skłodowska-Curie received the first of her Nobel prizes for her work on radioactivity, but initially only her husband, Pierre, with whom she collaborated, was considered for the award.[8] The situation was only resolved after Pierre wrote to the Nobel committee to ask them to consider Marie too (McGrayne 2001).

Issues of collaboration and credit feature strongly in the case studies on early European female astronomers presented by Gabriella Bernardi. These women mostly worked alongside male family members in home-based enterprises and businesses which illustrates the opportunities available to women within artisan, craft and guild traditions. This setting not only facilitated participation in science but also gave girls and women access to an astronomical and mathematical education not commonly considered appropriate for the female sex. Despite the extent of their contributions, these women were subject to the 'Matthew/Matilda' effect with credit going to male collaborators. The culture of public and private is implicated here, as it was acknowledged that women *did* science, but it was not accepted for them to act in the masculine, public sphere. Discussing some productive and successful astronomical dynasties, Bernardi demonstrates how it was the men who assumed professional, visible public position while women worked fully, but in the shadows. These women made observations, produced astronomical calendars, discovered comments, wrote astronomical texts, and undertook complex calculations as computers.[9]

Pioneering women in the history of science offer highly visible examples of women's education, practice and recognition across different disciplines, yet their stories are so often the exception rather than the norm and are tied to the specific conditions of gendered possibilities, limitations and perceptions during the period in which they lived. As Emily Winterburn's chapter on the lives of Caroline Herschel and her indirect female descendants points out, the educational models afforded to one generation of women did not necessarily extend to or benefit her successors. Women's historical engagement in science is interwoven with the story of education for women and girls, an often-contentious subject that centred on debates surrounding women's virtue, mental capacity and domesticity. Herschel's late eighteenth- and early nineteenth-century astronomical work stemmed from her informal training as an assistant to the prominent scientific men in her family, a grounding that led to her success in the public sphere. For the subsequent women of the Herschel family, formal education was more widely available, however evolving nineteenth-century expectations about women's roles as wives and mothers circumscribed the boundaries of their scientific educations and pursuits at a time when science was moving outside the home and into more formal environments. Drawing on the diaries and letters of the Herschel family, Winterburn's chapter sheds light on how, despite these barriers, women still found personally enriching ways to participate in science, even as their interests and labours often went unnoticed and uncredited.

Women, of course, are not just practitioners of science, but also a favourite object of investigation. The chapter by Amanda Caleb shows how, from the later nineteenth century, a traditional culture of scepticism concerning women's innate capacity for intellectual work was enhanced and placed on a privileged, scientific footing by Darwin's evolutionary theory. According to the many scientific and medical followers of Darwin, women's limited rationality when compared to men made them unsuited to abstract thought, something that precluded women's serious participation in higher education and scientific pursuits. Indeed, to the scientists and medical men espousing this view, the participation of women in science and scientific institutions may tarnish the image of these places as élite spaces of science. As Caleb demonstrates, women engaged with, and sometimes internalized, these exclusionary ideas. They also responded, at times using literature and poetry to protest and reveal the consequences, for women and men, of this hierarchical and gendered science of sex.

As Corinna Schlombs points out in her chapter on women and computing, gendered perceptions of hierarchies frequently shift over time, and not always in linear progressive ways. Beginning with the computing vision of Ada Lovelace in Victorian Britain and continuing through the nineteenth and early- to mid-twentieth centuries in America, women played an active role in developing modern computer programming, whether through the innovative design of programmes and systems, or through their work in telephone and radio technologies. Yet enduring societal disregard for women's historical and contemporary contributions to the field ensured that computing did not evolve into a woman's profession in the post-World War Two period, due in part to an international masculinization of the field that discouraged women's involvement. While such problems were exacerbated by the so-called software crisis of the 1960s, women's marginalization from academic computing, and the gendered division of labour that relegated women to low-paid data entry jobs and men to operating and programming, small numbers of women nonetheless remained and thrived in the field. By the 1980s, women began organizing their own creative and collective spaces to learn and share computing skills, including through the network surrounding computer scientist Anita Borg, thereby reshaping computing to suit their own needs and purposes. Challenging the historical gendering of computing technologies, Schlombs highlights how global twenty-first-century initiatives aim to bring women and girls into computing, identifies the work still to be done, and lays the groundwork for a shift in thinking that challenges the dominant gendered hierarchies in computing technologies.

The idea that there is a tension between femininity and mathematics and science is one that has received much attention in the scholarship of women in science; it is cited as one of the reasons why women have been under-represented in many scientific disciplines historically and is associated with the STEM gender gap today. However, we must be careful not to impose a Western understanding on other cultures and assume a uniformity than hides the complexity of women's experience in science. Carol Mukhopadhyay presents an anthropologically informed analysis of the culture of STEM in India which cautions against over generalization and frameworks that assume reasons for the gender gap are constant (even if the gap itself is). As Mukhopadhyay demonstrates, in India complex family requirements, networks, traditions and cultures have affected girls' and women's education and professional choices, and together provide a contrasting, social explanation for women's absence and presence in differing scientific disciplines. Most striking is the absence in India of essentialist beliefs about intrinsic intellectual differences between the sexes and, indeed, the 'Western gender-differentiated brain theory' was 'startling' to the individuals questioned as part of Mukhopadhyay's extensive research.

Women's contributions to science have historically been viewed and represented as marginal across a variety of contexts, yet their presence was vital to the formation of one of the most identifiable symbols in science, as evidenced in Brigitte Van Tiggelen and Annette Lykknes's chapter on women's contributions to the periodic system. Dominant narratives historically framed the periodic system through the accomplishments of individual male chemists rather than examining the development of the system over time, which included significant elemental discoveries by a range of women chemists in the nineteenth and twentieth centuries. Whether working with wet-chemical analyses, or the mixing of substances in their liquid phase, which originated in domestic spaces such as kitchens thereby allowing women a natural point of access, or through the mastery of analytical chemical work, including the determination of measurements such as atomic weights, which required specialist training in laboratories and universities, women chemists found innumerable ways to contribute to the identification and positioning of the periodic system. Looking beyond the more well-known female chemists such as Marie Skłodowska-Curie, Lise Meitner and Irène Joliot-Curie, this chapter reveals how the meticulous labour of overlooked women chemists in America and Europe enabled the separation and analysis of individual elements. Throughout their chapter, Van Tiggelen and Lykknes highlight the impact of female scientists on the periodic system while also questioning why the stories of these ground-breaking women have not been adequately acknowledged until recently.

Certain spheres of scientific activity across the late eighteenth to early twentieth centuries were especially populated by women and understood as complementary to femininity. Women found roles in science as writers, popularizers and translators of science, often writing for an amateur or child audience (Benjamin 1991; Fyfe 2000). For example, between 1806 and 1815, Margaret Bryan published well-received texts on astronomy and natural philosophy (including optics, hydrostatics, pneumatics and acoustics) (Ogilvie and Harvey 2003). Jane Marcet's *Conversations on Chemistry in which the Elements of that Science are Familiarly Explained and Illustrated by Experiments* was an enormous success when published in 1806, and inspired Michael Faraday to take up science (Phillips 1990, pp. 110–11). Mary Somerville's work is an excellent example of a form of writing that combined 'vision on a cosmic scale with a restrained poetic quality' to evoke the 'scientific sublime' (Neeley 2001, p. 8), thus making it both imaginatively engaging and scientifically informative. Central to the circulation of scientific ideas, postulates and findings across cultural and linguistic boundaries was, therefore, the ability to communicate them effectively on a visual and textual level, as the contributions to Part V of this *Handbook* on 'Scientific Communication and Representation' show. While the fundamental importance of translation was rarely overlooked by those involved in scientific pursuits, as Alison E. Martin's chapter shows, the agents responsible for transferring and transforming information from one language into a different language tended to remain near-invisible (Venuti 1995). To some extent, this invisibility served female translators well, given the dictates of female modesty in the public arena of publishing. Yet translation could also be used as a subtle way to display scientific knowledge, and, by strategic use of a translator's foreword or paratextual additions such as footnotes, could be employed to help women position themselves within scientific economies of knowledge. While women's translation activities brought texts into international circulation, a passion for collecting scientific artefacts also saw the movement of objects across cultural and geographical boundaries.

As the chapter by Andrea Gáldy and Anne E. Harbers on the Swedish royal collections clearly demonstrates, collecting practices also enabled women to partake in the display and representation of scientific knowledge. Focusing on the collections of Queen Lovisa Ulrika of Sweden (1720–1782), they investigate how the contents of her exhibition cabinets cast her as a woman of taste and status, who acted as a patron of the arts and literature, but also of science. The busts of Swedish scientists located in these rooms suggest a broader concern to promote her country's scientific achievements, in particular the work of scholars at Uppsala University, the oldest of Sweden's seats of learning and an institution with a clear focus on studies in the natural sciences. Her cabinet of natural history also, though, gave insights into her own personal interests, notably a preoccupation with mineralogy. While this collection primarily articulated and embodied the visual communication of expertise in a courtly environment, increasingly diverse audiences (which also included a small percentage of women) were beginning to visit such collections, and therefore they also served to stimulate intellectual curiosity in those who did not belong within the direct sphere of their curators. As such, then, these collections were key nodal points in networks of communication and exchange and the patronage of aristocratic women was essential to their success.

Being well-connected was also of importance to the Victorian traveller and botanical artist Marianne North (1830–90), as the chapter by Philip Kerrigan reveals. The daughter of a Member of Parliament who himself had been a particularly well-travelled amateur naturalist and botanist, North essentially inherited her father's scientific connections, and these may well have given her the vision to depart from traditional forms of botanical illustration. Travelling unchaperoned, North initially ventured to the United States and Canada, before heading to Australia, Tasmania and New Zealand. While her written outputs have tended to be cast as complicit in the British imperial project, Kerrigan argues that in her artistic contributions she looked beyond the conventional media of line and watercolour to work in oils and adopt a looser, more painterly technique. Taxonomic accuracy was still paramount to her, but she was concerned to convey a sense of how it was to paint 'in the field' and did not shy away from illustrating specimens with imperfections. Her subjects also included examples of conflict in the plant kingdom, notably parasitic plants and their survival strategies. A woman who was keenly aware of the dangers of what might be deemed 'unwomanly' behaviour in an environment adverse to female intervention in a more proficient and professional capacity, her work was extremely well received for its unique and striking representation of the natural world.

Successfully conveying a particular message—whether through a visual or textual medium—was key to women gaining public recognition in science. In a more contemporary setting, the chapter by Jordynn Jack signals the importance of phatic communication—communication used to create goodwill and maintain relationships—in facilitating scientific research pursued by women. Taking the example of the American zoologist Catherine Henley (1922–1999), Jack explores how in the mid-1940s she entered into a scientific field in which women were well-represented but under-acknowledged, with a mere handful of female figures reaching faculty positions in top research institutions. In particular, Jack investigates by using Bruno Latour and Steven Woolgar's notion of 'cycles of credit' how women had difficulties building a career for themselves in an environment that all too frequently relegated them to subordinate positions such as secretarial staff or laboratory technicians, thus placing them at the periphery of cutting-edge research. Conscious self-positioning and social awareness were essential in helping women to acquire intellectual capital in the world of science, while phatic communication also strengthened networks of exchange for women who otherwise lacked the components traditionally needed to launch a scientific career.

The extremely influential, but highly controversial, American science writer and public educator, Rachel Carson (1907–1964), is a fascinating example of how a woman acquired scientific credibility despite her apparent lack of academic credentials. As Ruth Watts notes, Carson did not originally sign up to study biology but English. However, it was precisely this understanding of the importance of language in enhancing scientific accessibility that was central to the career that she forged for herself. First gaining recognition through the radio scripts she wrote for the Bureau of Fisheries, and subsequently working as a civil servant authoring pamphlets on conservation and public resources, she honed her skills at making science accessible to a non-specialist audience through vivid portrayals of the ecology of the seashore. Now best known for *Silent Spring* (1962), written to alert its readers to the dangers of new synthetic pesticides, she adopted an overtly biocentric, rather than anthropocentric, perspective, to address contemporary concerns about humankind's unthinking contamination of nature. The chemical industry unsurprisingly sought to undermine the scientific credibility of a woman with no doctorate or university position. But the meticulous nature of her secondary research, the public recognition she had hitherto gained through her prize-winning writing, and her own concern to stay up to date with the latest scientific thinking enabled her writing to give new impetus to the burgeoning conservation movements in the second half of the twentieth century.

The visibility of women scientists in the public sphere today of course owes much to their representation through film and television. The ripple effects triggered by successful female figures in the world of science should not be underestimated, argues Amy C. Chambers, in her analysis of the spectrum of female scientific figures present in the media today. While women in STEM tended to be defined by their relationship to male counterparts, the media representation of women—memorably embodied in the iconic figure of Uhura from *Star Trek*—has increasingly improved in the last few decades, even if diversity is an issue still side-lined in favour of celebrating the privileged white woman. The intense over-simplification of earlier portrayals of women scientists on screen has since given way to their presentation as complex and capable figures driving plot and action. Nevertheless, women still tend to be characterized in roles associated with the 'soft' biosciences, while the 'hard' sciences of engineering and physics still appear to belong within the male domain. More intersectional approaches need to be adopted, Chambers suggests, if we can begin to rethink the expectations of what science constitutes, who scientists are, and how we envisage them.

While the term 'scientist' is frequently used in a wide spectrum of contexts—from academic and institutional research to industry and public policy—it pays to reflect on what it has come to represent both regarding routes into scientific professions and issues of gatekeeping. As the contributions to Part VI of this *Handbook*, 'Access, Diversity and Practice', illustrate, women working in the sciences have constantly queried what 'professionalization' might mean, particularly where it grants a status and authority which it has proven institutionally difficult for them to acquire. Taking 1833, the date of the alleged coinage of the word 'scientist' by the British polymath William Whewell, as a pivotal moment in women's engagement in science, Mary Orr investigates how Whewell responded in a review essay the following year to Mary Somerville's *On the Connexion of the Physical Sciences* (1834). Such position statements by men were, Orr argues, forms of critical reception and evaluation which had powerful potential to facilitate or block women's presence in the scientific arena, even if the 'Somerville effect' triggered women's conspicuous attendance at meetings of the British Association for the Advancement of Science. Taking the case study of Sarah Bowdich (Lee) (1791–1856), Orr demonstrates how women were already well established in the practice of science at the very time that the potentially exclusionary term 'scientist' came into being, while the outputs of Margaret Gatty (1809–1873) and Athénaïs Michelet (1826–1899) show how women successfully advanced scientific knowledge through their publishing activities outside institutional frameworks.

Demands for women's increased political and social action by the end of the nineteenth century were vital in guaranteeing their appointment to positions of authority, even if female professionalization was largely a process of élite formation which also emphasized distinctions between women. In her chapter on women's access to the medical profession in Britain and Ireland, Laura Kelly investigates how restrictions on women training and qualifying at university medical schools meant that they faced significant resistance in attempting to pursue an institutionally recognized medical career. The legal exclusion of women as 'outsiders' to the medical profession and the refusal by British licencing bodies to permit them to take exams saw some British women embrace the internationalization of the medical sciences and study and qualify overseas. Kelly charts how contemporary attitudes in late-Victorian and Edwardian periods in Britain to women's admission were based not only on proficiency and aptitude but also on gendered subjectivities and sexualized dynamics (notably with regard to the practice of dissection), which would only change at the watershed marked by the First World War.

It is common knowledge that, with the outbreak of war, employment opportunities emerged for women not just in the medical profession, but also in the defence industry. Less well known is the role that women played in aeronautics, as discussed in the chapter by June Barrow-Green and Tony Royle. Innovations in aeronautics, developments in theoretical aspects of aerodynamics and improvements to the strength and integrity of aircraft structures were the main areas in which women mathematicians helped to make significant advances. A key contributing factor was the gradual emergence of educational opportunities for women in the final quarter of the nineteenth century. Women could at last study on the Mathematical Tripos offered by Cambridge University, at that time the fulcrum of British mathematics, through the establishment of the women-only colleges of Newnham and Girton. Likewise, the founding of Bedford College, East London College and Royal Holloway in London enabled a pool of female talent in mathematics to emerge. Job openings at institutions such as the Admiralty Air Department, the National Physical Laboratory and the Royal Aircraft Factory ensured women could enter traditionally male-dominated domains, and therefore partially overcome the institutionalized prejudices of the era.

Despite this, the interwar period did not necessarily see women's entrance into science-based professions made significantly easier. Nina Baker argues in her chapter that the two World Wars could be seen as 'false entry' points, since women who had taken up jobs in these domains tended to be forced out of work when peace was restored. Particularly in the heroic, 'muscular' world of engineering—dominated historically by figures such as James Watt or Isambard Kingdom Brunel—the public persona of the engineer was defined by his ability to achieve in the face of physical adversity. This, coupled with embedded prejudicial structures and social conventions (notably the 'marriage bar' which carried the expectation that women would leave their employment once they got married), meant that women had difficulties establishing themselves in engineering, despite the first engineering degree conferred on a woman in the British Isles dating back to the very early years of the twentieth century. It was particularly in emerging fields of engineering post-World War Two—aeronautics, computing and bio-engineering—that women were more successful in gaining a foothold. While progression up the career ladder has subsequently proven easier, women are still frequently stymied at mid-career level by implicit institutional biases which prevent them from rising to the top.

An interesting comparison is offered by Claire Brock's chapter on the role of women in the surgical sciences in the 1920s and 1930s. The interwar period was characterized by more positive public responses to the notion of women working as surgeons, primarily influenced by the significant contribution they had made during World War One. While voices of dissent queried the rationale behind training women in surgery, given that they were then likely to leave the profession when they married, these were outweighed by the sense that surgery was no longer simply 'men's business'. Taking the examples of the careers of Louisa McIlroy (1874–1968), appointed University of London Chair in Obstetrics and Gynaecology in 1921, Louisa Martindale (1872–1966), President of the Medical Women's Federation and a pioneer in cancer treatment, and the orthopaedic specialist Maud Forrester-Brown (1885–1970), Brock reveals how these women were pivotal in encouraging public confidence in the work of women surgeons, and also enabled a body of experience to grow that later formed the legacy on which other women surgeons would build.

Bringing us into the twenty-first century, the contribution by Wendy DuBow investigates the still largely homogeneous character of the US technology work force, which primarily comprises white or Asian American men.

While just over half the professional workforce is made up of women, they only constitute one quarter of the computing workforce. Growing demands for transparency, together with the requirement that employers publish their demographic characteristics, has made the acute lack of diversity in the technology industry even more visible. It in part reflects the fact that women and racial/ethnic minorities are under-represented at degree level, which in turn is due to the choices made at secondary school level, where teachers, parents and careers advisors still tend to guide girls, and girls or boys of colour, towards non-technical areas of study. DuBow argues that this lack of diversity has tended to be tackled through initiatives geared towards individuals, rather than addressing the urgent need for systemic change across the board: only by implementing larger-scale strategies can changes be made that bring diversity to the field of technology, and with them greater stability and economic potential for women and people of colour. DuBow's chapter serves as an important reminder not only of the centrality of gendered issues to the world of science, but also of racial and ethnic concerns that have still to be addressed in relation to women's involvement in academic training and industry.

This *Handbook* therefore traces significant shifts in the self-representation of women in science from the early modern period to the present day. It charts women's social, political and intellectual activism in seeking to gain acknowledgement, authority and, more concretely, appointment—in scientific professions. Above all, its contributions reflect the importance that researchers are only starting to assign to histories of gender and work. The legacy of women's involvement in scientific pursuits—whether as amateur or professional—continues to influence their understanding of their role in the sciences today.

Notes

1. Women are often, literally, simply a footnote in landmark texts on the history of science. Just to take examples from the British context, Shapin and Schaffer's *Leviathan and the Air Pump* (2017, p. 30) briefly mentions the visit of Margaret Cavendish, Duchess of Newcastle, to the Royal Society of London in 1667, but only in a footnote barely indicates her scientific interest and expertise, instead concentrating on how 'full of admiration' she was for what she saw. The masculine perspective of Andrew Warwick's detailed text on the training of mathematicians at Cambridge University is revealed by its title, *Masters of Theory* (2003). Even a recent history of the Cavendish Laboratory at Cambridge (Longair 2016) scarcely mentions women at all despite the female researchers in its past; for a discussion of women researchers at the Cavendish in the late nineteenth century see Gould (1997).
2. For example, see Richmond (2001) on women in the early history of genetics and Sanz-Aparicio (2015) on women in the emergence of x-ray crystallography.

3. For example, see Hartley and Tansey (2015) and Iliff (2008).
4. Indeed, one only has to look at the titles of scholarship on Lovelace to recognise this romantic appeal. For example, Woolley's 1999 biography *The Bride of Science: Romance, Reason and Byron's Daughter*. Despite her limited legacy to science, in Britain women in science are celebrated every year, on the second Tuesday of October, with 'Ada Lovelace Day'. Romantic themes, even with sexual allusions, are popular in titles of works on women scientists, for example Robyn Arianrhod (2012) *Seduced by Logic: Emilie Du Chatelet, Mary Somerville and the Newtonian Revolution*; and Brenda Maddox (2003) *Rosalind Franklin, Dark Lady of DNA*.
5. Ormerod is widely understood as a pioneering technological scientist and economic entomologist whose science and science communication activities made her instrumental in establishing and defining the discipline of economic entomology; see Clark (2004).
6. Examples include Patricia Phillips (1990) *The Scientific Lady: A Social History of Woman's Scientific Interests 1520–1918* and Londa Schiebinger (1991) *The Mind has No Sex? Women in the Origins of Modern Science*.
7. It is not clear if Zehui He was the first woman elected an academician of the Chinese Academy of Science, but she was certainly one of very few women. As well as Skłodowska-Curie and He, in the history of nuclear science the physicist Lise Meitner (1878–1978) must be remembered as the co-discoverer of nuclear fission; see Sime (1997).
8. Data to March 2021 reveals that 4 women and 114 men have won the Nobel Prize in Physics; 7 women and 186 men have won the Nobel Prize in Chemistry; and 12 women and 111 men have won the Nobel Prize in Medicine (Nobel Prize 2021).
9. Before computers came into use, women were often employed as computers to carry out the 'tedious' calculations needed by science. For example, early women mathematics graduates in the UK were employed at the Greenwich laboratory in the nineteenth century and later, in the USA, at NASA (Grier 2007; Shetterley 2017).

References

Arianrhod, Robyn. 2012. *Seduced by Logic: Emilie Du Châtelet, Mary Somerville and the Newtonian Revolution*. Oxford: Oxford University Press.

Benjamin, Marina. 1991. "'Elbow Room': Women Writers on Science, 1790–1840." In *Science and Sensibility: Gender and Scientific Enquiry 1780–1945*, ed. Marina Benjamin, 27–59. Oxford: Basil Blackwell.

Catalyst. 2020. "Women in Science, Technology, Engineering, and Mathematics (STEM)." https://www.catalyst.org/research/women-in-science-technology-engineering-and-mathematics-stem/. Accessed March 13 2021.

Clark, J. F. M. 2004. "Ormerod, Eleanor Anne (1828–1901), Economic Entomologist." *Oxford Dictionary of National Biography.* https://www.oxforddnb.com/view/10.1093/ref:odnb/9780198614128.001.0001/odnb-9780198614128-e-35329. Accessed February 6 2021.

Fidecaro, Maria, and Sutton, Christine. 2011. Zehui He: Following a Different Road. *Cern Courier.* https://cer.courier.com/a/zehui-he-following-a-different-road/. Accessed March 13 2021.

Findlen, Paula. 1993. "Science as a Career in Enlightenment Italy: The Strategies of Laura Bassi." *Isis*, 84, no. 3: 441–69.

Fox Keller, Evelyn. 1985. *Reflections on Gender and Science.* New Haven: Yale University Press.

Fyfe, Aileen. 2000. "Young Readers and the Sciences." In *Books and the Sciences in History*, ed. Marina Frasca-Spada and Nicholas Jardine, 276–90. Cambridge: Cambridge University Press.

Gould, Paula. 1997. "Women and the Culture of University Physics in Late Nineteenth-Century Cambridge." *British Journal for the History of Science*, 30, no. 2: 12–49.

Grier, David Alan. 2007. *When Computers were Human.* Princeton, NJ: Princeton University Press.

Hartley, J. M., and E. M. Tansey. 2015. "White Coats and No Trousers: Narrating the Experiences of Women Technicians in Medical Laboratories, 1930–90." *Notes and Records of the Royal Society of London*, 69, no. 1: 25–36.

Iliff, Rob. 2008. "Technicians." *Notes and Records of the Royal Society of London*, 62, no. 1: 3–16.

Jones, Claire G. 2016. "Women and Science." *Routledge History of Feminism.* https://www.routledgehistoricalresources.com/feminism/essays/women-and-science. Accessed March 13 2021.

Longair, Malcolm. 2016. *Maxwell's Enduring Legacy: A Scientific History of the Cavendish Laboratory.* Cambridge: Cambridge University Press.

Maddox, Brenda. 2003. *Rosalind Franklin: Dark Lady of DNA.* London and New York: Harper Perennial.

McGrayne, Sharon Bertsch. 2001. *Nobel Prize Women in Science: Their Lives, Struggles, and Momentous Discoveries.* Washington, DC: Henry Joseph Press.

Merchant, Carolyn. [1976] 1999. "Isis' Consciousness Raised." In *History of Women in the Sciences: Readings from Isis*, ed. Sally Gregory Kohlstedt, 11–22. Chicago: University of Chicago Press.

Morrison-Low, Alison D. 1991. "Women in the Nineteenth-century Scientific Instrument Trade." In *Science and Sensibility: Gender and Scientific Enquiry, 1780–1945*, ed. Marina Benjamin, 89–117. Oxford: Basil Blackwell.

Mozens, H. J. [1913] 1991. *Woman in Science: With an Introductory Chapter on Woman's Long Struggle for Things of the Mind.* Notre Dame: University of Notre Dame Press.

Neeley, Kathryn A. 2001. *Mary Somerville: Science, Illumination, and the Female Mind.* Cambridge: Cambridge University Press.

Nobel Prize. 2021. "Women Who Changed the World." https://www.nobelprize.org/prizes/lists/nobel-prize-awarded-women/. Accessed March 14 2021.

Olgilvie, Marilyn Bailey. 2000. "Obligatory Amateurs: Annie Maunder (1868–1947) and British Women Astronomers at the Dawn of Professional Astronomy." *British Journal for the History of Science*, 33, no. 1: 67–84.

Ogilvie, Marilyn, and Joy Harvey. 2003. *The Biographical Dictionary of Women in Science: Pioneering Lives From Ancient Times to the Mid-20th Century*. Abingdon: Routledge.

Phillips, Patricia. 1990. *The Scientific Lady: A Social History of Women's Scientific Interests 1520–1918*. London: Weidenfeld and Nicolson.

Pycior, Helena M., Nancy G. Slack, and Pnina G. Abir-Am, eds. 2006. *Creative Couples in the Sciences*. New Brunswick, NJ: Rutgers University Press.

Richmond, Marsha L. 2001. "Women in the Early History of Genetics: William Bateson and the Newnham College Mendelians, 1900–1910." *Isis*, 92, no. 1: 55–90.

Rippon, Gina. 2019. *The Gendered Brain: The New Neuroscience that Shatters the Myth of the Female Brain*. London: Bodley Head.

Rossiter, Margaret W. 1993. "The Matthew/Matilda Effect in Science." *Social Studies of Science*, 23, no. 2: 325–41.

Royal Astronomical Society. 2021. "Women and the Royal Astronomical Society." https://women.ras.ac.uk/women-and-the-ras/history-of-women-at-the-ras. Accessed March 13 2021.

Sanz-Aparicio, Julia. 2015. "The Legacy of Women to Crystallography." *Arbor*, 191, no. 772. https://doi.org/10.3989/arbor.2015.772n2002. Accessed March 13 2021.

Schiebinger, Londa. 1991. *The Mind has No Sex? Women in the Origins of Modern Science*. Cambridge, MA: Harvard University Press.

Shapin, Steven, and Simon Schaffer. 2017. *Leviathan and the Air Pump: Hobbes, Boyle and the Experimental Life*. Rev. ed. Princeton NJ: Princeton University Press.

Shetterley, Margot Lee. 2017. *Hidden Figures: The Untold Story of African American Women Who Helped win the Space Race*. New York: William Collins.

Shteir, Ann B. 1997. "Gender in 'Modern Botany' in Victorian England." *Osiris*, no. 12: 21–38.

Sime, Ruth Lewin. 1997. *Lise Meitner: A Life in Physics*. California: University of California Press.

Toole, Betty Alexandra. 2004. "Byron, (Augusta) Ada [Married Name (Augusta) Ada King, Countess of Lovelace] (1815–1852), Mathematician and Computer Pioneer." *Oxford Dictionary of National Biography*. https://www.oxforddnb.com/view/10.1093/ref:odnb/9780198614128.001.0001/odnb-9780198614128-e-37253. Accessed February 16 2021.

UNESCO. 2021. "Women in Science." http://uis.unesco.org/en/topic/women-science. Accessed March 13 2021.

Venuti, Lawrence. 1995. *The Translator's Invisibility: A History of Translation*. London: Routledge.

Warwick, Andrew. 2003. *Masters of Theory: Cambridge and the Rise of Mathematical Physics*. Chicago: University of Chicago Press.

Woolley, Benjamin. 1999. *The Bride of Science: Romance, Reason and Byron's Daughter*. London: MacMillan.

PART II

Strategies and Networks

CHAPTER 2

The Natural Philosophy of Margaret Cavendish, Duchess of Newcastle: Nature, Self-Knowing Matter, and the Dialogic Universe

Brandie R. Siegfried

Over the course of her two-decade career as a philosopher of nature, Margaret Lucas Cavendish, Duchess of Newcastle (1623–1673), published twelve volumes, the majority of which were committed to refining a complex, multi-faceted theory of matter, perception, and human nature.[1] Indeed, Cavendish was the first English woman to develop and publish several volumes on her own theory of the natural world. She did so with astonishing literary versatility, hoping to win over her readers by means of literary pleasure. With sometimes startling reorientations of perspective, her poems, plays, fictional dialogues, scientific treatises, epistolary fiction, dramatic orations, and science fiction, all form a rich body of work deserving of further scholarly attention in the history of science.[2]

The Scientific Revolution

Seventeenth-century natural philosophy—the science of nature—included collective endeavours that would come to be known as the 'scientific revolution' and, as part of that effort (and despite restrictive attitudes towards women and publication), Cavendish engaged with the ideas of other major philosophers of the day, including Pierre Gassendi, René Descartes, Constantijn Huygens, Robert Boyle, Joseph Glanvill, Robert Hooke, Thomas Hobbes, Walter Charleton, and other contemporaries. As did many of her peers, she also

---

B. R. Siegfried (✉)
Formerly of Brigham Young University, Provo, UT, USA

took up classical authors such as Lucretius, Heraclitus, Aristotle, and Plato (to name but a few) and provided critical evaluations of their respective philosophies of the natural world. Cavendish was quite aware that her advances into the realm of science would be seen as highly unusual. Even late in her career, having published several volumes, she was still addressing the issue. In the preface to *Observations upon Experimental Philosophy* (1666), she wrote:

> It may be, the world will judge it a fault in me, that I oppose so many eminent and ingenious writers: but I do it not out of a contradicting or wrangling nature, but out of an endeavor to find out truth, or at least the probability of truth, according to that proportion of sense and reason nature has bestowed upon me. (2001, p. 9)

She knew she would encounter a certain amount of resistance, but embraced the possibility of intellectual combat:

> I am as ambitious of finding out the truth of nature, as an honorable dueler is of gaining fame and repute; for as he will fight with none but an honorable and valiant opposite, so I am resolved to argue with none but those which have the renown of being famous and subtle philosophers. (2001, p. 10)

Suspecting that many of her contemporaries would not respond publicly to a woman, she nevertheless sent copies of her works to be archived in the major universities of England for future readers.[3] Her brio defined her subsequent fame.

Of male philosophers and other readers who might be dismissive of her efforts, she wrote, 'They will perhaps think me an inconsiderable opposite [an inconsequential fencing partner], because I am not of their sex, and therefore strive to hit my opinions with a side-stroke, rather covertly, than openly and directly' (2001, pp. 9–10). She was partly right. Prominent philosopher Henry More wrote to Anne Conway that Cavendish 'is affrayed some man should quit his breeches and putt on a petticoat to answer her in that disguise [...] but I believe she may be secure from anyone giving her the trouble of a reply' (Conway 1992, p. 237). The issue had less to do with Cavendish's science than with the breaking of protocol: publishing her ideas in the public forum of print—as opposed to circulating manuscripts privately among coteries of family, friends, and neighbours—was considered inappropriate behaviour for a woman in the seventeenth century. Indeed, this meant that even family associates such as Kenelm Digby, John Evelyn, Thomas Hobbes, and Walter Charleton responded courteously but cautiously to gifts of her printed volumes. Even so, Digby, Charleton, Constantijn Huygens and Joseph Glanvill each engaged in correspondence with her, discussing various topics of interest. These men were not corresponding with Cavendish merely out of a commitment to courtesy or due to hopes of additional patronage, as some scholars have implied; rather they each shared particular scientific interests with her.

For instance, Digby (1602–1665), a natural philosopher and courtier, shared with Cavendish an interest in embryology and the related question of how one form of matter grows into another. Both leaned toward Aristotle's theory of epigenesis against the prevailing theory of pre-formation (Cavendish added several caveats to her own theory, however). In epigenesis, the embryo—equipped with its own 'potential' for growth—acquires form in stages. The coming together of the human, for instance, occurs only gradually and can be substantially reoriented during that process to produce offspring that might be larger, stronger, and more intelligent than either parent (similar to how developmental biologists currently describe the importance of environment on an embryo's growth in stages). Preformation, in contrast, referred to the notion that the embryo was already fully organized from the beginning, in a sense 'preformed' before actual growth even began (similar to how genetic determinists currently think about genetic codes, though this was long before anyone knew of genes). Digby's work on embryonic development is held to be foundational for subsequent advances on the topic, and Cavendish's critiques and additions are part of that story.[4]

Charleton (1619–1707), another of Cavendish's correspondents, was a friend of Digby, an admirer of Hobbes, and a member of the Cavendish Circle. Famous as a physician and natural philosopher, he was not only a friend of the Newcastles, but he also held a deep interest in the revival of Epicurean atomism that was every bit as intense as Cavendish's early espousal of the Epicurean philosophy. In fact, like Cavendish, he followed Gassendi in pioneering the integration of mechanical philosophy (atomism) with traditional Aristotelianism and Christianity, though her work had a much lighter touch with respect to religion.[5] Cavendish would later abandon the classical form of atomism, thereby moving away from Charleton's life-long theories; nevertheless, she retained important aspects of particle theory as a component of her subsequent, more sophisticated notion of how matter relates intimately to matter within a continuous, infinite, and eternal nature.

Huygens (1596–1687), another with whom Cavendish exchanged letters, was a poet, composer, and diplomat of international repute, and was knighted by King James of England for his service between nations. His circle of friends included Descartes, John Donne, and Anna Maria van Schurman. Given his ongoing correspondence with Descartes, and Cavendish's many critiques of Cartesian theory, his interest in Cavendish is noteworthy (Akkerman and Corporaal 2004, pp. 2–21). Huygens discussed various topics of interest with Cavendish in the course of their friendship and enjoyed reading her work. Writing to his friend Utricia Swann, he pronounced Cavendish's first book, *Poems and Fancies*, a 'wonderful book, whose extravagant atoms kept me from sleeping a great part of last night in this my little solitude'. Moreover, curious to see what she was up to in her laboratory, Huygens visited her household on several occasions while she and William lived in Antwerp, often asking her for solutions 'upon diverse questions sometimes for an entire afternoon' (Whitaker 2002, p. 170).

Joseph Glanvill (1636–1680), a clergyman, philosopher, and theologian who disagreed with Cavendish on some things, but agreed on others, demonstrated a keen interest in Cavendish's ideas on how science pertained to religion. Cavendish did not believe in witches, as described in her *Philosophical Letters* (1664), whereas Glanvill certainly did, and wrote at length on the topic. However, the same interest in material phenomena that led her to be dismissive of witchcraft also led her to concur with Glanvill's stance against undue deference to the scholastic tradition in natural philosophy. Glanvill not only struck up a correspondence with Cavendish, but sent her a copy of his own book, *A Philosophical Endeavor towards the Defense of the Being of Witches and Apparitions* (1666) as well. In his earlier book, *The Vanity of Dogmatizing* (1661), he had praised Descartes while criticizing Cartesian physics and psychology, a not uncommon stance towards the French philosopher in England. This position may have gained traction in part from Cavendish's early critiques of Descartes in her books *Poems and Fancies* and *Philosophical Fancies*, both of which caused a stir in the 1650s.[6]

Of the four correspondents covered here, Glanvill was the one most likely seeking some form of substantial patronage—or at least, an advantageous social connection—in his discourse with the Duchess. However, such initial motivations were extremely common among male-to-male correspondents of differing social status and cannot be the grounds for dismissing his very real interest in Cavendish's ideas. Moreover, Digby, Charleton and Glanvill were active members of the Royal Society, and each played a role in helping to secure an invitation to that scientific body when Cavendish expressed interest in seeing its members at work. The interest was mutual, as illustrated by another member of the Royal Society, Nehemiah Grew who, as Katie Whitaker points out, 'would later pioneer the study of plant anatomy and become secretary of the Royal Society'. Grew studied closely the first edition of another of Cavendish's books on natural philosophy, *Philosophical and Physical Opinions* (1655a), and drew up 'a detailed, eight-page summary of Margaret's book', which was included among his overviews of other major treatises of the period (Whitaker 2002, p. 315). In short, Cavendish was relatively well known among the scientific virtuosi of her day, and though some (men and women alike) were disapproving of her public literary productions, many were nevertheless conversant with her ideas.

Whether toying with the implications of algebraic geometry and infinitesimals, or developing a theory of matter that could accommodate her notion of a non-mechanistic world vibrant with its own intelligence, Cavendish was an original thinker. Many of her ideas—thanks to advances in physics, chemistry, and biology—ring true for a twenty-first century audience in ways that they did not for readers of her own day. Despite her annoyance with gendered notions of propriety surrounding work in science, Cavendish persisted in contributing to the ongoing conversations on natural philosophy through her independent publications. From her first book on science, *Poems and Fancies* (1653) to her last, *Grounds of Natural Philosophy* (1668), Cavendish asserted her right to participate in the development and criticism of what was then considered a 'new' science of experiment, observation, and technological advancement (including the use of the microscope, telescope, and air pump).[7] She was a particularly productive, original, and intrepid seventeenth-century natural philosopher.

Cavendish's Early Particle Theory: 'Poems and Fancies' and 'Philosophical Fancies' (1653)

Cavendish's interest in science seems to have flourished despite the unusual circumstances of her early career. Born Margaret Lucas, she was only nineteen in 1642 when civil war broke out in England and having become a maid of honour to Queen Henrietta Maria, by 1644 she was forced to follow the Queen into exile in France when the King's army was defeated. In France, she was wooed by the royalist commander William Cavendish (then Marquis of Newcastle), who had recently lost the crucial battle of Marston Moor. Eventually, she and William settled in Antwerp, where they spent their exile in the house rented from the widow of painter and diplomat Peter Paul Rubens. William and his brother Charles were patrons and correspondents with an international group of mathematicians, political philosophers and scientists (now commonly referred to as the Cavendish Circle); their status in the world of science meant that Margaret had ready access to conversations about theories of classical atomism, movements in the new mathematics, news on astronomy, physiology and chemistry, and a host of other interesting topics. No passive listener, and an active reader of science and philosophy, Cavendish swiftly crafted her first two books of natural philosophy and in 1653 released both for publication.

Cavendish's first foray into science took the form of a book of poems titled *Poems and Fancies* (1653). In this, she followed the model of Roman poet and philosopher Lucretius (99–55 BCE), whose famous Latin poem *On the Nature of Things* (*De Rerum Natura*) expounded Epicurean atomism. Lucretius's work was being read and discussed by many writers in the seventeenth century, and so pausing to understand its influence helps to clarify Cavendish's intentions. Classical atomism posited that reality, at its most basic level, consists of tiny indivisible particles moving about in otherwise empty space or 'void'. Leucippus, Democritus, and Epicurus each propounded a theory of atoms. Though slightly different, each theory presumed an entirely naturalistic account of how the world works by means of impersonal processes, rather than the volition of deities. Lucretius's poem, which was so influential in the seventeenth century, followed the Epicurean model. Epicurean atomism presumed that: (i) matter exists in the form of innumerable, indivisible and unchangeable particles; (ii) empty space (void) allows matter to move; (iii) atoms fall downward in the void but occasionally swerve without cause; and

(iv) this spontaneous, chance swerve results in atoms colliding and becoming locked together, yielding compound structures or bodies. Lucretius's poem developed explanations and dramatizations of how all forms in nature are based on atoms, and how even human society—made up of peoples, structures, and practices—must give way to an atomic pattern of growth, maturity, decline, and dissolution into component atoms, a pattern attested by history.

Cavendish was deeply interested in atomism as a means of explaining how nature works, and in her early writing embraced several aspects of Lucretius's narrative. For instance, she seems to have accepted the concept of void (empty space in which atoms could move freely) and the likelihood that atoms were the fundamental building blocks of nature. Without denying the existence of God (she was at pains throughout her career to avoid being labeled an atheist), she nevertheless preferred the explanatory power of observation and reason over human claims about divine intention and direction.[8] Moreover, though she would later eschew some elements of atomism as her theory of physics evolved, she maintained throughout her career a strong affinity for the Epicurean stress on an epistemology of pleasure and probability rather than detachment and certainty.[9] However, even in her early work, she challenged Lucretius on several counts. She did not, for instance, believe that chance collisions of atoms could account for the variety and complexity of life, and instead posited a world in which atoms are active and purposeful. In *Poems and Fancies* she writes in the poem 'A World Made by Atoms' (2018, p. 81):

> Small atoms of themselves a world may make,
> For being subtle, every shape they take.
> And as they dance about, they places find;
> Of forms that best agree, make every kind.
> [...]
> So atoms, as they dance, find places fit,
> And there remaining, close and fast will knit.

Although in this early encounter with atomism Cavendish embraced the notion of 'void' as the space necessary for the 'dance' of movement, she later developed a more complex notion of how matter is always, in some sense, enmeshed with other matter (more on that below).

Even so, *Poems and Fancies*, and her other 1653 volume, *Philosophical Fancies*, contain the seeds of that later development, for she had already invoked the seventeenth-century notion of 'sympathy' (the idea that a fundamental power or energy coordinated a reciprocal affinity between things otherwise different by nature) to help account for how atoms could communicate and collectively generate new forms of life.[10] Others who relied on a theory of sympathy in their own work included early renaissance thinkers Marsilio Ficino (1433–1499) and Giovanni Pico della Mirandola, as well as several prominent seventeenth-century writers such as Neo-Platonists Ralph Cudworth and Henry More. Kenelm Digby and Thomas Browne also employed the concept of sympathy in their thinking about science and, like them, Cavendish considered sympathy a natural, material principle, rather than a magical non-material one. A sympathetic universe helped to counter Lucretius's depictions of dead atoms colliding accidentally into life, instead allowing Cavendish to set forth a model defined by sensitive matter's innate vital connections to other sensitive matter.

Indeed, Cavendish insisted that life did not spark from otherwise 'dead' atoms, as Lucretius's poem taught, but from materials that were always already 'vital' (or alive). She derived from this a theory of *purposeful* forms of matter in which all matter is 'self-knowing', as she explained it. To put it another way, Cavendish was not only interested in explaining the world geometrically, according to particles of particular shapes and sizes (as Lucretius and others did), but dynamically, according to sensible forces or energy. However, in contrast to some of her peers in the period—such as Cudworth and More, who believed that the vital energies that animated bodies were nonmaterial and metaphysical—Cavendish considered the vital force to be fully material, a force that her early conception of atomism helped to explain. Even what might be considered 'spirit' or 'soul', according to Cavendish, was really just a more refined or 'thin' form of vital matter. For a period in which many still adhered to gendered notions of spirit (masculine) and body (feminine), this was a startling and potentially paradigm-shifting assertion.[11]

Cavendish also departed from Lucretius on the issue of whether there are particles that constitute absolutes (bits of matter that cannot be divided any further). Rather, she took up the mathematical idea of infinitesimals (an indefinitely small quantity that approaches but never reaches zero) in relation to atoms, insisting in *Poems and Fancies* that atoms might be 'Thinner and less, and less still by degree' (2018, p. 128). That is, particles exist in a range of sizes approaching infinity, and all atoms may be subject to further division. Interestingly, her insistence that all particles can be divided ad infinitum led her to embrace Lucretius's notion of infinite worlds at both the macro and microscopic levels. Her poems 'Of Many Worlds in This World', 'A World in an Earring', 'It Is Hard to Believe That There May Be Other Worlds in This World,' and 'Several Worlds in Several Circles' all embrace the notion of a plurality of worlds, an opinion revived in late sixteenth- and early seventeenth-century science and embraced by Giordano Bruno, Johannes Kepler, René Descartes, Pierre Gassendi, Thomas Digges, and Galileo Galilei.

Poems and Fancies also features Cavendish's interest in what could be termed an early form of ecology, the branch of biology focused on the interactions among organisms and their environment. Part Two of *Poems and Fancies* is especially notable in this regard. Here, through a series of dialogue poems, Cavendish explores her sense that all things are connected. 'A Dialogue betwixt Man and Nature', 'A Dialogue between Earth and Cold', 'A Dialogue betwixt Birds', and several others in this vein, dramatize the interconnectedness of earth's creatures and their landscapes. Moreover, Cavendish's first book also seems to embrace a nascent form of environmentalism, giving voice to creatures whose habitats and lives bear the brunt of brutal and unthinking human behaviour. Two of her most famous poems, 'The Hunting of the Hare' and 'The Hunting of a Stag', give a rich description of the suffering of creatures who are hunted for sport rather than necessity. Other poems, such as 'Of the Knowledge of Beasts', 'Of Fish', 'Of the Ant', and 'Of Birds', posit that animals have intelligence—a position fully congruent with her theory that all matter is intelligent—and insists that humans should have humility in relation to creatures whose knowledge and skills in their own realm often outstrip those of humans. Nor is Cavendish shy about taking her proposition to its logical conclusion; in *Philosophical Fancies* she writes: 'Who knows, but Vegetables and Mineralls may have some of those rationall spirits, which is a minde or soul in them, as well as Man? [...] For had vegetables and Mineralls the same shape, made by such motions as the sensitive spirits create, then there might be wooden men, and Iron beasts.' Although the knowledge of other creatures may be different from that of humans, 'yet it is knowledge', she insists, and 'Vegetables and minerals may know / As Man, though like to Trees and Stones they grow' (2018, pp. 54–56).[12]

The poem 'Earth's Complaint' in *Poems and Fancies* more directly addresses the problems of how humans use the earth: they often destroy habitat for other creatures, she notes, while greedily searching for metals, precious stones, etc., for themselves. She castigates wealthy landowners who ignore the needs of the environment at their own peril, contrasting them with poor folk who seem to live in harmony with nature:

> Rich men, when they for to delight their taste,
> Suck out the juice from th' Earth, her strength they waste.
> For bearing oft, she'll grow so lean and bare,
> That like a skeleton she will appear. (2018, p. 137)

Epicureanism tended towards a radical egalitarianism since all creatures and things were made up of the same fundamental particles. Although Cavendish critiqued many elements of Epicurean atomism, her first books imaginatively explored what this egalitarianism might mean in the broad realm of nature, as well as within the sphere of human society (which is the focus of the final poems of the volume).[13]

Literary Forms and Cavendish's Later Science

After her first two books were published, Cavendish penned two more while still in exile: *The World's Olio* (1655) and *Nature's Pictures* (1656), in which she continued to develop her ideas on science, couching them in creative literary forms. Later, after the restoration of the monarchy in 1660, she and her husband William returned to England. After settling into a new life back in their native land, Cavendish began work on an even wider range of literary endeavours. She published *Orations and Plays* (1662) followed swiftly by two more books, *Sociable Letters* and *Philosophical Letters* (1664), epistolary genres that framed knowledge-making, politics, science, and society as dialogic endeavours; that is, processes that depend on an egalitarian method of argument based on *validity* rather than on a scholastic tradition based on *authority*. Women feature prominently in these works, displaying for her readers the possibility of a world in which half of humankind was no longer forced to remain silent—through lack of formal education and opportunity—on issues at the heart of science and society.

In the year of the great fire of London and a devastating plague (1666), Cavendish brought forth two more works, *Observations upon Experimental Philosophy* (her most detailed science compendium to date) and the *Blazing World* (her science fiction romance, appended to *Observations*).[14] It was the next year, 1667, well after the publication of *Observations upon Experimental Philosophy*, that Cavendish made her famous visit to the Royal Society. Before turning to that moment, a look at some of the major points made in *Observations* is useful for understanding why the Society was willing to make the unusual invitation to Cavendish in the first place, for the Royal Society did not admit female fellows, and this particular female had vigorously critiqued several of the Society's most prominent members the previous year.

OBSERVATIONS UPON EXPERIMENTAL PHILOSOPHY (1666)

Cavendish's interest in particle theory continued, but in the 1660s she more forcefully rejected the classical definition of atoms as irreducible elements of matter: particles were important parts of nature, but their divisibility meant they could not constitute the fundamental and overarching material intelligence of nature. In *Observations upon Experimental Philosophy* (1666) she wrote, 'there can be no atom, that is, an indivisible body in nature; because whatsoever has body, or is material, has quantity; and what has quantity, is divisible'. This did not mean that she had abandoned the work of particles in her theory of nature, for 'parts' are in fact crucial to nature's power of variability. She explains that 'nature is a self-knowing infinite body, divisible into infinite parts' (2001, p. 125). Cavendish also reemphasized her earlier refusal to subordinate matter to form, or body to spirit. As she would insist, 'form cannot be created without matter, nor matter without form; for form is no thing subsisting by itself without matter; but matter and form make one body' (2001, p. 203). In this regard, she carefully delineated her revised notion of nature, which is infinite in variability, divisibility, and expansiveness. Technically, this was not a new element added to her theory (as noted above, *Poems and Fancies* had already embraced each as fundamental to nature), but in *Observations*, nature's power in this regard, as opposed to the freedom of any given atom, was asserted more clearly and advanced as essential for understanding transformation. How does food become part of one's body? How does a seed become a tree? How does healthy tissue become diseased? These and other questions are at the heart of her thinking in *Observations*. In contrast to Descartes and the Neo-Platonists, who each asserted that body is inert and inanimate (and thus in need of vivification by an *immaterial* substance of 'mind' or 'spirit'), Cavendish asserted that all matter in nature is self-knowing and self-moving.[15] More particularly, she insisted that all things are made up of two kinds of matter: the 'inanimate' (or passive) and the 'animate' (or active). Animate matter, in turn, was characterized by two interactive traits: 'sensitive' (having perception) and 'rational' (having reason). These two attributes, taken together, are what make matter intelligent.

While in *Observations* Cavendish grants that each form of matter has its peculiar attributes, she also notes that 'no particle in nature can be conceived or imagined, which is not composed of animate matter, as well as of inanimate' (2001, p. 158). That is, since the animate and inanimate are so thoroughly mixed, there cannot be a separation of the two, another reason why she had to abandon the strict notion of atoms as ultimately indivisible quantities. As Cavendish puts it, 'As infinite nature has an infinite self-motion and self-knowledge, so every part and particle has a particular and finite self-motion and self-knowledge, by which it knows itself, and its own actions, and perceives also other parts and actions' (2001, p. 138). Thus, while nature has *infinite* self-motion and knowledge that embrace all her parts, those various parts each have only *finite* capacities. Cavendish conceptualized this relationship as an ongoing dialogue: nature is in constant dialogue with her parts, and the parts are in dialogue with other related parts. This dialogic model allowed Cavendish to explain how an orderly nature might nevertheless occasionally produce bodily disease, disability, famine, or other recognizable disorders. Even in nature, communication may go awry.

By rejecting the indivisibility of atoms in classical particle theory, Cavendish increased the overall coherence of her own philosophy of science since the complexity of any given particle's nature—which included inanimate, sensitive, and rational elements—could account for nature's overarching harmonies and the general predictability of interactions between entities (2001, pp. 129, 169 and 207–208). In short, she was willing to allow that there are particles in nature that form complex structures, but she also insisted that each particle is (i) a part of nature's greater body; and (ii) divisible and a complete blend of inanimate, sensitive, and rational matter. Any given particle could be split, but each new part would still contain all three essential elements of matter. This allowed nature an infinite capacity for creating new forms (Cavendish would not have been at all uncomfortable with Darwin's eventual theory of evolution; she simply would have insisted that nature might make new forms quickly as well as slowly). Moreover, for Cavendish, all parts of nature 'depend for their existence and properties on their *relation* to each other and to the whole of nature' (2001, pp. 126–127–emphasis mine). While glimpses of this holistic ecology can be caught in the earlier *Poems and Fancies*, it dominates the scene in her later *Observations upon Experimental Philosophy*.

Additionally, *Observations* included Cavendish's thoughts on both the value and limitations of scientific instruments such as microscopes, telescopes and air pumps, and the book gives special attention to the work of Robert Boyle (1627–1691) and Robert Hooke (1635–1703) in this regard. Boyle was famous for his improvements on Otto von Guericke's air pump, and the resulting 'Pneumatical Engine' became the central tool for a series of experiments on the properties of air. Boyle's subsequent book, *New Experiments Physico-Mechanical, Touching the Spring of the Air, and its Effects* (1660) was widely read, and 'Boyle's Law' came about when Boyle provided clarifications in response to a critique of that work. Hooke, whose book on microscopy had been published in 1665, the year before Cavendish's *Observations* came out, was also an extremely influential member of the Royal Society. Hooke's volume, *Micrographia, or, Some Physiological Descriptions of Minute Bodies Made by Magnifying Glasses, with Observations and Inquiries Thereupon* was a sensation. Filled with detailed illustrations, descriptions, and explanations of things viewed under his microscope—including revelatory images of small entities such as a fly, a flea, cork, petrified wood, and the crystals in flint—the volume dramatized the advantages of the magnifying instrument for advancing knowledge of 'minute bodies'.[16]

That Cavendish published her own ideas on science, despite strong social pressures to avoid the 'immodesty' of appearing in print, was bold enough; that she publicly critiqued Boyle and Hooke, two of the most consequential figures of the Royal Society, was equally striking. In the case of Hooke, her criticism had mostly to do with an assumption he made about the nature of reason versus the nature of experimentation. As Cavendish quoted in *Observations*, Hooke asserted that human reason is fallible, and 'the remedies can only proceed from the real, the mechanical, the experimental philosophy' (2001, p. 49). Rather than assume things about a flea based on speculation, Hooke preferred to examine one under the microscope and simply describe what he saw. Theory should only develop after amassing large quantities of such data. Cavendish agreed that reason is imperfect, but disagreed about the superiority of sense over reason, for the senses (such as sight) are also subject to imperfections of both perception and interpretation. In this regard, she felt that Hooke was being naïve on three counts. First, what Hooke referred to as 'our power over natural causes and effects' is limited at best, even when augmented by instruments such as the microscope or telescope. Humans are within nature, as Cavendish insists, and 'they must be as nature is pleased to order them; for man is but a small part' (2001, p. 49). Viewers are not somehow above nature, looking down with a god's eye view. This is not to say that microscopical studies are useless, but rather to question the sudden assumption that humans would now have increased power over nature. After all, Cavendish insists, human 'powers are but particular actions of nature, and therefore […] cannot have a supreme and absolute power' (2001, p. 49). According to Cavendish, Hooke's perception of what he is accomplishing is out of scale in relation to the depths of intricacy of any given part of nature, even with respect to something as small as a flea.

Secondly, Cavendish did not believe in what Hooke called 'a first groundwork, which ought to be well laid on the sense and memory', by which he asserts that 'speculative philosophy' must properly follow after experiments. Cavendish countered that every experiment must have had a rationale (a reason) underpinning its method in the first place, and that reason (whether strong or weak), would thus continue to guide both the process and the subsequent interpretation of the information gathered. 'And hence I conclude', she explains, 'that experimental and mechanic philosophy cannot be above the speculative part, by reason most experiments have their rise from the speculative' (2001, p. 49). Finally, she notes that the microscope cannot mend senses that 'be defective, either through age, sickness, or other accidents'. Someone with cataracts looking through a microscope, for instance, will see a distortion of the figure presented, no matter how fine the lens is. Similarly, a less than stellar intellect may quite easily misconstrue, and thus misrepresent to others, what is under the microscope. Elsewhere, she notes other potential problems: the observer might begin to think that surface views are more representative of the creature under observation than they really are. Even with anatomization, a theory of the forces that makes movement, digestion, breathing, and other life-sustaining processes possible, would require much more than the information garnered from the microscopic study of internal organs and tissues. Above all, she insisted, taking things out of their natural habitat to pin them under a lens distorts the possibility of knowing their true nature.[17] As the example of Hooke demonstrates, Cavendish's critiques of experimental philosophy in *Observations upon Experimental Philosophy* were vigorous, useful, and in several cases, seem to have anticipated issues of context and perspective more commonly addressed in later periods.

Current scholarship does not address how Hooke may have taken Cavendish's criticism. Was he, like Huygens, interested in her thinking? Or did he consider her criticism beneath him, not worth considering? Cavendish seems to have seen herself as a peer-respondent whose practical observations would be of use to Hooke and others. Male-to-male critiques were understood to be part of the process of refining knowledge, and lively responses to a given philosopher's position were not only common to the Royal Society but had always been a feature of the 'republic of letters' (the international intellectual community of correspondents) as well. Cavendish decided to participate via her published works. And while it was the case that occasionally criticism of one philosopher by another would become the intellectual equivalent of fisticuffs (as in the Hobbes-Wallis controversy), or arguments might arise as to who really conceived of a new idea (as in Newton's excoriation of Hooke for daring to suggest a hand in understanding the gravitational force), Cavendish does not seem to have stirred such passions. Indeed, good evaluative responses to one's experimental assays were expected to be rigorous, for without robust responses to one's theories of nature, large blunders were much more likely.[18] In this regard, it is worth noting that while Cavendish's critiques of various philosophers were certainly robust, and sometimes included a hint of humorous sarcasm, a dollop of playful parody, or a dash of hyperbole, they were not mean-spirited, and she often admits that she might be wrong. Indeed, she frequently augments her observations with humour, meant both to underscore her point and to lessen tensions arising from intellectual conflict. Moreover, she did not, as some have suggested, simply dismiss experimental science and the use of technical instruments. With regard to Hooke's work, for instance, there are several places in *Observations* where she engages with images and explanations from his *Micrographia* in order to expand upon her own theory of matter and transformation in nature. A fine example is 'Of Wood Petrified' where she is responding to the segment in Hooke's volume, 'Of Petrifyed Bodies'. There, she suggests how her own theory of matter helps to explain how one kind of material might be replaced incrementally by that of another without change in outward appearance (2001, pp. 90–91).[19]

The Blazing World

The Royal Society—a body of gentlemen devoted to expanding knowledge of the natural world and experimenting with the methods and devices suited to that endeavour—was not yet the authoritative institution it would come to be. The Society received its Royal Charter in 1662 and was sometimes lampooned, and at other times singled out for praise and emulation. In 1666, the year before her famous visit, Cavendish had a bit of fun with the Society's members: in *The Blazing World*, the fictional work appended to *Observations*. Here she refracted her evaluations of their experimental methods through the lens of a three-part philosophical romance, which included humorous parodies of herself as well as caricatures of key fellows of the Royal Society. Cast as animal men, these fellows were mirrored in the very creatures Cavendish had praised in her science treatises as having their own peculiar intelligence, a means of underscoring her contention that humans—even astute philosophers of science—are merely part of nature's larger landscapes. We learn that the Empress-heroine of the story assigns each animal a job suitable to its own inclinations: 'The Bear-men were to be her Experimental Philosophers, the Bird-men were to be her Astronomers, the Fly- Worm- and Fish-men her Natural Philosophers, the Ape-men her Chymists, the Satyrs her Galenick Physicians, the Fox-men her Politicians, the Spider- and Lice-men her Mathematicians [and so on]' (2016, p. 71). In other words, Cavendish's engagement with the natural philosophers and experimentalists of her day was not limited to *Observations* but continued in a lively new literary form—science fiction— meant to give pleasure to imaginative readers. Indeed, most of those who picked up *The Blazing World* would have immediately recognized the allusions to famous figures of the Royal Society. Given that the novel was appended to the science treatise in which Cavendish had explained the problems she saw with the methods employed by both Boyle and Hooke, the two-volume set basically provided two different genres: serious scientific disquisition on the one hand, and humorous science fiction on the other, for exploring problematic methodologies at the heart of the new science.

The Blazing World is especially notable as a proto-science fiction novel. It begins with a merchant and his crew kidnapping a 'young lady', setting out to sea with her, and getting blown to the North Pole. Captain and crew there freeze to death, but the warmth of the lady's beauty aids her in surviving not only the frigid turbulence, but also the shift from her own world to another, for the boat was 'driven to the very end or point of the Pole of that world, but even to another Pole of another world, which joined close to it' (2016, p. 61). In moving from one sphere to the next, the Lady not only encounters a new realm of experience, she also quickly ascends to the pinnacle of society, becoming an empress possessed of 'an absolute power to rule and govern all that world as she pleased' (2016, p. 70). As already noted, key questions from *Observations upon Experimental Philosophy* re-emerge in this fictional format, and the plot of *Blazing World* revolves around the newly appointed empress's engagement with science. She participates in discussions on telescopes, microscopes, algebraic geometry, and theories of state. She organizes learned societies, calibrates the kingdom's religion to better accommodate the needs of all, and ponders the possibility of building new worlds. As Sara Mendelson points out in her introduction to *Blazing World*, the 'overall effect [...] is to subvert contemporary notions about gender hierarchy in the spheres of rational thought and literary creation'. Moreover, by 'insinuating current scientific and philosophical concerns into her flights of fancy, Cavendish challenges the hegemony of male rationality and deflates masculine pretensions to intellectual superiority' (2016, pp. 22–23). *The Blazing World* was a realm in which a female protagonist could stumble yet excel.

Several of the fellows of the Royal Society felt the sting of Cavendish's criticisms, but others were eager to extend her an invitation in response to her query of interest. Several of those with whom she had corresponded directly, or with whom she had engaged indirectly in her published works, understood her attention to be a boon to the Society on a variety of levels (putting on a good face in response to her critiques, public advertisement, possible patronage, and so on). The visit was well attended and well documented: she arrived with an elaborate train, her philosophical correspondent Walter Charleton gave an oration for the occasion, and Robert Boyle, Robert Hooke, and others performed demonstrations for her, each duly noted in the Society's official record.[20]

Conclusion: Cavendish on Education for Women

Cavendish understood that women's lack of university education was a severe liability when broaching topics of science and philosophy, and that English society was not yet ready to broach the issue. As she wrote in a preface to *Poems and Fancies* in 1653, 'But I imagine I shall be censured by my own sex', and men 'will cast a smile of scorn upon my book because they think thereby women encroach too much upon men's prerogatives; for they [men] hold books as their crown, and the sword as their scepter, by which they rule and govern' (2018, p. 61). Even so, she was determined to move forward in her chosen field. Indeed, by the time of her death in 1673, Cavendish had published numerous volumes over the course of her twenty-year career as an author and philosopher of nature. Parallel to her ideas on matter and energy, on ecology and the environment, and on intelligence and perception, is a persistent observation about the lack of educational opportunities for women. In one of her early books, *Philosophical and Physical Opinions* (1655), Cavendish wrote a dedication, 'To the Two Universities.' Her point is worth quoting at length, for she urged professors to receive her book:

> without a scorn, for the good encouragement of our sex, lest in time we should grow as irrational as idiots, by the dejectedness of our spirits, through the careless neglects, and the despisements of the masculine sex to the effeminate, thinking it impossible we should have either learning or understanding, wit or judgment, as if we had not rational souls as well as men, and we out of a custom of dejectednesse think so too, which makes us quit all industry toward profitable knowledge, being imployed only in looe [low], and pettie imployments, which takes away not only our abilities towards arts, but higher capacities in speculations.

Similar protests limn her later works as well, and collectively help to explain why Cavendish would be quoted by an undisputed early feminist, Bathsua Makin (1600–1675).

Makin spent her early years as both a student and teacher at her father's school in London. She was well known as a poet and medical practitioner and returned to teaching and writing after raising her children. Importantly, she published the first rigorous defence of women's right to an education, *An Essay to Revive the Antient Education of Gentlewomen* (1673), where Makin declared of Cavendish, 'The present Duchess of Newcastle, by her own genius rather than any timely instruction, over-tops many grave grown-men' (Teague and Ezell 2016, p. 60). Obviously, Makin was not arguing that education was unnecessary for women since they might attain to Cavendish's stature through self-education. Rather, she claimed that the famous Duchess might have achieved even more, had Cavendish received the level of education accorded her male peers. The point, for both Cavendish and Makin, is clear: women's added perspectives and experiences (were women to receive a good education) would prove to be a collective benefit to all knowledge-making endeavours, including the study of the natural world and human nature.

NOTES

1. Several of Cavendish's volumes went through subsequent revised editions. For an overview of her publications, see introduction to *Poems and Fancies with The Animal Parliament,* ed. Brandie R. Siegfried (2018). For a thorough extended analysis of Cavendish's natural philosophy, see Sarasohn (2010).
2. Selections of Cavendish's literary skills are showcased in Bowerbank and Mendelson (2000); see also Cavendish (1994).
3. For an extended analysis of gender in relation to Cavendish's science and politics, see Walters (2014).
4. Digby's writing was extremely influential. His *Two Treatises*, which included 'On Bodies' and 'On the Soul', went through several printings between 1644–1669. 'On Bodies' is especially worth reading in relation to Cavendish's work on a range of similar topics. In *Observations upon Experimental Philosophy* [1666] (2001), Cavendish responds to Digby's 'On Bodies' in several places, including 'Of Natural Productions' (pp. 66–68), 'Of the Nature of Water' (pp. 91–95) and 'Of Chemistry and Chemical Principles'(pp. 227–239).
5. See Charleton's *The Darkness of Atheism Dispelled by the Light of Nature* (1654a), *Physiologia Epicuro-Gassendo-Charltoniana* (1654b), which followed Cavendish's 1653 volume of atomic poems, and *The Immortality of the Human Soul Demonstrated by the Light of Nature* (1657). For more on how Cavendish's science fits within similar questions of religion, see Siegfried and Sarasohn (2016).
6. Glanvill was also interested in her views on spirits, the pre-existence of the human soul, the possibility of a 'world soul', the likelihood of innate ideas, and the compatibility of science and scripture. Her responses can be traced in the essays included in Siegfried and Sarasohn (2016), in particular *Observations upon Experimental Philosophy* [1666] (2001) in the segments 'Of Natural Sense and Reason' (pp. 215–216) and 'Of the Rational Soul of Man' (pp. 221–224).
7. There is some argument as to whether the period saw a 'scientific revolution', or whether it was rather a scientific 'continuation' and 'evolution', especially given the many experiments and technological advancements by classical thinkers such as Archimedes.
8. For more on Cavendish's ideas about religion in relation to her theories of nature, see Siegfried and Sarasohn (2016).
9. Epicureanism taught that the highest form of pleasure consisted in living modestly, gaining knowledge of nature, and developing tranquility. On the question of detachment, Cavendish expressed special concern about painful experiments on living creatures and the moral implications of such.
10. For a survey and analysis of theories of sympathy in the seventeenth century see Lobis (2015).

11. A slightly different feminist strategy may be observed in the work of the French Cartesian Poullain de la Barre. René Descartes' proposed split between mind and body led de la Barre to further assert that female intelligence and capacity was equal to that of men. For a lengthy discussion of this topic in relation to women and science see Schiebinger (1991b).
12. *Philosophical Fancies* ends with a description of a possible alien world, where alternative ecologies are in play, and where other kinds of beings—such as 'a woman out of flowers' and 'a man out of metals'—are as sentient and 'natural' as humans are in ours.
13. For discussions of the ecological and environmental strands in Cavendish's work, see Merchant (1980); Schiebinger (1991); and Bowerbank (2004).
14. Further publications include *The Life of William Cavendish* (1667); *Plays Never Before Published* and *Grounds of Natural Philosophy* (1668); new editions of *Poems and Fancies* and *Blazing World* (also in 1668), and second editions of *Nature's Pictures* and *The World's Olio* (1671).
15. Descartes was not a strict Neo-Platonist, but he did share with that movement the sense that 'matter' or ('body') and 'mind' (or 'spirit') were radically distinct things.
16. For more on Hooke, see Chapman (2004).
17. Note that Cavendish engages Hooke, as well Descartes, Boyle, Power, and Glanvill, throughout *Observations upon Experimental Philosophy*. Of special note here is that in the segment 'Of Natural Senses and Reason' she works through the paradox of having to use reason to critique reason, a clear indication of her willingness to take seriously critiques of her own position.
18. This period's literature of natural philosophy reveals the extent to which all of the major philosophers developed what now appear to be blunders, even as in other instances they made significant discoveries. Cavendish was quite aware that her own work probably suffered from a few blunders (it did), which is why she preferred a stance of epistemological probability rather than certainty in her writing.
19. See *Observations upon Experimental Philosophy* ([1666] 2001, pp. 90–91) for both Cavendish's explanation of petrified wood and an example of her use of humour. There she jokes that 'if all creatures could or should be metamorphosed into one sort of figure, then this whole world would perhaps come to be one stone, which would be a hard world'.
20. For an appraisal of Cavendish's visit to the Royal Society, in the context of that several early fellows shared with her certain reservations concerning particular scientific methods, see Wilkins (2014).

References

Akkerman, Nadine, and Marguérite Corporaal. 2004. "Mad Science Beyond Flattery: The Correspondence of Margaret Cavendish and Constantijn Huygens." In *Early Modern Literary Studies*, no. 1.

Boyle, Robert. 1660. *New Experiments Physico-Mechanical, Touching the Spring of the Air, and its Effects*. London.

Bowerbank, Sylvia, and Mendelson, Sara. 2000. *Paper Bodies: A Margaret Cavendish Reader*. Toronto: Broadview Press.

Bowerbank, Sylvia. 2004. *Speaking for Nature: Women and Ecologies of Early Modern England*. Baltimore: Johns Hopkins University Press.

Cavendish, Margaret. 1653. *Philosophical Fancies*. London: J. Martin and J. Allestrye.

Cavendish, Margaret. [1653] 2018. *Poems and Fancies with The Animal Parliament*, ed. Brandie R. Siegfried. Toronto: Iter Academic Press.

Cavendish, Margaret. 1655a. *Philosophical and Physical Opinions*. London: J. Martin and J. Allestrye.

Cavendish, Margaret. 1655b. "To the Two Universities." In *Philosophical and Physical Opinions*. London: J. Martin and J. Allestrye.

Cavendish, Margaret. 1664. *Philosophical Letters*. London.

Cavendish, Margaret. [1666] 1994. *Margaret Cavendish: The Blazing World and Other Writings*, ed. Kate Lilley. London: Penguin.

Cavendish, Margaret. [1666] 2001. *Observations upon Experimental Philosophy*, ed. Eileen O'Neill. Cambridge: Cambridge University Press.

Cavendish, Margaret. [1666] 2016. *The Description of a New World, Called the Blazing World*, ed. Sara H. Mendelson. Ontario: Broadview Press.

Cavendish, Margaret. 1668. *Grounds of Natural Philosophy*. London.

Chapman, Allan. 2004. *England's Leonardo: Robert Hooke and the Seventeenth-Century Scientific Revolution*. Bristol: CRC Press.

Charleton, Walter. 1654a. *The Darkness of Atheism Dispelled by the Light of Nature*. London: W. Lee.

Charleton, Walter. 1654b. *Physiologia Epicuro-Gassendo-Charltoniana*. London: Thomas Heath.

Charleton, Walter. 1657. *The Immortality of the Human Soul Demonstrated by the Light of Nature*. London: Henry Herringman.

Conway, Anne. 1992. *The Conway Letters: The Correspondence of Anne, Viscountess Conway, Henry More, and their Friends 1642–1684, Revised Edition*, ed. Marjorie Hope Nicholson and Sarah Hutton. Oxford: Clarendon Press.

Del Soldato, Eva. 2020. "Natural Philosophy in the Renaissance." In *Stanford Encyclopedia of Philosophy*, ed. Edward N. Zalta. https://plato.stanford.edu/archives/fall2020/entries/natphil-ren/. Accessed November 30 2020.

Foster, Michael. 2009. "Digby, Sir Kenelm (1603–1665)." *Oxford Dictionary of National Biography*. https://doi-org.liverpool.idm.oclc.org/10.1093/ref:odnb/7629. Accessed November 30 2020.

Glanvill, Joseph. 1666. *A Philosophical Endeavor towards the Defense of the Being of Witches and Apparitions*. London: James Collins.

Glanvill, Joseph. 1661. *The Vanity of Dogmatizing*. London: Henry Eversden.

Gosse, Edmund William. 1911. "Huygens, Sir Constantijn." In *Encyclopedia Brittannica*, vol. 14: 22. Cambridge: Cambridge University Press.

Henry, John. 2004. "Charleton, Walter (1620–1707)." *Oxford Dictionary of National Biography*. https://doi-org.liverpool.idm.oclc.org/10.1093/ref:odnb/5157. Accessed on 30 November 2020.

Hooke, Robert. 1665. *Micrographia, or, Some Physiological Descriptions of Minute Bodies Made by Magnifying Glasses, with Observations and Inquiries Thereupon*. London: Royal Society.

Lobis, Seth. 2015. *The Virtue of Sympathy: Magic, Philosophy, and Literature in Seventeenth-Century England*. New Haven: Yale University Press.

Merchant, Carolyn. 1980. *The Death of Nature*. San Francisco: Harper and Row.

Sarasohn, Lisa T. 2010. *The Natural Philosophy of Margaret Cavendish: Reason and Fancy during the Scientific Revolution*. Baltimore: Johns Hopkins University Press.

Schiebinger, Londa. 1991a. "Margaret Cavendish, Duchess of Newcastle." In *A History of Women Philosophers*, vol. 3, edited by Mary Ellen Waithe. Dortrecht: Klewer Academic Publishers.

Schiebinger, Londa. 1991b. *The Mind Has No Sex? Women and the Origins of Modern Science*. Cambridge, MA: Harvard University Press.

Shapin, Steven, and Schaffer, Simon. 2017. *Leviathan and the Air Pump: Hobbes, Boyle, and the Experimental Science* (revised edition). Princeton: Princeton University Press.

Siegfried, Brandie R., and Sarasohn, Lisa T., eds. 2016. *God and Nature in the Thought of Margaret Cavendish*. Burlington, VT: Ashgate.

Teague, Frances, and Ezell, Margaret J.M., eds. 2016. *Educating English Daughters: Late Seventeenth-Century Debates*. Toronto: Iter Academic Press.

Walters, Lisa. 2014. *Margaret Cavendish: Gender, Science, and Politics*. Cambridge: Cambridge University Press.

Whitaker, Katie. 2002. *Mad Madge: The Extraordinary Life of Margaret Cavendish, Duchess of Newcastle*. New York: Basic Books.

Wilkins, Emma. 2014. "Margaret Cavendish and the Royal Society." *Notes and Records of the Royal Society of London* 68, no. 3: 245–60.

CHAPTER 3

Navigating Enlightenment Science: The Case of Marie Geneviève-Charlotte Darlus Thiroux D'Arconville and Gabrielle Émilie Le Tonnelier De Breteuil and the Republic of Letters

Leigh Whaley

> I feel the full weight of prejudice that excludes us [women] so universally from the science...[1]

This chapter highlights key strategies utilized by two leading Enlightenment polymaths, Gabrielle Émilie Le Tonnelier de Breteuil Du Châtelet-Lomont (1706–1749) and Marie-Geneviève-Charlotte Darlus Thiroux d'Arconville (1720–1805), to practise science in a predominantly unreceptive cultural context in which the ideal woman was passive and domestic.[2] At a time when women were excluded from formal educationand membership of scientific academies, the Marquise du Châtelet and Madame Thiroux d'Arconville worked creatively around existing cultural norms and institutions which enabled them to make significant contributions in their respective fields of mathematics and physics (Du Châtelet) and chemistry and anatomy (D'Arconville).

Barriers to a career in science during the Enlightenment included the lack of a formal education (women were barred from attending universities in France), and the contemporary construction of women and gender roles. Female exclusion from prestigious scientific institutions such as the Royal Academy of Sciences, a premier scientific establishment during the Enlightenment, ensured women's marginalization from professional science. Membership of

L. Whaley (✉)
History and Classics, Acadia University, Wolfville, NS, Canada
e-mail: leigh.whaley@acadiau.ca

© The Author(s), under exclusive license to Springer Nature Switzerland AG 2022
C. G. Jones et al. (eds.), *The Palgrave Handbook of Women and Science since 1660*, https://doi.org/10.1007/978-3-030-78973-2_3

the Academy guaranteed that one's scientific endeavours would be taken seriously rather than perceived as dilettantism. In addition, academicians met with other fellows exchanging the latest discoveries and scientific concepts. As members of this exclusive group, publication of one's research was ensured. Tradition rather than statutes banned women from the Academy and this tradition persisted until 1962 (Whaley 2003; Petrovich 1999; Schiebinger 1989). In the eighteenth century, women were limited to the position of observers at the public lectures which took place bi-annually (Arianrhod 2012). Du Châtelet once remarked that it was easier for women to gamble, than to enter the Academy of Sciences (Watts 2007, p. 71).

A less formal venue where scientific theories were debated was the café; however, even there, women were prohibited from entering unless they were servers. Thus, women were excluded from participating in important scientific exchanges in two key venues, the academy and the café. When Du Châtelet attempted to join her tutor, Maupertuis, at a café named Gradot located on the Quai du Louvre, which was popular with scientists, philosophers, and mathematicians, she was barred from entering. Rather, Du Châtelet was forced to wait in her carriage for him (Zinsser 2006). Women were forced therefore to be creative if they wanted to practise science. Several identifiable strategies were key to Du Châtelet's and D'Arconville's ability to contribute to the conversation and advancement of science. Their strategies included networking in the informal gatherings of their private homes and conducting experiments in private laboratories, often located in their residences.[3] The acquisition of an education and having access to a good library, and collaboration with key male scientists in informal networks and communities, were also crucial to navigating science. Both women had at least one male mentor who collaborated and encouraged their investigations and publications. In Du Châtelet's case it was Voltaire, while D'Arconville was both encouraged and assisted by Pierre-Joseph Macquer and Bernard de Jussieu. Anonymous publication, which involved translation and the production of one's own treatises, and writing one's own rather than a collaborative text, were also employed to a varying degree. Although Du Châtelet conducted experiments with Voltaire, she often wrote clandestinely.

Enlightenment Cultural Norms: The Nature of Woman

To fully understand the circumstances which faced women who desired to study and contribute to science, the culturally held norms concerning women's roles and education must be considered. This section will therefore briefly review the context in which these women attempted to practice science. The period of the cultural revolution of the eighteenth century, commonly known as the Enlightenment, was not a time for the progress of women's rights. The debate over the different natures of the sexes intensified during the so-called 'Age of Reason'. During the period of the 'High Enlightenment', a new corpus of literature was produced which demanded finer definitions and boundaries of sex differences than had been articulated in the past. Previous debates concerned women's moral value and were expressed in the light of Christianity. Enlightenment debaters stressed reason and secularism. As historian of science, Londa Schiebinger has articulated, the arguments concerning the nature of women which began to be stated in a so-called 'impartial' or 'scientific' manner during the eighteenth century, greatly impacted 'women's relationship to science' (1986, p. 43).

The late eighteenth century produced three general perspectives about women's nature and abilities: (i) women were mentally and socially inferior to men (Rousseau and Kant); (ii) women were capable of rational thought, but no questioning of contemporary gender roles (Voltaire); and (iii) women were potentially equal in both mental ability and contribution to society (Condorcet, d'Alembert and Helvétius) (Whaley 2003, pp. 117–138). Leading the debate was Jean-Jacques Rousseau, whose anti-feminist writings were highly influential. According to Rousseau, women's roles were to devote herself to motherhood and to the service of her male partner. Philosophes, including those who edited and contributed to the leading publication of the age, the *Encyclopédie*, echoed Rousseau's sentiments. These included men like Montesquieu and Diderot. In Germany, Immanuel Kant took up Rousseau's views. Women, Kant argued, had a different kind of mind than men, a mind in which abstract thinking was impossible. Writing that 'a woman who [...] carries on fundamental controversies about mechanics, like the Marquise de [sic] Châtelet, might as well even have a beard', Kant concluded that 'woman will learn no geometry' (Clack 1999, p.147).

When discussing views on the nature of woman and her role, the most influential thinker of the age, Rousseau, put forth a theory of complementarianism: 'Yet where sex is concerned man and woman are unlike; each is the complement to the other' (1974, p. 321). Rousseau presented a binary understanding of gender, maintaining that women were 'weak and passive and timid' while men were 'strong and active' (p. 328). Unable to think in abstract terms, there was no place for women in science. According to Rousseau (p. 327):

> The search for abstract and speculative truths, for principles and axioms in science, for all that tends to wide generalization, is beyond a woman's grasp; their studies should be thoroughly practical. It is their business to apply the principles discovered by men, it is their place to make the observations which lead men to discover those principles [...] the works of genius are beyond her reach, and she has neither the accuracy nor the attention for success in the exact sciences; as for the physical sciences, to divide the relations between living creatures and the laws of nature is the task of that sex which is more active and enterprising, which sees more things, that sex which is possessed of greater strength and is more accustomed to the exercise of that strength. Woman, weak as she is and limited in her range of observation, perceives and judges the forces at her disposal to supplement her weakness, and those forces are the passions of man.

Voltaire's attitudes towards women's participation in the male domain of science were more complex than those of Rousseau. He often made contradictory remarks. On the one hand, he celebrated Du Châtelet's intellectual capabilities. This is clear in his Epistle to Madame la Marquise du Châtelet at the start of his tragedy *Alzire*, performed in 1736, where he enthused over her 'genius for geometry' and downplayed his own writings in the arts. He echoed these positive sentiments in his letter at the start of their joint *Éléments de la Philosophie de Newton* where he credited her with kindling his interest in the sciences and lauded her 'vast powerful mind' (Voltaire 1738). On the other hand, he seemed to cling to the notion that science was a man's pursuit; Voltaire wrote in his Preface to her translation of Newton's *Principia*, that the translation should have been made by 'the most learned men of France', but 'surprisingly to the glory of her country, it had been achieved by a woman, Gabrielle Émilie de Breteuil, Marquise du Châtelet'. Questioning the female aptitude for mathematics, he continued, 'It is much for a woman to know simple geometry [...] We have seen two miracles; one that Newton wrote this work; the other, that a lady has translated and explained it' (Newton 1759). Moreover, echoing Rousseau, 'it is not good for a woman to abandon her duty to cultivate sciences' (Voltaire 1736). Her physical beauty was not designed for the roughness of science. Voltaire described Du Châtelet's 'lovely hands' that were not intended for 'a mathematician's compass or a physicist's lens, nor such charming eyes for observing the orbit of a planet' (Zinsser 2007, p. 79). Writing to Frederick II approximately a month after her death in 1749, Voltaire referred to the woman he had loved so intensely as 'a friend of twenty-five years, a great *man*, who had the defect to be a *woman*' (my italics) (Voltaire 1970, p. 179).

Unlike Voltaire's capriciousness, Condorcet's constant support for female scientists never wavered. Condorcet was one of the few Enlightenment thinkers who recognized that women were first and foremost human beings before they were wives and mothers (1847–1849, vol. 7). He criticized contemporaries such as Rousseau who focused on the physical differences between men and women. Such philosophers exaggerated these differences and consequently 'assigns to each sex their roles, their prerogatives, their occupations, their duties and almost their desires, pleasures and feelings' (1847–1849, vol. 9, p. 630). Condorcet blamed that lack of genius and great discoveries by women on their poor education which did not focus on the development of women's thinking faculties. Women, he alleged, were endowed with reason. Their minds had not been put to the most 'enlightened use' when they were developing. This was 'an obstacle to the advance of women in the arts and sciences' (Condorcet 1976, p. 26).

Scientific Education for a Woman

Despite the almost hegemonic anti-feminist ideology which produced numerous barriers to women's participation in science, there were a few exceptional women such as Du Châtelet and D'Arconville who made significant contributions to science.[4] With education as a prerequisite to scientific research and writing, what were the methodologies utilized by Du Châtelet and D'Arconville to gain access to scientific training? Although both women were from privileged, wealthy, and aristocratic backgrounds, that did not necessarily mean they received an education fit for scientific research and writing. Indeed, in the case of D'Arconville, the daughter of a government financier, or tax farmer, and king's secretary André-Guillaume Darlus d'Arconville, an academic education was non-existent.[5] Having lost her mother at age four and a half, the young Marie-Geneviève was raised primarily by a woman who D'Arconville described as 'a governess incapable of raising me'.[6] Marie-Geneviève's education was focused on domestic rather than academic skills in preparation for an arranged marriage at age fourteen to Louis-Lazare Thiroux d'Arconville, a wealthy councillor and later President of the Paris Parlement (the Supreme Court in France's *ancien régime*) and the Chambre des Enquêtes (French Chamber of Inquiries). By the age of twenty she had given birth to three sons in succession born in 1736, 1738, and 1739 (Candler Hayes 2012). D'Arconville later recounted that although her father loved her tenderly, he did not attend to her schooling. She did not learn to read until she was eight years old and only at that age because she had asked her father to be taught letters (Bret and Van Tiggelen 2011, preface).

Lacking an education as a child, D'Arconville was largely self-taught as an adult. At age twenty-two she fell ill from smallpox and almost died; this unfortunate experience changed the course of her life. After her bout with smallpox, she resolved to pursue an academic education starting with the study of English, Latin, Italian, and German (Gargam 2012). The expertise she developed in foreign languages, English in particular, would prove fruitful for her later publications, many of which were translations of scientific texts by English physicians. Once she had gained competency in foreign languages, D'Arconville began to frequent a leading centre of learning in Paris, the *Jardin du Roi* (Cunningham 2010, pp. 98–99). At the *Jardin du Roi* she studied natural sciences and interacted with prominent men of science. Founded in 1626 by Louis XIII, also known as the *Jardin Royal des Plantes*, or *Jardin Royal*, it was an important centre of scientific teaching, particularly in medicine and pharmacy. Here, the King's physicians taught courses to future doctors and apothecaries. Science courses were open to lay persons, including women, from 1740. To reach as broad an audience as possible, the courses were free, paid for by the monarchy, and delivered in French rather than Latin. There were often 500 to 600 persons in attendance (Cunningham 2010; Schiebinger 1995). The courses focused on botany, anatomy, and chemistry and, although rigorous, they lacked examinations and diplomas (Laissus and Torlais, 1986 p. 300).

D'Arconville's experience at the *Jardin du Roi* was a unique opportunity for her to learn outside the home. Indeed, she took advantage of this new educational opportunity by studying physics, anatomy, botany, and natural history and, most importantly given her keen interest in chemistry, she had an opportunity to study with chemist François-Guillaume Rouelle who was a *démonstrateur en chimie* at the *Jardin* from 1742 and delivered courses there until 1768 (De La Porte 1835; Bardez 2011, p. 39). Rouelle taught chemistry to a number of famous eighteenth-century chemists, most notably Antoine-Laurent de Lavoisier (Lemay and Oesper 1954). The role of demonstrator was officially to reinforce the material taught by the lecturers, who at this time were chemists Louis-Claude Bourdelin, court physician to Marie-Antoinette, Paul-Jacques Malouin, and Pierre-Joseph Macquer (Shaw 1759, p.iv).

Du Châtelet fared much better when it came to education. Although many details about Du Châtelet's education remain unknown, Du Châtelet expert, Theodore Besterman, concluded that her education was 'unique in the eighteenth century' (1967, pp. 178–179). She was fortunate to receive the best education a girl could hope for at this time mainly due to the efforts of her father, the baron Louis-Nicolas le Tonnelier de Breteuil, a wealthy nobleman, who was Chief Protocol for Louis XIV (Barber 2006, p. 5). Émilie had five older brothers, the younger of which was destined for the Church rather than the army. He required a first-rate education, so Émilie was able to share in his lessons and thus receive a much superior education than was the norm for girls of the nobility who typically would study religion and manners (Zinsser 2016). Rather than sending the young Émilie to a convent, which was the custom for the daughters of the nobility, she was tutored at home focusing on Latin and literature (Kawashima 2004). Voltaire recounted that Du Châtelet 'knew by heart the beautiful pieces of Horace, Virgil and Lucretius; all of Cicero's philosophical works were familiar to her'. Further, he wrote that she also read English and Italian, but that her preference was for 'physics and metaphysics' (Voltaire 1965, pp. 20–21). Her father provided a conducive home environment for learning, allowing her to receive instruction in astronomy at age ten from Bernard Le Bovier de Fontenelle, a friend of Émilie's father who was perpetual secretary of the Royal Academy of Sciences (Bodanis 2006).

Du Châtelet stressed the importance of an education in several of her writings. In the Preface to her major work, *Institutions de Physique* (translated as *Foundations of Physics*), a text dedicated to her son, she explained:

> I have always thought that the most sacred duty of men was to give their children an education that prevented them at a more advanced age from regretting their youth, the only time when one can truly gain instruction. You are, my dear son, in this happy age when the mind begins to think and when the heart has passions not yet lively enough to disturb it. (2009, p. 116)

On her own education and contemporary prejudice about women and education, she wrote, 'I am convinced that many women are either ignorant of their talents because of the flaws in their education, or bury them out of prejudice and for lack of a bold spirit. What I have experienced myself confirms in me this opinion' (2009, p. 49). Further, because women had so few options open to them education was crucial:

> Undeniably, the love of study is much less necessary to the happiness of men than it is to that of women. Men have infinite resources for their happiness that women lack. They have many means to attain glory, and it is quite certain that the ambition to make their talents useful to their country and to serve their fellow citizens, perhaps by their competency in the art of war, or by their talents for government, or negotiation, is superior to that which one can gain for oneself by study. But women are excluded, by definition, from every kind of glory, and when, by chance, one is born with a rather superior soul, only study remains to console her for all the exclusions and all the dependencies to which she finds herself condemned by her place in society. (2009, p. 357)

Although Du Châtelet's early education was provided by her father, according to her chief biographer, she was mostly an autodidact, similar to D'Arconville (Zinsser 2018). As an adult, she sought to improve her understanding of physics and mathematics by employing several tutors. Nevertheless, neither Du Châtelet nor D'Arconville was educated for a public scientific career in the same systematic way as their male counterparts.

Personal Networks of Correspondents and Researchers

Perhaps the most important strategy employed by both women was to cultivate friendships with learned men and to develop a personal network of correspondents and researchers.[7] Personal relationships with tutors and mentors facilitated scientific opportunities otherwise closed to women. In addition, these relationships facilitated access to good libraries which was key to scientific research. With prestigious academies, the sources of knowledge, power, and networking, closed to women, the *Jardin du Roi* provided not only educational instruction for D'Arconville, but also the venue where she formed key friendships which would determine the future of her scientific writings (Bardez 2011, p. 21). The community of the *Jardin du Roi* was composed of a multi-disciplinary group of scientists including anatomists, physicians, botanists, chemists, and pharmacists led by Macquer, Jussieu, Lavoisier, Antoine François de Fourcroy, Bernard-Germain de Lacépède, François Poulletier de la Salle, Rouelle, and Jean-Baptiste Sénac.[8] These men composed a 'Republic of Science' referring to 'the loosely knit community of scientists', most of whom were also members of the Academy of Sciences in the late seventeenth and eighteenth centuries (Hahn 1971).

D'Arconville's social and intellectual life revolved around many of these famous personalities, the greatest scientists of the Enlightenment, in particular chemists Lavoisier and Fourcroy, in addition to Macquer under whom she studied and with whom she corresponded. Poulletier de la Salle, the discoverer of cholesterol, corresponded with D'Arconville aspects of putrefaction.[9] With her community of teachers and mentors, D'Arconville created what Margaret Carlyle called a 'private research team' (Carlyle 2011, p. 80). This collaborative effort would eventually lead to D'Arconville's translation and correction of Shaw's *Chemical Lectures* (1759) and Alexander Monro's (primus) *The Anatomy of the human bones, and nerves* (1759), in addition to the production of her own major work, *Essai pour servir à l'histoire de la putréfaction* (1766).

With the assistance of the scientists at the *Jardin*, and botanist Jussieu in particular, D'Arconville initially studied plants (what she called 'agriculture') because 'my interest in chemistry led me to acquire an introductory knowledge of agriculture, botany in particular, because these subjects were analogous to chemistry' (D'Arconville 1800–1805, pp. 197–199). Jussieu, professor of botany and director of the *Jardin du Roi* from 1722 to 1775, was instrumental in the development of D'Arconville's knowledge of organic chemistry; he also invented a method of plant classification that was particularly helpful to her studies (D'Arconville 1766, pp. xxvi–xxvii).

Jussieu's magnanimous sharing of his knowledge of botany, and his generosity in supplying D'Arconville's arboretum at her country home at Crosne (today a suburb of Paris) with exotic plants, led to her own observations about plant decay (D'Arconville 1800–1805, p. 190). It was here that D'Arconville, in her private laboratory, discovered that certain substances which impeded putrefaction existed in the gastric juices of plant roots, except in myrtle. In this flowering plant, she noted that the 'flesh took more than six months to decompose and was thus of a superior substance' (1766, pp. xxvi–xxvii). D'Arconville's examination of plant decomposition led to more general studies in chemistry. In addition to her work with Rouelle, she studied chemistry with Jean-Baptiste Sénac, the King's first physician, and with Macquer who was the key figure amongst these eminent scientists (Bardez 2009, p. 258). A former student of Rouelle, Macquer accomplished pioneering work in the field of chemistry as one of the first to distinguish the discipline from pharmacy, and by developing the practical arts such as dyes and porcelain. Author of a respected *Dictionnaire de Chimie*, Macquer did much to enhance the scientific career of D'Arconville. He encouraged her work and introduced her to the scientific community of Paris including important figures such as the naturalist Bernard-Germain Lacépède and the chemist Antoine Fourcroy.

Within D'Arconville's community of scientists, her name and works were both well-known and well-received. The greatest chemist of the age, Fourcroy, praised D'Arconville for her experimental work on putrefaction, placing it on the same level as studies on decomposition by Stephen Hales, John Pringle, David MacBride, and Antoine Baumé. Notably, Fourcroy commented that 'a French woman [D'Arconville] has distinguished herself through the production of a large number of intellectual works, including the translation of Shaw's chemistry course, which has led to much experimental research on the same topic'. When discussing in his own works the different conditions under which decomposition occurs, Fourcroy included D'Arconville's reasons for this phenomenon (1801, pp. 96–102). Likewise, Du Châtelet cultivated the friendship and mentorship of the greatest mathematicians and philosophers of the age. The most renowned and arguably most influential of these mentors was Voltaire. Du Châtelet made Voltaire's acquaintance for the first time at her father's salon in 1729 when she was thirteen years old. They became friends a few years later in May 1733 after she had given birth to her third child (Zinsser 2006, pp. 64–65). Describing herself as his 'secretary', Du Châtelet was initially Voltaire's assistant and collaborator (Besterman 1958, p. 100). Not only did Voltaire and Du Châtelet conduct experiments together, but they also co-authored the *Élements de la Philosophie de Newton*. She soon surpassed him with her original contributions to physics.

Voltaire's connections with the savants of the Enlightenment were key to Du Châtelet's scientific endeavours. Voltaire introduced her to astronomer and mathematician Pierre Louis Moreau de Maupertuis, a member of the Academy of Sciences and author of a *Discourse on the Different Shapes of Stars* (1732). Maupertuis' work introduced the French to Newtonian theories (Sutton 1995). Du Châtelet soon became his student, introducing her to the latest scientific and mathematical discoveries. Her letters, written to Maupertuis in the 1730s, clearly portray her keen interest in mathematics and their teacher–student relationship. She wrote of 'an extreme desire to study', her knowledge of Newton's philosophy and her interest in becoming a geometrist (Besterman 1958, p. 3). These letters also demonstrate her frustration and impatience with her tutor's neglect of her. As a rising star at the Academy, Maupertuis had more important priorities than tutoring his pupil (Besterman 1958, p. 236). In the light of his neglect, Du Châtelet turned to other tutors to satisfy her thirst for knowledge. These tutors included mathematician Alex-Claude Clairaut, who was Maupertuis' protégé, Johann Bernoulli II, youngest son of Johann Bernoulli, who created an equation for kinetic energy, and the Swiss mathematician Samuel de König, all of whom were major scientists at the time. Under their tutelage, she came to an understanding of the theories of Leibniz and Newton.

If the men of the *Jardin du Roi* composed D'Arconville's network, many members of the Academy of Sciences were participants in Du Châtelet's own private *Académie* at her country home, the Château de Cirey-sur-Blaise, 'a chateau located on the border of Champagne and Lorraine' (Besterman 1958, p. 12).[10] With the financial help of Voltaire and the moral support of her husband, she hosted a 'brilliant salon' to which many of the leading mathematicians and scientists, including Maupertuis, Samuel König, and Bernoulli II, named '*les Émiliens*' by Voltaire, of the day conversed (Voltaire 1965, p. 20; Zinsser, 2006, p. 116). Du Châtelet's academy was more than a social gathering of élites. It was the venue where colleagues of the international Republic of Letters such as Italian polymath Francesco Algarotti, author of *Il newtonianismo per le dame ovvero dialoghi sopra la luce e i colori* (1737), discussed and wrote their major works. Du Châtelet's network was European in scope. As well as those centred around the Bernouilli family, others were members of

Frederick the Great's Prussian Royal Academy of Sciences in Berlin including Algarotti, mathematician Leonhard Euler, König, physician Julien Offray de La Mettrie, Maupertuis, and Voltaire (Project Vox 2019).

The château at Cirey was a centre of study and scientific research. Voltaire describes a well-stocked library and a 'beautiful physics laboratory'. He provided details of the considerable accomplishments of Du Châtelet. Initially focusing on the study of Leibniz, 'Émilie developed a part of his system in a very well-written book entitled, *Institutions de Physique*'. However, her preference was for Newton, and she soon began 'translating the entire *Principes mathématiques*' (Voltaire 1965, pp. 20–21). Through her personal network, her *Institutions du Physique* was translated into Italian. One of her correspondents, Father Francois Jacquier, who was Professor of Physics and Astronomy and corresponding secretary of the Academy of Rome, oversaw the Italian edition which was published in 1743. Du Châtelet made her correspondence connection with Jacquier through the intervention of Johann Bernouilli II (Gardiner 1985, p. 184).

Translations, Commentaries, and Anonymity

A strategy commonly used by women scientists and humanists in early modern Europe was the translation of major texts. Defending the practice of translation, and alluding to the difficulties women faced when producing a treatise of their own, Du Châtelet wrote in the Preface to her translation of Mandeville's *The Fable of the Bees*:

> Translators are the entrepreneurs of the Republic of Letters and they should at least be praised for perceiving and knowing their limitations and for not undertaking to produce works themselves, and thus attempting to carry a burden under which they would succumb. Besides, if their work does not require creative genius, which no doubt holds the first rank in the empire of the fine arts, it calls for an application for which they must be even more grateful, as they can expect less glory from it. (Du Châtelet 2009, p. 46)

Further, referring to her own lack of originality, and intellectual capability, 'nature has refused me the creative genius that discovers truths', she defended as a means of 'rendering with clarity the truths others have discovered, and which the diversity of languages renders useless for most readers' (2009, p. 49).

Translation provided a vehicle through which women could disseminate their own ideas. These translations were much more than simple transliterations. Du Châtelet and D'Arconville not only translated the original texts, but they also provided extensive original commentaries and corrective notes to key texts. In the case of Du Châtelet, it was Newton's *Principia*, which to this day, is the standard French translation of that seminal work in the history of science. Only recently have scholars given Du Châtelet the credit she is due for her translation of Newton's *Principia* which had previously been attributed to Voltaire (Zinsser 2016, "Betrayals"). Du Châtelet's translation, published as *Principes Mathématiques de la philosophie naturelle, par feue Madame la Marquise Du Chastellet*, included her original commentary which explained complex Newtonian theories to French readers. The authors of a widely used university-level physics textbook underscore the importance of Du Châtelet's text for students today: 'A major work by de Breteuil was the translation of Newton's *Philosophiae Naturalis Principia Mathematica* into French. Her translation and commentary greatly contributed to the acceptance of Newtonian Science in Europe' (Larson et al. 2008, p. 344). Du Châtelet's translation remains the only complete French edition of Newton's major work (Fara 2004, p. 14). She completed her translation and commentary in 1749 shortly before her death in childbirth, but the text was not published until 1759 (Zinsser 2001).

For her part, D'Arconville translated and provided detailed notes and comments to Peter Shaw's *Chemical Lectures*. Macquer, her tutor, had advised her to focus on the links between chemistry and medicine. Specifically, he advised her to centre her research on infections in wounds and gangrene. Tissue decay, prevalent in wounded soldiers, involved decomposition or putrefaction; at the time this was attributed to the principle of infection, but the science of its origin was still ignored. Without the comprehension of the complications of war wounds, researchers attempted to identify substances which could act as antidotes and this search was the principal rationale for Macquer's request to D'Arconville to translate Peter Shaw's *Chemical Lectures*. Her resulting French translation of the work was published in 1759 as *Leçons de chymie, propres à perfectionner la physique, le commerce et les arts*; she added to this extensive notes and commentaries, transforming the work into her own. In addition to making revisions to Shaw's work, d'Arconville provided a ninety-four-page preface, 'Discours Préliminaire', itself a noteworthy contribution to applied chemistry. In this preface, D'Arconville wrote that the work included a history of chemistry, that it explained the types and goals of chemistry, and she included comments on chemistry's relationship to medicine and pharmacy stressing the importance for physicians to understand the intricacies of pharmacy in order to successfully practice medicine (Shaw 1759, p.vi). She provided a survey of the contributions made by chemists such as Joachim Becher, Hermann Boerhaave, Georg-Ernst Stahl, Nicolas Lémery, and Etienne-François Geoffrey. She also referred to her own experiments, which would later appear in 1766, in the *Essai pour server à l'histoire de la putréfaction*.

In addition to her translation of Shaw, D'Arconville translated, illustrated, and annotated Alexander Monro's (primus) classic textbook *The Anatomy of the human bones, and nerves* (1759).[11] This work was intended to serve as a commentary on dissections and demonstrations. It contained original descriptions of the cranium and D'Arconville was the first woman to make illustrations of the human skeleton.The translation of *The Anatomy of Human*

Bones has been attributed to Jean Joseph Sue, a Professor of Anatomy at the Royal College of Surgery; however, recently historian of science, Londa Schiebinger, has provided convincing evidence to dispute Sue's contributions. She concludes that the introduction and illustrator were of the same voice, quoting the author: 'The plates were drawn under my eyes, and there were many that I had redone many times in order to correct the slightest fault' (1986, p. 77). In addition, Schiebinger, who has conducted extensive research on this issue, states that the introduction to the translation can be found re-printed in other works by D'Arconville. What is intriguing is that the drawings of the female pelvis by D'Arconville are much wider than those of the male pelvis, while the female skull is much smaller than its male counterpart, thus stressing the differences between male and female (Schiebinger 2003, p. 309). D'Arconville succeeded in the publication of original works because she made sure they were published anonymously.[12] And at the same time, the right people knew of their authorship (Bret and Van Tiggelen 2011, p. 10). Moreover, D'Arconville's decision to remain anonymous would assure that her gender would remain unknown and she would exercise control over her work (Candler Hayes 2012, pp. 381–382).

Du Châtelet also went to great lengths to keep her anonymity when submitting her *Dissertation sur la Nature et Propagation du Feu* (Essay on the Nature and Propagation of Fire) to the Academy of Sciences essay contest, and when it was subsequently published. The first edition of Du Châtelet's major solo work, *Foundations of Physics* (*Institutions de Physique*), was published anonymously in 1740 and, similar to D'Arconville, everyone who mattered knew that it was her book.[13] In *Foundations of Physics,* Du Châtelet boldly challenged contemporary beliefs about 'forces vives', or living forces, meaning the 'force calculated as the product of mass and the square of distance—modern notation mv2' (Hutton 2004, p. 522). Perhaps she overstepped the boundaries for a woman scientist as she quarrelled with the perpetual secretary of the Paris Royal Academy of Sciences, Jean-Jacques Dortous de Mairan, who took issue with her findings, claiming that she had misunderstood Newton. Ironically, this attention from the leading scientist at the time only validated Du Châtelet's standing as a serious physicist (Zinsser 2007, pp. 169–196). The first edition of *Foundations* was published anonymously; however, the second edition, published in Amsterdam and entitled, *Institutions physiques de Madame la marquise du Châstellet adressés à Mr. son fils,* had her name in the title (Reichenberger 2018, p. 2). Similar to the Academy essay, she wrote her two-volume translation of Newton's *Principia* in secret, yet her name is on the title page as 'par madame la marquise du chastellet' (Zinsser 2007, p. 242).

Although Du Châtelet never became a member of the Paris Academy of Sciences, she came as close a woman could get to that in eighteenth-century France by submitting her own treatise on the Academy's topic of the Nature and Propagation of Fire for the 1736 essay prize.[14] There are several points of interest here which pertain to the question of women in science. First, not only was Du Châtelet the first woman in France to submit an essay to an

Academy competition, but it was actually published in the Academy's journal, which was remarkable for a non-prize-winning paper (Zinsser 2007, p. 152).[15] These were major successes for a woman in science. The story of Du Châtelet's journey from assistant and co-researcher with Voltaire, to independent scientist and published author, is informative of gender issues for a woman in science.

As recounted in detail by Du Châtelet experts, the study of fire and its qualities was a joint venture with Voltaire (Kawashima 2005; Zinsser 2007, pp. 152–162). They conducted experiments together; they read and discussed the data collaboratively. Their intention was to submit a joint essay to the Academy. However, once the experiments began to yield results, and Du Châtelet found that her conclusions differed from those of Voltaire, she felt compelled to write up her original conclusions and submit an essay on her own, anonymously (Voltaire 1970, p. 164). This choice was problematic because of her gender. Firstly, she did not want to compete publicly with her lover as she did not want to hurt his feelings or wound his pride. So, she wrote at night, and hid her work from him. She explained to Maupertuis the reasons behind hiding her own work from that of Voltaire: 'The work of M. de Voltaire, which was almost finished before I had begun mine, gave me some ideas and a desire to follow a similar course. I began working without knowing if I would send my essay, and I said nothing to M. de Voltaire because I did not want to suffer shame, in his eyes, for undertaking an enterprise which I feared he would disapprove of' (Voltaire 1970, p. 165). Secondly, for a woman to submit an independent work to the Academy was in direct contradiction of the gender norms at this time. Women's role was to support men, not to challenge them. Frederick the Great, a close friend of Voltaire, encapsulates this view in his comments about Du Châtelet, writing to Voltaire that, 'She wants to steal from us men, all the advantages of which our sex is privileged [...] You should oppose the progress of women' (Voltaire 1970, p. 276).

Thirdly, Du Châtelet suffered from feelings of insecurity and incompetence. She felt that her essay was not well-organized (Voltaire 1970, p. 164). She told Maupertuis that she did 'not want the prize' (Besterman 1958, p. 272). In a self-effacing final paragraph of her *Dissertation*, she wrote to the Academy's referees that she 'had already imposed too much on the patience of this respectable body to which I dare present this weak essay', but that her 'love for the truth will stand in place of talent, and that the sincere desire to contribute to its knowledge will obtain forgiveness for my errors' (Du Châtelet 2009, p. 102). At the same time, Du Châtelet's genuine desire to be useful, to contribute to knowledge, and to find the truth about the nature of fire, was too strong a motive for her not to submit her own essay. When neither Voltaire nor Du Châtelet won the prize, she told him about her submission and wrote, 'It seems to me that the refusal by the Academy that I shared with him became a source of pride for him' (2009, p. 102). Keiko Kawashima explains this contradiction in Du Châtelet's emotions in terms of gender expectations. By writing an independent essay and submitting it to the Academy, Du Châtelet was deviating from all eighteenth-century French gender norms (2005, pp. 32–33).

Voltaire was very supportive of Du Châtelet and her pursuit of scientific truths, but at the same time, as noted above, he never challenged contemporary gender norms. On the positive side, he wrote to Maupertuis urging the Academy to publish both of their essays. Moreover, he was keen that the Academy acknowledged that Du Châtelet's conclusions were independent and different from his own (Voltaire 1970, p. 15). Her essay, he wrote, 'is full of things that would make physicists proud' and, perhaps referring to the Academy's sexism, that she would have won the prize had it not been for the Academy's 'absurd and ridiculous belief' in the 'swirling chimera' or she-goat, a reference to a female Greek mythological creature, part snake, goat, and lion, that breathed fire (Voltaire 1970, p. 162). At the same time, in his review of her essay in the *Mercure de France* when referring to Du Chatelet's essay, he remarked that 'It might be very estimable if it were from the hand of a philosopher solely occupied with his own research: but, that a woman occupied with other things—domestic cares, governance of a family, and many business affairs—should have composed such a work, I do not know of anything so glorious in this century' (Voltaire 1739, pp. 1327–1328). This comment on the surface seems very positive and lacking in gender bias, which it would have been had Voltaire not discussed woman's legitimate role as domestic caretaker. As stated above, Voltaire's views were laudable yet remained wedded to the traditional Enlightenment concept of woman.

Du Châtelet submitted her essay for publication anonymously as she had done in the Academy contest. Members of the Academy knew the essay was written by her as they referred to the author as 'a young woman of high status' rather than 'Mr la M. du Chastellet', the name under which the Academy had originally planned to publish the essay (Voltaire 1970, p. 15). Submitting the essay signed would have been a step too far, as 'writing women' before the nineteenth century required great courage to publish. They were challenging all respectability and gender norms (Zemon Davis 1981).

While D'Arconville never submitted any of her work to the Paris Academy, she did produce an original scientific text, *Essai pour servir à l'histoire de la putrefaction* (1766), describing in great detail the results of over 300 experiments on meat conservation and decomposition conducted in her private laboratory. She discovered thirty-two substances that either retarded or produced putrefaction. The text is composed of D'Arconville's original research on the action of strong and weak acids on human and beef bile.

Unlike Du Châtelet, D'Arconville worked alone, although the impetus for writing up her experiments originated with her mentor, Macquer, to whom she dedicated her *Essai*, describing him as her 'friend and master' (1766). Macquer advised her to focus her work on the links between chemistry and medicine. Specifically, he asked her to centre her research on infections in wounds, gangrene, and the necrosis of tissues. These problems, prevalent in wounded soldiers, involved decomposition or putrefaction attributed to the principle of infection but the science of their origin had been ignored. Without any comprehension of the reasons for the complications of war wounds, researchers attempted to identify substances which could act as antidotes. Macquer advised her to verify Sir John Pringle's recent conclusions about the science of putrefaction. Pringle was general physician to the armies in England and the results of his work had been published by the Royal Society of London.[16] Pringle had presented seven lectures to the Royal Society between 1750 and 1752 where he provided a list of septic substances that favoured decomposition, and others known as antiseptics, which fought decomposition. According to Pringle, cinchona bark possessed a great power not only to preserve animal substances but to prevent them from corruption as well as re-constituting them to their original state once they had putrefied (Pringle 1752, p. 314).

D'Arconville's comments concerning Pringle's experiments are lengthy and substantive indicating her sophisticated knowledge of chemicals. With respect to Pringle's observations on septic and anti-septic substances, she noted the power of quinaquina in the preservation of animal matter. However, her results were not always the same as Pringle's, especially those experiments using chamomile. Although she intended that her work served as a corrective to his conclusions, she denied being his rival stating, '[t]he superiority of Mr. Pringle's talents and the goal we both propose exclude any rivalry'. Further, she added that 'the goal is not reached if students are jealous of masters' (Pringle 1752, p. iv). In other words, although D'Arconville was confident in challenging Pringle's conclusions, she hesitated to question his authority as the teacher. Considering herself in the role of the student, she puts herself in the subordinate position to her male mentors, not unlike Du Châtelet when it came Voltaire.

Conclusion

As Du Châtelet's and D'Arconville's contributions demonstrate, scientific research in eighteenth-century France was not limited to the formal institutional framework of the French Academy. Their lives are a testament to the creativity of women who were determined to participate in the Republic of Science. These exceptional women, using a variety of strategies, practised science at a time when science was a male fiefdom. Their tactics ranged from securing an education to cultivating key personal relationships with leading men of science who would serve as mentors and collaborators. D'Arconville achieved her goal to be a vital part of the scientific community through collaboration and dialogue with her male counterparts and through her own publications and translations in the fields of chemistry and anatomy. Translating major works, and publishing their own works usually anonymously, were other methods of navigating science. These methods proved to be successful as both women made significant contributions to the Republic of Science in eighteenth-century France.

Notes

1. From preface to Marquise Du Châtelet's English translation of Bernard Mandeville's *The Fable of the Bees*. See Du Châtelet (2009, p .48).
2. Du Châtelet and D'Arconville contributed substantially to the humanities as well as to the sciences. D'Arconville wrote prose, poetry, a play, and philosophical treatises, while Du Châtelet wrote a treatise on happiness and the Bible.
3. In a recent study of Madame Dupiéry's contributions to astronomy, Isabelle Lémonon (2015) suggests that women's largely overlooked contributions to scientific knowledge have been focused in the domestic sphere rather than in the academy.
4. Other exceptional women of science in eighteenth-century France include Nicole-Reine Etable de Labrière Lepaute, Louise du Pierry, and Marie-Anne Pierrette Paulze Lavoisier.
5. Her mother, Françoise Gaudicher, was the daughter of a notary from Angers (De la Porte 1835, pp. 14–39).
6. Thiroux-d'Arconville Darlus. M.G.C. 1800–1805. "Histoire de mon enfance" in *Pensées et réflexions morales* (3), pp. 259–260. University of Ottawa Archives and Special Collections. Her nephew, a physician and botany professor, who wrote a short biographical notice, recalled that his brilliant aunt had received the education of a housekeeper (Hippolyte Bodard de la Jacopière 1810, p. xxvi).
7. Margaret Carlyle (2011, p. 80) confirms D'Arconville's creation of a 'personal research team'.
8. D'Arconville corresponded with several men of science including her mentor, chemist Pierre-Joseph Macquer between 1770 and 1778. Bibliothèque Nationale de France (BNF), Département des manuscrits, Fr. 12,305, fols. 19–25.
9. Similar to D'Arconville, Poulletier worked on bile from 1754 to 1755. Bardez (2009, p. 259).
10. Madame du Châtelet and Voltaire lived at Cirey, the Du Châtelet Estate, from 1734 to 1749.
11. Alexander Monro (primus), (1697–1767), was a Scottish anatomist and professor of Anatomy at Edinburgh University.
12. Married women in pre-revolutionary France were not only required to have their husband's permission to publish, but if they did, their work was owned by their husband. These legal obstacles may have also been factors in the decision to publish anonymously. Robin Craig notes that most women who published were single, separated, or widowed (2006, p. 552).
13. Judith Zinsser writes that Du Châtelet 'assumed she could publish the Institutions anonymously' and that three drafts of the Preface demonstrate Du Châtelet depicting herself as a male parent (2006, p. 171).

14. For a detailed discussion of the essay, its contents and Du Châtelet's strategies, see Kawashima (2005).
15. Zinsser (2007, p. 323) notes that there are three published versions of the essay that are extant, two by the Academy in 1739 and 1740; and a revised edition in 1744.
16. Pringle (1752) made an important contribution to the understanding of decay processes in the cause of disease. His understanding of putrefaction led him to advocate proper hygiene, latrines, and ventilation in army hospitals.

References

Arianrhod, Robyn. 2012. *Seduced by Logic: Émilie du Châtelet, Mary Somerville and the Newtonian Revolution*. New York: Oxford University Press.

Barber, W.H. 2006. "Madame Du Châtelet and Leibnizianism: The Genesis of the *Institutions de Physique*." In *Émilie du Châtelet, Rewriting Enlightenment Science and Philosophy*, edited by Judith P. Zinsser and Julie Candler Hayes, Part 1. Oxford: Voltaire Foundation.

Bardez, Élisabeth. 2009. "Au fil de ses ouvrages anonyms, Madame Thiroux d'Arconville, femme de lettres et chimiste éclairée." *Revue d'Histoire de la Pharmacie*, no. 363: 259.

Bardez, Élisabeth. 2011. *Madame d'Arconville et les Sciences: raison ou résonance?* Paris: Hermann.

Besterman, Theordore. 1958. *Les lettres de la marquise Du Châtelet*, Vol. I. Genève: Institut et Musée Voltaire.

Besterman, Theodore. 1967. *Voltaire*. New York: Harcourt.

Bodanis, David. 2006. *Passionate Minds: The Great Love Affair of the Enlightenment*. New York: Crown.

Bret, Patrice and Van Tiggelen, Brigitte. 2011. *Madame D'Arconville (1720–805): Une Femme de Lettres et de Sciences au Siècle des Lumières*, with a preface by Élisabeth Badinter. Paris: Hermann.

Candler Hayes, Julie. 2012. "From Anonymity to Autobiography: Mme d'Arconville's Self-Fashionings." *The Romantic Review* 103 ,no. 3-4: 381–397.

Carlyle, Margaret. 2011. "Femme de sciences, femme d'esprit: 'le traducteur des Leçons de Chymie." In *Madame d'Arconville (1720–1805): Une Femme de Lettres et de Sciences au Siècle des Lumières*, edited by Patrice Bret and Brigitte Van Tiggelen with a preface by Élisabeth Badinter, 71–92. Paris: Hermann.

Clack, Beverley, ed. 1999. *Misogyny in the Western Philosophical Tradition: A Reader*. Basingstoke: Palgrave.

Condorcet, Nicolas de Caritat. 1847–1849. *Oeuvres*, Vols. 7 and 9. Paris: Firmin Didot Frères.

Condorcet, Nicolas de Caritat. 1976. "Reception Speech." *In Condorcet: Selected Writings*, edited by K.M. Baker, 26. Indianapolis: Bobbs-Merrill.

Craig, Robin. 2006. "Les Droits d'auteur: French Women Writers and the Legal Borders of Eighteenth-Century Authorship." *Historical Reflections/Réflexions historiques*, no. 32: 543–558.

Cunningham, Andrew. 2010. *The Anatomist Anatomis'd: An Experimental Discipline in Enlightenment Europe*. Farnham: Ashgate.
D'Arconville, Marie Geneviève-Charlotte. [1760] 1800–1805. *Pensées et réflexions morales*. Avignon.
D'Arconville, Marie Geneviève-Charlotte. 1766. *Essai pour server à l'histoire de la putréfaction*. Paris: Didot le jeune.
De Fourcroy, Antoine-François. 1801. *Système de connaissances chimiques et leurs applications aux phénomènes de la nature et de l'art*. Paris: Baudoin.
De la Porte, H. 1835. *Notices et observations à l'occasion de quelques femmes de la société du dix-huitième siècle*. Paris: H. Fournier.
Du Châtelet, Emilie. 2009. *Selected Philosophical and Scientific Writings*, edited by Judith P. Zinsser, trans. Judith P. Zinsser and Isabelle Bour. Chicago: Chicago University Press.
Fara, Patricia. 2004. "Emilie du Châtelet: The Genius Without a Beard." *Physics World*, no. 6: 14–15.
Gardiner, Linda. 1985. "Women in Science." In *French Women and the Age of Enlightenment*, edited by Samia I. Spencer, 181–193. Bloomington: Indiana State University Press.
Gargam, Adeline. 2012. "La Chair, l'os et les éléments: l'heureuse fécondité de la traduction scientifique au XVIIIe siècle: le cas de Marie-Geneviève Thiroux d'Arconville." In *Les Rôles transfrontaliers joués par les femmes dans la construction de l'Europe*, edited by Guyonne Leduc, 59–76. Paris: L'Harmattan.
Hahn, Roger. 1971. *The Anatomy of a Scientific Institution: The Paris Academy of Sciences, 1666–1803*. Berkeley: University of California Press.
Hippolyte Bodard de la Jacopière, P.H. 1810. *Cours de botanique médicale comparée*. Paris: Méquignon l'Aîné.
Hutton, Sarah. 2004. "Emilie Du Châtelet's Institutions De Physique as a Document in the History of French Newtonianism." *Studies in History and Philosophy of Science Part A* 35: 515–531.
Kawashima, Keiko. 2005. "The Issue of Gender and Science: A Case Study of Madame du Châtelet's Dissertation sur le feu." *Historia Scientiarum: International Journal of the History of Science Society of Japan* 15, no. 1: 23–43.
Kawashima, Keika. 2004. "Birth of Ambition: Madame Du Châtelet's Institutions de Physique." *Historia Scientiarum, International Journal of the History of Science Society of Japan* Series 2, 14, no. 1: 49–66.
Laissus, Y., and Torlais, J. 1986. *Le Jardin du Roi et le Collège royal: dans l'enseignement des sciences au XVIII siècle*. Paris: Hermann.
Larson, Ron, Hostetler, Robert P., and Edwards, Bruce H. 2008. *Essential Calculus: Early Transcendental Functions*. Boston: Houghton Mifflin.
Lemay, P., and Oesper, R.E. 1954. "The Lectures of Guillaume François Rouelle." *Journal of Chemical Education* 31, no. 7: 338–343.
Lémonon, Isabelle. 2015. "Gender and Space in Enlightenment Science: Madame Dupiéry's Scientific Work and Network." In *Domesticity and the Making of Modern Science*, edited by Donald L. Opitz, Staffan Bergwick, and Brigitte Van Tiggelen, 41–60. Basingstoke: Palgrave.
Newton, Isaac. 1759. *Principes mathématiques de la philosophie naturelle [traduit du latin] par feue madame la marquise Du Chastellet [Avec une préface de Roger Cotes et une préface de Voltaire]*. https://gallica.bnf.fr/ark:/12148/bpt6k1040149v/f13. item. Accessed February 26 2019.

Petrovich, Vesna C. 1999. "Women and the Paris Academy of Sciences." *Eighteenth-Century Studies* 32, no. 3: 383–390.
Pringle, John. 1752. *Observations on the diseases of the Army*. London: A. Millar.
Project Vox. 2019. "Gabrielle Émilie Le Tonnelier de Breteuil, la Marquise Du Châtelet." *Duke University Libraries*. http://projectvox.org/du-chatelet-1706-1749/. Accessed December 13 2020.
Reichenberger, Andrea. 2018. "Émilie Du Châtelet's Interpretation of the Laws of Motion in the Light of 18th Century Mechanics." *Studies in History and Philosophy of Science* Part A, 69: 1–11.
Rousseau, Jean-Jacques. [1762] 1974. *Émile, Or Education*, trans. by Barbara Foxley and Peter D. Jimack. London: Dent.
Schiebinger, Londa. 1986. "Skeletons in the Closet: The First Illustrations of the Female Skeleton in Eighteenth Century Anatomy." *Representations*, no. 14: 42–82.
Schiebinger, Londa. 1989. *The Mind has no Sex? Women and the Origins of Modern Science*. Cambridge, MA: Harvard University Press.
Schiebinger, Londa. 1995. *Nature's Body: Gender in the Making of Modern Science*. Boston: Beacon Press.
Schiebinger, Londa. 2003. "Skelettestreit." *Isis* 94, no. 2: 307–313.
Shaw, Peter. 1759. *Leçons de chymie propres à perfectionner la physique, le commerce et les arts*, trans. and preface by Marie Geneviève-Charlotte D'Arconville. Paris: J. Thérissant.
Sutton, Geoffrey V. 1995. *Science for a Polite Society*. Boulder: Westview.
Voltaire. 1736. "À Madame la Marquise du Châtelet". In *Alzire, ou les Américains*. University of Toronto Libraries, Eighteenth Century Collections Online. https://guides.library.utoronto.ca/primary. Accessed February 17 2019.
Voltaire. 1738. M. Voltaire to the Marchioness du CH**. In *The elements of Sir Isaac Newton's Philosophy*, edited by John Hanna. University of Toronto Libraries, Eighteenth Century Collections Online. https://guides.library.utoronto.ca/primary Accessed February 17 2019.
Voltaire. 1739. "Extrait de la Dissertation de Madame L.M.C.D.C. Sur la Nature du Fer." *Mercure de France* (June): 1327–1328.
Voltaire. 1965. *Mémoires pour server à la vie de M. de Voltaire, écrits par lui-même, suivis de Lettres à Frédéric II*, edited by J. Brenner. Paris: Mercure de France.
Voltaire. 1970. *The Complete Works of Voltaire/Oeuvres Complètes de Voltaire*, Vol. 95, edited by Theodore Besterman. Geneva: Institut et Musée Voltaire.
Watts, Ruth E. 2007. *Women in Science: A Social and Cultural History*. London: Routledge.
Whaley, Leigh. 2003. *Women's History as Scientists: A Guide to the Debates*. Santa Barbara: ABC-CLIO.
Zemon Davis, Natalie. 1981. "Gender and Genre: Women as Historical Writers, 1400–1820." In *Beyond their Sex: Learned Women of the European Past*, edited by Patricia H. Labalme, 153–182. New York: New York University Press.
Zinsser, Judith P. 2001. "Translating Newton's *Principia*: The Marquise Du Châtelet's Revision and Additions for a French Audience." *Notes and Records of the Royal Society of London* 55, no. 2: 227–245.
Zinsser, Judith P. 2006. *La Dame d'Esprit: A Biography of the Marquise du Châtelet*. New York: Viking.
Zinsser, Judith P. 2007. *Emilie Du Châtelet: Daring Genius of the Enlightenment*. New York: Penguin.

Zinsser, Judith P. 2016. "Émilie Du Châtelet, Part 1 and 2." *History of Women Philosophers and Scientists*. https://www.youtube.com/watch?v=gI54d64YoHY. Accessed December 10 2020.

Zinsser, Judith P. 2016. "Betrayals: An Eighteenth-Century Philosophe and Her Biographers." *French Historical Studies* 39, no. 1: 3–33.

Zinsser, Judith. P. 2018. "Gabrielle Émilie le Tonnier de Breteuil : Marquise du Châtelet (1706–49)." In *Women and Science: Changing the World One Idea at a Time*. http://womeninscience.history.msu.edu/Biography/C-4A-0/marquise-du-chtelet/. Accessed December 10 2020.

CHAPTER 4

'A Valuable Gift': The Medical Life of Margaret Mason, Lady Mount Cashell

Alexis Wolf

In 1823, *Advice to Young Mothers on the Physical Education of Children* was published anonymously, its authorship attributed only to 'a Grandmother'. The language of the title highlights the central message of female enfranchisement found throughout *Advice*: that of an older and more knowledgeable woman mentoring younger women in need of both technical medical guidance and maternal confidence. The text offers itself as a handbook to help young mothers 'prevent' illnesses rather than 'to cure' them (p. vi). It combines medical knowledge with prudence, providing practical, how-to guidance for pregnancy, childbirth, and maternal care from infancy to adolescence. In the preface, the author acknowledges that the text's medical advice includes 'many known truths and many old remarks'. Its importance, however, lies in its 'utility', offering the reader 'such a book as [she] would herself, at the age of twenty, have received as a valuable gift' (p. xi).

While the contents of *Advice* may not be entirely ground-breaking, its message of women's medical empowerment may be read as more revolutionary. The anonymous author states in the preface that 'she is anxious to diffuse amongst her own sex, a species of knowledge which may enable mothers to educate their children with better prospects of health and happiness'. She encourages women to accept personal responsibility for 'the welfare of their offspring' rather than feeling alienated and incapable (p. iv). Women,

A. Wolf (✉)
University of Lancaster, Lancaster, UK
e-mail: a.wolf@lancaster.ac.uk

© The Author(s), under exclusive license to Springer Nature Switzerland AG 2022
C. G. Jones et al. (eds.), *The Palgrave Handbook of Women and Science since 1660*, https://doi.org/10.1007/978-3-030-78973-2_4

whether as 'observing mothers' or 'attentive nurses', she insists, are advantageously positioned to understand and oversee the health of infants and children. Were they to combine this first-hand experience with 'a moderate degree of scientific knowledge' through practical training and study, their skills would surely surpass those of any male physician (p. vii). The author acknowledges that the care of trained medical practitioners may be necessary at different stages of life. In the case of childbirth, *Advice* suggests the employment of 'well-instructed' female midwives who have 'knowledge of the organization of the tender being' (p. 31). Properly trained women midwives are 'as capable of exercising the obstetric arts as a professional man' (p. 4).

The text proved popular. Several new editions were printed in 1835, including one in London revealing the author as 'Margaret Jane Moore, Countess of Mountcashell' [*sic*]. However, by the time of the second printing, she was known to her family and friends as 'Mrs Mason'.[1] Mason was indeed a grandmother to many children, yet she was also a political activist, children's book author, novelist, and self-proclaimed and practising physician.[2] Mason's life and writings offer an important case of women's engagement with medicine in the early nineteenth century and provide a lens through which to trace the transnational dimension of women's diverse involvement in medicine in Britain and Italy. Recognizing her accomplishments as a physician and author allows for a reconsideration of the forms of medical knowledge acquired by and exchanged between women. Such practical engagements with medicine are illuminated by Mason's apprenticeship as a physician under renowned surgeon Andrea Vaccà in Pisa, the publication of her domestic medical text, as well as evidence of her ongoing medical practice as an expatriate physician tending to the poor. Perhaps most importantly, Mason's writings intervened in scientific debates surrounding maternal health. Her *Advice* pushed for women's professionalization and education as a means of reconstituting and reimagining their traditional roles as midwives, a position which had been usurped by the male medical establishment throughout the second half of the eighteenth century.

A Physician's Life

Mason's medical and literary life was shaped in large part by the relationships of her early years. She was born in 1772, the second of twelve children and the eldest daughter of the aristocratic, Anglo-Irish Kingsborough family of County Cork, Ireland. Her mother, Caroline Fitzgerald, Lady Kingsborough, did not attend personally to her children's health or education, rather leaving them, as Mason later wrote, to 'the care of hirelings from the first moment of [...] birth' and 'subjecting them to the discipline of governesses & teachers' before the age of three. Her early education was an irregular one, which she described as 'not calculated to improve my good qualities or correct my faults'.[3] She was not close with her mother, who provided 'only a negative model' of adult womanhood (Denlinger 2018).

All of this changed with the arrival of Mary Wollstonecraft as governess to the Kingsborough family. Wollstonecraft, the philosopher, novelist, and early women's rights author who would write, soon after, her influential treatise *A Vindication of the Rights of Woman* (1792), was in the formative stages of her career. She disapproved of Lady Kingsborough's parenting style, viewing her aristocratic employment of wet nurses, governesses, and teachers as a form of parental negligence. Wollstonecraft caricatured Lady Kingsborough in her early novel, *Mary, A Fiction* (1788), in the figure of Eliza. 'After the mother's throes', Eliza 'felt very few sentiments of maternal tenderness; the children were given to nurses, and she played with her dogs' (p. 9). Eliza's inattention results in 'milk-fevers', 'consumption', and the death in infancy of all but two of her children. Echoes of Wollstonecraft's criticisms of Lady Kingsborough can be found in Mason's later reference in *Advice* to 'the great influence which the conduct of a mother, during the time of pregnancy and nursing, is likely to have on the health of her offspring' (1823, p. xi). Wollstonecraft's influence can also be detected in Mason's later preference for the values and practices of the 'middle rank of life' rather than those of the 'higher sphere' into which she was born.[4] Wollstonecraft drew on her experiences of re-educating the young Margaret King, as Mason was then known, and her sister Mary, to write her didactic manual, *Original Stories from Real Life* (1788). This text offers a fictionalized representation of the governess and her two pupils. It is composed of a series of short instructive tales touching on moral, physical, and charitable conduct for young women, borrowing from aspects of Rousseau's *Émile* (1762). This formative educational agenda can be traced to Mason's focus in *Advice* on 'the strict connexion [*sic*] between mind and body' as necessary components of physical and moral education (1823, p. xi).

Wollstonecraft was unceremoniously dismissed due to her tension with Lady Kingsborough as much as for her powerful influence over her young pupil (Todd 2005, p. 105). Wollstonecraft's radical and holistic approach to intellectual and personal development made a lasting impact on Mason, who afterwards endeavoured to 'correct those faults she had pointed out & to cultivate my understanding as much as possible'.[5] However, she 'wanted advice' to cultivate her 'exalted views', and was soon pressured by her family into an aristocratic marriage expected of a young woman of her station. She married Stephen Moore, 2nd Earl of Mount Cashell (1770–1822) in 1791. The marriage produced seven living children, whose education Mason made 'an object of importance', despite the Earl's unwillingness to invest in materials and tutors. Despite her conventional familial role as a young aristocratic wife, Mason was radical in her views. She attended political and literary salons that supported Irish independence from Britain and published anonymous poetry and pamphlets that engaged with the Irish Rebellion of 1798 and subsequent Union Crisis of 1799–1800. Mason travelled to the Continent with her family and a companion, Katherine Wilmot, during the Peace of Amiens of 1801–1803. Mason examined the extensive collections in the Cabinet of Natural

History in Turin and socialized with professors of anatomy in Florence, experiences which may have secured Italy as a place of scientific opportunity in her mind (Wilmot 1924, pp. 108 and 130). It was during this Grand Tour that Mason met her second partner, George Tighe. Mason chose to abandon her previous marriage in order to begin an expatriate existence with Tighe, also gaining intellectual freedom. In doing so, she was forced to make the heartbreaking choice of relinquishing access to her seven children. In the years that followed, Mason wrote children's books published by the Juvenile Press of William Godwin, Wollstonecraft's widower, including the successful *Stories of Old Daniel: or Tales of Wonder and Delight* (1808).[6]

It was during this period that Mason began to seriously pursue a medical education. Anecdotal evidence suggests she may have disguised herself as a man to gain entry to lectures at the University of Jena, aided in doing so by her tall stature (Stocking 1995, p. 135).[7] Mason and Tighe eventually settled in Pisa, where she adopted the name of Mrs. Mason, borrowed from the fictional governess of Wollstonecraft's *Original Stories*. She gave birth to two daughters, Laurette and Nerina, on whom she would test her scientific theories of childrearing. Mason formed a friendship and working relationship with the renowned doctor Andrea Vaccà Berlinghieri, Professor of Surgery at the University of Pisa. Vaccà's reforming surgical practices attracted pupils from across Europe. His own medical text on the treatment of venereal disease, published in 1800, was widely reprinted and translated throughout the first half of the nineteenth century. Mason would maintain a lifelong personal connection to the doctor, discussing and receiving counsel on difficult medical cases. Mason's expatriate medical pursuits seem to embody Wollstonecraft's earlier call to action in her *Vindication of the Rights of Woman* that women practise as physicians, regulate farms, manage shops, and stand 'erect, supported by their own industry' (1792, p. 340).

The idea that Mason may have employed a masculine disguise to study medicine is both romantic and intriguing. However, the transnational differences in women's medical education in the period indicate that such extremes may have been unnecessary. Women in Britain were not welcome as medical students in universities, though they could obtain some medical training as nurses or midwives at lying-in hospitals or through private tuition from a male practitioner (Wilson 1995, p. 201). By contrast, some women received medical training and other scientific qualifications in European universities in the eighteenth and early nineteenth centuries. Dorothea Erxleben (1715–1762) received a medical degree in Germany in 1754 and wrote a treatise arguing for women's rights to university education. In Napoleonic France, women received formal medical training, embracing professionalization while upholding society's view of the feminine connection between the moral and physical body (Burton 2007). Women occupied important roles as medical professionals in France under these systems, as seen in the example of Marie Boivin (1773–1841), a midwife, hospital director, and gynaecological author who invented a widely used vaginal speculum (Ogilvie 1986, p. 43). Boivin published *Mémorial de l'art des accouchements* in 1812, a practical handbook for midwives and medical students which contained 133 instructive plates. The text was based in part on the teachings of her own mentor, Marie-Louise Lachapelle (1769–1821), herself a midwife and medical author who served as the Head of Obstetrics at the largest public hospital in Paris (Oakes 2002, p. 208). Italy, where Mason settled, had a particularly established tradition of formal scientific and medical training for women stemming from the Middle Ages. During the eighteenth century, numerous women graduated from Italian universities. Anna Morandi Manzolini (1716–1774) held the Chair of Anatomy at the University of Bologna, where she also created wax anatomical models (Alic 1886, p. 104). Maria Dalle Donne (1778–1842) subsequently became the first woman to receive a doctorate in medicine in Italy in 1799 and published scientific papers on female reproduction, fertility, and midwifery. She later served as Director of the Department of Midwifery at the University of Bologna (Windsor 2002).

Mason apparently never received a formal medical qualification, despite her association with the University of Jena and later connection to the University of Pisa through association with Vaccà.[8] Instead, her letters indicate that from 1815 onwards she studied personally with the Professor through an informal mentoring arrangement which took place beyond the University and continued over the course of a decade. Considerable evidence remains of her uses for the medical knowledge she acquired from this relationship as well as from her independent study. A personal manifesto written in 1818 outlines Mason's thoughts on the connection between mind and body, which would inform her treatments of moral and physical health. She states that one must 'take voluntary Exercise both of mind & body' to avoid 'Mind[s] overrun with Weeds' & 'Bodies with diseases'.[9] Other documents show the depth of her knowledge in administering and manufacturing remedies. A 'Memorandum on the Children' appears to be the first written example of her medical advice, outlining methods of care for her two young daughters during her absence.[10] Her prescriptions draw on regionally available remedies, including the use of 'a syrup of squills', a coastal Mediterranean plant of the lily family, to be administered for a common cough or difficult breathing, guidance which would eventually be printed in *Advice to Young Mothers*.

Many aspects of Mason's familial and sociable medical writings were later revised for publication in *Advice*, which suggests that informal literary practices contributed to women's ability to identify as medical practitioners in the period. Correspondence, manuscripts, and receipt books blurred the boundary between public and private care and discourse while creating networks of medical knowledge. Letters exchanged with the Shelleys in 1819 show Mason prescribing a course of 'a few drops (7 or 8) of sulphuric ether on sugar' as well as 'bathing the feet or hands (or both) in warm water with mustard sufficient to smart a little' to treat Percy Shelley's 'nervous difficulty of breathing'. Mason asserts herself to be 'as good a physician for Mr S[helley] as any one, were not the first requisite wanting – I mean the confidence of a patient'.[11]

She leaves the couple to judge whether she has 'more medical skill than other old women who boast of the cures they have made' based on the efficiency of the prescription. In 1822, Mason wrote to Mary Shelley about the ill health of Mary's friend Marianne Hunt. She advised 'almond milk [...] with the yolk of an egg & a few drops of lemon juice' for the patient's night sweats and loss of appetite. Mason provided a hand-drawn diagram on how to apply a linen cloth over the treatment area for Hunt's comfort in case of acute pain. She recommended 'these things because both as an invalid & as a Physician I have had experience of their advantage', asserting her empathy as well as her expertise.[12]

Mason's identification as a physician with the ability to aid others further increased with the publication of *Advice*. In June 1823 she wrote to Mary Shelley while the latter was en route to Britain. Mason encouraged her to use the book as a tool for self-education on maternal healthcare, writing 'I hope you have by this time received the "Grandmother's Advice" as it may be of use to you in your long journey' and that 'should little Percy get the Hooping [*sic*] Cough, that book would give you full instructions how to manage him'.[13] Mason reassures her young friend that the book's medical guidance is sound; she has 'lately cured another child (about a year old) with the mixture of Kermes Mineral', the corresponding description of which is printed in *Advice* (p. 226). Mason also exchanged letters with prominent physicians in Italy and abroad, including George Parkman of Boston in 1825. Mason sent Parkman a copy of *Advice to Young Mothers* with a request for comments in anticipation of a potential American edition of the text, an interaction which points to her sense of agency as a physician engaged in medical discourse.[14]

Mason offered free medical care to the Pisan poor from her home, the Casa Silva (Denlinger 2018). Evidence of her practising public medicine in this way can be found in an Italian-language novel written by her daughter, Laurette, many years after Mason's death.[15] The novel, entitled *Una Madre: Racconto Del Secolo XIX* (A Mother: A Story of the Nineteenth Century), depicts Mason offering medical treatment to the children of impoverished Italian mothers. The fictionalized version of Mason pays a visit to a young patient. The boy's mother cries out 'come, come! [...] if anyone can save him now, it's certainly her' (Tighe 1857, pp. 16–17).[16] Mason takes the child's pulse, examines his tongue and eyes, and evaluates his fever before diagnosing that worms are the cause of his ailment. She promises to send medicine which will cure the child's malady. The scene illustrates Mason's medical reputation as a physician among the poor of Pisa and indicates that her dissemination of medical assistance to young mothers extended beyond the pages of her book.

'IT IS A MISTAKE TO SUPPOSE THE AID OF MEN NECESSARY'

A robust body of historical scholarship has outlined the trends of medical professionalization that increasingly excluded women's knowledge and practice of traditional medicine and midwifery in Britain from the mid-eighteenth century (Lay et al. 2000; Moscucci 1993). In the centuries prior, childbirth was historically managed by the woman midwife, passed on through oral tradition and apprenticeship, and largely relegated within the closed domain of women's collective culture. Shifting societal preferences for male practitioners professionally trained in midwifery came at the expense of female midwives, transforming 'what was a vaguely defined traditional craft' into 'a nascent profession' (Herrle-Fanning 2000, p. 29). The rise of the 'man-midwife' in the eighteenth century was supported by prominent medical authors and teachers such as William Smellie (1697–1763). Midwifery institutions in London trained large numbers of male pupils in obstetric practice, including in the use of forceps for difficult births. The advent of lying-in hospitals and charities contributed to childbirth evolving as a distinct branch of institutionalized medicine (Wilson 1995, p. 3). The intervention of the man-midwife in childbirth often led to a continued association with the family as physician, an arrangement which established a trajectory of male control over matters of maternal health.

Male practitioners and medical authors criticized women's traditional midwifery, labelling such practices as dangerous and unscientific, while also outlining new standards of institutional practice and professional expectation. William Buchan's influential and much reprinted *Domestic Medicine: or, A Treatise on the Prevention and Cure* (1774) states that 'almost one half of the human species perish in infancy, by improper management or neglect' (p. vi). Buchan attributes the widespread infant mortality of the eighteenth century to 'the superstitious prejudices of ignorant and officious midwives', a crisis which could only be remedied by the employment of skilled midwives, whether male or female (p. 582). Women in the period were gaining training and qualifications to answer these demands, as well as publishing literature which combated the 'ignorant midwife' theory (Marland 1993, p. 2). Skilled women midwives set up schools to further advance women's ability to access medical qualifications. Margaret Stephen (fl. 1765–1795), for instance, ran a school of midwifery in London in the 1790s, served as midwife to Queen Charlotte, and published a treatise entitled *Domestic midwife; or, the best means of preventing danger in child-birth* (1795) which argued that her female pupils were 'as well qualified as men' to engage in the medical practice of midwifery (p. 4).

Through her writings, Mason can be seen as participating in the contemporary transnational and gendered debates surrounding midwifery. In one sense, Mason's utilization of a language of maternal medical mentorship in *Advice* harkens to an older tradition rooted in the oral transmission of female-centric knowledge, yet her concerted interest in the Enlightenment exchange of medical knowledge, as well as her insistence on proper training and qualification, connects her writing with the increasing professionalization of maternal health and midwifery in the period. The ideological frame with which she advocates for women's right to train and practice as medical agents, then, also links her text to Wollstonecraft's desire for women to have both equal rights and professional lives.

Mason engaged in international social debates and professional dialogue surrounding women's ability to serve as medical practitioners. One example can be seen in annotations to the proofs of a forthcoming essay entitled 'The State of the Poor in Italy'.[17] The unnamed author attributes a high occurrence of 'acquired deformity' among the Italian peasantry to 'their extreme carelessness in the science of midwifery, that generally being practiced by women'. A cross next to this assertion corresponds to a note at the bottom of the page in which Mason denounces the author's claim: 'This is not true—for the Midwives are very skilful & it is a mistake to suppose the aid of men necessary' [n.d.]. Mason knew first-hand the level of skill possessed by Italian midwives; she worked alongside them as a physician to the poor of Pisa. In Italy, 'a new midwife' had been created in the second half of the eighteenth century as the state 'opened schools of obstetrics for both surgeons and midwives' (Filippini 1993, p. 163). Mason's assertion of the capabilities of midwives, as well as her suggestion that attempts to discredit them were untruthful, opposes the dominant narratives being promoted by male-authored medical texts of the period.

Yet Mason's affirmation of the skill of the Italian midwives also draws out her complex view of the changing expectations facing the woman midwife. In *Advice*, Mason essentially repeats Buchan's language of the 1770s. Mason derides the 'common sort of female midwives', whose 'ignorance' and 'temerity' are 'the cause of many fatal accidents'. Instead, a proper midwife should have 'regular instructions', having gone through 'professional examinations'. A female midwife qualified as such would be 'as capable of exercising the obstetric arts as a professional man'. Any woman under her care would have no 'recourse to the assistance of the other sex at the time of labour' (pp. 3–4). As it was not yet '*the fashion*' or 'general custom' to 'employ female midwives', Mason recommended that her readers locate 'the best female practitioner' possible, who could operate independently while a male physician remained elsewhere in the house in case of difficult births.

Mason's modern vision of the professional future of women's midwifery builds on a notable, if small, body of British texts disseminating female-centric scientific medical knowledge throughout the eighteenth and early nineteenth centuries. While women's medical publishing was in a developing stage in the period, there nevertheless existed a 'population of professional women' drawing on a 'range of approaches and strategies that their authors utilised in order to break new ground in publicly acknowledging their skills and abilities as practitioners and writers' (Blackwood 2017, p. 90). The majority of these texts were written with a view to assist both the midwife and the mother, blurring the boundaries between the professional and the laywoman reader. Jane Sharp's *The midwives book, or the whole art of midwifry discovered* (1671) occupies the earliest part of the eighteenth-century tradition to offer instruction to other female practitioners and calls out for higher standards and better training for women's midwifery education. Sarah Stone's *Complete practice of midwifery* (1737) actively resisted the rise of the male-midwife by recommending itself to 'All female practitioners in an art so important to the lives and well-being of the sex' (titlepage, n.d.). Margaret Stephen's *Domestic midwife* pre-emptively defends her treatise against male readers and medical practitioners who may find an issue with it. The short and straightforward text, she states, is a portable object intended to extend the thorough learning of female pupils in her midwifery school, as well as being useful and accessible to 'any woman who is, or may be a mother' (1795, p. 3). Stephen directly attacks the male establishment for excluding women midwives from institutional medical establishments. She writes unapologetically that she 'intends to continue my lectures as usual to women entering upon the practice of midwifery, until the men who teach that profession render them unnecessary, by giving their female pupils as extensive instructions as they give the males' (pp. 5–6). Martha Mears' *Pupil of Nature, or Candid Advice to the Fairer Sex* (1797) contrasts with these earlier examples by employing the language of sensibility to 'allay the fears of pregnant women' and to 'inspire them with a just reliance on the powers of nature' while simultaneously praising Smellie and other eminent male-midwives for the 'zeal and ability they have displayed in combatting prejudice and error' (p. 3).[18] *An essay to instruct women how to protect themselves in a state of pregnancy* (1798), by Mrs. Wright, a midwife, is closely aligned with the aims and language of Mason's *Advice*, despite its brevity at thirty-six pages. The author uses her expertise to mentor and educate expectant mothers directly, while also outlining treatments for children's ailments.

Women habitually exchanged vital medical knowledge among themselves in other ways as well, often by drawing on less formalized channels including familial and sociable networks of manuscript circulation, or by embedding medical advice under the guise of more acceptably feminine domestic pursuits. The extant manuscript and published texts of domestic author Maria Eliza Rundell illustrate several of these tactics. Rundell was known as the author of *A New System of Domestic Cookery: Formed upon Principles of Economy, and Adapted to the Use of Private Families*, first published in 1806 and frequently reprinted throughout the nineteenth century. The early sections of *Domestic Cookery* contain day-to-day recipes for familial use as well as for entertaining. The final section of the text includes recipes for medicinal purposes under the heading 'Cookery for the Sick'. This practice aligns with long-standing traditions of women's engagement in domestic medicine through the composition, circulation, and publication of receipt books (DiMeo and Pennell 2013; Leong 2018). Another text attributed to Rundell, an unpublished manual on women's general health, pregnancy, and childbirth, which was nominally addressed to her daughter Harriet but may have been more widely circulated, reveals the depths of her maternal medical knowledge.[19] The manuscript contains many recipes with health-restoring properties in the style of Rundell's *Domestic Cookery*, but it also includes a lengthy treatise on enlightened parenting, as well as essays and instructions on women's health more generally, such as at 'the turn of life'. The manuscript is comprehensive in instruction and wide-ranging in scope, addressing topics including lying-in, the immediate care of a newborn, and methods of weaning, in a style similar to Mason's text.[20]

By the time *Advice* was published in 1823, Mason had borne ten children, and had been practising medicine for the poor of Pisa for more than eight years (McAleer 1958, p. 182). The text runs to more than 350 pages. It is a practical manual drawing on the diversity of her own experience, organized into essays separated by age group and into subject headings as well as containing an appendix of medicinal treatments. The early sections discuss pregnancy and childbirth, followed by instructions for children up to two years of age, and from two years of age upwards and beyond. It is written without the inclusion of 'technical terms' with 'a view to real *utility*' for the laywoman (p. xi). The general purpose of the text is to inspire confidence in young mothers' abilities to feed, clothe, and care for the day-to-day moral and physical health of their children. The process of motherhood begins, Mason insists, with a pregnant woman's self-reliance and personal healthfulness. A woman 'who desires to produce an offspring, well constituted in body and mind, should pay the strictest attention to her own conduct, both physical and moral' (p. 1). She has the ability to stave off the need for medical intervention through her own moderation by avoiding 'a great quantity of animal food', 'fermented liquors in abundance', and 'a sedentary life', which will lead to 'the necessity of bleeding and other medical aid' (pp. 1–2).

Class concerns, as well as matters of fashion, remain at the forefront of Mason's advice concerning nursing and the care of infants. The text speaks to women 'who are not rich enough to have the advice of a really skilful physician'. Drawing on her work as physician to the poor families of Pisa, Mason counsels a calm and holistic approach. Nursing mothers 'should observe with attention the state of the child: as long as it looks well and does not appear to suffer from indigestion, there is nothing to dread' (p. 18). She acknowledges that many wealthier women are encouraged to give up suckling their own children at the first sign of difficulty, particularly those of the middle and upper classes with the financial means to hire a nurse. This is a mistake; no pain or inconvenience 'should prove sufficient obstacles to mothers really determined to follow the dictates of nature in this matter' (p. 17). Contrary to popular opinion, nursing will not require women to forego socializing or living 'the life of a recluse'; there is no reason 'why a lady, who suckles a healthy child, should not enjoy an hour or two in the diversions she has been accustomed to, even in the crowded assemblies of the metropolis' (p. 17). Here, Mason is perhaps thinking of her own experience as a young mother in Ireland, where she participated in salon culture while nursing her first seven children. Mason underscores the importance of doing what is best for the child as a first priority rather than bending to the fashion for wet nurses. The text is interjected, occasionally, with abrasive maxims that recall Wollstonecraft's criticism of Mason's own mother, Lady Fitzgerald, and which take on a shaming tone directed at the upper classes. 'A woman who cannot submit to a little trouble, which lasts but a short time, for the benefit of her child', she writes, 'does not deserve to have a child' (p. 21). Mason's intended readership, then, were all young mothers disposed to accept moral and physical responsibility for their offspring, regardless of their class position in society.

The physical health of children is treated in all aspects, including guidance on sleeping patterns, clothing to be worn at particular times of the year, and specific instructions on how to produce remedies for illnesses. Lengthy sections on diet spare no attention to detail, outlining strict instructions and suggestions for how to prepare and serve each meal, as well as guidelines for monitoring children's 'evacuations' for proper digestion (p. 134). The text also addresses the proper way a mother may help a daughter navigate puberty in a section entitled 'Cautions respecting the treatment of young females at a critical time of life'. Mason states that this difficult transition can pass without illness or medicine if attention is paid to 'exercise and amusement, early rising, good nourishment, sufficient sleep, and tranquillity of mind' (p. 165). She cautions mothers 'in the middle and lower ranks' not to keep their daughters too confined to needle-work or sedentary tasks during this stage of development, but rather to 'employ them in the more active business of the household' (p. 166). Young females should 'be informed' about the specific delicate state of their 'human frame' at this stage of life, trained to take stewardship over their own well-being through avoiding foods difficult to digest, or exposing themselves to wet feet or cold (p. 169). Such instruction and control over their own bodies will allow them to remain 'healthy females [...] as they advance in years'. In this way, Mason's physical advice frequently intersects with the moral. Sensibility or excessive 'proofs of feeling' should be discouraged in the young as 'a quality which tends to injure the health and debilitate the mind', and is described as a 'moral malady' that leads to errors in judgement (p. 325). Here, as elsewhere, Mason reminds mothers that 'the body influences the mind', and that parents 'who sincerely wish to see their children healthy [...] will never play upon their feelings' (pp. 325–326). Instead, youthful minds should be encouraged towards good health and self-reliance, a blessing 'which few would disregard, if it were represented to them in the important light it deserves' (p. 349).

Conclusion

Shortly after the publication of *Advice*, Mason wrote to Mary Shelley that she had been motivated to write the text by the conviction that 'no one but myself *would* & *could* execute such a work well', as well as by 'a sincere wish to do good by imparting to others the knowledge I had obtained in thirty years of study, observation & experience'.[21] Mason stated that she had already given away forty copies and would consider all of her effort repaid if only twenty children benefitted from it.

The wide circulation of the book, both before and after her death in 1835, would greatly surpass Mason's modest hopes. The text found devoted readers, including Frances Rossetti, scholar and mother to artists and authors Maria, Dante, William Michael, and Christina. Rossetti's heavily annotated copy of the 1823 edition of *Advice* indicates years of steady use. She eventually passed the volume on to the family of her second son, William Michael, where it continued to be used until at least 1905. The book went through several editions in the 1830s and 1840s in America and Britain as well as being printed in an Italian translation. A new British edition was praised by newspapers including *The Sheffield Iris* which viewed it as 'a valuable little book', helpful 'to all young mothers as containing a fund of useful advice, in training up children, and guarding their health' (3 November 1835).

A contemporary review of the 1835 edition of *Advice* illuminates some of the possibilities and prejudices faced by women medical practitioners as well as by mothers interested in acquiring medical knowledge to tend to their own children. Following the success of earlier editions, the 1835 printing of the text included the author's full name and aristocratic title, 'M.J., Dowager Countess Mount Cashell', as a selling point. The author's class is particularly noted by the *Medical Quarterly Review*, along with the comment that her mentor, the late 'eminent surgeon' Andrea Vaccà had 'no reason to be ashamed' of the book of his pupil (p. 423). The reviewer describes the text as 'a hygienic treatise of very extraordinary merit' in its particular medical methodologies of care, despite the author's misguided desire to 'teach mothers to treat the diseases of their children', a thing which simply 'cannot be done' (p. 426).

This negative view of a mother's diagnostic capabilities stands in stark contrast with the appraisal of Vaccà himself, who proclaimed that 'a woman of good sense who studies that book will want no physician for her children', and that 'there is a great deal [in] that book of which many physicians are ignorant'.[22] At the same time, the reviewer's focus on the female author's 'noble' background underlines the intersectional problematics of both class and gender inherent in women's medical literature of the eighteenth and early nineteenth centuries. While an aristocratic woman of significant training might be accepted as a knowledgeable resource for practical maternal health advice, a wider attainment of self-confident medical skill held by the masses of young mothers was far more dangerous, whether in a capacity to lead to mistreatment, or even to undermine the professionalization of such care under solely male practitioners. In *Advice*, Mason aimed to achieve precisely that level of medical autonomy for young women, as well as an increase in the general health of children thanks to an improved and independent medical competency among mothers across class and national boundaries. Mason argued that by 'enabl[ing] young mothers to direct the physical education of their children with success', 'the happiness of mankind would be much augmented' (pp. 349–50).

Through *Advice*, Mason was able to share medical knowledge with untold numbers of young women, disseminating rational and empowering guidance. At the same time, her text advocates for the woman physician and the woman midwife as knowledgeable and valid medical practitioners, and encourages mothers to respect their role in society. Mason's medical life, both as a medical author and self-proclaimed physician, provides a significant example of British women's transnational involvement in medical sciences in the early nineteenth century. It also illustrates the ways in which women with medical interests engaged in the literary dissemination of medical knowledge in the period.

Acknowledgements I am grateful to the Birkbeck School of Arts and the Wellcome Trust for the award of a joint Institutional Strategic Support Fellowship in 2018–19, which enabled me to complete this chapter. I would also like to extend my thanks to Elizabeth Denlinger and Charles Carter of the Carl H. Pforzheimer Collection of Shelley and His Circle at the New York Public Library for their assistance with my research.

Notes

1. The author's highest title was used for the publication to increase the work's prestige, despite the fact that it was no longer one to which she was entitled. She used several names each connected to distinct stages in her life. She was born Margaret King to the Kingsborough family. She was imparted the title of Countess as Lady Mount Cashell upon her first marriage to Stephen Moore, Earl of Mount Cashell. Following the dissolution of their marriage, she assumed the name 'Mrs Mason', using the fictionalized name of her governess Mary Wollstonecraft in *Original Stories from Real Life* (1791).
2. Denlinger (2018) notes that the pseudonymous title of 'a Grandmother' 'must have been freighted with sadness, as Mrs Mason had never seen the grandchildren of her first marriage'.
3. New York Public Library (NYPL), Carl H. Pforzheimer Collection of Shelley and His Circle, Mount Cashell-Tighe-Cini Family Papers, S'Ana 0763, 1.
4. NYPL, Mount Cashell-Tighe-Cini Family Papers, S'Ana 0763, 5.
5. NYPL, Mount Cashell-Tighe-Cini Family Papers, S'Ana 0763, 2.
6. Mason wrote several children's books which assisted in providing financial support for her family's life in Italy. The success of *Stories of Old Daniel* provoked a sequel, *Continuation of the Stories of Old Daniel* and *Simple Stories, in words of one syllable, for little boys and girls*. For more on her juvenile literature, see Markey (2010).
7. Claire Clairmont related a story of Mason's incognito attendance at Jena to Edward Augustus Silsbee nearly four decades after her death in 1875. See Stocking (1995).

8. It is possible that further evidence pertaining to Mason's medical career exists within a private collection of some of her family's papers held at Palazzo Cini, Pistoia, Italy, which is not currently accessible to scholars.
9. NYPL, Mount Cashell-Tighe-Cini Family Papers, S'Ana, 1120.
10. NYPL, Mount Cashell-Tighe-Cini Family Papers, S'Ana, 0764.
11. University of Oxford, Bodleian Library, Abinger Collection, MS. C 45, fol 35r.
12. University of Oxford, Bodleian Library, Abinger Collection, MS. C 46, fol 25v.
13. University of Oxford, Bodleian Library, Abinger Collection, MS. C. 46, fol 83r.
14. NYPL, Mount Cashell-Tighe-Cini Family Papers, S'Ana, 0782a.
15. Laurette Tighe shared in her mother's literary pursuits. She assisted her mother in running the *Accademia dei Lunatici*, a liberal literary society, between 1827 and 1832. Laurette published several Italian-language novels under the pen name 'Sara', including *Una Madre* (1857). In the text she explains that the character of Eleonora Melin is a representation of her mother, designed to honour her 'noble and beautiful' character.
16. Author's own translation. The original text reads: 'venga, venga!' gridò con voce soffocata, 'se qualcuno può ancora salvarlo, è certo lei'.
17. NYPL, Mount Cashell-Tighe-Cini Family Papers, S'Ana, 0886.
18. Herrle-Fanning (2000) notes that much of Mears' *Pupil of Nature* was copied unattributed from Thomas Denman's *Introduction to the Practice of Midwifery* (1788), rather than offering her own experiences of midwifery. Denman was a contemporary of Mears, and a leading man-midwife in London in 1790s.
19. Wellcome Library (London), Archives and Manuscripts, MS. 7106.
20. While the manuscript is dedicated to Rundell's daughter, about whom nothing is known, its elegantly scripted presentation and shared organizational methodology with *Domestic Cookery* indicates that it may have been circulated more widely in fair-copy. If so, it raises interesting questions about the contrasting public and private literary personas adopted by women with medical skill in the period. The mere existence of the manuscript in either capacity suggests that women's agency over their own maternal health, and as purveyors of that knowledge, found channels in which to thrive at the outset of the nineteenth century.
21. University of Oxford, Bodleian Library, Abinger Collection, MS. C. 47, fol 45r.
22. University of Oxford, Bodleian Library, Abinger Collection, MS. C. 46, fol 113v.

REFERENCES

Alic, Margaret. 1886. *Hypatia's Heritage: A History of Women in Science from Antiquity to the Late Nineteenth Century*. London: The Women's Press.
Blackwood, Ashleigh. 2017. *Managing Maternity: Reproduction and the Literary Imagination in the Eighteenth Century*. Doctoral thesis, Northumbria University.
Boivin, Marie A. 1812. *Mémorial de l'art des accouchements*. Paris: L'Hospice de la Maternité.
Buchan, William. 1774. *Domestic Medicine: or, A Treatise on the Prevention and Cure of Diseases by Regimen and Simple Medicines*. London: W. Strahan and T. Cadell; Edinburgh: J. Balfour and W. Creech.
Burton, June K. 2007. *Napoleon and the Woman Question: Discourses of the Other Sex in French Education, Medicine, and Medical Law, 1799–1815*. Lubbock: Texas Tech University Press.
Cieślak-Golonka, Maria and Morten, Bruno. 2000. "The Women Scientists of Bologna: Eighteenth-Century Bologna Provided a Rare Liberal Environment in Which Brilliant Women Could Flourish." *American Scientist* 88, no. 1: 68–73.
Dalle Donne, M. 1800. *Theses ex Universa Medicina Deprompte*. Bologna: Ex Typographia Thomae Aquinatis.
Denlinger, Elizabeth Campbell. 2018. "King [Married Names: Moore; Tighe], Margaret Jane, Countess Mount Cashell, [Known as Mrs. Mason], (1772–1835), Writer and Political Activist." *Oxford Dictionary of National Biography*. https://doi-org.liverpool.idm.oclc.org/10.1093/odnb/9780198614128.013.109696. Accessed on 13 November 2020.
DiMeo, Michelle and Pennell, Sara. 2013. *Reading and Writing Recipe Books, 1550–1800*. Manchester: Manchester University Press.
Donnison, Jean. 1977. *Midwives and Medical Men: A History of Interprofessional Rivalries and Women's Rights*. New York: Schocken.
Erxleben, Dorothea Christiana. 1754. *Dissertatio Inavgvralis Medica Exponens Qvod Nimis Cito Ac Ivcvnde Cvrare Saepivs Fiat Cavssa Minvs Tvtae Cvrationis*. Halae: Magdeburgicae.
Filippini, Nadia Maria. 1993. "The Church, the State and Childbirth: The Midwife in Italy During the Eighteenth Century." In: *The Art of Midwifery: Early Modern Midwives in Europe*, edited by Hilary Marland, 152–175. London: Routledge.
Herrle-Fanning, Jeanette. 2000. "Figuring the Reproductive Woman: The Construction of Professional Identity in Eighteenth-Century British Midwifery Texts." In: *Body Talk: Rhetoric, Technology, Reproduction*, edited by Mary M. Lay, Laura J. Gurak, Clare Gravon, and Cynthia Myntti, 29–48. Madison: University of Wisconsin Press.
Lay, Mary M., Gurak, Laura J., Gravon, Clare, and Myntti, Cynthia, eds. 2000. *Body Talk: Rhetoric, Technology, Reproduction*. Madison: University of Wisconsin Press.
Leong, Elaine. 2018. *Recipes and Everyday Knowledge: Medicine, Science, and the Household in Early Modern England*. Chicago: University of Chicago Press.
Markey, Anne. 2010. "The English Governess, Her Wild Irish Pupil, and Her Wandering Daughter: Migration and Maternal Absence in Georgian Children's Fiction." *Eighteenth-Century Ireland* 25: 161–176.
Marland, Hilary. 1993. *The Art of Midwifery: Early Modern Midwives in Europe*. London: Routledge.

Marland, Hilary and Rafferty, Anne Marie. 2002. *Midwives, Society, and Childbirth: Debates and Controversies in the Modern Period*. London: Routledge.

McAleer, Edward C. 1958. *The Sensitive Plant: A Life of Lady Mount Cashell*. Chapel Hill: University of North Carolina Press.

Mears, Martha. 1797. *The Pupil of Nature; or Candid Advice to the Fair Sex, on the Subjects of Pregnancy; Childbirth; the Diseases Incident to Both; the Fatal Effects of Ignorance and Quackery; and the Most Approved Means of Promoting the Health, Strength, and Beauty of Their Offspring*. London: Benjamin Crosby and Co. and Robert Faulder.

Moore, Margaret J.K. (Countess of Mount Cashell). 1808. *Stories of Old Daniel: Tales of Wonder and Delight*. London: Juvenile Library.

Moore, Margaret J.K. (Countess of Mount Cashell). 1820. *Continuation of the Stories of Old Daniel*. London: Mary Jane Godwin and Co.

Moore, Margaret J.K. (Countess of Mount Cashell). 1822. *Simple Stories, in Words of One Syllable, for Little Boys and Girls*. London: J. Harris and Son.

Moore, Margaret J.K. (Countess of Mount Cashell). 1823. *Advice to Young Mothers on the Physical Education of Children, By a Grandmother*. London: Longman and Co.

Moscucci, Ornella. 1993. *The Science of Woman: Gynaecology and Gender in England, 1800–1929*. Cambridge: Cambridge University Press.

Oakes, Elizabeth H. 2002. *International Encyclopedia of Women Scientists*. New York: Facts on File.

Ogilvie, Marilyn Bailey. 1986. *Women in Science: Antiquity Through the Nineteenth Century*. Cambridge, MA: MIT Press.

Rousseau, Jean-Jacques. 1762. *Émile, ou De l'éducation*. Amsterdam: Jean Néaulme.

Sharp, Jane. 1671. *The Midwives Book, or, The Whole Art of Midwifry Discovered*. London: Simon Miller.

Smellie, William. 1752–64. *A Treatise on the Theory and Practice of Midwifery*. London: D. Wilson.

Stephen, Margaret. 1795. *Domestic Midwife; or, the Best Means of Preventing Danger in Child-Birth*. London: S.W. Fores.

Stocking, Marion Kingston, ed. 1995. *The Clairmont Correspondence: Letters of Claire Clairmont, Charles Clairmont, and Fanny Imlay Godwin, 1808–1879*. Baltimore: Johns Hopkins Press.

Stone, Sarah. 1737. *A Complete Practice of Midwifery: Consisting of Upwards of Forty Cases or Observations in That Valuable Art, Selected from Many Others, in the Course of a Very Extensive Practice*. London: T. Cooper.

Tighe, L. 'Sara'. 1857. *Una Madre: Racconto Del Secolo XIX*. Turin: Tipographia Ecomonica.

Todd, Janet. 2003. "Ascendancy: Lady Mount Cashell, Lady Moira, Mary Wollstonecraft and the Union Pamphlets." *Eighteenth-Century Ireland* 18: 98–117.

Todd, Janet. 2005. *Daughters of Ireland: The Rebellious Kingsborough Sisters and the Making of a Modern Nation*. New York: Ballantine Books.

Vaccà Berlinghieri, Andrea. 1800. *Traité des maladies vénériennes*. Paris: P.P. Alyon.

Wilmot, Catherine. 1924. *An Irish Peer on the Continent, 1801–1803, Being a Narrative of the Tour of Stephen, 2nd Earl of Mount Cashell, as Related by Catherine Wilmot*. London: Williams & Norgate.

Wilson Adrian. 1995. *The Making of Man-Midwifery: Childbirth in England 1660–1770*. Cambridge, MA: Harvard University Press.

Windsor, Laura Lynn. 2002. *Women in Medicine: An Encyclopedia*. Santa Barbara: ABC-CLIO.

Wollstonecraft, Mary. 1788. *Thoughts on the Education of Daughters: With Reflections on Female Conduct, in the More Important Duties of Life*. London: J. Johnson.

Wollstonecraft, Mary. 1788. *Mary, A Fiction*. London: J. Johnson.

Wollstonecraft, Mary. 1791. *Original Stories from Real Life with Conversations, Calculated to Regulate the Affections, and Form the Mind to Truth and Goodness*. London: Printed J. Johnson.

Wollstonecraft, Mary. 1792. *A Vindication of the Rights of Woman*. London: J. Johnson.

Wright, Mrs. 1798. *An Essay to Instruct Women How to Protect Themselves in a State of Pregnancy from the Disorders Incident to That Period, or How to Cure Them Also, Some Observations on the Treatment of Children*. London: Barker, Lee. Hurst and Kirby.

CHAPTER 5

Janet Taylor (1804–1870): Mathematical Instrument Maker and Teacher of Navigation

John S. Croucher

When Janet Taylor, born the daughter of a country clergyman, passed away on Wednesday 26 January 1870, her death certificate recorded her occupation simply as 'Teacher of Navigation'. But she was far more than this. A mathematician, astronomer, author, instrument maker and inventor, with her own business and nautical academy, Mrs. Taylor's singular contribution to the maritime community and to the welfare of mariners was widely acknowledged. She was awarded gold medals (the highest award possible) by the kings of Prussia and Holland—and even by the Pope—and was eventually awarded a Civil List pension by the British government in recognition of her work. Indeed, Mrs. Taylor's story is a remarkable one, especially for a woman in science and technology in the nineteenth century.

Janet Taylor's story is one of gender, but it is also one of class, in a society in which gender and class were both significant markers. Her beginnings were modest, but her career in maritime and nautical pursuits testifies to an intellectual curiosity and hunger for answers, leading her into fields that were unusual for a woman in her time. Britain in the mid-nineteenth century was the richest country in the world and, while railways opened up the country, shipping opened up empires. There were many opportunities to meet the great hunger for knowledge and for devices to make sea navigation safer. Inspired by her scholarly father in the wonders of navigation, Mrs. Taylor became a prodigious

J. S. Croucher (✉)
Macquarie University, Sydney, NSW, Australia
e-mail: john.croucher@mq.edu.au

© The Author(s), under exclusive license to Springer Nature Switzerland AG 2022
C. G. Jones et al. (eds.), *The Palgrave Handbook of Women and Science since 1660*, https://doi.org/10.1007/978-3-030-78973-2_5

author of nautical treatises and textbooks, born of a fascination, particularly, for measuring longitude by the lunar distance method. She conducted her own Nautical Academy in Minories in the East End of London, was a sub-agent for Admiralty charts, ran a manufacturing business for nautical instruments—many of which she designed herself—and embarked on the business of compass adjusting at the height of the controversies generated by magnetic deviation and distortions on iron ships. Mrs. Taylor was also the mother of eight children and stepmother to a further three, with both she and her husband changing their names on their marriage: he from George Taylor Jane to George Taylor; she from Jane Ann Ionn to Janet Taylor.

Learning About Mrs. Taylor

Tracking down Mrs. Taylor and her activities is not without difficulty; as discussed in Anne Barrett's chapter elsewhere in this collection, women's records are often deemed less important to keep or get lost in the archive, hidden away in the papers of men. In the absence of many personal records, such as a diary, Mrs. Taylor's story had to be pieced together mostly from public records including her publications and their reviews and discussion in relevant journals.[1] Other sources evidencing Mrs. Taylor's role include her correspondence with the Admiralty in the Archives of the United Kingdom Hydrographic Office, Taunton, and correspondence with the Astronomer Royal, the archives of which are held by Cambridge University.[2] There are also Mrs. Taylor's letters to various nautical magazines, principally in defence of her own works. With respect to biographical sources, the secondary literature includes a small number of brief biographical notes or cameos, including one from 1871 (Anderson 1871, pp. 228–240).[3] Another source offers biographical information of a local kind in an unpublished typed manuscript by J. H. Lambert (1960).[4] Of the more 'orthodox' secondary literature, there are entries in two dictionaries of biography (Boase 1901, p. 94; Fisher 2004).[5] There are also family records including those held by the descendants of Frederick Peter Ionn, Mrs. Taylor's youngest brother, in particular a large parchment chart with linen backing, written in 1869.[6] On this parchment, under the heading 'Mems', Frederick set down his memories of his family, including a section on his sister. After her death in 1870, a note was added to record her passing.[7]

Mrs. Taylor is also included in two biographical dictionaries focusing on particular professions. The first was published in 1966 by Professor Eva Taylor in *The Mathematical Practitioners of Hanoverian England 1714–1840*, detailing over 2000 mathematicians who contributed to the profession in the designated period, including a small handful of women (1966, pp. 101–102 and 461–462).[8] The second was the entry by Dr. Gloria Clifton, then Curator of Navigational Instruments and Head of the Royal Observatory, in the landmark *Directory of British Scientific Instrument Makers 1550–1851* (1995, p. 275). Professor Taylor refers to the navigational warehouse that Mrs.

Taylor ran, initially with her husband George Taylor, a former naval lieutenant. From here she sold a wide range of nautical and mathematical instruments and, from 1835, acted as a sub-agent for Admiralty charts. Professor Taylor described this business as one of 'the most socially interesting warehouses', and unique in being run by a woman, 'and not simply a widow carrying on her husband's business' (1966, pp. 101–102). Mrs. Taylor is also included in Mary Brück's *Women in Early British and Irish Astronomy*, which states that in 'addition to her teaching, Janet Taylor engaged in every aspect of the navigational business' (2009, p. 53).

Mrs. Taylor's Early Life

Janet's father, the Reverend Peter Ionn, was curate and schoolmaster at the Free Grammar School in Wolsingham, County Durham, where he taught, among many other things, the subject of navigation, prompted by his own education at the Free Grammar School in Penrith.[9] With an active and eager intellect, Janet (then Jane Ann) absorbed what she could from her teacher-father. When she was nine years old, a scholarship for the daughters of clergymen, naval or military officers and 'orphans of gentlefolk', became available to improve their chances of good marriages through better education and accomplishments. The scholarship was to the Royal School for Embroidering Females, in Ampthill, Bedfordshire, established under the patronage of Queen Charlotte, the wife of George III. Although the normal earliest entry age of the students was fourteen, Janet was so outstanding that she was accepted as a pupil at the age of only nine. It is reported that the Queen said, 'Let her come—not to work at the tapestry, but at her figures; and let her be educated with the others until she is of age to take the place of one of them'.[10] When Queen Charlotte died on 17 November 1818, the school closed and Janet, now aged fourteen, obtained a position as governess to the family of the Vicar of Kimbolton, the Reverend John Huntley.[11]

On 2 May 1821, Janet's father Peter died suddenly when Janet was sixteen years old. She inherited some money from his estate and went to live with her brother Mathew Seymour Ionn who had opened a linen draper's shop at 44 Oxford Street in London. According to her other brother, she went to live with Seymore as he needed a 'housekeeper' (Croucher and Croucher 2016, p. 253). With the failure of the business in 1829, it appears that Janet moved to Antwerp, in Belgium, to be with her sister, Isabella, who was expecting her first child.[12] Soon after, on 30 January 1830, Janet married in The Hague the widower George Taylor Jane, a brewery manager and former naval lieutenant.[13] She was aged twenty-five and he was forty-one. Interestingly, on the marriage he changed his name to George Taylor and she became 'Mrs Janet Taylor' for the rest of her life. The couple set up residence at 6 East Street, Red Lion Square, near Oxford Street in London. Janet, as she was now known, also inherited George's three children from his previous marriage. With George, she had a further eight children of her own. Janet was a determined woman and one with great plans: to write books, to design nautical instruments and to teach navigation. George supported her fully. As a former naval man and a publican, he had understood dealings with men. But what was most telling was that he was a 'Dissenter', brought up outside the Church of England. They educated girls with boys and, indeed, dissenting backgrounds have been identified as creating the conditions for women to enter science, so Janet's fearsome intelligence and determination were perhaps not such a surprise to him (Watts 2007, pp. 79–98).

JANET TAYLOR—AUTHOR

Mrs. Taylor's entry into publishing began in 1833 at the age of twenty-nine, and she went on to produce a number of major works of importance to the maritime community, each of which went into many editions. There were seven editions of her first book, *Luni-Solar and Horary Tables* (or *Lunar Tables*) alone, appearing between 1833 and 1854. Her *Principles of Navigation Simplified: with Luni-Solar and Horary Tables*, first published in 1834, went through three editions, while *An Epitome of Navigation and Nautical Astronomy* went to twelve editions between 1842 and 1859. Her *Planisphere of the Fixed Stars with Book of Directions*, with its beautiful plates of star charts, was first published in 1846 and reached its sixth edition in 1863. And then there were the *Diurnal Register for Barometer, Sympiesometer, Thermometer and Hygrometer* (seven editions); *A Guide Book to Lt Maury's Wind and Current Charts* (seven editions); and the *Handbook to the Local Marine Board Examinations for Officers of the British Mercantile Marine Board* (twenty-seven editions).[14] This was a remarkable volume of work and, almost without exception, her contributions were highly praised and well received. Indeed, Taylor's texts were reviewed in a wide variety of journals, technical journals, newspapers and magazines. A singular exception to positive receptions was the *Nautical Magazine's* 1835 review of her first book, *Luni-Solar and Horary Tables*, which prompted a strong response from Taylor, and a counter-response from the reviewer in the succeeding issue. This exchange also seems to have led to a revision of her formula (Croucher and Croucher 2016).

An essential ingredient of success in the maritime world in the nineteenth century was to gain the imprimatur of 'the big three': The Admiralty, Trinity House and the East India Company. It was the Admiralty that oversaw the navy and in 1795 established its own Hydrographic Office, with Captain Francis Beaufort in charge from 1829.[15] The Corporation of Trinity House, first incorporated by Royal Charter in 1514, was the general lighthouse authority for Britain and responsible for the examination and licensing of pilots. And then there was the East India Company that had been granted a charter of incorporation by Queen Elizabeth I on 31 December 1600, giving it a monopoly of Indian and Far Eastern trade. It was the major provider of survey charts used on British merchant ships. It was critical for Mrs. Taylor to gain the endorsement and approval of at least one of these bodies or her work would simply not be accepted by the seafaring world as having integrity or indeed use. The worth of Mrs. Taylor's tables, with an abridged or simplified method for clearing lunar distances, was firmly acknowledged by grants of £100 from each body.[16] Moreover, as further evidence of their value, they also allowed her to dedicate her works to them.

Mrs. Taylor's books were key teaching tools in her Nautical Academy. Established first in 1835, the Nautical Academy would continue for over thirty years at Minories.[17] It was 'much patronised by officers of the East India Company and of the Navy' (E.C.M. 1933). In the 1861 census, at which time Taylor, now widowed, was living at 1 Grove Park Terrace, Camberwell, she was described as 'Janet Taylor (widow) – authoress and instructress in Navigation and Nautical Astronomy'.

Mathematical Instrument Manufacturer—And Inventor

By 1845 Janet and George opened a business premises at 104 Minories where they sold a wide range of nautical and mathematical instruments, illustrated by an advertisement in the *Mercantile Marine Magazine*, October 1854, declaring that the firm manufactured 'every description of nautical and mathematical instruments', many of which she designed herself.[18] Taylor also became a sub-agent for Admiralty charts.[19] Indeed, Mary Brück (2009, p. 55) observes that

> [c]learly Janet Taylor was not only an accomplished mathematician and geometer but—what was exceptional among intellectual women of the early nineteenth century—a successful businesswoman in a man's world.

A distinct chapter in Taylor's life is as an inventor. In March 1834, still before her thirtieth birthday, Mrs. Janet Taylor of East Street, Red Lion Square, Middlesex, lodged an application for a British patent for 'A Mariner's Calculator', claiming 'improvements in instruments for measuring angles and distances, applicable to nautical and other purposes'. The application, No. 658, was granted in September 1834. Between 1617 and 1852 only seventy-nine patents were awarded in the category 'Compasses and Nautical Instruments' (Woodcroft 1854, pp. 529–237). The patents were awarded to renowned leaders in the field such as John Hadley in 1734,[20] Edward Troughton in 1788[21] and Edward Dent in 1844.[22] But quite intriguing is that, during this 235-year period, only one of these was to a woman: Mrs. Janet Taylor, aged thirty.

The Mariner's Calculator was an ingeniously clever concept, combining several nautical instruments in one. The specifications written by the young Mrs. Taylor included a lengthy three-page technical description of precisely how the device operated, along with two large diagrams (Woodcroft 1854,

p. 532). Detailed descriptions are also found in Taylor's own texts. She delivered a prototype of her new device to the Admiralty for assessment, and it was given to their own Hydrographer, Captain Francis Beaufort.[23] Beaufort was a man of high reputation and esteem, and a worthy one, in ordinary circumstances, to undertake an assessment of the Mariner's Calculator. On this occasion, however, the timing could not have been worse, as he was in the throes of a great personal crisis with the imminent death of his wife of over twenty-one years, Alicia, from breast cancer (Friendly 1977, pp. 268–269). In the midst of this great upheaval in his life, Beaufort finally delivered his report to the Admiralty, as noted in the Admiralty Minute Book. It was not favourable. It was not that he thought the instrument would not work, but in the 'clumsy fingers of seamen', he thought not. He also considered it would encourage 'slovenliness' (perhaps because it would do too much of the hard work).[24] By the time Patent Number 658 for 'The Mariner's Calculator' was formally granted on 27 September 1834, however, its fate had effectively been sealed. Without the Admiralty's backing, there was little chance that the other two organisations would support it either. The fact that Janet Taylor's name still sat in the register of patents for 'Compasses and Nautical Instruments', alongside men like John Hadley, would have been cold comfort. Indeed, it may have left her in dire financial straits, as she had gambled much of her capital on its success. It is known that she had inherited money from her father, which in turn he had inherited largely from her mother.[25] Eventually, Mrs. Taylor left nothing at her death, having to 'sacrifice' everything to meet the demands of her creditors.[26]

A particular question in the legacy of Mrs. Taylor is whether her Mariner's Calculator, in which she had placed so much faith, was given a fair assessment. Among her many qualities, Mrs. Taylor was undoubtedly a brilliant mathematician, and an extremely gifted author, astronomer, businesswoman and inventor. How could she have been so wrong with her invention—or was she? Alongside Captain Beaufort, another that expressed reservations about her instrument was the nautical historian, Charles H. Cotter, who wrote in his *History of the Navigator's Sextant* (1983, p. 152):

> Mrs Taylor claimed that [...] this instrument has an advantage over the sextant; for the observer can rest it steadily against the shoulder, and take the most correct observations with the greatest of ease [...] How far this claim was justified is not known, for it is probable that only a very few of these clumsy, yet ingenious, instruments were made.

In other words, Cotter viewed the Mariner's Calculator as exceptionally clever, but totally impractical. A reconstruction of the instrument was commissioned to test this question. It was undertaken under the supervision of the late Ron Robinson, one of England's leading compass adjusters and a specialist in the restoration of period nautical instruments (Croucher and Croucher 2011). Robinson's judgement was strikingly similar to Beaufort's. He said, 'to get a true sense of it, imagine giving something like the Mariner's Calculator to someone like a coal miner, with fingers like sausages, in poor light and under seagoing conditions'. The science was sound, but the instrument would be unworkable in practice by men with hands like these. Where Cotter suggested that the instrument was 'clumsy, yet ingenious', perhaps it would be fairer to say that the instrument was ingenious, but impractical in the 'clumsy' hands of its potential users. And so, it remains, it seems, an inspired, curious, white elephant of the history of nautical invention. The Mariner's Calculator was not a success, but this meant that it became a rarity. There is now only one of Mrs. Taylor's instruments known to be in existence. It is not in the possession of the National Maritime Museum at Greenwich, but is in private hands, having been purchased at considerable cost at auction in the 1990s.[27]

Another of Mrs. Taylor's many nautical instruments has found a special niche, again based on its appearance and provenance, rather than for its practicality. In the National Maritime Museum at Greenwich, London, a sextant entered by Mrs. Taylor in the Great Exhibition of 1851 is held for posterity. It is a magnificent, and flamboyantly ornate, sextant, or rather 'quintant', that appears to combine the principal aspirations of the exhibition in one: Arts and Manufacturing. At the centre of the instrument is the Prince of Wales crest, in the German of the Hanovers, '*Ich Dien*' (I serve') which also appears on its box. The elegance of Mrs. Taylor's creation, however, was lost on the Exhibition jury who did not deem it worthy of an award. It was seemingly assumed that Janet Taylor had exhibited a sextant intended for show rather than use, and her device was a regarded as pretty, rather than given a fair assessment as a scientific instrument: evidently, one was seen as incompatible with the other.

JANET TAYLOR—COMPASS ADJUSTER

Britain in the mid-nineteenth century was heavily reliant on shipping, not only for military engagements, but also for imperial ambitions and responsibilities. The method of construction of these ships was also changing significantly, using more and more iron. Weighing less than wooden ones, iron ships could be much larger and carry more cargo. From the early decades of the nineteenth century, ships had iron straps around their wooden hulls. But, by the 1860s, plates of iron encased the wooden hulls—the 'ironclads' as they were known—and then, in time, ships were constructed completely with iron hulls. These had an Achilles heel however: iron severely affected the ship's compasses. This now provided the next great challenge for navigators and the problem was far more complex than simply a mechanical one. As remarked by the historian of science, Alison Winter (1994, p. 72):

> The iron ships were testaments to the role of science in Britain's dominion over the world's magnetic and oceanic currents. But the fate of the entire industry hinged on the navigational problems raised by the use of iron in the construction of the ship, because the magnetic character of the hull caused

compass deviations which made them unreliable. And when the ships began to go astray, misled by errant compasses, the lack of a reliable means of correcting compasses threatened to produce a consequent shipwreck of public confidence in the scientists and engineers whose status had rested on the success of the enterprise.

The nautical world divided between 'applied' and 'theoretical' approaches to addressing the 'compass problem' and Mrs. Taylor became a key player in the middle of it. She ran a large compass adjusting business, based on the application of techniques for correcting magnets to counteract the effects of iron advocated by the Astronomer Royal, Professor George Airy.[28] Despite these techniques, the problem was still far from solved in the eyes of the public, and of the Admiralty Compass Committee which challenged Airy. After three years of deliberation, the Committee decided upon recommending a standard compass design for all naval compasses and the adoption of deviation tables. The compasses were to be checked regularly for deviation by 'swinging', but they were *not* to be corrected; this was a decision contrary to Airy's recommendation. Rather, a 'deviation table' was to be kept, to keep track of the deviation for each particular compass, so that adjustments could be made accordingly, as explained in an Admiralty publication in 1859. This 'tabular' approach was in contrast to the 'mechanical' approach of Airy, that relied on compensating magnets. Cotter refers to these approaches as 'two distinct schools of thought' (Cotter 1977, p. 300). The Royal Navy adopted the 'tabular' approach, but the mercantile marine followed Airy.

While the navy was now using the tabular method, merchant seamen meanwhile clamoured to have their compasses corrected according to Airy's system and this was 'almost universally adopted in iron ships of the mercantile marine' (Cotter 1977, p. 300). However, the introduction into the maritime community of the technical experts necessary to perform Airy's exacting method was not straightforward, as evidenced by continued shipwrecks and loss of life. For those with an entrepreneurial bent, however, there was now an opportunity in the field of compass adjusting for merchant ships. A new profession was born—and Mrs. Janet Taylor was one who joined it.

By 1854, when controversy erupted over the sinking of the steamship *Tayleur*, whose compasses had reportedly been adjusted according to Airy's methods, Taylor's firm had adjusted 'hundreds' of iron vessels and 'nearly all the largest steam ships and sailing vessels which have left the Thames'.[29] The issue of whether the compasses were, in fact, 'properly adjusted', became a *cause célèbre* in the nautical world in 1854, leading to attacks on compass adjusters in general, as well as specific criticism of Airy's methods. In both cases, Mrs. Taylor became a vocal participant (Croucher and Croucher 2018). Indeed, Mrs. Taylor wrote to Airy to advise him of her extensive practical application of his system since 1843 without a failure. She also expressed a preference for Airy's system over the 'tabular' method used by the Admiralty, writing to him that: 'I have seen with much regret the efforts made to do away with your system of magnets, knowing the difficulties and accidents that must arise if a 'Table of Errors' is alone to be relied on'.[30] Airy thanked her for her letter which, he remarked 'may prove of considerable value'.[31] So comprehensive and compelling was her unsolicited report that he immediately passed it on directly to Captain Beaufort for his perusal.

Airy was finally vindicated in 1862 by the findings of the Liverpool Compass Committee, as reported in the *Journal of Navigation* (Cotter 1977, p. 226). As a result, Airy's system of mechanical correction became generally accepted by the nautical community: first by merchant seamen, and gradually by the Compass Department of the Admiralty, the two methods complementing each other in time. For Janet Taylor, compass adjusting was not just a business opportunity but an intellectual challenge, and she proved her mettle in the maritime community, establishing a professional correspondence at the highest levels, with those like the Astronomer Royal.

JANET TAYLOR—A CIVIL LIST PENSION

On 10 January 1860 Mrs. Taylor was recognised with the award of a Civil List pension of £50 per year. This was awarded in consideration of 'her benevolent labours among the seafaring population of London and of the circumstances of difficulty in which she is placed by the death of her husband'.[32] This was not without precedent; the mathematician and astronomer Mary Somerville had been awarded £200 a year in 1835, increased to £300 a year in 1837 (Patterson 1983). Civil List pensions were formally granted in the name of the sovereign on the recommendation of the Prime Minister. They could be awarded in recognition of personal services to the Crown, by the performance of duties to the public, or by useful discoveries in science and attainment in literature and the arts. Mrs. Janet Taylor's circumstances of difficulty were true enough, but the decline of her resources was not only attributable to George's death, but also to difficulties with her business and debt. As for the award of £50—the lowest amount awarded—was this a fair acknowledgement of Mrs. Taylor's contribution? How does it sit alongside the award to Mrs. Somerville? A comparative analysis concludes that while both Mrs. Taylor and Mrs. Somerville made notable contributions to science, they were from 'different sides of the tracks', to use a contemporary metaphor (Croucher and Croucher 2012). They both challenged and defined mid-nineteenth-century ideals of womanhood. However, the contrast between their awards was ultimately not so much about gender—in which they can both be seen to defy the masculine focus of science and navigation—but about class: the differences in their awards reflect their situation in distinct social spheres. Mrs. Somerville and Mrs. Taylor appear to have come from very different ends of the middle-class scale. Mary Somerville was the daughter of an Admiral; Mrs. Taylor a daughter of a country curate, albeit a clever one. The Somervilles moved in elevated circles. Mrs. Somerville's second husband, her cousin Dr. William Somerville, was a Fellow of the élite Royal Society. Mrs. Taylor's husband was a brewery manager. The Somervilles' house in Hanover Square was in walking distance of the Royal Institution (Somerville 1874, p. 106). The Taylors lived in the heart of the East End, within walking distance of the Tower of London and the great docks on the Thames.

The difference in class is also evident in the scientific community, broadly between the 'high' end of theoretical scientific pursuits, and the 'low' end of applied work. Mrs. Somerville can be seen to reflect the 'West End' science group with an emphasis on theory, described by Hannah Gay as the 'gentlemanly science' of Westminster, as distinct from the applied 'East End' science of workers such as Janet Taylor (1997, p. 155).[33] Indeed, Taylor described herself as seeking to contribute to 'the practical science'.[34] In this regard, Mrs. Somerville was an 'insider', allied to the gentlemen scientists, with 'easy access to the foremost scientific minds in the nation' (Patterson 1983, p. 145). In contrast, Mrs. Taylor was very much an 'outsider', seeking acknowledgement and recognition from them. Women who were acknowledged in the Civil List pensions for literary contribution were generally given £50 or £100 (Patterson 1983, p. 156). Mrs. Somerville's award of £300 put her on a par with men like Airy himself. It seems fair to argue, then, that the award to Mrs. Somerville was more about class than it was about gender.

It was widely agreed that Mrs. Taylor was done an injustice by being denied a Civil List pension for so long and then, when it finally arrived, being awarded such a paltry sum. As the Bishop of Durham, writing to Prime Minister Melbourne, explained,

> if her works be of the value they are said to be and if her character corresponds to her talent (neither of which have I any reason to doubt, but I never vouch for what I do not myself know) she is more deserving of a pension even than Mrs Somerville; because the works of Mrs Taylor are of such practical usefulness in a branch of knowledge most important to this country.[35]

REMEMBERING MRS. TAYLOR

Janet Taylor died on 26 January 1870 and was interred in the burial plot of her brother-in-law, Canon Matthew Chester, who was married to her sister Elizabeth, at St Helen Auckland near Durham. Just ten days after her death, on 5 February 1870, an obituary was published in *The Athenaeum* (Y.L.Y. 1870):

MRS JANET TAYLOR

The past week's obituary records the departure of a remarkable person. Mrs. Janet Taylor was a mathematician of the first class: as such to be commemorated by the side of Mrs. Somerville; less universally cultivated but no less admirable in exposition than the latter named lady. In any event, little is known of her in the outside world. But her logarithmic tables we have been assured on fair authority, are correct and complete in no ordinary degree. And it was her singular occupation to prepare young men for the sea, by her tuition in the higher branches of mathematics.

A more quiet, a more singular union of rare powers of will and knowledge, especially in a woman, than her do not occur to us. She lived at the extreme east end of London, among her pupils and clients. We believe that she was as gentle and simple in herself as she was deeply versed in the abstract sciences which she professed. Perhaps some surviving relative or friend may be able to throw light on the life and labours of one who was as extraordinary from her acquirements of knowledge as from her social reticence.

While the obituary opens with acknowledging that Mrs. Taylor was 'a remarkable person' and a mathematician 'of the first class', and compares her, in class terms, with Mrs. Somerville, the dominant thread is framed in terms of gender: that she was 'gentle' and 'simple', 'socially retic[ent]' and that it was her 'singular occupation to prepare young men for the sea'. Presented in this way, the description of Mrs. Taylor speaks of the constructions of femininity of its time and clearly points to a tension that was not erased, even as her expertise and contributions to science and technology were clearly acknowledged.

Notes

1. This chapter is written by a descendent of Janet Taylor's older brother. Taylor did not keep a diary, despite what has been asserted by Laura Rose (2014, p. 105). Based on the internal references in the will and the family details, the document referred to by Rose is not of Janet Taylor, but of Jane Ann Ionn, Mrs. Taylor's niece, the daughter of Janet's oldest brother, William Ionn, who named his daughter after his sister, Jane Ann Ionn (who became Janet Taylor after her marriage).
2. In particular, see the UK Hydrographic Office Archives, Taunton, Somerset (Minute Books); and Royal Greenwich Observatory Archives, Papers of George Biddell Airy, University of Cambridge (RGO 6/687).
3. Anderson included a fair amount of detail of Mrs. Taylor's early life, and the narrative, being written so close to the time of her death, bears the hallmarks of family informants, probably Janet's youngest brother, Frederick Peter Ionn. Anderson also had insights as to her physical appearance that only someone who had met her, or spoken to someone who had, could know.
4. J. H. Lambert .1960. "Wolsingham from early times to 1938." 1960. Typed unpublished manuscript, Durham County Records Office, Durham.
5. A principal reference listed by Fisher is the short booklet on Mrs. Taylor published by Lieutenant-Commander Ken Alger (1982).
6. A scant few family papers have been located in Australia, England and France. Some glimpses are also recorded in correspondence with Taylor's daughter-in-law, Rachel Henning, published in Adams (1966).

7. Ionn Family Papers, France. See also Croucher and Croucher (2016, pp. 252–253 and 152–153).
8. This entry ran to almost a full page, while many other biographies spanned only a couple of lines; Mrs. Taylor was singled out for special mention, even though her life's work continued for many decades after 1840.
9. He was the curate of the Church of St Mary and St Stephen and schoolmaster of the Free Grammar School at Wolsingham. His appointment as schoolmaster on 13 May 1783, and curate after his ordination on 25 September 1785, is recorded in the Durham Diocesan Records, Durham University Library, DJR/EA/CLO/3/1785/21. An account of Peter Ionn as a schoolmaster is given by one of his former pupils (Nicholson 1926, p. 295). That her father taught navigation to his pupils is referred to in Edward Maltby to Lord Melbourne, 5 February 1838, Royal Archives, Windsor Castle, Melbourne Papers, Reference: RA MP 9/15.
10. Lambert reported (somewhat quaintly and most likely an erroneous transcription from his source) that the Queen supposedly said, 'Let her come, not to work at tapestry, as at her age it would spoil her figure. But let her be educated with the four until she is of age to take the place of one of them'. J. H. Lambert .1960. "Wolsingham from early times to 1938." 1960. Typed unpublished manuscript, Durham County Records Office, Durham, p. 170. The Ionn Family Papers include the following comment that 'her forte seemed to be figures', which is more accurate.
11. The appointment as governess is recorded by Janet's brother in his chart of 'Mems' (Croucher and Croucher 2016, p. 253).
12. From the 'Mems', we learn that Matthew's business failed and that he went north to Hastings in 1829. On 19 February 1829, their sister Isabella married in London, with her siblings as witnesses, then moved to Antwerp. Given Janet's marriage in The Hague on 30 January 1830, and the birth of Isabella's child on 1 January, it can be assumed that Janet moved to be with her sister in 1829 after Matthew's business failed.
13. They were married at the British Legation in The Hague by the Chaplain of the English Church, Rotterdam: Gemeente Achief 'S–Gravenhage Holland.
14. The 25th edition, 1865, is still available through Bibliobazaar print-on-demand.
15. On 1 October 1846 Beaufort was made a rear-admiral, and on 29 April 1848 a KCB for his civil services as hydrographer (Laughton 2008).
16. Admiralty: extract from UK Hydrographic Department Archives, Taunton, 'Minute Book No. 2, 1831–1837', p. 185; UK Hydrographic Department Archives, 'Letter Book No. 6, 1834–1836', p. 173; *Nautical Magazine*, July 1835, p. 438. To place this amount of money

into the context of the value of the currency of the time, it represented, for example, nearly three-months full salary for Beaufort himself. Trinity House: Court Minutes, Trinity House, p. 67, 2 July 1835. The 'handsome pecuniary award' was announced, with congratulations, in the *Nautical Magazine*, July 1835, p. 438.

17. For example, see the reference to her teaching in Janet Taylor to Captain Beaufort, 6 September 1836, UK Hydrographic Department Archives, letter no. T179. Other references are included in Taylor's books: the dedication page of the third edition of *Lunar Tables* (1836), gives her address as 'Nautical Academy, 103 Minories'. The title page of the first edition of *An Epitome of Navigation and Nautical Astronomy* (1842) has at the bottom 'Nautical Academy and Navigational Warehouse'. An example of an advertisement of the academy's programme was included in the back page of *Directions to Planisphere of the Stars* (1846). A crest of the academy is evident on her letters from mid-1845, for example Janet Taylor to (Beaufort?), 26 July 1845, UK Hydrographic Department Archives, letter no. T206. A description is given of her teaching in Alger and Croucher (2005).

18. An advertisement in the Mercantile Marine Magazine, October 1854, p. 45 suggests the wide range of instruments being produced: 'Manufacturer of every description of nautical and mathematical instruments'. See also Taylor (1966) and Clifton (1995). In 1856 a review appeared of 'Mrs. Janet Taylor's New Sea Artificial Horizon' in *Mercantile Marine Magazine and Nautical Record*, February 1856, p. 71.

19. She was appointed a sub-agent in 1835. In the 1841 Census, at which time Taylor was living in Minories, she was listed as 'chart-seller'. The principal chartseller was Robert Brettel Bate. See Robert Bate to Captain W Parry, 31 January 1829, UK Hydrographic Office Taunton, MLP 62, I.VII.

20. For a quadrant: Patent No. 550 (Woodcroft 1854, p. 530). Hadley had presented his findings to the Royal Society (Hadley 1731).

21. For a method of framing for nautical instruments: Patent No. 1644 (Woodcroft, p. 530).

22. For a ship's compass: No. 10,277. Woodcroft (p. 532). Dent patented another compass in 1850: No. 13,176.

23. We know this because it was Captain Beaufort who delivered a report on it in May: Minute Book, 21 May 1834, Hydrographic Office, Taunton.

24. Beaufort's report is recorded in Minute Book No. 2, May 1834, extracts 1831 to 1837, p. 114 and p. 21, Hydrographic Office, Taunton.

25. The Reverend Peter Ionn died on 2 May 1821, aged 59. His Will, dated 14 April 1818, is held at the University of Durham library, Durham Probate Records, 1822. The standing of his in-laws and the modest

inheritance through his wife is recorded in family records held in private hands in France.

26. This is based on deductions from two sources: first, the London Gazette of 11 May published a notice of adjudications and first meetings of creditors, among them listing, for 11am on 31 May, 'Janet Taylor, The Grove, Camberwell'. Under the Bankruptcy Act 1571, Commissioners of bankrupts could be appointed to allow a bankrupt to legally discharge their debts to creditors by an equitable and independent distribution of assets and then begin trading again with outstanding debts wiped out. The notices in the London Gazette were published by the Commissioners to inform creditors. Secondly, in a letter from Taylor to her children, she referred to having to 'sacrifice' her beautiful things: Janet Taylor to Deighton and Rachel Henning, 17 August 1869, in the possession of the author.

27. The National Maritime Museum wanted to acquire it, but at the hammer price of £15,000 it was beyond their budget.

28. The advertisement also noted: 'The Deviations in the Compasses of Iron Ships found and corrected'. Janet engaged in correspondence with the Astronomer Royal, Professor George Airy on the subject: Janet Taylor to George Airy, Astronomer Royal, 30 November 1854, Airy Papers, RGO 6/687, Section 22, 173–175, Cambridge University; George Airy to Janet Taylor, 11 December 1854, Airy Papers, Sect. 24, 194; Janet Taylor to George Airy, 16 December 1854, Airy Papers, Section 24, 196–203.

29. Janet Taylor to George Airy, 30 November 1854. Airy Papers, University of Cambridge.

30. Janet Taylor to George Airy, 30 November 1854. Airy Papers, University of Cambridge.

31. George Airy to Janet Taylor, 6 December 1854. Airy Papers, University of Cambridge.

32. 'Return of all Pensions granted and charged upon the Civil List, in accordance with the Act 1 Vict. c. 2, with the Grounds upon which such Pensions have been granted', *Accounts and Papers of the British House of Commons*, vol. xxxiv, 1861.

33. Gay acknowledges however that 'it would be false to claim any rigid dichotomy of a West End devoted to pure science and an East End devoted solely to applied science' (1997, p. 155).

34. Janet Taylor to George Airy, 29 November 1837, Airy Papers, University of Cambridge.

35. Edward Maltby to Lord Melbourne, 5 February 1838, Royal Archives, Windsor, Melbourne Papers, Reference No. RA- MP/9/15.

REFERENCES

Adams, David, ed. 1966. *The Letters of Rachel Henning*. Sydney: Angus and Robertson.
Alger, Ken and Croucher, John S. 2005. "An Exceptional Woman of Science." *Bulletin of the Scientific Instrument Society* 84: 22–27.
Alger, Ken. 1982. *Mrs. Janet Taylor: Authoress and Instructress in Navigation and Nautical Astronomy (1804–1870)*. London: City of London Polytechnic: LLRS Publications.
Anderson, W. 1871. *Model Women*, 2nd edition. London.
Boase, Frederic. 1901. *Modern English Biography*, vol. III. Truro: Netherton and Worth.
Brück, Mary. 2009. *Women in Early British and Irish Astronomy*. London: Springer.
Clifton, Gloria C. 1995. *Directory of British Scientific Instrument Makers 1550–1851*. London: Zwemmer and National Maritime Museum.
Cotter, Charles H. 1983. *A History of the Navigator's Sextant*. Glasgow: Brown, Son and Ferguson.
Cotter, Charles H. 1977. "John Thomas Towson (1804–1881): His Contribution to Navigation." *Journal of Navigation* 30, no. 2: 220–231.
Cotter, Charles H. 1977. "The Royal Society and the Deviation of the Compass." *Notes and Records of the Royal Society of London* 31, no. 2: 297–307.
Croucher, John S. and Croucher, Rosalind F. 2018. "Compasses and Sinking Ships: Mrs. Janet Taylor's Contribution to the Compass Adjusting Controversy Mid-19th England." *International Journal of Maritime History* 30, no. 2: 234–251.
Croucher, John S. and Croucher, Rosalind F. 2016. *Mistress of Science: The Story of the Remarkable Janet Taylor, Pioneer of Sea Navigation*. Stroud: Amberley.
Croucher, John S. and Croucher, Rosalind F. 2012. "Mrs. Janet Taylor and the Civil List Pension: A Claim to Recognition by Her Country." *Women's History Review* 21: 253–280.
Croucher, John S. and Croucher, Rosalind F. 2011. "Mrs. Janet Taylor's 'Mariner's Calculator': Assessment and Re-Assessment." *British Journal for the History of Science* 44, no. 4: 493–507.
E.C.M. 1933. "The Beginnings of Organised Instruction in Navigation." *Nautical Magazine*, 104.
Fisher, Susanna. 2004. "Taylor [née Ionn] (1804–1870)." *Oxford Dictionary of National Biography*. https://doi-org.liverpool.idm.oclc.org/10.1093/ref:odnb/49543. Accessed on 13 December 2020.
Friendly, Alfred. 1977. *Beaufort of the Admiralty: The Life of Sir Francis Beaufort*. London: Hutchinson.
Gay, Hannah. 1997. "East End, West End: Science Education, Culture and Class in Mid-Victorian London." *Canadian Journal of History* 32: 153–183.
Hadley, John. 1731. "The Description of a New Instrument for Taking Angles." *Philosophical Transactions (1683–1775)* 37: 147–356.
Laughton, J.K. rev. Roger, N.A.M. 2008. "Sir Francis Beaufort (1774–1857)." *Oxford Dictionary of National Biography*. https://doi-org.liverpool.idm.oclc.org/10.1093/ref:odnb/1857. Accessed on 13 December 2020.
Mrs. Janet Taylor's New Sea Artificial Horizon. 1856. *Mercantile Marine Magazine and Nautical Record*. February: 71.

Nicholson, W. 1926. "Recollections of My Youth." In: *Records of Wolsingham*, edited by Thomas Valentine Devey, 295. Newcastle-upon-Tyne: Northumberland Press.

Patterson, Elizabeth Chambers. 1983. *Mary Somerville and the Cultivation of Science, 1815–1840*. The Hague: Martinus Nijhoff.

Rose, Laura. 2014. *Poppy's with Honour*. Bloomington: AuthorHouse.

Somerville, Mary. 1874. *Personal Recollections, from Early Life to Old Age, of Mary Somerville*. Boston: Roberts Brothers.

Taylor, Eva G.R. 1966. *The Mathematical Practitioners of Hanoverian England 1714–1840*. Cambridge: Cambridge University Press.

Taylor, Janet. 1842. *An Epitome of Navigation and Nautical Astronomy*. London: G. Taylor.

Taylor, Janet. 1846. *Directions to Planisphere of the Stars*. London.

Taylor, Janet. 1836. *Lunar Tables*. London: G. Taylor.

The Admiralty. 1859. *Practical Rules for Ascertaining and Applying the Deviations of the Compass Caused by the Iron in a Ship*. London: Hydrographic Office.

Watts, Ruth. 2007. *Women in Science: A Social and Cultural History*. London: Routledge.

Winter, Alison. 1994. "'Compasses All Awry': The Iron Ship and the Ambiguities of Cultural Authority in Victorian Britain." *Victorian Studies* 38: 69–98.

Woodcroft, Bennet. 1854. *Subject Matter Index of Patents and Inventions from 2 March 1617 to 1 October 1852*. London: Queens Printing Office.

Y.L.Y. 1870. "Mrs. Janet Taylor." *The Athenaeum*, 5 February, no. 2206: 199.

CHAPTER 6

Early Female Geologists: The Importance of Professional and Educational Societies During the Late Nineteenth and Early Twentieth Centuries

Cynthia V. Burek

The importance of professional and amateur societies to women trying to educate themselves and follow academic careers was paramount in late-Victorian society and during the early years of the twentieth century. These societies operated at three different levels: international, national, and local with each catering for a different type of educated female. Some women chose to belong to more than one type of society according to their limited time and financial constraints. This chapter analyses four different societies to understand both the role that they played in the development of early female geologists, and to illuminate the contributions of women themselves to the societies of which they were members. As large international societies were restricted in access to all but the wealthiest and independent female geologists, the main focus here will be on the local level. The Geological Society of London represents the British national professional body; the British Federation of University Women represents the national networking organization for graduate women; the Chester Society of Natural Science represents a local amateur society; and Cambridge University's Sedgwick Club represents one of the first student geological societies to admit women. The significance of

C. V. Burek (✉)
Department of Biological Sciences, University of Chester, Chester, UK
e-mail: c.burek@chester.ac.uk

these societies to women will be illustrated by case studies of the careers of Gertrude Elles, Agnes Arber, Maria Ogilvie Gordon, Ethel Wood, Ethel Skeat and Caroline Coignou. However, to understand these women's experiences it is necessary first to detail the social attitudes and legislative context which provided the backdrop to their geological careers.

SOCIAL CONTEXT AND ATTITUDES TO SCIENTIFIC WOMEN

Late Victorian and early twentieth-century British society was a time when women were slowly allowed greater freedom to assume a wider role in politics, education, industry, voluntary work, and life in general. The relevant legislation arguably starts with the 1832 Reform Act as this prompted discussions as to women's place in society and connected to the growing debate on women's suffrage. The Education Act of 1870—the first of a series of acts which introduced compulsory education in England and Wales for children aged from five to thirteen years—increased the need for trained teachers. In 1849 Bedford College became the first higher education college for women in England; by the 1860s, other colleges were being founded and some established universities were beginning to open their doors to admit women. These included University College, London, which allowed women to sit degrees from 1878, and the new colleges for women at Cambridge: Girton College, established in 1869, and Newnham College, established in 1871 (although women at these colleges were not eligible to be awarded University degrees until much later). As the twentieth century dawned, the new 'red-brick' civic universities began to appear and offer more opportunities for higher education to women, such as Owens College in Manchester and Mason College in Birmingham, but not all of these offered geology courses. The situation in Scotland was slightly different.

The Married Women's Property Acts of 1870 and 1882 paved the way for married women to own their own property including land and, more importantly, money and items such as books, furniture, and laboratory specimens and equipment. The importance of this to early female geologists becomes clear below in sections looking at Maria Ogilvie Gordon, Ethel Wood, and Dame Ethel Shakespear. The 1888 County Council Act gave women rate payers the right to vote in Council and Borough elections and began to encourage the appointment of women to positions of influence. In 1919 the Sex Disqualification (Removal) Act made it illegal for institutions to discriminate membership on the basis of sex and it was this piece of legislation that is of paramount importance as it made a real difference for professional women. Up until this point women scientists, especially geologists, had been banned from fellowship of many professional societies such as the Geological Society, the Chemical Society, and the Antiquities Society. Some societies, for example the Linnean Society, had allowed women to participate and become members or fellows previously to the 1919 Act but, in contrast the Linnean, most of these tended to be the less élite, more amateur and/or local societies, as shall be demonstrated below in relation to the Chester Society for Natural Science. The Geological Society of London was and remains the only professional society for geologists in the UK. The new enabling legislation of 1919 was significant in allowing female participation for the first time in these professional societies. The barrier had finally been removed. However, it took further legislation with the 1975 Sex Discrimination, and time for society's attitudes towards female participation to change, to make any significant difference. Women were still regarded as fragile, lacking brain capacity and requiring chaperoning especially outside the home. This last point limited fieldwork for early female geologists, something which was a key requirement of the profession.

INTERNATIONAL GEOLOGICAL SOCIETIES

International Earth Science societies that were open to female geologists during the period under consideration were the International Geological Congress and some specialized societies, for example the International Union of Geodesy and Geophysics and, from 1899 to 1914 when it was disestablished, the International Association of Academies. These international societies had limited general appeal however, as travel, money and transport often presented a barrier to participation. Some women did engage through correspondence with international societies though, and many preferred to publish their work in international journals if they could, such as Maria Ogilvie Gordon. Some aristocratic and/or independently wealthy women were able to afford to attend international geological meetings; one example is Rachel Workman, an American who was wealthy in her own right, who attended the 1910 International Congress in Stockholm and, in 1911, married Alexander MacRobert, to become Lady MacRoberts of Douneside, Aberdeenshire. Catherine Raisin, a single lecturer and expert on serpentinites, from Bedford College, London, also attended the Stockholm International Geological Congress. Raisin took part in the discussion on Pre-Cambrian geology using her petrological experience from the English Midlands which she had earlier described in an account published in the *Manchester High School Magazine* in 1902. Rachel Workman would meet Catherine Raisin on future International Geological Congresses such as in Toronto three years later (Burek 2009). Rachel MacRoberts corresponded with both Catherine Raisin and Maria Ogilvie Gordon, forming her own small geological network while ensconced in the highlands of Scotland. Maria Ogilvie Gordon also attended overseas conferences, presented papers to international meetings, and received tremendous accolade for her work on the structure of the Dolomites and Alps. She really valued this as she felt that few at home understood her work in the Dolomites. In a letter dated 1929 to Julius Pia, the curator of the Vienna

Natural History Museum, she wrote that 'In my own country I never count at all. I am made to feel a complete outsider' (Pia 1939, p. 184). Despite this, Gordon seemed to feel the lack of any network or support from fellow geologists while working overseas too. On the award of the Geological Society of London's Lyell medal in 1932, she spoke of deriving special pleasure in receiving final recognition from the geologists of her homeland as 'my work had to be done outside Great Britain and was humanly of so isolated a character [...]' (Gordon 1932, p. 60). Perhaps her lasting mark of esteem was to have a fossil fern from the Dolomites named after her (*Gordonopteris lorigue*) in 2005, many years after she had died (Wachtler 2016).

These geologists were exceptions and they individually gained tremendously from the networking and exposure to international ideas, as is shown in Rachel Workman's letters to her fiancée, Sir Alexander MacRobert, written from the Swedish International Geological Congress held in August 1910:

> Everyone is most awfully nice, but I had to overcome the usual annoyance men have when women are about on scientific expeditions. Now that they have found that I am not a drag on them or bore them with talk they are very pleasant.[1]
>
> Everyone is most awfully kind to me, and I am having a splendid time. It is most inspiring to meet so many eminent geologists and to hear their discussions etc. in the field.[2]

International conferences did not feature widely in the lives of most early female geologists. However, international meetings and societies were important for a few early women as they afforded them international exposure for their work, but it was the national and local networks which really mattered to many early women geologists.

NATIONAL GEOLOGICAL SOCIETIES

National scientific societies were much more accessible to early female geologists as the financial and travel burdens were far less arduous. While these national societies covered a wide range of disciplines, there were also societies for educated, professional women wanting to mix freely with others of a like mind but outside their own discipline. There were two main national geological societies for the UK, each of which catered for a different audience: the Geological Society of London (GSL) and the Geologists' Association (GA). The GSL, from 1874 based in Burlington House, Piccadilly, London, represented the professional geologist. By contrast, the GA concentrated on amateur geology, fieldwork, and regional geological discussion. The latter admitted women in from its initiation with the first female members joining in 1859 (Green 2008).

At the GSL, papers of the highest calibre were read, and rigorous discussion and debate took place on important geological ideas of the time. The location and seating arrangements were fashioned on the Houses of Parliament with members sitting opposite each other in true debating but also antagonistic style. This was not something that most women had ever been exposed to and, furthermore, it was considered intimidating as most regional societies were 'lecture style' or in the round. For the early female geologist, the GSL was out of bounds until they were allowed to attend lectures from 1904 or permitted entry on special occasions to use the library. This was the case with Margaret Crosfield: only after she wrote letters to the President of the GSL was she allowed to use the facilities (Burek 2021a, b). Female geologists could not become fellows of the Society until 1919 when the effects of the Sex Disqualification (Removal) Act of 1919 were anticipated. Admission of women to the Society had been discussed several times before 1919 but always failed on the vote, as the fellows felt the presence of women would have a 'dampening down' affect, by distracting the male fellows and disrupting the deep intellectual conversations. Charles Lyell wanted to introduce women to the society as visitors in March 1860 but, after consulting with T. H. Huxley, he received a letter on the seventeenth of that month in which the following statement was made, '[…]the Geological Society is not, to my mind, a place of education for students, but a place of discussion for adepts, and the more it is applied to the former purpose the less competent it must become to fulfil the latter' (Herries-Davies 2007, p. 185). Thus, the valuable face-to-face networking resulting from membership was denied to women. Instead, they had to resort to letters being read at committee meetings to have their opinions heard and were also denied access to the often-heated debates on contentious issues. As Rachel, Lady MacRoberts, states in a letter to her husband 'there is no doubt one needs to be in touch with the current thought and to attend lectures and hear and discuss things and keep abreast of the times.'[3]

Although women could submit papers and have them read by male fellows, they themselves were not able to attend to hear the discussion or to answer questions. Caroline Coignou, a scholar at Newnham College, Cambridge, found a new species of trilobite on a fieldtrip with her professor. She wrote her findings up as a paper to be read at the Geological Society in 1890 by Professor McKenny Hughes. Coignou's paper was well received; that early female work was valued cannot be denied as they received medals and grants, as shown in Table 6.1.

However, male fellows had to receive these accolades on their behalf. Funds were often awarded to encourage further fieldwork as was made clear in congratulations speeches; for women, these monies were received by a male fellow and he was expected to give it to the female recipient (Burek 2021a, b). Miss Catherine Raisin was awarded the Lyell Geological Fund in 1893 and received it via Professor Bonney:

Table 6.1 Medals and grants awarded to early female geologists by the geological society

Date	Name	Medal	Fund	Received by
1893	Catherine Raisin		Lyell £24 16s 3d	Professor Bonney
1898	Jane Donald		Murchison £28 14s 3d	Mr Newton
1900	Gertrude Elles		Lyell £19 6s 0d	Professor McKenny Hughes
1903	Elizabeth Gray		Murchison £22 15s10d	Unknown
1904	Ethel Wood		Wollaston £34 6s10d	Professor Marr
1906	Helen Drew		Daniel Pidgeon Fund (undisclosed amount)	Sent
1907	Ida Slater		Daniel Pidgeon Fund (undisclosed amount)	Sent
1908	Ethel Skeat		Murchison £25 8s 4d	Herself
1919	Gertrude Elles	Murchison		Herself
1919	Eleanor Mary Reid		Murchison (undisclosed amount)	Unknown
1920	Ethel Wood Shakespear	Murchison		Herself
1920	Marjorie Chandler		Daniel Pidgeon Fund	Sent
1927	Marjorie Chandler		Wollaston	Unknown
1932	Maria Ogilvie Gordon	Lyell		Herself

In asking you to forward to Miss Raisin one moiety of the balance of the proceeds of the Lyell Geological Fund I am performing a very pleasing duty... Miss Raisin's excellent work both in the field and with the microscope... The Council in making this award, wish it to be regarded as an acknowledgement of past work and at the same time as an encouragement for the future. (Hudlestone 1893, p. 42)

For Miss Gertrude Elles, seven years later in 1900, the same thing happened: 'the proceeds of the Lyell Geological Fund, awarded to Miss Gertrude L. Elles of Newnham College, to Prof T. McKenny Hughes F.R.S. for transmission to the recipient [...]'. She is then informed by the President that 'this Award from the Lyell Geological Fund will show her that her work is valued and will encourage her to continue it.' Elles replies via Professor McKenny Hughes that 'I am bound to regard it as an incentive to future work [...] I will strive my very utmost to make the work which I may do in the future worthy of the confidence' (McKenny Hughes 1900, pp. 48–49). In this way, early female geological research was often undertaken in isolation without the benefit of a wider audience. Frequently the financial awards received from the Society were used for fieldwork research and, indeed, recipients were encouraged to use the money for this purpose. When they had the opportunity, women tended to undertake their research fieldwork in pairs for safety reasons, often partnering with someone they had met through networking while at university. There was safety in numbers when doing fieldwork in isolated areas; women working alone could be subject to hassle from quarrymen and, indeed, Catherine Raisin on occasion would usher her students away from a quarry when male students or working men appeared. This situation has not changed today when most universities have 'working alone' policies. The advance of technology in the form of mobile phones can add to security precautions, providing reception is available.

The oft-called quartet of female palaeontologists: Margaret Crosfield, Ethel Skeat, Ethel Wood and Gertrude Elles, all from Newnham College, Cambridge, are an example of women working together. These four women had studied together at Cambridge from 1891 to 1894 and maintained their friendships throughout their lives. Gertrude Elles, in particular, helped identify graptolites for Ethel Skeat (later Mrs. Woods) and Margaret Crosfield when they worked together on the stratigraphy of north east Wales. This resulted in a paper in the Quarterly Journal of the Geological Society (Woods and Crosfield 1925). Several other significant pieces of research were produced. Perhaps the most important of these was the eighteen-year collaboration between Gertrude Elles and Ethel Wood (later Dame Shakespear) on the *Monograph on British Graptolites* 1901–1918 published by the Palaeontological Society. This work was acknowledged by the President of the Geological Society, George Lamplugh, in his Murchison Medal presentation and address to Ethel Shakespear in 1920:

> In collaboration with Dr. Gertrude Elles and Prof. C Lapworth, you have produced a Monograph of British Graptolites which is, and will remain, of the utmost service to all students of the Palaeozoic rocks […] Your results in this field […] have provided us with a record of the Graptolite succession of Silurian times which has become the standard of reference wherever the rocks of this age are studied […] It has been the privilege of the Society to publish most of the papers in which these far-reaching stratigraphical researches are embodied […] In handing you this [Murchison] Medal and Award on behalf of the Council, I ask you to accept also my personal congratulations on your success in advancing geological science. (Lamplugh 1920, p. 44)

Margaret Crosfield and Ethel Skeat (later Mrs. Wood) also produced some highly regarded fieldwork papers published by the Geological Society, especially on the area around Carmarthen, South Wales (Skeat and Crosfield 1896). This research was acknowledged and used by the Geological Survey geologist, Thomas Crosby Cantrill, an early mapper of the South Wales coalfields (Elles 1953). Another close association developed between two slightly later Newnham female geologists, Ida Slater and Helen Drew. These two

Table 6.2 The first female fellows of the geological society 1919–1922

Year	Date	Title	First names	Surname
1919	21 May	Miss	Margaret Chorley	Crosfield
		Miss	Gertrude Lilian	Elles
		Mrs	Maria Matilda Ogilvie	Gordon
		Miss	Mary Sophia	Johnston
		Mrs	(Mary) Jane Donald	Longstaff
		Lady MacRobert	Rachel	Workman
		Miss	Mildred Blanche	Robinson
		Mrs	Ethel Gertrude Skeat	Woods
	25 June	Miss	Catherine Alice	Raisin
		Mrs	Margaret Flowerdew	Romanes
	3 December	Miss	Mary Kingdon	Heslop
		Miss	Dorothy Margaret	Woodhead
1920	21 January	Miss	Florence Annie	Pitts
		Mrs	Eleanor Mary	Reid
	23 June	Dr	Ethel Mary Reader Wood	Shakespear
	1 December	Miss	Lucy	Ormrod
1922	22 February	Miss	Irene Helen	Lowe
	22 March	Miss	Edith	Goodyear
		Muir	Helen Margarueite	Muir-Wood
	14 June	Dr	Marie Carmichael	Stopes
	6 December	Miss	Agnes Irene	McDonald

women are included below in the discussion of the Sedgwick Club. Table 6.2 lists the number of women who were elected to a fellowship of the Geological Society in the first years after the 1919 Sex (Disqualification) Removal Act. Numbers are small and do not represent the rush that was anticipated by some of the male fellows; indeed, only twenty-one women were admitted as fellows between 1919 and June 1922, representing a tiny fraction of the overall fellowship.

Networking Societies

Opportunities for networking and comradeship were vitally important to early female geologists. These allowed women operating in a male dominated discipline to communicate with other professional or educated women at the same intellectual level. Most of these women were also experiencing the same difficulties in progressing their careers and advancing their education and research, for example due to financial problems, discrimination in their research, and lack of facilities and employment opportunities. Sharing ideas, discussing good practice, and overcoming problems of discrimination, financial inequality and not being taken seriously, took place during these meetings. Friendship became important as well as networking for mutual support.

A good example of one of these networking societies is the British Federation of University Women (BFUW, now the British Federation of Women Graduates) founded in 1907 in Manchester. An informal meeting of seventeen graduate women, all working in Manchester, took place at Manchester High School and it was decided to form a group of graduate women who would help each other through friendship and networking. At that meeting a proposal was put forward by Miss Caroline Coignou and Miss Ida Smedley to form a Federation of University Women (Morley 1949). The Federation was to act as a means of communication and of united action in matters affecting the interests of women. Although no list survives of all the participants, it is likely that Marie Stopes would have attended as she was at Manchester University researching palaeobotany and was a member of the staff, as a demonstrator, at the time. Stopes felt she needed the support of other educated women in a department dominated by men and, certainly, she joined the organization and later applied for grants to support her research. At a second meeting in May 1907, thirty-four women met formally in the Board room of Manchester High School to constitute the organization (Coignou 1928). Its aims were initially fourfold, as quoted in Sondheimer (1957): to promote women's work on public bodies; to work for the removal of sex disabilities; to facilitate the intercommunication and cooperation of university women; and to afford opportunity for the expression of a united opinion by university women on subjects of special interest to them.

By 1909 local associations of the BFUW had been set up in London, Cardiff, Leeds, Liverpool, Bangor, Cambridge, St. Andrews, and Sheffield (Morley 1949). The first annual general meeting was held in Sheffield in June 1910 and, by the end of 1913, membership reached 789. This coincides with the rise of the new civic 'red-brick' universities at the beginning of the twentieth century such as Birmingham (1900), Manchester (1903), Liverpool (1903), Leeds (1904), Bristol (1905) and Sheffield (1909). These grew out of former engineering or technical institutions in Victorian and Edwardian industrial cities. All had become universities before World War One. Following the end of the conflict, in 1919, an international Federation was set up spearheaded by Virginia Gildersleeve, Rose Sidgwick and Caroline Spurgeon. Its mission was to encourage friendship across the world between educated women and to ensure that another war of such dimensions never happened again. This was the International Federation of University Women (today Graduate Women International) made up of the British Federation of University Women, American Association of University Women, and the Canadian Federation of University Women.

University women had particular problems as, like most women, they were undervalued with lower salaries than men and few paid posts. Dr. Catherine Raisin at Bedford College, London, constantly complained about this in her letters to University governors during the early 1900s. In particular, Raisin objects that she is awarded only one-year contracts at a time and that this gives her no security. Women were in the minority and at the bottom of the academic hierarchy, so the need for informal, supportive networks was considerable (Dyhouse 1995). The BFUW's role in providing this was crucial. Several early female geologists joined BFUW branches and took key roles; of the twenty first female fellows of the Geological Society, women who had been admitted up to June 1922, at least nine were members of the BFUW. However, there were several more who had studied geology at university but chose not to seek election to the fellowship of the Geological Society. This may have been because fees were quite high compared to salaries paid to women. If you were not undertaking research within geology, but were teaching at primary or secondary school level, then the incentive to join was small. Caroline Coignou, who had studied geology at Newnham College, Cambridge, is one example of a non-joiner. Others include Sedgwick Club members Elizabeth Dale (Girton College, Cambridge) and Agnes Robertson (later Dr. Agnes Arber, Newnham College, Cambridge). Dr. Ethel Skeat is an exception to this, but she was still researching while teaching at the Queens School in Chester. Table 6.3 shows these early female fellows of the Geological Society and the roles they played within the BFUW.

Many of these women found the friendship offered by the local branches invaluable, especially the single women. Here they met for discussions on issues such as equal pay, voting rights, pensions, and work opportunities. The early minutes of the meetings of the London group illustrate the range of topics covered and the actions the women undertook to help each other. Within the minutes for 9 November 1918 is a resolution that still has a contemporary echo to it: 'It was agreed to put the following resolution before the general meeting at 3 o'clock. "that equal pay for equal work is the only principle consistent with justice to men and women alike"'.[4]

These concerns included the right to proper remuneration, job security and the right to be interviewed for professorial appointments. Catherine Raisin, an eminent geologist and head of the geology department at Bedford College, London, became a life member of the BFUW in 1917; she attended the London branch dinners on a regular basis, taking time out from a very heavy workload to talk to women in other disciplines. Raisin acted as a role model for many women, including Dr. Doris Reynolds. Reynolds, one of Raisin's graduate students and an expert on granite, recalled many years later the effect

Table 6.3 Early female geologists and membership of the British Federation of University Women (BFUW)

Name	Branch	Dates	Membership	Role
Catherine Raisin FGS (Fellow of the Geological Society)	London	1912	Life member 1917	Helped with the sixth annual general meeting (AGM) held at Bedford College, London, in June 1915
Caroline Coignou	Manchester Leeds	1907–1913 1913–1932	Life member	Representative for Leeds at AGMs; voting delegate at International Federation of University Women (IFUW) Paris conference 1921; served on 1923 Committee for International Relations; left legacy to British Federation in 1936
Gertrude Elles FGS	Cambridge	1913	Life member 1919	Elected a member of the national executive for one year in 1914
Ethel Skeat, later Mrs Wood FGS	Cambridge	1919		A delegate for Cambridge to the national AGM in 1920
Ethel Woods, later Dame Shakespear FGS	Birmingham	1922	Member until 1939	President of Birmingham branch and representative to national AGM in 1925
Maria Ogilvie Gordon FGS	London	1922		
Irene Helen Lowe FGS	London	1917	Life member until death in 1946	
Helen Drew	Bristol	1923–1926		

(continued)

Table 6.3 (continued)

Name	Branch	Dates	Membership	Role
Mary Kingdon Heslop FGS	London Newcastle	1922–1925 1928–1931		
Edith Goodyear FGS	London	1914	Life member	Represented London at national AGMs and served on London Committee 1928–1931
Isabel Ellie Knaggs FGS	Cambridge London	1917–1920 1920–1925		
Marie Stopes FGS	London	1917–1938	Life member from 1917	
Mary Foley	London	1916–1924	Died 1925	Honorary Secretary of London branch 1917–1924

that Catherine Raisin had had on her at Bedford College (Burek 2007; Lewis 2020).

Caroline Coignou (1865–1932) was also an active life member of the BFUW at branch, national and international level. She was an only child, born to immigrant French parents, and attended Manchester Grammar School for Girls where she was one of six girls to study geology in the sixth form. Coignou gained a place at Newnham College, Cambridge, with the highest mark in botany in the Cambridge local exam of 1883—an examination which was open to both boys and girls. She received a Clothworkers' Scholarship to read the Cambridge Natural Science Tripos, including geology, botany and chemistry. On graduating, Coignou became one of the 'Steamboat Ladies' who, from 1903–1907, collected their degrees from Trinity College, Dublin, as at that time neither Oxford nor Cambridge universities granted women degrees even when they had passed the relevant exams (Parkes 2007). Coignou became an assistant mistress at Pendleton School, Salford, from 1890–1894, before returning to her old school, Manchester High School for Girls, where she remained until 1910.

While an assistant mistress at Pendleton, Coignou published an article in the *Quarterly Journal of the Geological Society* (Coignou 1890). A fossil was later named after her (Miller 1973, 1976). Geology was not widely taught in schools and often teachers combined or included it within teaching chemistry, geography, botany, or physics. Other openings for female geologists in the UK were few and far between, unlike the States where the First World War had opened up opportunities for women in the oil fields when the men left for war (Gries 2018, 2021). A few openings for qualified female geologists existed in the mining and chemical industries, but most went into teaching if they did not marry. Women were not permitted to work as geologists in the Geological Survey of Great Britain until 1943. For her teaching, Coignou specialized in botany, nature studies, geology, physiography (now recognized as landscape and physical geography), chemistry and hygiene. Ethel Skeat at Queens School in Chester covered a similar range of subjects and included geology wherever she could. She supplemented this by running geology fieldtrips for the girls. At Manchester Grammar School for Girls, Coignou also enjoyed taking the pupils on geological field excursions. One of these was a five-day trip for the girls to Cambridge in 1901 which illustrates the value of networking from her Newnham College student days. While away, the Manchester High School party had several invitations including:

> An invitation to dine at Clough Hall, Newnham on the Friday evening. The evening was passed pleasantly in renewing acquaintance with old High School girls [...] Saturday morning was filled in by visits to the Balfour Laboratory over which [...] Miss Klaussen showed the party. The original research work which was going on in the building proved a source of great interest to the girls. A short time was spent in the Geological Museum, a building which seems rather inadequate to contain such a fine collection of specimens.
>
> In the afternoon, Professor Hughes, at whose house the party had lunched, was kind enough to accompany his guests to Trinity College... Through Professor Hughes' influence, Dr. Perowne's unique collection of amber was seen. This collection includes many large pieces of amber, of shades ranging from honey colour to dark sherry colour, and all are very beautiful. (McNichol 1901)

It came as no surprize that in 1902 Coignou received the coveted Frances Mary Buss Memorial Scholarship which enabled her to travel to the USA for six months as an exchange teacher. This trip had a tremendous influence on her and broadened her scope within teaching. Coignou now had an international network of fellow teachers.

Coignou's keen involvement in setting up a group of like-minded individuals was understandable. She had intelligent parents but no siblings of her own age and was quite isolated. She missed her group of friends at Newnham College who, like her, were interested in science subjects. Coignou took an active part in the Manchester branch of the BFUW and later represented it at national Annual General Meetings. In 1910 she gave up teaching to move to Leeds to become an examiner for schools for the West Riding of Yorkshire education department, a position she held for eight years, becoming inspector of secondary schools from 1918 to 1927. During this period, her free time was occupied with BFUW matters. By 1913 Coignou was a member of the Leeds branch and represented it on the national executive. In 1920–1921 she travelled as a British voting delegate to the Paris International Federation of University Women conference. In 1923 she was elected to the BFUW Committee of International Relations, reflecting her keen interest in international matters following her exchange visit to America. After her retirement, Coignou spent time networking at Crosby Hall, run by the BFUW in London, which was one of a network of residential clubs for women academics (Goodman 2011). 'But to tell the whole story of the Federation of University Women and its wonderful international development during the past twenty-one years is for another pen than mine' she wrote towards the end of her life, adding that for 'friends and members not only of the British Federation of University Women but of the International Federation of University Women all over the world, Crosby Hall stands today dedicated to the encouragement of learning and the promotion of friendship between the women of all nations' (Coignou 1928, pp. 9–10). The importance of a safe residence for female geologists travelling alone to consult material in London, such as types of fossils in the Natural History Museum or key mineral specimens in the Geological Survey, was essential. Coignou contributed financially to the upkeep of the Yorkshire room at Crosby Hall and clearly loved visiting. She used her time there to visit geological museums and attend meetings. In her will she left a small legacy of £163.8s.3d to the BFUW which shows how much she valued the facilities and friendship she derived from it. During her life, she had believed passionately in women's education, led her female students on geological fieldtrips, and did all in her power to deliver a scientific education to her pupils while never giving up on learning herself. The friendships and support she received from colleagues in the BFUW helped her achieve her aims and potential.

Local Societies

Local societies were a key mechanism for women to network and follow their geological interests. These societies adopted various practices according to the local situation and membership. Two case studies will be presented here: one of the branches of the amateur Society of Natural Science, and an early student geology society, the University of Cambridge's Sedgwick Club. These local societies were important in the early days for networking and learning at a time when women were mostly not readily admitted into learned societies or indeed university lectures. Certainly, Bedford College, London, had expanded its science facilities, especially geology, under Catherine Raisin, and both Cambridge and Oxford universities had alternative geological and other science facilities. Royal Holloway College as well as University College in London taught geology and had science laboratories for women to use. But considering the small number of female students that went to these universities, most women aspiring to learn about geology had to resort to other means and this was often the local natural science society.

Chester Society of Natural Science (CSNS) was a local amateur society originally set up by Charles Kingsley in May 1871. Kingsley was well known as the author of *The Water Babies* (1863) and a keen geologist and naturalist. Initially the Society was dominated by presidents who were geologists, although all branches of natural science were covered by the Society. When Charles Kingsley moved to Westminster in 1873, Professor McKenny Hughes, Woodwardian Professor of Geology at Cambridge, became its next president for sixteen years (Burek and Hose 2016). It is fortuitous that McKenny Hughes did become president because he was a keen advocate of female education and encouraged women to attend field trips, present papers and specimens, and lecture. The Society accepted women members from its initiation and, by attending lectures, fieldtrips, exhibitions and social events, these women used the CSNS to extend their horizons. When the Grosvenor Museum in Chester, which had a lecture theatre, was opened in 1886, ladies were welcomed to the occasion and from then on women participated in future events held there. Female membership of the Society remained between 28 and 42% throughout this time as Table 6.4 demonstrates. However, there were no female honorary members.

Ethel Skeat, one of the Cambridge quartet of palaeontologists, was a member of the Chester Society of Natural Science (CSNS) from 1904 until her departure to get married in 1910. She moved from teaching in Penarth, South Wales, to Queen's School, Chester, in 1904, following her former roommate and friend Beatrice Clay. Clay had become headmistress of Queens and invited Ethel to fill the science teacher vacancy. Within a year, Ethel Skeat gave her first geological talk to the CSNS on 'Jurassic shorelines, or a fragment of world history.' This was followed in 1909 by another on the 'Bernese Oberland'

Table 6.4 Chester Society of Natural Science (CSNS)—Women's Membership to 1910

Year	Total membership	% Women	Number of women	Honorary members
1871	356	41.8	142	0
1872	454	41.6	189	0
1873	502	39.4	198	0
1884	616	28.3	174	0
1890	661	33.7	223	0
1901	901	41.8	377	0
1910	1028	35.8	369	0

and just before she left in 1910, a third on 'Life's failures'—a look at extinctions (Burek and Hose 2016). Both ladies remained members of the Society for many years and Clay gave lectures on English literature when the Society widened its interests to include Arts and Humanities. The CSNS gave women in the Chester area the opportunity to listen to lectures on a vast range of subjects and to attend field excursions. The Society was also important for giving qualified women the chance to perform publicly, especially as teachers may have been used to addressing audiences—for example, their pupils— but other women were not. This forum allowed the amateur as well as the professional to gain experience in public speaking, especially to a mixed audience, something most women had never done. Women typically had neither the confidence, nor the opportunity, to do so elsewhere.

As well as delivering public lectures in April 1905, Ethel Skeat gave a demonstration to school children at the Grosvenor Museum on a recent exploration in Greenland, using materials lent to her by the Danish Authorities (at the time Greenland was a colony of Denmark). As she was a teacher at the Queen's School in Chester, this was seen as appropriate but also highlighted her own research. Later that year, in May 1905, like many early successful university women, she was awarded a ScD by Trinity College, Dublin, specifically in recognition of her contribution to geological research. This was followed with other accolades. In 1908 for example, Skeat was awarded the Murchison Fund of the Geological Society of London (£25.6.4d) as documented in Table 6.1. The CSNS must have impressed her as in 1925–1926 she gave her collection of graptolites from North Wales to the Grosvenor Museum Chester, which was run by the Society.

Societies such as the CSNS, and the Woolhope Society in Bristol, grew up during the last years of the nineteenth century and gave scientific women the opportunity to educate themselves. The membership fees were not cheap however and were often aimed specifically at the middle classes. It is ironic that the CSNS developed from a set of lectures to some young men of Chester given by Charles Kingsley. They were later published as a book: *Town Geology* (Kingsley 1873). A need was identified and fulfilled as the book proved popular to many readers not just men. The formal lectures deemed suitable only for men initially were replaced by a natural history society catering for all.

The Sedgwick Club at Cambridge University was founded in 1880 in memory of Professor Adam Sedgwick (Woodwardian Professor of Geology 1818–1873) and it still meets today. At the first meeting on Saturday 13 March 1880 in rooms at St John's College, Cambridge, it was resolved that a club should be formed to promote the study of geology by the reading and discussion of papers. However, it was an exclusive club limited to ten members and women were not allowed. In fairness however, the first female students had only just arrived in Cambridge. It was not until 26 January 1896 that a proposal was put forward for women to be admitted as members. Mr. W. A. Brend B.A. of Sidney College stated:

> [...]that the Club existed with the object of promoting the study of geology, and that as there were now ladies working at the Woodwardian Museum who had shown great ability in the pursuit of that science he thought that the Club would benefit by admitting them as members.[5]

Mr. E. H. Cunningham-Craig B.A. from Clare College, Cambridge, seconded the motion and added 'that as a matter of common justice whatever advantages the Club offered should be open to students of the university, whether men or women on exactly the same footing'. Elections were held and the proposal passed unanimously on 21 February 1896. The Club rules were amended allowing up to six women and twelve men to be members. However, at that time Newnham and Girton were not yet part of Cambridge University so a special phrase mentioning those colleges specifically had to be inserted into the rules.

The first ladies admitted to a Sedgwick club meeting were Miss Gertrude Elles, Miss Ethel Wood and Miss Louisa Jebb as members of Newnham College, and Miss Stephen, Vice-Principal of Newnham College, attending as a guest. Gertrude Elles was to become a stalwart member of the Club during her lifetime, reading many papers and becoming a strong advocate of female participation in club events (Tubb and Burek 2021). She herself became secretary and then president. Elles often proposed female members as well as later acting as chaperone. Louisa Jebb was an agriculture student interested in soils and geological applications. Jebb returned home to Shropshire to run the family farm using her geological learning to the advantage of it; later in life, in 1919, she set up the Women's Land Army and received an OBE (Order of the British Empire) for her efforts. The first paper read by a woman was delivered by Elles on 03 March 1896, on 'Structure of Graptolites', in Mr. Brend's rooms in Sidney Sussex College. It was also at this meeting that Gertrude Elles and Louisa Jebb proposed and seconded the admission of Ethel Skeat as a member. Elles had completed her geology degree the previous year in 1895. The location is unusual as female students were normally chaperoned and it was considered highly irregular to go into male colleges, especially a male room. However, because Mrs. Seward and Miss Stephen were also present at that meeting as guests, they were regarded as chaperones. Gertrude Elles was supported by her good friend and research collaborator Ethel Wood who had also completed her degree the previous year. Thirteen men were also present at this meeting. The following meeting on 05 May 1896 was attended by guest Mrs. Marr, the wife of the lecturer who was based at St. John's College, as well as Miss Jebb. Attendance was sporadic at the fortnightly meetings after that until January 1897 when Miss Elles and Miss Wood again attended together. Attendance at the meeting for the first few years shows a pattern which emerges as the women often attended

Table 6.5 Female attendance at the Sedgwick Club, University of Cambridge

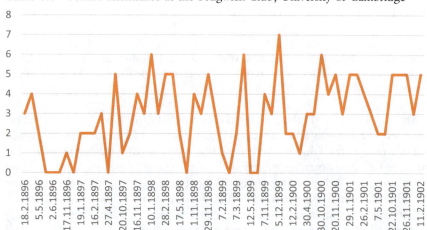

in twos and partnerships in friendship and research flourished, as illustrated in Table 6.5.

The freedom this club gave to its female geologists was significant as it allowed them to mix freely with the male geology students and to hear up-to-date information from their own and visiting lecturers. The willingness of the male lecturers too, in the guise of Professor McKenny Hughes and Dr. Marr, also encouraged the free mixing of the sexes at a time when other institutions were much more rigid in their attitudes towards the mixing of students, even in supervised situations. This was a unique opportunity for female students in Cambridge, envied and valued by many in other disciplines even within Cambridge itself. It was valued tremendously by all as shown by a quote from Miss Ball from 1908–11 who spoke of 'those glorious years, which culminated in an unforgettable Sedgwick Club trip to Church Stretton' (Ball 1908, pp. 76–78).

The networking centred around the Sedgwick Club resulted, inevitably, in romantic entanglements and several marriages. At least four early geological marriages were the result of the mixing of the sexes at these initial meetings. In 1906 Ethel Wood married fellow student Gilbert Shakespear, and in 1915 Margaret McPhee married former geology student James Romanes. Ethel Skeat also met her husband during her study years at Cambridge and in 1910; many years after she had left, she married Henry Woods, the palaeontology lecturer who had first started his geological training in 1887 at St. John's College. We have already met her when discussing the Chester Natural Science Society. Agnes Robertson had met her future husband, Edward Arber, when he was a lecturer in the geology department. In 1895 he came up to Trinity College, Cambridge, specializing in botany and geology. He graduated in 1899 when Professor McKenny Hughes nominated him as a demonstrator in palaeobotany at the Woodwardian, later Sedgwick, Museum. He held this post for the rest of his life and was involved in curating the palaeobotanical collections, lecturing, and demonstrating in fossil botany. As student and honorary members, respectively, Agnes and Edward often attended the same Sedgwick Club meetings, for example on 15 May 1900 to listen to a talk by fellow geology student Mr. Coles on South African Diamonds, and on 14 December 1900 when Mr. Walker gave a talk on the morphology of shorelines. There is evidence of a further four meetings at the Club during 1901 and 1902 when Agnes and Edward listened to a variety of talks given by Gertrude Elles, Dr. John Marr and Mr. Downing. In 1902 Agnes gained a first-class pass in Part II of the Natural Science Tripos and then spent a year as an assistant to the botanist Ethel Sargant. Between 1903 until her marriage, she held a Quain Scholarship at University College, London, and published papers on palaeobotany and plant anatomy, culminating in the award of a doctorate. Agnes and Edward became engaged in September 1906 and married on 5 August 1909, Edward's thirty-ninth birthday. She then left London—and her scholarship—to live in a rented house on Huntingdon Road, Cambridge. Mrs. Arber collaborated with her husband, a palaeobotanist in the geology department, in his field work, drawing fossil figures for and proofreading some of his geological and palaeobotanical books. She also held a research fellowship at Newnham College and had a demonstrator role at the Balfour Biological Laboratory for Women at Cambridge (Haines 2011). Thus, she was able to continue her research until she became a mother. Agnes Arber became a plant morphologist and anatomist and historian and philosopher of botany; she was elected a Fellow of the Royal Society (FRS) in 1946, the first female botanist and only the third woman overall to be so honoured (Packer 1997). Edward Arber was sadly killed during World War One, leaving behind a four-year-old daughter, Muriel. After his death, Arber wrote to Mr. Wordie of the Sedgwick Club in May 1919, to thank him for the copy of a resolution passed by the Sedgwick Club and conveyed her gratitude for the kind expressions regarding her husband's connections with it. This case study shows the resilience of some women in carrying on with their careers after marriage, providing they had the support of family, especially husband. Ida Slater, a gifted research palaeontologist trained at Newnham College, Cambridge, was unable to carry on after her marriage and we lose her altogether. For some women, marriage opened up research opportunities for them, as is the case with Annie Greenly who helped her husband Edward map Anglesey and produce the definitive text *Geology of Anglesey* in 1919.

By 02 February 1897, there is a separate entry in the Minutes Book under members termed 'Lady members'. On 16 February1897, Gertrude Elles gave her second paper and was supported by her friend Ethel Wood. Again, the talk was held in one of the male colleges, this time Mr. Thomas' rooms in Sidney Sussex College. Table 6.6 shows numerous talks given by female geologists, providing an excellent opportunity for women to present papers on many

Fig. 6.1 The Sedgwick Club, University of Cambridge, 1900. Photograph reproduced by kind permission of the Sedgwick Museum of Earth Sciences, University of Cambridge

different topics in a learned but friendly environment, before venturing further afield in front of often less-accepting audiences sceptical of women's expertise and authority. The friendship, networking, and support that the Sedgwick Club gave to all of its geological student members, male and female, was second to none at this time (Fig. 6.1).

Conclusions

The importance of national and local societies to early female geologists is demonstrated by the comments from the women themselves. The work they did—often in glorious isolation—was recognized by the initiated but not publicized or applauded by those at large. These women needed the companionship, friendship, support, networking opportunities and experience of other women in similar situations, to feel confident to carry on. Several early female geologists have been looked at in detail at the various social levels to illustrate this hypothesis. Each society offered a different experience and some women

Table 6.6 Papers read by women members at the Sedgwick Club, University of Cambridge

Date	Presenter	Title of paper	Location	Attendees & guests Female	Male
03 March 1896	Miss Elles	Structure of Graptolites	Sidney Sussex College	4	13
16 February 1897	Miss Elles	The Vestergothia District	Sidney Sussex College	2	15
10 January 1898	Miss Dale	The Glacial Geology of the Neighbourhood of Buxton	2, Trumpington Street	6	17
03 May 1898	Mrs. Hughes	A Tour through Russia	Secretary's Room	5	19
05 December 1899	Miss Elles	Graptolites	Mr. Arber's Rooms	7	9
30 October 1900	Miss Dale	On Recent and Fossil Species of Sequoia	2, Trumpington Street	6	4
06 November 1900	Miss Elles	Notes on the Geology of the St. David's District	College House, Grange Road	5	5
12 November 1901	Miss Elles	The Tectonic Structure of England & Wales	Miss Elles' Rooms, College House	5	3
10 November 1903	Miss Slater	The Probable Origin of some Types of Valleys	R.H. Rastall's Rooms, 19 Sidney Street	8	9
03 May 1904	Miss Elles	The Highland Lochs	Combination Room, Newnham College	7	11
08 November 1904	Miss Slater	Some Aspects of Vertebrate Palaeontology	Mr. Slater's Rooms, Emmanuel College	2	10
14 February 1905	Miss Elles	The Southern Uplands of Scotland	Mr. Pocock's Rooms, Sidney Sussex College	2+	13
15 May 1906	Miss Elles	The Carboniferous of the Bristol District	Miss Elles' Rooms	3+	1+

(continued)

Table 6.6 (continued)

Date	Presenter	Title of paper	Location	Attendees & guests Female	Male
23 October 1906	Miss Baker	Volcanoes	Mr. Rastall's Rooms	3+	4+
19 February 1907	Miss Watney	The Geology of the Rhone Valley	The Taylor Library, Sidney Sussex College	2	5+
03 November 1908	Miss Elles	Spitzbergen	Miss Elles' Rooms, Newnham College	5	14
26 January 1909	Miss MacPhee	Firth of Clyde	Newnham College	5+	8+
26 October 1909	Miss Haddon	Grand Canyon of Colorado	Lecture Room Newnham College	5	2+
17 May 1910	Miss Elles	The Assynt District	Lecture room Newnham	9	3+
6 March 1911	Miss Elles	Geology of the Church Stretton District	Miss Elles' Rooms Newnham	5+	1+
6 June 1911	Miss HB Leach	Hidden Coalfields	Lecture Room Newham	7	7
20 February 1912	Miss Jewson	The Origin of Earthquakes	Lecture Room Newnham College	6	8
5 March 1912	Miss Elles	Geology of the Bristol Area	Miss Elles' Rooms Newnham College	7	12
19 November 1912	Miss Haddon	Geology and Evolution	Mr Kennedy's Rooms, Trinity College	6	13
22 April 1913	Miss Rolland	Earth Movements and Rock Types	Lecture Room Newnham College	6	9
28 November 1919	Miss Chandler	Peat	Miss Elles' Rooms Newnham College	4	7

belonged to more than one, for example, Gertrude Elles belonged to three. So, what did these societies offer them? The Geological Society gave them the professional recognition they sought; The British Federation of University Women gave them friendship and networking from like-minded women; The Sedgwick Club gave them a geological network unlike any other at that time; and the Chester Society for Natural Science provided an opportunity for women to hone their presentation and listening skills in a local, less financially demanding way. Both the societies and the women themselves paved the way and opened the doors for other women of science to follow, albeit a trickle at first. Today we need to recognize their contributions and continue to work towards the equality that is still lacking in this particular discipline.

Notes

1. Rachel Workman letter to Sir Alexander of Douneside, 1910. MacRobert Archive (MRA), Tarland, Aberdeenshire, Scotland. MS 2, 9/5/4.
2. Rachel Workman letter to Sir Alexander of Douneside, 1910. MRA, MS 2, 9/5/7.
3. Rachel Workman letter to Sir Alexander of Douneside, 1910. MRA, MS 2, 9/5/50.
4. Minute Book 1916–1920, London Association of the British Federation of University Women Archive, Women's Library, London School of Economics, 5 LAU/1.
5. Sedgwick Club Minutes, Sedgwick Museum of Earth Sciences, University of Cambridge, GB 590 SGWC, Book VI, January 1893–November 1896.

References

"A geological Expedition in Derbyshire." 1902. *Manchester High School Magazine* IV, no. 12: 109–112.

Ball, M.D. [1908] 2010. "Newnham Scientists." In *A Newnham Anthology*, edited by Ann Phillips, 76–78. Cambridge: Cambridge University Press.

Boylan, P.J. 2009. "The Geological Society and Its Official Recognition, 1824–1828." In: *The Making of the Geological Society of London*, edited by C.L.E. Lewis and S.J. Knell, 319–330. London: Geological Society.

Burek, Cynthia V. 2007. "The Role of Women in Higher Education—Bedford College and Newnham College." In: *The Role of Women in the History of Geology*, edited by Cynthia V. Burek and B. Higgs, 9–38. London: Geological Society.

Burek, Cynthia V. 2009. "The First Female Fellows and the Status of Women in the Geological Society of London." In: *The Making of the Geological Society of London*, edited by C.L.E. Lewis and S.J. Knell, 373-407. London: Geological Society.

Burek, Cynthia V. 2021a. "Female Medal and Fund Recipients of the Geological Society of London: A Historical Perspective." In: *Celebrating 100 Years of Female*

Fellowship of the Geological Society: Discovering Forgotten Histories, edited by Cynthia V. Burek and B. Higgs, London: Geological Society.

Burek, Cynthia V. 2021b. "Margaret Chorley Crosfield: The First Female Fellow of the Geological Society." In: *Celebrating 100 Years of Female Fellowship of the Geological Society: Discovering Forgotten Histories*, edited by Cynthia V. Burek and B. Higgs. London: Geological Society.

Burek, Cynthia V. and Hose, Thomas A. 2016. "The Role of Local Societies in Early Modern Geotourism: A Case Study of the Chester Society of Natural Science and the Woolhope Naturalists' Field Club." In: *Appreciating Physical Landscapes: Three Hundred Years of Geotourism*, edited by Thomas A. Hose, 95–116. London: Geological Society.

Burek, Cynthia V. and Kölbl-Ebert, M. 2007. "The Historical Problems of Travel for Women Undertaking Geological Fieldwork." In *The Role of Women in the History of Geology*, edited by Cynthia V. Burek and B. Higgs, 115–122. London: Geological Society.

Coignou, Caroline. 1890. "On a New Species of *Cyphaspis* from the Carboniferous Rocks of Yorkshire." *Quarterly Journal of the Geological Society of London* 46: 421–422.

Coignou, Caroline. 1928. "Crosby Hall". *Manchester High School Magazine* XXX: 9–10.

Dyhouse, Carol. 1995. "The British Federation of University Women and the Status of Women in Universities, 1907–1939." *Women's History Review* 4, no. 4: 465–485.

Elles, Gertrude L. 1953. "Obituary—Margaret Chorley Crosfield." *Proceedings of the Geological Society of London* 1502: 131–132.

Fraser, Helen and Cleal, Christopher. 2007. "The Contribution of British Women to Carboniferous Palaeobotany During the First Half of the 20th Century." In *The Role of Women in the History of Geology*, edited by Cynthia V. and B. Higgs, 51–82. London: Geological Society.

Gledstone, M. 1949. "The First Forty Years: The British Federation of University Women 1909–1949." *The University Workers' Review*, no. 39: 7–9.

Goodman, Joyce. 2011. "International Citizenship and the International Federation of University Women Before 1939." *History of Education* 40, no. 6: 701–721.

Gordon, M.O. 1932. "Anniversary Meeting—Lyell Medals." *Proceedings of the Geological Society of London*, no. 88: 59–60.

Gordon, Maria, Mathilda, Ogilvie. 1934. "Geologie von Cortina d'Ampezzo und Cadore." *Jahrbuch der Geologischen Bundesanstalt* 84, no. 1–4: 59–215.

Green, C.P. 2008. "The Geologists' Association and Geoconservation: History and Achievements." In *The History of Geoconservation*, edited by Cynthia V. Burek and C.D. Prosser, 91–102. London: Geological Society.

Gries, R.R. 2018. Anomalies-Pioneering women in Petroleum Geology:1917-2017. Revised Ed. Steuben Press, Longmont. Colorado.

Gries, R.R. 2021. Buried discoveries of early female petroleum geologists. 245-260. In: Celebrating 100 Years of Female Fellowship of the Geological Society: Discovering Forgotten Histories, edited by Cynthia V. Burek and B. Higgs, 33-53. London: Geological Society.

Haines, Catherine M.C. 2011. "Agnes Arber (née Robertson): Botanist, Historian of Natural History, Biographer." *Archives of Natural History* 36, no. 2: 363–364.

Herries Davies, Gordon L. 2007. *Whatever Is Under the Earth: The Geological Society of London 1807–2007*. London: Geological Society.

Hudlestone, W.H. 1893. Award of the Lyell Fund. *Proceedings of the Quarterly Journal of the Geological Society* 49: 42.

Kingsley, Charles. 1873. *Town Geology*. New York: D. Appleton and Company.

Lamplugh, G.W. 1920. Award of the Murchison Medal. *Proceedings of the Quarterly Journal of the Geological Society*. London, 75: 44.

Lewis, Cherry. 2020. "Two for the Price of One: Doris Livesey Reynolds (1890–1985)." In: *Celebrating 100 Years of Female Fellowship of the Geological Society: Discovering Forgotten Histories*, edited by Cynthia V. Burek and Bettie Higgs. London: Geological Society.

McKenny Hughes, T. 1900. Award of the Lyell Fund. *Proceedings of the Quarterly Journal of the Geological Society* 56: 48–49.

McNichol, M. 1901. "A School Journey to Cambridge." *Manchester High School Magazine* III: 48–49.

Miller, J. 1973. "*Coignouina decora* sp. Nov. and *Carboncoryphe habnorum* sp. Nov. (Trilobita) from a Visean fissure deposit near Clitheroe, Lancs." *Geological Magazine*, no. 110: 113–124.

Miller, J. 1976. "Synonymy of the Carboniferous Trilobites *Namuropyge* and *Coignouina*." *Palaeontology* 26, no. 14: 917–919.

Morley, E.J. 1949. "Looking Backward—The British Federation of University Women 1909–1949." *The University Workers' Review*, no. 39: 307.

Packer, Kathryn. 1997. "A Laboratory of One's Own: The Life and Works of Agnes Arber, F.R.S. (1879–1060)." In *Notes and Records of the Royal Society of London* 51, no. 1: 87–104.

Parkes, S.M. 2007. "Steamboat Ladies (act. 1904–1907)." *Oxford Dictionary of National Biography*. https://www.oxforddnb.com/view/10.1093/ref:odnb/9780198614128.001.0001/odnb-9780198614128-e-61643. Accessed on 18 February 2021.

Pia, Julius. 1939. "Maria Matilda Ogilvie Gordon." In *Mitteilungen des Alpenlandischen geologischen Vereines*, no. 5: 173–186.

Robinson, E. 2007. "The Influential Muriel Arber: A Personal Reflection. In: *The Role of Women in the History of Geology*, edited by Cynthia V. and B. Higgs, 287–294. London: Geological Society.

Skeat, Ethel G. and Crosfield, Margaret C. 1896. "The Geology of the Neighbourhood of Carmarthen." *Quarterly Journal of the Geological Society of London*, no. 52: 523–541.

Sondheimer, J.H. 1957. *History of the British Federation of University Women 1907–1957*. London: British Federation of University Women.

Tubb, Jane and Burek, Cynthia V. 2021. "Gertrude Elles: The Pioneering Graptolite Geologist in a Woolly Hat. Her Career, Achievements and Remembered by Her Family and Colleagues." In: *Celebrating 100 Years of Female Fellowship of the Geological Society: Discovering Forgotten Histories*, edited by Cynthia V. Burek and B. Higgs. London: Geological Society.

Wachtler, Michael. 2016. "Early-Middle Triassic (Anisian) Ferns from the Dolomites." In: *Fossil Triassic Plants from Europe and Their Evolution Vol. 2: Lycopods, Horsetails, Ferns*, edited by Michael Wachtler and Thomas Perner, 101–103. San Candido, Italy: Dolomythos Museum.

Woods, E.G. and Crosfield, M.C. 1925. "The Silurian Rocks of the Central Part of the of Clwydian Range." *Quarterly Journal of the Geological Society*. London, 81: 170–194.

PART III

Making Women Visible: Institutions, Archives and Inclusion

CHAPTER 7

Where Are the Women? How Archives Can Reveal Hidden Women in Science

Anne Barrett

The importance of finding women in science from history is inestimable. Not only does it add to 'Herstory', but it may encourage present girls and women to take up work in STEM by giving them an understanding of the variety of roles and different ways of being a female scientist, or a woman involved professionally with science. One way to discovery in history is through archives. Archives make us aware that many scientists are never acknowledged by celebrity, even when it is their work that brings the breakthroughs, the inspirational teaching, and new developments in science, technology, and medicine, that ensure knowledge and applications continue. Peer pressure and the cult of celebrity can obscure the real opportunities available if an individual's nascent interest in, and concentration on, science can be harnessed early in an educational career. There are many levels of valuable work in STEM, something which could helpfully be borne in mind by families, schools, universities, careers advisors, and government. So how can archives help in the recruitment and retention of girls and women into STEM, against what they sometimes feel are gender odds in favour of boys and men?

Archives reveal women who have studied science, become researchers, teachers and provided technical or administrative support in a scientific milieu, and who have helped make scientific institutions more comfortable places

A. Barrett (✉)
Imperial College London, London, UK
e-mail: a.barrett@imperial.ac.uk

© The Author(s), under exclusive license to Springer Nature Switzerland AG 2022
C. G. Jones et al. (eds.), *The Palgrave Handbook of Women and Science since 1660*, https://doi.org/10.1007/978-3-030-78973-2_7

for everyone. 'What is past is Prologue', wrote Janet Watson in her inaugural lecture as Professor of Geology at Imperial College in 1977.[1] With this phrase in mind, understanding the past achievements of women should inform the careers of women in STEM and spur on each successive generation. In recent years, archives have been mined to produce various university studies on women in science in general or women in specific subjects. In response to an absence of women, archives have also been created by, for example, various universities and by countrywide initiatives focusing on their own specific national experience.

Harvard University Medical School asked the question: 'If there is no documentation of women leaders in medicine, how do we write the history of their achievements?' The low number of women who had collections in the archives as contrasted to men—twenty women as opposed to 900 men—led to the founding of the Archives for Women in Medicine in 2005. Harvard had a long history of women being interested in attending their Medical School (since 1845) and tuition was given by women lecturers from 1910, but it was World War Two and a shortage of available men that made the difference to women being formally admitted as students, albeit in 1945. The caveat then was a ten-year trial to assess if women could make the grade as physicians. The Archives for Women in Medicine was created 'to address that gap in documentation, collect the papers of Harvard's first-generation women leaders in medicine and science, provide access to collections, and promote and celebrate women's achievements in science and medicine' (Harvard University 2005). They also regard this initiative as a way to inspire the coming generations of women to train in medicine. In 2018 the programme expanded and is now part of the Harvard Archives for Diversity and Inclusion (Harvard University 2021).

Iowa State University holds a large archive of Women in Science and Engineering (WISE) which includes alumni as well as women not directly associated with the University. The archive interprets holdings widely and, except for medicine, collects broadly to include women's organizations. Books are included in the range of materials, along with their manuscript preparation papers, and women who are significant within their collecting policy are well documented. The entries in the catalogue are easily located and there is a biographical entry and text about the papers to a general level (Iowa State University 2021).

There are also nationwide initiatives to identify women. To reveal women in Australia, the National Foundation for Australian Women and the University of Melbourne have partnered to produce The Australian Women's Archive Programme. The resulting Australian Women's Register 'is a rich and growing source of information about Australian women and their organisations' (Australian Women's Register 2020). It contains over 7000 entries with references to over 4000 archival resources, circa 9000 published resources and over 1200 digital resources. The Register can be searched to find a large range of resources including on women, organizations, archives, publications, and digital resources. This database includes women's achievements in all areas of life, but a search under 'Explore Themes' and then 'Women's Occupations' reveals subject by subject, named individuals. A search under 'S' and then 'Scientists' (for example) reveals names and specialities cross-referenced to disciplines, for example 'Environmental Science', with hyperlinks to the archival collections where applicable. If a name is known a search can be made under that name and the same for a subject. The Register is already a very comprehensive, informative resource which will grow as women are added to it.

In Canada, the International Network of Women Engineers and Scientists (INWES) Research and Education Institute was founded to ensure standards of education in the teaching of STEM subjects. One of its objectives has been the creation of the Canadian Archive of Women in STEM (2019), in partnership with the University of Ottawa, which holds the Archive of the Canadian Women's Movement. This partnership highlights how important university archives are in keeping the record of women in science, engineering, mathematics, and technology. The Institute has identified a problem that affects all scientists' archives: namely that scientists typically do not believe their papers are worth keeping. This may be exacerbated for women of course, especially if they suffer from 'imposter syndrome'. In response, the project aims to raise awareness of the importance of making provision for the deposit of an archive; to identify archives of women in STEM in Canada; and to develop a website about the archives. It seeks to educate people to look after their own archives and to understand that they have a value to history and to further generations as inspiration. This is essential given the digital lives people now live; students and early career scientists need to understand that they must be responsible for the retention of their material. If they are not, it will disappear, along with their voice. Scientists neither live nor work without influences from a range of sources—societal, social, disciplinary, and work—and this is something archives can also reveal. However, in order to do so, it is essential that scientists are educated to see that their manuscript papers, their communications, and other interests, form the material saved for cataloguing. Their published output is not indicative of the whole person, important though it is to their career. This applies to everyone and is a message that should begin with students in the academic environment. This chapter now presents case studies from Imperial College London, the UK university concentrating on science, technology, and medicine, to illustrate what can be revealed about women's role and situation in science from an analysis of the archives.

CASE STUDY: IMPERIAL COLLEGE AND ITS DEVELOPMENT

Imperial College has always been thought of as a male domain. This is clear from its outset in the nineteenth century, except that women were there, as the archives reveal. Despite this, as late as the mid-twentieth century there is an example of a woman who, on her arrival at Imperial, suggested to her mother that her acceptance had been a mistake, and that this was a college for men.[2] But she remained and prevailed as discussed below. She is one of the women who rose through the scientific ranks with international accolades: Marjorie de Reuck, Chemical Engineer.

The medical schools, which joined Imperial to form the Imperial College School of Medicine in 1997, had eighteenth, nineteenth, and twentieth-century origins. One joined in 1988: St. Mary's Hospital Medical School Paddington, and four in 1997: Chelsea and Westminster Medical School, Charing Cross Medical School, The National Heart and Lung Institute, and The Royal Postgraduate Medical School. In their early days, these schools were very restrictive and ambivalent as to women entering the medical profession. But again, women overcame the obstacles, and one even rose to become the first woman president of their professional societies, as will be discussed below. Interestingly, the Imperial College Business School had its origins in the Mechanical Engineering Department as the Industrial Sociology Unit, and then the Management School. The Industrial Sociology Unit had four women lecturers in the 1960s and 1970s, one of whom, Joan Woodward, became the second female professor at Imperial in 1970. The other three also became professors, and two moved to other organizations: Dame Sandra Dawson as the first woman master of Sydney Sussex College Cambridge 1999–2013; Dorothy Wedderburn to Bedford College, later Royal Holloway and Bedford New College, as Principal 1981–1990; and Dot Griffiths OBE as Professor in Imperial College Business School and Provost's Envoy for Gender Equality.[3]

There were far fewer higher education colleges and universities in the early to mid-nineteenth century, therefore the constituent colleges of Imperial College were well placed to develop a social agenda to expand scientific teaching to enable industry and individuals to thrive. Through the founders of its constituent colleges, the Imperial College of Science, Technology, and Medicine was in direct communication with relevant members of Government from its inception in 1907. These colleges included The Royal College of Chemistry (RCC); the Government, later Royal School of Mines (RSM); the Royal College of Science (RCS); and the City and Guilds College (C&G). This communication and exchange reached from the 1840s to the early twentieth century when these institutions were at the forefront of government scientific higher education.[4]

Imperial College's examples of hidden women begin in the mid-nineteenth century. To launch the RCC, a high-profile patron was found in Prince Albert; however, in the absence of government funding, subscriptions were required, and these subscribers included women. The registers held by Imperial College Archives reveal subscriber names and addresses, the amount paid, membership status and sometimes a little more information. One woman's subscription was Lady Davy (1780–1855) a woman subscriber connected with London science through her husband, Sir Humphrey Davy (1778–1829), a renowned chemist. Lady Davy, who became a life member, paid twenty guineas. A Mrs. Cludde of Orleton Wellington Salop is also subscribed as a life member; however, no other details are recorded.[5]

The Royal School of Mines (RSM) opened in 1851. As the first government funded higher education institution in the UK, it was obliged in return to run a programme of public lectures. These were the 'Occasional Lectures' which were held every week across a range of subjects. They were the same as those given for matriculated (registered full-time) students and were therefore uncompromisingly scientific.This is in contrast to many of the popular science lectures given by other institutions or by itinerant lecturers in the earlier nineteenth century. The RSM archives reveal that women did attend the 'Occasional Lectures', but mostly with a male companion, typically a husband or father. Miss Annabella Homeria Pollock (d.1873) accompanied her widowed father, Field Marshall Sir George Pollock (1786–1872) to the geology course in 1852. The following year a Miss Wollaston (d.1873), who had also taken the geology course, is recorded as attending with them under the name of Lady Pollock. Women attending lectures alone are also recorded; for example, a Mrs. Wilkinson, who attended geology lectures 1854–1855. Geology proved popular with women, as the chapter by Cynthia Burek elsewhere in this volume demonstrates. Unmarried women also attended lectures alone, notably a Miss Garrett who took natural history 1861–1862. She later became known as Elizabeth Garrett Anderson (1836–1817), the celebrated campaigner for women's right to enter medical school. Garrett was the first woman to be entered on the British Medical Register in 1865; she achieved this through gaining the Licence of the Society of Apothecaries which brought with it the naming of licence holders on the Medical Register. However, when it was realized that a woman had discovered and made use of this loophole, it was quickly closed to ensure no other women could follow her example (Elston 2004). Garrett's career developed and, after medical training in Paris, she co-founded the London School of Medicine for Women, the first Medical School for women in Britain (Crawford 2002, pp. 76–97). Garrett was also an active suffragist, one of just two women to be elected to the first London School Board in 1870, and the first female mayor in England, taking Aldeburgh in 1908. Some women attended lectures for longer periods, such as Lady Maria Forester (1824–1894) who attended physics courses from 1862 to 1864. These courses were given as series of lectures, but did not have an exam or any certification attached to them, although originally it was suggested that they might have.[6]

In the 1890s, the attention of the Department of Science and Art turned to the need for science teaching in schools, which in turn required scientific education for school teachers. To this end, it created the system of Science Teacher in Training Scholarships, to assist school science teachers in attending the constituent colleges in South Kensington. The purpose was to improve the standard of science teaching throughout the British Isles and these courses were as rigorous as those followed by matriculated science students. The scholarships enabled:

> Country teachers to gain experience of the latest discoveries and newest experiments. They attend on the understanding that they do their work thoroughly. They are required to take notes and afterwards write a full report of the lecture and take part in practical work. Their reports are examined and marked by the lecturers [...] The main purpose of these courses is to enable teachers to give practical instruction the sciences. (Becker 1874, p. 163)

Although few women were offered these scholarships, in 1873 Mary Anne Gahan from the Roscrea YMCA Institution, County Tipperary, achieved this distinction (Phillips 1990, pp. 231–232). A less usual archival source describing the experience of being a recipient of a Science Teacher Scholarship is to be found in H.G Wells' novel *Love and Mr Lewisham* (1900). This text is semi-autobiographical and includes a fictionalized account of his experience as a Science Teacher in Training. In the story, both he and a girl he has a friendship with attend classes in South Kensington. His expressed attitude may not be objective, as he did not enjoy his time at the College, apart from the biology courses he took with T.H. Huxley, an anatomist and Darwinist and one of the leading figures in science at the time. Wells spent much time on student matters, and on launching what is a good source of archival information, *The Science Schools Journal*, for the RCS, thus neglecting his studies. Wells did though maintain a long friendship, mostly by correspondence, with Elizabeth Healey, a fellow student.[7]

WOMEN AT IMPERIAL COLLEGE, LONDON: MATRICULATED STUDENTS

One of the first students at the City & Guilds College (which opened to students in 1885), was a Miss A. Grace Heath who took Chemistry 1885–1887 and gained the Associateship of the City and Guilds Institute (ACGI) in 1887. Subsequently she taught at the North London Collegiate School. Her case is interesting as she was taught by Henry Edward Armstrong, Professor of Chemistry at the City and Guilds College, who was known to oppose women science teachers teaching boys, and women joining professional science societies, due to women's 'mental disabilities' (Armstrong 1904, p. 14). Sara Phoebe Marks, later Hertha Ayrton, was a student in 1884 of electro-technics, studying with Professor Edward Ayrton, her future husband, at Finsbury Technical College which became the feeder to the City and Guilds College when it opened to students in 1885. The first woman Associate of the Royal College of Science (ARCS) was Chemistry and Biology student Elizabeth Healey, the friend of H.G. Wells mentioned earlier, who attended the College from 1881 to 1886. Healey taught science at Battersea Polytechnic Institute and was the author of *The Educational Systems of Sweden, Norway, and Denmark, with Special Reference to the Education of Girls and Adults: Being the Report Presented to the Trustees of the Gilchrist Educational Trust on a Visit to Scandinavia* (1892). This was completed as a Gilchrist Travelling Scholar, made possible by Gilchrist, an educational trust that had been operating since 1865. Later in life, Healey was appointed MBE (Member of the British Empire). The first woman to gain Associate status of the RCS in Physics was Marion Whiteford Acworth, née Stevenson, in 1893. In that year, she married Joseph Acworth, a chemist who had also attended the RCS. They subsequently worked together and published papers jointly, a particular interest being photographic processes. Marion extended her horizons to flight in 1927 when she published *The Great Delusion: A Study of Aircraft in Peace and War* under the pseudonym of Neon. In this she described her opposition to the development of military aircraft, as she believed civilian aircraft could be adapted for warfare. This, of course, was in the early days of the aviation industry.[8]

A chemistry teacher, Martha Annie Whiteley (1866–1956) was appointed to the staff of the Royal College of Science (RCS) in 1904. She was the first female academic at the RCS, and indeed Imperial College. Her mentor was her tutor, Professor William Tilden, with whom she had worked as a student. Whiteley became an Assistant Professor but apparently refused further promotion; despite this, both she and her work were revered at the RCS, Imperial College and beyond, as witnessed by her appointment to the OBE (Order of the British Empire) for her World War One work on lachrymatory gases and incendiary devices (Fara 2018, pp. 167–180). The then Rector Sir Alfred Keogh (1857–1936) remarked that 'Here you have no ordinary woman. I know of no one more likely to inspire women students to great things in science than Dr. Whiteley'.[9] He especially recognized her powers of engendering support for her science and causes, and for her assistance to others; indeed, in 1912 he asked Whiteley to form the Imperial College Women's Association for the support of women students. This Whiteley did, although her support extended to male students and colleagues too.

Scientific women continued to join constituent colleges after they amalgamated to form Imperial College in 1907, albeit in low numbers. Some female students came from overseas, for example Mariam Oommen, a lecturer in Chemistry at the Women's Christian College, Madras University, India. To attain higher degrees, Mariam attended the Royal College of Science between 1927 and 1930 and was awarded a Ph.D. and Diploma of Imperial College (DIC), before returning to resume lecturing in India. Whilst at the RCS, Mariam worked closely with Martha Annie Whiteley. Some women joined Imperial College as refugees fleeing wars and political situations, before and during World War Two: Charlotte (Lotte) Kellner of Berlin was awarded a free place at Imperial in 1933 to continue her work on infrared spectroscopy. She received grants from the Academic Assistance Council (ACC) and the Jewish Professional Committee and was supported by Imperial physics professors, Herbert Dingle and G.P. Thomson. The AAC tried to persuade her away from science, but Dingle obtained a Rockefeller Foundation Grant for her, and employment as a Demonstrator in Physics from 1938. Kellner later became a lecturer at Imperial College from 1945. Professor James Watson Munro (1888–1968, Zoology and Applied Entomology) supported Russian entomologist, Nadia Waloff (1909–2001). Waloff later gained a lectureship in Munro's department. Dr. Gertrude Kornfeld (1891–1955) was supported by G.P. Thomson Professor of Physics. Kornfeld had emigrated to Britain in 1933 and became a British Federation of University Women (BFUW) German Scholar Residential Fellow at Crosby Hall, Chelsea. She was also a Research Fellow in the Department of Astronomy, 1934–1935.[10] Some female scientists had to overcome stereotypical views such as student Lettice Digby whom the Registrar notes as being away from College on nursing duties, when in fact she was working as an Assistant Bacteriologist at a hospital in Richmond, Surrey.

IMPERIAL COLLEGE WOMEN SUPPORTED BY MEN IN SCIENTIFIC CAREERS

There are many examples of men supporting women at the constituent colleges and many women (most who cannot be included here for lack of space) have mentioned their male mentors when discussing their careers. There has been much scholarship in recent years on the importance of male supporters and mentors in gaining access to science for women, especially access to scientific education and institutional facilities and, certainly, the archive of Imperial College evidence this.[11] One notable female scientist who was supported by her husband is electrical engineer and physicist Hertha Ayrton (1854–1923). Following their marriage in 1885, Professor William Edward Ayrton (1847–1908) of the City and Guilds College, who had taught Hertha, would not work with his wife for fear that her work would be accredited to him, even though she was already the holder of several patents for scientific instruments. While a professor at the Finsbury Technical College, Ayrton had given his wife access to his laboratories and student assistants and it was here that she carried out her first important research, on the hissing of the electric arc, which led to her being elected a member of the Institution of Electrical Engineers in 1899 (Jones 2009, pp. 67–92). Hertha Ayrton was also the first woman to be nominated, in 1902, to a fellowship of the Royal Society of London, although this was unsuccessful as the Society argued that a married woman was ineligible and so they could not consider her application (Mason 1995). Hertha did, however, receive the Royal Society's Hughes Medal for Original Research in 1906 and exhibited at the Society's prestigious scientific soirée, as described elsewhere in the handbook in the chapter by Claire Jones. William Ayrton described his wife as a 'genius' to her cousin, Dr. Philip Hartog, saying 'you and I are able people, but Hertha is a genius'.[12] Both her scientist cousins, Philip Hartog and Marcus Hartog, supported her applications to professional societies, as did Silvanus Philips Thompson, physicist at the City and Guilds College. Hertha also worked for women's suffrage and was the inspiration for the main character, a young woman scientist, in her stepdaughter Edith Ayrton Zangwill's suffrage and science novel *The Call* (1924).

Letitia Chitty (1897–1982) BA, MA; MICE, FRAeS was supported throughout her career by two men, both of whom took up posts at Imperial College and invited her to join them there[13]: Civil Engineer A.J. Sutton Pippard (1891–1969) and Mathematician and Engineer Richard V. Southwell (1888–1970). Other men in the engineering profession also supported her, by requesting her to work with them, and respecting her knowledge and expertise. It was war time opportunity that gave Chitty her access to a career in academic civil engineering, via a male mentor. As a student of Mathematics at Cambridge University, she obtained permission to take a break in her studies to assist the World War One effort in 1918. She joined the Admiralty Air Department and carried out stress analysis on RAF experimental aircraft. This led her to her future career in civil engineering, and the investigation of stresses in other structures too, particularly arch dams. A.J. Sutton Pippard was a colleague of Chitty's at the Admiralty Air Department. Recognizing her outstanding abilities, even though she had not completed her studies, he was keen to continue supporting her career. Chitty returned to Cambridge after the War, changing to the Engineering Tripos, and becoming the first woman to gain first-class honours in the Cambridge University Mechanical Engineering Tripos.[14]

Following her graduation, Chitty returned to the Air Ministry and worked with Richard V. Southwell who later became Rector of Imperial College and a great supporter of her career. Chitty acted as his Research Assistant working on the Airship Stressing Panel; her analytical skills were essential to the safe build of these aircraft following the disaster of Airship R 38, then the largest airship built, which broke up over Hull on 24 August 1921, crashing into the Humber Estuary, due to stresses exerted during manoeuvres. There were only five survivors out of the crew of forty-nine. Throughout her career, Chitty applied her analytical skills to the design as well as stresses and frames of aircraft, becoming the first female Fellow of the Aeronautical Society in 1934. When Pippard became a Professor of Civil Engineering at Imperial College, he ensured that Chitty also joined the Civil Engineering Department in 1934. It was Pippard whose insistence eventually gained her membership of the Institution of Civil Engineers (ICE). This was difficult because, although she had undertaken the right work to the highest of standards, Chitty had not followed ICE's acknowledged career path and so did not have a research degree. Nevertheless, during World War Two Chitty had been sought after by the Admiralty to investigate stresses in submarine hulls from explosions, and by the Ministry of Supply; in 1943, she was promoted to a lectureship at Imperial in the Civil Engineering Department. In addition, she was awarded an Institution of Civil Engineers 'Telford Premium'—awarded for the best published paper in the field—three times in the course of her career. This prize was created by the bequest of the first President of the Institution of Civil Engineers (ICE), Thomas Telford (1757–1834). Despite this initial award, the ICE did not seem able to aggregate it with her research output enough to overcome their requirement for a thesis. Pippard's papers in the Imperial Archives reveal how he battled with the Institution to gain membership for her—and succeeded. He gathered others behind him, including another Imperial Professor, Cecil L. Fortescue, and one other member of the ICE, and together they made a particular case for her acceptance on the sole basis of her work. Thus, in 1947, she became only the third Woman Associate Member of the Institution of Civil Engineers. Chitty went on to become the first woman member, in 1958, of the Technical Committee of the ICE. In 1969 she was the first female full member of the Institution to receive the Telford Gold Medal for her publications. This was awarded for her publications throughout her career, culminating in her work on dams, and for her major presentation on her research at the ICE conference in 1968.

The career of Elen Elaine Austin (1895–1987) F.R.G.A., F.R.Met.Soc. saw several 'firsts' for women, as well as support from one man in particular.[15] A specialist in dynamical meteorology, and an early worker on climatology, particularly winds and seas, Austin was one of the first women graduates to work in a technical position at the Air Ministry Meteorological Office. In 1921, she was also the first woman to be seconded from there to work at Imperial College. Her roles included Demonstrator; Lecturer; Personal Assistant to Professor of Meteorology, Sir (William) Napier Shaw (1854–1945); Shaw's co-author on *Manual of Meteorology 1926–1941*; and Principal Scientific Officer at the Meteorological Office. After graduating in Mathematics and Natural Sciences in 1917 from Newnham College, Cambridge, Austin joined the Meteorological Office in 1918. Here she came to the attention of the Meteorological Office Director (from 1905 to 1920) Sir William Napier Shaw. He had effectively professionalized the office and introduced modern working methods, including employing more women. Shaw became Professor of Meteorology at Imperial College in 1920, and Austin was seconded by the Meteorological Office, part of the Air Ministry, to work as his personal assistant from 1921 to 1924. She undertook advanced study in Dynamical Meteorology whilst working for him at Imperial College and continued to work on their four-volume *Manual of Meteorology* after returning to the Meteorological Office. There she also prepared handbooks on the weather over the oceans during World War Two. Retiring as Principal Scientific Officer in 1955, she published *A Bibliography of the Works of Sir Napier Shaw, FRS, 1854–1931*, and chose to take a less senior role to continue working for some years. The Elaine Austin Centenary Memorial Prize in Aeronautics was created in her name at Imperial College in 2007. A photograph from The Graphic (18 April 1925) shows Elaine on the extreme right-hand corner, standing behind the men; beneath the caption are the names of the sixteen men in the photograph, and last, her initials, EEA, in her handwriting.[16]

Women working and publishing in science were vulnerable to mistakes of gender identity, for example Constance Elam Fligg Tipper (1894–1995). Tipper worked as a top academic researcher for thirty-two years before she was recognized officially with any academic title. In 1949 she was made a Reader and gained full membership of the Faculty of Engineering at Cambridge University. Tipper worked in metallurgy and crystallography and was the first person to apply scanning electron microscopy (SEM) to metallurgical work. Indeed, she created the discipline of strength of materials and became renowned for her 'Tipper Test' for fatigue in metal. She came to particular notice for her work on the US Liberty ships in World War Two, which suddenly broke up in cold weather when crossing the Atlantic, with catastrophic results. Tipper discovered that this was due to a weakness called 'brittle fracture' (Tipper 1962). Recognizing her great ability, Cambridge University Professor John Baker (1901–1985) put her forward to take on the role of testing for the Admiralty Ship Welding Committee. Baker was just one of her supportive male colleagues who championed her throughout her career; others were Professor Geoffrey Ingram Taylor (1886–1975), and Professor Sir Harold Carpenter (1875–1940) for whom she worked as research assistant 1917–1929 at the Royal School of Mines.

Tipper's use of initials on research publications, instead of a gender-identifying first name, almost led to an embarrassing experience for her at the Royal Society Dining Club. Women were not allowed at the Club dinners, but a dinner was usually held for speakers after delivering the Bakerian Lecture. In 1923, Tipper gave the Bakerian Lecture on the topic of crystal distortion in metals, along with her co-author, Professor Geoffrey Ingram Taylor. Tipper managed to discover the ban in time to avoid the situation and the Committee sent her a box of chocolates as a consolation. It was not until 1973 that she eventually received an invitation and attended the Dining Club for a commemorative Bakerian Lecture. In 1963 Tipper was awarded an Imperial College Fellowship. On her 100th birthday, in 1994, Newnham College marked the event of their esteemed alumna by planting a tree in the College grounds.[17]

The first female professor at Imperial College was Helen Archbold Kemp Porter (1893–1987). Following an undergraduate degree at Bedford College, London, Helen Archbold, as she then was, attended Imperial College under the tutelage of Martha Annie Whiteley, holding one of the places Whiteley had persuaded her head of department, Professor Jocelyn F. Thorpe, to create for women. This provided a positive culture to work in and facilitated a move into a newer area of research with another mentor, Professor Vernon Blackman. Archbold repaid them both by seizing any opportunities that arose from working with them. In order to work on the problem of cold storage of apples in transit, with Blackman, Helen surreptitiously took refresher evening courses in biology and so was able to undertake her move into biochemistry and that area of study to advantage. Archbold worked rigorously with her mentors and colleagues at Imperial, Cambridge, Rothamsted Experimental Laboratories, and as a visiting lecturer at Swanley Horticultural College. Another male mentor enabled her to further develop her work, Professor F.G. Gregory (1893–1961), Head of Plant Physiology at Imperial. After a year in America, she moved to Bangor University in Wales where her research was recognized by the Nuffield Foundation which provided a grant, so that she could set up her own research laboratory at Imperial College. This was a huge achievement for her in 1949. Archbold's breakthrough work on plant metabolism came when she utilized a new method of research chromatography and radioactive tracers. Following this, she was elected to a Fellowship of the Royal Society in 1956, to Reader in Enzymology, and to Principal Scientific Officer at the Research Institute in Plant Physiology at Imperial in 1957. This led to the Headship of the Plant Physiology Unit and to promotion to the first woman professor at Imperial College in 1959 (Northcote 1991).[18]

The case studies presented briefly above evidence the importance of archives for identifying and tracing scientific women whose careers are otherwise typically overlooked and forgotten. At Imperial College, it is clear from archival records that women had followed academic careers in science at the institution and its constituent colleges from the late nineteenth century at least. As well as the individuals discussed above, in more recent years at Imperial two more notable women must be mentioned: Janet Vida Watson was awarded a Personal Chair in Geology in 1974, and Anita Bailey was appointed to the Kodak Chair of Interface Science in 1976.

WOMEN OF SCIENCE AND FAMILY CONNECTIONS

Family connections and social relations played a large role in creating opportunities in science in the late nineteenth and early twentieth centuries. This was especially so for women who often depended on a male mentor for access and career development. This process can be identified from the archives of Imperial College London, where working at the College was at times very much a 'family business' (Barrett 2017).

The first Director of the Royal College of Chemistry (RCC) in 1845, Professor August von Hofmann (1818–1892), a German chemist of renown, married Helen Moldenhauer in 1846. They were later joined at the RCC by Hofmann's sister Hannah, who besides assisting domestically, also undertook assaying work. At the Royal School of Mines (RSM), the Reeks clan made it a family affair, with three of the four children of Trenham Reeks (1823–1879), the Registrar and Curator of the Museum, working or studying at there: Margaret (1855–1937), Maria Ellen (1858–1929), and Trenham Howard (1860–1938). Margaret was the longest serving, from 1883 to 1928; she sometimes worked as an art teacher and possibly a sculptor, but her main job was as a Draughtsman at the RSM, later successfully lobbying for a change of title to Technical Artist, which is in fact what she was for the College. She drew the illustrations, teaching charts, diagrams, and explanatory pictorial images, and made models for the professors' lectures, according to their requirements. Her sister, Maria Ellen, was a sculptor, but she sometimes helped Margaret with illustration, as the workload was often too high for one person, albeit efficient and experienced, to manage. Despite this, Margaret also made time to undertake work on illustrating crystals, working in particular with the Natural

History Museum and the Imperial Institute. She put this work to good use in writing a student's guide to the drawing of crystals: *Hints for Crystal Drawings* which was published in 1908. A second but very different book followed in 1920, when Margaret was asked to write the history of the RSM. She was clearly well placed to do this having been born into the Institution due to her father having been the first Registrar since before her birth. This resulted in a book of two parts, one part her history, and the other a register of past students; the *Royal School of Mines History and Register 1851–1920* is a useful reference work for tracing biographical details of the staff and for identifying student careers in mining. There being no set retirement age, and still being 'fully up to the work', Margaret worked into her seventies.[19]

The Hill family worked in the same department at the Royal School of Mines and, together, members of the family provided technical support to the Imperial College Geology Department for 116 years. Mr. Edward (Ted) John Hill began work in 1925 as a 'Laboratory Boy' rising to the role of Departmental Superintendent 1960–1971. His wife, Violet (Peggy) Hill took a temporary job as a 'Laboratory Woman' in the department in 1943 (this was in wartime therefore fewer men would have been available for laboratory work) however she remained post-war, developed her skills, and became Technician in Charge of Ore Polishing and Chief Technician, 1966–1968. Peggy reports that the Head of Department once remarked to her that she was the first woman to work in the Geology Department 'in a technical capacity'.[20] Other families, including women, can be identified across scientific disciplines. These include the three Conrady sisters, whose father was Alexander Eugen Conrady (1866–1944), Professor of Optical Design in the Physics Department from 1917 to 1931. Two of these, Doris and Irene, took degrees in chemistry at Imperial and became teachers. The other, Hilda Gertrude Conrady (1902–2003, later Gertrude Kingslake) studied physics from 1920 to 1924. One of her tutors was her father. She followed her father's specialism of Optical Engineering and, on marriage to fellow student Rudolf Kingslake (1903–2003), also an optical engineer, the couple moved together to the Optical Institute, University of Rochester, New York. Hilda held several roles there. The couple developed the Institute together, researched optics, and published both collaboratively and individually. Hilda later became the Institute's historian.[21] In medicine, several women associated with Imperial's constituent medical schools were the first women in their sphere of activity to take on roles or be appointed to boards. One of the most well-known of these is Dame Janet Maria Vaughan (1899–1993). Vaughan's specialism was haematology, including anaemias, pathology, blood transfusion, bone and bone marrow pathology, and radiological effects. She became the first female Councillor of the Royal College of Physicians and, in 1938, she created the National Blood Transfusion Service. Vaughan was also interested in medical education and worked at the Royal Postgraduate Medical School and The Hammersmith Hospital, London (Owen 1995). Vaughan's attitude to being a woman and a medical scientist is reflected in her remark that she would 'like to be remembered as a scientist. That I have been able to throw light on fascinating problems. But as a scientist who had a family. I don't want to be thought of as a scientist who just sat thinking. It's important that you have a human life' (Dacie 1993).

Other women revealed by the archives include Professor Dame Margaret Elizabeth Harvey Turner-Warwick (1924–2017), a specialist in thoracic medicine, who became the first woman President of the Royal College of Physicians in1989, serving until 1992. She worked at the Brompton Hospital, and the National Heart and Lung Institute, in London. Turner-Warwick's strategy as a woman was pragmatic and clear; in her memoir she wrote that it was to 'Systematically ignore gender prejudice because it is irrelevant to medicine and other management situations' (Turner-Warwick 2005, p. 23). Finally, the Imperial College Archives reveal that the medical profession was also changed by the achievements of another woman, Professor Averil Olive Mansfield (b.1937), a specialist in vascular surgery and training; surgery to the carotid artery for the prevention of stroke; and women in surgery. Mansfield served as the first female Professor of Surgery in the UK, the first female Chairman of a department of surgery, and the first elected Chairman of the Court of Examiners of the Royal College of Surgeons of England. She was also the founding Chairwoman of Women in Surgical Training (WiST). Professor Mansfield worked in London at The Hammersmith Hospital, the Royal Postgraduate Medical School, and St Mary's Hospital Paddington (Barrett 2017).

Collegiality and Visibility

These examples of women and their relationships with men at Imperial College London have described a situation that was necessary for women to gain access to science—or at least institutionally-based science—in the past. However, the importance of men and women working together today to support and develop STEM, and to promote science for all to be a part of, cannot be underestimated. At Imperial, the setting up of a day nursery was one outcome to assist women to work, but also families as men, too, could use the crèche for their children, as could students. One eminent physicist with an international career in spectroscopy, Lady Anne Thorne, had four children between 1961 and 1968 and strove to ensure these facilities were provided. It took five years, but with the help of colleagues, it was achieved in 1970 (Gay 2007). In 2021 it is called the Early Years Education Service and still serves the College community.

During the 1950s and 1970s there were developments in student support designed to help women. A system of senior tutors for female students was created in 1958, initiated to counter the male-dominated culture at the College. In addition, many women students had not been away from home before and may well have been the first person in their family, male or female, to attend university. The quotation about Imperial being a 'men's college' from the introduction was made by Marjorie de Reuck, née Gratwick. She, having four brothers, understood her situation well, and very quickly found her feet, though she joined the cause for support for female students. Reuck's career centred on physics, chemical engineering and, from 1966, thermodynamics. She retired as Professorial Research Fellow after a career based around work on the IUPAC (International Union of Pure and Applied Chemistry) Thermodynamic tables which were of huge importance to industries.[22]

The many women in senior positions at Imperial College, both administrative and academic, are visible through their roles, but there are other highly visible ways of revealing the women: two new Student Halls of Residence are now named after women, including Porter Hall after Professor Helen Kemp Porter, the biologist/biochemist discussed above. Another mark of visibility–literally–is women's portraiture. At Imperial, the role of 'Provost's Envoy for Gender Equality' was created in 2014, following the establishment of an Academic Opportunities Committee in 1998. Both initiatives created a senior, visible, and accountable position to challenge the continued over-representation of men. Much of this chapter has focused on Imperial College as a case study of a scientific institution which has evolved and continues to evolve to accommodate new ways of working and ideas. This may be an inspiration to women and other organizations.

Conclusions

Table 7.1 details where to look in archive collections, from a university/institutional standpoint, to reveal the women, and provides a useful starting point for finding both student and staff records. Information can be found in a range of printed works which can identify the person and then lead to their manuscript record. Frequently in printed works, the women are easily identified by their title, the prefix Miss/Mrs, or by their first name published in full, whereas men are often represented solely by initials. However, this is not always the case, so caution is needed—and, of course, 'Dr' and 'Professor' can hide gender when used without a first name. Personal records can be retrieved from these sources, with the proviso that relevant legislation as to the use of personal information and an organization's privacy and reproduction policies must be adhered to. When writing about living people, they should have the opportunity to give permission and view the text to discuss content. Other useful sources include audio and visual interviews such as 'Voices of Science' by The British Library (2009–2013). One of the most poignant interviews in this series is with Professor Dame Julia Polak (1939–2014), formally Director of the Tissue Engineering and Regenerative Medicine Centre at Imperial College. She was a pathologist who required a heart and lung transplant herself, so she eventually undertook the pathology on her own lungs. This is something not many would wish to do, but such was her dedication to scientific discovery and courage in the cause of science (Polak 2007). Even before the development of digitized records and systems enabling searching across material which anyone

Table 7.1 Finding women in the university archive: primary sources and strategies

Checklist for finding women in the Archives

- College Calendars – that is the institutional publication laying out the organization's purpose and general information of interest for applicants, similar to a prospectus, but with detail of departments and staff, governing body members, committees and award holders. Also awards, scholarships and grants, and their criteria for applicants
- Governing Body Minutes
- Board of Studies Minutes
- Annual Reports
- Departmental histories
- Departmental Reports
- An organisations internal and published histories
- Lists of theses
- Manuscript Student Registers
- Annual Lists of Students
- Student Year Books
- Alumni Registers
- Student files (not current)
- Staff files (not current)
- Student and staff newspapers and other informational publications
- Memorials - names on boards, for example Student Union Officers, War Memorials, plaques for names of award holders, buildings, Deans
- Correspondence
- Photographs
- Films / videos
- Ephemera eg. Student Union election material, Student representatives, organisations events
- References in published catalogues of scientists
- References in scientist's papers – both manuscript and published
- Organisational website

An organisations social media output

create and post, verification of sources and background, with critical analysis applied, was essential. Arguably this is even more imperative now and adds an extra dimension to archival research: the need for training researchers to use caution and ever greater critical analysis in considering the huge amount of material available online before use.

Professor Janet Watson's phrase in the introductory paragraph, 'What is past is Prologue' sums up the work of women in STEM as described in this chapter. How we know so much about them is because their achievements are stored—'hidden'—in the archives, awaiting release by research. Archives therefore have a vitally important role in revealing scientific women and understanding their work, their negotiations and support networks, their struggles for acknowledgement, and the profound gains and contributions they made during their careers in STEM.

NOTES

1. Imperial College Archives, London, 'Inaugural Lectures Series, 1977'.
2. Remarked in a private conversation with the author, Anne Barrett.
3. Imperial College Archives: Personal papers, official files, and student records.
4. Imperial College Archives: registers, memos, minutes, correspondence, prospectuses, personal manuscript papers, published reports, and essay.
5. Imperial College Archives: Subscribers' Registers. Royal College of Chemistry, C2/2.
6. Imperial College Archives: Student Registers and Governing Body Minutes (1851).
7. Imperial College Archives: student registers; *The Science Schools Journal*; and RCS reports. Rare Book and Manuscript Library, University of Illinois, H.G. Wells papers (correspondence, 1845–1946).
8. London School of Economics, Archives and Special Collections: Pamphlet Collection, 'Elizabeth Healey Report'. Imperial College Archives: student registers.
9. Imperial College Archives: Martha Annie Whiteley Biographical Files, including correspondence, lecture notebooks, publications, government files recording war work, and award of OBE.
10. Imperial College Archives: staff and student files; war work files. See also Brinson (2006).
11. For example, see Slack et al. (1996) and Abir-Am and Outram (1987).
12. The Institution of Engineering and Technology (IET) Archives: biographical information on Hertha Ayrton, the first woman member of the Institution of Electrical Engineers (IEE), including her research on the electric arc and the Ayrton Flapper Fan.
13. B.A. Bachelor of Arts; M.A. Master of Arts; Member Institution Civil Engineers; FRAeS Fellow Royal Aeronautical Society.
14. Newnham College, University of Cambridge, College Archives: *Newnham College Register 1871–1923*, vol. 1, 1963.
15. FRGS Fellow Royal Geographical Society; Fellow Royal Meteorological Society.
16. For the photograph, see: History of Meteorology and Physical Oceanography Special Interest Group. 2011. Newsletter 3, p.15. https://www.rmets.org/sites/default/files/hisnews1103.pdf. Accessed on 14 March 2021.
17. Imperial College Archives: Staff Lists, Royal School of Mines; papers of Sir Patrick Linstead; Tipper letter to Lady Linstead (27 December 1963). Trinity College Cambridge, Special Collections: Catalogue of the papers and correspondence of Sir Geoffrey Ingram Taylor (1886–1975).

18. Imperial College Archives: student and staff files and registers; department papers and correspondence; College calendars; Board of Studies (Senate) minutes.
19. Imperial College Archives: Margaret Reeks staff file; departmental correspondence; staff and student files.
20. Imperial College Archives: Peggy Hill quote from 'A History of the Geology Department Imperial College 1958–1988' by Mary Pugh, one of the internal histories of the Geology Department; staff lists; committee minutes.
21. Imperial College Archives: student records; student and staff files; departmental and biographical papers.
22. Imperial College Archives: staff and student files; minutes.

References

Abir-Am, Pnina G., and Dorinda Outram. 1987. *Uneasy Careers and Intimate Lives: Women in Science 1789–1979*. New Brunswick, NJ: Rutgers University Press.

Armstrong, Henry E. 1904. "Record of Professor Henry E. Armstrong, Ph.D., LLD, FRS." *Reports of the Mosely Educational Commission to the United States of America*. London: Co-operative Printing Society, 7–25.

Australian Women's Register. 2020. https://www.womenaustralia.info/. Accessed on 16 February 2021.

Barrett, Anne. 2017. *Women at Imperial College: Past, Present and Future*. London: World Scientific Europe.

Brinson, Charmian. 2006. "Science in Exile: Imperial College and the Refugees from Nazism—A Case Study." *The Leo Baeck Institute Year Book*, 51, no. 1: 133–152.

British Library. 2009–2013. "Voices of Science." https://www.bl.uk/voices-of-science/about-the-project. Accessed on 16 February 2021.

Canadian Archive of Women in STEM. 2019. "INWES-ERI Project." https://owhn-rhfo.ca/the-canadian-archive-of-women-in-stem/. Accessed on 16 February 2021.

Crawford, Elizabeth. 2002. *Enterprising Women: The Garretts and Their Circle*. London: Francis Boutle.

Dacie, Sir John. 1993. "Dame Janet Vaughan". Munks Roll, Royal College of Physicians. https://history.rcplondon.ac.uk/inspiring-physicians/dame-janet-maria-vaughan. Accessed on 17 February 2021.

Elston, M. A. 2004. "Anderson, Elizabeth Garrett (1836–1917), Physician." *Oxford Dictionary of National Biography*. https://www.oxforddnb.com/view/10.1093/ref:odnb/9780198614128.001.0001/odnb-9780198614128-e-30406. Accessed on 15 February 2021

Fara, Patricia. 2018. *A Lab of One's Own: Science and Suffrage in the First World War*. Oxford: Oxford University Press.

Gay, Hannah. 2007. *The History of Imperial College London, 1907–2007: Higher Education and Research in Science, Technology, and Medicine*. London: Imperial College Press.

Harvard University. 2005. "Archives for Women in Medicine 2005." Centre for the History of Medicine at Harvard Countway Library. http://collections.cou

ntway.harvard.edu/onview/exhibits/show/women/archives-for-women-in-medicine. Accessed on 16 February 2021.

Harvard University. 2021. "Archives for Diversity and Inclusion." Centre for the History of Medicine at Harvard Countway Library. https://countway.harvard.edu/center-history-medicine/collections-research-access/adi. Accessed on 16 February 2021.

History of Meteorology and Physical Oceanography Special Interest Group. 2011. Newsletter 3. https://www.rmets.org/sites/default/files/hisnews1103.pdf. Accessed on 14 March 2021.

Iowa State University. 2021. "Women in Science and Engineering Digital Collection." University Library. https://digitalcollections.lib.iastate.edu/women-in-science-and-engineering. Accessed on 16 February 2021.

Jones, Claire G. 2009. *Femininity, Mathematics and Science, 1880–1914*. Basingstoke: Palgrave.

Mason, Joan. 1995. "Hertha Ayrton and the Admission of Women to the Royal Society of London." *Notes and Records of the Royal Society of London*, 49: 125–140.

Napier Shaw, Sir William, with the assistance of Elaine Austin. 1926–1941. *Manual of Meteorology*. Cambridge: Cambridge University Press.

Northcote, D. H. 1991. "Helen Kemp Porter. 10 November 1899–7 December 1987." *Biographical Memoirs of Fellows of the Royal Society*, 37: 400–409.

Owen, Maureen E. 1995. "Dame Janet Maria Vaughan, D.B.E., 18 October 1899–09 January 1993." *Biographic Memoirs of Fellows of the Royal Society*, 41: 482–498.

Phillips, Patricia. 1990. *The Scientific Lady: A Social History of Woman's Scientific Interests 1520–1918*. London: Weidenfeld and Nicolson.

Polak, Julia. 2007. "Conversations with Pathologists." https://conversations.pathsoc.org/index.php?option=com_content&view=category&id=23&Itemid=194. Accessed on 16 February 2021.

Reeks, Margaret. [1908] 2009. *Hints for Crystal Drawing with a Preface by John W. Evans with Illustrations by the Author*. Charleston, SC: BiblioLife.

Slack, Nancy G., Helena M. Pycior and Pnina G. Abir-Am, eds. 1996. *Creative Couples in the Sciences*. New Brunswick, NJ: Rutgers University Press.

Tipper, Constance Fligg Elam. 1962. *Brittle Fracture Story*. Cambridge: Cambridge University Press.

Turner-Warwick, Margaret. 2005. *Living Medicine: Recollections and Reflections*. London: Royal College of Physicians.

Wells, H. G. [1899] 2005. *Love and Mr Lewisham*. London: Penguin Classics.

Zangwill, Edith. 1924. *The Call*. London: Allen and Unwin.

CHAPTER 8

'A Very Worthy Lady': Women Lecturing at the Royal Geographical Society, 1913–C.1940

Sarah L. Evans

In *The Lost City of Z*, a film made in 2017 about the trials and tribulations of Percy Fawcett (a British explorer active in early-twentieth century South America), one scene stands out for the ways in which it communicates popular ideas about lecture spaces in early twentieth century learned societies.[1] In this scene, set in 1911, a triumphant Fawcett, freshly returned from a dangerous and pioneering expedition in the Amazon in search of the titular lost city, addresses the Royal Geographical Society of London (RGS) on the scientific results of his expedition. The scene provides an entertaining caricature of such lectures, showing us row upon row of sombrely dressed white men, whose sartorial composure is undermined by their braying shouts in favour or against Fawcett's claims of a lost, technologically advanced, South American civilisation. The space of the lecture resembles nothing so much as the UK House of Commons on a particularly rowdy day—Fawcett, getting into the spirit of things, provides witty retorts to dryly sarcastic objections from the audience, against a backdrop of jeers and hoots—and there is no sense from what is depicted of a carefully sourced and cited scientific paper being read to the audience, as was more frequently the case. However, the film's wider (mis)representations of scientific lectures given at learned societies aside, one aspect of the scene is particularly striking and provides compelling evidence as to popular modern understandings of the extremely gendered nature of such

S. L. Evans (✉)
Royal Geographical Society (with IBG), London, UK
e-mail: s.evans@rgs.org

© The Author(s), under exclusive license to Springer Nature
Switzerland AG 2022
C. G. Jones et al. (eds.), *The Palgrave Handbook of Women and Science since 1660*, https://doi.org/10.1007/978-3-030-78973-2_8

lectures and such spaces. This is the physical set-up of the scene, as one main hall with wooden rows/pews of seating, and above, a gallery to which the female attendees are confined—with the explicit corollary that they silently observe proceedings and are not able to contribute to them. The gendered separation of the space is reinforced by Nina Fawcett asking to sit with her husband downstairs, and being told that 'It's men only, I'm afraid, madam'.

As with much of the film, the caricature presented here contains grains of truth, but nonetheless obscures a far more complex and nuanced reality about women's access and participation in such lecture spaces in the early to mid-twentieth century (Evans 2017). The lecture hall as a particular space of scientific knowledge dissemination and reception has increasingly been considered within the history of geographical thought and practice, and within the wider history of science. Increasing attention is being paid to the ways in which these spaces were gendered (and classed and raced) and to the extent to which women could access and participate in them. These considerations are part of a wider flowering of interest in the feminist historiographies of geography and in the histories of women's geographical work, itself set within the vast literature considering women and their places in the history of science. The discussion in this chapter builds on the groundwork laid in two related strands of the literature: work by feminist historical geographers like Alison Blunt and Avril Maddrell into the reception of women's geographical work; and the recent literature on aurality and orality within the reception of geographical work more broadly (1994; 2009; Alberti 2003; Livingstone 2005; Keighren 2006; Finnegan 2017).

This chapter will explore how a particular group of women participated in the set of lecture spaces around and within the RGS in the early twentieth century. The discussion will demonstrate how they were able to disseminate the scientific results of their expeditionary work through these networks and will consider how that work was received—and how that reception both was, and was not, gendered. This was a space in which gender was visibly present, made so by the embodied presence of both speaker and audience. Women's history is very frequently a process of recovery and reconstruction, dismantling pre-existing popular understandings, adding detail, nuance, and complexity, as well as demonstrating the presence of women and their achievements in spaces from which they might otherwise be considered completely absent (Russ 2018). This chapter will demonstrate that far from being silent observers, women could and did speak within these lecture spaces, both at the time represented in *The Lost City of Z*, and in the decades before and after. As has increasingly been acknowledged within the feminist historiography of geography and of science more broadly, it is important to look at the interaction of women—as marginalised historical actors both in their own time and within subsequent histories of the discipline—within an é{lite, even hegemonic, institution like the RGS, in order to write more accurate histories of that institution and of the discipline's broader development. It is also vital to assert the historical fact of women's presence within these spaces, even if that presence was frequently a marginal one. Examining these women as a cohort, and not isolated individuals, allows us to reveal their similarities to one another and to their male peers, colleagues, and relatives, in terms of their highly privileged class and raced positions and their role as agents of empire and participators in the RGS, a deeply colonial organisation. The women discussed here—at least those whose own words are recorded—were overwhelmingly white, well-educated, and from upper-class or upper-middle-class backgrounds. Noting this allows us to begin to consider additional excluded others whose contributions are left out of the history of science.

This chapter focuses on a subset of women who either contributed to a lecture, or lectured on their own, to the RGS about the results of expeditionary work undertaken while on projects supported in some way by the Society. It explores how these women gained access to the podium, the nature of the work they discussed, and how this was received by the gatekeepers who invited lectures and so shaped contributions through review and feedback, and by audiences and wider RGS networks. It should be emphasised that this particular group of women is but one subset of those women who lectured to the Society during this period and so this chapter is part of a much larger and longer story about women's engagement. However, their experiences have important things to say about women's interactions with the RGS as a set of overlapping networks and the ways in which their work was received within those networks. This is particularly the case for women who received formal recognition of the validity of their work through it being supported by the Society; indeed, support for an expedition was often dependent on demonstrating not only appropriate expertise and experience, but also sociability (Evans 2016).

The RGS was founded in 1830, with one of its key aims being the promotion of geographical science (which remains its key objective today).[2] The Society can be understood as a series of interlinked networks and spaces encompassing groups including its governing bodies (the Council and Committees), the Society's staff, elite Fellowship groupings like the Geographical Club, the London-based Fellowship who attended lectures, the wider Fellowship throughout the UK and overseas, the members of other learned societies with whom the Society might collaborate, and university-based academics with a connection to the Society. While these spaces were deeply gendered masculine, women's access to them varied considerably, with some spaces open to women from the mid-nineteenth century and others remaining closed to them until the early twentieth century or much, much later.[3] One of the spaces to which women had the greatest access from the earlier years of the Society's existence was that of the Society's lectures.

Women had the right to attend RGS lectures from 1853 (albeit as guests of male Fellows), with women regarded as helping to bring a convivial element to the lectures and, more importantly, to the spaces of socialising that frequently preceded and followed them. The first woman to lecture to the Society was Isabella Bird in 1897, although Alexandra Tinné's earlier paper in 1889 was read to the Society by her son (Maddrell 2009). The Fellowship debates of 1892/93 over the admission of women had been at least partly sparked by Bird's earlier refusal to lecture before a Society that would not admit her as a Fellow (Middleton [1965] 1982; Bell and McEwan 1996; Blunt 1994; Maddrell 2009; Evans et al. 2013). By 1913, when women were granted the permanent right of admission to the Fellowship of the Society,[4] a number of women had given lectures to the Society. It was at least in part by demonstrating their expertise in this way that women's ability to produce geographical knowledge was acknowledged and their right to admission accepted. Fellowship opened up new opportunities for women, for example access to the Society's library and training opportunities and admission to lectures in their own right rather than as guests, and this helped in small ways to shift the culture towards greater access for women who were not Fellows.[5] 1913 thus marks an important moment in the history of women's relationships and interactions with the RGS.

Reflecting this, this chapter considers a group of women's lectures to the Society which took place between 1913 and 1940 (when the outbreak of the Second World War marked an interruption in the Society's support of expeditions, as it turned its efforts and energies towards supporting the British war effort). These lectures all presented findings from RGS-supported expeditions which took place between 1913 and 1939 and had female team members. A substantial number of the women taking part in RGS-supported expeditions during this period were participating alongside male family members, particularly husbands (eighteen of the twenty-three expeditions where a woman participated 'indirectly' [i.e. without being the one to make the application to the RGS for support], saw a woman participating in this way) (Evans 2016). This is both a clear continuity with the pre-Fellowship period, in the specific case of the RGS, and part of a much longer and wider tradition of women collaborating with male family members to produce scientific work, across many disciplines, including women undertaking exploratory travel and fieldwork alongside their husbands (Hoe 2012).

During this period every lecture given at the RGS was printed in the Society's *Geographical Journal* (*GJ*). This analysis draws on these printed lectures as well as upon RGS administrative correspondence with potential speakers and discussants. Between 1913 and 1939, there were nineteen lectures from sixteen RGS-supported women-participating expeditions, out of a total of thirty-two (Evans 2016). Several of the remaining sixteen expeditions disseminated their work through other Society outputs, such as short articles in the *GJ* or book reviews and mentions in the Monthly Record, where Society announcements appeared. Of these nineteen lectures, eleven were read by men, seven by women, and one was split in two and read jointly by a man and a woman. Of the seven lectures given solely by women, two were about the results of a mixed expedition (rather than from a women-only one), and in both instances these women—Katherine Routledge and Gertrude Caton-Thompson—had been responsible for the majority of the scientific work (see

Table 8.1 Women's Royal Geographical Society (RGS) lectures on their RGS-supported expeditionary work, 1913–1940

Name	Date	Title
Katherine Routledge	1917	'Easter Island'
Gertrude Caton-Thompson and E. W. Gardner (read by GCT)	1928	'Recent Work on the Problem of Lake Moeris'
Gertrude Carton-Thompson and E. W. Gardner (read by GCT)	1932	'The Prehistoric Geography of Kharga Oasis'
Meta McKinnon Wood (half lecture with A. V. Coverley Price)	1933	'Professor J. W. Gregory's Expedition to Peru, 1932'
Freya Stark	1936	'Two Months in the Hadhramaut'
Gertrude Caton-Thompson	1938	'Climate, Irrigation, and Early Man in the Hadhramaut'
Freya Stark	1938	'An Exploration in the Hadhramaut and Journey to the Coast'
Olive Murray Chapman	1940	'Primitive Tribes in Madagascar'

Table 8.1 detailing women's RGS lectures on their RGS-supported expeditionary work, 1913–1940). All of the lectures listed appeared first in *The Geographical Journal*; the citations for these are available in the reference list.[6]

In Routledge's case, in March 1917, her authorship of the work was made explicit by her husband William Scoresby Routledge in a brief speech before the lecture, in which he explained that his wife had asked him

> to say a few words as to why she is giving [the lecture] and not I. The point is this. You have lately elected ladies to be Fellows of the Society; and as I think a most worthy lady has most worthily carried out work of a character suitable for a lecture, it seems to me it would be much better for her to give an account rather than that I should do so. (Routledge, William S. 1917, pp. 340–341)

It was highly unusual for anyone other than the President of the Society to speak before the lecture. Scoresby Routledge can be seen here to be using his own status to give his wife explicit permission to transgress social conventions against women giving scientific lectures. Importantly, he also uses the Society's own policies—admitting women to the Fellowship and so recognising women's capacity for geographical knowledge production—to bolster and validate this decision.

Despite this, it is clear that male speakers dominated RGS lectures, especially after taking into account the large number of lectures from men-only expeditions during this time, where the speakers were necessarily male. This was the case even where subsequent publications by the expeditionary team were co-authored by women participants, such as in the case of Stella and Edgar Barton Worthington on the Cambridge East African expedition in 1930, where Stella was lead author on the subsequent monograph (1933). Stella Worthington was mentioned favourably by the President after her husband's lecture, with particular emphasis on her dedication to the expedition and its aims, to the extent that she had given up completing her degree in geography at Newnham College, Cambridge (Goodenough 1932). However, she does not speak in the printed lecture or discussion.[7]

Women involved in these expeditions were sometimes invited however to give a prepared comment in the post-lecture discussion. The handful of women in question generally spoke after lectures given by their husbands about their joint expeditionary work. For example, both Mollie Courtauld and Phyllis Wager were invited to comment on their experiences on the 1935 Wager Greenland expedition after the lectures given by their respective husbands in late 1936 and early 1937. In her comment Wager focused on the domestic roles played by her sister-in-law and herself, and the relationships they built with the Inuit women who were also part of the expedition, but does not mention the women's contributions to the botanical and geological work of the expedition:

> It is a very beautiful country indeed. My sister-in-law and I certainly did the drudgery work, such as washing up, cooking, and baking of bread, but we enjoyed it all very much. I thought the winter might be a little trying during the dark months, but I did not find it so. When the men were away we spent hours in the Eskimo house, talking with them, and they came over to our house, generally just at meal-times so that they could share the meal with us; and we had an arrangement whereby the Eskimo came to tea every Sunday. We used to fill them up with biscuits and jam, and then we had to entertain them afterwards. The gramophone came in very useful for that, and when the snow melted from our large door-step we used sometimes to dance outside, once in a very heavy snowstorm. (1937, p. 424)

Wager focuses on these social and domestic elements, possibly as the part that she was most interested in, or possibly in keeping with gendered notions of modesty and appropriate expeditionary work for a woman. Her comment is highly evocative of their day-to-day life at base camp, and as such is an important resource for reconstructing these aspects of their expeditionary experiences. We also have a sense here of Wager's privileged class and racial position, in that her female Inuit colleagues' own voices are absent from the text, even as highly personal details of their experiences are shared. For example, in his own comment to the lecture, the expedition's doctor, Dr. E. C. Fountaine, mentions that Widimina, one of these women, had asked him for a cure for cystitis and shared details of a traditional remedy she had been using to treat the condition.

Another example of a woman contributing to the discussion after her husband's lecture is that of Inezita Hilda Baker, who gave an extended comment about her experiences of the 1933 Oxford Expedition to the New Hebrides (Baker 1935). Her comment was unusual in terms of its length and that it was illustrated by slides, serving as an appendix to the lecture itself, or even as a small lecture in its own right. Like Wager, Baker also discusses how she developed relationships with local people who were part of the expedition team or working closely with them. She also describes at length her adventures with Tom Harrisson, another team member, after they were stranded on a different island to the rest of the team, after having travelled there with the ambition of climbing an unexplored mountain while in pursuit of geological specimens. It is probably for these latter details that Baker was invited to give this extended comment, as well as for the slides that she showed. Adventurous narratives and lantern slides were both a popular draw for audiences throughout this period, if sometimes disparaged as being sensationalist (Hayes 2018).

Such sensationalism can also be seen in the correspondence recording preparations for a lecture given by Rosita Forbes in 1921 on her Kufara expedition. Arthur Hinks, RGS Secretary, commented in a letter to David Hogarth (who had read and commented on Forbes' lecture) that 'whatever the defects [in the paper] I am sure she will create a sensation on our platform'.[8] There had been a public dispute between Forbes and Hassanein Ahmed Bey, the Oxford-educated Egyptian diplomat and writer with whom she collaborated on the expedition (not supported by the RGS, and so not included in the set of lectures discussed here), as to who had been the originator of the expedition and so responsible for its findings. This dispute had racial as well as gendered dimensions; Hassanein Bey argued that his contributions were downplayed by Forbes and in popular reporting about the expedition because he was not English, in spite of his strong credentials, network connections, and participation. Hassanein Bey later gave subsequent lectures to the Society and was awarded its Founders Medal in 1924, partly in recognition of his contributions to the Kufara expedition.

Hogarth had been dismissive about Forbes' contributions to the leadership of this expedition, or to the knowledge produced by it, and this is likely to be the reason for Hinks' comment on the possible 'defects' of Forbes' paper. However, as Forbes was young and thought to be beautiful and engaging, with a reputation both for adventurous travel and for potentially disreputable sexual morals (according to the gendered social conventions of the time), she was still likely to attract a good audience for the Society. There are parallels here with the treatment of Wilhelmina Elizabeth Ness around this time, with Hinks' citing the fact that she was 'particularly nice-looking' as a reason to help her on her travels.[9] Women's beauty and sociability could open doors for them, but to the detriment of their intellectual or scientific accomplishments—and without the woman in question having necessarily consciously used such a strategy for advancement.

Particularly high-status women could also play the role of invited discussant. Gertrude Bell, who commented on a lecture on the Balkans given by her friend and colleague David Hogarth in 1913, was introduced by the RGS President, Lord Curzon, as 'one of our greatest authorities on the Near East, who has written excellent books on the subject' and as 'one of our recently elected Lady

Fellows' (Curzon, p. 337). The lecture took place on 10 March, a few weeks after Bell had been elected in the first cohort of women admitted in February 1913, and around the same time that she was awarded the Gill Memorial Prize in recognition of her expertise and work. Both of these can be seen as marks of her high status within the Society at the time. We now turn to considering the seven lectures given by women about their own work, beginning with the process of preparing a lecture.

Arthur Hinks played an important role in organising the Society's programme of lectures in this period, both in his capacity as Secretary of the RGS and editor of the *GJ*. It is illuminating to reconstruct some of these processes through analysis of Hinks' correspondence with invited speakers; potential discussants; potential guests; and other members of RGS staff. For example, it was Hinks who first made the suggestion that the archaeologist Gertrude Caton-Thompson give a lecture to the RGS on her Fayum expedition's findings, during their correspondence in May 1928 about the grant the RGS had made to her: 'I am glad to hear that my letter of April 3 has at last reached you and to learn of your success in this season's work. We shall evidently expect a paper from you and Miss Gardner at an Evening Meeting next session'.[10]

As a paper would need to be sent out to discussants prior to a lecture, the processes of preparing the text for a lecture, along with any accompanying illustrative slides, and of preparing it for subsequent publication in the *GJ*, ran alongside one another. Before this could happen however, the paper generally had to pass a process of peer review, either by Hinks or his colleagues at the RGS, or by other experts to whom Hinks had passed the paper for comment. An example of the reviewer comments for one of Evelyn Cheesman's later papers (not about a project supported by the RGS, although Cheesman had taken part in the RGS-supported St. George expedition in 1924) survives in the RGS-IBG archives, as well as the instructions sheet sent to the reviewer, Henry Balfour.[11] This gives some idea of the criteria by which papers were judged, including originality, whether the Society should accept it for publication, what alterations were needed, and whether it could be adapted for an evening meeting lecture.[12]

The role of Arthur Hinks in particular, and other RGS staff members in general, as gatekeeper for the reception of knowledge at the Society and through its lecture programme, is made explicit in the case of Violet Cressy-Marcks (Maddrell 2009). In 1937 there was a 'protracted correspondence' between Hinks and Cressy-Marcks as to the possibility of Cressy-Marcks giving an evening lecture, which Hinks did his best to obstruct. Maddrell demonstrates how the RGS positioned Cressy-Marcks as not quite one of their own, for apparent reasons of social snobbery and class-based anxiety about protecting the status of the Society, with the result that she was not given much space within the Society for dissemination of her work and ideas (2009). The complexity of this classism, in spite of Cressy-Marcks' access to ample funds, is elucidated by the fact that Cressy-Marcks was not only able to fund her own travels, but left a bequest to the RGS, somewhat pointedly earmarked as for *women's travel*.

In the case of Caton-Thompson's 1928 paper, Hinks himself read it and offered comments. His suggestions shed light on the expectations and norms governing an RGS evening lecture at this time, particularly his comment that while the 'great amount of material' they had amassed was to be congratulated, it made for 'rather stiff reading'.[13] Hinks made a number of suggestions for improving the paper and making it more accessible for the general audience (as was his usual practice when commenting on papers), suggesting that they added a 'brief introductory geographical paragraph on the present geography of the Fayum', on the basis that 'one cannot assume that the reader is familiar with the essential facts'. He also added that:

> there are places where the argument seems to me to need a little elucidation, especially for the lecture audience. One must not assume that they understand the implications of the technical terms. For the lecture I think that it will be necessary to summarise large sections in plainer language, omitting the details. An audience cannot follow a detailed argument unless they are familiar with the matter, and will be quite prepared to accept the conclusions without treading in all the steps.[14]

Not only did Hinks have to ensure that Caton-Thompson submitted the paper to him in good time so that it could be checked, and provision made for any accompanying slides and illustrations, he also had to corral potential discussants and extend invitations to anyone within the Society's networks with an interest or expertise in the topic, to both the lecture and if necessary the Geographical Club dinner beforehand. For Caton-Thompson's lecture, this task was complicated by the fallout of the Fayum dig dispute in 1927 and 1928.

The first two seasons of Caton-Thompson's Fayum work (1924–1925 and 1925–1926) had been funded by Flinders Petrie's British School in Egypt (with Egypt then under British colonial control as a British protectorate) (Drower 2006). In 1926 this came to an end, with Petrie moving his base of operations to Palestine. Concessions to excavate were 'given to an approved institution, not to an individual' (Caton-Thompson and Gardner 1934, p. 6). Caton-Thompson and Gardner therefore needed to find another 'body of standing' to act as sponsor, as well as financial support (Caton-Thompson 1983, p. 102). Gardner had just been appointed Lecturer in Geology at Bedford College, and would not have access to leave of absence until the 1927–1928 season, while Caton-Thompson had 'much to write up and lecture about' (Ogilvie and Harvey 2000, p. 483). They therefore did not return to the Fayum until October 1927, planning to move to a fresh dig site adjacent to their previous sites in order to systematically continue their work, and having secured formal sponsorship from the Royal Anthropological Institute as well as funding from a number of bodies, including the RGS (Caton-Thompson 1983).

In the meantime, Caton-Thompson had been unable to formally renew the concession. They were however unconcerned about the delay, 'tranquil in the tradition which forbids appropriation of another person's work without inquiry as to their intentions to continue it', and content that once they had secured a suitable sponsor, the concession would be forthcoming (Caton-Thompson 1928, p. 109). However, in the meantime,

> owing to alleged sensational discoveries ... an American expedition had secretly applied for, and been virtually accorded the N. Fayum concession. Prolonged negotiations with the Dept. of Antiquities, so devoid of prehistorians as to be unable to verify the authenticity of the Fayum discoveries, resulted in acknowledgment of our moral right to continue the work in which we had led the way, but left undefined the area to be assigned to us. The positions of the sites coveted by the Oriental Institute of Chicago were widespread: no attempt was made from that quarter to alleviate our position; and on arrival in Egypt in November we found ourselves re-allotted a restricted concession within the area we had already exhausted both prehistorically and geologically. (Caton-Thompson 1928, p. 109)

Whilst Caton-Thompson and Gardner were able to secure some significant findings in going over their previous ground, thanks largely to some fortuitous rains, this was a serious breach of professional etiquette on the part of the rival team.[15] It had repercussions within the London circle of the RGS, due to Kenneth Sandford's involvement with the debacle, and to the social conventions around lectures given at the RGS, discussed further below. This can be seen from a letter from Caton-Thompson to Hinks in November 1928, immediately prior to that in which Hinks gave feedback on the paper:

> Thinking of what we said about Dr. Sandford's invitation to the Moeris lecture, and my readiness that he should come if he cares to, I should, perhaps, have made it clearer that it would be pleasanter for both of us not to meet <u>at dinner</u> [emphasis original]. Miss Gardner feels quite as hurt as I do about his conduct, and would, I know associate herself with this request. Neither can I think Dr. Sandford would wish to meet us in any but a purely professional way.[16]

Hinks, previously unaware of the extent of the dispute, saw Caton-Thompson and Gardner privately to discuss the matter, and Caton-Thompson also supplied him with a copy of her article from *Man*, which set on record her version of events.[17]

In the subsequent correspondence between Hinks and the various concerned parties, as he tried to ascertain the truth of the matter and work towards a resolution, there is an interesting exchange between Hinks and Sandford which invokes various gendered norms and discourses around expeditionary work and geographical capability. Sandford makes multiple references in the statement he placed on record for Hinks to Professor Breasted, the leader of the rival team, being considered the 'right man for the job' to carry out excavations in the area, in preference to Caton-Thompson 'even if available'.[18] Sandford also complains of the 'fierce' public 'attacks' that Caton-Thompson has made against him in the press, and laments that 'allowing her a woman's privilege'—presumably of chivalrously not raising to the bait, implicitly not treating her as an academic equal—'has not been the best policy'.[19] In response, Hinks rebukes Sandford for implying that Caton-Thompson was unqualified for the work that she had planned:

> I do not feel that I am convinced by some of your arguments, particularly about letting Miss Caton-Thompson "loose" as you call it in an area which she had made her own for two years, and which in her opinion someone had stolen from her. But I do not want to enter into the details. I have known Miss Caton-Thompson for a good many years, and have naturally a slight prejudice on her side, but am most anxious that the Society should do the proper thing.[20]

The 'proper thing' turned out to be Hinks engineering an invitation for Sandford to the Kosmos Club, another dining society associated with the RGS, so that he would not be present at the Geographical Club dinner, on the grounds that it 'would be a pity if anything would remind Miss Caton-Thompson and Miss Gardner of the controversy just before they read the paper'.[21] In any case, although Hinks had hoped that 'the lecture by you and Miss Gardner might be the occasion of improving the situation rather than the reverse',[22] the feud continued for many years, with Caton-Thompson requesting not to be seated near Sandford at subsequent dinners.[23]

The feud also had its professional, scientific dimensions, as the various academics concerned also disagreed with each other's interpretations of the data, provoking a series of papers in the *GJ* throughout the 1920s and 1930s from Caton-Thompson, Gardner, Sandford, A. J. Arkell, H. L. Beadnell, and John Ball, on the precise prehistoric levels of Lake Moeris and other related geological and archaeological phenomena. This is all discussed at length in correspondence between Hinks and the various authors over this period, with Sandford commenting in 1933 that 'I see Beadnell and Miss Caton-Thompson have been having a little discussion in this month's Journal, & find that a good bit of the battle takes part [?] over my presumably prostrate body',[24] and requesting the right to reply, to Hinks' apparent annoyance. Hinks commented that 'I am anxious to do everything reasonable in this matter consistent with the interests both of the combatants and the readers of the Journal, who may to tell the truth be getting a little tired of Pliocene and Pleistocene Tufas'.[25] Here, we can note both the importance of the close intertwining of the professional and the social in sustaining this debate, and the mediating role played by Hinks as RGS Secretary.

The foregoing discussion also shows how collaborative, and potentially contested, the process of dissemination, or at least of getting the speaker to the stage, paper in hand, could be. However, it also makes clear just how closely embedded a woman like Gertrude Caton-Thompson could be in the Society's networks and practices, even as she dealt with gendered constraints that her male colleagues did not face. What are also striking throughout is the unspoken yet ever present colonial underpinnings of the dispute—throughout all the discussion of stolen dig sites, moral right to the work and to interpret the site, and questions of professional etiquette, no mention is made that Egypt, as a British protectorate, is under colonial and not local control. It was this colonial dispossession which made possible the archaeological work, the authority to undertake it, and was the context for the disputes that surrounded it.

While we can reconstruct these processes of preparation from surviving correspondence, it is rather more difficult in these particular seven and a half cases of women delivering lectures to reconstruct the actual lecture, including excavating traces of the embodied gendered nature of this performance, as Keighren does for some of Ellen Churchill Semple's lectures (2010). Whilst the printed versions of the lectures survive, it is clear from some of the correspondence around Forbes' 1921 paper that speakers did not necessarily read the whole of the printed paper, and in some cases owing to length were actively encouraged not to.[26] Similarly, it is unclear whether Hinks' suggested amendments for Caton-Thompson's paper, quoted above, were made to the printed version, or simply to that read by Caton-Thompson on the night, for the RGS evening lecture audience. Caton-Thompson does not discuss the experience of giving the lecture in any of her subsequent publications, or in the surviving correspondence with Hinks. Freya Stark does discuss her nervousness about lecturing at a range of venues, including the RGS, in a letter to her mother:

> I must send you a hasty line to say that all went well. I know today by the general feeling of lightness what a burden it was. The hall was crammed – 800 people: Iveaghs, Lady Halifax, Goschens, and lots of friends there. The Admiral was in the Chair and Lord Wakefield on my other side at dinner, and he, Mr. Perowne, and Violet Leconfield made speeches after, all full of nice things – dreadful to listen to when one is perched on a platform. But the audience was charming, and laughed at all my jokes. (Stark [1938] 1976, p. 235)

The lecture spaces in and around the RGS can be characterised as constituting and representing a particular public and audience, and as operating as a kind of semi-public space. Whilst the lectures were not, as far as can be seen from the available archival evidence, open to all-comers and so interpreted as fully public events, there were other dimensions which suggest that they can be seen as semi-public spaces. RGS lectures were often publicised in the press. Some surviving letters mention this, such as the correspondence between Hinks and Forbes in 1921 about her forthcoming lecture, which had been advertised in the Times,[27] as well as Wilhelmina Elizabeth Ness's later complaints to Hinks that her own lecture had been insufficiently publicised; Ness drew explicit comparison with the usual publicity obtained for male speakers.[28] Furthermore, the audience was often an interested general audience rather than specialists, at least for the main evening lectures. That general audience was made up largely of Fellows and their invited guests, although the Society would also invite other interested parties and experts, and speakers were also encouraged to invite guests who could contribute to the discussion.[29] However, as lists of attendees have not been preserved it is difficult to precisely reconstruct audiences for given lectures in great detail. Nevertheless, it is possible to sketch the general outlines of likely audiences during this period. As discussed above, the RGS can be envisaged as a series of overlapping and interlinked networks. One of the most important of these networks was the Fellowship of the Society, made up of its Fellows and members (and for the purposes of this chapter, that proportion of the Fellowship, based in and around London, who regularly attended evening and afternoon lectures). At that time, as has been the case for much of the Society's history, Fellowship was dependent on demonstrating an interest in geography rather than necessarily having produced original geographical knowledge. The evening lectures in particular were considered important social occasions for the largely upper-middle-class and upper-class Fellowship, so that the programme of lectures fulfilled social as well as geographical and educational purposes.

Considering other dimensions of the 'lecture itself', particularly its physical aspects, it is worth noting that the 'RGS lecture hall' was in fact made up of a number of different physical spaces during the period in question. After its foundation in 1830 the RGS moved between numerous different homes, most notably Whitehall Place from 1854 and Savile Row from 1871 (Mill 1930). None of these premises had lecture capacity, so throughout its early history other locations were used for the Society's evening meetings. As for many other scientific societies during this period, the question of securing their own lecture facilities remained a perennial preoccupation for the RGS (Mill 1930; Naylor 2002).

The Society moved to Lowther Lodge, its present site, in 1913, having purchased the building in 1912. The building had formerly been a private home and as such had no large lecture space, so the Society continued to make use of other venues, including the theatre at Burlington Gardens until 1920, and the Aeolian Hall until 1930 (Mill 1930; Hayes 2018). Between 1929 and 1930 the Society embarked on an ambitious remodelling and extension of Lowther Lodge, including the construction of the Hall (what is now the Ondaatje Theatre) and the Ambulatory. This created on site capacity for lectures, and for refreshments and social circulation after lectures, an important part of these events (Higgitt and Withers 2008). The mobility of the Society's evening meetings up to 1930 means that it is not always clear where a pre-1930 lecture took place, although this was sometimes recorded in the printed lecture or, occasionally, in the evening minute books. As a result, it is difficult to envisage the interaction between speaker and physical setting, an important element of the practice of lecturing. However, post-1930 this becomes easier.

As well as multiple lecture locations, there were contrasting forms of lecture during this period, which were aimed at different audiences and governed by particular discursive norms. The main evening meetings were aimed at a general audience and often included adventurous or entertaining material alongside their scientific content. There were also smaller afternoon meetings in the Map Room, which were run by the Research Department and aimed at specialists (Mill 1930). As noted above, a written paper was generally submitted to the Society prior to the lecture, and, at least in the case of evening meetings, sent out for comment to discussants who would be called on to read a prepared response after the lecture. The paper would subsequently be printed in the *GJ*, along with the prepared comments from discussants. It is unclear whether afternoon meetings followed this precise format, as their printed versions do not generally contain individual comments. It is likely however, that additional discussion took place which, because it was extempore and not pre-prepared, was not recorded.

As already noted, there were strong social elements to the lectures, including the serving of refreshments before and after lectures; this sociability was particularly important for the evening lectures. One particularly important aspect of socialising around the lectures involved the Geographical Club. This was a men-only dining club composed of the Council and senior Fellows, who would invite the speaker and other important guests to dinner before the lecture then given to the wider Fellowship. Women were not generally permitted to attend, although an exception was made in the case of women speakers. These dinners formed an important venue for intellectual discussion and informal networking among the social elite of the RGS (Mill 1930). The topics covered in the whole set of lectures given by women on the results of their RGS-supported expeditionary work (whether evening or afternoon lectures) range quite widely, as does the tone adopted by the speaker in question. Caton-Thompson's is quite dryly scientific, while Olive Murray Chapman is rather self-deprecating. There are potential class dimensions to this, with Caton-Thompson's self-confidence likely emanating from her upper-class background as well as from her educational accomplishments and expertise. The earlier example of Mary Kingsley is potentially instructive here. While Kingsley gave her own lectures in northern manufacturing towns, and particularly to the Liverpool Geographical Society, she was more sensitive about lecturing in the home counties, due to anxieties about her accent (Blunt 1994; Maddrell 2009).

The invited comments on these lectures tend to conform to the polite conventions—of praise and congratulation for the lecturer—which governed this particular set of lecture spaces, at least as represented in the *GJ* (very different from the rowdy scenes depicted in *The Lost City of Z*, with which this chapter opened, although we must acknowledge the uncrossable gap between the printed version, and the lived experience of the lecture itself which is lost to us). Little attention is explicitly drawn to their gender, and the women lecturing are generally positioned as high achieving equals. This contrasts with the treatment of earlier writers like Kingsley, whose femininity was emphasised by reviewers in order to downplay her achievements, and is potentially reflective of the later women's formal intellectual achievements and qualifications, with many of them having studied for degrees, an opportunity barred to their earlier counterparts like Kingsley (Blunt 1994; Evans 2016). Katherine Routledge was praised for her scientific achievement and hard work on Rapanui, while Caton-Thompson and Gardner were praised in similar terms in the comments to their joint paper on Lake Moeris (co-authored but read by Caton-Thompson). In both cases there is intellectual engagement with, and significant criticism of, their arguments, in ways that frame them as intellectual and professional equals to the male discussants making these comments. This contrasts to the way that, in earlier decades, women's geographical work was often described as 'suggestive', in the sense of being thought-provoking, but not endowed with a sense of authority (Maddrell 2009). Meanwhile, Freya Stark, in comments to her paper on the Wakefield expedition and in particular her exploration of the Hadhramaut coast, is effusively praised for her courage and personal qualities, and her contributions to British imperial interests in the Middle East, as well as for her 'lively, entertaining and humorous narrative' (Perowne 1939, p. 14). While these comments have echoes of the 'pluck in the face of adversity' tropes used to characterise earlier women travellers, as well as of the strong discursive links between gender and nationality used to describe Kingsley, Stark is clearly being positioned as an asset (both to the RGS, and to Britain), and not an eccentric oddity.

As noted above, these examples are a handful of the lectures given by women to the Society during the period in question (1913–1939), and there were a number of such lectures, beyond those reporting back on RGS-supported expeditions. Moreover, the women discussed here—particularly Caton-Thompson and Stark—enjoyed high status and recognition at the Society, with Caton-Thompson serving on Council during the 1930s, and Stark described by the President as the Society's 'valued friend' (Goodenough 1939, p. 14). Further research in this area could compare the treatment of these high-status, high-achieving women, with that experienced by other women lecturing to the Society, and other female Fellows of the Society during this period. Women remained a minority of the Fellowship throughout this period, making up c.5% of all Fellows in 1918, rising to c.16% by 1939.

Conclusions

A number of points emerge from the foregoing discussion. The first is while the lecture spaces of the RGS were strongly male dominated during this period (1913–1940), women were both visibly present and frequently heard, both on the stage and in the audience. It is important to stress this, and to make such women visible within the histories we write and tell, giving a fuller, more nuanced, and more accurate picture of the pasts that we are thereby reconstructing. There is also considerable scope for further research to uncover more about these women, whether by investigating lectures given by women to the RGS beyond the particular dataset considered here, or by seeking to more clearly map out the composition of audiences at RGS lectures during this period, and women's participation within these. While the latter task is complicated by the fact that attendee lists were either not compiled or do not survive, the discussions recorded in the *GJ*, along with reference in surviving correspondence to particular individuals being invited to attend or mentioning their own attendance, provide potential avenues for research in this direction. Groups we do know of to date, as mentioned above in reference to this particular set of lectures, include family members of speakers; those with a professional interest in the topic in hand; colleagues of speakers; and students interested in the topic. In this, there are continuities with the pre-Fellowship period, as across the question of women's access to the resources of the Society.

Secondly, in the case of these specific lectures, the reception given to the female lecturers is complex, being both gendered, and non-gendered. At least as recorded in the printed lectures, the women are often engaged with as scientific colleagues and implicit equals, at least in the public space of the lecture, something that for Stark and Caton-Thompson at least is probably reflective of their high status within the RGS and its networks. Within more private spaces, and within private correspondence, we can see a more ambiguous picture, where gendered expectations and norms become more pronounced. Even in the public space, such apparent transgression of norms seems to have needed to be noted and authorised, as in Scoresby Routledge's introduction to his wife's lecture.

There is similar ambiguity in the levels of access to the platform itself. Women tended to speak—that is, to give the main lecture—only when there were no men to speak for them (either because the expedition had only had female European team members, or because any male team members were not considered to have led or significantly directed the scientific or geographical work of the expedition). As we have shown, there were also other opportunities to contribute to the dissemination of expeditionary knowledge through speech, in the form of invited comments. Here, however, there was a gender difference, in that their invited comments tended to focus on the everyday, domestic side of their expeditions, rather than on scientific content, except in very rare cases.

Finally, discussing these women and their experiences together, as a cohort—as their numbers permit us to do—helps us to understand their similarities, both to one another, and to the male colleagues, partners, relatives, and friends with whom they collaborated and worked. This aids us further in viewing them intersectionally, considering the ways in which they were both frequently marginalised by their gender, but enabled by other aspects of their identities.

NOTES

1. Gray, James, dir. *The Lost City of Z*. 2017. Culver City, CA: Amazon Studios; New York, NY: Bleeker Street.
2. RGS-IBG Strategic Plan 2017–2021, available from https://www.rgs.org/about/the-society/strategy/ [Accessed 7 July 2019].
3. The first woman to sit on the RGS Council, Wilhelmina Elizabeth Ness, was elected in 1930, around the same time that Elizabeth Fea joined the staff of the Society. The first female Director was Dr. Rita Gardner in 1995, as Director of the newly merged Royal Geographical Society (with IBG), while the first President was Professor Judith Rees in 2013.
4. During the Fellowship debates of 1892–1893, twenty-two women had been admitted to the Fellowship, but as a result of these debates, no further women were admitted until 1913, as a means of resolving the dispute. See Bell and McEwan, 'The admission of women fellows'; Maddrell, *Complex locations*; and Evans, Keighren, and Maddrell, 'Coming of age?'
5. This was a period when women who were not Fellows, guests of Fellows, or known students were excluded from the Society's premises for fear of militant suffragette activity. See RGS, Council Minutes vol. 9, Minutes of Council, 23 March 1914, p. 76.
6. Full citations for each lecture are available in the References list.
7. It is worth noting here that the University of Cambridge did not formally award degrees to women until 1949, even if a woman had completed her studies and sat her exams, so Worthington would not have been awarded the full degree even had she not left Cambridge to join the expedition.
8. Arthur Hinks to David Hogarth, 4 May 1921, Correspondence Block 9 Rosita Forbes (Mrs. McGrath), 1921–1930, RGS-IBG Archives.
9. Hinks to de Bunsen, November 23 1920, Correspondence Block 9 Wilhelmina NESS (Mrs. Patrick Ness), 1921–1930, RGS-IBG Archives.
10. Arthur Hinks to Gertrude Caton-Thompson, 3 May 1928, Correspondence Block 9 Gertrude Caton-Thompson 1925–1940, RGS-IBG Archives; Caton-Thompson's colleague, friend, and frequent collaborator, Elinor Wight Gardner, a geologist with whom Caton-Thompson collaborated on several RGS-supported expeditions. See Evans, Mapping Terra Incognita.
11. Arthur Hinks to Henry Balfour, 1 November 1932, Correspondence Block 9 Lucy Evelyn Cheesman, RGS-IBG Archives.
12. See Newman (2019) for a fuller discussion of reviewing practices at the Society and in its journal, including reference to the first item published by a woman, a review by a Miss Wilkins in 1840.
13. Arthur Hinks to Gertrude Caton-Thompson, 7 November 1928, Correspondence Block 9 Gertrude Caton-Thompson 1925–1940, RGS-IBG Archives.

14. Arthur Hinks to Gertrude Caton-Thompson, 7 November 1928, Correspondence Block 9 Gertrude Caton-Thompson 1925–1940, RGS-IBG Archives.
15. See Caton-Thompson, G., and Gardner, E. W. (1929) Recent Work on the Problem of Lake Moeris. *The Geographical Journal* 73(1): 20–58.
16. Gertrude Caton-Thompson to Arthur Hinks, 4 November 1928, Correspondence Block 9 Gertrude Caton-Thompson 1925–1940, RGS-IBG Archives.
17. Gertrude Caton-Thompson to Arthur Hinks, 9 November 1928, Correspondence Block 9 Gertrude Caton-Thompson 1925–1940, RGS-IBG Archives. See Caton-Thompson, Recent excavations in the Fayum.
18. Kenneth Sandford to Arthur Hinks, undated statement [12 November 1928], Correspondence Block 9 Dr. K. S. Sandford letters 1925–1939 RGS-IBG Archives.
19. Kenneth Sandford to Arthur Hinks, 14 November 1928, Correspondence Block 9 Dr. K. S. Sandford letters 1925–1939 RGS-IBG Archives.
20. Arthur Hinks to Kenneth Sandford, 15 November 1928, Correspondence Block 9, Dr. K. S. Sandford letters 1925–1939. RGS-IBG Archives.
21. Arthur Hinks to Kenneth Sandford, 13 November 1928, Correspondence Block 9 Dr. K. S. Sandford letters 1925–1939. RGS-IBG Archives.
22. Arthur Hinks to Gertrude Caton-Thompson, 7 November 1928, Correspondence Block 9 Gertrude Caton-Thompson 1925–1940, RGS-IBG Archives.
23. Gertrude Caton-Thompson to Arthur Hinks, 10 June [1932], Correspondence Block 9 Gertrude Caton-Thompson 1925–1940, RGS-IBG Archives.
24. Kenneth Sandford to Arthur Hinks, 11 December 1933, Correspondence Block 9 Dr. K. S. Sandford letters 1925–1939, RGS-IBG Archives.
25. Arthur Hinks to Kenneth Sandford, 2 March 1933, Correspondence Block 9 Dr. K. S. Sandford letters 1925–1939, RGS-IBG Archives.
26. Arthur Hinks to Rosita Forbes, 18 May 1921, Correspondence Block 9, Rosita Forbes (Mrs. McGrath), RGS-IBG Archives.
27. Arthur Hinks to R. B. Burney, 26 May 1921, Correspondence Block 9, Rosita Forbes (Mrs. McGrath), 1921–1930 RGS-IBG Archives.
28. In 1930, Ness became the first woman to serve on the Society's Council, shortly after this incident, apparently for similarly pragmatic reasons to those which helped drive the 1913 permanent admission of women. In March 1928, Ness wrote to Arthur Hinks that:

I, as well as other people, have noted with regret & some surprise, that, since all papers however good or bad that are read before the Society find some mention in the daily papers, the fact that a woman has once more lectured (after a lapse of years) at an evening meeting has been carefully omitted. With regard to your recent question as to whether I would help some geographical expeditions financially, I regret to say, that at the moment I am not inclined to do so, though, I need hardly add, that since the £250 is promised for the Blue Nile Expedition it will be forthcoming if required at some future date not too remote.

Ness was unwilling to countenance such unequal treatment, and instead asserts her right to be taken seriously, using the means available to her as an independently wealthy woman who could withdraw funding from the RGS. In this use of leverage to combat gender discrimination, there are some parallels with the earlier example of Bird's refusal to lecture to the RGS (i.e. withholding something they wanted) which, as discussed above, was instrumental in the debates around women's access to the Fellowship. Less than two years after this letter was sent, Ness joined the Council of the RGS. No discussion of the decision to elect Ness to Council is recorded in the minutes, or in letters between Ness, Hinks, and other Council members. Nonetheless, it seems likely that in awarding her this honour, the other Council members were seeking to ensure her continued good relationship with the Society and thus a continued stream of funding, particularly since Ness had been a key contributor to the Building Fund.

See Wilhelmina Elizabeth [Mrs. Patrick] Ness to Arthur Hinks, March 22, 1928, Correspondence Block 9 Mrs. Patrick Ness 1921–1930, RGS-IBG Archives.

29. Arthur Hinks to Gertrude Caton-Thompson, 2 November 1928, Correspondence Block 9 Gertrude Caton-Thompson 1925–1940, RGS-IBG Archives.

References

Alberti, Sara J. M. 2003. "Conversaziones and the Experience of Science in Victorian England." *Journal of Victorian Culture* 8(2): 208–230.
Baker, Zelda. 1935. [Untitled Comment]. In: J. R. Baker [Baker, Z.], and W. Goodenough. "Espiritu Santo Discussion". *The Geographical Journal* 85(3): 230–233.
Bell, Morag, and McEwan, Cheryl. 1996. "The Admission of Women Fellows to the Royal Geographical Society, 1892–1914: The Controversy and the Outcome." *The Geographical Journal* 162: 295–312.
Blunt, Alison. 1994. *Travel, Gender and Imperialism*. New York: The Guilford Press.
Caton-Thompson, Gertrude. 1928. "Recent Excavations in the Fayum." *Man* 28: 109–113.

Caton-Thompson, Gertrude. 1983. *Mixed Memoirs*. Gateshead and Tyne & Wear: Paradigm Press.
Caton-Thompson, Gertrude, and Gardner, Elinor W. 1929. "Recent Work on the Problem of Lake Moeris." *The Geographical Journal* 73(1): 20–58.
Caton-Thompson, Gertrude, and Gardner, Elinor W. 1932. "The Prehistoric Geography of Kharga Oasis." *The Geographical Journal* 80(5): 369–406.
Caton-Thompson, Gertrude, and Gardner, Elinor W. 1934. *The Desert Fayum* Bedford Place. London: The Royal Anthropological Institute.
Caton-Thompson, Gertrude, and Gardner, Elino W. 1938. "Climate, Irrigation, and Early Man in the Hadhramaut." *The Geographical Journal* 93(1): 18–35.
Coverley Price, A. V., and McKinnon Wood, Meta. 1933. "Professor J. W. Gregory's Expedition to Peru, 1932." *The Geographical Journal* 82(1): 16–38.
Curzon, George N. 1913. [Untitled Comment]. In: "The Balkan Peninsula: Discussion." *The Geographical Journal* 41(4): 336–340.
Drower, Margaret S. 2006. "Gertrude Caton-Thompson 1888–1985." In: Getzel M. Cohen and Martha Sharp Joukowsky (eds.) *Breaking Ground: Pioneering Women Archaeologists*, 351–379. Ann Arbor: The University of Michigan Press.
Evans, Sarah L. 2016. "Mapping Terra Incognita: Women's Participation in Royal Geographical Society-Supported Expeditions 1913–1939." *Historical Geography* 44: 30–44.
Evans, Sarah L. 2017. "Lost Histories." *Geographical* 89(5): 22–23.
Evans, Sarah L., Keighren, Innes, and Maddrell, Avril. 2013. "Coming of Age? Reflections on the Centenary of Women's Admission to the Royal Geographical Society." *The Geographical Journal* 179(4): 373–376.
Finnegan, Diarmid A. 2017. "Finding a Scientific Voice: Performing Science, Space and Speech in the 19th Century." *Transactions of the Institute of British Geographers* 42(2): 192–205.
Goodenough, William. 1932. [Untitled Comment]. In: W. Goodenough, G. A. S. Northcote, R. E. Dent, C. W. Hobley, E. B. Worthington, and J. S. Gardiner "The Lakes of Kenya and Uganda: Discussion." *The Geographical Journal* 79: 293–297.
Goodenough, William. 1939. [Untitled Comment]. In: W. Goodenough, S. Perowne, L. Wakefield, and L. Leconfield. "An Exploration in the Hadhramaut and Journey to the Coast: Discussion." *The Geographical Journal* 93(1): 14–17.
Hayes, Emily. 2018. "Geographical Light: The Magic Lantern, the Reform of the Royal Geographical Society and the Professionalization of Geography c.1885–1894." *Journal of Historical Geography* 62: 24–36.
Higgitt, Rebecca, and Withers, Charles W. J. 2008. "Science and Sociability: Women as Audience at the British Association for the Advancement of Science, 1831–190." *Isis* 99: 1–27.
Hoe, Susanna. 2012. *Travels in Tandem: The Writing of Men and Women Who Travelled Together*. Oxford: The Women's History Press.
Keighren, Innes M. 2006. "Bringing Geography to the Book: Charting the Reception of Influences of Geographic Environment." *Transactions of the Institute of British Geographers* 31: 525–540.
Keighren, Innes M. 2010. *Bringing Geography to Book: Ellen Semple and the Reception of Geographical Knowledge*. London and New York: I. B. Tauris.
Livingstone, David N. 2005. "Text, Talk and Testimony: Geographical Reflections on Scientific Habits. An Afterword." *British Journal for the History of Science* 38(1): 93–100.

Maddrell, Avril. 2009. *Complex Locations: Women's Geographical Work in the UK 1850–1970*. Oxford: RGS-IBG Book Series/Blackwell.
Middleton, Dorothy. [1965] 1982. *Victorian Lady Travellers*. Chicago: Academy Chicago Publishers.
Mill, Hugh. 1930. *The Record of the Royal Geographical Society, 1830–1930*. London: Royal Geographical Society.
Murray Chapman, Olive. 1940. "Primitive Tribes in Madagascar." *The Geographical Journal* 96: 14–25.
Naylor, Simon. 2002. "The Field, the Museum and the Lecture Hall: The Spaces of Natural History in Victorian Cornwall." *Transactions of the Institute of British Geographers* 27: 494–513.
Newman, Ben. 2019. "Authorising Geographical Knowledge: The Development of Peer Review in The Journal of the Royal Geographical Society, 1830–c.1880." *Journal of Historical Geography* 64: 85–97.
Ogilvie, Marilyn, and Harvey, Joy. 2000. *Biographical Dictionary of Women in Science: Pioneering Lives from Ancient Times to the Mid-Twentieth Century*. London and New York: Routledge.
Perowne, Stewart. 1939. [Untitled Comment]. In: W. Goodenough, S. Perowne, L. Wakefield, and L. Leconfield. "An Exploration in the Hadhramaut and Journey to the Coast: Discussion." *The Geographical Journal* 93(1): 14–17.
Routledge, Katherine. 1917. "Easter Island." *The Geographical Journal* 49(5): 321–340.
Routledge, William S. 1917. Untitled Comment. In: W. S. Routledge, H. Read, T. A. Joyce, A. P. Maudslay, B. Thomson, H. Balfour, J. W. Evans, H. O. Forbes, and H. Howorth. "Easter Island: Discussion." *The Geographical Journal* 49 (5): 340–349.
Russ, Joanna. [1983] 2018. *How to Suppress Women's Writing*. Austin: University of Texas Press.
Stark, Freya. 1936. "Two Months in the Hadhramaut." *The Geographical Journal* 87(2): 113–124.
Stark, Freya. 1938. "An Exploration in the Hadhramaut and Journey to the Coast." *The Geographical Journal* 93(1): 1–14.
Stark, Freya. 1976. *Letters: Volume Three the Growth of Danger 1935–39*, edited by Lucy Moorehead. Wiltshire: Compton Russell.
Wager, Phyllis. 1937. [Untitled Comment]. In: H. Balfour, E. C. Fountaine, W. A. Deer, A. Courtauld, L. R. Wager, and E. Munck. "The Kangerdlugssuak Region of East Greenland: Discussion." *The Geographical Journal* 90(5): 422–425.
Wilkins, Miss. [no first name given] 1840. "Rise in die Steppen des dulichen Russlands, &c. Journey Through the Steppes of Southern Russia, undertaken by Dr F Goebel, accompanied by Dr C. Claus and Mr A. Bergmann." *Journal of the Royal Geographical Society* 10: 537–543.
Worthington, Stella, and Worthington, Edgar Barton. 1933. *Inland Waters of Africa: The Result of Two Expeditions to the Great Lakes of Kenya and Uganda, with Accounts of Their Biology, Native Tribes and Development*. London: Macmillan & Co.

CHAPTER 9

Women at the Royal Society Soirée Before the Great War

Claire G. Jones

In the novel *The Call* (1924), a narrative based on the life of physicist and electrical engineer Hertha Ayrton, the heroine's mother exclaims of the Royal Society ladies' soirée that she is always 'so alarmed at those scientific functions. I think I'll get electrocuted or something' (Ayrton-Zangwill [1924] 2018, p. 28). Although an amusing aside, this remark provides an insight into the special character of the soirée as an entertainment that challenged boundaries as much as it communicated and displayed science. Becoming a highlight of the London social season soon after its establishment in 1876, the annual Royal Society ladies' soirée mixed science with society and provided an active experience reflective of a scientific experiment itself. Guests were anything but passive vessels to be enlightened by expert exhibitors, and the roles of exhibitor and audience blurred as participants inhabited both of these roles at once. This was certainly the case at the ladies' soirées of 1887 and 1888 when a large electrical eel was provided by the Zoological Society of London, and guests were invited to present themselves for shocks in the Principal Library. This recalled the eighteenth-century tradition of science as spectacle when thrilling electrical shocks were part of polite entertainment and women were 'essential protagonists of electrical soirées' (Bertucci 2007, p. 90). At these two soirées at least, women did not seem 'alarmed' by the experience and proved themselves as enthusiastic participants as the men.

C. G. Jones (✉)
University of Liverpool, Liverpool, UK
e-mail: C.G.Jones2@liverpool.ac.uk

This chapter explores women's presence—as both exhibitors and guests—at the Royal Society annual soirées from 1872 to 1914 when the events ceased temporarily on the outbreak of World War One.[1] Then, as today, the Royal Society was one of the world's most prestigious scientific institutions and its oldest, having been founded in 1660 and gaining its Royal Charter two years later in 1662 (Hall 1992, p. 1). From its earliest days, the Society sought to advance learning based on experiment, observation, and demonstration rather than through authority and dogma, an aim reflected in the activities that characterized the annual soirées. Soirées or conversaziones were not unique to the Royal Society; many learned societies held them, both locally and in the metropolis, and they were popular especially with the middle and upper classes and élite urban networks (Alberti 2003). The nomenclature of these events as '*soirées*' or '*conversaziones*' is muddled and the terms were used interchangeably, both by the Royal Society and other institutions. 'Conversazione' derives from the Italian and was used from the early eighteenth century to denote a gathering particularly for the purposes of learned conversation; 'soirée' derives from the French word for evening and connotes a formal social occasion held late in the day. The mix of these two terms reflects the character of the Royal Society events well and, as soirée was the term chosen for the first Royal Society Soirée Committee in 1872, it will be the name preferred throughout this chapter.

THE LADIES' SOIRÉE: CONTEXT AND DESCRIPTION

Soirées were a feature of Royal Society culture long before the first one was held at the Society's London premises in April 1872. Prior to that, the soirée was hosted by the President—usually a man of wealth and privilege—and held at his expense at his home (Bluhm 1958, p. 61). These events were exclusively for men, mostly Royal Society fellows, until the Society introduced a second ladies' soirée in 1876. From then on, two annual soirées were held: one in April or May for fellows and male guests, and the second in June which was open to women too. As *The Times* explained, the first of these was colloquially named 'the black one' as 'it was exclusively confined to the sombre sex' with exhibits that 'would appeal to the specialist' whereas 'Ladies' Night' was 'generally supposed to be less severely scientific' (*The Times* 1900). However, this distinction does not fully hold and there is little evidence that content was especially 'dumbed down' for the ladies. Even at the first ladies' soirée, on 14 June 1876, there seemed little hesitation in exposing the ladies to strongly technical displays. Exhibits that evening included various kinds of microscopes, electro-magnets, nautical instruments, prisms showing the spectra of gases, and a device for the distribution of time-signals on the railway. One item demonstrated that night may have attracted particular feminine interest as it was destined to have a close connection to the emancipated 'New Woman' of the 1890s: 'The Type-Writer, a Machine to supersede the Pen in Letter-writing, or Manuscript work of any kind' was exhibited by the Remington

Sewing-Machine Company. It was not uncommon for exhibits from the men's soirée to be shown again at the following ladies' event; for example, in 1888 Eadweard Muybridge presented his electrical lantern show illustrating bipedal locomotion at both of that year's soirées. Repeating exhibits provided opportunity for women whose work was displayed in their absence at the men's soirée to exhibit in person in June; for example, astronomer Annie Maunder's photographs, taken on the British Astronomical Association solar eclipse expedition to India, were featured at both soirées in 1898.

Despite the deep scientific and technical nature of many of the exhibits at the ladies' soirée, the Society vocalised a desire to include displays 'appropriate' for women and non-specialists. Herbert Rix, who served the Royal Society in an administrative capacity for some eighteen years, described exhibits at the June soirée at which ladies attended as 'generally of a more popular character'.[2] He may have had in mind displays linked to everyday experience, such as 'a number of flies sucking sugar, greatly magnified under the microscope' which featured in 1888, or Francis Galton's 'Finger prints as a means of Identification' exhibited in 1891. Active exhibits, which provided the fun of participation alongside education, were particularly welcomed. Galton's instrument for measuring an individual's reaction time proved popular at the June soirée in 1888 and John Perry's 'spinning tops' exhibit was actively sought for 1891 as 'very suitable for the ladies'.[3] In contrast, Karl Grossman's model of a crater showing its perlitic (glass-like) structure was turned down as 'too technical a character' for the ladies' soirée.[4] The diversity of exhibits included at both annual soirées evidences an inclusivity of science now mostly lost due, in part, to increasing specialization and the narrow definition of contemporary science. The soirées showcased artworks and photographs of flora, landscapes and architecture; artefacts from other cultures; ancient and medieval items and jewellery; and, at the first ladies' soirée in 1876, specimens of 'a new kind of Art-Pottery at the Lambeth Pottery Works' exhibited by Mr. H. Doulton. These combined to make an eclectic mix of exhibits alongside technology, experimental science, inventions, and natural curiosities such as, in June 1890, the eggs of a large python and mummy heads of priests from Upper Egypt. The inclusiveness of late nineteenth- and early twentieth-century science provided scope for scientific women such as archaeologists, artists, and photographers to be admitted as specialists and exhibitors at the soirée, something which will be illustrated in more detail below. The display of cultural artefacts and ethnological representations also contributed to a narrative of hierarchy and empire which both echoed and reinforced the colonial rhetoric and scientific racism of the time (Coombes 1994; Wintle 2008).

The ladies' soirée, in particular, mixed science with entertainment and élite social networking. The event was held at the Royal Society's Burlington House where rooms were decorated with flowers and guests were entertained with music and refreshments, including wines and ices. Science and spectacle often fused together, as in 1891 when the National Telephone Company, using an

'Edison Loud-speaking telephone' and 'Bell's receivers', transmitted performances of light opera from the Savoy Theatre London and Princes Theatre, Birmingham. Ambitions were even higher the following year when telephonic communication was made with the Paris Opera House and enough telephones were provided for ten persons to listen simultaneously. Thrilling lantern shows, exhibited at set times in a darkened room, were also shown at the ladies' soirée. These were often of an astronomical or biological character, for example Ray Lankester's slides of jelly fish displayed on a gigantic scale in 1888, or Norman Lockyer's regular demonstrations of astronomical phenomena. Of less obvious connection to science was the 1894 exhibit by the Post-Master General, 'Freaks of the Post Office and its clients' which included examples of 'remarkable letters and envelopes' and a collection of Christmas cards. As the latter exhibit suggests, attention was paid to the entertainment of the guests and perceived 'feminine' sensibilities. Reviewing the 1880 ladies' soirée, the committee were mortified to learn that the new electric light, screened by yellow glass, was thought by some of the ladies to 'make them look bilious'.[5]

Despite occasional problems with lighting, such was the appeal of the ladies' soirée that competition for invitations was intense and there were occasional complaints about overcrowding. Indeed, numbers of guests grew steadily and, by 1887, over six hundred guests were present, with carriages blocking the roadway outside. Fielding letters from would-be guests became a tiresome job for the President, who found himself compelled to decline some scores of applications and, in 1890, to issue a circular to prevent a number of applications being made to him for extra invitation tickets.[6] Those seeking admission were sometimes not averse to underhand tactics. When, in 1902, it became clear that an alteration had been made to one invitation card and 'Not transferable' had been altered, the new Society secretary, Robert Harrison, felt it necessary to investigate. He wrote to the bearer of the invitation, Miss Dorothea Beale, mathematician, suffragist, and Principal of Cheltenham Ladies' College, as 'no doubt she did not do this herself, and so he thinks she should be acquainted with it'. Beale's opinion on the matter was sought as 'cases have been brought before the President and have caused him some annoyance'.[7] It is unsurprising that demand for invitations was fierce—the ladies' soirée was acknowledged as one of the high points of the London season. Indeed, as the opening lines of the 1888 *Times* report of the ladies' soirée gushed, 'Last night a distinguished and brilliant company gathered in the Rooms of the Royal Society at Burlington House', then going on to list the guests before moving on to the secondary matter of a description of the exhibits (*The Times* 1888).

Who were the women attending the ladies' soirée? Mostly those from privileged backgrounds who were the wives or daughters of scientists and fellows. A woman was rarely eligible by dint of her own scientific standing alone, unless she was an exhibitor herself, or connected to a man who could take her as his 'plus one' or put her name forward for an invitation. Each Royal Society fellow could bring a lady, and fellows and committee members could nominate other persons of standing to be sent an invitation. The soirée committee meeting of 18 May 1909 revised the system of direct invitations to ladies so that most received only a single invitation, with an exception being made for widows of fellows and elderly ladies. As well as general over-demand for invitations, this may also be connected to the steady increase, in the 1890s and 1900s, of university women who were now taking bachelor and higher degrees in science and undertaking research alongside Society fellows. Despite the single ticket only policy, qualified women could often be disappointed. For example, when Miss A Porter of the Zoological Research Laboratory, University College London, lobbied for an invitation to the 1907 ladies' soirée, she received a terse 'no' in response.[8] The ladies' soirée was a powerful vehicle for securing patronage and influence and officers invited any visiting distinguished person or foreign dignitary deemed worthy. In 1903, an invitation was extended to Monsieur and Madame Curie who were visiting England to speak at the Royal Institution—or rather Pierre to speak and Marie to watch, as she was disqualified from lecturing due to her sex (Ogilvie 1991, p. 67).

The inequity of women of science being denied access in their own right to this most prestigious of scientific events—even when they had acknowledged scientific credentials—did not pass without comment. This was especially so in the first decade of the twentieth century when more and more women were accessing a university education, and when the campaign for women's suffrage was provoking wider debate about women's rights and role. In response, spaces were emerging where respectable women could meet in public on their own, such as tearooms, department stores and, from the 1890s onwards, social clubs. Founded in 1903 for female professionals, the Lyceum Ladies' Club, whose membership included women scientists, was located at 128 Piccadilly, just down the road from Burlington House (Black 2012, pp. 195–96). Despite these pressures, the fellowship of the Royal Society remained exclusively male until 1945, when the first two women were elected, despite the passing of the 1919 Sex Disqualification (Removal Act) and challenge presented by the nomination to a fellowship of Hertha Ayrton in 1902. Ayrton's bid was unsuccessful, ostensibly because she was a married woman, but women's relationship with the Society was increasingly subject to scrutiny (Mason 1991). This came to a head on 16 June 1914, on a day coinciding with the summer soirée, when an anonymous correspondent mounted a blistering 'Complaint against the Royal Society, The Handicap of Sex' in *The Times*. The ladies' soirée was the main target with the author objecting to the fact that on the one night when the Society's doors were opened to the public and women 'mingle with the hoary-headed scientists', it was the *wives* of scientists and not *woman scientists* who were granted admittance:

> But any Amelia or Leonora whom chance married to a scientific man is eligible as his "lady". Women high up in scientific positions, women with international reputations, women who would themselves bear the magic title of F.R.S. if they

could disguise from the world the fact of their sex—such women are shut out from the concourse of their intellectual fellows, shut out from the opportunities of meeting and talking with their scientific colleagues, unless they know by chance some bachelor Fellow, or one whose wife does not care to show off her diamonds, who will take her *incognito* as his "lady".

The author was in fact Dr. Marie Stopes—writing with her usual forthrightness—who was paid £3 for her contribution (Jones 2009, p. 201). Remembered now primarily for her work on birth control, for sixteen years prior to this Stopes forged a significant career in geology and palaeobotany. This saw her travel widely for research, accept government commissions, become the first woman on the science teaching staff at Manchester University, and publish nearly forty academic papers and books (Chaloner 2005). Despite her strident criticism, Stopes was one of the first women to receive a major research grant from the Royal Society—for a year-long research trip to Japan based at the Imperial University in Tokyo in 1907–1908—and had demonstrated her resulting fossils at the ladies' soirée in 1909.

Women Exhibitors

Excluded from the fellowship of the Royal Society, exhibiting at the annual soirée was one way that women could embody scientific authority and navigate entry into the élite networks of London science. An examination of soirée programmes from 1872 to 1914, alongside Royal Society correspondence and minutes, and secondary sources including news reports, reveals some thirty women to have exhibited, or had their work exhibited, at the annual soirée. Nine of these women had work shown at the exclusively male event, but they were not there in person and their contribution was exhibited for them by a male fellow. Although women at the June soirée mostly exhibited their science themselves, this was not always the case and there are a handful of women whose work was exhibited for them by a man, probably due to difficulty in gaining invitations to this prestigious event. These women were typically young university researchers working in a department alongside a male scientist, usually their senior, who exhibited their research for them. The work of a significant number of female exhibitors centred on the art of seeing, observation, and representation in science. These were scientific illustrators and artists who were working within a tradition of feminine involvement in botany, natural history, and sketching and drawing, all of which had roots in eighteenth-century ideas of acceptable female pursuits (Shteir 1997). At the end of the nineteenth century, this cultural affinity extended to palaeobotany—the focus of at least two female exhibitors—as the discipline became strongly occupied by women (Fraser and Cleal 2007). At this time too, photography became accessible to women due to easier processes and equipment, and the growth of photographic clubs and associations (Denny 2009, p. 811). This added a new technique to the art of observation in science. Women with other scientific interests were represented at the soirées too, including astronomy and archaeology. These disciplines retained a strong presence in the field and so were more accessible to women, unlike wholly laboratory-based science which often excluded women as it moved from the domestic space into an institutional setting.

The first soirée hosted in the Royal Society rooms in 1872 was an exclusively masculine affair, with the exception of the exhibition of a series of fifty-four photographs of sixth-century Irish architecture by 'Miss Stokes' exhibited by J.H. Lamprey. This was no diminutive 'Miss', but rather Margaret McNair Stokes (1832–1900) who had taken on the completion of the third Earl of Dunraven's two-volume *Notes on Irish Architecture* upon his death in 1871. In the preface to volume one, Dunraven's son stresses how unfinished the text was with only 'rough notes and fragments of manuscript' and states that only Stokes had the knowledge to complete it (Dunraven 1885/1887). Stokes had accompanied Dunraven on research trips and was an expert archaeologist, artist, and photographer. Well connected to Dublin intellectual circles, she was elected an honorary member of the Royal Irish Academy in 1876, only the second woman to have achieved this. Stokes took a scientific rather than aesthetic approach to her illustrations, taking rubbings to identify architectural designs and inscriptions precisely, and then taking photographs only when the sun was at the right angle to properly reproduce them (Cunningham 2018). Lamprey, who was the conduit for her photographs at the soirée, was part of a movement to standardise scientific observation and develop techniques for producing systematic, comparable images; Stokes's technique echoes this and points to the growing significance of photography as an evidentiary tool in science. Lamprey was assistant secretary of the Ethnological Society of London and, in 1869, had produced the first guidelines for creating systematic photographic images (Sera-Shriar 2015, pp. 158–59).

The work of a female illustrator of a different kind, the botanical illustrator Marianne North (1830–1890), whose art is examined by Philip Kerrigan elsewhere in this volume, featured at a number of soirées, both the 'black' and the ladies'. North's work was displayed at the men's soirées of 1874 (Brazilian landscapes and flora) and 1877 (paintings of vegetation in California, Japan, Ceylon, and the Malayan Islands). She also exhibited herself at the ladies' soirées of 1879 (flora of India), 1882 (views of the landscape and the coast of Ceylon), and 1886 (studies of plants and curious nests). As well as an inveterate global traveller, North was a respected botanical artist and plant hunter with six species of plant registered to her name. The most notable of these was the pitcher plant which she found in Borneo and named *Nepanthes northeana* (Middleton 2004). In 1882 North endowed a special gallery at Kew Royal Botanic Gardens which now displays some 800 of her botanical paintings. North was joined at the ladies' soirée in 1886 by Mrs. Anna Lee Merritt (1844–1930), an American painter and printworker who practised as a professional artist for most of her life. Merritt had studied mathematics, and anatomy at the Women's Medical College in Philadelphia, although her contribution that night was a portrait of the astronomer and inventor Warren de la Rue FRS. At the time of the soirée, Merritt had already exhibited at the Royal Academy of Arts in London and would go on to exhibit at the 1893 World's Columbian Exposition in Chicago (Heller and Heller 2013, p. 379). Another female scientific artist and traveller exhibited her sketches of landscape near an active volcano, and of an extinct crater, at the ladies' soirée in 1889, both products of a recent expedition to Fiji and Hawaii. Constance Frederica Gordon-Cumming (1837–1924) was a Scottish painter and draughtsperson, as well as a prolific writer of mostly travel memoirs. She had travelled to Fiji in 1875, in the company of the new governor, Sir Arthur Gordon, and his wife. Gordon-Cumming painted her water-colours on site and regarded them as 'representational' rather than 'art', so therefore of scientific interest. A number of her paintings are now held by the Cambridge University Museum of Archaeology and Anthropology (Laracy 2013, pp. 85–88).

Two other established female illustrators and artists exhibited at the Royal Society too. The botanical illustrator Marian Ellis Rowan (1848–1922) exhibited water-colours of Australian wildflowers at the ladies' soirée of 1895. Rowan was fresh from exhibiting at the World's Columbian Exposition two years earlier. She specialized in flowers, birds, and insects, recording them with scientific precision. Many of the plants Rowan found and illustrated in this way were classified by the Australian government botanist Sir Ferdinand Mueller (Hazzard 1988). Both soirées of 1902 featured the geological sketches of Adela Breton (1849–1923); Breton presented these herself at ladies' night, but they were exhibited on her behalf at the men's soirée by the Rev. H.H. Winwood. Breton was an English artist and explorer who was respected internationally for her scholarship on Mayan/Mexican archaeology. As her geological interest grew, she began producing detailed scale drawings and water colours of geological sites and phenomena, often adding labels with technical details. Like Stokes, Breton took a strong interest in the debate about accurate scientific observation but was critical of photography which she felt could distort and gave no information about colour. The sketches illustrated at the Royal Society were of canyons, glaciers, and waterfalls in the US and British Columbia. Breton's scientific focus and precision is illustrated by her response when finding herself in the midst of an earthquake in San Francisco in 1891. To record this phenomenon, she centred her pencil on a blank sheet of paper so that, as the quake progressed, the pencil moved over the page showing the changes in motion taking place, timing the process as she did so. In this way, Breton produced an image that foreshadowed that of a modern seismograph (McVicker 2005, pp. 13–32).

Three other female scientific artists—albeit not as well known—also exhibited at the Royal Society soirée, although their work was shown on their behalf and they did not exhibit in person. These were women who practised their art in the amateur, private sphere and did not enter the public arena in the same way as the women already discussed. At the men's soirée of 1881, Miss M.A.

Hicks' painting of 'The Meteor of 7 June, observed at Exeter', was exhibited by the amateur astronomer and mathematician Henry Perigal. Hicks is likely the sister of mathematician William M. Hicks, who was a member of the London Mathematical Society alongside Perigal. This connection suggests the importance of familial and friendship networks in mediating women's access to science and the soirée at this time. This is illustrated again by the contributions of Ellen Martha Busk (1846–1889) to the ladies' soirée of 1885, and Caroline Lawes to the men's soirée of 1891. Busk's portrait of her father, the naturalist George Busk FRS, was exhibited by the Linnaean Society of London of which he was a prominent member. Busk was an artist, working in oils rather than the usual water-colours deemed more appropriate for women, and had exhibited at the Royal Academy in 1878 (Harms and Scott 2013, pp. 145–48). Lady Caroline Lawes (1822–1895) was the wife of Sir John Bennet Lawes, founder in 1843 of the Rothamsted Experimental Station, one of the first ever agricultural research institutions. Her coloured drawings of drain-gauges were exhibited by her husband at the men's soirée of 1891. Lawes was an amateur water colourist and seems to have had the job, as so many wives and daughters of scientific men, of putting her artistic skills at the service of her male relative. Lawes' drawings were not repeated as an exhibit at the later ladies' soirée, most likely as they were deemed not of a sufficiently 'popular' character.

The contributions of the last two women exhibiting visual representations at a soirée during the period 1872–1914 can certainly be placed at the 'popular' end of the spectrum. These women were photographers and indicative of the increasing number of women, not only from privileged backgrounds, taking up this new technology from the 1880s onwards. As well as enjoyed as an amateur pastime, photography gave some women opportunity to set up as professionals with their own studios (Denny 2009, p. 811). Mrs. F.W.H. Myers, née Evelyn Tennant (1856–1937) largely fits this profile. She married Cambridge don Frederic William Henry Myers—remembered as the co-founder of the Society for Psychical Research (SPR)—in 1880 and had taken up photography as a hobby when her children were young. Myers later established her own studio at her Cambridge home, with her own developing and printing facilities, and practised as a portrait photographer (Oberhausen and Peeters 2016). At the ladies' soirée of 1890 she exhibited platinotype photographs including portraits of Arthur Balfour and William Gladstone. Many of Myers' portraits are now held by the National Portrait Gallery in London. Less is known about the Miss Rhodes who exhibited 'Stereoscopic Views of Victoria Falls' at the ladies' soirée of 1906. We do know, again, that she was connected to the Royal Society through friendship networks, particularly with G.W. Lamplugh FRS who had suggested her stereoscopic photographs as a possible exhibit to the then Society Secretary, Robert Harrison.[9] Stereoscopic photography was an early form of three-dimensional photography that became somewhat of a craze after the London Stereoscopic Company produced compact cameras, complete with chemicals, packed in a portable wooden case. The technique comprised taking photographs of a scene from two slightly different angles and mounting them together; when seen through a stereoscopic viewer, these give the impression of one three-dimensional image (Metherell 2018).

Two years earlier in 1904, photographs of a completely different kind were exhibited at the ladies' soirée by 'Mrs D.H. Scott'. This was the botanist, palaeobotanist and pioneer filmmaker Henderina Scott, née Klaassen (1862–1929). Scott was married to 'father of palaeobotany' Dukinfield Henry Scott FRS and the couple were well-known in London's scientific circles. Scott published her research into living and fossil plants in various scientific journals and collaborated with her husband and the botanist Ethel Sargant. In 1905, Scott and Sargant were among the first women admitted as 'Lady Fellows' of the Linnaean Society. On the evening of the soirée, Scott demonstrated what she called 'animated photographs' of the growth and movements of plants; she showed, in slow motion time-lapse photography, the opening of buds, pollination by a bee, the unravelling of a shoot and other manifestations of plant activity. Scott used a Kammatograph which was an early camera and projector in one which used glass plates with miniature images arranged in a spiral to project movement. This 'filmless camera' was used with a lantern projector for showing. This is one of the earliest examples of this technology being used to record botanical phenomena and Scott was invited to demonstrate this novel work to the Royal Horticultural and Botanical Societies, as well as at the Royal Society. However, she practised her filmmaking and science in the domestic sphere in a private capacity, rather than as a public professional; as a result, her work, celebrated at the time, is now forgotten (Jones 2016, pp. 89–93).

Female botanists contributed several exhibits to Royal Society soirées in the 1890s onwards, most of them researchers at university colleges in London. At the ladies' soirée in June 1892, Margaret O. Mitchell and Frances G. Whitting contributed 'a new Order of Algae' (a kind of seaweed). They did not exhibit their find themselves, instead this was done on their behalf by George Murray FLS from the Botanical Department at the British Museum. These women, both former students of Newnham College, Cambridge, were researchers alongside Murray at the British Museum. Whitting later moved to a lecturing role at Kings College London (B. A. 1892). If Whitting and Murray were scientific researchers, the 'Miss York', whose 'Dried Specimens of Cape Plants' were exhibited at the 1894 ladies' soirée, was most likely a collector and amateur botanist. Her specimens were exhibited on her behalf by the Director of the Royal Gardens at Kew, Sir William Turner Thiselton-Dyer. York may have been connected to the family of a colonial administrator, as many wives and daughters of colonial officials collected plants in the eighteenth and nineteenth centuries, becoming only footnotes in history with little known about them (Horwood 2010).

One female collector with scientific interests who is well known provided an exhibit for the men's soirée of 1879. Baroness Burdett-Coutts contributed 'Assegai' (spears) from the Battlefield of Isandula where, in January that year, Zulu fighters had defeated a substantial British force and shocked Victorian society. Angela Burdett-Coutts (1814–1906) had inherited a large fortune which she used for philanthropy, often in the field of science. In 1861 she had endowed two Oxford scholarships in geology and natural science and had given money to other scientific organizations including Kew Gardens (Healey 2004). The usefulness of soirées in cultivating patronage such as this is clear (even if Coutts was not at this men's soirée in person). Individuals of influence, including royalty, colonial officers, and archbishops, were sought out regularly for invitations to the ladies' soirée in particular.

The ladies' soirée of 1894 featured a reversal of roles when a woman exhibited the scientific art created by a man—a situation that again underlines the importance of status and family connections. This was especially so in relation to the Royal Society, although this was being challenged from end-of-century as science was infiltrated by commerce, and by non-élite women (and men) taking advantage of increasing opportunities for a scientific education. Lady Mariabella Fry (1833–1930) was the granddaughter of Luke Howard FRS, the meteorologist and chemist, and author of *On the Modification of Clouds* (1803). This text gave us our nomenclature of clouds and informed the landscape paintings of William Turner and John Constable. Lady Fry exhibited water sketches of clouds which had recently been found among Howard's papers; these later came into the ownership of the Royal Meteorological Society (Burton 2004).

The ladies' soirée of 1903 featured separate exhibits by another two female botanists and university researchers: Edith Saunders and Lettice Digby. Edith Rebecca Saunders (1865–1945) exhibited specimens of her experimental work on cross-breeding and blended heredity in plants. This was a repeat of the same exhibit shown on her behalf by a male fellow at the men's soirée in May. At the time, Saunders held teaching posts at Newnham and Girton Colleges and was Director of the Balfour Biological Laboratory for Women at Cambridge. She enjoyed a long and significant research collaboration with William Bateson who headed a school of genetics at Cambridge which comprised primarily women from Newnham College (Richmond 2001). Saunders co-authored with Bateson four reports to the Royal Society, between 1902 and 1908, on the application of Mendelism to understanding heredity, as well as publishing widely under her own name alone. With Scott and Sargant, Saunders was among the first women to be elected to a fellowship of the Linnean Society in 1905, becoming its first female vice-president in 1912–1913. She was later active in the British Association for the Advancement of Science (Alexander 2020). Lettuce Digby (1877–1972) exhibited cell-phenomena connected to apogamy (asexual development in plants without fertilization), alongside J.B. Farmer FRS and Mr. J.E.S. Moore. Aged only 26 years at the time, Digby had been a student of Farmer's at the Royal College of Science in South Kensington, which merged into Imperial College London in 1895, and then worked as a researcher at the Jodrell Laboratory at Kew (Sarbadhikari 1946). Digby went on to publish widely and in 1914 co-authored a study with Farmer on the dimensions of chromosomes (Bretland-Farmer and Digby 1914). Another female collaborator of William Bateson exhibited at the ladies' soirée of 1914; Florence Margaret Durham (1869–1949) showed a 'fertile canary hybrid' and 'a rat of a new colour'. Durham was a physiology lecturer at Newnham specializing in the inheritance of coat colour in mice, a subject on which she had co-written a 1908 report to the Evolution Committee of the Royal Society with Bateson and Saunders (Crowther [1952] 2009, p. 281).

The first decade of the twentieth century witnessed strong interest in heredity and, by association, eugenics. At the men's soirée of 1908, Karl Pearson, the eugenicist and mathematician who had taken up directorship of the Francis Galton Laboratory for National Eugenics (formally the Eugenics Record Office) the previous year, credited Amy Barrington alongside himself for an exhibit comprising specimens of the hair of a chestnut horse. Barrington (1857–1942), a former student of Girton College, Cambridge, and then a lecturer in teacher education at Bedford College, was one of a team of female researchers employed by Pearson at the Biometrics Laboratory at UCL (Walsh 2014, p. 102). These women collected and measured biological and physical data to produce biometric tables that informed eugenic theory. Barrington was among several women with which Pearson co-wrote scientific papers (Jones 2009, p. 177). Another female collaborator with Pearson was Alice Lee (1858–1939). At the ladies' soirée of 1895, the UCL maths department exhibited 'A series of Diagrams calculated and prepared by Miss Alice Lee' to illustrate time-decay caused by a Hertzian oscillator, a device for transmitting electrical waves (although it is not clear if Lee was present in person). In 1895, Lee had just begun attending Pearson's lectures in statistics at UCL after graduating in mathematics from Bedford College, London, the previous year. She stayed on at Bedford in a teaching role and began a Ph.D. with Pearson on the connection of skull capacity to intellectual ability. After controversy (her examiners were sceptical that she had completed the work herself and did not like results which contradicted existing thinking) Lee was awarded the Ph.D. in 1901 (Love 1979).

Unlike Lee, Dorothea Bate (1878–1951) did not have university qualifications, however her expertise as a palaeontologist was well known at the time. At the ladies' soirée of 1903, Bate exhibited remains of pygmy elephant and hippopotamus which she had discovered during her recent research expedition to Cyprus. That expedition had been partly funded by the Royal Society which had awarded her a grant of £30. Despite her scientific standing, Bate had always to scavenge for research funds. She was associated with the Natural History Museum from 1898, yet was not paid or made a formal member of staff until 1948 when she was in her late 60s. This illustrates how precarious scientific women's situation could be; excluded from salaried professional positions, apart from teaching (which could be seen as an extension of the feminine nurturing role), women typically pursued their science in a private capacity. This was certainly the case with several of the women scientists discussed in this chapter. Despite this, Bate's scholarship was accepted and her finds were sought for display at Royal Society soirées another three times before 1914. In 1905 her remains of fossil mammals from Crete were shown at the men's soirée in May and she exhibited them herself at the ladies' soirée the following month. Bate was invited to exhibit fossil finds from Cyprus again at the ladies' soirée in 1910 (Shindler 2005). And, at the ladies' soirée in 1911, she exhibited 'the peculiar Goat-like Animal *Myotragus balearicus Bate*' with photographs of the locale in Majorca where she had made the discovery.

Palaeobotany was a growing discipline from the 1880s to 1914, driven by fossil plants being revealed in new coal seams and due to the expansion of the railways. At the ladies' soirée in June 1906, a research team from UCL, including two women, exhibited 'Fossil Plants from the English coal Measures'. Winifred Brenchley (1883–1953) had studied botany at UCL and would soon take up the Gilcrist studentship for university women at the Rothampsted Experimental Station. She was awarded a DSc in 1911, became a fellow of UCL in 1914, and was elected to a fellowship of both the Linnean and Royal Entomological Societies (Jenkinson 2004). Margaret Jane Benson (1859–1936) had achieved first-class honours in botany at UCL in 1891 and, in 1893, became head of the newly established Department of Botany at Royal Holloway, University of London. She established herself as one of the most significant palaeobotanists of the early twentieth century (alongside Marie Stopes), publishing around twenty original papers in this area (Creese and Creese 1998, p. 29). As noted earlier, Stopes exhibited 'The Microscopic Structure of Fossil Plants from Japan' at the ladies' soirée of 1909. On that occasion, she was joined by another female exhibitor, Mrs. Gadow, who displayed 'Specimens of natural history from Mexico' with her husband, Dr. Hans Gadow FRS. This was the second time Mrs. Gadow had exhibited, having shown 'Ethnological Specimens from Southern Mexico' at the ladies' soirée in 1905 (on this occasion her husband had a separate exhibit). Clara Maud Gadow, née Pagett (1857–1949) was part of an influential scientific family and one of the three daughters of Sir George E. Paget, Regius Professor of Physics at Cambridge. All sisters married Cambridge scientists, notably Rose Pagett (1860–1951) who researched at the Cavendish Laboratory and married its Director, the celebrated physicist J.J. Thomson (Peterson 1984, pp. 680–88). These exhibits were the product of expeditions the couple had made to Southern Mexico in 1902 and 1904. Hans Gadow published an account of these travels which recalled the hair-raising episodes experienced while collecting botanical and zoological specimens, including 'my wife' struggling to capture snakes which had invaded the couple's temporary abode (Gadow 1908, p. 82). Clara is referred to throughout anonymously, no doubt to preserve her privacy as a woman; this reticence to include women in public narratives (middle-class ones at least) is another reason why the contributions of women to science—which is typically a collaborative process—are often lost to the historical record.

Women with technical interests were represented at the soirées too. The ladies' soirée of 1898 featured photographs taken by the astronomer Annie Maunder, née Russell (1868–1947). A former student of Girton College, Cambridge, Maunder met her husband, Edward Maunder, while working as a 'lady computer' at the Greenwich Observatory where he led the solar section. After marriage, they worked together and separately on astronomical projects (Ogilvie 1991, p. 129). Maunder had taken the photographs, of the sky in the neighbourhood of the constellation of Monoceros, as one of the astronomers on the British Astronomical Association's expedition to India to view the solar eclipse of January 1898. She had designed her own equipment, and borrowed a telephoto lens, to enable her to photograph the longest possible extension of the coronal streamers present during the eclipse (Ogilvie 2000, p. 79). Two years later, in 1900, Maunder again exhibited at the ladies' soirée, this time showing photographs of the Milky Way and, in 1902, she exhibited photographs of the 'The Corona of 1901'.

At the ladies' soirée of 1888, one of the most popular exhibits were the 'Voice Figures' created by the Welsh soprano Megan (Margaret) Watts-Hughes (1842–1907). Hughes had invented a device she called an 'eidophone' to visualize the forms created by the human voice while singing. This consisted of a flexible elastic membrane, tightly stretched over a receiver, into which the voice was directed; sand, powders, or thick coloured liquids were then applied which produced various shapes according to the sound vibrations (Watts-Hughes 1891). According to *The Times*, which picked out the exhibit for special merit, the voice figures were 'exquisitely beautiful' and 'strikingly like those of natural flowers, ferns, trees, etc.' (*The Times* 1888). Hughes published a book on her invention, the 'eidophone', in 1904 and exhibited her voice figures again at the ladies' soirée of 1906 (Watts-Hughes 1904). Another woman—more anonymous than the celebrity soprano discussed above—was also singled out for special attention in *The Times*, this time in its report of the ladies' soirée of 1900:

> Another exhibit which attracted general attention, and which deserves special recognition, was that of the enlarged models of gnats (mosquitoes) and of human blood corpuscles infested by the malaria parasite, the work of Miss Delta Emett, shown by Professor Ray Lankester. The workmanship and truth to nature of these models reflect the greatest credit upon Miss Emett's manipulative skill and accuracy of observation [...] The immense practical and scientific importance of Miss Emett's work must be evident. (*The Times* 1900)

Little is known about Delta Emett, except that she is listed as a 'maker of museum models' at the Natural History Museum, at the time when Ray Lankester was director of the natural history department and keeper of zoology at the British Museum.[10] Scholarship is only just beginning to recover the many technicians and assistants—generally individuals of lower middle or working class and/or women—who are not typically on the public record, in spite of the importance of their work to the collaborative process of science (Hartley and Tansey 2015). That a woman technician was recognized by *The Times* in this way is intriguing.

Hertha Ayrton—an extract from whose fictionalized biography began this chapter—exhibited at the ladies' soirée four times, in 1895, 1899, 1904, and 1908. Ayrton (1854–1923) was an electrical engineer and physicist who won the Royal Society's Hughes Medal for original research in 1906. Although from a modest background, she had studied mathematics at Girton College, Cambridge and then, in 1884, embarked on evening classes in electro-technics at the new City and Guilds Technical Institute at Finsbury, marrying her professor there, William Ayrton FRS, in May the following year. Collaboration with her husband gave Ayrton access to the laboratories at Imperial College, something that she lost on his death in 1908, forcing her to set up her own, much less-well equipped, laboratory in the drawing room of her London home. This illustrates again the precariousness of women in disciplines requiring institutional laboratory facilities to pursue their research. At the 1895 ladies' soirée, Ayrton exhibited sepia drawings of the electric arc and, at the 1899 ladies' soirée, demonstrated live experiments on the hissing of the arc. Arc lights were used for searchlights and streetlights; to strike the arc a high voltage was established between two carbon rods a short distance apart, but the result was unstable. Ayrton's experiments were designed to find a way to lessen the noise, sputtering and hissing. Reporting on the soirée, the *Daily News* wrote that what 'astonished the lady visitors [...] was to find one of their own sex in charge of the most dangerous-looking of all the exhibits – a fierce arc light enclosed in glass. Mrs Ayrton was not a bit afraid of it' (Sharp 1926, p. 143). Ayrton's last two contributions to the ladies' soirée were connected to the second of her major researches, into the formation of ripple marks and water vortices; in 1904 and 1908, she demonstrated less 'fierce' experiments utilizing water tanks, sand, and coloured liquid.

Conclusions

Women may not have been eligible for a fellowship of the Royal Society, but the soirée gave them opportunity to be visible as experts, even if a woman embodying scientific authority was not without its tensions. This is clearly evidenced by the *Daily News*'s report of Ayrton's 1899 soirée demonstration, which conveys surprise that she was not at least a little afraid of her experiments. The number of individual women exhibiting between 1872 and 1914 is small at just thirty (although a handful of women exhibited more than once), and it must be noted that not all women were present in person when their work was exhibited. Nevertheless, it is clear that these women were known and respected within their own scientific networks, and that history of science has mostly failed to see them, looking as it has—until recently at least—through its own exclusively masculine prism. One reason for this is the public/private and amateur/professional distinction; this distinction, less rigid in the years around 1900, when applied through modern eyes can distort accurate appreciation of women's significance. Women pursued science then, but often in a private capacity with no institutional affiliation, such as Scott, Ayrton, Gadow, and Maunder. These women's access to institutional science and networks was smoothed by their husband's connections and, in the case of Gadow, by her father's too. The husbands of Scott, Gadow, and Ayrton were active and respected fellows of the Royal Society. Annie Maunder's husband, Edward, was secretary, and then vice-president of the élite Royal Astronomical Society (RAS) (Crommelin 1928). The RAS did not admit women as fellows until 1916, still twenty-nine years before the first female fellows of the Royal Society were elected. As the new century approached, the impact of the opening of higher education to women was evidenced by female researchers achieving a presence at the soirée. However, they still needed the support of a male friend or mentor as intermediary and often exhibited with senior male collaborators and colleagues.

Women exhibitors at the soirée are more visible in the biological sciences, natural history, and art-related disciplines; as outlined above, these had a long tradition of female participation and so did not clash with ideas of feminine suitability and propriety. It must be remembered too, that the nineteenth-century conception of what counted as science was inclusive, as is clearly evidenced by the eclectic nature of exhibits at the soirée, especially the ladies' soirée. This broad understanding embraced scientific illustration, photography, collecting, archaeology, science writing and scientific translation, and this supported women's involvement in science. Additionally, these branches of science did not transfer to an institutional setting at the end of the nineteenth century and so were more easily accessed by women pursuing their interest in science in the domestic sphere. The women who attended the ladies' soirée, as guests and/or exhibitors, were illustrative of an excitement and fascination with science which characterized the later Victorian and Edwardian periods (Broks 1996). This curiosity with science extended to all classes, although the Royal Society soirée, with its mix of science with élite social networking, catered mostly for the privileged. Notwithstanding, the women at the soirée before the Great War—whether exhibitors, guests, or both—illustrate that there was clear feminine engagement with science, on a variety of levels, to 1914 (Fig. 9.1 and Table 9.1).

9 WOMEN AT THE ROYAL SOCIETY SOIRÉE BEFORE THE GREAT WAR 187

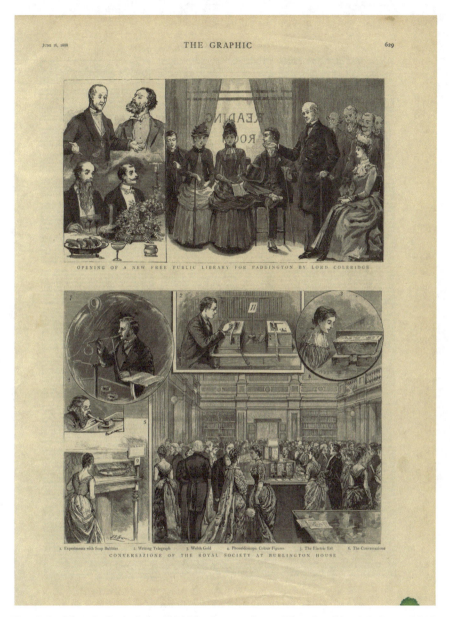

Fig. 9.1 The Ladies' Soiree 1888. Image from *The Graphic*, 16 June 1888, reproduced by kind permission of The Royal Society of London

Table 9.1 Women exhibitors at Royal Society Soirées 1872–1914

Year	Men's Soirée (women did not exhibit in person)	Ladies' Soirée
1872	**Margaret McNair Stokes** Archaeologist, Artist, Photographer	
1873		
1874	**Marianne North** Botanical Artist	
1875		
1876		*First annual ladies' soirée*
1877	**Marianne North** Botanical Artist	
1878		
1879	**Baroness Angela Burdett-Coutts** Collector	**Marianne North** Botanical Artist
1880		
1881	**Miss M. Hicks** Astronomical Artist	
1882		**Marianne North** Botanical Artist
1883		
1884		
1885		**Ellen Martha Busk** Portrait Artist
1886		**Marianne North** Botanical Artist **Anna Lee Merritt** Artist
1887		
1888		**Margaret Watts-Hughes** Instrument inventor
1889		**Constance F. Gordon-Cumming** Artist and draughtsperson
1890		**Evelyn Myers** Photographer
1891	**Lady Caroline Lawes** Technical Artist	
1892		**Frances G. Whitting*** Botanist **Margaret O. Mitchell*** Botanist **Not exhibiting in person*

(continued)

Table 9.1 (continued)

Year	Men's Soirée (women did not exhibit in person)	Ladies' Soirée
1893		
1894		**Miss York*** Botanic Collector *Not exhibiting in person
1895		**Hertha Ayrton** Physicist and Electrical Engineer **Marion Ellis Rowan** Botanical Artist **Alice Lee** Mathematician and Eugenicist **Lady Mariabella Fry** Exhibiter of artwork of family member
1896		
1897		
1898		**Annie Maunder** Astronomer and Photographer
1899		**Hertha Ayrton** Physicist and Electrical Engineer
1900		**Annie Maunder** Astronomer and Photographer **Delta Emett*** Model-maker *Not exhibiting in person
1901		
1902	**Adela Breton** Geologist and Artist	**Adela Breton** Geologist and Artist **Annie Maunder** Astronomer and Photographer
1903	**Edith Saunders** Botanist	**Dorothea Bate** Palaeontologist **Edith Saunders** Botanist **Lettice Digby** Botanist
1904		**Hertha Ayrton** Physicist and Electrical Engineer **Henderina (Rina) Scott** Botanist, Palaeobotanist, Filmmaker

(continued)

Table 9.1 (continued)

Year	Men's Soirée (women did not exhibit in person)	Ladies' Soirée
1905	**Dorothea Bate** Palaeontologist	**Dorothea Bate** Palaeontologist **Clara Maud Gadow** Zoologist and Botanist
1906		**Winifred Brenchley** Botanist and Palaeobotanist **Margaret Jane Benson** Botanist and Palaeobotanist **Miss Rhodes** Stereoscopic photographer **Margaret Watts-Hughes** Instrument inventor
1907		
1908	**Amy Barrington** Biologist and Eugenicist	**Hertha Ayrton** Physicist and Electrical Engineer
1909		**Marie Stopes** Geologist and Palaeobotanist **Clara Maud Gadow** Zoologist and Botanist
1910		**Dorothea Bate** Palaeontologist
1911		**Dorothea Bate** Palaeontologist
1912		
1913		
1914		**Florence Margaret Durham** Biologist and Geneticist

Acknowledgements Sincere gratitude is extended to the Royal Society archivists for their generous help and expertise, without which this chapter would not have been possible.

NOTES

1. This chapter draws on holdings of the Royal Society Library and Archives (RS) including Conversazione Programmes 1872–1889, 1890–1900, 1901–1908, and 1909–1914; soirée committee minutes; and correspondence 1870–1914.
2. Herbert Rix to Editor of *The Builder*, 12 June 1891. RS, NLB/4/444.
3. Herbert Rix to Prof. John Perry, 05 June 1891. RS, NLB/5/420. This exhibit centred on gyroscopes and their laws of behaviour.
4. Herbert Rix to Karl Grossman, 18 May 1894. RS, NLB/9/439.
5. RS Soirée Committee Minutes, 22 January 1880.

6. Herbert Rix to William Marcet, 01 June 1889. RS, NLB/3/402; Soirée Committee Minutes, 06 February 1890.
7. Robert Harrison to Dorothea Beale, 05 July 1902. RS, NLB/24/760.
8. Robert Harrison to Miss A Porter, 14 June 1907. RS, NLB/35/583.
9. Robert Harrison to Miss Rhodes, 15 June 1906. RS, NLB/33/8.
10. Natural History Museum (London, England), Library and Archives, Tring Correspondence, PX4048.

References

Alberti, Samuel. 2003. "Conversaziones and the Experience of Science in Victorian England." *Journal of Victorian Culture* 8, no. 2: 208–30.
Alexander, Christine. 2020. "Celebration of Edith Rebecca Saunders." University of Cambridge. https://www.gen.cam.ac.uk/department/history-of-the-department/celebration-of-edith-rebecca-saunders. Accessed on 08 August 2020.
Ayrton-Zangwill, Edith. [1924] 2018. *The Call*. London: Persephone Books.
B. A. 1892. "Phycological Memoirs." *Nature* 46 (26 May): 75–76.
Bertucci, Paola. 2007. "Sparks in the Dark: The Attraction of Electricity in the Eighteenth Century." In *Endeavour*, edited by Donald L. Opitz, 31, no. 3: 88–93. Amsterdam: Elsevier.
Black, Barbara. 2012. *A Room of His Own: A Literary-Cultural Study of Victorian Clubland*. Athens: Ohio University Press.
Bluhm, R.K. 1958. "A Note on the Origin of the Royal Society's Conversaziones." In *Notes and Records of the Royal Society of London* 13, no. 1: 61–63.
Bretland-Farmer, John, and Digby, Lettice. 1914. *On Dimensions of Chromosomes Considered in Relation to Phylogeny*. London: Royal Society.
Broks, Peter. 1996. *Media Science Before the Great War*. Basingstoke: Palgrave.
Burton, Jim. 2004. "Howard, Luke (1772–1864)." *Oxford Dictionary of National Biography*. https://www.oxforddnb.com/view/10.1093/ref:odnb/9780198614128.001.0001/odnb-9780198614128-e-13928. Accessed on 15 August 2020.
Chaloner, William. 2005. "The Palaeobotanical Work of Marie Stopes." In *History of Palaeobotany: Selected Essays*, edited by A.J. Bowden, C.V. Burek, and R. Wilding, 127–35. London: Geological Society.
Coombes, Annie. 1994. *Reinventing Africa: Museums, Material Culture and Popular Imagination in Late Victorian and Edwardian England*. New Haven: Yale University Press.
Creese, Mary, and Creese, Thomas. 1998. *Ladies in the Laboratory? American and British Women in Science, 1800–1900: A Survey of Their Research*. MD Scarecrow Press.
Crommelin, Andrew Claude de la Cherois. 1928. "Obituary: Edward Walter Maunder." *The Observatory* 51: 157–59.
Crowther, James. [1952] 2009. *British Scientists of the Twentieth Century. Vol. 9*. London: Routledge.

Cunningham, Bernadette. 2018. "Margaret Stokes: Antiquarian Scholar with an Artist's Eye." *The Royal Irish Academy*. https://www.ria.ie/news/library-library-blog/margaret-stokes-antiquarian-scholar-artists-eye. Accessed on 25 April 2020.

Denny, Margaret. 2009. "Royals, Royalties and Renumeration: American and British Women Photographers in the Victorian Era." *Women's History Review* 18, no. 5: 801–18.

Dunraven, Edwin. 1885/1887. *Notes on Irish Architecture*, edited by Margaret Stokes. London: George Bell and Sons.

Fraser, Helen, and Cleal, Christopher. 2007. "The Contribution of British Women to Carboniferous Palaeobotany During the First Half of the 20th Century." In *The Role of Women in the History of Geology*, edited by C. V. Burek and B. Higgs, 51–82. London: Geological Society.

Gadow, Hans. 1908. *Through Southern Mexico: Being an Account of the Travels of a Naturalist*. London: Witherby.

Hall, Marie Boas. 1992. *The Library and Archives of the Royal Society 1660–1990*. London: The Royal Society.

Harms, Elree, and Scott, Shirley. 2013. *A Gallery of Her Own: An Annotated Bibliography of Women in Victorian Painting*. London: Routledge.

Hartley, J.M., and Tansey, Elizabeth. 2015. "White Coats and no Trousers: Narrating the Experiences of Women Technicians in Medical Laboratories, 1930–1990." *Notes and Records of the Royal Society of London* 69, no. 1: 25–36.

Hazzard, Margaret. 1988. Rowan, Marion Ellis (1848–1922). *Australian Dictionary of Biography*, Vol. II. Melbourne: Melbourne University Press.

Healey, Edna. 2004. "Coutts, Angela Georgina Burdett-, suo jure Baroness Burdett-Coutts (1814–1906)." *Oxford Dictionary of National Biography*. https://www.oxforddnb.com/view/10.1093/ref:odnb/9780198614128.001.0001/odnb-9780198614128-e-32175. Accessed on 15 August 2020.

Heller, Jules, and Heller, Nancy. 2013. *North American Women Artists of the Twentieth Century: A Biographical Dictionary*. London: Routledge.

Horwood, Catherine. 2010. "Intrepid Lady Plant Collectors." *HerStoria* 6: 31–33.

Jenkinson, D.S. 2004. "Brenchley, Winifred Elsie (1883–1953)." *Oxford Dictionary of National Biography*. https://www.oxforddnb.com/view/10.1093/ref:odnb/9780198614128.001.0001/odnb-9780198614128-e-51673. Accessed on 15 August 2020.

Jones, Claire. 2009. *Femininity, Mathematics and Science, 1880–1914*. Basingstoke: Palgrave.

Jones, Claire. 2016. "The Tensions of Homemade Science in the Work of Henderina Scott and Hertha Ayrton." In *Domesticity in the Making of Modern Science*, edited by D. Opitz, S. Bergwik, and B. Van Tiggelen, 84–104. Basingstoke: Palgrave.

Laracy, Hugh. 2013. *Watriama and Co: Further Pacific Islands Portraits*. Canberra, Australia: ANU Press.

Love, Rosaleen. 1979. "Alice in Eugenics-Land: Feminism and Eugenics in the Careers of Alice Lee and Ethel Elderton." *Annals of Science* 36, no. 2: 145–58.

Mason, Joan. 1991. "Hertha Ayrton (1854–1923) and the Admission of Women to the Royal Society of London." *Notes and Records of the Royal Society of London* 45, no. 2: 201–20.

McVicker, Mary Frech. 2005. *Adela Breton: A Victorian Artist Amid Mexico's Ruins*. Albuquerque: University of New Mexico Press.

Metherell, Colin. 2018. "Early 3D: The British Contribution to Early Stereoscopic Photography." *The Stereoscopic Society*. http://www.stereoscopicsociety.org.uk/WordPress/early-3d/. Accessed on 15 August 2020.

Middleton, Dorothy. 2004. "North, Marianne (1830–1890)." *Oxford Dictionary of National Biography*. https://www.oxforddnb.com/view/10.1093/ref:odnb/9780198614128.001.0001/odnb-9780198614128-e-20311. Accessed on 15 August 2020.

Oberhausen, Judy, and Peeters, Nic. 2016. "Eveleen Myers (1856–1937): Portraying Beauty: The Rediscovery of a Late-Victorian Aesthetic Photographer." *The British Art Journal* 17, no. 1: 94–102.

Ogilvie, Marilyn Bailey. 1991. *Women in Science: Antiquity Through the Nineteenth Century*. Cambridge, MA: MIT Press.

Ogilvie, Marilyn. Bailey. 2000. "Obligatory Amateurs: Annie Maunder (1868–1947) and British Women Astronomers at the Dawn of Professional Astronomy." *British Journal for the History of Science* 33, no. 1: 67–84.

Peterson, Mildred Jeanne. 1984. "No Angels in the House: The Victorian Myth and the Paget Women." *The American Historical Review* 89, no. 3: 677–708.

Richmond, Marsha. 2001. "Women in the Early History of Genetics: William Bateson and the Newnham College Mendelians, 1900–1910." *ISIS* 92, no. 1: 55–90.

Sarbadhikari, P.C. 1946. "Botany at University College of Science, Calcutta." *Nature* 158 (27 July): 125.

Sera-Shriar, Efram. 2015. "Anthropometric Portraiture and Victorian Anthropology: Situating Francis Galton's Photographic Work in the Late 1870s." *History of Science* 53, no. 2: 155–79.

Sharp, Evelyn. 1926. *Hertha Ayrton: A Memoir*. London: Edward Arnold.

Shindler, Karolyn. 2005. *Discovering Dorothea: The Life of the Pioneering Fossil-Hunter Dorothea Bate*. London: HarperCollins.

Shteir, Ann. 1997. "Gender in 'Modern Botany' in Victorian England." *Osiris* 12: 21–38.

The Royal Society Conversazione. 1888. *The Times*, 07 June, 6.

The Royal Society Conversazione. 1900. *The Times*, 10 May, 11 and 22 June, 4.

Walsh, Brendan. 2014. *Knowing Their Place: The Intellectual Life of Women in the 19th Century*. Stroud: The History Press.

Watts-Hughes, Margaret. 1891. "Visible Sound-Voice Figures." *The Century Illustrated Monthly Magazine* 42, no. 1: 37–40.

Watts-Hughes, Margaret. 1904. *The Eidophone: Geometrical and Natural Forms Produced by Vibrations of the Human Voice*. London: Christian Herald.

Wintle, Claire. 2008. "Career Development: Domestic Display as Imperial, Anthropological, and Social Trophy." *Victorian Studies* 50, no. 2: 279–88.

CHAPTER 10

Career Paths Dependent and Supported: The Role of Women's Universities in Ensuring Access to STEM Education and Research Careers in Japan

Naonori Kodate and Kashiko Kodate

One should not be dissatisfied with the inadequacy of a social institution and wait vainly for the moment when a chance is given. If one strikes it, the back door will open, and if one tries to break it in, one will be given a small desk there. Then one should be grateful for what has been given and devote all your spirit to learning.

Sechi Katō[1]

The low ratio of female researchers in STEM domains has been widely debated across the world. In many advanced economies, gender equality in STEM education is now firmly ensconced in the public policy agenda. While the issue is often presented as binary, i.e. that of freedom of individual choice in liberal society or something that society needs to address and enforce, it is more complex than it first appears when it comes to achieving results. One reason for this is that individual choice of subjects and the education environment are intricately intertwined and shaped by other factors such as family (e.g. parents' influence, education levels, socio-economic status), teachers' advice, gender stereotypes, social norms, the labour market and school system of that

N. Kodate (✉)
School of Social Policy, Social Work and Social Justice & UCD Centre for Japanese Studies, University College Dublin, Dublin, Ireland
e-mail: naonori.kodate@ucd.ie

K. Kodate
Japan Women's University, Tokyo, Japan

particular point in history. Many of these factors are rooted in history and social changes are often piecemeal and path-dependent (Collier and Collier 1991; Kodate and Kodate 2015). Therefore, it is critically important to understand who the major players are and how those local agents make sense of the meaning of education for women at a given point in time, in that locality. With a better understanding of these, it could be possible to achieve the goal of creating an environment where every person's skills and talents are supported and can flourish. With the exception of the USA, many Western nations have adopted co-education as the standard norm for higher education institutions. Yet Japan still maintains a number of women-only higher education institutions (Kodate et al. 2010). Recently, in some European countries such as Germany and Austria, women-only environments have been re-introduced for combatting the underrepresentation of female engineers (Freitag et al. 2012). This chapter examines the extent to which women-only universities matter(ed) in ensuring STEM education in Japan, and their past and future roles in a changing higher education landscape. This local landscape of higher education in Japan, and its impact on women in STEM, is the primary theme of this chapter.

Great efforts have been made in Japan in the last 20 years to address gender inequality in academia, and STEM in particular. The initiatives and policy change originated in the bursting of the economic bubble in the early 1990s, when the Japanese economic model began to be called into question. The male-breadwinner model of lifetime employment, involving long work hours and male-dominated workplaces, has been challenged. The situation has been compounded by dramatic demographic change, particularly the low fertility rate and rapid pace of ageing. Given the country's traditional view of gender roles—men work outside the home, while women stay at home—the government was originally hesitant to adopt proactive steps (Yamamoto and Ran 2014). However, when strong evidence pointed to the need to engage both men and women in scientific enquiry and the labour force to sustain its economy, the government began to act.

The evidence base was provided by the Japan Inter-Society Liaison Association Committee for Promoting Equal Participation of Men and Women in Science and Engineering (EPMEWSE), which was formed from major academic societies in STEM fields. The EPMEWSE conducted and published a large-scale survey in 2004. One of its major findings was a clear gender gap in the treatment of men and women in STEM. Gender inequality in the rank of positions was found, but even greater inequality was identified in the allocation of basic resources necessary for research and development, such as the number of subordinates and level of funding. This issue of a gender gap in the allocation of research resources was more acute in universities and public research institutions, compared with that in corporations. The report also noted that female researchers and engineers have far fewer children than their male counterparts, which is considered 'almost abnormal' (EPMEWSE 2004, p. 47).

After the publication of these results in a White Paper by the Bureau of Gender Equality ('Danjo Kyōdō Sankaku Kyoku'), in 2006 the government launched 'Support for best practice for the promotion of female researchers' as part of the process of Competitive Research Grants for the Promotion of Science. Ten universities were awarded under this scheme: Hokkaidō, Tōhoku, Tokyo University of Agriculture and Technology (TUAT), Waseda, Ochanomizu, Tokyo Women's Medical University, Japan Women's University (JWU), Kyoto University, Nara Women's University and Kumamoto University. Four out of the ten universities were women-only institutions. In the era of gender equality, there was a revival of women-only education institutions in Japan, which contrasts with European countries in particular. This can be partly explained by the relatively low ratio of women entering large research-intensive national universities in Japan. The ratio of female undergraduate students at seven former *Imperial Universities* (national universities) ranges from 33.1% at Osaka University to 19.0% at the University of Tokyo, as of 2016. In contrast to this, parity has been achieved at universities such as Harvard in the USA (50%) and almost achieved at Oxford in the UK (46%). It is known that in Japan, national universities generally provide better facilities and equipment integral to training for skills and the knowledge base required for certain types of STEM research (e.g. engineering and physics).

However, the revival of women-only universities should not be ascribed only to the very small proportion of women pursuing scientific research and STEM-related careers in national universities as there are other historical and institutional reasons for this. To address this, the following discussion is organised into four parts. The first section deals with the historical developments of female education in Japan, particularly the establishment of different types of women-only higher education institutions in the first half of the twentieth century. The second will then focus on the significance of women-only universities, as they created the only 'pipelines' for pioneering women in STEM by offering them access to further research environments. Hokkaidō (Imperial) University and Rikagaku Kenkyūjo (RIKEN, the Institute of Physical and Chemical Research) are singled out as uniquely 'open' institutions in the country during the interwar period. The third section will briefly consider the trend in favour of co-education as a relatively recent phenomenon in Japan. While competition among different higher education institutions is increasing nationally and globally, the mission of women-only universities is ever more accentuated in STEM domains. The final section will outline the present landscape of Japanese higher education, with a focus on women-only institutions.

The Establishment of 'Universities' and Women-Only Education Institutions in Late Nineteenth-Century Japan

In the late nineteenth century, the Japanese social system, including education, was reformed on the lines of 'Enriching the country, strengthening the military' ('Fukoku Kyōhei') and 'Promoting new industries' ('Shokusan Kōgyō'). Both 'Fukoku Kyōhei' and 'Shokusan Kōgyō' became national slogans to underpin various public policies which aimed to modernise the country based on the Western models.[2] A number of universities were founded with the clear mission of importing and translating Western ideas into Japan to catch up with the West. In 1877, the University of Tokyo was established with four faculties (Law, Humanities, Science and Medicine) and became the Imperial University in 1886 by the enactment of the Order of the Imperial University. At this stage, it was the only Imperial University in Japan, absorbing the former Imperial College of Engineering. The Order was a catalyst that changed the landscape of higher education in Japan. Prior to 1886, the university ('daigaku') was not a major player in higher education in Japan. Instead, many publicly established colleges such as Sapporo Agricultural College (later Hokkaidō University) and Tokyo Vocational School (later Tokyo Institute of Technology) existed to provide education to students. People also studied at private schools such as Keio Gijuku (later Keio University) and Meiji Law School (later Meiji University). The Order set the scene for the establishment of seven Imperial Universities (including Tokyo) across the country as Japan's top-level education organisations, while simultaneously other types of higher education institutions began to be established and foster professionals and policymakers. The establishment of these Imperial Universities was an integral part of the grand scheme to foster national leaders and élite civil servants for modern Japan (Yoshimi 2011). With the increasing industrialisation of Japan, the demand for educated and skilled workers also heightened. Higher education and vocational training were both deemed necessary and this led to the founding of state-funded vocational schools and the promotion of women's education through a separate single-sex schooling system.

The mainstream view of female education's role was to foster 'good wives, wise mothers' ('Ryōsai Kenbo'). The term 'good wives, wise mothers' was coined in 1875 by a pioneer in women's education, Masanao Nakamura, and began to represent the ideal for womanhood in the late 1880s and early 1900s (Sievers 1983). This concept fitted very well with 'Enriching the country, strengthening the military' and implied there was no need for women to enter 'university'. The Women's Normal School (Kōtō-shihangakkō) was opened as a teachers' college for women in 1875, and it not only prepared women for teaching positions but also became the national centre for girls' education (MEXT, n.d.). The meaning of education for women was seen as completely different from that for men, and the education system did not assume women's access to university. In 1882, its preparatory school was transformed into

Higher School for Girls (Kōtō-jogakkō). A few amendments were made to the 1886 Middle School Order in 1891, and the Higher School for Girls was then classified as being of the type Ordinary Middle School (Jinjō-chūgakkō). In 1899, a separate Order for Higher Schools for Girls (Kōtō-jogakkō) was enacted. It stipulated that the girls eligible to enrol in this school were those who had completed the higher grades of primary school (and were over twelve years of age), and each prefecture had the responsibility for planning and providing this type of school for its residents. In the same year, the Vocational School Order was also passed. With the 1910 amendments, these schools were permitted to establish home economics-based specialised courses, as the primary aim of education for women was to produce homemakers.

The Private School Order and the Professional School Order were issued in 1901 and 1903 respectively. The former was designed to regulate Christian schools established by many Christian educators, while the latter was applied to certain 'professional schools' for women (Senmon gakkō) established at the turn of the century. Up to 1919, universities meant only the imperial universities stipulated by the 1886 Order of Imperial University. However, the 1919 Order of University granted a 'university' status to colleges such as Osaka (Prefectural) Medical College, Keiō, Waseda, Dōshisha and Meiji.

While the significance of female education was recognised and laws and infrastructure were gradually updated, single-sex education became embedded in the landscape of the Japanese higher education system. This era saw a nascent form of women-only universities of various types. Although there are different ways in which these educational institutions can be categorised based on the gender of the founders, founding principles and the focus of the curriculum, for the purpose of this chapter, four elements were used to categorise the women's universities. These were: teacher training college, 'vocational' school (with strong focus on homemaking), professional school (with some emphasis on higher learning) and Christian school; see Table 10.1 detailing examples of four types of higher education institutions for women between 1870 and 1930.

The first type of women-only universities (Type A) was teacher training colleges (Normal Schools), established by the government. The Tokyo Women's Normal School (TWNS) was established in 1875. It was subsequently transformed into its current form as Ochanomizu University in 1949. The two Women's Higher Normal Schools (Ochanomizu and Nara), with their strong orientation towards teaching qualifications, played a major role as the highest level of educational institution for women. Women were not granted an official entitlement to study at any of the Imperial Universities until the end of World War Two, and access to further education for women was considerably restricted. After the enactment of the 1919 Order of University, two more teacher training colleges were established in 1929 (Liberal Arts and Science Universities in Tokyo and Hiroshima), and these opened their doors to women. There was a movement among Ochanomizu Alumnae to upgrade TWHNS to university status in the 1920s, but this was not supported

Table 10.1 Four types of higher education institutions for women between 1870 and 1930 (selected examples)

Type A Teacher Training College	Type B Vocational School	Type C Professional School	Type D Christian School
1875: Tokyo Women's Normal School (Ochanomizu Daigaku) 1908: Women's Higher Normal School in Nara (Nara Joshi Daigaku) 1929: (co-ed) Liberal Arts and Science Universities in Tokyo and Hiroshima (⇒ closed in 1962, became Tsukuba Daigaku and part of Hiroshima Daigaku respectively)	1875: Atomi School (Atomi Gakuen Joshi Daigaku) 1881: Wayō Sewing School (Tokyo Kasei Daigaku); Tōyō Jojuku (Jissen Joshi Daigaku) 1901: Japan Women's University (Nihon Joshi Daigaku) 1886: Kyōritsu Girls' Vocational School (Kyōritsu Joshi Daigaku) 1903: Women's School of Commerce (Kaetsu Daigaku [co-ed]) 1908: Ōtsuma Sewing School (Otsuma Joshi Daigaku) 1917: Gunze Girls' School (co-ed in 1924, closed)	1900: Women's Institute for English Studies (Tsudajuku Daigaku); Tokyo Women's Medical School (Tokyo Joshi Ika Daigaku) 1901: Japan Women's University (Nihon Joshi Daigaku) 1907: Tokyo Women's School of Pharmacy (Meiji Yakka Daigaku [co-ed]) 1924: Osaka Prefectural Professional School (Osaka Joshi Daigaku ⇒ closed in 2014) 1925: Imperial Women's Medical and Pharmaceutical College (Tōhō Daigaku [co-ed])	1870: Joshigakuin (⇒ TWCU) 1875: Kōbe Home (Kōbe Jogakuin Daigaku) 1881: Tokyo Anglo-Japanese Girls' School (Aoyama Gakuin Joshi Tanki Daigaku ⇒ closed in 2019) 1884: Tōyō Eiwa Jogakuin (⇒ TWCU) 1916: Sacred Heart (Seishin Joshi Daigaku) 1918: Tokyo Woman's Christian University (TWCU, Tokyo Joshi Daigaku)

by the government. Despite this, Type A universities have been the primary driving force behind education and training of women in STEM. Type B, the second type of women-only education institutions, was vocational schools. In 1886, Kyōritsu Girls' Vocational School was founded with the goal of attaining women's independence and autonomy (Abe 2018). However, Abe cautioned that this 'women's independence' did not mean that the School's ethos was out of line with 'good wives, wise mothers', and this is reflected in the curriculum with a strong focus on sewing, embroidery, knitting and home economics (2018, p. 13). This strong emphasis on homemaking and family life was found even in sericultural institutes which provided intensive courses for training female supervisors in the silk-reeling industry. In 1917, Gunze Girls' School was founded by Gunze Co. (silk manufacturing company, established in 1896) for meeting the rising demand for women with technical education (Kiyokawa 1991). Female workers in the School learned about the importance of a family life, and quit the job when they got married (Kiyokawa

1991; Sugimoto 2006). The third type (Type C) is privately-owned professional schools (Senmon Gakkō), with emphasis on mathematics and science education. Japan Women's University (JWU) was established in 1901 by Jinzō Naruse, a Christian church minister. JWU also emphasised the importance of science education from the onset of its foundation.

JWU invited distinguished professors such as Nagayoshi Nagai (pharmacologist and first President of the Pharmaceutical Society of Japan) to give lectures on natural science subjects. In order to match the high standards of lectures given by those professors, in 1906, German-style laboratories and Leitz microscopes were installed in a new building designed for carrying out scientific experiments (Kodate and Kodate 2015). JWU, just as Ochanomizu, attempted to upgrade its status to university in the late 1920s, providing a curriculum with science subjects. However, that movement did not continue into the 1930s (Yamamoto 2012). Students who wanted to pursue research needed to look elsewhere. Some of them later went on to study at Imperial Universities, once their doors opened (Fig. 10.1).

The other two women-only universities stand out in that they were established by women themselves, and these pioneering women had access to

Fig. 10.1 Autopsy class at Japan Women's University (JWU). Image provided courtesy of Naruse Memorial Hall, JWU

science. In 1900, the Women's Institute for English Studies (Joshi Eigaku-juku) was founded by Umeko Tsuda and later became Tsuda University. Tsuda studied biology and education at Bryn Mawr College in Philadelphia, and continued her study at St. Hilda's College, Oxford (Takahashi 2002). She was one of the rare Japanese women at that time who had access to both scientific knowledge and English language outside of Japan. Unlike many other women-only colleges, the Women's Institute for English Studies placed a strong value on women's independence and professional career. Similarly, in healthcare-related fields, in 1900, Yayoi Yoshioka (who was a female doctor) decided to establish a women-only medical school (later known as Tokyo Women's Medical University, TWMU) after hearing that a woman had been denied admission to Saiseigaku-sha (precursor to Nippon Medical University). It is important to note that Tokyo Women's Medical School (TWMS) was a private entity, unlike teacher training colleges.

Several other private schools were established for training female doctors, dentists, pharmacists and nurses in the first half of the twentieth century. The Imperial Women's Medical and Pharmaceutical College (IWMPC, later renamed Tōhō University) also played a significant role as a hub of medical and scientific learning. Prefecture-based (publicly-owned) Schools of Medicine for Women were founded in some areas later with the view to making up for the shortage of male doctors during World War Two. While Tōhō and Prefectural Medical Schools became co-ed or merged with another co-ed institutions in the post-war era, TWMU remained a single-sex university. In addition, although several Women's Schools of Pharmacy were established and became popular, particularly in the 1930s, the compatibility of female pharmacists with homemaking activities was favourably seen, and used to promote the qualification (Amano 1986; Kimura 2019).

The fourth type of higher education institution for women was initiated by several individual educationalists and founded as private schools. They were often set up under Christian guiding principles, and maintained a Christian ethos. As in other East Asian countries, Christianity in Japan played a great role in providing women with access to higher education in those early days (Takahashi 2009). In 1870, the Presbyterian Mission Female Seminary (renamed Joshigakuin in 1890) was founded in Tokyo by Julia Sarah Carothers, a missionary from Ohio, USA. In 1884, Tōyō Eiwa Jogakuin was also founded by a Methodist missionary from Canada, named Martha J. Cartmell. The 'high-school' section of Tōyō Eiwa Jogakuin later merged with its parent institution, thereby forming Tokyo Woman's Christian University (TWCU) in 1918. The establishment of a woman's higher education institution was announced at the World Missionary Conference in Edinburgh in 1910. Inazō Nitobe, an economist, diplomat and educator, was a member of the Religious Society of the Quakers, and firmly believed in men and women's equal rights to education. Like JWU, TWCU used the word *Daigaku* (university) in its name, although it was classified as a professional school (Senmon Gakkō) by the government until it was upgraded to a university in 1948.

However, TWCU was unique in that its second College President, Tetsu Yasui, created the Department of Mathematics in 1927, with her belief that 'the very thing that women need is mathematics' (TWCU 1968, p. 87). Like Tsuda, Yasui studied education and psychology abroad, more specifically at Hughes Hall, Cambridge, and Oxford University in the UK. Yasui observed how women were educated in the UK, and came to firmly believe that women should study mathematics, not home economics (Aoyama 1949). As a result of the establishment of the Department of Mathematics, in 1937, TWCU became the first women-only college that was accredited by the then Ministry of Education, where the graduates were automatically granted the mathematics teacher qualification. At that time, the only other institution that had the same accreditation in Japan was Tokyo Academy of Physics (founded in 1881), which was a men-only professional school and later became Tokyo University of Science. A string of alumnae from TWCU went on to study mathematics at Imperial Universities such as Hokkaidō and Osaka (Yamamoto 2010). It is worth noting that mathematics teachers from TWCU were able to teach at boys' schools, when women were not granted the equal right to higher education (TWCU 1968). This attests to the fact that the government in the interwar period already recognised a strong track record of women-only colleges in science education (mathematics, in this case), and did not necessarily discriminate against female teachers.

Women-Only Higher Education Institutions and Pioneers in STEM

The year 1913 was the first year when female students were admitted to one of the Imperial Universities. The then President of Tōhoku Imperial University, Masatarō Sawayanagi, a former civil servant in the Ministry of Education, decided to admit three female students in STEM to the university (Yukawa 1994). Two of them, Chika Kuroda (Chemistry) and Raku Makita (Mathematics), were graduates of the TWHNS, and the third student, Ume Tange (Chemistry), was a graduate of JWU. Professor Nagai (the Imperial University of Tokyo and JWU) strongly supported the entrance of Kuroda and Tange. However, this decision was controversial inside and outside Tōhoku Imperial University. The admission of these three female students to a male-only Imperial University would not have been possible without support and commitment from President Sawayanagi and other distinguished male professors, including Tsuruichi Hayashi (Mathematics) and Rikō Majima (Chemistry) at Tōhoku Imperial University. However, even President Yanagisawa did not believe in the notion of women working outside the home as professionals after graduation. Newspapers from that time reflect how Japanese society in 1913 was not ready to accept that women should or could study at the university level as equals of men.[3] The founder of JWU, Jinzō Naruse, wrote that 'what we need to consider seriously is whether co-education is good or not'. Women's opinions were also divided on the issue (Yukawa 1994, 131). While a prominent female educator and founder in 1886 of Kyōritsu Girls' Vocational School, Haruko Hatoyama, praised the courageous decision made by Tōhoku, another female educationalist, Ayako Tanahashi criticised it, because 'women's debt to the nation is to raise their children' (Yukawa 1994, p. 131). Regardless of opposition, two of the three first students went on to conduct further research, and completed their doctoral degrees in Science and Agriculture respectively. Kuroda was awarded her doctoral degree by Tōhoku University in 1929, which made her the second female doctor in Japan after Kono Yasui (doctoral degree in Science, Tokyo Imperial University, 1927). Yasui completed her degree while working at the TWHNS. In contrast, Tange went on to the United States, and earned her Ph.D. at Johns Hopkins University in 1927. She became the first Japanese female scientist with a Ph.D. obtained overseas. Subsequently, she earned another doctorate in agriculture from Tokyo Imperial University in 1940 (Fig. 10.2).

Female scientists with doctoral degrees, including Chika Kuroda, Ume Tange, Michiyo Tsujimura, and Sechi Katō (Fig. 10.3), continued working at RIKEN. Others had the opportunity to continue their research activities at various Imperial Universities, while simultaneously fulfilling their duties as teachers. Thus, if a woman wished to become a researcher at that time, she had to accept one of two occupations. For many of them, it meant giving up marriage and a family life.

After 1913, women began to be granted access as non-regular students or guest auditors to universities such as Hokkaidō Imperial University, Kyoto Imperial University and private institutions including Tōyō, Waseda, Keiō, Dōshisha and Meiji University. However, access for women to science education remained very restricted, and women-only universities connected female students with research intensive (male) universities such as Hokkaidō and Tōhoku Imperial University. Tōhoku Imperial University admitted two female

Fig. 10.2 Dr. Ume Tange, 1940. Image courtesy of Naruse Memorial Hall, Japan Women's University (JWU) and the Ochanomizu University Digital Archives

Fig. 10.3 Three pioneering women in STEM at RIKEN. From left to right: Dr. Michiyo Tsujimura, Dr. Sechi Kato, Dr. Chika Kuroda. Images provided courtesy of Hokkaidō University Archives

students in mathematics as auditors in 1922. From 1925, the university codified female admission, and almost half of all the female full-time university students prior to World War Two were in Tōhoku Imperial University. This action was emulated by Hokkaidō, Kyūshū, Osaka and Nagoya Imperial, but Kyoto and Tokyo officially admitted female undergraduate students only after World War Two.

In terms of colleges from which these women graduated, twenty were from TWHNS, followed by JWU (eleven), TWCU (nine) and Nara (eight). Out of these sixty-two, approximately half (thirty) specialised in mathematics, followed by fifteen in biology. Chemistry had nine, while physics had four. Half of the students were graduates from Higher Normal Schools in Tokyo and Nara. Women's options for studying science were limited, and women-only universities with a strong emphasis on mathematics and science education provided the bridge between education and research in STEM (Fig. 10.4).

THE SPECIAL CASES OF HOKKAIDŌ UNIVERSITY AND RIKEN

Among the former seven Imperial Universities in Japan, Hokkaidō is an interesting case for examining women in STEM, given its more consistent open-door approach towards women since 1918. Although Tōhoku University became the first university to admit female students in 1913, it only admitted

Fig. 10.4 Five female students at the Department of Physics and Mathematics, Hokkaidō University, 1939. Front row, left to right: Yurie Yokota, Akiko Soeya. Back row, left to right: Hisako Nakayama, Yoshi Kobayashi, Yuki Uenaka. Image provided courtesy of Hokkaidō University Archives

three postgraduate students (in total) in STEM areas between 1918 and 1945 (Tōhoku University, n.d.). The next female undergraduate student admitted to physics on the record was Ms. Michiko Toshima in 1942. In 1924, a mathematics student at Tōhoku who wanted to specialise in physics was asked by the university to change her major to mathematics. The main reason for this was that there was a reluctance on the part of the faculty to mix male and female students in lab-based experiments (Yamamoto 2012, p. 50).

While Kyūshū had been granted Imperial University status in 1911, following Tokyo (1886), Kyoto (1897) and Tōhoku (1907), the first faculty which officially opened the door to two female students in 1925 was that of Law and Literature (Kyūshū University Editorial Committee for Centennial Publication 2017). Following the admission of three female students by Tōhoku University in 1913, discussions were also held at Kyūshū University. One of the primary drivers for admitting women was purely practical, that is, to fill the placements unoccupied by male students. As a result, the Faculties of Law and Agriculture decided to allow female students to study there. However, the Faculty of Engineering was divided and hesitant on the grounds that allowing female students could lead to co-education, which some faculty members refused to accept. Their reluctance became clear when Hitoe Baba applied to audit classes in applied chemistry. Although it was officially announced in 1922 that female students were allowed to audit classes, Baba's application took seven months before approval was granted, and by that time, the module that the applicant wanted to audit was no longer offered. Baba never reapplied. Two years later, a different student, Chiyoko Hirano, applied and was allowed to sit in a mathematics module and became the first auditing student at the Faculty of Engineering at Kyūshū University (Kyūshū University Office for the Promotion of Gender Equality, n.d.) Osaka and Nagoya were established as Imperial Universities later in the 1930s. Subsequently, Osaka admitted three postgraduate female students (all in mathematics) at the postgraduate level between 1939 and 1941 (Yamamoto 2018, p. 51). While the demand for students in STEM at Hokkaidō Imperial University was greater than that in other regions, the record suggests that these women sought out Hokkaidō as the place for their further learning and training.

Sechi Katō was the very first student admitted as a 'special course' student to Hokkaidō Imperial University in 1918, meaning she was not given full enrolment status. She was a graduate of TWHNS, from the physics and chemistry section. Sapporo Agricultural College (founded in 1876) was part of Tōhoku Imperial University between 1907 and 1918, when it was separated as an Imperial University. President of the university, Shōsuke Satō, made a remark in July 1918 that 'this university's door is not closed to women' in reference to the twenty-nine students from TWNHS who were paying a visit to the university campus. Katō was accompanying the group of students as an alumna of TWHNS, and she had been teaching at Hokusei Jogakkō (an all-girl's school, also established by American missionary Sarah Clara Smith) based in Sapporo since April of that year. Having listened to the President's speech, she applied to the university, only to find that her application was rejected due to strong opposition to admitting girls to the university. She appealed directly to the President and was eventually admitted as a 'special course' student to agricultural science. Although President Satō's remark was genuine, there was strong opposition within the faculty. In order to overcome this, Satō created a separate panel of eight professors to handle Katō's reapplication. The compromise reached was that she was allowed to undertake all the coursework but would not be awarded the degree (hence the title, 'special course' student). Later, Katō became the third female doctor in science in Japan, receiving her Doctorate from Kyoto University in 1931, and became the first female Chief Researcher at RIKEN in 1951. She also raised two children while working at RIKEN, with support from her own mother.

Satō's support for equal access for women was demonstrated by his subsequent action and speeches. While Nitobe and Yasui made efforts to establish the TWCU, Satō was involved in the selection of the Dean for TWCU, and provided assistance (Yamamoto 2008). Satō also gave a speech which distinguishes his ideas regarding education for women from those in the mainstream—returning to the theme of fostering 'good wives, wise mothers'. At a ceremony at another missionary girls' school based at Hakodate (Iai Joshi Women's Academy) in 1917, he presented the following rarely expressed view:

In the past, in Japan we have held on to the belief that education for girls means the nurturing of good wives and wise mothers, but today's situation has created a need to broaden their mission, namely that there is a great need for girls to have professional training, and in the future, girls will need to engage in a highly professional career. (Cited in Yamamoto 2008, p. 25)

In 1920, at Hokkaidō Imperial University, Michiyo Tsujimura accepted an unpaid assistant job in an agro-chemistry laboratory, an arrangement common for both men and women at the time, thereby becoming the first female staff member of Hokkaidō University. Yasu Honma followed in Kato's footsteps, registering herself as a 'special course' student in Agrobiology, and became an (unpaid) assistant researcher in 1923 in the same Faculty of Agriculture. Both Tsujimura and Honma were TWNHS graduates. Tsujimura left Hokkaidō for Tokyo Imperial University, and subsequently joined RIKEN. She became the first Japanese female to receive a Doctorate in Agriculture (Tokyo Imperial University) in 1932.

Originally proposed by renowned chemist Jōkichi Takamine in 1913, RIKEN was funded jointly by the government, industry and the Imperial Household, and later became a publicly funded, national science research institute in 1958. Before the end of World War Two, RIKEN was the only place for women in STEM. When women were still not allowed to study as equals of men at Imperial Universities, RIKEN provided female researchers with a place for conducting research alongside their daily jobs as teachers and lecturers at TWHNS or women-only colleges. Katō reflected on RIKEN's environment, and highlighted the level playing field for men and women at RIKEN (Katō 1967) (Fig. 10.5).

Hokkaidō University founded its Faculty of Science in 1930. From its inception, despite certain special eligibility criteria (e.g. graduate of a teacher training college), women were allowed to enter on an equal basis as men. Another TWNHS alumna, Fuji Yoshimura, was admitted as Hokkaidō University's first female formal, full-time undergraduate student specialising in botany. Between 1918 and 1945, Hokkaidō Imperial University admitted twenty-four female students in STEM. Twenty-two of these graduated from the Faculty of Science, and one from the Faculty of Agriculture. Out of the twenty-four, six were alumnae of TWNHS, eight of JWU, four of TWCU and two of Nara Women's. Two others qualified as secondary school teachers, one each from Osaka Prefectural Women's Professional School and Tokyo Women's School of Pharmacy.

Women-only higher education institutions such as JWU and TWNHS provided basic training for these female students in STEM. Yurie Yokota and Akiko Someya had known each other from JWU when they entered the Faculty of Science, and in subsequent years paved the way for their juniors. Someya pursued her research career as an assistant at Hokkaidō University. While these personal connections and exchanges of information were invaluable for women in STEM to succeed, there were a number of obstacles. For

Fig. 10.5 Drs. Tsujimura and Kato visited by Hokkaidō students at RIKEN in 1936. Image provided courtesy of Hokkaidō University Archives

them, the search for a place for learning required both space (institution open to women) and people (where male and female mentors were present).

Pioneers' Chosen Environment and Slow-Moving 'Pipelines'

In 1978, a special article in science magazine *Shizen* (*Nature*) entitled 'Free world of women scientists' commemorated the sixtieth anniversary of RIKEN. It covered a panel discussion among five women pioneers in STEM (Drs. Sechi Katō, Kazuko Kubo, Sumi Nishida, Fumiko Fukuoka and Mizu Wada), with Drs. Yoshio Suge (physicist, former member of the Board of RIKEN, 1966–1970), Shinji Fukui (mechanical engineer, Director, 1975–1980) and Yukio Miyazaki (physicist, Deputy Director, 1976–1980). The panel discussion touched upon the excellent 'liberal' research environment, salary differences between men and women, and the importance of male mentors dedicated to supporting them along the way (*Shizen* 1978, pp. 52–57). It also highlighted the stark difference in facilities and experimental infrastructure between women-only higher education institutions and Imperial Universities at that time. Just as in the case of President Satō and male and female founders and educators of women-only institutions in earlier years, active supporters for the

Fig. 10.6 Dr. Toshiko Yuasa, 1978. Image courtesy of the Ochanomizu University Digital Archives

pioneering women scientists were primarily men. However, they were few and far between.

In the face of these barriers, some female students in physics and engineering, such as Toshiko Yuasa (1909–1980), sought opportunities abroad (Fig. 10.6). After studying at TWHNS, Yuasa entered the Department of Physics at Tokyo Liberal Arts and Science University (predecessor of Tokyo University of Education, currently Tsukuba University) in 1931, as the first Japanese woman to specialise in physics. At that time, Tokyo Liberal Arts and Science University was the only university located in the Kantō (Greater Tokyo) area, among five universities across the country, that accepted female regular students. She conducted research on nuclear spectroscopy. Inspired by the discovery of artificial radioactivity in France, Yuasa decided to go there. Her decision was supported by her father (who was an engineer) (Tsugawa and Kanomi 1996). There was a shortage of equipment and staff in Japan for conducting experimental physics, and it was particularly difficult for a female associate professor at TWHNS. In 1939, Yuasa passed the examination given by the French government and went to Paris in 1940. Her advisor at the Collège de France was Professor Jean Frédéric Joliot-Curie, who discovered artificial radioactivity with his wife, Professor Irène Joliot-Curie. In 1943, Yuasa obtained her degree in physics (Doctorat en Science) for her research on the beta-ray spectra from artificial radioactive nuclei. Upon her return from France, she became a professor at TWHNS (1945 to early 1949), but once again found her research environment to be inadequate. The Occupation Forces in Japan prohibited any research in experimental nuclear physics. She left Japan, relying on her mentor Professor Joliot's help to return to France. She became *Maître de Recherche* (chief researcher) at CNRS in 1957 and continued her research at l'Institut de Physique Nucléaire de l'Université de

Paris, Orsay. She remained in Paris until her death in 1980 (Yagi and Matsuda 2007).

In the interwar period, Japan faced a number of challenges, one of which was a shortage of educated and skilled men. This prompted the development of initiatives to encourage educational institutions to produce graduates with numerical and scientific skills, regardless of gender. Between the late 1920s and mid-1940s, various science courses were established for women at several universities. Major initiatives included the Imperial Women's College of Science and Kyoto Prefectural University Women's College. In 1943, Tsuda College created thirty places each for physics, chemistry and mathematics. However, at the end of World War Two, a new era arrived for higher education in Japan. A new university system was introduced by the Fundamental Law of Education and the School Education Law. In April 1948, prior to the nationwide launch of the reformed university system, eleven private colleges and one prefectural college were upgraded to university status. They were Kokugakuin, Sophia, JWU, TWCU, Tsuda College, Sacred Heart, Dōshisha, Ritsumeikan, Kansai, Kwansei Gakuin, Kobe College and Kobe University of Commerce. Five of them (JWU, Kobe College, Sacred Heart, Tsuda and TWCU) were women-only colleges, and seven out of the eleven were established on Christian principles.

In the following year, TWHNS and Nara Women's Higher Normal School became the two national women-only universities—as Ochanomizu University and Nara Women's University respectively. The three women with doctoral degrees (Kono Yasui, Chika Kuroda and Toshiko Yuasa) became professors in the Faculty of Science at Ochanomizu University. At the same time, the Imperial Women's College of Science became co-educational and was renamed Tōhō University in 1950. Tsuda College dropped its physics and chemistry courses, but retained mathematics. Additionally, in the late 1950s, JWU had a plan to reorganise the Faculty of Home Economics into the Faculty of Science (which later materialised in 1992) (Kodate and Kodate 2015).

These reforms officially granted women equal access to higher education for the first time in Japan. However, on the flipside, the idea of 'good wives, wise mothers' ('Ryōsai Kenbo') was never questioned, and more women-only institutions were established in the following decades to serve this purpose. They primarily fit in with Types B and D (Table 10.1). Therefore, the landscape of higher education actors was maintained overall, and female students continued choosing either co-ed liberal arts colleges or women-only universities, instead of opting for former Imperial Universities. This trend only started to change in the last three decades. Women-only universities remained the main 'pipelines' for women in STEM after World War Two, as demonstrated by looking at the members of the Science Council of Japan (SCJ).

The SCJ was established in January 1949 as a special advisory body to the Prime Minister, independent of the government, with a mission to enhance and promote the field of science. It has 210 council members, and another 2000 affiliate members are selected from academic communities ranging from humanities, social and natural sciences to engineering. The first female member, who was elected for the 12th SCJ in 1981, was an earth scientist, Katsuko Saruhashi (only one woman was elected). She was a member for one term (1981–1985). In the following term (thirteenth, 1985–1988), three female scholars were elected, two of whom were social scientists (Yasuko Ichibangase and Etsuko Yasukawa). Masako Hayashi was the only natural scientist (architecture, JWU graduate). The number of women in the SCJ did not increase in the following terms. The number of female members selected from the natural sciences stayed at one (0.5%) from 1981 until 1997. Even the total number remained negligible, fluctuating between one and three (1.5%), until the SCJ changed its selection rules in 2005 (Kurokawa 2006). The aim of the new rules was to increase the proportion of both female and younger members selected from academic societies. As a consequence, the number of members selected from the natural sciences quadrupled in 2005, from four to seventeen. The total number of female members also tripled (from thirteen to forty-two, accounting for 20%). Since the first female member was elected in 1981–2011, there have been forty-one female members in STEM fields. Of this number, nine members obtained their first degrees from the University of Tokyo and Ochanomizu University, while three are graduates of the University of Kyoto. Two female members came from each of the following: Nara Women's University, JWU, Hokkaidō University and Kyūshū University. In terms of doctoral degrees, more than half (twenty-two members) obtained their degrees from the University of Tokyo, followed by four from the University of Kyoto, and two each from Hokkaidō, Kyūshū and Nagoya. Seven out of nine graduates from Ochanomizu went to the University of Tokyo. The path that their predecessors took since the early twentieth century has continued, when we zone in on these selected women scientists in leadership positions. This resulted from a combination of the following two factors: the persistence of girls' (or their parents') preference for leading women-only universities over co-educational institutions, and the success of those institutions with the clear mission of fostering female scientists.

When we look at the twenty-fourth SCJ term, 2017–2020, it becomes clearer that things have changed. The total number of women is up to sixty-nine, of which forty-five members belong to STEM domains including medicine, nursing and dentistry. The majority of female members in STEM are graduates from former Imperial Universities, national universities or co-ed private universities (e.g. Waseda, Keiō, Tokyo University of Science and Aoyama Gakuin). Although the record is incomplete (as not all the data about education profiles were available), only four studied at Ochanomizu at undergraduate and/or postgraduate levels. This signifies that the "pipeline" of female scientists is no longer the leading single-sex education institution, seventy years after equal access to large-scale national universities was granted. The level playing field seems to have begun producing results. However, when it comes to nurturing female students, women-only universities are still providing 'choice' for certain types of students. As outlined in Table 10.1, women's universities in Types A and C are particularly successfully highlighting their strengths and unique vantage point of nurturing women STEM professionals and female leaders in society. This could also represent the situation where gender equality is far from being achieved in wider society.

THE CONTEXT OF SUPPORT FOR JAPANESE WOMEN IN STEM IN THE PRESENT DAY

As of 2020, there are 795 universities in Japan, eighty-six of which are national; ninety-four local (prefectural and municipal); and 615 private (MEXT 2021). Approximately one tenth of the total (seventy-six universities) are women-only universities. As described below, Ochanomizu and Nara Women's are both national universities, while Fukuoka Women's and Gunma Prefectural Women's University are run by the local government. The remaining seventy-two are private universities. In recent years, there has been a trend for these women-only universities to open their postgraduate programmes to men, and now half of them have become co-educational at that level.

Just as many higher education institutions in the world are subject to ranking in various league tables, those in Japan are no different. Out of the top 150 universities published in the Japan University Rankings 2019 (Times Higher Education), nine (Ochanomizu, Fukuoka Women's, Tsuda, Showa Women's, Tokyo Woman's Christian [TWCU], Nara Women's, JWU, Doshisha Women's College, Osaka Jogakuin University) are women-only universities and six of these benefited from the above-mentioned government schemes. Outside this league table, Tokyo Women's Medical University and Mukogawa Women's University also were previously awarded the funding. At Mukogawa Women's University, a new Department of Architecture is scheduled to be opened in 2020 (Mukogawa Women's University Institute for Education 2019). While Nara Women's University is planning to create one National University Corporation in 2021 with co-ed Nara University of Education, it aims to found a new Department of Engineering in 2022. Therefore, several women-only institutions are still seen as viable and thriving in the Japanese higher education context.

In terms of more traditional STEM subjects (e.g. biology, chemistry, mathematics), two national (Ochanomizu and Nara Women's) and three private universities (Tsuda, TWCU and JWU) offer modules and degrees. Ochanomizu, Nara Women's and JWU have a faculty of science. In recent years, Ochanomizu and Tsuda have been particularly successful in creating collaborative networks through a type of government funding scheme, teaming up with industry, research institutes and engineering-heavy universities such as the Shibaura Institute of Technology (SIT) and the University of Electro-communications (UEC). Ochanomizu led the three-year Program for Supporting Research Activities of Female Researchers (2014–2016), with a focus on engineering disciplines, together with SIT and the National Institute for Materials Science (NIMS). In contrast, Tsuda University is a partner in the scheme 'Women at the Frontier of Communication' (selected as part of the Japan Science and Technology-funded 'Initiatives for the Implementation of Diverse Research Environment: Collaborative Type'), led by the UEC. Since 2016, Tsuda University and UEC have worked with their industry partner, NTT Science and Core Technology Laboratory Group, and this collaboration is set to continue until 2022.

Women-only universities are active in terms of degree programmes. In 2016, the two national women-only universities (Ochanomizu and Nara Women's) founded a postgraduate joint programme called Human Centred Engineering. Ōtsuma Women's University founded its Faculty of Social Information Studies in 2017. Tsuda University also launched its new College of Policy Studies, as a multidisciplinary programme with a strong focus on data sciences. In April 2015, Ochanomizu University established the Collaborative Organisation for Research in Women's Education of Science, Technology, Engineering and Mathematics, in collaboration with Nara Women's University. Women-only universities constitute a long-established, yet niche position in the Japanese higher education landscape.

Table 10.2 shows the proportion of female students enrolled in four-year undergraduate programmes by discipline and subject. STEM subjects were individually classified. This clearly indicates that agriculture achieved the overall average split (fifty-five male/forty-five female). In terms of subject, biology has the highest proportion of female students (40%), followed by chemistry and geology. The proportions for physics and engineering subjects remain very low.

Table 10.2. Proportion of female undergraduate students in Japan by discipline and subject (MEXT 2018)

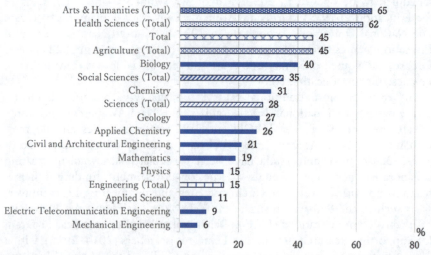

Despite this, a large number of female students in STEM graduate from larger universities such as Tokyo Institute of Technology, Waseda and TUAT. Tokyo Institute of Technology (established in 1881) is the largest national institution for higher education in Japan dedicated to science and technology, consisting of more than 9000 undergraduate and postgraduate students. It is composed of three Schools (Science, Engineering and Bioscience and Biotechnology). As of 2018, out of 4828 undergraduate students, 637 (13.2%) were female. Waseda University (founded in 1882) is a major private university, having more than 50,000 (undergraduate and postgraduate) students and has three Schools dedicated to sciences and engineering (Fundamental Science and Engineering, Creative Science and Engineering, and Advanced Science and Engineering). The latest figures show that 1716 female students are enrolled as STEM students in one of the three Schools, accounting for 22.4%. TUAT (originated in agricultural and sericultural vocational schools which were established in 1874) is another national, but smaller university, consisting of two faculties (agriculture and engineering) and has roughly 5000 students. In line with Table 10.2, near-parity has been achieved in agriculture, with 674 (48.2%) female students enrolled in the Faculty of Agriculture. On the other hand, 563 (23.2%) female students are registered in the Faculty of Engineering at the undergraduate level.

While the student ratio is one key barometer of gender equality in STEM domains, the importance of role models has been stressed by many studies (Hill et al. 2010; Drury et al. 2011). Since 2003, the Japan Association of National Universities (JANU) has been conducting and publishing an annual survey concerning equal participation of men and women in eighty-six national universities (JANU 2018). The 2018 results are shown in Table 10.3.

Table 10.3 Proportion of female academic staff members in national universities in Japan by rank (JANU 2018)

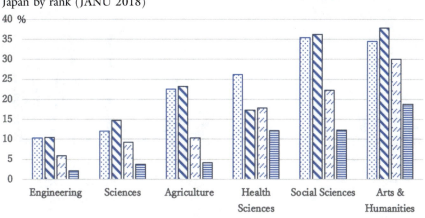

At Tokyo Institute of Technology, there are seventeen female professors (4.6%), thirty-nine female associate and assistant professors (11.2%) and twenty-eight female assistants (8.8%), as of 2018 (Tokyo Institute of Technology 2019; n.d.). At TUAT, there are eight female professors (5.2%), forty-one female associate and assistant professors (21.0%), and six female assistants (12.2%), as of 2018 (Tokyo University of Agriculture and Technology 2020a, 2020b). At Waseda, the latest data available (2015) show that there were fourteen female academic staff, out of 316 in total, in sciences and engineering, accounting for 4.4% (Waseda University Office for Promotion of Equality and Diversity 2015; n.d). Yukari Matsuo (professor in physics at Hosei private university, and a member of the Science Council of Japan) pointed out that the low ratio of female researchers and academic staff in private universities has been neglected simply because many private universities do not cover those domains (Matsuo 2016). At the very top position, female presidents account only for 11.3% (eighty-five out of 752 universities surveyed), as of 2018.

Table 10.4 shows the proportion of female academic staff by rank in 7 former Imperial Universities, in contrast to one private and two national women-only universities (Japan Women's University 2018; Ochanomizu University 2016; n.d.). While this table contains all disciplines, and the size of the seven universities and the three women-only universities differ greatly, the contrast is stark in all ranks.

If the goal is to achieve parity in numbers, there is no question that more female students and academic staff members need to be affiliated to the

Table 10.4 Proportion of female staff members by rank in seven former Imperial Universities and one private and two national women-only universities (JANU 2018)

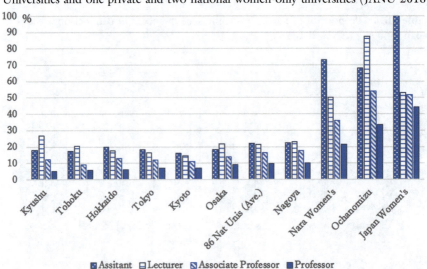

large universities. In fact, given the shrinking young population in Japan, the competition among universities, particularly private ones, has become fierce in recent years. In the post-COVID-19 pandemic era, it is anticipated that the landscape of higher education not only in Japan but also in the world will change dramatically. Women-only universities will face new types of challenges including greater openness and accessibility to students from diverse backgrounds (e.g. cultures, ages, races, religions, sexual orientations, abilities, and ethnicities). However, when it comes to fostering female STEM researchers, the track record of an educational institution and the quality of its research environment is a key factor. Whether or not they can retain the reputation and the environment in the coming decades remains to be seen.

Conclusions

This chapter has demonstrated that women's opportunities in STEM have been influenced by the continuity of major actors in Japanese higher education, including women's universities. Even after World War Two, no major shift occurred in the hierarchy of higher education institutions, with former Imperial Universities at the helm. This continued until the Japanese economic model began to be challenged in the 1990s. Well into the twenty-first century, competition in the global higher education market started to shake the educational environment; as a result, the perception of what is the required skillset for the young and future generations has started to change. Furthermore, concerns about the declining population add to fierce competition among higher education institutions and there is now a new dimension to the urban–rural divide in Japan. From this perspective, it is interesting to review the historical roles of women-only institutions in producing women in STEM, and how they have been maintained over many years in Japan. Although a recent revival of these institutions may be short-lived under the current competitive climate, women-only institutions in Japan have kept their reputation for introducing and connecting students to the external world, be it former Imperial Universities or RIKEN, carving out and harnessing their niche positions in the higher education sector.

The other side of the story of women-only institutions in Japan is about the role of male mentors in connecting women in STEM to further education, training and the professional research world. While increasing female representation is crucial, supporting male mentors for women in STEM is as well (Kodate et al. 2014). If staff and students are aware of the mission of women-only institutions, provide good mentors and role models, and are determined to continue playing their connecting roles with larger entities and the outside world, even under this rapidly changing higher education landscape, women-only institutions, with a strong track record in nurturing female leaders and researchers, can remain a credible force.

Acknowledgements The authors are grateful to those who provided assistance while we carried out this research, particularly Ms. Mihoko Yamamoto (Hokkaidō University) and Dr. Takako Kodate (Tokyo Woman's Christian University). This chapter is dedicated to Mrs. Toyoe Andō (1936–2020) and all women who continue to strive for a better world.

Notes

1. Sechi Katō, cited in Yamamoto (2017, p. 65). Translated by the authors.
2. The concept, 'Fukoku Kyōhei', is believed to originate in the Chinese text from the Warring States period (5th to third centuries BCE). Faced with Western imperialism in China and other parts of Asia in the latter half of the nineteenth century, 'enriching the country, strengthening the military' became known as a slogan for the industrialisation and the westernisation of society, including education.
3. While an article in *Osaka Mainichi Shimbun* (22 August 1913) praised the decision of Tōhoku Imperial University's President Sawayanagi, another article in *Yomiuri Shimbun* on the same day expressed reservations, particularly in relation to training female university students in law and politics (Yukawa 1994).

References

Abe, Tsunehisa. 2018. "Memorandum Regarding the History of Kyoritsu Women's Education Institution." ["Kyōritsu Joshigakuen no Rekishi ni tsuite no Oboegaki." In Japanese]. *Kyoritsu Kokusai Kenkyu* 35: 1–20.

Amano, Masako, ed. 1986. *Fundamental Problems of Higher Education for Women* [*Joshi Kōtō Kyōiku no Zahyō*. In Japanese]. Tokyo: Kakiuchi Shuppan.

Aoyama, Nao. 1949. *Tetsu Yasui: Biography* [*Yasui Tetsu Den*. In Japanese]. Tokyo: Tokyo Women's University Alumnae Association.

Collier, Ruth Berins, and David Collier. 1991. *Shaping the Political Arena: Critical Junctures, the Labor Movement, and Regime Dynamics in Latin America*. Princeton: Princeton University Press.

Drury, Benjamin J., John Oliver Siy, and Sapna Cheryan. 2011. "When Do Female Role Models Benefit Women? The Importance of Differentiating Recruitment from Retention in STEM." *Psychological Inquiry* 22, no. 4: 265–69.

Freitag, Daniela, Birgit Hofstätter, and Anita Thaler. 2012. "Women-Only Engineering Education: A Promising Austrian Model Initiative." *GIEE 2011: Gender and Interdisciplinary Education for Engineers*. Rotterdam: Sense Publishers. 297–308.

Hill, Catherine, Christianne Corbett, and Andresse St. Rose. 2010. *Why so Few? Women in Science, Technology, Engineering, and Mathematics*. Washington, DC: American Association for University Women.

JANU. 2018. "Report on the 14th Follow-Up Survey on the Implementation of Gender Equality Promotion at National Universities." ["Kokuritsu daigaku ni okeru danjo kyōdō sankaku suishin no jisshi ni kansuru dai 14 kai tsuiseki chōsa hōkokusho." In Japanese]. https://www.janu.jp/gender/201801houkoku_01.pdf. Accessed on 4 February 2021.

Japan Inter-Society Liaison Association Committee for Promoting Equal Participation of Men and Women in Science and Engineering (EPMEWSE). 2004. "Diverse Visions of Scientists and Engineers in the 21st Century: For the Promotion of Gender Equality (Report on the MEXT-Commissioned Project, FY2003)." https://www.djrenrakukai.org/doc_pdf/EPMEWSEreport.pdf. Accessed on 4 February 2021.

Japan Women's University. 2018. "University Full-Time Faculty Male-Female Ratio by Job Rank (Commuting Type)." ["Daigaku sennin kyōin shokkai betsu danjyo hiritsu (tsūgaku)." In Japanese]. https://www.jwu.ac.jp/content/files/unv/info_disclosure/3_kyouin/2018/danjobetsu.pdf. Accessed on 4 February 2021.

Katō, Sechi. 1967. "Joy and Excitement as the First Female Student at Hokkaidō University" ["Hokudai saisho no joshi gakusei to shite no kangeki", in Japanese]. *Sapporo Dōsōkai-shi (Sapporo Alumni Magazine)* 2: 55.

Kimura, Yuuka. 2019. "Research on Women's Colleges of Pharmacy in Pre-War Japan: Examining Factors That Affect Female Students to Enter These Colleges" ["Senzenki Nihon no Joshi Yakugaku Senmon Gakkō ni kansuru Kenkyu: Joshi Seito no Shingaku Yōin ni Kansuru Kentō wo Chushin ni." In Japanese]. *Bulletin of Waseda University Graduate School of Education* 26, no. 2: 77–87.

Kiyokawa, Yukihiko. 1991. "The Transformation of Young Rural Women into Disciplined Labor Under Competition-Oriented Management: The Experience of the Silk-Reeling Industry in Japan." *Hitotsubashi Journal of Economics* 32, no. 2: 49–69.

Kodate, Naonori, and Kashiko Kodate 2015. *Japanese Women in Science and Engineering: History and Policy Change*. London: Routledge.

Kodate, Naonori, Kashiko Kodate, and Takako Kodate. 2010. "Mission Completed? Changing Visibility of Women's Colleges in England and Japan and Their Roles in Promoting Gender Equality in Science." *Minerva* 48, no. 3: 309–30.

Kodate, Naonori, Kashiko Kodate, and Takako Kodate. 2014. "Paving the Way and Passing the Torch: Mentors' Motivation and Experience of Supporting Women in Optical Engineering." *European Journal of Engineering Education* 39, no. 6: 648–65.

Kurokawa, Kiyoshi, ed. 2006. "How Far Have We Come? Equal Participation of Men and Women in Academia." ["Dokomade susunda danjo kyōdō sankaku." In Japanese]. Tokyo: SCJ Foundation.

Kyūshū University Editorial Committee for Centennial Publication. 2017. *Kyushu University Centennial History, Vol. 1*. Fukuoka: Kyushu University.

Kyūshū University Office for the Promotion of Gender Equality. n.d. "History of Kyushu University and Women." ["Kyushu Daigaku no rekishi to jyosei." In Japanese]. http://danjyo.kyushu-u.ac.jp/activity/history1.php. Accessed on 4 February 2021.

Matsuo, Yukari. 2016. "Challenges for Fostering Female Researchers in STEM at Private Universities." ["Shiritsu Daigaku ni okeru Rikei Jyosei Kenkyūsha Ikusei no Kadai." In Japanese]. *Gakujutsu no Dōkō [Trends in the Sciences]*. Tokyo: SCJ Foundation. 50–51.

Ministry of Education, Culture, Sports, Science and Technology (MEXT). n.d. "Diagram of school system." ["Gakkō. Keitōzu." In Japanese]. https://www.mext.go.jp/b_menu/hakusho/html/others/detail/1318188.htm. Accessed on 10 October 2021.

Ministry of Education, Culture, Sports, Science and Technology (MEXT). 2018. "School Basic Survey (Heisei 30)." ["Gakkō kihon chōsa." In Japanese]. https://www.e-stat.go.jp/stat-search/files?page=1&toukei=00400001& tstat=000001011528. Accessed on 4 February 2021.
Ministry of Education, Culture, Sports, Science and Technology (MEXT). 2021. "Handbook of Education and Science Statistics." ["Monbu kagaku tōkei yōran." In Japanese]. https://www.mext.go.jp/b_menu/toukei/002/002b/1417059_00006.htm. Accessed on 10 October 2021.
Mukogawa Women's University Institute for Education. 2019. "Women's university statistics / university basic statistics." ["Jyoshi daigaku tōkei / daigaku kihon tōkei." In Japanese]. http://kyoken.mukogawa-u.ac.jp/statistics/. Accessed on 10 October 2021.
No title. 1913. *Osaka Mainichi Shimbun*, 22 August.
No title. 1913. *Yomiuri Shimbun*, 22 August.
Ochanomizu University. n.d. "Collaborative Organization for Research in Women's Education of Science, Technology, Engineering, and Mathematics." http://www-w.cf.ocha.ac.jp/cos/. Accessed on 4 February 2021.
Ochanomizu University. 2016. "Collection of Role Models in Women Researchers in Engineering." ["Kōgakukei jyosei kenkyūsha rōru moderū shū." In Japanese]. http://www.cf.ocha.ac.jp/igl/renkei/rolemodel_contents/. Accessed on 4 February 2021.
Sievers, Sharon L. 1983. "Flowers in Salt: The Beginnings of Feminist Consciousness in Modern Japan." Redwood City, CA: Stanford University Press.
Shizen (Nature). 1978. "Special Issue on RIKEN: Tracing Its 60-Year History." Tokyo: Chūō Kōron Publishing.
Sugimoto, Seiko. 2006. "Japan's Modern Silk Industry and Christian Spirit." ["Nihon no Kindai Seishigyō to Kirisutokyō Seishin." In Japanese]. *Senri Ethnological Reports* 62: 71–91.
Takahashi, Yūko. 2002. *Social History of Umeko Tsuda* [*Tusda Umeko no Shakai-shi*. In Japanese]. Tokyo: Tamagawa Daigaku Shuppan.
Takahashi, Yūko. 2009. "A Japanese American Enterprise: Umeko Tsuda's Bryn Mawr Network and the Founding of Tsuda College." In *China's Christian Colleges: Cross-Cultural Connections, 1900–1950*, edited by Daniel H. Bays and Ellen Widmer. Stanford: Stanford University Press. 271–86.
Tōhoku University, n.d. http://tumug.tohoku.ac.jp/100th/rekishi.html. Accessed on 4 February 5 2021.
Tokyo Institute of Technology. n.d. "Faculty/Students." ["Kyōshokuin/gakusei." In Japanese]. https://www.titech.ac.jp/about/disclosure/pdf/facts_2_student_181 112.pdf. Accessed on 4 February 2021.
Tokyo Institute of Technology. 2019. "Tokyo Institute of Technology Student/Faculty Male/Female Ratio." ["Tokyo Kōgyō Daigaku gakusei/kyōshokuin no danjo hiritsu." In Japanese]. http://www.gec.jim.titech.ac.jp/about/doc/womenraito.pdf. Accessed on 4 February 2021.
Tokyo University of Agriculture and Technology. 2020a. Number of Students/Graduates/Degrees Awarded ["Gakuseisū / sotsugyōsei – shūryōsha / gakui-juyosū." In Japanese]. https://www.tuat.ac.jp/campuslife_career/campuslife/number/. Accessed on 4 February 2021.

Tokyo University of Agriculture and Technology. 2020b. No title. https://www.tuat.ac.jp/documents/tuat/outline/data/shokuinkoubo/h30_syokuin.pdf. Accessed on 4 February 2021.

Tokyo Woman's Christian University. 1968. *Fifty Years of Tokyo Woman's Christian University* [*Tokyo Joshi Daigaku Gojūnenshi*. In Japanese]. Tokyo: TWCU.

Tsugawa, Akiko, and Satoko Kanomi. 1996. *Opening: Tracing the Paths of Japanese Female Scientists* [*Hiraku: Nihon no josei kagakusha no kiseki*. In Japanese]. Tokyo: Domesu Publishing.

Waseda University Office for Promotion of Equality and Diversity. n.d. "Data." https://www.waseda.jp/inst/diversity/information/data/. Accessed on 4 February 2021.

Waseda University Office for Promotion of Equality and Diversity. 2015. "8 Number of Full-Time Faculty Members and Female Ratio by Faculty (FY2015)." ["8 Gakujutsuin betsu sennin kyōinsū to jyosei hiritsu (2015 nendo)." In Japanese]. https://www.waseda.jp/sankaku/figure/pdf/8_faculty_students.pdf. Accessed on 4 February 2021.

Yagi, Eri, and Matsuda, Hisako. 2007. "Toshiko Yuasa (1909–80): The First Japanese Woman Physicist and Her Followers in Japan." *AAPPS Bulletin (The Association of Asia Pacific Physical Societies)* 17, no. 4: 15–17.

Yamamoto, Mihoko. 2008. "Sato Shosuke's Concept of Women's Higher Education: On the Admission of Women to Hokkaido Imperial University." ["Sato Shosuke no joshi koto kyoikuron: Hokkaidō Teikoku Daigaku ni okeru jyosei no nyugaku wo megutte." In Japanese]. *Hokkaidō Daigaku Bungaku Shokan Nenpō (Annual Report of Hokkaidō University Archives)* 3: 18–42.

Yamamoto, Mihoko. 2010. "Tokyo Woman's Christian University Students Who Went on to Study at Hokkaidō Imperial University." ["Hokkaidō Teikoku Daigaku he shingakushita Tokyo Jyoshi Daigakusei tachi." In Japanese]. *Hokkaidō Daigaku Bungaku Shokan Nenpō (Annual Report of Hokkaidō University Archives)* 5: 76–103.

Yamamoto, Mihoko. 2012. "About Alumnae of Japan Women's University Who Pursued Science at Hokkaidō Imperial University's Faculty of Science." ["Hokkaidō Teikoku Daigaku ni shingaku shita Nihonjyoshi Daigakko Sotsugyosei tachi." In Japanese]. *Hokkaidō Daigaku Bungaku Shokan Nenpō (Annual Report of Hokkaidō University Archives)* 7: 42–58.

Yamamoto, Mihoko. 2017. "Brief Note: 'Science Is the Most Beautiful Dress That a Woman Could Wear'. Inspirational Messages from Sechi KATO, a Pioneer Woman in the Study of Physical Chemistry" ["Shōkai: 'Kagaku ha Josei ni totte Nanimono nimo Suguru Bifuku dearu': Josei Kagakusha no Senkusha Kato Sechi no Ayumi." In Japanese]. *Hokkaidō Daigaku Bungaku Shokan Nenpō (Annual Report of Hokkaidō University Archives)* 12: 53–67.

Yamamoto, Mihoko. 2018. "Female Student's Progression to the Imperial Universities' Graduate Schools Between 1918–1945." ["1918 kara 1945 nen ni okeru Teikoku Daigaku Daigakuin he no Josei no Shingaku Jyokyo (1): Kagaku Senko no Shingakusha ni chakumoku shite." In Japanese]. *Hokkaidō Daigaku Bungaku Shokan Nenpo (Annual Report of Hokkaidō University Archives)* 13: 48–61.

Yamamoto, Masahiro, and Weina Ran. 2014. "Should Men Work Outside and Women Stay Home? Revisiting the Cultivation of Gender-Role Attitudes in Japan." *Mass Communication and Society* 17, no. 6: 920–42.

Yoshimi, Shun'ya. 2011. "What is a university? [Daigaku to wa nani ka. In Japanese]." Tokyo: Iwanami Shoten.

Yukawa, Tsugiyoshi. 1994. "Admitting Women to Universities in Taisho Era: The Case of Tohoku Imperial University in 1913 and Its Aftermath." ["Taishō-ki ni okeru jyosei no daigaku no monko kaihō: Taishō 2 (1913) nen no Tōhoku Teikoku Daigaku no jirei to sonogo no tenkai." In *Japanese*]. *Kyoikugaku Kenkyu (Japanese Journal of Educational Research)* 61, no. 2: 129–38.

CHAPTER 11

Internationalism and Women Mathematicians at the University of Göttingen

Renate Tobies

The focus of this chapter is the beginning of (official) women's study of mathematics in Germany, which started at the little University of Göttingen, then situated in the largest German federal state, Prussia. It will demonstrate the decisive role that the mathematician Felix Klein (1849–1925) played as one of the foremost promoters of women studying at the university level as well as the supporting role played by David Hilbert (1862–1943) in these developments. The discussion will be embedded in an international framework.

Non-German Women Paving the Way in Germany

Although women were not legally permitted to enrol in German universities during the nineteenth century, the first woman to earn a doctoral degree in mathematics nevertheless did so at the University of Göttingen in 1874: the Russian Sofia Kovalevskaya (1850–1891). The life and work of Kovalevskaya—and the circumstances of her doctorate *in absentia*—have received sufficient scholarly attention. Kovalevskaya's Ph.D. examination records were first published in Tollmien (1997; see also Tobies 2001; Kaufholz-Soldat and Oswald 2020). It should be stressed that Kovalevskaya's career, until she became a full professor in Stockholm, had been assisted by mathematicians

R. Tobies (✉)
History and Philosophy of Natural Sciences,
Friedrich Schiller University, Jena, Germany
e-mail: renate.tobies@uni-jena.de

© The Author(s), under exclusive license to Springer Nature Switzerland AG 2022
C. G. Jones et al. (eds.), *The Palgrave Handbook of Women and Science since 1660*, https://doi.org/10.1007/978-3-030-78973-2_11

from Sweden, Germany, France, and Italy (Coen 2012, pp. 477, 509–15). Felix Klein, too, then a young professor at the University of Erlangen (in the German federal state of Bavaria), immediately recognized the significance of Kovalevskaya's thesis, '*Zur Theorie der partiellen Differentialgleichungen*' (1875) and praised it in a letter to his main collaborator at that time, the Norwegian Sophus Lie (1842–1899) (Tobies 2019, p. 368; Tobies 2021, p. 411).

Kovalevskaya's career, however, was an exception, and it was not until 1895 that the next women mathematicians completed their doctorates at German universities. Marie Gernet (1865–1924) would become the first German to do so, in 1895 at the University of Heidelberg (in the German federal state of Baden) and under the direction of Leo Königsberger (1837–1921), with whom Kovalevskaya had begun her studies in 1869. Marie Gernet became a teacher at the first German (private) secondary school for girls in Karlsruhe (Baden) where it was possible to take the *Abitur*, the examinations required for entrance to German universities. The next German woman to complete a doctorate in mathematics at a German university (Erlangen) was the famous Emmy Noether (1882–1935), who is discussed later in this chapter, but this was not until 1908.

In the meantime, women's study was further developed in other countries. Felix Klein, a corresponding member of the British Association for the Advancement of Science since 1873, a foreign member of the London Mathematical Society since 1875, and a participant in other foreign institutions, saw and recognized these new trends and supported mathematically gifted students regardless of their sex, religion and nationality. From the early stages of his career, he co-operated with a number of international colleagues in the field of geometry who likewise supported women mathematicians. These included Gaston Darboux (1842–1917) in France, Luigi Cremona (1830–1903) in Italy, Arthur Cayley (1821–1895) in the United Kingdom, Hieronymus G. Zeuthen (1839–1920) in Denmark and James Joseph Sylvester (1814–1897) in the United Kingdom and in the United States as well.[1]

From an international perspective, between the year 1874 (the year of Kovalevskaya's doctorate) and 1895, five other female mathematicians received a doctoral degree. The next was the Russian Elizaveta Fedorovna Litvinova (1845–1919)—a friend of Kovalevskaya's—at the University of Bern (Switzerland) in 1878, since Swiss universities officially allowed women's studies earlier than German universities. Incidentally, a German female mathematician also received her doctoral degree from the University of Bern, in 1907, when not all German universities had opened their doors to women. The latter, Annie Reineck (1880–1978), is notable because she was able to continue her occupation as a teacher in Switzerland after she married, whereas in Germany she would have lost her position because of the so-called *Beamtinnenzölibat* which demanded female teacher celebacy. With the support of her husband, Reineck-Leuch took on a leading position in the *Schweizerischen Verband für Frauenstimmrecht*.[2]

Back to the five women, mentioned above, who earned degrees between 1874 and 1895: Litvinova was followed by the Englishwoman Charlotte Angas Scott (1858–1931), from the University of London, in 1885. She had previously studied at the University of Cambridge. The latter university did not allow women to pursue a doctoral degree or obtain a formal research position, although a remarkable number of women did earn undergraduate degrees in mathematics there, first from Girton College in 1873 and then from Newnham College in 1875.[3] On these grounds, some of them tried their luck abroad. Recommended by the mathematician Arthur Cayley, Charlotte Angas Scott became chair of the mathematics department at the newly founded Bryn Mawr College for women in Pennsylvania (United States), where she supervised female doctoral students and maintained contact with Göttingen. For the same reason, Grace E. Chisholm (1868–1944), who graduated from Girton in 1892, would also go to Göttingen, which means to Felix Klein.

Klein, who was invited to succeed Sylvester at Johns Hopkins University, Baltimore, in 1883, knew of the developments taking place in North America where the next three women would receive a doctoral degree in mathematics. Winifred Harington Edgerton (1862–1951) was the first to do so, with a thesis submitted at Columbia University in 1886. She was followed by Ida Martha Metcalf (1857–1952) from Cornell University in 1893; and one year later, likewise from Cornell, the Canadian Annie Louise MacKinnon (1868–1940) (Fenster and Parshall 1994, p. 235; Green and LaDuke 2009). MacKinnon went on to conduct post-doctoral research in Göttingen as discussed below, after which she continued her career in the United States where there were better job opportunities. In 1895, the first female Danish mathematician received her doctoral degree in Copenhagen: Thyra Eibe (1866–1919) was supervised by one of Klein's collaborators in the field of geometry, the aforementioned Zeuthen, and by Johan Ludvig Heiberg (1854–1928), a historian of mathematics and philologist (Hoyrup 1993). In 1895, too, Felix Klein passed his first female doctoral students at the University of Göttingen: the Englishwoman Grace E. Chisholm (later Chisholm-Young), as mentioned above, and the American Mary F. Winston (1869–1959). Yet Klein had to fight for this, as will be described below.

It should be noted that it was not in Germany alone that foreigners paved the way for native women interested in science and mathematics. The situation was similar in France.[4] A well-known case is that of the Polish woman Maria Skłodowska-Curie (1867–1934), who earned a *Lizenziat* in physics (1893) and mathematics (1894) at the Sorbonne in Paris, completed her doctorate, and earned two Nobel prizes. Remarkably, the Russian mathematician Sofia Kovalevskaya had already become a member of the French Mathematical Society (*La Société Mathématique de France*) as early as 1882, before she was made a professor in Stockholm. Kovalevskaya's Finnish student Ebba Louise Nanny Lagerborg (later Cedercreutz; 1866–1950) would also complete her *Lizenziat* at the Sorbonne and become a member of the French Mathematical Society in 1890 (Rahikainen 2009).

Felix Klein, Fighting for the Right of Women to Study

In the autumn of 1893, Felix Klein made it possible for the first women to enrol at the University of Göttingen. These women were at first exclusively foreigners. In order to understand this situation, it is necessary to examine the context in closer detail.

The Humboldtian university reform in 1810 had provided a decisive impetus for mathematical research in Germany, and this attracted an increasing number of foreign students during the last third of the nineteenth century. Both women and men wished to study at the centre of scholarly activity. As early as Klein's time in Bavarian universities (Erlangen 1872–1875, and Munich 1875–1880), young men came to study with him from Scandinavia and Italy. With his move to Leipzig (Saxony) in 1880, American, Austrian (Czech, Hungarian), British, French, Italian and Russian students came to learn from him, and when, in 1886, he became a professor in Göttingen, the number of foreign students further increased. Between 1886 and 1895, ten or so Americans earned a doctoral degree under Klein's supervision. The significance of these numbers becomes clear when we learn that, throughout the 1880s and 1890s in Leipzig and Göttingen, more Americans studied mathematics under Klein and his successor at Leipzig (the Norwegian mathematician Sophus Lie) than under any professor of mathematics in the United States. It goes without saying that the subsequent development of mathematics in that country was emphatically influenced by this contact (Parshall and Rowe 1994). Moreover, two of Klein's German students, Oskar Bolza (1857–1942) and Heinrich Maschke (1853–1908), became professors of mathematics at the University of Chicago and maintained close contact with Klein in Göttingen.

While women in many other countries were already able to take qualifying examinations and even to study at university (Costas 2002), the ministerial decrees allowing women to matriculate in German states were first issued between 1900 and 1909. It should be noted that every German federal state had its own laws (Tobies 1997, pp. 18–25). However, because foreign women, like men, wanted to study where the highest standards of scholarship could be expected, for some time they attempted to gain access to German universities even while official status as students could not yet be granted to them.

Sandra L. Singer (2003) has written a detailed book about North American women at German-speaking universities during the period of 1868–1915. Before her, in a pioneering study, Margaret Rossiter underscored Felix Klein's special role in promoting American women mathematicians (1982, pp. 40–41). Singer made good use of Rositter's findings and drew upon additional archival sources. An outline of the achievements of the first women in the American mathematical community can be found in the work of Della D. Fenster and Karen H. Parshall (1994). Up until the beginning of the 1890s, Berlin was regarded as the centre of mathematics in Germany, with professors there such as Karl Weierstrass (1815–1897), Ernst Eduard Kummer (1810–1893), and Leopold Kronecker (1823–1891). But women could not study at the University of Berlin, as Kovalevskaya experienced (she was promoted and taught privately by Weierstrass). In the 1890s, Felix Klein achieved a turning point in Göttingen, where he created the basis for an international centre for mathematics, science and technology. Yet it took some time before women could participate in university courses.

In July 1891, when Felix Klein was asked by the American Ruth Gentry (1862–1917) if she could attend his lectures and seminars, he regrettably had to turn her away.[5] Decisions of this sort fell to the conservative *Kurator* of the University of Göttingen, the legal scholar Dr. Ernst von Meier (1832–1911). A staunch opponent of women studying, he also forbade Christine Ladd-Franklin (1847–1930), who had already completed her Ph.D. requirements at Johns Hopkins, from attending Klein's lectures. She had come to Göttingen in the fall of 1891 with her husband Fabian Franklin (1853–1939), who studied under Klein. In a letter to Klein, von Meier dismissed the desire of women to study with the following curt remark: 'This is worse than social democracy, which only wants to do away with differences in property. You want to abolish the difference between the sexes.'[6]

A new situation arose in the run-up to the 1893 World's Fair in Chicago. Klein's former student Heinrich Maschke, who was by then a full professor at the University of Chicago, asked in a letter dated 8 April 1893:

> One of our students of mathematics, Miss Mary F. Winston, is applying for a scholarship, on the basis of which she intends to go to Germany next year. She has [...] talent, thinks independently, and is certainly above average. [Oskar] Bolza and I have encouraged her [...] to go to Göttingen and have just as forcefully discouraged her from going to Berlin in order to keep her away from the stiff atmosphere there. Now the question remains whether female doctoral or post-doctoral students may enrol at Göttingen or whether, if that is not the case, you think you might exert your influence to make an exception. (quoted in Tobies 1991/1992, p. 153)

In order to accomplish this and other goals, Klein sidestepped the prescribed order of command (evading the conservative university *Kurator*) and communicated directly with the influential official Friedrich Althoff (1839–1908) at the Prussian Ministry of Culture.

The ministry in Berlin was far more open to the idea of women studying than the *Kurator* in Göttingen ever was. A farsighted official, Althoff recognized the signs of the time and, on 20 May 1892, he assembled a new file with the title 'The Request of Persons of the Female Gender to Matriculate and Attend Lectures at the Royal State Universities.' The file opened with excerpts from newspapers about the ability of women to study in foreign countries. Developments abroad, in other words, influenced the decisions made at the Prussian Ministry of Culture. Thus, Klein received the following promising message from the ministry in Berlin, on 6 July 1893, shortly before he would first travel to the United States to take part in the Mathematical Congress organized in connexion with the World's Fair:

> With respect to women studying, I can confidentially say that, as I know from Mr. Althoff, the Ministry will not hinder the matter, although it will not especially encourage such questions. Regarding their [women's] participation in lectures, this custom will also become more entrenched than limited; and if American women come to study in Germany, they will not have difficulties here. Mr. Althoff is of the opinion that, without asking, you could just arrange for your numerous female American admirers to come over. (Tobies 1991/1992, p. 154)

Klein's attitude was reinforced in the United States, where he not only confirmed Mary F. Winston's outstanding talent as a mathematician, but also witnessed women in positions that a Swiss delegate to the World's Fair described as follows:

> The Americans find nothing unusual in the fact that, for instance, a woman is the director of a national bank, as in Texas, or that women have found a place on the supervisory committees of universities or in the national department of education, and this is not to mention professorial positions, of which there are many for women [...]. Not only have universities been made available to women but also preparatory secondary education, be it in connexion with schools for boys or in parallel institutions, as in Boston [...]. America knows no difference in the practice of scientific careers between men and women [...]. At the University of Chicago, there are six female professors. (Tobies 1991/1992, p. 152)

Mary Frances Winston, an honorary fellow of mathematics at the University of Chicago, participated in the International Congress of Mathematicians in Chicago which was held 21–26 August 1893. While still in America, Klein wrote to Althoff in Berlin, saying that he would like the minister to make all the necessary preparations to ensure that Winston, 'despite the existing legal regulations of the matter,' would be admitted as a visiting student during the winter semester of 1893 to 1894. Anticipating the negative attitude of the University of Göttingen's *Kurator*, Klein added that Althoff should speak with the Minister of Culture and explain the matter in such a way that von Meier would not feel left out (Tobies 2019, p. 370; Tobies 2021, p. 413).

Having returned from Chicago to Göttingen, Klein proposed to the Ministry of Culture in Berlin that Winston, Grace Chisholm, and the American Margaret Eliza Maltby (1860–1944), be allowed to enrol in university. Despite another negative vote by the university *Kurator*, the ministry approved the application of these women within six days. The *Kurator* resigned from his position in a huff, and his successor was welcoming to women. All three women participated officially in lectures and seminars and completed their doctoral theses by 1895—Chisholm and Winston under Klein, Maltby in the field of physical chemistry under the supervision of Walther Nernst (1864–1941).

The female students were not officially matriculated but rather possessed the status of auditors (every professor had to be asked individually for permission, which ultimately had to be granted on an individual basis by the Ministry). In the meantime, additional female students had arrived at Göttingen. Klein personally helped them to receive permission to attend courses at the University. In the autumn of 1894, for example, he wrote the following to the Prussian minister of culture Robert Bosse (1832–1901):

> Your Excellency!
>
> In addition to the two women, Miss Chrisholm and Miss Winston, who for the last year have been studying mathematical subjects at the local university and whose diligence and abilities I have repeatedly commended, there are now two new applicants, Miss MacKinnon and Miss Maddison, who are likewise requesting permission from the appropriate instructors to participate, as of next semester, in lectures on mathematics, physics, and astronomy. I have examined the qualifications of both women and am thus able to support their applications in every respect.[7]

The above-mentioned Annie L. MacKinnon gave five presentations in Klein's seminars during her time in Göttingen (1894–1895). She went on to teach mathematics at Wells College in the United States; her courses included spatial geometry, analytic geometry and differential and integral equations. Encouraged by Klein, she continued to conduct further research; in a letter to him dated 2 January 1897, for instance, she remarked:

> As promised, I am writing to you now during the Christmas vacation about the work on number theory that I told you about last summer. I find that I have both the time and desire to undertake such a study and, following your suggestion, I would like to work on it for a year in order to see what I can do with it.[8]

Ada Isabel Maddison (1869–1950) was a British woman who, like Scott and Chisholm, had studied with Arthur Cayley in Cambridge. After graduating from Girton College in 1892, she attended Bryn Mawr College in the United States, where she won a fellowship for studying abroad. After returning to Bryn Mawr, Maddison completed her Ph.D. and also translated Klein's address to the Royal Academy of Science in Göttingen on 'The Arithmetizing of Mathematics' which was published in 1896 in the *Bulletin of the American Mathematical Society*.

On 1 November 1895, the mathematician Arthur Schönflies (1853–1928), an associate professor of descriptive geometry at Göttingen, wrote the following remarks: 'We now have nine women studying mathematics, and yesterday they formed a club; they will meet once a week for coffee' (Tobies 1991/1992, p. 157). These meetings can be interpreted as the formation of the first women's network of mathematicians. In the same year, because of growing demand for access, the Prussian ministry decided that universities only needed to provide this with a list of the female participants enrolled in courses as auditors, and so not required to take exams, rather than students proper.

Charlotte Angas Scott, who supervised doctoral theses at Bryn Mawr (the first was Ruth Gentry's in 1894, the second Ada Isabel Maddison's in 1896), arranged for further students to pursue studies at the University of Göttingen. On 19 March 1897, she wrote the following to Klein: 'I expect to send two of my best students to Göttingen next year. Both have been awarded a College Fellowship, and both are eager to study under your direction for a year, if this is agreeable to you.'[9] Thus in the fall of 1897, Emilie Norton Martin (1869–1936) and Virginia Ragsdale (1870–1945) arrived in Göttingen along with other American students. Having benefitted from their time with Klein and David Hilbert, both went on to complete their doctorates under Scott's supervision at Bryn Mawr. It should be underscored that Ragsdale contributed to Hilbert's 'sixteenth problem' (the Ragsdale conjecture). In 1900, Hilbert had presented twenty-three unsolved mathematical problems in an influential and now famous speech to the International Congress of Mathematicians in Paris. Hilbert's speech, which was delivered in German, was translated by Klein's doctoral student Mary F. Winston (married name, Newson) and published under the title 'Mathematical Problems' in the *Bulletin of the American Mathematical Society* in 1902.

Such developments continued. Male and female mathematicians, who reported enthusiastically about the stimulating atmosphere in Göttingen, encouraged additional students and colleagues to study at the university there. They arrived from North America, Russia, Denmark, Norway and elsewhere. Women from Poland and, finally, from around Germany, came there as well. Klein's first women students from the German-speaking region were Frieda Hansmann (b. 1873 in Northeim, Prussia) in the summer semester of 1895, who would later earn a doctoral degree in chemistry from the University of Bern; the second was Elsa Neumann (1872–1902), who attended Klein's lecture on technical mechanics in the summer semester of 1896 and would become, in 1899, the first woman ever to be awarded a doctoral degree from the University of Berlin (her field was physics; see Vogt 1999). Klein kept a record of the students who attended his courses. These records extend from his time as a university lecturer (the summer semester of 1871) until the year 1920, even though he had already become a professor emeritus in 1913.[10]

DAVID HILBERT IN KLEIN'S FOOTSTEPS

David Hilbert became a full professor at the University of Göttingen in 1895, at the instigation of Klein, and was also a staunch supporter of women's right to study. Both men pursued the goal of promoting mathematical research in a broader, all-encompassing sense. On the basis of their open-minded approach to all areas of mathematics and its application, they were able to strengthen Göttingen as an international centre for mathematical research. Klein's female doctoral students (Chisholm and Winston) had already completed their dissertations when Hilbert arrived in April of 1895. By that point, Klein had supervised almost fifty doctoral students altogether, these two women included (Klein 1923, pp. 11–13). Hilbert, previously at the University of Königsberg, did not have any doctoral students until 1898. While in Göttingen, he would go on to supervise sixty-nine in all, including six women (Hilbert 1935, pp. 431–33). As early as the summer of 1895, Klein arranged for Hilbert and himself to conduct a joint research seminar in which women also took part.

Hilbert's female doctoral students worked in his specific fields of research; one was American, three Russian and two were German women from Jewish families.[11] His first female doctoral student, Anne Lucy Bosworth (1868–1907) from the United States, was already a professor of mathematics at the Rhode Island College of Agriculture and Mechanic Arts when she came to Göttingen. She earned her degree with Hilbert in 1899 in the field of the foundations of geometry. However, Bosworth met the engineer and mathematician Theodore Moses Focke (1871–1949) in Göttingen, married him in 1901, and did not continue her academic career.

Hilbert's three Russian doctoral students belonged to a group with an excellent education in mathematics, physics, and the German language because they had each attended the famous advanced courses for women in St. Petersburg (the so-called Bestuzhev Courses), which had been established in 1878 (Borisovna 2003). These courses were taught by some of the best mathematicians from St. Petersburg, and these professors of mathematics also had close contact with Klein. As of 1895, sixteen well-educated women from here came to study at the University of Göttingen, including Helene von Bortkiewicz (1870–1939) and her friend Alexandrine von Stebnitzky (b. 1868), who were born to Polish officer families. Neither completed a doctorate in Göttingen, but they attended lectures in Klein's seminars (co-taught with Hilbert) on differential calculus and on number theory in the summer and autumn of 1895.[12] Helene von Bortkiewicz was the sister of the statistician and later professor of economics Ladislaus von Bortkiewicz (1868–1931), who had completed his doctoral degree in Göttingen under the direction of Wilhelm Lexis (1837–1914).

The first Russian woman to earn a doctoral degree under Hilbert's supervision was Ljubov Nikolaevna Zapolskaya (Sapolsky, Sapolski) (1871–1943). Having successfully completed the Bestuzhev Courses in 1894, she began her studies in Göttingen in October 1895. She attended Klein's lecture on number theory and on the theory of the top. Participating in the research seminars on number theory, Zapolskaya gave two lectures on this topic: '*Theorie der höheren Congruenzen*' on 12 February, 1896, and '*Zwei Methoden der Erzeugung eines cubischen Abelschen Körpers*' on 13 May 1896.[13] It was in this field of algebraic number theory that she wrote her doctoral thesis: '*Über *die Theorie der relativ-cubischen Abelschen Zahlkörper*' (oral doctoral examination on 29 June 1900). Zapolskaya received a post-doctoral degree from the University of Moscow in 1905 and worked as a teacher at secondary schools; from 1919 to 1923, she was a lecturer at a pedagogical institute in Ryazan, offering courses in higher mathematics, and she would go on to hold similar positions elsewhere (Makeev 2011).

Born in Simbirsk, Nadeschda Nikolaevna von Gernet (1877–1943) was educated at the women's Bestuzhev Courses in St. Petersburg, too, and she started at the University of Göttingen in the summer of 1899 where she gave a presentation on '*Die Reihe von Lagrange*' in Klein's seminar on the theory of functions.[14] She went on to complete her doctoral thesis on the calculus of variations in 1901, supervised by Hilbert. Von Gernet later became a lecturer at her alma mater in St. Petersburg, and held further positions at other academic institutions, from where she in turn sent students of her own to Göttingen. An active researcher, von Gernet published a book in 1913 on the calculus of variations. After earning her degree, she returned regularly to Göttingen in the summers before the outbreak of the First World War, participating in the meetings of the Göttingen Mathematical Society, which Klein had founded in 1892.

The third of Hilbert's Russian doctoral students, Vera Lebedeva (1880–1970), arrived in Göttingen in 1903, likewise with a diploma from the Bestuzhev Courses in St. Petersburg. Working in Hilbert's field, the theory of integral equations, she defended her thesis in 1906. She also met her future husband there, the Romanian Alexandru Myller (1879–1965), who completed his own doctorate in the same year and was also supervised by Hilbert. Both became professors at the University of Iasi in Romania—he in 1910, she in 1918. She continued to publish in German and French journals, and together they created an influential school of mathematical thought. With her appointment, in fact, Vera Myller-Lebedeva became the second female full professor of mathematics in all of Europe. It is still remarkable that they were able to hold professorships simultaneously at the same university. This was possible in Romania at the time, but not elsewhere (Abele et al. 2004, pp. 133–47; Lykknes et al. 2012).

Hilbert's female doctoral students from Jewish German families, Margarete Kahn (1880–1942) and Klara Löbenstein (1883–1968), contributed to the theory of curves topology with their doctoral dissertations, became secondary school teachers, and suffered a hard fate because of the Nazi dictatorship, although Löbenstein was able to emigrate to Argentina (König et al. 2014). Klein gradually allowed Hilbert to assume the leading role in pure mathematical research, while he himself turned his focus to applied mathematics, university administration and educational reform.

New Regulations for Women in Higher Education and Secondary Schools

Although it was not until 1908 that women could attend Prussian universities as more than mere auditors, the presence of women in university classrooms became an increasingly normal occurrence in Göttingen. In 1896, Klein offered the following response to a question about the ability of women to participate in higher education:

> I am all the more pleased to answer this question as the opinion prevailing in Germany, which is that the study of mathematics must be virtually inaccessible to women, essentially blocks all efforts directed toward the development of women's higher education. In this regard, I am not referring to extraordinary cases, which as such do not prove very much, but rather to our average experiences in Göttingen. Though this is not the place to enter more deeply into the matter, I would simply like to point out that during this semester, for instance, no fewer than six women have participated in our higher mathematics courses and *practica* and, have continually proven themselves to be equal to their male classmates in every respect. The nature of the situation is that, for the time being, these women have been exclusively foreigners: two Americans, an Englishwoman, and three Russians, but certainly no one would wish to assert that these foreign nations possess some inherent and specific talent that we are lacking, and thus that, with suitable preparation, our German women should not be able to accomplish the same thing. (Kirchhoff 1897, p. 241)

Klein's conclusion was that the infrastructure for educating girls should be enhanced in Germany. New secondary schools for girls were needed, where mathematics and the sciences would be taught. Within the framework of the educational reform movement, Klein voiced his opinion on numerous committees, in many publications and lectures, and in speeches held as a member in the Upper House (*Herrenhaus*) of the Prussian Parliament (*Landtag*); he was the only representative of the University of Göttingen in this chamber. The result was new educational and professional opportunities for women.

Klein had high regard for Thekla Freytag (1877–1932), who was the first woman in Prussia to fight for the right to pass the examination for secondary school teachers (for mathematics and scientific subjects), and he wrote about all of the obstacles that she had had to overcome to do so in 1905 (Lorey 1909; Tobies 2017). In 1908, Prussia enacted new laws, which allowed the (full-time and official) enrolment of women and the establishment of new types of secondary schools for girls, where it was possible to prepare for the end of school *Abitur* examinations. This led to even more women entering higher education. At that time, the preferred career goal of both female and male students of mathematics was to become a secondary school teacher of mathematics and other subjects (Abele et al. 2004). Klein's youngest daughter, Elisabeth (1888–1968, married name Staiger), reaped the benefits of his efforts and was able to study as an officially matriculated student (Tobies

2008). Because she became a widow in 1914, and so was not subject to the ban on married women teachers, she worked as a secondary school teacher of mathematics, physics and English, and she ultimately became the principal of a school for girls (though she was demoted when the Nazis came to power in 1933). With the right of women to matriculate and with the new opportunity of becoming secondary school teachers (and even a principal at a secondary school for girls), the number of female German students of mathematics began to increase. The files of the women (and men) who had studied mathematics (and at least two other subjects) at Prussian universities and became teachers are kept in the BBF archive in Berlin.[15] Because it was long obligatory for female civil servants to remain 'celibate,' these teachers as a rule remained unmarried or had to leave their positions if they did marry.

With the reform of mathematical and scientific education, which even during his lifetime was known as 'Klein's Reform,' the number of female students increased in general, and many new career options were made available to women and men (Tobies 2011; 2012a). A teaching programme in applied mathematics, initiated by Klein in 1898, also enabled women to become, for instance, an actuary at an insurance company or an industrial mathematician (Abele et al. 2004; Tobies and Vogt 2014). A notable example is Iris Runge (1888–1966), the eldest daughter of the Göttingen mathematician Carl Runge (1856–1927) and sister-in-law of another mathematician, Richard Courant (1888–1972); she became an impressive industrial mathematician at OSRAM and Telefunken (Tobies 2012b).

A post-doctoral degree, the *Habilitation*, was required for a professorship at German universities. Until a law was passed on 21 February 1920, this qualification was restricted to men. Cordula Tollmien (1990; 2021) has already discussed in detail how Emmy Noether needed three attempts to earn this degree, even though the Göttingen professors of mathematics supported her (see also Rowe and Koreuber 2020). Noether was denied in 1915 and 1917 because she was a woman. In 1919, Klein personally supported her *Habilitation*. Encouraged by Albert Einstein (1879–1955), Klein wrote a letter to the Prussian Ministry of Culture on 5 January 1919 in which he explained that Emmy Noether was, at that time, the most productive mathematician at the University of Göttingen (Tobies 1991/1992, p. 172). In 1919, before the official law was passed, Noether became the first woman mathematician to achieve this post-doctoral degree. In 1922, she received the title *professor*, but without pay. As of 1923, she earned income from a teaching assignment in algebra, and every semester she had to reapply for this teaching contract. Noether was never offered a full professorship in Prussia, where no woman was appointed to this role until 1945 (Noether's first full professorship was at Bryn Mawr, after her emigration). Nevertheless, it was Klein (who died in 1925) and other mathematicians in Göttingen who created the conditions that would make it possible for women to attain faculty positions at universities. Throughout all of Germany before 1945, there were only two women who became full professors: Margarete von Wrangel (1877–1932) at the Agricultural University of Hohenheim (in Württemberg) and Mathilde Vaerting (1884–1977) at the University of Jena (Thuringia), both in 1923.

THE FIRST FEMALE MEMBERS
OF THE GERMAN MATHEMATICAL SOCIETY

Whereas the London Mathematical Society had been established in 1865 and the French Mathematical Society in 1874, the German Mathematical Society, *Deutsche Mathematiker-Vereinigung* (DMV) was not formed until 1890. The first women to become members of the DMV were foreigners.

The aforementioned Charlotte Angas Scott became the first female member of the DMV in 1898, when Felix Klein was its president (Toepell 1991, p. 354). She already belonged to the London Mathematical Society (Oakes et al. 2005), had joined the New York Mathematical Society in 1891, and was a founding member of the American Mathematical Society (AMS) in 1894, where she served as the first woman on its council. At that time, she and her first (female) Ph.D. student were two of nine women among an AMS membership of 250. Scott again served on the AMS Council from 1899 to 1901, and in 1905 she became the vice-president (Kenschaft 1987, p. 105). She was one of only four women—the others were the Italian Iginia Massarini (b. 1887), Vera von Schiff from St. Petersburg and Charlotte Wedell (1862–1953) from Denmark—to attend the inaugural International Congress of Mathematicians in Zürich in 1897. Felix Klein, then president of the DMV, participated and delivered an invited lecture (Curbera 2009, p. 16; Mihaljević and Roy 2019). It is noteworthy that one of these four women, Charlotte Wedell, had just studied under Klein in Göttingen and had presented a lecture in his seminar ('*Die Gauss'sche Summen*') on 13 January 1897.[16] We could call her an intellectual grandchild of Klein, given that his former doctoral student, Adolf Hurwitz (1859–1919) at the *Polytechnikum* in Zürich, had supervised Wedell's thesis: '*Applications de la théorie des fonctions elliptiques à la solution du problème des Malfatti.*' At that time, the *Polytechnikum* was not permitted to grant doctoral degrees in mathematics, and so Wedell officially completed her doctorate at the University of Lausanne in 1897.

Returning to the female members of the DMV: the next after Scott was Hilbert's above-mentioned doctoral student Nadeschda N. von Gernet, who was a member from 1901 to 1938. It should be noted that her teacher, Hilbert, served as president of the organization during the year of her appointment. In 1904, the American Helen Abbot Merrill (1864–1949) became a DMV member (Toepell 1991, p. 254). She studied at the University of Göttingen from 1901 to 1902 and earned a doctoral degree from Yale two years later (Singer 2003, p. 93). Elizabeth Buchanan Cowley (1874–1945) followed as the next female American member of the DMV in 1908 (Toepell 1991, p. 75). In that year, she received her Ph.D. from Columbia and embarked upon further studies at the Universities of Göttingen and Munich.

In 1907, she and Ida Whiteside (b. 1883) had published an article together in the journal *Astronomische Nachrichten*, for which they were awarded a prize by the German Astronomical Society.

The first Italian woman to join the DMV did so in 1905: Laura Pisati had earned her doctoral degree in Rome in 1903, taught at the Technical School 'Marianna Dionigi,' and was also—as of 26 February 1905—a member of the *Circolo Matematico di Palermo* (Toepell 1991, p. 291; Jones 2009, p. 91). Her article '*Sulla estensione del metodo di Laplace alle equazioni differenziali lineari di ordine qualunque con due variabili indipendenti*' (1905) was long a fixture in scholarly bibliographies (Ganzha et al. 2008). On account of her tragic premature death, Pisati was unable to deliver her lecture, 'Essay on a Synthetic Theory for Complex Variable Functions,' at the Fourth International Congress of Mathematicians in Rome in 1908, where she would have been the first woman to have spoken at this event. A memorial for Pisati was held during Section I of the Congress in Rome (Curbera 2009, p. 44).

Although Emmy Noether accompanied her father to the Congress in Rome, she did not give a talk there. Having just finished her doctorate under Paul Gordan (1837–1912) at the University of Erlangen, she would deliver her first lecture at the annual meeting of the DMV in Salzburg one year later. In that same year, 1909, she became the first German woman to be granted membership to the DMV (Toepell 1991, p. 276).

In subsequent years, a number of other female German mathematicians earned a doctoral degree and became members of the DMV (Tobies 2006). Additional non-German women joined as well. Here the Austrian Hilda Geiringer (1893–1973), who came to Berlin after completing her doctorate in 1917 at the University of Vienna, may be singled out. Her supervisor Wilhelm Wirtinger (1865–1945) had enjoyed Klein's inspiring seminars as a post-doctoral student at the end of the 1880s (Tobies 2019, p. 306) and recommended her to go to Germany. Geiringer herself was in contact with Klein as well, and she was made a member of the DMV in 1921.[17] She would become, in 1927, the first woman in Germany to earn a post-doctoral degree in applied mathematics at the University of Berlin (Siegmund-Schultze 1993). Olga Taussky (1906–1995), another Austrian who completed her doctorate with Klein's former student Philipp Furtwängler (1869–1940) in Vienna, became a member in 1930 when she was called upon in Göttingen to edit chapters on (algebraic) number theory in Hilbert's collected works (in doing so she would correct many errors in Hilbert's papers). Born to a Czech Jewish family, she continued her research in the United Kingdom and the United States, accompanied by her Irish husband Jack Todd (1911–2007).

Finally, mention should be given to the British mathematician Dorothy Wrinch (1894–1976), a biochemist who applied mathematical principles. She joined the DMV from 1933 to 1937 (Toepell 1991, p. 424; Senechal 2013). At that same time, however, some of the female German-speaking members of the association were forced to emigrate (including Noether, Geiringer and

Taussky) when the Nazis came to power (Siegmund-Schultze 2009; König et al. 2014).

Female Contributors to the Journal Mathematische Annalen and to the Encyklopädie Project

Klein's network was based on his scientific desire to acquaint himself with as many mathematical schools as possible. He collaborated as a peer reviewer for several journals of mathematics and he organized international exchanges of scholarship and bibliographical material. He wanted to maintain the highest international standards for his journal *Mathematische Annalen* and for the book projects under his purview. Klein was one of the chief editors of *Mathematische Annalen* from 1876 to 1924, and it is worth noting when women authors first published in this journal. The first female contributors were Winston in 1895; Scott in 1899; Lebedeva in 1907, 1909 and 1911; Emmy Noether in 1915, 1916 (four times), 1917, 1920 (two times), 1921, 1922 and 1923; and Tatyana T. Ehrenfest-Afanasyeva in 1916.

All of these women's papers were closely associated with Klein's and Hilbert's fields of research. Scott's contribution, for instance, concerns her important proof of a theorem by Max Noether (1844–1921): '*A Proof of Noether's Fundamental Theorem*' (1899). Mary F. Winston, Klein's American doctoral student, had attended his course on hypergeometric functions and in the research seminar she presented novel results in a lecture entitled '*Die gewöhnlichen Kugelfunctionen als Specialfälle der hypergeometrischen Function.*'[18] Klein accepted a short paper by her (completed October 1894) to be published in *Mathematische Annalen* (Winston 1895). Shortly thereafter, she finished her doctoral thesis: '*Über den Hermite'schen Fall der Lamè'schen Differentialgleichungen.*' It is notable that Klein cited her results in a lecture titled 'The Mathematical Theory of the Top,' which he delivered at Princeton University in October of 1896, and that Charles Hermite (1822–1901), whose equations were used by Klein, had written the following, endorsing the importance of the findings, as early as 27 January 1896: 'Le résultat concernant les formues pour le mouvement d'un corps pesant de révolution est d'une bien haute importance.'[19]

Although Klein's British doctoral student Grace Chisholm, whose life and work have been discussed in detail, did not publish an article in 'Klein's journal,' it should be pointed out that Klein encouraged her to write an elementary book on geometry and the first English textbook on set theory; the latter she wrote together with her husband, William Henry Young (1863–1942) (Mühlhausen 2020). Klein also kept this couple in mind when he hoped to produce an English version of his famous encyclopaedia, the six-volume *Encyklopädie der mathematischen Wissenschaften mit Einschluss ihrer Anwendungen* (edited from 1898 to 1935). Because of the First World War, however, an English translation was never produced, though an incomplete French version did eventually appear.

At Klein's instigation, one female author did contribute to the *Encyklopädie*: the Russian mathematician Tatyana Afanasyeva (1876–1964), who had also attended the courses in St. Petersburg and from 1902 had studied in Göttingen. She took part in Klein's lectures on the encyclopaedia of mathematics (1902–1903) and, in November of 1902, she gave a presentation in Klein's seminar on the principles of mechanics entitled '*Der Begriff des unendlich Kleinen und die Ableitung der Differentialgleichungen der Bewegung in allgemeinen Koordinaten bei Lagrange.*'[20] Klein quickly recognized her talents and commissioned her and the Austrian physicist Paul Ehrenfest (1880–1933) as authors. The two had met one another in Göttingen and married on 21 February 1904. Together, they wrote the famous contribution on statistical mechanics for the fourth volume of the great *Encyklopädie*.

Emmy Noether's co-operation with Klein and Hilbert on the general theory of relativity has been discussed at length, recently by Rowe (2019). Her important paper on the topic, which includes the Noether theorems named after her, was outlined by Klein in the *Göttinger Gesellschaft der Wissenschaften* in 1918, published in the *Göttinger Nachrichten*, and ultimately submitted as her post-doctoral thesis.

That Klein was an influential role model for promoting women is clear and demonstrated by the achievements of his former students. Hilbert followed in Klein's footsteps to support mathematically-gifted women, but he was not alone; a number of Klein's students became the first mathematicians at their respective institutions to supervise female doctoral students. Examples include Adolf Hurwitz and Heinrich Burkhardt (1861–1914) in Zürich, Wilhelm Wirtinger and Philipp Furtwängler (1869–1940) in Vienna, Georg Pick (1859–1942) in Prague, Virgil Snyder (1869–1950) at Cornell University and Max Winkelmann (1879–1946) at the University of Jena. In 1919 in Prague (and under Pick's supervision), Saly Ruth Ramler (1894–1993) became the first woman there to earn a doctoral degree in mathematics (Bečvářová 2020). In Jena, Winkelmann supervised the doctoral studies of Dorothea Starke (1902–1943; married name Werner), the first woman to earn such a degree at that university; he also hired her to work as an assistant at his research institute (Bischof 2014).

Over time, the number of foreign women who earned a doctoral degree in mathematics at German universities fell as the number of German women to do so rose. Up to 1906, seven foreign women (four Russians, two Americans, and an Englishwoman) had defended a mathematical dissertation in Germany (all in Göttingen). From 1907 to 1945, only three foreign women did the same, two from Great Britain (at the Universities of Marburg and Göttingen) and one from Denmark (at the University of Freiburg). The cause of this regression was above all the First World War. Afterwards, many nations, most notably the United States and Russia, established new research centres for mathematics, so that women were then more likely to pursue doctoral research in their home country. Nevertheless, Göttingen remained an important international centre for research up until 1933, to which point it continued to attract both male and female mathematicians from abroad.

Altogether, in Germany it was mainly non-mathematicians, especially psychologists and allied specialists, who rejected the idea that women could be capable mathematicians. In contrast, most mathematicians appreciated women's mathematical achievements, given that such results could be evaluated objectively. Yet it would be a long time before male professors would allow women to join the highest professional ranks. Of course, their marginal position in the field meant that many important results—Emmy Noether's key theorems in theoretical physics being a striking example—took a long time to achieve general recognition (Kosmann-Schwarzbach 2011). Today, female professors of mathematics are a welcome part of the profession, but they still remain in the minority (Flaake et al. 2006; Stiller 2019).

Acknowledgements This chapter was translated by Valentine A. Pakis.

Notes

1. Florence Nightingale (1820–1910) is regarded as Sylvester's most distinguished female British student.
2. Annie Reineck-Leuch completed her doctorate with distinction with a thesis entitled *Die Verwandtschaft zwischen Kugelfunktionen und Bessel'schen Funktionen* (Tobies 1997, pp. 41, 137).
3. See Davis's archive of female mathematicians, which includes a chronological list of graduates from the University of Cambridge (1873–1940): http://www-history.mcs.st-and.ac.uk/Davis/Indexes/xCambridge.html. Accessed on 8 August 2019.
4. This is based on a lecture given by Catherine Goldstein at the University of Würzburg in October 2015.
5. As a rare exception, Gentry had previously (in 1890 and 1891) been allowed to attend lectures in Berlin by Lazarus Fuchs (1833–1902) and Ludwig Schlesinger (1864–1933).
6. The original in German reads: '*Das ist schlimmer als die Sozialdemokratie, die nur den Unterschied des Besitzes abschaffen will. Sie wollen den Unterschied der Geschlechter abschaffen!*'. UBG (*Handschriftenabteilung der Niedersächsischen Staats- und Universitätsbibliothek Göttingen*), Cod. MS. Felix Klein, Personalia 22 L.
7. UBG, Cod. Ms. Klein, I C2: fols. 95–96.
8. UBG, Cod. MS. F. Klein 10: No. 905.
9. UBG, Cod. Ms. Klein 11: No. 947. See also Parshall 2015.
10. UBG, Cod. Ms F. Klein 7 E.
11. The doctoral files of these women are published in Tobies (1999).

12. Library of the Mathematical Institute of the University of Göttingen. Protocols (scientific notebooks) of Felix Klein's mathematics seminars 1872–1912, vol. 12, pp. 226–36, 311–22 and 330–39 Online: http://www.uni-math.gwdg.de/aufzeichnungen/klein-scans/klein/ Accessed on 13 December 2020.
13. Klein Protocols, vol. 13, pp. 8–17 and 31–39.
14. Klein Protocols, vol. 15, pp. 113–31.
15. BBF (Bibliothek für bildungsgeschichtliche Forschung), Berlin. Personal records of Prussian teachers.
16. Klein Protocols, vol. 13, pp. 124–31.
17. While writing a popular book on mathematics, which made use of Klein's educational reform ideas and his conceptual coupling of precise and approximate mathematics, Geiringer wrote two letters to Klein, dated 7 November and 3 December 1921 (UBG, Cod. MS Felix Klein 9: No. 307 and No. 308). Klein had read the manuscript of her book (Geiringer 1922, pp. 93–95) and sent comments to her.
18. Klein Protocols, vol. 12, pp. 29–32.
19. UBG, Cod. F. Klein 10: No. 632.
20. Klein Protocols, vol. 19, pp. 16–21.

References

Abele, Andrea, Neunzert, Helmut, and Tobies, Renate. 2004. *Traumjob Mathematik: Berufswege von Frauen und Männern in der Mathematik*. Basel: Birkhäuser.

Bečvářová, Martina. 2020. "Women and Mathematics at the Universities in Prague." In *Against All Odds: Women's Ways to Mathematical Research Since 1800*, edited by E. Kaufholz-Soldat and N. M. R. Oswald, 73–111. Cham: Springer Nature.

Bischof, Thomas. 2014. *Angewandte Mathematik und Frauenstudium in Thüringen: Eingebettet in die mathematisch-naturwissenschaftliche Unterrichtsreform seit 1900 am Beispiel Dorothea Starke*. Jena: Geramond.

Borisovna, V. O. 2003. *Духовное пространство Университета: Высшие женские (Бестужевские) курсы (1878–1918 гг.): Исследования и материалы*. Монография: St. Petersburg.

Coen, Salvatore, ed. 2012. *Mathematicians in Bologna 1861–1960*. Basel: Birkhäuser.

Costas, I. 2002. "Women in science in Germany." *Science in Context* 15: 557–76.

Curbera, Guillermo P. 2009. *Mathematicians of the World, Unite! The International Congress of Mathematicians—A Human Endeavor*. Wellesley, MA: CRC Press.

Fenster, D. D., and Parshall, K. H. 1994. "Women in the American mathematical research community." In *The History of Modern Mathematics*, vol. 3, 229–61, edited by E. Knobloch and D. E. Rowe. Boston: Academic Press.

Flaake, K., Hackmann, K., Pieper-Seier, I., and Radtke, S. 2006. *Professorinnen in der Mathematik. Berufliche Werdegänge und Verortungen in der Disziplin*. Bielefeld: Kleine Verlag.

Ganzha, E. I., Loginov, V. M., and Tsarev, S. P. 2008. "Exact solutions of hyperbolic systems of kinetic equations: Application to Verhulst model with random perturbation." *Mathematics in Computer Science* 1, no. 3: 459–72.
Geiringer, Hilda. 1922. *Die Gedankenwelt der Mathematik*. Berlin: Verlag der Arbeitsgemeinschaft.
Green, Judy, and LaDuke, Jeanne. 2009. *Pioneering Women in American Mathematics: The Pre-1940 PhD's*. Providence, RI: American Mathematical Society and London Mathematical Society.
Grinstein, Lousie S., and Campbell, Paul J. 1987. *Women of Mathematics: A Bio-Bibliographical Sourcebook*. Westport: Greenwood Press.
Hag, K. and Lindquist, P. 1997. "Elizabeth Stephansen: A pioneer." *Det Kongelige Norske Videnskabers Selskab: Skrifter* 2: 1–23.
Hilbert, David. 1902. "Mathematical Problems," trans. Mary F. Winston Newson. *Bulletin of the American Mathematical Society* 8, no. 10: 437–79.
Hilbert, David. 1935. *Gesammelte Abhandlungen*, vol. 3. Berlin: Julius Springer.
Hoyrup, Else. 1993. "Thyra Eibe - the first female mathematician in Denmark." *Normat* 41, no. 2: 41–44.
Jones, Claire G. 2009. *Femininity, Mathematics and Science, 1880–1914*. Basingstoke: Palgrave Macmillan.
Kaufholz-Soldat, Eva, and Oswald, Nicola M. R., eds. 2020. *Against All Odds: Women's Ways to Mathematical Research Since 1800*. Cham: Springer Nature.
Kenschaft, Patricia C. 1987. "Charlotte Angas Scott, 1858-1931." *College Mathematics Journal* 18, no. 2: 98–110.
Kirchhoff, A., ed. 1897. *Die akademische Frau: Gutachten hervorragender Universitätsprofessoren, Frauenlehrer und Schriftsteller über die Befähigung der Frau zum wissenschaftlichen Studium und Berufe*. Berlin: Steinitz.
Klein, Felix. 1923. *Gesammelte Mathematische Abhandlungen*, vol. 3. Berlin: Julius Springer.
Klein, Felix. 1896. "The Arithmetizing of Mathematics," trans. Isabel Maddison. *Bulletin of the American Mathematical Society* 2: 241–49.
König, Y.-E., Prauss, C., and Tobies, R. 2014. *Margarete Kahn, Klara Löbenstein: Mathematicians—Assistant Headmasters—Friends*. Jewish Miniatures 108. Berlin: Hentrich & Hentrich.
Kosmann-Schwarzbach, Yvette. 2011. *The Noether Theorems. Invariance and Conversations Laws in the Twentieth Century*. Trans. Bertram Schwarzbach. New York: Springer.
Kovalevskaya, Sophia. 1875. "Zur Theorie der partiellen Differentialgleichungen." *Journal für die reine und angewandte Mathematik* 80: 1–32.
Lorey, W. 1909. "Die mathematischen Wissenschaften und die Frauen: Bemerkungen zur Reform der höheren Mädchenschule." *Frauenbildung: Zeitschrift für die gesamten Interessen des weiblichen Unterrichtswesens* 8: 161–78.
Lykknes, Annette, Opitz, Donald L., and Van Tiggelen, Brigitte, eds. 2012. *For Better or for Worse? Collaborative Couples in the Sciences*. Basel: Birkhäuser.
Makeev, N.N. 2011. "Lubobj Nikolaevna Zapolskaya." *Vestnik permskovo universiteta* 3, no.7: 82–87.
Mihaljević, Helena M., and Roy, Marie-Françoise. 2019. *A Data Analysis of Women's Trails Among ICM Speakers*. https://arxiv.org/pdf/1903.02543.pdf. Accessed on 13 December 2020.

Mühlhausen, E. 2020. "Grace Chisholm Young, William Henry Young, Their Results on the Theory of Sets of Points at the Beginning of the Twentieth Century, and a Controversy with Max Dehn." In *Against All Odds: Women's Ways to Mathematical Research Since 1800*, edited by E. Kaufholz-Soldat and N. M. R. Oswald, 121–32. Cham: Springer Nature.

Oakes, Susan M., Pears, Alan R., and Rice, Adrian C. 2005. *The Book of Presidents 1865–1965: London Mathematical Society*. London: London Mathematical Society.

Parshall, Karen H. 2015. "Training women in mathematical research: The first fifty years of Bryn Mawr College (1885–1935)." *Mathematical Intelligencer* 37: 71–83.

Parshall, Karen H., and Rowe, David E. 1994. *The Emergence of the American Mathematical Research Community 1876–1900: J.J. Sylvester, Felix Klein, and E.H. Moore*. Providence, RI: American Mathematical Society.

Pisati, L. 1905. "Sulla estensione del metodo di Laplace alle equazioni differenziali lineari di ordine qualunque con due variabili indipendenti." *Rendiconti del Circolo Matematico di Palermo* 20: 344–74.

Rahikainen, A. 2009. "Cedercreutz, Nanny (1866–1950)." In *Biographical Dictionary for Finland*. Sweden: Swedish Literary Society.

Rossiter, Margaret W. 1982. *Women Scientists in America*. Baltimore: Johns Hopkins University Press.

Rowe, David E. 2019. "On Emmy Noether's role in the relativity revolution." *The Mathematical Intelligencer* 41, no. 2: 65–72.

Rowe, David E., and Koreuber, Mechthild. 2020. *Proving It Her Way: Emmy Noether, a Life in Mathematics*. Springer Nature.

Scott, Charlotte A. 1899. "A Proof of [Max] Noether's Fundamental Theorem." *Mathematische Annalen* 52: 593–97.

Senechal, Marjorie. 2013. *I Died for Beauty: Dorothy Wrinch and the Cultures of Science*. New York: Oxford University Press.

Siegmund-Schultze, R. 1993. "Hilda Geiringer-von Mises, Charlier series, ideology, and the human side of the emancipation of applied mathematics at the University of Berlin during the 1920s." *Historia Mathematica* 20: 364–81.

Siegmund-Schultze, Reinhard. 2009. *Mathematicians Fleeing from Nazi Germany*. Princeton: Princeton University Press.

Singer, Sandra L. 2003. *Adventures Abroad: North American Women at German-Speaking Universities, 1868–1915*. Westport: Praeger.

Stiller, S. 2019. "Mathematik-Professorinnen und –professoren an Universitäten in Deutschland. Das Problem der Zahlen." *Mitteilungen der Deutschen Mathematiker-Vereinigung* 27, no. 2: 52–53.

Tobies, Renate. 1991/1992. "Zum Beginn des mathematischen Frauenstudiums in Preußen." *NTM–Schriftenreihe für Geschichte der Naturwissenschaften, Technik und Medizin* 28, no. 2: 151–72.

Tobies, Renate, ed. [1997] 2008. *Aller Männerkultur zum Trotz': Frauen in Mathematik, Naturwissenschaften und Technik*. Frankfurt am Main: Campus.

Tobies, Renate. 1999. "Felix Klein und David Hilbert als Förderer von Frauen in der Mathematik." *Prague Studies in the History of Science and Technology* 3: 69–101.

Tobies, Renate. 2001. "Femmes et mathématiques dans le monde occidental, un panorama historiographique." *Gazette des mathématiciens* 90: 26–35.

Tobies, Renate. 2006. "Biographisches Lexikon in Mathematik promovierter Personen." In *Algorismus, Studien zur Geschichte der Mathematik und der Naturwissenschaften*, vol. 58, edited by Menso Folkerts. Augsburg: Dr. Erwin Rauner.

Tobies, Renate. 2008. "Elisabeth Staiger: Oberstudiendirektorin in Hildesheim." *Hildesheimer Jahrbuch für Stadt und Stift Hildesheim* 80: 51–68.
Tobies, Renate. 2011. "Career paths in mathematics: A comparison between women and men." In *Foundations of the Formal Sciences VII: Bringing Together Philosophy and Sociology of Science*, edited by K. François, B. Lowe, and T. Muller, 229–42. Milton Keynes: College Publications.
Tobies, Renate. 2012a. "German graduates in mathematics in the first half of the 20th century: Biographies and prosopography." In *Biographies et prosopographies en histoire des sciences*, edited by L. Rollet and P. Nabonnand, 387–407. Nancy: Presses Universitaires.
Tobies, Renate. 2012b. *Iris Runge: A Life at the Crossroads of Mathematics, Science, and Industry*. Trans. Valentine A. Pakis. Basel: Birkhäuser.
Tobies, Renate. 2017. "Thekla Freytag: 'Die Mädchen werden beweisen, dass auch sie exakt und logisch denken können....'" In *Scriba Memorial Meeting: History of Mathematics*, edited by G. Wolfschmidt, 330–79. Hamburg: Tredition.
Tobies, Renate. 2019. *Felix Klein: Visionen für Mathematik, Anwendungen und Unterricht*. Berlin: SpringerSpektrum. Engl. revised and extended ed.
Tobies, Renate. 2020. "Internationality: Women in Felix Klein's Courses at the University of Göttingen." In *Against All Odds: Women's Ways to Mathematical Research since 1800*, edited by Eva Kaufholz and Nicola Oswald, 9–38. Cham: Springer Nature.
Tobies, Renate. 2021. *Visions for Mathematics, Applications, and Education*. Cham: Birkhäuser.
Tobies, Renate, and Vogt, Annette B., eds. 2014. *Women in Industrial Research*. Stuttgart: Franz Steiner.
Toepell, M. ed. 1991. *Mitgliedergesamtverzeichnis der Deutschen Mathematiker-Vereinigung 1890–1990*. Munich: Institut für Geschichte der Naturwissenschaften.
Tollmien, C. 1990. "Sind wir doch der Meinung, daß ein weiblicher Kopf nur ganz ausnahmsweise in der Mathematik schöpferisch tätig sein kann Eine Biographie der Mathematikerin Emmy Noether (1882–1935) und zugleich ein Beitrag zur Geschichte der Habilitation von Frauen an der Universität Göttingen." *Göttinger Jahrbuch* 38: 153–219. (Tollmien, Cordula. 2021. *Die Lebens- und Familiengeschichte der Mathematikerin Emmy Noether (1882-1935) in Einzelaspekten*. Bd. 1 und 2. Hamburg: tredition.)
Tollmien, C. 1997. "Zwei erste Promotionen: Die Mathematikerin Sofja Kowalewskaja und die Chemikerin Julia Lermontowa." In *Aller Männerkultur zum Trotz': Frauen in Mathematik und Naturwissenschaften*, edited by Renate Tobies, 83–129. Frankfurt am Main: Campus.
Vogt, Annette B. 1999. *Elsa Neumann: Berlins erstes Fräulein Doktor*. Berlin: Verlag für Wissenschafts- und Regionalgeschichte Dr. Michael Engel.
Winston, M. F. 1895. "Eine Bemerkung zur Theorie der hypergeometrischen Function." *Mathematische Annalen* 46: 159–60.

PART IV

Cultures of Science

CHAPTER 12

Astronomy, Education and the Herschel Family: From Caroline to Constance

Emily Winterburn

INTRODUCTION

The astronomer Caroline Herschel (1750–1848) did a lot with a little. She grew up with the lowest of expectations. While better educated than her mother—she was at least sent to school to learn to read and write—Caroline's childhood education was severely limited. Her hours were filled learning the myriad of skills needed to run the home of an eighteenth-century musician, but only cursory attention was ever given to her academic schooling in contrast to her brothers. She could cook, clean, make clothes and raise children. Yet much to her bitter disappointment was never taught French or calculus. As an adult she was taught music and then astronomy, primarily for the convenience of her brother who would then be saved the bother of hiring, training and paying an assistant. Hers was an education of utility, each step designed to make her more useful to the family around her, yet she made it her own, and stood firm, crucially whenever the question of credit and acknowledgement arose. She was supported in this by her brother, William Herschel, and several prominent men of science who saw to it that her work was noticed, celebrated and accepted. Between Caroline and her male supporters, credit was secured for each of her comet discoveries, in some cases even resulting in her own words appearing in the prestigious academic journal of the Royal Society.[1] Her work as assistant was acknowledged in the Royal pension her

---

Present Address:
E. Winterburn (✉)
Coop Academy Nightingale, Leeds, UK

brother obtained for her. Her independent organisational work revising John Flamsteed's cataloguewas published and promoted for her by her brother, and by her friend and supporter the Astronomer Royal Nevil Maskelyne and by the Royal Society (Herschel 1798). Hers is a story of success against the odds. But what if the odds had been a little more in her favour? What of those women who succeeded her? Without the obstacles she had to overcome surely theirs would have been a quicker, simpler, smoother rise to the top?

The story of the Herschel women paints a rather more complex view of what constitutes an obstacle to participation and credit in science. Their story tells us about the opportunities available to some nineteenth century women that were not available to women like Caroline. It also shows us new obstacles, however. As with so many stories about women in science, the Herschels' story is one of finding a way. Theirs is a story that tells of science in the nineteenth century on the one hand moving out of the home and away from women's traditional reach, and yet, on the other, of women still finding ways to participate in science.

For the men within this multigenerational family, success evolved and changed in line with broader changes taking place within the scientific world. William Herschel succeeded as a gentleman amateur, John Herschel as a Cambridge man within a network of Cambridge men and his sons trained and succeeded in scientific professions as science moved beyond the home and into field, lab and observatory work. Caroline's brother William had been a pioneer himself who like Caroline had taken a circuitous route to astronomy. In time he became known throughout the world for his many achievements: the first named discover of a planet, builder of the biggest telescope in the world at the time, a surveyor searching for structure among the fixed stars thereby creating a whole new field within astronomy, but initially he, like Caroline, had been an outsider. To English philosophical society William was an outsider not by gender but by class; he was a jobbing musician and economic refugee with no experience of the conventions needed to express his ideas and have them accepted. He made sure that his son, John, faced no such obstacles. Unlike his father John was schooled and coached in all subjects but especially maths from a young age ensuring his outstanding success at the University of Cambridge. After that he dabbled, travelled, networked and experimented, publishing as he went, thereby steadily creating a reputation in science beyond that of famous son. A few years in South Africa completing his father's work for the southern hemisphere and many positions on prestigious committees ensured that John Herschel became well established as the archetypal man of science of his day. His sons too were able to make very respectable if less high-profile careers for themselves in science thanks to the education and cultural capital of their family background. Indeed, the changing face of science during the nineteenth century (of which John Herschel was an instrumental part), meant that his sons were able to train for and get secure, scientific jobs for life as engineers and university teacher/researchers, positions that barely existed in William's time. The combination of their family's interest and understanding of science, and the changing scientific job market worked to ensure William's grandsons had the chance to work as he had done in a field they enjoyed, but without the insecurity and isolation experienced by himself and his peers.

For the women of the Herschel family, there is a different story. This chapter will explore how the history of women in science is one imbedded in the broader story of women's education, and the status of women as capable, functioning human beings. Caroline's story spans a period in which women's capacity for education was being questioned on biological grounds, while at the same time, the desirability of educating women was also being openly debated (Schiebinger 1989). Centuries after philosopher François Poullain had declared 'the mind has no sex' (Poullain 1673), many claimed to have found biological differences that would explain why men rather than women dominated the intellectual and public sphere. As Caroline rose to fame, these arguments were part of a lively public discourse. Mary Wollstonecraft pointed out it would be hard to compare intellectual ability without first offering parity in education and opportunity (Wollstonecraft 1792). Arguments for why society should bother to educate women had to some extent more homogeneity. Decades earlier, Jean-Jacques Rousseau had posited that the purpose of a girl's education had to be to create good and virtuous wives and mothers so they might carry out their civilising role on the men around them (Rousseau 1762; Cohen 2006). This continued to be a major argument for educating women throughout the eighteenth and nineteenth centuries (Jordan 1991). While some feared an academic education for girls might give them impossible dreams, making the dreary drudge of domestic duty unbearable, others considered an academic education to be essential in preparing girls to become both a stimulating intellectual companion to their husbands and a suitably well-informed teacher of their sons. Though not everyone was convinced, opportunities in education for women (and men) did begin to increase in the nineteenth century.

Caroline's successors benefited then from being born into a time when education was becoming more widely available to all. Early nineteenth-century female education nevertheless was very mixed. Lower-class girls, alongside lower-class boys, were starting to gain some form of academic education in Britain, initially through the Sunday school movement, where they could go for a few hours a week to be taught basic reading, writing arithmetic, and knowledge of the Bible and later in state run schools (Fyfe 2004). Higher-class girls had a more varied experience. While some had families who prized academic curiosity in their daughters and provided governesses, tutors and access to well-stocked libraries, others were actively discouraged from learning too much. In some upper middle-class families, a daughter's education was deliberately devoid of academic content, with the learning of 'accomplishment' a preferred means of making daughters more desirable in the competitive marriage market. They were taught singing, dancing, art and needlework, while anything too academic was frowned upon (Delamont 1989). In those families who did prize academic education for their daughters, it was often unclear what the purpose of that education was and that purpose almost certainly varied from family to family. Women from all types of families were routinely trained to be good wives and mothers, whether that meant academic education or not. Men on the other hand were generally educated with a view to taking some form of work outside the home. Just as we rightly assume 'some form of work outside the home' is not a single thing, requiring a single type of education, so too should we problematise the concept of good wives and mothers.

The Herschels were one of a number of scientific families who dominated British science in the nineteenth century. The Herschels, the Darwins and the Airy family for example all knew one another, had one another's children to stay and discussed schools, tutors and when to teach their children what—and all had daughters. It has been observed elsewhere that the daughters of clergymen and those from 'intellectual families' had an advantage (Delamont 1989, p. 66), and certainly it is difficult to think of a high-profile pre-twentieth century British woman in science who did not come from such a family. These were, after all, the families with libraries, and with an interest in and understanding of learning. Discussions within and between these families suggests even in such a small group of families there were differences of opinion as to what to teach daughters and why, hence the need to confer.

The Herschel women who came after Caroline had a very different experience of education to Caroline. They were growing up in a time when the academic education for women of their class was not unheard of. They were from that slim strata of society—as daughters of clergymen and scientific men—who historians have long recognised as standing alone for their valuing of academic learning for their daughters. They were, as all girls were in the nineteenth century, being educated primarily with a view that they should become good wives and mothers, however what that meant deserves closer inspection. How that could translate into a scientific life—if one resulting in limited credit and documentation—deserves further study. The example of the Herschel family women gives us the opportunity to do just that.

TRAINING FOR LIFE AS A SCIENTIFIC COMPANIONATE WIFE

For John Herschel's wife, Margaret Brodie Stewart (later Herschel), the purpose of her academic education seems to have been primarily to help her be a good, intellectually thorough, evangelical, proselytising Christian. Margaret was the youngest daughter of an evangelical minister of the Scottish church called Alexander Stewart. Among his contemporaries, Alexander was considered to be a radical. He was a minister but also a Gaelic scholar and follower of the radical thinker James Haldane. Alexander wrote about his faith with such zeal that even many evangelists found him too extreme. His book, *The Free Presbyterian* was for example described by one evangelical scholar as 'Sandemanian' (Beaton 1985, p. 179). Stewart's interpretation of Christianity was strongly spiritual. Belief was of utmost importance, entirely separate from politics and other worldly concerns and it was this separation that allowed space for an exploration of science and its place in his and his family's religious world view.

Margaret's mother, Emilia Stewart (née Calder) was also from a religious family. Like her husband Alexander Stewart, Emilia's father, Rev. Charles Calder was a minister in the Church of Scotland. Both the Calders and the Stewarts were part of a movement within the Church of Scotland, and evangelical Christianity more broadly, who were trying to make sense of the emerging dominance of science in society. They were looking at these new ideas and discoveries in astronomy, geology and biology, and trying to understand them and unite them with their religious teachings. Aileen Fyfe points out that these two differing ideologies—evangelicalism and science—did not fully come together until the 1840s with the emergence of the Religious Tract Society (2011). In order, however, for this series to emerge, there needed to be discussion, development and refinement of how to present this union of science and religion. This was the movement in which Margaret's father and to an extent the rest of their family were involved.

Margaret was the youngest of four. We know little of her education pre-1827, when she married John Herschel, but correspondence between the siblings both around this time and later show theirs was a family that prized academic learning, independent thinking and strong religious conviction. She and her sister were taught at home, by family members who encouraged questioning and curiosity. She was taught academic subjects such as mathematics, languages and some astronomy to a reasonably high standard. Though never explicitly stated, this education seems to have been designed to train the girls—like any other middle-class nineteenth-century girl—to become good wives and mothers. The content of that education, however, tells us what kind of wife and mother they hoped they might become.

Until her marriage to John Herschel, there is no indication that Margaret had any ambitions for herself other than to become a good, evangelical companionate wife and mother as she had been trained to become. Not long after her marriage however, things began to change. She began to comment on her admiration for the intellectual women in John's circle, making names for themselves in print, showing themselves to be the intellectual equal of any man. She looked to Mary Somerville, a regular visitor to the Herschel home in the early days of their marriage, as John assisted with her book, *Mechanisms of the Heavens*. She saw too Maria Edgeworth—whose writing on the education of children she would come to rely on heavily—hold her own in discussions with men on intellectual subjects. As she wrote to Edgeworth on one occasion:

> in the abundance of your charity pray come and infuse some of your own clear sighted and right thinking metaphysics into Herschel and Mr Whewell who are just now deep in profound dissensions on these subjects, and sadly want some

fresh ideas on cause & effect – My poor brains are quite addled with it all, but there is enough of wickedness left in me to desire to see the gentlemen floored by a women, for we claim you, however nearly you may be allied in intellect to the other sex.[2]

Looking at these women, Margaret was able to see a new model for how an educated woman might use her academic learning. Following her education from her family, she could already see that an academic education might help her better understand the world created by their god. Through these women she could see too that an academic education might allow a woman a voice, and an opportunity to participate in the creation of new knowledge.

Surrounded by these intellectual women, Margaret began to look for advice. She actively sought out a friendship with John's aunt Caroline, asking about her life and how was she had created her role in science.[3] Margaret's own education might have been slowed down by the birth of her first child a year into her marriage, but the arrival of that baby girl only made her more interested in how to educate her girls like these intellectual women. As her daughters grew older, she even asked Caroline directly for advice on how to educate them so they might match the achievements of their great aunt. Caroline's response was cautious, keen to highlight the perils of her path rather than encourage emulation:

> Now a few words to my dear Niece and her Darlings your description of young ladies progress they are making under the tuition of Madam Garlach; and we ought not to forget the example of their excellent mama before them. It is delightful and would tempt me almost to wish I were some 10 years younger that I might know them to be happily married to deserving and good men. But I trust the caution is not wanted here that it is safest rather to conceal some than showing too much approbation of their abilities.[4]

Margaret was undeterred. Like Caroline she wished she had had more opportunity to learn and was aware—especially among her husband's friends—of gaps within her education but unlike Caroline she felt she nevertheless had some control over that situation. No one had withheld that knowledge from her, she just had not yet found time to learn it.

A few years into her marriage, when Margaret had been blessed not only with three small children but also a plethora of servants to care for them, she began to make time to return to her studies. As she told her mother,

> After tea there is singing & arranging French sentences with a box of letters till bed time at 8 o'clock, & then comes the only hours I have to read or write, learn German or copy for Herschel. (Warner 1991, p. 87)

After a while, she also found some time in the morning too:

for instance I am just now learning Algebra from him [John Herschel], & reading every morning at Breakfast some pages of a MS Vol on the subject which I dare say he will publish some day, having first tried its effect on learners of the lowest capacity. (Warner 1991, p. 100).

She began to discuss the education she might give her daughters, comparing possible strategies with other mothers. When discussing the introduction of French to the curriculum of their daughters (both called Caroline), Margaret wrote to her friend Mary Maclear (wife of Thomas Maclear, then Astronomer Royal at the Cape):

I am inclined to add that any change from the old style, is synonymous with improvement, it is not improbably that both the Carrys may escape the horrors of French grammar which so sorely puzzled their Mothers. (Warner 1991, p. 69)

These conversations suggest a tradition of sorts already existed for girls' education at least among certain families. These women recognised in each other comparable backgrounds and levels of education. They had both been academically educated as children and considered this the norm, the standard against which they should devise an education for their daughters. Opportunities outside the home for educated women in the 1830s were just as limited as they had been two decades earlier for Margaret. Most could aspire to be wives to the kind of men who sought an intellectual companion as a wife, a few, though not many, might consider writing or running their own school, while those who fell on hard times could use their academic education and good family to get themselves work as a governess or school teacher. The most secure of these was to find a rich, successful husband to whom they could provide—with their academic education—intellectual companionship and an education for their own children. Unsurprisingly then, this was at least initially the primary ambition these women had for their daughters.

Historians have long recognised the way in which women have traditionally lost out in the 'great' men account of history. To rebalance this, attempts were for a time made to dig out stories of 'great' women only to find that their stories were almost always intertwined and interdependent on the surrounding men in their lives. While it is fair to surmise that this is equally true of those 'great' men, it nevertheless leaves us with a problem. How do we decide which women to study? How do we find out about them? What constitutes a historically recognisable female role in science? One possible avenue—at least as a way of recovering some of the stories of rich, white women in science—is to recognise the validity of the role of companionate wife in nineteenth century life. Back in 1989, Jeanne Peterson pointed out that historians had for too long conflated the modern concept of the trophy or 'incorporated' wife 'tied to her husband's career as a status symbol and decorative companion but who dares not enter into his professional life intellectually or politically' (Peterson 1989, p. 186), with that of Victorian marriages. By studying a range of Victorian marriages within an extended intellectual family, Peterson found this was not the case. Instead she discovered the phenomenon Barbara Caine later described as that of the 'companionate' wife (Caine 2005):

> wives were assistants, colleagues, and partners in the work that men did. Their husbands took the public credit for the task performed – these were not 'dual careers', nor was there any ideology of equality. These were 'single-career families', but both husband and wife partook of that single career. (Peterson 1989, p. 166)

Within this context, without projecting back our modern concepts of what constitutes a successful female life in science, we could conclude that in fact Margaret Brodie Stewart was able to live up to her better-known aunt-in-law. Her family's ambitions for her were very different from that of Caroline's. Margaret was expected to marry, though the family's ambitions regarding what type of man that should be are unclear. More concrete, is their expectation that she should become a good, evangelical, free-thinking Christian, and in that she seems to have more than met those expectations. In terms of how her own ambitions shifted this imposed ideal, she was perhaps more compromised. Early on in her marriage she looked to that group of contemporary women—Maria Edgeworth and Mary Somerville—and seen in them a new aspirational model for what an educated woman might become. In them, she saw women expressing their own ideas in print, arguing their position in male dominated company and like William before them, pursuing knowledge for its own sake, rather than as a means of making themselves more useful to others.

Around the middle of the nineteenth century, as Margaret was beginning to put into practise her beliefs in how best to educate her daughters, many debates raged over the purpose of education in general and who should be taught what and why. Matthew Arnold (Arnold 1869) and Thomas Henry Huxley (Huxley 1866) were debating the purpose of male education and the relative value of classics and science as a means of training the next generation of young men. At the same time, the purpose of female education was being debated and those philosophies were being put into action (Donnelly 2002; Jordan 1991). Emily Davies (later founder of Girton College Cambridge) was a central figure in this debate, arguing forcefully and uncompromisingly in favour of women having equal access to the same education as men (Davies 1866). On the other side of this debate were the 'separatists' such as Anne Jemima Clough who felt women should have an equal but different education to men, arguing that existing education favoured a male outlook and that a female centred education would and should therefore be different but equal (Delamont 1978). While the focus on all these debates was on university education and the select few who were able to attend, many of the ideas touched on broader ideas regarding the purpose of education and this influenced how girls' education was designed and discussed.

For the women within the Herschel family, like many middle-class girls in the mid-nineteenth century, these discussions were of only passing interest. The Herschel girls were educated entirely at home. There was—as yet—no possibility of them attending university, and so no schools that could train them for that goal were ever discussed. While the boys had their careers decided for them at the age of eleven, and were then sent off to school to prepare for the appropriate training college or university, the girls were left to develop their interests in their own time. They continued to be taught at home, but how was that home learning designed? With what aim or purpose was a family like the Herschels teaching its daughters?

Navigating the Marriage Market and Compromised Potential

If Margaret succeeded in her chosen path as companionate wife and mother, how well did her daughters fare? Caroline (born 1830), Isabella (born 1831) and Louisa (born 1834) were all taught at home alongside their brothers until each brother reached the age of eleven and was sent away to school. When the boys left, the girls' education carried on without interruption suggesting very little gendered difference in aim or design of this infant training. All the children, Margaret told her brother Charles in 1840, were:

> getting on admirably with their lessons under their excellent German governess & as Willie [then aged seven] is working hard at German I have deferred his Latin & Greek for a little while - Ancient history, & natural history are their favourite studies and Caroline [aged ten] shews a good deal of talent for music.[5]

While the girls' education continued as a means of developing their interests and 'talent', concern for the boys' futures began to occupy the minds of their parents. When an old family friend of Margaret's—Eneas Mackintosh—offered to find William a place in the East India Company she jumped at the chance, telling her friend Maria Edgeworth that since 'it has been determined that our Eldest boy is to follow the same path in India [as one of her brothers had before] - we have two more to turn into active useful Herschels of some right sort in England'.[6]

Margaret's phrasing here, in her praise for work in India is interesting in light of William and Caroline's eighteenth-century education and their respective descriptions. Caroline always claimed her training and her desire for more education was to enable her to become more 'useful', in contrast William always claimed to have been driven by a love of learning. Here, two generations later, that gendered language appears reversed. For this generation, the boys were to be trained to be useful, the girls meanwhile were kept at home, allowed to develop their love of learning.

We get a good idea of what the Herschel children were taught day by day from the diary of Louisa, then aged ten from 1844. In it she describes the roles of the different governesses, as well as their mother and father, in the teaching of their various lessons. Miss Agar (one of the family's two governesses at the time) taught them elementary mathematics (the older, more advanced children were taught by their father) and took them for walks; Miss Rufenacht (their other governess) taught geography, German, French and natural history. Other subjects were taught by their parents, including music, religion and literature from their mother, Latin and advanced mathematics from their father. In addition, they spent most evenings translating between English, French, German and Italian or occasionally reading plays with their mother. Sometimes they would play games such as dominoes or chess. On one occasion Louisa complained about having to play cricket with her brother William, back from Clapham Grammar School for the holidays. An outside tutor was brought in for dancing and painting. In the summer they did gardening and periodically they were taken to see lectures and shown experiments by their father.[7]

As Kathryn Hughes' study of governesses shows, this was a pretty standard division of labour within upper middle-class Victorian homes (Hughes 1993). Large families often employed more than one governess to allow children of different ages to be taught separately. Nursery governesses would be employed to teach boys and girls reading and writing from aged four to eight. A preparatory governess would then be appointed for girls aged between eight and twelve. She might typically teach 'the basics of English grammar, history, "use of globes", French, perhaps Italian or German, piano, singing, drawing' and whatever else she might know well (Hughes 1993, pp. 60–61). The teaching of arithmetic, Hughes tells us, remained weak until the end of the century and science was almost unknown in most homes. Latin was taught only if the governess was the daughter of a clergyman or if the child's father took on the responsibility.

There was a couple of areas in which the Herschel education did deviate from that described by Hughes however. The Herschels were much stronger in their teaching of arithmetic and science than the families typically studied by Hughes and others. In this, the Herschels were following their own experiences; John had been given additional tutors in mathematics from the moment he started school while Margaret's correspondence with her brothers suggests both mathematics and science were key components of her and her sister's schooling. They were also following the late eighteenth-century advice of their friend Maria Edgeworth. In *Practical Education*, Edgeworth asserted that 'Natural history is a study particularly suited to children: it cultivates their talents for observation, applies to objects within their reach, and to objects which are every day interesting to them' (Edgeworth and Edgeworth 1798, p. 338). She defended the teaching of accomplishments on the grounds that they act as 'resources against ennui, as they afford continual amusement and innocent occupation. This is ostensibly their chief praise; it deserves to be considered with respect' (Edgeworth and Edgeworth 1798, p. 522). She concluded, however, that the study of nature would serve the same purpose and had the additional advantage of making your girls stand out in the marriage market. Following this advice, the Herschels evidently chose to teach both.

By 1844, when Louisa's diary tells us their daily routine, the Herschels had established a gender-free education for their young infants. They were given a solid grounding in maths, English and various languages. They were taught some history and natural history generally through activity and excursions and these academic pursuits were balanced out with lots of sport, dance, music and painting. From this starting point, the boys were sent to Clapham Grammar School, and then onto (in age order) Haileybury (the East India Company training college for its civil servants), Cambridge as preparation for a life in academia, and Addiscombe (the East India Company training college for its engineers). In all three cases, the education they had received at home was considered a good grounding, preparing them to meet the challenges of first grammar school and then later these different institutions of higher education. The girls received the same preliminary education yet there was no possibility (at least in the 1840s) of imagining they might go onto any form of training for a career in the civil service, academia or engineering. Instead, their education continued at home, building on this same base but preparing them for a life closer in spirit to that of the leisured gentleman amateur of their grandfather and great aunt's generation than the paid professions of their brothers.

In April 1852 the Herschels' eldest daughter Caroline rejected her first marriage proposal. The proposal was from the young and wealthy but otherwise unknown William Peareth. Margaret's approval of this decision was expressed first to Caroline's sister Bella: 'Of course his great inferiority in talent (though we need not say so) - & his extreme youth (not yet 21) were sufficient objections in Carry's eyes & would be in ours'.[8] Then to her husband John noting 'that it seldom did well, that the wife should be the superior of the two (& they all acknowledge Carry's pre-eminence)'.[9]

That both mother and daughter objected to this man on the basis of his 'great inferiority in talent' gives us some indication of the aspirations they shared for her future. Education then for these girls was then to some extent about preparing them for their role as a particular kind of wife of a particular kind of husband. Their daughters they ensured through their education were highly educated. Their husbands however were expected to be educated to an even higher standard. Why might that have been? In part, it would have been to ensure they remained or even rose in class status. They should marry men who came from similarly highly academic, financially secure families. However, it may also have been to allow their daughters' access to intellectual work. To prepare them—as Margaret's probing questions to Caroline on her training seemed to imply—for a life of intellectual companion and assistant as Caroline had been to her brother, William.

A few months later Caroline married Alexander Hamilton Gordon, a Scottish soldier, and the son of the Tory Prime Minister, the 4th Earl of Aberdeen, George Hamilton Gordon, a distant relative of Margaret's (Mosley 1999, p. 10). John expressed some regret at this proposal too, suggesting he had higher or at least different aspirations for his eldest daughter, though he did not then spell out what they might have been.[10] Why this match was considered more suitable than the last for Margaret is not made clear, though the family link may have made it more difficult to turn down. As for Isabella and Louisa, the other two eldest daughters, Isabella never married, preferring instead to stay at home, educating her younger sisters and becoming archivist and keeper of the family archive. Louisa meanwhile left home in 1858 to marry Reginald Dyke Marshall, the son of a Leeds linen manufacturer.

While these eldest daughters of John and Margaret Herschel were educated to a high standard, with aspirations that they might use this to select the perfect intellectual marriage companion, the reality of the marriage market showed this a more difficult undertaking that they had originally imagined. Instead, they were forced to make compromises. These daughters had to choose between marriage or a life in science, while at the same time the family took stock looking to see what other options might be available to the girls within this generation.

A Scientific Education Becomes More Mainstream

The Herschels had so many children—twelve over a span of twenty-five years in total—that to consider them all together, as though society and family aspirations didn't change at all over time would be misleading. The first three daughters grew up initially in Colonial South Africa and then in the family's large home, Collingwood, in Hawkhurst, Kent. Their early education took place in the 1830s and 1840s when their parents were working out what to teach and how and while around them arguments raged over the status of science in society and the value of providing academic education for girls. Around them, the number of schools for girls rose dramatically with literacy levels for women doubling between 1780 and 1840. In the 1840s, evangelicals began publishing books on science designed to educate the working class about this new discipline in a controlled and moral way. Science was becoming a mainstream concern, and as such it made sense to teach it to one's daughters, especially as it now has religious backing. These, however, were changes that took place as the elder daughters were growing up. It would not impinge immediately on their education, but rather form a backdrop that could then inform the teaching of their younger siblings.

These eldest children were the first recipients of an education their young mother was assembling drawing on advice from her mother, John's aunt Caroline, scientific wives and mothers the family knew and a good deal of reading. For these eldest children, the Herschels were trying out new ideas, testing out alternatives to their own experiences (changing for example how they taught

French following a conversation with fellow mother, Mary Maclear in South Africa) (Warner 1991, p. 69). The daughters who followed were able to reap some of the benefits of those tentative experiments in how to teach.

The Herschels' middle daughters—Maria Sophia (born 1839), Amelia (born 1841), Julia (born 1842), Matilda Rose (born 1844) and Francisca (born 1846)—all grew up more or less together. By now the family was determinedly scientific with an unashamed determination to raise academically trained daughters. To set the tone, many were named after important women in the Herschels' lives and assigned godparents that would support their parents' aspirational aims. Maria Edgeworth became godmother to Amelia; the photographer Julia Margaret Cameron became godmother to Julia Mary and Mary Somerville became godmother to Matilda Rose (Somerville and McMillan 2010, p. 220). In thanking Maria Edgeworth for performing this duty, Margaret told her 'I only pity my poor little girl on the high nature of the hopes & expectations which may be formed of her'.[11]

In February 1854 the family's governess of 10 years, Miss Rufenacht, left. Bella, now aged twenty-two, stepped in to act as tutor. Margaret wrote to John: 'Bella is everything in the house just now, Housekeeper, Secretary, Tutor, Hostess & Student - I can only call her incomparable'.[12] Bella took the children on expeditions to learn botany: 'The children had their botany lesson al fresco yesterday' she announced in one letter.[13] She then had them studying their finds back at the house: 'Fancy [Francisca, then aged eight] is busy looking at mosses through her microscope & has found some wonderful little seed vessels "just like dust" she says'.[14] Bella also encouraged the children to think about the scientific properties of the everyday objects they saw around them, just as Edgeworth had earlier recommended (Edgeworth and Edgeworth 1798, p. 429). When Bella thought her own knowledge lacking, she turned to her father. At this time, John was away in London working as Master of the Mint. 'I want to know all about air!' Bella wrote on one typical occasion:

> & about water & steam & heat, & weight, & pressure, & resistance & attraction &c &c &c &c ... There's a modest little want for you! The immediate reason, is that one day in talking to the school-children I plunged into the subject of clouds & dew & then found that I could tell them a great deal that they could quite understand about the atmosphere, the properties of air, & of water - with ample references of course to tea-kettles, roasted chestnuts &c. - but still I feel on rather unsafe ground about it all.[15]

While her initial request for help was so she might teach better, she quickly went on to admit it was also to satisfy her own intellectual curiosity warning:

> I don't promise that I will stop at the "useful" though I suppose I must at the intelligible - for I confess to the selfish desire of knowing its contents for myself, as well as for the benefit of the children.[16]

Bella's offhand confession to her 'selfish desire' implies she thought he might approve of her seeking scientific knowledge purely out of interest. While her great aunt had needed to justify her learning in terms of its usefulness to others, Bella was allowed simply to want to learn.

Bella was not left to be sole governess to her sisters for too long. Within the year, the family had appointed Miss Karth, 'a French Alsastian' fluent in both French and German (*Newsletter of the Japan Herschel Society* 1993–1994). At the same time, John, their father was now involved in the various parliamentary committees tasked with reviewing and advising on teaching at Cambridge. The committees were set up to enquire into 'the state, discipline, studies and revenue of our University of Cambridge' including finding out and evaluating the merits of what was taught and how.[17] Herschel and his fellow committee members discussed the current situation with regard to professorships, private tuition and subjects offered by the University for Examination. They interviewed members of the colleges about precisely what took place:

> What is the mode of instruction pursued within your college? What amount of oral examination is given at the college lectures? [...] What are the elementary treatise on subjects of mathematics given or applied which have during the last years been published by members of the University not being fellows of the college chiefly for the use of students and of them what are read in the University?[18]

They also presented recommendations: 'The system of teaching the physical science should not remain barren of application', and 'the importance of experimental and illustrative lectures' should not be underestimated. Cambridge, with the help of John Herschel, was changing, transforming itself from a classics-led institution preparing aristocratic men for the professions into a training ground for upper middle-class professional scientists. Back home, the education the Herschels designed for their daughters similarly shifted, increasingly focusing as we have seen from Bella's lessons to focus on the application and the physical sciences and the importance of experiment.

These middle daughters, like their older sisters, were expected to marry, and so looked for husbands through whom they might be able to live interesting lives. Of these daughters, the first two did as was expected of them, marrying men of a similar social standing to their own family, and becoming wives and mothers with no obvious outlet for intellectual occupation beyond that of teacher to their children. They did not marry scientific men as such, but rather a neighbour and a diplomat. Maria Sophia married a neighbour, Henry Hardcastle (a member of the same family her brother William also married into). Curiously, the correspondence between Maria and Henry's fathers leading up to their marriage is all about money. They begin by establishing that the marriage is not in any way an arrangement influenced by the wishes of their parents, as Maria's father John says, 'for our young people seem to have settled the matter so decisively between themselves, and so very much to their own satisfaction'. He then goes on to discuss money, stating that 'I cannot make any immediate provision for her. At my death she will come into a portion'.[19] Henry's father, Joseph replies, 'I propose to settle £500 a year upon Henry and his future wife, by a bond which I shall put into your hands, and I propose further to secure £10,000 to be put in trust at my death for them'.[20] This was a local family, who they were friendly with, and although this might not have been a magnificent opportunity for their daughter to utilise her carefully honed intellect, it was a marriage that both families agreed would make their children happy. Amelia meanwhile married a diplomat, Thomas-Francis Wade, a British diplomat and expert in Mandarin-English translation, and this made her Lady Wade. They travelled extensively, visiting Amelia's various family members stationed in India and living for a time in China.

The next three Herschel daughters, Julia, Matilda Rose and Francisca, led slightly more scientific lives. Julia—whose godmother and namesake was the photographer Julia Margaret Cameron—like her sisters before her was educated entirely at home. She was taught by her parents, her sisters and a succession of governesses, the distribution of teaching responsibilities changing depending on the skills of each newly arrived employee. As she got older, she in turn took over teaching her younger siblings. As her youngest sister Constance would later describe, reflecting back on her education of the 1860s (when Julia would have been in her early twenties):

> One of each of my three sisters (who were still at home) took the morning lesson for two days in the week. Isabella taught me geography, Julia history and literature and Rose drawing, (*Newsletter of the Japan Herschel Society* 1993–1994)

For some of the daughters, staying in the family home, teaching younger relatives and archiving the family archive was sufficient. This was the life favoured by Isabella and Francisca, but Julia had other plans. In 1876, when Julia was thirty-four years old, she announced that she planned to marry Jack Maclear, the son of the astronomer couple the Herschels had been so friendly with in South Africa in the 1830s. Margaret Herschel was not happy with the match. She encouraged Julia to put off marrying him until he had accumulated what she considered to be a reasonable 'nest egg'. She complained to John that Maclear did not seem interested in becoming richer. Julia's attraction to Maclear was in part, so her mother told her brother, because: '[S]he is intensely interested in all the scientific results of this "Challenger" cruise & is wonderfully neat handed in collections &c &c so that he has indeed got a treasure'.[21] Her mother saw this is as a failing, regarding Jack Maclear as Julia's intellectual inferior. For Julia, it was an opportunity to become involved in a major scientific research project.

The Challenger 'cruise' of which Margaret refers to here was the expedition of HMS Challenger that set out to explore the oceans and is now generally considered to be the founding expedition of the discipline of oceanography. Jack Maclear was commander, under the ship's captain Sir George Nares, and part of a crew of around 240 men, who set sail from Portsmouth in 1872 to systematically collect specimens from all the oceans of the world. They collected specimens of water, of sea creatures and of the ocean floor and their aim was to study them, catalogue them, draw them and send them back to England for further study. Findings from this expedition are still used today to help marine biologists and oceanographers study changes in the oceans over long periods of time.

As Margaret told John (Jr), '[Jack's] intellectual capacity is, I imagine bounded by his professional acquirements & the scientific branch in which, it is generally admitted, he is a careful observer'.[22] In other words, Jack would do better out of the marriage than her daughter. After initially listening to her mother's reservations (Julia broke off the engagement in 1877 and travelled to India to visit her sister and two brothers), Julia married Maclear in 1878. Through her marriage to Jack Maclear, Julia had found a way to participate in contemporary science, albeit without having her contribution documented or acknowledged.

Julia's story tells us one way in which this generation of upper middle-class nineteenth-century women saw were able to participate in science. Her mother may not have understood it, and her contribution may be hard to uncover and entirely overlooked by generations of historians but that is not because she did not pursue participation in science, only that she did not prioritise recognition for that involvement. Her younger sisters Matilda Rose, Francisca and Constance meanwhile found other ways to engage with the scientific world of their age.

THE BEGINNINGS OF ACCESS TO THE SCIENTIFIC WORLD WITHIN THE UNIVERSITY OF CAMBRIDGE

Constance Herschel was sixteen when her father died. Matilda Rose was by then twenty-seven years old, Francisca was twenty-five. While John had been away in London—fulfilling his role as Master of the Mint—when his older children were small, he was back at home by the time Constance's education began. As a child, Constance had been taught maths and Latin by her father, and he had encouraged her to take a scientific interest in the natural world around her. Her sisters, as we have seen, taught her other subjects: geography, history, literature and drawing. (Sir William Herschel's Biographer—Lady Constance Lubbock 1994) Her mother focused on her religious education, while governesses taught her European languages, elementary mathematics and other subjects depending on their interests and expertise. When John died in 1871, Constance and her mother had to make a decision about her education. Girton College had opened in 1869 becoming the first women's college at the University of Cambridge and it was decided that she would go there. To ensure she was ready, she was sent first to London to stay with her sister and attend lectures at University College, Gower Street, which had already started to admit women a year earlier in 1868 (*Newsletter of the Japan Herschel Society* 1993–1994). Then, in October 1874 Constance began her studies in Physiology, Comparative Anatomy and Chemistry at Cambridge.

While at Cambridge, Constance met a scholar called Henry Melvill Gwatkin, giving her sister Matilda a way into scientific research without the necessity of marriage. Matilda was at this time in India, staying with her brother when Constance wrote with a suggestion for her:

> There is a gentleman her (a Mr Gwatkin – such a nice man!) who is collecting snails tongues to classify them. He has got almost all possible British ones but he is very anxious to get some outlandish ones. If you ever come across any shell beats land, freshwater or marine which look odd it really would be a good act to put them in a bottle of spirits – mind it is the tongue he wants not the shell (thought that is important too to know the beast by) so empty shells are no good & the creature must be enspirited while fresh. Do go in for snails – it will be a novel occupation I am sure, and among Indian univalves there is an unexplored field open to you.[23]

Whether or not Matilda did ever go in for snails is unclear, but what this exchange shows is the shift in scientific power that came with entry to a prestigious institution. Constance was now part of this institution. She was working and dining alongside these various men who took the importance of their thoughts and actions in scientific research for granted. She was not the wife of someone who was interested in snails' tongues, dutifully taking notes, assisting and easing social networks, she was a fellow scholar and, in that role, could insert herself into the process, offering roles to potential collaborators.

Curiously, their father and their brothers had been working like this for years, using the status and connections they acquired at Cambridge to help out the other boys in the family, allowing them to become involved in and contribute to contemporary science. John had used his Cambridge connections to find Alexander work after graduation, first at the London school of Mines and later at Anderson College in Newcastle. He had put researchers in touch with his son John too in India, suggesting he was well placed to take eclipse observations for them from that part of the world. Alexander meanwhile had encouraged his brothers to contribute to his scientific research, sending him samples of meteorites from India and sending them books to support their research in return.

Matilda's career as a researcher on snail tongues—if she ever took up the offer—was short-lived. Unlike her brothers, there was little possibility of any work she produced being published under her name or leading to any form of recognition. Instead, she married a friend of her brother's, fellow civil servant in India, William Waterfield, remaining in India until he retired and then returning to England. Constance similarly began independently. She took her final exams at Cambridge in 1878 and then remained in college as a technician, helping the students that followed her pursue their research. Back home she was treated as one of the family's scientists, arguing with Alexander about how best to organise the family laboratory, but to the outside world her options were still very limited. She had a degree, though not formally. Cambridge would allow women to study and take exams from 1869 but it would be another nearly 80 years before they would be issued with actual degrees. She was still forbidden from joining most professional societies and it was still extremely rare for women to be published in scientific journals. She could pursue science, but as an independent female researcher there would be no credit or recognition for her work. There would be no professional position for her in any institution. There would be no way for her work to contribute to any wider discussion within the scientific community.

After a few years as a technician at Cambridge Constance decided to move on, transferring her interests to areas in which she would be able to make more of an impact. She married her old childhood friend and neighbour, Neville Lubbock, now an archaeologist and travelled the world with him sharing in his work. After he died, she like many of her aging siblings returned home and there she worked with her sisters Isabella and Francisca, both of whom had rejected marriage in favour of archiving and preserving their family papers, to study her family history. In 1933, she published *The Herschel Chronicles*, still regarded as an essential starting point for any researcher looking to better understand the life and work of William and Caroline Herschel (Lubbock 1933).

Unexpectedly, it was Francisca who would become the only woman in this generation to gain recognition for her work from a professional society. Where Caroline two generations before had been given 'honorary' fellowship of the Royal Astronomical Society (RAS) alongside Mary Somerville when the society grudgingly agreed to acknowledge their work in 1828, Francisca was awarded full fellowship of the same society in 1916. It is unclear why she was chosen, but she was. She was listed alongside ten other women 'balloted for and duly elected' to become the first female fellows of the society in 1916 (Bowler 2016, pp. 4, 18). Sue Bowler (2016) editor of the Royal Astronomical Society's current journal, *Astronomy and Geophysics*, suggests that:

> While little remains in the academic record of the work of these women Fellows, the fact of their election to the Society in itself demonstrates that their work met the standards of other RAS Fellows of their time. They are thinkers, observers, teachers, writers and curators – just as their fellow (male) Fellows were.

Just as plausibly however, these women may simply have been chosen because they each had sufficient sympathetic friends within the society willing to put their names forward for ballot. As Bowler points out, without the remaining records we may never know.

CONCLUSIONS

The women within the Herschel family who came after Caroline appear, on the surface to have had more opportunities to pursue science than her, yet to have achieved much less, but is that fair? Much of how we conclude this question depends on how we see success and how we measure achievement. Margaret was successfully trained to become a companionate wife, sufficiently well-educated that she could converse with her husband and design a thorough and rigorous education for her sons. She and John then set out to train their daughters to become the female equivalent of a gentlemen amateur, the pinnacle of scientific respectability in the early nineteenth century. Sadly however, by the time their education was complete, times had moved on. Now science was in the hands of the professionals. Just as eighteenth-century women struggled to gain the education needed to participate in science, and were then prevented from getting credit for any work that they did thanks to their lack of access to institutions and publications, now nineteenth century women were being prevented from participating in science for even more formal reasons. Where once they could participate in science but not get credit—when that science was in the home, and publication the domain only of men—now they were not even able to participate as science moved out of the home and into the male-only space of the laboratory. The Herschels' had trained their daughters to become 'grand amateurs' (Chapman 1998), once the most respected scientific position, only to find the scientific world had changed their criteria.

What their story tells us about women in science in the nineteenth century is that although science was moving out of the home, women were still finding ways to participate in science. There was field work (such as the collection of snails' tongues) and the organisation of collected samples (Julia's work on the Challenger expedition) that could spill over out of the laboratory and into the home. But there is too another story that should be told, and that is why women who grew up enjoying science, and seeking out opportunities to study science—women such as Constance—would choose to give it up. What was it about that journey that was so off putting as to make an entirely unknown new path seem like an attractive and preferable alternative?

When writing the story of women in science, those women who battled against the odds and found ways of participating that allow us with hindsight to fit their achievements into a 'great men of science' narrative are always fascinating. However, to truly understand the history of women in science, those who worked unnoticed, who didn't insist on getting credit, those who gave up, and those nearly made it must be studied too to truly understand why women were—and still are—not better represented in the sciences.

Notes

1. See, for instance 'Account of the Discovery of a New Comet'.
2. Margaret Brodie Herschel to Maria Edgeworth, 'April 1841'. JHS John Herschel-Shorland's private collection (JHS papers), BFR 5.
3. Caroline Herschel, Notebook 'Autobiography begun in 1840', British Library (BL): microfilm M/588 (4).
4. Caroline Herschel to Margaret Brodie Stewart, Hannover, April 5 1840. BL, Eg.3762 55–56.
5. Margaret Brodie Herschel to Charles Stewart, 28 January 1840. JHS papers, BFR 7.
6. Margaret Brodie Herschel to Maria Edgeworth, 16 November [1843]. JHS papers, BFR 5.
7. Margaret Louisa Herschel's diary, JHS papers, ARM.
8. Margaret Brodie Herschel to Isabella Herschel, 25 April 1852. JHS papers, BFR 6.
9. Margaret Brodie Herschel to John F. W. Herschel, undated. JHS papers, BFR 6.
10. John F. W. Herschel to Margaret Brodie Herschel, 25 August 1852. Herschel Collection at the Harry Ransom Centre, University of Texas, folder L0543.14.
11. Margaret Brodie Herschel to Maria Edgeworth, 15 April 1841. JHS papers, BFR 5.
12. Margaret Brodie Herschel to John F. W. Herschel, 9 February [1854]. JHS papers, BFR 6.
13. Isabella Herschel to Margaret Louisa Herschel, c. 1854. JHS papers, BFR 6.
14. Maria Sophia Herschel to John F. W. Herschel, 4 March [1854]. JHS papers, BFR 6.
15. Isabella Herschel to John F. W. Herschel, undated [1854]. JHS papers, BFR 6.
16. Isabella Herschel to John F. W. Herschel, undated [1854]. JHS papers, BFR 6.
17. Papers relating to the Cambridge Commission. 1854–1855. BL: microfilm M/588(7).
18. Papers relating to the Cambridge Commission. 1854–1855. BL: microfilm M/588(7).
19. John Herschel to Joseph Alfred Hardcastle, 28 July 1865. Quoted in Sheppard (2020).
20. Joseph Alfred Hardcastle to John Herschel, 3 August 1865. Quoted in Sheppard (2020).
21. Margaret Brodie Herschel to John Herschel (Jr), 1 June 1876. JHS papers, TEU.
22. Margaret Brodie Herschel to John Herschel (Jr), 1 June 1876. JHS papers, TEU.

23. Constance Herschel to Matilda Rose Herschel, 3 February 1881. Archives of Diana Ladas (photocopies with Lubbock descendants).

References

Arnold, Matthew. 1869. *Culture and Anarchy: An Essay in Political and Social Criticism*. London: John Murray.
Beaton, Rev. Donald. 1985. *Some Noted Ministers of the Northern Highlands*. Glasgow: Free Presbyterian Publications.
Bowler, Sue. 2016. "Observers, writers, teachers..." *Astronomy & Geophysics* 57, no. 4: 4.18–4.19.
Caine, Barbara. 2005. *Bombay to Bloomsbury: A Biography of the Strachey Family*. Oxford: Oxford University Press.
Chapman, Allan. 1998. *The Victorian Amateur Astronomer*. Chichester: Wiley.
Cohen, Michele. 2006. "'A little learning'? The curriculum and the construction of gender difference in the long eighteenth century." *British Journal for Eighteenth-Century Studies* 29: 321–335.
Davies, Emily. 1866. *The Higher Education of Women*.
Delamont, Sara. 1978. "The contradictions in ladies' education." In *The Nineteenth Century Woman: Her Culture and Physical World*, edited by Sara Delamont and Lorna Duffin, 134–163. London: Routledge.
Delamont, Sara. 1989. *Knowledgeable Women: Structuralism and the Reproduction of Elites*. London: Routledge.
Donnelly, James F. 2002. "The 'humanist' critique of the place of science in the curriculum in the nineteenth century, and its continuing legacy." *History of Education* 31, no. 6: 535–555.
Edgeworth, Maria, and Richard Lovell Edgeworth. 1798. *Practical Education*. London: Printed for J. Johnson.
Fyfe, Aileen. 2004. *Science and Salvation: Evangelical Popular Science Publishing in Victorian Britain*. Chicago: University of Chicago Press.
Herschel, Caroline. 1796. "Account of the Discovery of a New Comet. By Miss Caroline Herschel. In a Letter to Sir Joseph Banks, Bart. K. B. P. R. S." *Philosophical Transactions of the Royal Society of London* 86: 131–134.
Herschel, Caroline. 1798. *Catalogue of Stars: Taken from Mr Flamsteed's Observations*. London: Peter Elmsly.
Hughes, Kathryn. 1993. *The Victorian Governess*. London: Bloomsbury.
Huxley, Thomas Henry. 1866. *Science and Culture*. London: Macmillan.
Jordan, Ellen. 1991. "'Making good wives and mothers'? The transformation of middle-class girls' education in nineteenth-century Britain." *History of Education Quarterly* 3, no. 4: 439-62.
Lubbock, Constance. 1933. *The Herschel Chronicles*. Cambridge: Cambridge University Press.
Mosley, Charles, ed. 1999. *Burke's Peerage and Baronetage*. 106th ed. London: Fitzroy Dearborn.
Peterson, M. Jeanne. 1989. *Family, Love and Work in the Lives of Victorian Gentlewomen*. Bloomington: Indiana University Press.

"Portrait: A Note by Lady Lubbock to Her Children." 1993–1994. *Newsletter of the Japan Herschel Society.* https://www.ne.jp/asahi/mononoke/ttnd/herschel/n-text/newsindex-e2.html. Accessed on 9 March 2021.

Poullain de la Barre, François. 1673. *De l'Égalité des deux sexes.* Paris: Jean Du Puis.

Rousseau, Jean-Jacques. 1762. *Emile ou De l'Education.* Amsterdam: Jean Néaulme.

Schiebinger, Londa. 1989. *The Mind Has No Sex?* Cambridge, MA: Harvard University Press.

Sheppard, Martin, ed. 2020. *Diaries and Letters of Joseph Alfred Hardcastle MP.* https://www.pakenham-village.co.uk/History/josephAlfredHardcastleAndNetherHall/josephAlfredHardcastleAndNetherHall.htm. Accessed on 28 January 2021.

"Sir William Herschel's Biographer—Lady Constance Lubbock." 1994. *Bulletin of the William Herschel Society* 45: 7–8.

Somerville, Mary. 1831. *Mechanisms of the Heavens.* London: John Murray.

Somerville, Mary, and Dorothy McMillan. 2010. *Queen of Science: Personal Recollections of Mary Somerville.* Edinburgh: Canongate Books.

Warner, Brian, ed. 1991. *Lady Herschel: Letters from the Cape 1834-1838.* Cape Town: Publication of the Friends of the South African Library.

Wollstonecraft, Mary. 1792. *A Vindication of the Rights of Woman.* London: Printed for J. Johnson.

CHAPTER 13

Domestic Astronomy in the Seventeenth and Eighteenth Centuries

Gabriella Bernardi

As the focus of this volume reflects, the beginnings of modern science—in Europe at least—is generally taken to be the period from around 1660 onward. Since then, history of science has investigated not only the development of the scientific method, but also the evolution of the scientific institutions which became so important as professional networks for men of science. That year, 1660, saw the founding of one of the oldest and most prestigious of scientific institutions—the Royal Society of London—an influential organization which is still, of course, active today. In common with nearly all such élite scientific institutions, women were not admitted as fellows for most of its history with the first two women being elected in 1945.[1] This does not imply, however, that women were not involved in science throughout this time, and this is true especially for astronomy which is among the most ancient of scientific disciplines. Indeed, history testifies to the presence of women active in astronomy since antiquity, and even reveals women participating in formal astronomical organizations and networks. This chapter uses examples of scientific women active in astronomy to understand how a woman could contribute to the scientific endeavour, and to describe the difficulties that she had to overcome to do so. These biographic case studies, then, recover certain lesser-known women of science and illustrate the importance of family networks and domestic settings in facilitating women's access to science. These examples of

G. Bernardi (✉)
INAF-Astrophysical Observatory of Turin, Turin, Italy

women astronomers also provide an insight into why, and how, history has tended to construct science erroneously as an almost exclusively masculine project.

ELISABETHA KOOPMAN HEVELIUS: STELLAR CARTOGRAPHER

One way that women negotiated access to science was through collaboration with a husband or male relative. An early example of a collaborative marriage is that of the astronomer Johannes Hevelius (1611–1687) and Elisabetha Koopmann Helvelius (1647–1693). Elisabetha, Helvelius's second wife and thirty-six years younger than her husband, was reputedly praised by the French mathematician and astronomer, François Arago (1786–1853) as being the first woman who, to his knowledge, was unafraid to face the demands of making astronomical observations and calculations. Elisabetha worked with her husband, a wealthy merchant in Danzig, to run an astronomical observatory and take observations. Her passion for astronomy is witnessed by her supposed 'marriage proposal' to her future husband, made one night while the couple were doing astronomical observations: 'To remain and gaze here always, to be allowed to explore and proclaim with you the wonder of the heavens; that would make me perfectly happy!' (O'Connor and Robertson 2008).

Elisabetha was Polish; her father, Nicholas Koopmann, was a prosperous local merchant in Danzig, a largely German-speaking city that was part of Poland at the time. She had dabbled in astronomy since childhood. At the age of sixteen, Elisabetha became the second wife of Johannes Hevelius. Their vast age difference was not uncommon at the time, but this couple was peculiar because they shared the same passion for astronomy. Their common interest seems to have made their marriage a happy one, despite the notable age difference. Moreover, women could take advantage from such unions because they granted them access to creative and intellectual pursuits through what could be termed a kind of 'conjugal apprenticeship'—such learning was otherwise obstructed by a society in which female education was almost completely disregarded. Elisabetha gave birth to a son and three daughters, but this did not stop her from becoming responsible for the couple's private observatory and serving as assistant to the many astronomers who came to visit. The Observatory's structure comprised three adjacent houses with each of their roofs connected to create a viewing facility, one of the finest and best equipped observatories in Europe: the *Stellaeburgum* or 'village of stars'.

The couple's close collaboration is perfectly depicted in a famous image from one of their publications, *Machinae Coelestis* (1673, p. 222). In this engraving, Elisabetha and Johannes are depicted in the act of making an observation with a large brass sextant (Fig. 13.1).

This image witnesses the preference of astronomers, at the dawn of the telescope era, for observations made with the naked eye. Nonetheless, Johannes built the most powerful telescope of the time; this telescope, which was

Fig. 13.1 Johannes Hevelius and wife Elisabetha observing the sky with a brass sextant. Engraving from *Machinae Coelestis: Pars Prior* (1673) by Johannes Hevelius. Library of Congress: Public Domain

constructed without a tube, was 150 feet (forty-six metres) long and used for the study of the surfaces of the Moon and planets. Elisabetha was clearly able to perform calculations and handle the complex instrumentation on the Observatory roof. In addition, she knew Latin better than her husband as is evidenced from the letters she sent to other scientists, although it is not clear where and when Elisabetha learned this language as the study of Latin was unusual for girls. However, this skill was useful to Johannes Hevelius, especially after 1664 when he became a fellow of the Royal Society of London, as Elisabetha helped him to keep in contact with other European astronomers by writing his letters in Latin.

Johannes Hevelius published his first star map in 1673, in the first volume of *Machinae Coelestis*. In this book Johannes describes his wife as his most valuable scientific collaborator, with great skills both as an observer and in mathematical work (Cook 2000). Elisabetha's considerable degree of scientific autonomy is proven by her continuation of this activity after the death of her husband in 1687. In the next three years, she published three books—discussed below—that originated from the completion of research initiated with her husband. The couple's lifelong work, which also included an improvement of Kepler's planetary ephemerides (calculated positions of celestial objects over time) was put at serious risk on 27 September 1679 when the Observatory was accidently set on fire. Not only were many fine brass instruments destroyed, but also astronomical data and most of the copies of *Machinae Coelestis*. In addition, the couple's private library and typographic records were almost completely lost. Fortunately, their research was ultimately preserved thanks to the miraculous salvation of a small leather-bound notebook containing the results of thousands of calculations from their observations, and to the help of some powerful patrons, including the king of France, Louis XIV, and of Poland, John III. This allowed the couple to rebuild the Observatory and recommence their programme of astronomical observation and research.

This major astronomical project was eventually to be completed by Elisabetha who, as indicated above, continued their joint research alone and published the results three years after the death of her husband. This research is composed of three separate works: *Prodromus Astronomiae* (1690) which outlines the methodology and technology used in creating the star catalog and provides examples of the use of sextant and quadrant; an atlas of constellations named *Firmamentum Sobiescianum sive Uranographia* (1690); and the *Catalogus Stellarum* (1687). The *Catalogus* is the most extensive stellar catalog obtained without the use of the telescope still in existence and contains the exact location of around 1564 stars. The *Firmamentum* takes its name from its dedication to the king of Poland, John III Sobieski, and comprises details of seventy-three constellations.

Elisabetha Hevelius died on 23 December 1693 at the age of forty-six. She left a complete set of the couple's published works to each of their three daughters. One of them, Catharina, received a special edition of the *Catalogus Stellarum* which had originally been planned as a gift for the king of France, Louis XIV. This was probably a sign of gratitude to her as the person who, at the age of thirteen, had saved the manuscript of the fixed-star catalog from the disastrous fire of 1679.

Maria Margarethe Winkelmann-Kirch and the Kirchin Dynasty

In contrast to Elisabetha Hevelius—who practiced astronomy in the private, home observatory that she ran with her husband—Maria Margarethe Winkelmann-Kirch worked within a setting that was not completely domestic, and which can be understood as halfway to an official scientific institution. This astronomical observatory, which was created on the instruction and with the support of the Berlin Academy of Science, was managed along business lines and featured contributions from Maria and other female family members. However, Maria became known especially for her scientific achievements. In 1709, the philosopher and mathematician Gottfried Wilhelm Leibniz, president of the Berlin Academy of Science, presented Maria to the Prussian court with these words:

> There is [in Berlin] a most learned woman who could pass as a rarity. Her achievement is not in literature or rhetoric but in the most profound doctrines of astronomy [...] I do not believe that this woman easily finds her equal in the science in which she excels [...] She favors the Copernican system (the idea that the sun is at rest) like all the learned astronomers of our time. And it is a pleasure to hear her defend that system through the Holy Scripture in which she is also very learned. She observes with the best observers, she knows how to handle marvelously the quadrant and the telescope. (Helly and Reverby 1992, p. 63)

Maria Margarethe Winkelmann-Kirch was the wife of the astronomer and mathematician Gottfried Kirch (1639–1710) who was one of the most famous German astronomers of his time. Like Johannes Hevelius, he supervised the studies of his three sisters, and received great help from his second wife, Maria, who had already benefitted from an astronomical education before her marriage. As was the case with Elisabetha, Maria's learning cannot be attributed to any enlightened attitude towards female education, but rather to a favourable familial environment. She was educated by her father, a Lutheran pastor, who believed in equal education for both sexes; when both her parents died when she was still a child, her uncle took over her education and introduced her to astronomy. Soon Maria became assistant to Christoph Arnold Sommerfeld, known as the 'peasant astronomer', who inspired her interest in astronomy even further. It is probably thanks to becoming part of Sommerfeld's astronomical circles that Maria gained opportunity to meet her future husband, Gottfried Kirch (who had worked for a short time, in 1674, at the Hevelius observatory). Maria and her husband lived in Saxony, moving in 1700 to Berlin where Kirch accepted tenure as astronomer. They had seven children, including daughters, all of whom were tutored from the earliest age to join the family business of astronomy. This was not unusual; Londa Schiebinger has noted that between 1650 and 1720 women constituted around fourteen per cent of German astronomers and attributes this to the strength of the artisan, craft tradition in that country (Schiebinger 1987, p. 177).

In Berlin, Maria and Gottfried worked at the production of almanacs, including books of observations and calendars which played an important role in astronomy, business, and society at this time. In that period, German protestant states introduced a calendar identical to the Catholic Gregorian calendar except for the date of Easter. This event had enough political importance to justify a specific 'calendar tax' to fund the work of qualified astronomers, and although the official appointment and recipient of this was Gottfried, the astronomical work was assigned to the care of his wife. Another important achievement of Maria's was the discovery of a comet on 21 March 1702. Maria and her husband alternated observing shifts with each other, giving her the opportunity to become the first woman documented to have discovered a comet. Comets were objects of utmost importance then and their discovery could change the life of an astronomer. Unfortunately, however, custom at the time was to incorporate all the Observatory's findings under the authorship of its male head, Gottfried, therefore the official assignment to Maria happened only a few years later. A similar fate happened to Maria's other works, except for some treaties concerning the observation of the northern polar lights of 1707, a short text on the conjunction of the Sun to Saturn and Venus of 1709, and the forecast of the conjunction of Jupiter and Saturn in 1712, all of which were published under her own name (Alic 1986, p. 121). Maria's work on the calendars, however, was published under the name of her husband and, later, under the name of their son Christfield (1694–1740) who was appointed astronomer at the Berlin Observatory in 1716 (Ogilvie 1991, p. 110). When her husband died in 1710, Maria asked the Royal Berlin Academy of Sciences if she, with her son, could continue to produce the calendars, as she had been doing during Gottfried's illness. She was quickly refused, however. As the secretary of the Academy, Johann Jablonski, noted in a letter to Leibniz:

> That she be kept on in an official capacity to work on the calendar or to continue with observations simply will not do. Already during her husband's lifetime the society was burdened with ridicule because its calendar was prepared by a woman. If she were now to be kept on in such a capacity mouths would gape even wider. (Schiebinger 1987, p. 187)

This quotation is significant in the stress laid on a woman inhabiting an 'official' role. Women may practice science, but they did so in a private, amateur capacity. To pretend to be an official, or public position was to contravene the norms of womanhood.

This negative opinion of women and their formal participation in astronomy, shared among everybody in the Academy except for Leibniz, prevented Maria from gaining an official appointment and the economic independence that would have come with it. She was undoubtedly disappointed and aware of the injustice. Around the same time, she was moved to write in the foreword of one of her publications that a woman could become 'as skilled as a man at observing and understanding the skies' (Helly and Reverby 1992, p. 65). Two years later, in 1712, Maria moved to the private observatory of Baron Bernhard Friedrich von Krosigk, where she continued the training of her children. Later, in 1714, she moved to Danzig for two years, finally returning to Berlin in 1716 when her son Christfried took up an appointment at the Berlin Academy Observatory. Once again, Maria's role was that of assistant, this time to her son; she resumed the actual, but unofficial, preparation of the calendars together with her two daughters—although she refused to stay in the shadows when guests visited the Observatory. She died after few years in Berlin, at the age of fifty.

Maria's work was carried on by her two daughters, Christine (1696–1782) and Margaretha (1703–1744), both of whom continued their astronomical observations after the death of their mother. The daughters' main activities were calculations for planetary ephemeris and for the compilation of annual calendars (almanacs); yet again, they had to carry out their work in the shadow of their brother Christfried who was director of the Berlin Academy Observatory. These women produced the *Almanac* and *Ephemeris* for the Academy—important publications which were 'among the sources of revenue of this learned body' (Mozens 1999, p. 174). Eventually Christine, in 1776 and at the end of her career, received official recognition for her professional contributions to astronomy and was awarded a salary of 400 thalers, the German silver coin, from the Academy (Bernardi 2016, p.105). She continued working until late in life, thus becoming the first female astronomer paid for her professional activity, a record that until recently was attributed to Caroline Herschel.

Maddalena and Teresa Manfredi: The Ephemeris Calculators

Female astronomers working in southern Europe shared similar educations and backgrounds to the women discussed so far and this supported their access to astronomy. These women's scientific learning can also be traced back to a favourable family environment in a society in which teaching girls academically, especially scientifically, was not usual. Despite some extremely rare cases from Italian universities in the seventeenth and eighteenth centuries, across southern and northern Europe, women were very rarely given an academic education.

In the *Introductio in Ephemerides* (*Introduction to Ephemerides*), produced by the Italian astronomer Eustachio Manfredi and preserved in Bologna, we can read at the end of the manuscript the sentence:

> I started the ephemeris on December 1712 in Bologna. With many interruptions they continued in the following years with the help of my two sisters Maddalena and Teresa, and of Mr. Giuseppe Nadi, and yet some more from Mr. Cesare

> Parisij [...] The table of longitudes and latitudes was calculated by my sister Maddalena in 1702 or 1703. (Manfredi 1750)

These lines reveal again the practice of the family guild or workshop in action. Although the two Italian sisters Maddalena (1673–1744) and Teresa (1679–1767) never signed their works, their contribution was significant. The women's co-operation in the family enterprise consisted in astronomical observations and mathematical calculations for the Ephemerides of their brother, the astronomer Eustachio.

The Manfredi family was classed among the petty bourgeoisie of Bologna. The brothers went to the University of Bologna and one of them, Eustachio, became first a public reader there and then, in 1699, professor of mathematics. The sisters were educated in a convent of tertiary nuns, but they also learned Latin, mathematics and astronomy within the family and from the circle of friends who frequented their house. This was possibly due to the *Accademia degli Inquieti*—literally 'Academy of the Restless'—which Eustachio founded and devoted to literature and the experimental sciences. Maddalena and Teresa applied their physical and mental strength to the administration of the family business, to their scientific collaboration, and to the production of literary works for the market of the educated bourgeoisie of Bologna. At the beginning of 1700, they all moved into the palace of Earl Luigi Ferdinando Marsili, which housed an astronomical dome, and here both brother and sisters engaged in helping with astronomical calculations. This latter collaboration is witnessed directly by Eustachio Manfredi himself in a note at the end of the introduction to *Effemeridi dei moti celesti* (*Ephemerides of celestial motion*, 1715). Here, he states that the table of longitudes and latitudes had been computed by Maddalena in around 1702 or 1703. When Eustachio was appointed astronomer of the Academy and moved to the new headquarters of Via Zamboni, his sisters followed him and, while remaining purposely in the shadows of their brother, they gained some fame thanks to the literature they wrote. According to the Italian writer Ilaria Magnani Campanacci, the lives of the two sisters Manfredi were representative of a female genius which acted as a 'complement', rather than an 'alternative', to the commitments of a woman; they worked not with a competitive attitude, but in close and supportive collaboration with the more recognized intellectual figure of their academic brother (Campanacci 1988).

As indicated above, the realization of *Ephemerides of celestial motion* was made by Eustachio Manfredi with the important assistance of his sisters Teresa and Maddalena, but it is possible that a third sister, Agnese Manfredi, also collaborated with this work. The project required an observational part, carried out with the use of a five-metre-long telescope and a sextant with telescopic sights, as well as a computational part. The calculations were quite long and tedious but around then, for the first time, techniques were being developed that made them accessible to non-specialists such as Agnese, who could then be employed to perform the calculations. The exceptional quality of *Ephemerides of celestial motion* is proven by the regard it enjoyed for decades among the manufacturers of all Europe; it was acknowledged as the most extensive and complete tool for the astronomical characterization of astronomical sites, thus greatly contributing to the reputation of the local university. According to the writer and historian Giovanni Fantuzzi, the Manfredi sisters acquired a great knowledge of tables and astronomical calculations and those included in the *Ephemerides* were, if not all, 'mostly due to the diligence and the study of these two calculators' (Fantuzzi 1786, vol. 5, p. 188). The term 'calculator' defines a specific role in astronomy and was also used later in other scientific and business disciplines. Before computing machines became available, the term 'calculator' identified people, frequently women of the family of an astronomer, who performed mathematical calculations. This task was long and tedious but, not unlike today, it often required a good deal of creativity to devise the most efficient method. Today, we would call these techniques an 'algorithm' designed to perform calculations at the required accuracy.

Maria Clara Eimmart: Images of the Sky Before Photography

Another type of mechanical aid that was missing in this era was photography. The only method astronomers had to record images of their observations was to draw them and, in some cases, it required a special ability that was beyond the reach of the average scientist. This is another field in which women found their way into astronomy; one of the most distinguished representatives in this area is Maria Clara Eimmart (1676–1707).

Comets, lunar phases, phases of Mercury, Venus, Mars, Jupiter, and Saturn—these are the subjects depicted by Maria Clara Eimmart in her short life. She was an astronomer and engraver born in 1676 at Nuremberg, the largest centre of astronomical observation in Germany. She worked on the roof of her father's house and became one of the first and most talented designers of astronomical tables. As common to other scientific fields, appendices with illustrations accompanied the astronomical treatises of the time. For example, the work *Ichonographia nova contemplationum de sole in desolatis antiquorum philosophorum ruderibus concepta* (New Contemplation of the Sun), published in 1701 and authored by Clara's father, Georg Christoph Eimmart, includes illustrations by his daughter. It is sometimes argued that this work was entirely produced by Maria Clara yet published under her father's name; however, there is no evidence to support this claim. Nonetheless, the scientific purpose of these drawings is evident not only from their accuracy, but also from their accompanying details which include the day of observation.

Clara's father was a successful painter and an amateur astronomer. Since his main business was profitable, Georg was able to buy several astronomical instruments and build a private observatory, which later became the Astronomical Observatory of Nuremberg. Maria Clara was his assistant at their observatory, which soon grew into a meeting point for several astronomers including her future husband, Johann Heinrich Müller, whom she married in 1706. Müller was one of her father's first students; he later became a professor of physics at the high school in Nuremberg and, finally, professor of astronomy at Altorf. In this environment, Maria Clara studied Latin, French and mathematics, specializing in botanical and astronomical illustration, becoming one of the first, and most talented and prolific, authors of astronomical drawings. For example, between 1693 and 1698, she produced 350 illustrations of the phases of the Moon, as well as several images depicting different comets and planets including Mercury, Venus, Mars, Jupiter, and Saturn. The total number is not known as some of these images have been lost, including those illustrating the total eclipse observed from Nuremberg in 1706. Despite this, her illustrations were highly regarded all over Europe. Some of them, drawn with pastels on blue sheets, were given as a gift to a scientific collaborator of her father, the same Earl Marsili who hosted the Manfredi family in his palace; these illustrations are still preserved in Bologna, at the Observatory's Museum. After the death of her father in 1705, Clara inherited the family observatory which, according to the law at the time, passed to her husband upon their marriage. It was then purchased by the City of Nuremberg and her husband was appointed as the new director. Unfortunately, Maria Clara died young in childbirth, in 1707.

Marie-Jeanne Amélie Harlay: Unofficial Assistant Astronomer

Another case with parallels to the 'Kirchin dynasty', but even more closely connected with academia, is that of the Frenchwoman Marie-Jeanne Amélie Harlay Lefrancais de Lalande (1768–1832). Her husband was the astronomer Michel Jean Jérôme Lefrancais de Lalande and, through him, she became a niece by marriage of the famous astronomer Joseph Jérôme Lalande, director of the Astronomical Observatory of Paris between 1795 and 1801. The latter wrote in his *Bibliographie Astronomique* (1803) that his niece helped her husband in his observations and drew results from calculations. He cited her contribution also in *Astronomie des dames*, where he wrote 'She has reduced the observations of ten thousand stars and prepared a work of three hundred pages of timetables, a gigantic work for her age and her sex, which are published in my *Abrégé de navigation*' (De La Lande 1817, p. 7). In the *Bibliographie Astronomique*, her famed uncle continued with more words of appreciation, stating that Marie-Jeanne had completed the tables for finding the time at sea based on the altitude of the Sun and stars; and that the National Assembly had ordered them to be printed in June 1791 (1803, pp. 703–704).

Marie-Jeanne's astronomical contributions began when she helped her husband with his astronomical observations and performed the mathematical calculations required for their interpretation. She therefore followed a common route of women accessing science, namely in a domestic setting, via the family network and interests/business of its male members. Marie-Jeanne became so expert that she also helped the astronomer Cassini in 1791 (Jean-Dominique, Comte de Cassini, also known as Cassini IV) then director of the Paris Observatory, in his observations at the College of France. Madame de Lalande also collaborated with Joseph Jérôme Lalande on the preparation of the *Histoire Céleste Francaise* (French Celestial History), the largest and most complete star catalog of the time. This star catalog included the positions and magnitudes of more than 47,000 stars; the amount of work implied by this publication can be appreciated by understanding that, just for the computational part, at least thirty-six operations were needed for each star. This text was published in 1801 under the name of Joseph Jérôme Lalande, but Marie-Jeanne is listed among the collaborators; this and her other works made her very well known throughout Europe. The mathematical and astronomical skills of his niece probably prompted Lalande to recognize the problem of female education. He referred to this subject in *Astronomie des dames* and reflected that what women were missing 'is just the educational opportunities and examples which they can emulate; we see them standing out enough, despite the obstacles of education and prejudice, to believe that they have just as much talent as most of the men who get a reputation in science' (De La Lande 1817, p. 7).

Marie-Jeanne also dedicated herself to another astronomical task: the writing of timetables for the Navy. These tables were common at the time as they were needed by sailors who used them to determine the time at sea by calculating the altitude of the Sun and of the stars. Marie-Jeanne's tables were published in 1793 with the title *Table horaires pour la marine*. This text consisted of 300 pages contained within Joseph-Jérôme Lalande's *Abrégé de navigation*, and it earned for her as author a medal of the Lycée des Arts, which was awarded to scholars and distinguished artists. Madame Marie-Jeanne de Lalande also enjoys a curious link with the last case study of this chapter, the astronomer Caroline Herschel. This information was once again provided by Jérôme Lalande in his *Bibliographie Astronomique*. Here, speaking of Marie-Jeanne's daughter, he wrote that his great-niece was born on 20 January 1790, 'a day when we saw in Paris for the first time the comet discovered by Miss Caroline Herschel; the child was then given the name of Carolina' (De La Lande 1803, p. 697).

CAROLINE LUCRETIA HERSCHEL—THE COMET DISCOVERER

Anglo-German astronomer Caroline Lucretia Herschel (1750–1848) is arguably the best-known of the female case studies in this chapter; certainly, she is a complex figure of domestic astronomy and the subject of the chapter by Emily Winterburn elsewhere in this volume. Caroline wrote a diary that was used by her nephew, John Herschel, in 1876 to write her biography. Filled with anecdotes of life and astronomy, its pages provide an interesting insight into Caroline's experiences, education, and access to astronomy:

> My father wished to give me something like a polished education, but my mother was particularly determined that it should be a rough, but at the same time a useful one; and nothing farther she thought was necessary but to send me two or three months to a sempstress to be taught to make household linen. (Herschel 1876, p. 20)

These few lines give a clear impression that Caroline's scientific life started on an uphill path. She was the eighth of the ten children of Isaac Herschel and Anna Ilse Moritzen, born on 16 March 1750 in Hanover, then one of the German states of the Holy Roman Empire, now Germany. Her father treated his sons and daughters equally with respect to their education. He was a gardener who became a military musician (an oboist) in the Hanover military band. Despite his poor education, Isaac Herschel did his best to educate his sons and daughters, bringing them closer to his interests, namely music, philosophy, and astronomy. On the other hand, as recalled by Caroline's words, her mother did not wholly support her husband's efforts:

> My mother would not consent to my being taught French, and my brother Dietrich was even denied a dancing-master, because she would not permit my learning along with him, though the entrance had been paid for us both; so all my father could do for me was to indulge me (and please himself) sometimes with a short lesson on the violin, when my mother was either in good humour or out of the way. (Herschel 1876, p. 20)

It seems that Caroline's mother opposed an academic education for her daughters because, as women, their main use for education would be to deal with housework and home management. In the case of Caroline though, the attitude of her mother has at least one possible explanation. At the age of three Caroline contracted smallpox and this had left her left eye slightly disfigured, plus she was short of stature at just 4ft 3inches. Given these difficulties, she was unlikely to attract a husband, especially as she was also without a good dowry. It is possible that a life at home as a domestic helper was the only possible future that Anna Ilse could envision for her daughter. However, this fate was subverted when Caroline's life became tightly connected with that of her celebrated brother, Friedrich Wilhelm Herschel, better known as renowned astronomer William Herschel (1738–1822). William began his career as a musician and, in 1766, moved to England to become organist and choirmaster of the Octagon Chapel in Bath, a fashionable resort of the wealthy in the eighteenth century. He asked his sister to join him there in 1772, when Caroline was aged twenty-two years. Bath proved welcoming to both siblings and became the setting for their future astronomical careers.

Once Caroline was settled in England, William began to give her English and singing lessons—very successfully as she won some acclaim for her recitals given locally and in Birmingham. At the same time, William pursued his interest in astronomy and, as Caroline wrote in her diary, was using the garden as a place to erect a twenty-foot telescope which made use of exceptionally large mirrors; he was also taking advantage of his roomy new home to create workshops and a place on the roof for observing (Herschel 1876, p. 41). To support these activities, Caroline received geometry and spherical trigonometry lessons to equip her to perform astronomical observations, first with William, and later by herself. William's research developed to the point that he needed larger mirrors for his telescopes and, as these could not be found 'off-the-shelf', he extended his studies to include the construction of these instruments, soon becoming among the best telescope-makers of Europe. Caroline found herself spending long hours helping him in this complex task. These efforts were rewarded in 1781 when William became famous thanks to the discovery of Uranus made with his new homemade telescope (Fig. 13.2).

Initially named *Georgium sidus* or 'Georgian star', in honour of George III, King of England, it earned the astronomer a salary of 200 pounds a year from the King. Although not a huge sum, it was enough to allow him to devote himself full time to astronomy. The musical careers of William and Caroline then ceased, and she became a kind of assistant-astronomer for William, helping him both in his observations and in the manufacturing of the instruments. In this way, the family enterprise of the Herschels was in part funded by building and selling telescopes. It was a completely artisan activity, and the assembly and usage instructions included with the telescopes were not printed at the time but written accurately by hand by Caroline. Soon William gave her a telescope with which Caroline began to observe on her own, conducting systematic surveys of the sky, and in particular seeking for those comets that had so fascinated her as a child.

In 1782, the Herschels moved to Datchet, near Windsor, which provided a much larger home for their astronomical activity. Here Caroline was asked to abandon her own observing activity to help her brother. William had begun a nebula survey project and needed to stay at the eyepiece without interruption, ready to call out a description of any nebula that came into view, while Caroline sat at a desk at an open window nearby, writing down his description and recording other data of the observation. During the day, Caroline worked on the results of the previous night—a task which required long calculations and which she conducted with extraordinary accuracy. Caroline, who never learned multiplication tables because she studied them too late in her life, used a table on a sheet of paper that she kept in her pocket when she worked. However, this task left much less time for her own observations, to which she could dedicate herself only when her brother was not at home. Caroline was particularly interested in the deep sky objects that she could observe thanks to the Herschel telescopes, which were the best equipment on the market. Despite the pressures on her time which obstructed her personal observations, by the end of 1783 Caroline had discovered fourteen objects, including galaxies and open clusters, which were included in William's catalog.[2] However, Caroline became most famous for 'her' comets. As noted earlier, the discovery of a comet was highly regarded and could determine the career of an astronomer; Caroline's chance to become a comet hunter coincided with the siblings' move

Fig. 13.2 Caroline Herschel recording her brother William's observations on the night he discovered the planet Uranus. Engraving by Paul Fouché from *Astronomie populaire: description générale du ciels* by Camille Flammarion (Paris, 1890, p. 73). Public Domain

in 1786 to Slough, to a new home which they called Observatory House. Here William gave Caroline a small telescope with a focal length of about seventy centimetres which was capable of thirty magnifications. It was with this relatively small telescope that, during the night of 1 August 1786, at the age of thirty-six, Caroline discovered her first comet. This comet is classified as C/1786 P1 Herschel and sometimes called 'the first lady's comet' as coined by the novelist Frances Burney.[3] It was of magnitude 7.5, which means that it could be observed only with a telescope; moreover, the sky conditions were not optimal, so Caroline had to wait the next night to confirm the discovery. This comet became brighter in the following few days and could be observed with the naked eye when it occurred later that year on 17 August.

Caroline's fame as the first woman to officially discover a comet was ensured. In the following years, at the request of William, King George III recognized her role of assistant to her brother by assigning to her a salary of fifty pounds per year. It was a modest sum, but an important symbol because it made her probably the first woman in Britain, and the second in Europe (after Christine Kirch) to have her professional work in astronomy officially recognized and remunerated. William's request to the King emphasized the industry, dedication, and expertise that Caroline brought to her vital role as his assistant, something that he could not manage without. It should also be pointed out that in addition he stressed that his sister was cheaper to employ than a regular assistant, suggesting that Caroline's recognition was not solely a consequence of her astronomical expertise and discovery of a comet but was a more complex affair (Hoskin 2002). William's marriage to Mary Pitt, in 1788, had caused a drastic change in Caroline's habits including a move away from the family home. Although eventually time helped to settle down the tense relationship between Caroline and William's wife, Caroline's salary was probably in part compensation for William's marriage and his sister's exit from the conjugal home. However, this move had the silver lining of making her independent from her brother. Their mutual collaboration continued, but on a more autonomous footing, from both the economic and the scientific point of view. Indeed, this new status supported the flourishing of Caroline's astronomical career, especially her success as a comet hunter. She discovered her second comet (35P / Herschel-Rigollet) on 21 December 1788 and, during the next ten years, discovered another eight. This was a record for a woman and was only surpassed in 1980 by the astronomer Carolyn S. Shoemaker. Only six of Caroline's comets are officially ascribed to her however; she was significantly involved in identifying the other two and shares their discovery with another.[4] It seems Caroline was proud of these discoveries as, after her death, all the documents relating to her comets were found neatly cared for, sorted, and saved.

There is no doubt that Caroline was known and respected as an astronomer in her own time. Indeed, Nevil Maskelyne, the Astronomer Royal was impressed with the telescope Caroline used to make her discoveries as evidenced by his correspondence held at the Royal Astronomical Society:

> She shewed me her 5 feet Newtonian telescope made for her by her brother for sweeping the heavens. It has an aperture of 9 inches, but magnifies only from 25 to 30 times, & takes in a field of 1° 49' being designed to shew objects very bright for the better discovering any new visitor to our system, that is Comets, or any undiscovered nebulae. It is a very powerful instrument, & shews objects very well. (Hoskin and Warner 1981, pp. 29–30)

In addition to her activity as a comet sweeper, Caroline embarked on a new project to check and correct the star catalog produced by John Flamsteed (1646–1919); this was presented in 1798 to the Royal Society of London as the *Index to Flansteed's Observations of the Fixed Stars*, together with a list of 560 other stars that had been ignored. This publication marks the temporary end of her work as a researcher after the death of her brother William. For the next twenty-five years Caroline devoted herself to the education of the son of William and Mary, John Herschel, who was born in 1792 and who was destined to become an astronomer as well. Nevertheless, Caroline's fame earned her much attention. In 1799 she was the guest of the Astronomer Royal, Nevil Maskelyne, at the Royal Observatory, and she was also the guest of members of the Royal Family on several occasions. She received awards for her science, the most prestigious being honorary fellowship of the Royal Astronomical Society (RAS), awarded to her and to another female scientist, the mathematician and astronomer Mary Somerville (1780–1872), in 1835. The Royal Astronomical Society did not admit women routinely, as ordinary fellows, until 1916 (RAS 2021).

Upon William's death in 1822, Caroline returned to Hanover and pursued independently some research projects in support of her nephew John. The outcome of this project was a catalog of 2500 nebulae, for which the Royal Astronomical Society awarded her its gold medal in February 1828 stating.

> That a Gold Medal of this Society be given to Miss Caroline Herschel for her recent reduction, to January, 1800, of the Nebulae discovered by her illustrious brother, which may be considered as the completion of series of exertions probably unparalleled either in magnitude or importance in the annuals of astronomical labour. (Hoskin 2011, p. 194)

On 9 January 1848, at the age of ninety-six, Caroline died in Hanover, two months before her birthday.

Conclusions

These case studies of early women astronomers reveal strong parallels in women's access and participation in astronomy in the seventeenth and eighteenth centuries in Europe. A lack of access to an academic education can be identified as a key obstacle to scientific women. This problem exacerbated when the development of modern science nurtured the formation of the first professional research institutions and moved science from a home setting to an institutional one. However, with its emphasis on observation of the night sky, undertaken in the field, women astronomers were shielded to an extent from this as they could still practice in a domestic setting. Here, the strength of the craft tradition, and families combining to pursue the 'family business', provided a gateway for women to learn and contribute to astronomy.

Indeed, this setting was conducive to collaboration with male family members or friends who couldprovide the necessary scientific instruction.

Arguably Caroline Herschel can be considered as the most significant, and perhaps last, of these female 'domestic astronomers'. In the second half of the nineteenth century, science in general, including astronomy, started to become more and more complex, requiring increasing education and resources, and sophisticated facilities such as laboratories and observatories which could be provided only by professional institutions. This caused the gradual disappearance of non-academic practitioners from the forefront of scientific research, rendering scientific women relatively invisible in the scientific panorama, along with home-based science. Although their number is small, in their own time these women were recognized and commanded the respect of their peers. However, history has been slow to understand their contributions to astronomy, but to do so—and therefore reveal the hidden diversity of the past—is important to resolving the present masculine bias in science.

NOTES

1. The first two women were elected as Fellows of the Royal Society in 1945: they were the crystallographer Kathleen Lonsdale and microbiologist Marjorie Stevenson.
2. The fourteen objects discovered by Caroline Herschel are classified in the New General Catalogue (NGC) of Astronomy as the following: two galaxies, NGC 253 and NGC 205; and twelve open clusters: NGC 7789, NGC 189, NGC 225, NGC 659, NGC 7789, NGC 6819, NGC 6866, NGC 752, NGC 7380, NGC 2360, NGC 2548, NGC 6633. In addition, the historian of astronomy Michael Hoskin (2006) reported of another open cluster discovered by Caroline with a catalog number of IC (Index Catalogue) 4665.
3. Caroline's discovery of an unknown comet caught the interest of people including the Court at Windsor where Fanny Burney was then a lady-in-waiting. Burney wrote in her diary in 1786 'The comet was very small, and had nothing grand or striking in its appearance; but it was the first lady's comet, and I was very desirous to see it' (Phillips 1990, pp. 161, 186).
4. In January and April 1790 Caroline identified 'C/1790 A1 Herschel' and 'C/1790 H1 Herschel'. On 15 December 1791 came 'C/1791 X1 Herschel' and on 7 October 1793 'C/1793 S2 Messier'. In this latter case, according to the official denomination, Charles Messier had already sighted the comet without Caroline knowing. The seventh and the eighth comets are '2P/Encke' discovered on 7 November 1795 and 'C/1797 P1 Bouvard-Herschel' discovered on 14 August 1797.

References

Alic, Margaret. 1986. *Hyfvpatia's Heritage: A History of Women in Science from Antiquity to the Late Nineteenth Century*. London: The Women's Press.

Bernardi, Gabriella. 2016. *The Unforgotten Sisters: Female Astronomers and Scientists before Caroline Herschel*. New York: Springer.

Campanacci, Ilaria Magnani. 1988. "Matematiche e poetesse nella Bologna del '700". *I Campanacciani*. http://www.icampanacciani.it/matematiche-e-poetesse-nella-bologna-del-700. Accessed on 21 February 2021.

Cook, Alan. 2000. "Johann and Elizabeth Hevelius, Astronomers of Danzig", *Endeavour* 24, no. 1: 8–12.

De La Lande, Jérôme. 1803. *Bibliographie Astronomique avec l'Histoire de l'Astronomie depuis 1792 jusqu' à 1802*. Paris: De L'Imprimerie de la République.

De La Lande, Jérôme. 1817. *Astronomie des dames*. Quatrième édition. Paris: Ménard et Deseene, Fils.

Fantuzzi, Giovanni. 1786. *Notizie degli scrittori bolognesi*. Bolognia: Stamperia di S. Tommaso d'Aquino.

Helly, Dorothy O. and Reverby, Susan. 1992. *Gendered Domains: Rethinking Public and Private in Women's History: Essays from the Seventh Berkshire Conference on the History of Women*. Ithaca: Cornell University Press.

Herschel, John. 1876. *Memoir and Correspondence of Caroline Herschel*. London: John Murray

Hoskin, Michael. 2002. "Caroline Herschel: Assistant Astronomer or Astronomical Assistant?", *History of Science* 11, no. 4: 425–444.

Hoskin, Michael, ed. 2003. *Caroline Herschel's Autobiographies*. Cambridge: Science History Publications.

Hoskin, Michael. 2006. "Caroline Herschel's Catalogue of Nebulae", *Journal for the History of Astronomy* 37, no. 128: 251–253.

Hoskin, Michael. 2011. *Discoverers of the Universe: William and Caroline Herschel*. Princeton: Princeton University Press.

Hoskin, Michael and Warner, Brian. 1981. "Caroline Herschel's Comet Sweepers", *Journal for the History of Astronomy* 12, no. 1: 27–34.

Manfredi, Eustachio. 1750. *Introductio in Ephermerides*. Naples: Constantini Pisarri S. https://archive.org/details/bub_gb_QG2L1dq0-VgC. Accessed on 12 February 2021.

Mozens, John Augustine. [1913] 1999. *Women in Science: With an Introductory Chapter on Woman's Long Struggle for Things of the Mind*. Notre Dame, IN: University of Notre Dame Press.

O'Connor, J. J. and Robertson, E. 2008. "Catherina Elisabetha Koopman Hevelius", *MacTutor History of Mathematics Archive*, University of St Andrews. https://mathshistory.st-andrews.ac.uk/Biographies/Hevelius_Koopman/. Accessed on 18 January 2021.

Ogilvie, Marilyn Bailey. 1991. *Women in Science: Antiquity through the Nineteenth Century*. Cambridge, MA: MIT Press.

Phillips, Patricia. 1990. *The Scientific Lady: A Social History of Woman's Scientific Interests 1520-1918*. London: Weidenfeld and Nicolson.

Royal Astronomical Society (RAS). 2021. "Women and the Royal Astronomical Society", https://women.ras.ac.uk/women-and-the-ras/history-of-women-at-the-ras. Accessed on 26 February 2021.

Schiebinger, Londa. 1987. "Maria Winkelmann at the Berlin Academy: A Turning Point for Women in Science", *ISIS* 78, no. 2: 174–200.

Wertheim, Margaret. 1997. *Pythagoras' Trousers: God, Physics, and the Gender Wars*. London: W.W. Norton.

CHAPTER 14

Darwin and the Feminists: Nineteenth-Century Debates About Female Inferiority

Amanda M. Caleb

In a letter to Prime Minister William Gladstone in 1870, Queen Victoria famously wrote, 'Let woman be what God intended, a helpmate to man, but with totally different duties and vocations' (Rappaport 2003, p. 428). At the same time, Charles Darwin was completing his work on *The Descent of Man, and Selection in Relation to Sex*, published the following year, in which he claimed (1871, pp. 329–330):

> ...although men do not now fight for the sake of obtaining wives, and this form of selection has passed away, yet they generally have to undergo, during manhood, a severe struggle in order to maintain themselves and their families; and this will tend to keep up or even increase their mental powers, and, as a consequence, the present inequality between the sexes.

Unlikely bedfellows, religion and science perpetuated the belief in separate spheres and female inferiority and, as the century progressed, science became the more dominant narrative that provided evidence of female inferiority, including smaller brains and an evolutionary shift that focused bodily energy in reproductive organs. As George Romanes sums up, 'the man has always been regarded as the rightful lord of the woman, to whom she is by nature subject, as both mentally and physically the weaker vessel' (Romanes 1887, p. 656).

A. M. Caleb (✉)
Geisinger Commonwealth School of Medicine, Scranton, PA, USA
e-mail: acaleb@som.geisinger.edu

Such claims regarding female inferiority were reactionary as part of the larger discussion of women's roles in the changing economic and social landscape, resulting in two well-repeated concepts of the age: separate spheres and the 'Woman Question'. The former is a product of the industrial revolution and of Enlightenment claims of biological determinism. The rise of factories led to men working outside the home and engaging with economics—and only explains the public role of men, not why women were confined to the home. That reasoning comes from biological determinism, which argued that a sex's biological makeup determined their abilities, thus introducing arguments that women were not suited, physically or intellectually, for work outside the domestic sphere. However, the industrial revolution also ushered in a need for a larger workforce within the factories and placed a greater demand on working-class families, which resulted in the employment of working-class women. Thus, the 'Woman Question' emerged in response to the need to shelter women from the public sphere even as women were working within that sphere. It was also a response to the first UK Reform Act (1832), which explicitly excluded women from voting; previous to that legislation, gender was not explicitly stated in voting laws, although very few women qualified to vote under the property requirements (Heater 2006, p. 107). The 'Woman Question' asked essentially—what were women capable of doing, and what should they be allowed to do?—revealing itself to be a question of biology and a question of social expectations. While many preferred women to continue in their roles as companions to men, or 'Angels in the House', others, primarily women, were looking for more opportunities for women in the public sphere, including through higher education and voting rights.

These historical contexts are relevant to understanding Darwin's theory on sexual difference, born of the 'intersection of evolutionary theory and social, psychological and medical theory', and very much a part of, and a response to, the Woman Question (Beer [1983] 2009, p. 196). Given the timeline of the UK women's movement, including meetings of the Langham Place Circle in the 1850s and the creation of the National Society for Women's Suffrage in 1867, Darwin's two major publications fit squarely within this developing period of nineteenth-century feminism, as does the authority that evolutionists asserted in this arena. For instance, T.H. Huxley advocated for woman's access to education because he did not think it would make a difference in the struggle for existence with men. Evelleen Richards concludes that his 'Darwinian barriers against the complete equality' of the sexes were:

> ...ratified by Darwin in the *Descent*, and by other prominent evolutionists who joined forces with anthropologists, psychologists, and gynaecologists to force a formidable body of biological theory which purported to show that women were inherently different from men in their anatomy, physiology, temperament, and intellect—that women, like the 'lower' races, could never expect to match the intellectual or cultural achievements of men or obtain an equal share of power or authority. (2017, p. 377)

Contrary to the claim that 'scientists never engaged in a conscious conspiracy against women' (Russett 1989, p. 190), this positioning by Huxley and others indicates a deliberateness to their actions and writings with regard to marginalizing women. In fact, it was by this very conscious and Foucauldian act of dividing and classifying women as Other that Darwinian scientists were able to scientifically prove and maintain Victorian gender divisions.

The arguments regarding women's intellectual capabilities, made explicit in *Descent*, are responses to the social changes occurring in Britain and part of a politically-motivated biological science designed to keep women out of the public sphere. Yet at the same time, these very claims were contextualized and redefined by feminists to assert women's social value, the need for equal access to education, and the right for women to control their own bodies; one account of feminism went as far as to claim that 'Darwin was the originator of modern feminism' (Walsh 1917). *Descent*, then, functioned in two ways: as a political response *to* and a political response *of* the women's movement, which reflected a vying for authority over the female body by the scientific community and feminists alike.

BIOLOGICAL DETERMINISM OF SEXUAL DIFFERENCE: ARGUMENTS AGAINST THE WOMEN'S MOVEMENT

The very basis of Darwin's theories, as Fiona Erskine notes, 'include gender specialization' (1995, p. 97). Both natural and sexual selection require differentiation as part of the adaptive process, something which created an imperative within Darwin's theories to not only identify these differences, but also to emphasize the need to maintain them as part of the evolutionary process. First thinking about sexual selection in the 1830s, Darwin began refining the theory in 1856, coming to his full understanding of the theory in 1858 (Richards 2017, pp. 291–293). In *On the Origin of Species*, he describes the process as 'a struggle between the male for possession of the females', citing examples of how sexual selection works within animal species, including female selection of males based on aesthetic evaluation among birds (Darwin 1859, pp. 88–89). Yet Darwin's desire to explain male plumage and vocal abilities painted him into a proverbial corner in *Descent*: if the same process that explained sexual differentiation applied to humans, then women could be seen as more powerful than men (Levine 2006, p. 201).

Darwin's solution was to distinguish humans from animals in this process, based, again, on contemporary observation and evolutionary theory. Women, not men, are the ones who 'delight in decorating themselves with all sorts of ornaments', whereas 'Man is more powerful in body and mind than woman, and in the savage state he keeps her in a far more abject state of bondage than does the male of any other animal; therefore it is not surprising that he should have gained the power of selection' (1871, pp. 371–372). Darwin's shift of female to male choice among humans is problematic in a number of ways: as many male species are physically stronger than their female counterparts, this claim of strength does not fully explain the complete change in choice, and it weakens the link between animals and humans that he argued for in the rest of the work. Indeed, it presented an inconsistency in thinking about anthropological claims for promiscuity in savage societies, which weakens the theory of sexual selection. Evelleen Richards posits that 'these inconsistencies seemingly escaped Darwin, blindsided by his single-minded focus on substantiating his key theses of female choice in animals and male choice in humans, the driver of race divergence' (2017, p. 453). Rosemary Jann instead focuses on the issue of gender: 'the construction of a biological rationale for gendered behaviour required that he project a patriarchal model of the family, with its dominant males and dependent females, back into the no-(hu)man's land between biology and culture' (1996, p. 86). Although Richards and Jann come to different conclusions, collectively they introduce an important aspect of Darwin's evolutionary theory: the differentiation of race and gender that supported a view of white male superiority. Indeed, Darwin's claim that 'in utterly barbarous tribes the women have more power in choosing, rejecting, and tempting their lovers, or afterward changing their husbands' indicates his Victorian prejudice against female choice and therefore female power (1871, p. 372).

Through Darwin, the intersection of race and gender, used as a call for women's education and suffrage, was being replaced by 'the equation of woman-as-child-as-primitive', derived from 'the interlinked social, racial, and anthropological hierarchies of the later Victorian period' (Richards 2017, p. 295). Thus, Mary Wollstonecraft's arguments in *A Vindication of the Rights of Woman* (1792) about women being 'prepared for the slavery of marriage' (1891, p. 221), and reasserted decades later by John Stuart Mill's claim that 'there remain no legal slaves, except the mistress of every house' (1869, p. 147), were supplanted by biological and anthropological claims of female inferiority connected to both animal and primitive development. In responding directly to Mill, Darwin writes, 'I am aware that some writers doubt whether there is any inherent difference [between the sexes]; but this is at least probably from the analogy of the lower animals which present other secondary sexual characters', concluding that mental ability is a secondary sexual characteristic in humans (1871, p. 326). From animals, Darwin moves to a comparison of women to savages: although women have a maternal instinct that includes tenderness and selflessness (a nod to the 'Angel in the House'), he concludes that 'with woman the powers of intuition, of rapid perception, and perhaps of imitation, are more strongly marked than in man' and that 'some, at least, of these faculties are characteristic of the lower races, and therefore of a past and lower state of civilization' (1871, pp. 326–237). He expands upon this difference between the sexes, pairing these less evolved faculties with physiological evidence of a lower evolution for women; referring to Alexander Ecker and Herman Welcker's works, Darwin notes that 'in the formation of her skull, [it] is said to be intermediate between the child and the man' (1871, p. 317).[1] Darwin attributed women's stymied intellectual growth to sexual selection and men's competition with other men that results in ambition, which he associates with intellect: 'Man is more courageous, pugnacious, and energetic than woman, and has a more inventive genius' (1871, p. 316). These traits, products of both natural selection and sexual selection through struggle and combat, do not belong to women because they neither compete nor choose mates in Darwin's model. Thus, as Richards concludes:

> human females, even primitive ones, were exempt from the struggle for existence in important respects: locked out of the means of acquiring intelligence through the improving powers of natural selection, dependent on males for care, subsistence, and protection, and locked into a model of primitive family life that coincided with the Victorian ideal of middle-class patriarchy and the sexual division of labor. (2017, p. 452)

The selectivity of Darwin's application of sexual selection to women indicates a social influence on the scientific theory, one used to 'legitimate every view of women's abilities and appropriate roles' (Paul 2003, p. 226).

A further claim Darwin made—and one that moved from anthropological sexual selection to more explicit contemporary distinctions—was the noted achievements of men that distinguished them from women, an approach adapted from Francis Galton's *Hereditary Genius* (1869) and praised in *Descent*. He notes:

> The chief distinction in the intellectual powers of the two sexes is shewn by man attaining to a higher eminence, in whatever he takes up, than woman can attain—whether requiring deep thought, reason, or imagination, or merely the use of the senses and hands. If two lists were made of the most eminent men and women in poetry, painting, sculpture, music,—comprising composition and performance, history, science, and philosophy, with a half-a-dozen names under each subject, the two lists would not bear comparison. (1871, p. 327)

Such observations, of course, fail to consider social context, educational limitations, and historical gender bias—or rather, these considerations were purposefully ignored in this line of reasoning. Darwin did recognize the possibility of women achieving the same intellectual abilities as men: in *Descent* he explains, 'In order that woman should reach the same standard as man, she ought, when nearly adult, to be trained to energy and perseverance, and to have her reason and imagination excised to the highest point; and then she would probably transmit these qualities chiefly to her adult daughters' (1871, p. 329). While flawed in his reasoning about acquired traits, he does entertain the possibility that changes to the social and educational environment for women could produce evolutionary changes that raise them to the level of men. However, Darwin saw the limitations of such a proposal, claiming the sexual selection drive for men to provide for their families would perpetuate the divide. The assumption that only men can provide for their families is more social than evolutionary, as evidenced in his response to Caroline Kennard's question about female intellectual inferiority:

> there is some reason to believe that aboriginally (and to the present day in the case of Savages) men and women were equal in this respect, and this would greatly favour their recovering this equality. But to do this, as I believe, women must become as regular 'bread-winners' as are men; and we may suspect that the early education of our children, not to mention the happiness of our homes, would in this case greatly suffer.[2]

Darwin's private letter insinuates that the division between the sexes in Western society is socially constructed: he does not suggest that altering such divisions would be against evolutionary interests directly, but rather against social structures that benefit society as currently organized.

Ignoring or dismissing systemic social factors that affected women's access to education, the workforce, etc., was standard practice for those scientists who wished to perpetuate the theory of female inferiority, privileging biological evidence over all else. George Romanes, echoing Darwin in his well-read article of 1887, claimed female inferiority 'displays itself most conspicuously in a comparative absence of originality, and this more especially in the higher levels of intellectual work' and assured the public that 'the disabilities under which women have laboured with regard to education, social opinion, and so forth, have certainly not been sufficient to explain this general dearth among them of the products of creative genius' (1887, pp. 655–656). The particular attention to creativity, in addition to intelligence, moves the discussion of sexual difference and evolution to that of psychology, particularly the early days of evolutionary psychology, which focused on brain development and function in relation to natural and sexual selection.

Romanes's focus on the anatomical and physiological signifiers of intellectual inferiority created a space in which to legitimize centuries-old assumptions about women's capabilities. Thus, he claims:

> [because] the average brain-weight of women is about five ounces less than that of men, on merely anatomical grounds we should be prepared to expect a marked inferiority of intellectual power in the former. Moreover, as the general physique of women is less robust than that of men—and therefore less able to sustain the fatigue of serious or prolonged brain action—we should also on physiological grounds be prepared to entertain a similar anticipation. (1887, pp. 654–655)

By pairing differences in physique—which few denied—with intellectual abilities, Romanes commits an association fallacy, one that he confirms in his advocacy of women's education that reflects their inferior intellectual capabilities and stamina:

> The channels, therefore, into which I should like to see the higher education of women directed are not those which run straight athwart the mental differences between men and women which we have been considering. These differences are all complementary to one another, fitly and beautifully joined together in the social organism. If we attempt to disregard them, or try artificially to make of woman an unnatural copy of man, we are certain to fail, and to turn out as our result a sorry and disappointed creature who is neither the one thing nor the other. (1887, p. 671)

His argument against equal education echoes a sentiment expressed by those who discussed the 'Woman Question': what happens when higher education leads to the unsexing of women?

This concept of the unsexed woman has roots in evolutionary theory: if women are no longer driven to procreate, then both sexual and natural selection cease in a civilized society. On the surface, this implies women have a great deal of power in the evolutionary process; however, contemporary discussions of women's intellect and access to higher education indicated women were at the mercy of their reproductive organs and therefore not in control of their decision making. Herbert Spencer, in discussing the difference between the sexes, claimed it 'results from a somewhat-earlier arrest of individual evolution in women than in men; necessitated by the reservation of vital power to meet the cost of reproduction' (1873, p. 340). This vital power Spencer imagines is not just physical but also mental. He explains:

> This rather earlier cessation of individual evolution thus necessitated, showing itself in a rather smaller growth of the nervo-muscular system, so that both the limbs which act and the brain which makes them act are somewhat less, has two results on the mind. The mental manifestations have somewhat less of general power or massiveness; and beyond this there is a perceptible falling-short in those two faculties, intellectual and emotional, which are the latest products of human evolution—the power of abstract reasoning and that most abstract of the emotions, the sentiment of justice—the sentiment which regulates conduct irrespective of personal attachments and the likes or dislikes felt for individuals. (Spencer 1873, pp. 341–342)

Spencer builds upon contemporary theories regarding brain size and introduces an implicit argument about women's mental fitness to be involved in decision making: if their capacity to understand and deliver justice is less than man's, then a world with women in charge, or who could vote, would be an inherently unjust world.

Spencer's argument regarding female evolution echoed sentiments expressed by the medical community, particularly in relation to limitations on the vital energies of the female body and the reproductive imperative. Rather than focus on female intellect directly, members of the medical community focused on two threads against equal education among the sexes: that each sex was made differently and therefore for different purposes, and that female reproduction and issues of physical stamina would affect reproduction and their identity as women. Both approaches stem from the medical view of the female body as 'disease or disorder, a deviation from the standard health represented by the male' and defined by its relationship to reproduction (Moscucci 1990, p. 102). Notably, gynaecology was defined in 1867 as 'embracing the physiology and pathology of the non-pregnant state' (Barnes 1867, p. 368). Although medicine's focus on the female reproductive body predates Darwin's theories, the rhetoric used by the medical community to advocate against women's higher education echoes evolutionary concerns and the language of Darwin and Spencer, particularly in thinking about women's relationship to men. As noted by Sir Patrick Geddes and John Arthur Thompson, 'Darwin's man is as it were an evolved woman, and Spencer's woman an arrested man' (1889, p. 37). In both cases, women's identity is defined by her relationship to men, though the shift in language signals that 'woman is no longer a misbegotten man. She has become an unevolved man' (Tuana 1993, p. 44).

The view of woman as an unevolved man echoes the rhetoric used by the medical community in the second half of the century. Ornella Moscucci contends that 'the medical profession was very slow to notice Darwin's controversial theories [in the *Descent*]', and the reason is largely because of Spencer's extensive and ongoing writing about the differences between the sexes, in fields such as sociology, psychology, anthropology, and biology—all fields that influenced medical views of women (1990, p. 22). Spencer's (and implicitly Darwin's) influence can be seen in a rhetorical and functional shift regarding how medical professions described women's biological functions. Understandings of female intelligence and reproduction were modified to reflect Spencer's logic: John Gideon Millingen claimed that woman 'is less under the influence of the brain than the uterine system', reasoning that reproductive drives drove physiological and mental needs (1849, p. 157). Three decades later, a column entitled 'Influence on Women of Special Brain-Work' in *The Lancet* stated, 'In the ordinance of nature the female is endowed with a force tending to the reproduction from her arrested or suppressed organism of the perfect organism of the male', reasoning that 'Experience seems to show that special brain-*work*, properly so called, on the part of the mother, exhausts the energy of brain-development—or reproduction' (1881, p. 379). The medical community also adopted Spencer's argument about intellectual and reproductive energies as being in conflict with each other. Most succinctly, psychiatrist Henry Maudsley claimed, 'When Nature spends in one direction, she must economize in another direction', asserting that the educational demands for women would reduce their reproductive energies (1874, p. 467).[3] By implementing evolutionary language and theories in medical discourse, physicians reinforced their authority over female bodies and articulated this as part of an effort to protect and promulgate the species.

What unified these different fields of study was an argument against the women's movement, particularly access to education—and by extension science—and participation in politics, thus revealing the influence of society on supposedly objective fields of study. Concerns regarding women's education were couched in the language of harming themselves and their future offspring. As Maudsley explains, 'Inasmuch as the majority of women will continue to get married and to discharge the functions of mothers, the education of girls certainly ought not to be such as would in any way clash with their organization, injure their health, and unfit them for these functions' (1874, p. 214). In viewing mothering as a biologically determined function and imperative, Maudsley claims a scientific authority in how women are educated and its impact on the family and social organization. In an even further-reaching argument, physician Thomas S. Clouston, agreeing with others that advocated for limited energies, argued for the importance of the physical development over the intellectual, which 'gives a great chance of health and happiness to the individual, and infinitely more change of permanence and improvement to the race', an obligation that was 'woman's chief work to the future of the world' (Clouston, 1882, p. 20). He concludes by stating, 'The health we must have, for it is requisite for the life of the race; the culture we must have in such degree as is consistent with the health' (1882, p. 48). Here it is clear that women's health, defined by medicine through evolutionary imperatives, takes precedence over current fads of a society—which is ironic, given the culturally-driven tone of Clouston's work.

Gynaecologist Lawson Tait claimed ultimate authority regarding women's rights: 'The questions raised by the advanced advocates of women's rights are to be settled, not on the platform of the political economist, but in the consulting-room of the gynaecologist' ([1874] 1883, p. 91). Positioning these political discussions within the realm of gynaecology furthered a male authority regarding women's bodies and their social status, couched in evolutionary rhetoric regarding species survival. Thus, Tait claimed: 'I may own myself an advanced advocate of women's rights; at the same time I cannot help seeing the mischief women will do to themselves, and to the race generally, if they avail themselves too fully of these rights when conceded' ([1874] 1883, p. 91). Such mischief included pursuing higher education, which would 'leave only the inferior women to perpetuate the species [and] will do more to deteriorate the human race than all the individual victories at Girton will do to benefit' ([1874] 1883, p. 91).[4] Again, the evolutionary imperative reinforces the biological determinism that was thought to rule women's bodies: to go against this drive would be to go against nature and condemn humanity to degeneration.

The effectiveness of this rhetoric stems from the intersection of male-dominated fields that asserted power by way of evoking an ultimate secular authority: nature. But what was at stake in the debate regarding female education and suffrage was a different type of power: the ability to influence society (Endersby [2003] 2009, p. 85). Herbert Spencer recognized this social struggle, positing that if 'the mental natures of men and women [are] the same [...then] an increase of feminine influence is not likely to affect the social type in a marked manner. If they are not, the social type will inevitably be changed by increase of feminine influence' (1873, p. 340). Given the numerous claims that woman's brain is different to man's, Spencer already believed the second statement to be true, revealing the real concern regarding female education and suffrage: that women might shape society in an antithetical way to men's wishes and/or the species survival.[5]

QUESTIONING 'THE BOOK OF DARWIN': FEMINIST ARGUMENTS FOR EQUALITY

Accounts of American women's challenge to *Descent* and its subsequent applications in the nineteenth century are well documented (Hamlin 2015; Hayden 2013; Deutscher 2004). However, responses from British women have received less attention. This perceived gap in scholarship is largely because of how British women contested Darwin's claims of female intellectual inferiority; these challenges generally fall into two categories: direct disagreement with the application of Darwin's theories by Maudsley and others, and literary responses that reassert female choice. The reason for a less direct engagement by British women (as opposed to American women) is ripe for speculation; however, looking at evolutionist and feminist Constance Naden's 'The Evolution of the Sense of Beauty' ([1885] 1891), one of the few direct responses to Darwin, offers some insight. In this essay, Naden mocks the near-Biblical authority of Darwin and the disciples that follow 'the book of Darwin, in the book of the Chronicles of the Descent of Man' ([1885] 1891, p. 80). In the context of Naden's essay—though admittedly delayed in its reply to *Descent*— it is possible that British feminists chose not to attack that which was held up to such a high standard but instead focused on those who applied Darwin's work to their own. As *Descent* does not directly propose educational or political limitations to women, but rather identifies the biological factors that create the condition for such legislation, there were more outspoken and vulnerable individuals to challenge.[6]

There are, however, some broad references to Darwin's work that did attract feminists' ire and revision, namely his evidence of small female skulls and his listing of accomplished men—though neither are uniquely Darwin's and, therefore, could be seen as a broader critique of male scientists who supported a theory of female inferiority. In an 1874 address to the London Anthropological Society, Emma Wallington acknowledged the smaller size of women's brains, but contended size 'is not always a criterion of mental ability', supporting her claim by listing twenty-five women who excelled in the sciences, mathematics, and medicine, fulfilling her claim that 'an ounce of fact is of more worth than any quantity of theory' (1874, pp. 558–559). Although Wallington's remarks were dismissed by the male-only membership of the Society, her willingness to take on anthropology directly—a field that helped fuel Darwin's arguments and perpetuate the belief in female intellectual inferiority—encouraged other women to do the same in related domains.

A discipline worthy of attention was psychiatry, particularly given Maudsley's outspoken comments about the ill effects of higher education on women and therefore the human race. Trailblazing physicians Elizabeth Garrett Anderson and Sophia Jex-Blake both responded to Maudsley's essay, addressing the environmental factors that impact women's health when pursuing education. Garrett Anderson argues that such strains may not be unique to women, and that even if women face additional strains through education:

> [this is better than] develop[ing] only the specially and consciously feminine side of the girl's nature [...] it is difficult to believe that study much more serious than that usually pursued by young men would do a girl's health as much harm as a life directly calculated to over-stimulate the emotional and sexual instincts, and to weaken the guiding and controlling forces which these instincts so imperatively need. (1874, pp. 588–590)

In comparing male educational pursuits to feminine education and stimulation (which includes novel reading, ballroom dancing, and so on), Garrett Anderson directly challenges the social environment that dictates educational needs for the benefit of the individual and society and that perpetuate separate spheres at the cost of the physical and intellectual development of women. In essence, she raises suspicions about why women need to be educated differently, which Sophia Jex-Blake addresses explicitly in her response to Maudsley's article: 'I think nothing is more curious than the placid way in which a certain class of male writers constantly assume that no labour is severe except that usually allotted to men, and that it is only when women venture to invade *that* field that they are likely to be overtasked' (1874, p. 457). Jex-Blake calls out the medical community for their selective application of how intellectual pursuits affect women's physiology and implicitly acknowledges Spencer's concern about women's influence on society.

This critique of Spencer's view of women's intellectual inferiority is made manifest in Josephine Butler's social reform work. In an 1874 essay advocating for the abolition of the Contagious Diseases Acts, she criticizes Spencer's claim that women lack a sense of justice, writing, 'The depressed condition of woman has prevented the free exercise of her judgment; her natural sense of justice has been, in secret, outraged almost to extinction; she has not been permitted to exercise or express it in any open or legitimate manner, nor encouraged to bring it to bear on any large or public questions. No wonder it has become enfeebled' ([1874] 2004, p. 62). In shifting the argument about justice from a biological development to a social restriction, Butler introduces the cultural contexts that are often misrepresented or summarily dismissed in Spencer's evolutionary arguments about women. To drive her point home, she critiques the male sense of justice, claiming it 'has been impaired, well-nigh to extinction, by a different process. It has been warped and corrupted by the almost exclusive possession of power in one direction, and by the privilege he has assumed to himself of forming a judgment on all that concerns one half of the human race, irrespective of any judgment which that half of the human race may have formed concerning their own interests' ([1874] 2004, p. 62). Such a scathing attack on male justice reveals the very hypocrisy of the claim of inherent male strength, and the injustice of its application to women.

In a similar vein, Edith Simcox delves into the cultural considerations of female intelligence to probe Romanes's understanding and application of evolutionary theory to women's mental capacities. She questions his claim about the evolution of male intelligence and female morals, calling them 'vague metaphysical notions about an *Ewigweibliche* [the eternal feminine]' and asserting that 'the psychological and other distinctions of sex are among the after-thoughts of the primeval mother nature' (1887, p. 391). Here she notes both the cultural construct of the eternal feminine—which contradicts an evolutionary view of adaptation and development—and emphasizes the gender of nature itself, which is a subtle reminder of not only the power of creation, but also the feminine design of life. Simcox then replaces Romanes's biological history with a cultural history, noting 'there is a difference between things practically useful under given material conditions and things belonging to the eternal and immutable "nature of things"' and asserting that 'the brains of both men and women exercise themselves habitually upon such stuff as the customs of their age and race set before them' (1887, pp. 392, 394). In other words, historical cultural factors, for example, the division of household duties, play a role in current practices that impact educational access—not the intelligence of women. In inserting cultural evolution into the equation, Simcox reclaims evolution from the masculine biological to an ungendered holistic view of adaptation.

Literary responses sought the same goals of reclaiming evolutionary rhetoric and theory, while focusing more directly on female choice as reflecting female intellectual superiority (Kohlstedt and Jorgenson 1999, pp. 278–279). There is significant research on the impact and integration of Darwin's ideas, particularly those from *Origin*, in Victorian literature, but there is far less scholarship on feminist literary responses to *Descent*. This is largely because the literary response was minimal, but no less significant, to that of the female medical and scientific community. Although there is little by way of literary responses to *Descent* and 'gendered science' in the 1870s (Murphy 2006, p. 215), the literary responses that do emerge in the two decades following the publication of *Descent* engage with it by reclaiming female choice in an effort to advocate for gender equality.

While male critics of *Descent* were trying to dismantle the idea of female choice among any species—a blatant assertion of sexism into science—feminists claimed female choice for all female species, as evident in the poetry of Mathilde Blind.[7] Blind appears to be the first, and one of the only, literary feminists alluding to *Descent* in the 1870s, publishing 'The Song of the Willi' in the magazine *Dark Blue* six months after *Descent* appeared. Although not an outright criticism—James Diedrick identifies her as 'implicitly refuting

Darwin's claim' of female inferiority through a sisterhood 'more powerful than the men they return to enthrall' (2016, p. 85)—Blind creates a space for female desire and choice, female physicality that matches that of men, and an intellectual equality, if not superiority.[8] The voice in the dramatic monologue, an affianced dead woman, describes two dances with her betrothed, the first while alive and the second as a supernatural spirit. In the first dance, the man refers to the speaker as 'my chosen', promising to 'dance into death with thee' (1871, st. vi). Implied in the positioning of her death—immediately following the dance—is that male selection and the subsequent dance led to her demise, as 'pallid the cheeks grew erst flushing with pleasure' (1871, st. xi). In death, male selection is reversed: the speaker, incredulous that her betrothed can sleep peacefully while she is dead and wandering the Earth, invokes another dance with her betrothed, who 'walketh like one in a trance' (1871, st. vi–xv). The dance, an echo of birds' mating rituals, is performed by a sisterhood of other dead women doing the same (Birch 2013, pp. 81–83). Female mental superiority, as represented through hypnotism, is coupled with female physical superiority, as the man, unable to keep up with her dancing and victim to her 'dank tresses [rolled] round thy face, round thy throat', willingly follows her into death (Blind 1871, st. xvii). While one could critique Blind's choice of describing women's superiority only in death (and a death that may have been caused by a worldly physical inferiority), this same death denies the reproductive imperative assigned to women through biological determinism and reasserts female choice.

Blind's critical response to *Descent* continues in her later work, including her 1889 *Ascent of Man*, which offers a model for gender equality. In his introduction to a posthumous edition, Alfred Russel Wallace writes that while 'her treatment of the subject, if not altogether satisfactory […] is undoubtedly poetical, and is imbued with modern ideas, though, as was perhaps inevitable, it deals more with the social and spiritual aspects of the subject than with those which are purely scientific, though these latter are by no means neglected' (1889b, pp. v–vi). That Wallace wrote the introduction indicates the text's significance to the scientific community, though his criticism of her representation of scientific ideas is warranted, particularly given the 'Lamarckian yearnings and a teleological drive' at the end of her work (Holmes 2012, pp. 52–53). Still, individual components of the volume engage with Darwin in ways that challenge his conclusions about women. For instance, Wallace describes 'The Leading of Sorrow' as showing 'the universality of sorrow and death [which is…] entirely opposed to that of Darwin and the present writer' (1889b, p. ix). In the midst of this biological struggle is a seemingly out-of-place story of a country-maiden who, jilted by a noble-man, dies from his sexually transmitted infection, while her lover 'preaches / Resignation—o'er his burgundy', deemed by the public to be 'Oracles of true philanthropy' (Blind 1889, ll. 317–318). Blind's description of a sexual double standard amidst biological struggle demonstrates the social struggle women face, one that is structured in ways that disadvantage them. The maiden, 'by right of nature / Sweet as odour of the upland thyme', is punished for both their sins before she dies (1889, ll. 243-244). A 'poor outcast', she is:

> howled at, hooted to the wilderness,
> To that wilderness of deaf hearts, blunted
> To the depths of woman's dumb distress. (1889, ll. 243, 245-248)

Her muted distress is unheard by those who are indifferent to women's struggle within a system that inherently favours men, which the speaker calls out as unjust and questions a society that looks to 'the frank ravening of the raw-necked vulture / As its beak the senseless carrion smites' rather than 'woman's nameless martyrdom'—a direct criticism of favouring biological struggles over socially engineered ones (1889, ll. 327-328, 332). Thus, the poem ends with Nature asking, 'redeem me from my tiger rages' by replacing them with 'the hero's deeds, the thoughts of sages, [... and] healing love of woman', a final call for gender equality that ascends the biological differences described by Darwin (1889, ll. 425, 427, 429).

On a lighter note, though no less poignant, Constance Naden's 'Evolutional Erotics' (1887) takes issue with Darwin's account of sexual selection, particularly female choice, through parody of his theories and those that strictly (and unquestioningly) adhere to them. In the most direct of these poems, 'Natural Selection', Chloe, the beloved of a Darwin-discipline geologist, chooses 'an idealess lad', who 'sings with an amateur grace / [and] dances much better' than the geologist (1887, ll. 17, 23-24). The blatant reference to Darwin's work on female choice among birds is echoed by the geologist's claim that 'we know the more dandified males / By dance and by song win their wives', reclaiming female choice as part of human sexual selection (1887, ll. 25-26). Naden undermines Darwin's exception for male choice among humans and negates male combat through the geologist's acknowledgement of his inferiority and reluctance to engage the dandy. As signalled in the opening lines of the poem, Chloe has no interest in the geologist's intellectual pursuits: she claims, 'But he ne'er could be true [.../] Who would dig up an ancestor's grave', to which he reflects, 'I loved her the more when I heard / Such filial regard for the Cave' (1887, ll. 5-8). The fitness of this match, then, is not evolutionary but social: the geologist has chosen Chloe for no obvious evolutionary benefit and certainly not for her intellectual compatibility (she fails to admire his research). However, the geologist conflates the biological with the social, concluding the poem by resigning himself to their relationship, 'for since Chloe is false, / I'm certain that Darwin is true!' (1887, ll. 31-32). John Holmes reads these final lines as a very obvious critique of 'Darwinians mov[ing] all too readily from anecdotal evidence to familiar stereotypes when describing sexual behaviour' (2012, p. 193). The criticism is certainly there and speaks to Naden's previous reproaches of Darwinians in 'The Evolution of the Sense of Beauty'. There is more at stake here, however. Patricia Murphy reads the final lines as Chloe's selection being destined for failure, but I would counter that such failure is only within the geologist's perception of his own selection (2006, p. 54). In other words, the failure is both his overlooking of his own Darwinian sexual selection and his reduction of woman's choice to that of female birds, assuming that Chloe must either be wrong or less evolved. This disconnect between the geologist and Chloe is evident throughout the poem, making it clear to the reader how wrong the geologist is in his theories and how right Chloe is to choose another.

The power of reclaiming female choice is perhaps best illustrated by Olive Schreiner's 'Life's Gifts' in which the speaker imagines a sleeping woman who is visited by Life and is given the option to choose Love or Freedom. She chooses Freedom, and Life tells her, 'If thou hadst said, "Love," I would have given thee that thou didst ask for; and I would have gone from thee, and returned to thee no more. Now, the day will come when I shall return. In that day I shall bear both gifts in one hand' (1889, p. 408). Life's award of both love and freedom reads as a feminist rebuke of a Darwinian nature that denies women choice and presumes a biological determinism for women to reproduce (Love). The allegory then ends with the cryptic line, 'I heard the woman laugh in her sleep' which could translate as disbelief or smugness, but given Schreiner's feminism, reads as a triumphant mockery of Darwin's *Descent* and the denial of woman's choice (1889, p. 408).

Conclusion

The Darwinian-charged debates about the 'Woman Question' hinged not only on evolution and sexual selection but also on authority. As scientists were 'establish[ing] science as an authority capable of transcending spheres, an authority relevant to the private life of the individual' (De Witt 2013, p. 15), feminists, too, claimed an authority: to move beyond the domestic sphere and into the public arena, including science. By the end of the century, however, their challenges to scientific views of female intellectual inferiority were limited by national concerns regarding declining birth rates, the fitness of British soldiers, and the pseudo-scientific claim of degeneration (Caine 1992, p. 250). Thus, a 'new' feminist emerged who 'yielded the concept of natural rights and equal opportunities […and] no longer challenged the implied assumption of sexual inferiority' (Erskine 1995, pp. 116–117). Many of these 'new' feminists appropriated biological determinism and gender roles in order to assert their value in society through scientific authority, most notably in their support of eugenics. This shift within Victorian feminism does not indicate its failings in challenging scientific authority; rather, it demonstrates the power and pervasiveness of scientific authority in the social and political spheres.

Acknowledgements Researching, writing, and revising this chapter was generously supported through a faculty research grant from Misericordia University.

Notes

1. Discussing Welcker's findings, Alexander Ecker claimed that 'the female character is [...] approaching that of a child; woman, in fact, holds an intermediate position between man and child' (1868, p. 352).
2. Charles Darwin, Letter to Caroline Kennard, 9 January 1882. *Darwin Correspondence Project*, University of Cambridge. https://www.darwinproject.ac.uk/letter/DCP-LETT-13607.xml. Accessed on 30 November 2020.
3. Elizabeth Garrett Anderson notes that Maudsley's essay is largely a paraphrase of Edward Hammond Clarke's *Sex in Education: Or, A Fair Chance for the Girls* (1873), a discussion of the American education system and thus not sufficiently adapted to the English system (1874, pp. 582-583).
4. The reference to 'Girton' is to Girton College, Cambridge, the first college for women at Cambridge, which had been established in 1869.
5. Spencer was 'definitely worried by the prospect that women, who generally love the helpless, would promote socialist legislation in order to benefit inferior members of society' (Francis 2007, p. 76).
6. In discussing the catalogue of Darwin's correspondence with women, Samantha Evans notes Darwin's encouragement of female botanists and dependence on women's research to aid his theories, arguing for a more malleable view of gender differences based on environmental changes (Evans 2017, pp. xix–xxvi).
7. For instance, Alfred Russel Wallace, in advocating for his vigour theory, proclaimed, 'The term "sexual selection" must, therefore, be restricted to the direct results of male struggle and combat. This is really a form of natural selection, and is a matter of direct observation' (1889a, p. 296).
8. While there is no definitive evidence that Blind read *Descent* immediately after its publication, an 8 March 1871 letter from Richard Garnett alerted her to a review of the work in the *Saturday Review* (Diedrick 2016, p. 71; Birch 2013, p. 79).

References

Anderson, Elizabeth Garrett. 1874. "Sex in Mind and Education: A Reply." *Fortnightly Review* 21: 582–594.

Barnes, Robert. 1867. "Midwifery and the Diseases of Children." In *A Biennial Retrospect of Medicine, Surgery, and Their Allied Sciences for 1856–66*, edited by Henry Power, Francis Edmund Anstie, Timothy Holmes, Thomas Windsor, Robert Barnes, and C. Hilton Fagge. London: The New Sydenham Society.

Beer, Gillian. [1983] 2009. *Darwin's Plots: Evolutionary Narrative in Darwin, George Eliot, and Nineteenth-Century Fiction*. 3rd ed. Cambridge: Cambridge University Press.

Birch, Katy. 2013. "'Carrying Her Coyness to a Dangerous Pitch': Mathilde Blind and Darwinian Sexual Selection." *Women: A Cultural Review* 24, no. 1: 71–89.
Blind, Mathilde. 1871. "The Song of the Willi." *Dark Blue* 1: 524–528.
Blind, Mathilde. 1889. *The Ascent of Man*. London: Chatto and Windus.
Butler, Josephine. [1874] 2004. "Some Thoughts on the Present Aspect of the Crusade Against the State Regulation of Vice." In *Josephine Butler and the Prostitution Campaigns: Diseases of the Body Politic*, vol. 3, edited by Jane Jordan and Ingrid Sharp, 62. London: Routledge.
Caine, Barbara. 1992. *Victorian Feminists*. Oxford: Oxford University Press.
Clouston, Thomas S. 1882. *Female Education from a Medical Point of View*. Edinburgh: Macniven and Wallace.
Darwin, Charles. 1859. *On the Origin of Species: By Means of Natural Selection, or the Preservation of Favoured Races in the Struggle for Life*. London: John Murray.
Darwin, Charles. 1871. *The Descent of Man and Selection in Relation to Sex*. London: John Murray.
Deutscher, Penelope. 2004. "The Descent of Man and the Evolution of Woman." *Hypatia* 19, no. 2: 35–55.
De Witt, Anne. 2013. *Moral Authority, Men of Science, and the Victorian Novel*. Cambridge: Cambridge University Press.
Diedrick, James. 2016. *Mathilde Blind: Late-Victorian Culture and the Woman of Letters*. Charlottesville: University of Virginia Press.
Ecker, Alexander. 1868. "On a Characteristic Peculiarity in the Form of the Female Skull, and Its Significance for Comparative Anthropology." *Anthropological Review* 6, no. 23: 352.
Endersby, Jim. [2003] 2009. "Darwin on Generation, Pangenesis and Sexual Selection." In *The Cambridge Companion to Darwin*, edited by Jonathan Hodge and Gregory Radick, 73–95. 2nd ed. Cambridge: Cambridge University Press.
Erskine, Fiona. 1995. "*The Origin of Species* and the Science of Female Inferiority." In *Charles Darwin's The Origin of Species: New Interdisciplinary Essays*, edited by David Amigoni and Jeff Wallace, 95–121. Manchester: Manchester University Press.
Evans, Samantha. 2017. *Darwin and Women: A Selection of Letters*. Cambridge: Cambridge University Press.
Francis, Mark. 2007. *Herbert Spencer and the Invention of Modern Life*. London: Routledge.
Geddes, Patrick and Thompson, John Arthur. 1889. *The Evolution of Sex*. New York: Scribner's.
Hamlin, Kimberley A. 2015. *From Eve to Evolution: Darwin, Science, and Women's Rights in Gilded Age America*. Chicago: University of Chicago Press.
Hayden, Wendy. 2013. *Evolutionary Rhetoric: Sex, Science, and Free Love in Nineteenth-Century Feminism*. Carbondale: Southern Illinois University Press.
Heater, Derek. 2006. *Citizenship in Britain: A History*. Edinburgh: Edinburgh University Press.
Holmes, John. 2012. *Darwin's Bards: British and American Poetry in the Age of Evolution*. Edinburgh: Edinburgh University Press.
Influence on Women of Special Brain-Work. 1881. *The Lancet* 120, 5 March: 379.
Jann, Rosemary. 1996. "Darwin and the Anthropologists: Sexual Selection and Its Discontents." In *Sexualities in Victorian Britain*, edited by Andrew H. Miller and James Eli Adams, 79–95. Bloomington: Indiana University Press.
Jex-Blake, Sophia. 1874. "Sex in Education." *The Examiner*, 2 May: 457.

Kohlstedt, Sally Gregory and Jorgensen, Mark R. 1999. "'The Irrepressible Woman Question': Women's Responses to Evolutionary Ideology." In *Disseminating Darwinism: The Role of Place, Race, Religion, and Gender*, edited by Ronald L. Numbers and John Stenhouse, 267–294. Cambridge: Cambridge University Press.

Levine, George. 2006. *Darwin Loves You: Natural Selection and the Re-enchantment of the World*. Princeton: Princeton University Press.

Maudsley, Henry. 1874. "Sex in Mind and Education." *Fortnightly Review* 21: 466–483.

Mill, John Stuart. 1869. *The Subjection of Women*. London: Longmans, Green, Reader and Dyer.

Millingen, John Gideon. 1849. *The Passions; or, Mind and Matter*. London: John and Daniel A. Darling.

Moscucci, Ornella. 1990. *The Science of Woman: Gynaecology and Gender in England. 1800–1929*. Cambridge: Cambridge University Press.

Murphy, Patricia. 2006. *In Science's Shadow: Literary Constructions of Late Victorian Women*. Columbia: University of Missouri Press.

Naden, Constance. [1885] 1891. "The Evolution of the Sense of Beauty." In *Further Reliques of Constance Naden: Being Essays and Tracts for our Times*, edited by George M. McCrie, 76–101. London: Bickers and Sons.

Naden, Constance. 1887. *A Modern Apostle: The Elixir of Life; The Story of Clarice; and Other Poems*. London: Kegan, Paul, Trench, and Co.

Paul, Diane B. 2003. "Darwin, Social Darwinism and Eugenics." In *The Cambridge Companion to Darwin*, edited by Jonathan Hodge and Gregory Radick, 214–239. Cambridge: Cambridge University Press.

Rappaport, Helen. 2003. *Queen Victoria: A Biographical Companion*. Santa Barbara: ABC-CLIO.

Richards, Evelleen. 2017. *Darwin and the Making of Sexual Selection*. Chicago: University of Chicago Press.

Romanes, George. 1887. "Mental Differences Between Man and Women." *Nineteenth Century* 21, no. 123: 656.

Russett, Cynthia Eagle. 1989. *Sexual Science: The Victorian Construction of Womanhood*. Cambridge, MA: Harvard University Press.

Schreiner, Olive. 1889. "Life's Gifts." *The Woman's World* 2: 408.

Spencer, Herbert. 1873. *The Study of Sociology*. London: Henry S. King.

Simcox, Edith. 1887. "The Capacity of Women." *The Nineteenth Century* 22: 391.

Tait, Lawson. [1874] 1883. *The Pathology and Treatment of Diseases of the Ovaries*. 4th ed. New York: William Wood and Co.

Tuana, Nancy. 1993. *The Less Noble Sex: Scientific, Religious, and Philosophical Conceptions of Women's Nature*. Bloomington: Indiana University Press.

Wallace, Alfred Russel. 1889a. *Darwinism: An Exposition of the Theory of Natural Selection with Some of Its Applications*. London: Macmillan.

Wallace, Alfred Russel. 1889b. "An Introductory Note." In *The Ascent of Man*, edited by Mathilde Blind, v–vi. London: T. Fisher.

Wallington, Emma. 1874. "The Physical and Intellectual Capacities of Woman Equal to Those of Man." *Anthropologia* 1: 558–559.

Walsh, Correa Moylan. 1917. *Feminism*. New York: Sturgis and Walton.

Wollstonecraft, Mary. [1792] 1891. *A Vindication of the Rights of Woman*. London: Walter Scott.

CHAPTER 15

Women, Gender and Computing: The Social Shaping of a Technical Field from Ada Lovelace's Algorithm to Anita Borg's 'Systers'

Corinna Schlombs

The person celebrated as the first computer programmer was a woman: Ada Lovelace. In the mid-nineteenth century, Lovelace showed that a computing machine could be instructed—or: programmed—to perform symbolic processes, and not just pre-specified numerical calculations. Despite the pioneering role of Lovelace and countless other women who programmed the first electronic computers in the mid-twentieth century, computer programming did not become a woman's profession. On the contrary, by the late twentieth century, computer programmers were widely thought of as young, Coke-guzzling and pizza-chomping men who binge-coded overnight, or as genial male millionaires creating platform-based economic ventures. At the same time, the participation of women in computing in most Western industrialized countries levelled off at less than 20%, raising concerns by institutions of higher education, national scientific funding agencies, inter-governmental agencies and other organizations that have sought to increase the participation of women in the field over the last decades. Addressing the participation of women in information and communication technologies, from mechanical calculators to telephone, radio and electronic computing, this chapter highlights the cultural and social factors—not just individual choice or personal aptitude—that shaped the gendering of the field.

C. Schlombs (✉)
Department of History, Rochester Institute of Technology, Rochester, NY, USA
e-mail: cxsgla@rit.edu

© The Author(s), under exclusive license to Springer Nature Switzerland AG 2022
C. G. Jones et al. (eds.), *The Palgrave Handbook of Women and Science since 1660*, https://doi.org/10.1007/978-3-030-78973-2_15

Scholars have in recent decades investigated questions of women, gender and computing. Initially, social scientists and historians studied reasons for the leaky pipeline, that is the observation that girls and women choose in lower numbers than boys and men to take computing courses or enrol in computing degrees, and that they more often leave the field (Margolis and Fisher 2002; McGrath Cohoon and Aspray 2006; Misa 2010).[1] More recently, historians have studied women in the field with a longer perspective. On the one hand, Janet Abbate has celebrated the niches that women have carved out for themselves in computing, from the early machines like ENIAC and Colossus to software service companies in the 1960s and later listservs (2012). On the other hand, Nathan Ensmenger has argued that the professionalization of computer programming in the wake of the software crisis of the late 1960s and the introduction of software engineering have led to a masculinization of the field, with women leaving computer programming (Ensmenger 2010a, 2015). Similarly, Mar Hicks has shown how job classifications in post-WWII Great Britain systematically discriminated against women programmers, stultified their careers and deprived the country of essential resources in a cutting-edge technology in which the country was to lose its lead within a short time (Hicks 2017). Together, these studies show how social perceptions of gender and computing have shaped the field over time.

A brief internationally comparative review sets the stage for this chapter by pointing to important cross-cultural differences in women's participation in computing; it highlights that computing need not be an inherently masculine or feminine field. Next, Ada Lovelace's achievements in early Victorian Britain introduce us to a time before today's disciplinary specialization and professionalization, point to the structuring forces of these social processes and complicate Lovelace's legacy as a hero for women in computing. In the modern period, the experiences of information and communication technology operators and users in the United States in the late nineteenth and early twentieth century, particularly those working with telephone and radio technology, reveal the gendering of the technologies, and how assumptions about men, women and technology shaped each other. Finally, the chapter discusses women's roles in electronic computing and programming in the mid- to late-twentieth century, between women's contribution to the field and the disadvantages and discrimination they sometimes faced. Gender and computing mutually shaped each other, rather than any inherent qualities of women, men and technology.

---

[1] For a more recent focus on gender and diversity, see Aspray (2016).

Electronic Computing in International Comparison: Not Inherently Masculine

A brief international comparison of the participation of men and women in computing reveals glaring differences between countries and global regions. These differences forcefully demonstrate that there is nothing inherently male or female in the ability to program; by nature, men and women are equally suited to program. Rather, it is the gendering of computer programming—that is social ideas about programming, women and men—that have turned computer programming into a masculine field in the United States and many European countries, and into a field more open to the participation of women in other parts of the world, such as Southeast Asian and Arab countries. The international comparison also demonstrates that the social shaping of gender and computing varied over time and by location. It raises the question of why computing has become a primarily masculine field in some countries, and a more feminine field in others.

In the United States, the participation of women in computing stands out among other scientific and engineering fields. Most fields like mathematics, physics and engineering had low female participation in the mid-twentieth century, with a slow but steadily rising rate of women receiving undergraduate degrees in these fields towards the end of the twentieth century. But the trajectory of women in computing was different: women worked in notable numbers in electronic digital computing from the inception of the field in the mid-twentieth century, and the rate of women receiving undergraduate degrees in computer science initially rose rapidly above engineering and the physical sciences to about a third of all graduates by the early mid-1980s. However, their rate dropped precipitously in the mid-1980s and again in the early 2000s, to level off at about one in five recipients of undergraduate degrees in computer science being a woman (Table 15.1).[2] Notably, this trajectory appears to be unique to the United States.

While most European countries today have similar rates of women receiving one in five degrees in computer science, they did not necessarily follow the same trajectory. In Germany, for example, the rate of women graduates with bachelors in computer science never reached above 20 per cent. It peaked in the early 1980s, at 18.7 per cent, hovered around 16 per cent for the remainder of the 1980s and dipped as low as 10.5 per cent by the late 1990s, only to rebound again to 16 per cent for the first decade of the 2000s.[3] In most European countries, the rate of women receiving bachelors, masters or doctoral degrees in Information and Communication Technologies varied between 15 and 21 per cent in 2016, including the Czech Republic, France,

---

[2] Similar trajectories and rates apply to graduate degree recipients and women programmers working in computing in the United States. For a closer discussion, see Hayes (2010).

[3] Data for Germany can be found at Catalyst (2018) and BMBF (2020). On the perception of computer science in Germany, see also Schinzel (2013).

Table 15.1 'Women as percentage of all Bachelors' recipients in the United States, by major field group, 1966–2015. Tabulated by the National Science Foundation, National Center for Science and Engineering Statistics; data from the Department of Education, National Center for Education Statistics: Integrated Postsecondary Education Data System Completions Survey

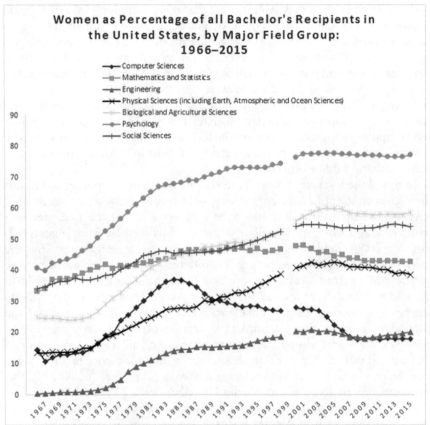

Finland, Hungary, Ireland, Italy, Norway, Poland, Slovenia and the United Kingdom. While some European countries saw female participation in tertiary degrees as low as 5.63 per cent in Belgium, 8.86 per cent in Switzerland, 14.2 per cent in Spain and 14.32 per cent in Austria in 2016, Southern and Eastern European countries led the statistics with women receiving 23.45 per cent of tertiary degrees in Portugal, 27.11 per cent in Serbia, 33.19 per cent in Romania, 38.76 per cent in Greece, and 40.05 per cent in Bulgaria, with Sweden being the only Nordic exception in this lead group at 28.17 per cent (Unesco Institute for Statistics 2021).

By contrast, globally, the participation of women in computing is comparatively high in Southeast Asian and Arab countries—parts of the world not usually known for their recognition of women's rights. For example, women comprised 42.39 per cent of information technology graduates with bachelor, masters or doctoral degrees in Iran in 2016, 43.61 in Saudi Arabia, 73.56 in Oman, 46.58 in Qatar, 57.78 in the United Arab Emirates, 46.32 in India, 44.98 in Sri Lanka and 44.55 in Malaysia in 2015 (Unesco Institute for Statistics 2021). However, closer studies have shown that this comparatively high participation of women fits with overall patriarchal values rather than indicating a sign of women's emancipation. In Malaysia, for example, women perceive programming as a suitable field for themselves because it allows them to meet their families' expectations of a degree with secure job offers and high income, and because it is perceived as safe indoor office work, as opposed to jobs that require outside contacts or even outdoor work, such as computer networking or civil engineering (Lagesen 2008; Mellström 2009).[4]

As this international comparison demonstrates, computing does not necessarily need to be gendered masculine. However, if it is—as has been the case in the United States and most Western industrialized countries in the past half century—it excludes the experiences and contributions of women, often along with those of other minorities such as people of colour and differing abilities. The result is social inequalities, with more than half of the population being deprived of opportunities for lucrative professional careers. Also, computing technologies designed and created by homogenous teams often serve some users—usually male, white and able-bodied ones—better than others, for example when biased online searches connect women of colour with pornography, or artificial intelligences label black persons' faces as gorillas. (Noble 2018). Ultimately, who becomes a computer programmer thus is a question of social justice as well as of technological supremacy and economic competitiveness. The remainder of this chapter will therefore explore how, historically, information and communication technologies have been gendered masculine in the West, and in the United States in particular.

ADA LOVELACE—FIRST COMPUTER PROGRAMMER?

In recent years, Ada Lovelace has been celebrated as the first computer programmer and a role model for girls to pursue careers in science and technology. Previously, critics had dismissed Lovelace as mathematically incompetent, psychologically unstable and addicted to opiates.[5] Yet, in the 1980s, the US military named a programming language in her honour; the 2010s saw a veritable wave of about a dozen children's and juvenile books remembering Lovelace; and when the *New York Times* added 'overlooked' obituaries for

---

[4] Refer to two chapters in this handbook: Kate Shaw on Women in Science in the Arab World, and Carol Mukhopadhyay On India.

[5] For a critical review of this slighting by biographers and computer historians, and an attempt to neutrally recognize Lovelace's contributions, see Misa (2016). Similarly, Krämer highlights the uniqueness of Lovelace's biography and her accomplishments, warning against too quick of a categorization (2015).

women in the wake of the me-too movement in 2018, Lovelace's was one of the first to be published (Hayes 2018; Doudna 2018; Lew 2018; Sanchez Vegara 2017; Robinson 2016; Stanley 2016; Bodden 2017; Labrecque 2017; Loh-Hagen and Bane 2018; Stone and Priceman 2018; Wallmark 2016).[6] Still, without diminishing Lovelace's accomplishments, it may be an overstatement to call her the first programmer when the first digital computer was still a century in the future. Yes, Lovelace wrote a program—that is a set of instructions to be performed by a machine—with nestling loops and decisions for a machine that was only described in fragments and still in flux (Misa 2016). But instead of hailing Lovelace's technological firsts, this chapter emphasizes that Lovelace lived before our socially embedded notions of the sciences, scientists and scientific careers—and their often implicitly assumed masculinity—emerged. Consequently, Lovelace did not distinguish between the sciences and humanities the way we do today, and although she is now often romanticized as a model for young girls, she did not pursue a scientific career herself. Instead, she cast herself as an assistant, a socially acceptable role for Victorian women.

Born in December 1815 as the only daughter of the Romantic poet Lord Byron and his wife, Lady Annabella Milbanke, Lovelace enjoyed a privileged yet extraordinary upbringing. Her parents separated a month after her birth, and Lovelace never again saw her father, who died at the age of thirty-six in self-imposed exile from Britain. The reasons for the separation were subject to wild speculation then and now, with only one-sided accounts remaining (Moore 1977; Woolley 1999; Markus 2015; Stein 1985).[7] Yet, during a time when women could not even own property, filing for divorce and maintaining custody over her daughter posed legal and social challenges to her mother. Lovelace grew up in privileged and financially prosperous conditions thanks to her noble birth and inheritances from her mother's family. At a time when girls were typically barred from attending schools and universities, Lovelace's mother arranged for the socially appropriate education at home for daughters, through governesses and tutors. She herself had received an extensive education, including instruction in mathematics, which was untypical for the time. Thus, Lovelace was taught in a wide range of subjects, from languages and music to math and science, in a highly regimented schedule enforced by reward tickets for successful lessons that was a common method of disciplining at the time.

Popular writing today often dwells on Lovelace's early interest in flying machines as an example of her attraction to technology and engineering. Yet, Lovelace pursued a wide range of subjects. For example, she studied engineering and mathematics, learned languages including French and Italian,

---

[6] For the *New York Times* obituary, see Miller (2018, p. B5).

[7] Speculations range from bipolar disorder and domestic violence to womanizing and an incestuous affair, with only Lovelace's mother's account being preserved, and her father's diary burnt after his death.

played and composed music, and practiced appropriate 'feminine' skills such as embroidery.[8] Indeed, it was Lovelace's knowledge of the French language, in addition to her technical understanding, that led her to engage with machine algorithms in the first place. More importantly, she approached the object of her learned interest from different angles at the same time. For example, when she took an intense interest in flying machines as a young teenager, she implored her mother for books on birds with trans-sections which she intended to use as animal models to build wings for her own size. At the same time, she metaphorically explored flying to overcome emotions related to her separation from her mother.[9] Thus, Lovelace would not have thought of calculating machines as a merely technical subject. Indeed, not until the midtwentieth century were the natural and technical sciences on the one hand, and the humanities on the other, seen as mutually exclusive approaches (Snow 1959). This broader, in today's sense interdisciplinary, way of thinking allowed Lovelace's prescient remarks on the machine's larger symbol-processing capabilities. Reading today's disciplinary distinctions into Lovelace's work risks the fallacies of ahistorical misunderstanding of her accomplishments.

At the age of seventeen, Lovelace was introduced to the royal court, and she now moved to London's social circles which included notable scientists. Thus, she met Mary Somerville, a highly regarded populariser of mathematical and astronomical knowledge who became a social and scientific mentor to her, and Charles Babbage, an independently wealthy gentlemanly scientist who pursued his wide-ranging interests in mathematics, astronomy, economics, theology, mechanical calculators and science reform. In addition to immersing herself in the busy social season with opera visits and dances, Lovelace also sought the company of her learned friends. Invited to one of Babbage's now infamous soirées with a demonstration of his mechanical calculating machine, the Difference Engine, her astute questions displayed her acute understanding of the mechanism. Babbage and Lovelace stayed in loose contact over the next decade, while Lovelace married William King—a man nine years her senior to whom she was introduced by Somerville's son—and gave birth to three children within three years.

From late 1842 to the summer of 1843, Lovelace devoted the better part of a year to translating a French article on Babbage's newly envisioned machine, the Analytical Engine, into English and adding original notes that further demonstrated the machine's capabilities. By this time, Babbage's work on his first machine had stalled; this was the Difference Engine which was devoted to one single purpose, to calculate nautical tables, and for which he had received

---

[8] Thanks to Dominique Molee for pointing this out.

[9] Biographies with a focus on Lovelace's mathematical accomplishments typically also discuss her early interest in flying machines, see Toole (1998, pp. 20–29), Hollings et al. (2018, pp. 17–21), Essinger (2014, pp. 47–56) and Warrick (2007). On Lovelace's integration of mathematical, scientific, musical, literary, and philosophical thought, see also Hartmann (2015).

government funding.[10] Indeed, a meeting between Babbage and the British prime minister had failed to secure additional funding and the machine was to be relegated to a museum. In this situation, Lovelace decided to intervene and support Babbage's visionary work. With encouragement from Charles Wheatstone, the British inventor of telegraph technology, she translated the only published description of Babbage's Analytical Engine, a French article in a Swiss journal by Italian military engineer Luigi Menabrea. When Lovelace showed the translation to Babbage in the early winter of 1843, he encouraged her to go beyond merely translating Menabrea's work and so she resolved to add notes under her initials.

Lovelace wrote seven notes which, together, were twice as long as the original article. The seventh and last note, on the calculation of Bernoulli numbers, is commonly seen as the most important. Bernoulli numbers are a set of rare numbers that had not been calculated manually; Lovelace was aware of their mathematical significance through a correspondence course on calculus with London University professor Augustus de Morgan. Babbage provided the general formula for the Bernoulli calculation, while Lovelace developed the step-by-step algorithm. Both corresponded, often daily, to ensure Lovelace's correct understanding of Babbage's envisioned design for the Analytical Engine, and to proofread. The preceding six notes served to prepare the seventh. For example, the Analytical Engine was to be run by punched cards, after the model of the Jacquard loom, a machine fitted to a loom to simplify the manufacture of complex textile patterns. In her notes, Lovelace explained how to use loops and decision conditions to reduce the number of punched cards needed, how to use the machine's 'storehouse' (memory, in today's terms) and how to solve analytical calculations with variables and complex problems. The notes included some remarks about the capabilities of the Analytical Engine which, according to some analysts, demonstrated that Lovelace's understanding of the machine surpassed even Babbage's own. While Babbage primarily saw the machine as a universal calculator, Lovelace pointed out that the machine could perform analytical operations, including symbol processing such as musical representations, and solve complex problems that could not be calculated manually, for example the three-body problem (Menabrea 1843).[11] However, Lovelace cautioned, in what has been read as an objection against artificial intelligence, the machine could not perform any operation unless it was first instructed to do so by a human operator.

Without diminishing Lovelace's important contribution, it can hardly be said that the publication of her article—the preparation of which took the

---

[10] Multiple reasons stalled the completion of the Difference Engine: Babbage continued to change the design, extraordinary needs for precision, conflicts with its master craftsmen, and financial shortfalls. See Swade (2000) and, for a broader historical analysis, Schaffer (1994).

[11] For analysis of the notes, see Misa (2016), Essinger (2014), Fuegi and Franci (2003) and Huskey and Huskey (1980).

better part of a year—amounted to a scientific career. Indeed, hardly any of Lovelace's learned contemporaries can be said to have pursued a scientific career, and the term 'scientist' was established only over the course of the nineteenth century. The gentlemanly natural philosophers of her time were expected to be independently wealthy, assuming that their financial disinterest allowed for objective observation of natural experiments (Shapin and Schaffer 1985). Babbage himself fits the model of the gentlemanly natural philosopher, although he worked towards science reform that aimed at more specialization and professional engagement. Babbage was independently wealthy through inheritance from his banker father, and he published in a wide range of fields from astronomy and mathematics to theology, in addition to building the calculating machine. While women were typically relegated to assistive roles, at least two of the women in Lovelace's circles derived their income from their scientific pursuits. Mary Somerville, Lovelace's mentor, derived her family income from the sales of her popular books, and Caroline Herschel, the aunt of Babbage's close friend John Herschel, received a stipend from the British government for her work assisting her older brother William Herschel in his astronomical observation. While Lovelace needed no income from her mathematical pursuits, she chose a similarly assistive role.[12] Although, in their correspondence, she and Babbage addressed each other like familiar colleagues, and she was not above occasionally commanding him around, she also called herself his 'fairy,' signalling her supportive role.

Lovelace's privileged social background and her extraordinary education enabled her to compose and publish the notes for which she is today celebrated as the first programmer. A closer look also reveals insights into the historical construction of the sciences and scientific careers. In the early nineteenth century, Lovelace and her contemporaries perceived natural phenomena and engineering problems from different perspectives which, today, would be considered truly interdisciplinary. Also, Lovelace can hardly be considered a professional scientist or engineer. Although she pursued a rigorous education and sought out advanced tutoring in mathematics as an adult, she divided her time between her mathematical and musical skills and her obligations to run a complex household, oversee the education of her three children and respond to her mother's needs. As we will see in the next section, by the late nineteenth and early twentieth century, the invention and operation of information and communication technologies became a career from which individuals derived an income as well as a social position—and which also took on gendered associations.

---

[12] Like her mentor Somerville, Lovelace added supplementary explanations to her translation. See Pohlke (2015).

Telegraph, Telephone and Radio Technology and Their Operators: Gendering

Social expectations about men, women and technologies in the United States shaped who developed and operated information and communication technologies, and what the social place of the technologies and the opportunities for the operators were from the nineteenth century onwards. Thus, in the two decades after Samuel Morse first demonstrated his electric telegraph technology in 1844, mostly young men operated telegraphs, playing pranks on each other and forming a competitive work culture. For many of the young telegraph operators, who included Thomas Edison and Andrew Carnegie among their ranks, these first jobs offered opportunities for social advancement and, in some cases, distinguished careers. By contrast, when Alexander Bell's company introduced telephones into the offices and homes of people in the last two decades of the nineteenth century, prank-playing young men were soon dismissed in favour of young women who appropriately provided a personal service to their customers. Many of them worked only for a few years, often under taxing conditions for a minimal pay, before marrying and leaving the workforce. Radio again became a young men's game in the 1910s, with male hobbyists competing in their attics to see how far they could listen at a given night. In view of changing ideas about masculinity from physical strength to technical control, mastering radio technology promised to prepare the young hobbyists for successful careers. Information and communication technologies from the mid-nineteenth to the early twentieth century were gendered; however, whether they became gendered masculine or feminine depended on larger social expectations of men and women and their work. Technically, telegraph and telephone technology were very similar in that both relied on electric currents signals and, in principle, both technologies could use the same wire infrastructure. Indeed, Western Union, which by then held monopoly control over telegraphy services, initially competed with Bell by offering its own line of telephones. Only corporate agreements between Western Union and the Bell company, reinforced by federal regulations, ensured that telegraphy and telephony became separate communication systems (John 2010). With regard to the gendering of their workforce, both companies also went separate ways, albeit for different reasons.

In the United States, mostly young men served as telegraph operators in the decades leading up to the Civil War, sending messages in Morse code from one station to the next, and copying incoming messages. Although women sometimes worked as telegraph operators in rural areas, mostly young men took on positions as telegraph operators in larger urban offices.[13] Skilled young

---

[13] For a labor history of telegraphy see Gabler (1988). After the Civil War, increasing numbers of women also worked in telegraphy; see Jepsen (2000). In Britain, by contrast, increasing employment of women followed the nationalization of the telegraph system through the 1868 and 1869 Telegraph Acts. See Bruton and Hicks (2020).

men could tinker with the apparatus or business organization, and improvements easily allowed them to climb the corporate ladder, offering many young men from farming backgrounds a chance at social uplift, with an office job in an urban centre. The young men quickly developed their own work culture, including languages, songs and rituals that revolved around competition. New operators were harangued as 'slugs' or 'plugs' for their lack of speed, and often introduced to the culture with harassing pranks. During downtimes, operators used telegraph lines to play chess at a distance, chat privately, develop friendships, and in rare cases—if a women operator was at the other end of the line—even romantic relationships. Telegraph work culture thus was characterized by competition and boyish pranks that could be seen as distinctly masculine.

Telephony, by contrast, became gendered female only three decades later, despite the technological similarities between the two. Bell, also an ardent eugenicist and believer in oral education for the deaf and hard-of-hearing—that is lip-reading and voicing rather than signing—originally developed voice transmission over electric lines when tinkering with communication aids for the deaf.[14] He initially demonstrated his new apparatus at the 1876 Centennial exhibition, and then toured the United States, presenting it at shows as a technological curiosity. A year later, he was granted a patent, and he founded the Bell telephone company. Given the similarities to telegraph technology, it is not surprising that the Bell company initially hired young men as telephone operators.

Yet, these young men soon proved utterly unsuited to operating the new technology, and young women quickly replaced them. Protected by patents from market competition, the Bell company envisioned telephony as a tool for business communication, and thus marketed it at rather high prices. Telephone devices were installed in offices and businesses as well as in the homes of well-to-do citizens—thus invading the protected domestic sphere. Clients, who included many men of higher ranks, needed to pick up the receiver and interact with the device if they wanted to place a call. In the early years of telephony, connecting calls was still a precarious activity, particularly if it involved going through multiple switchboards, and call quality was not satisfying. Young male telephone operators playing pranks on telephone clients—with whom they directly interacted, different from telegraph offices, where operators were not necessarily in customer-facing positions—did no good in this situation; young women proved better suited for these jobs.

Young women provided appropriate service to Bell's customers, fitting well into the tradition of the countless housemaids—and increasingly secretaries and sales assistants—that telephone customers were used to. The young women were cheerful and helpful, as they had been raised, and the Bell

---

[14] Ironically, this resulted in the exclusion of deaf and hard of hearing individuals from telephony for almost a century, until TTY modems were developed in the United States in the 1960s.

company later trained its operators to serve customers with a personable voice. In the early days of telephony, operators typically called each number in the morning to ensure that the battery was sufficiently charged, and they often developed a personal relationship with their clients. Male customers sometimes developed a flirtatious relationship with their operators, even giving them pet names. In other instances, operators served as babysitters, with telephone receivers being left near a sleeping baby's crib, and the operator notifying the mother when the baby awoke. Operators also discretely educated their customers to interact appropriately with the new technology, for example by not shouting into the line, and by picking up the receiver and greeting 'hello' to indicate they were listening (Schmitt 1930, p. 18).[15] By the early 1910s, the Bell and other phone companies firmly structured and controlled telephone operations according to the principles of scientific management, prescribing the exact wording to shave seconds of every single call, although the company upheld the idea of personal service provided by the women telephone operators.

For the women, working as telephone operators was a welcome opportunity. In the 1880s, telephone operators were among the first women to take on salaried work in urban centres, as were many women working as typists, file clerks and secretaries. In contrast to housemaids and sales assistants, these were white-collar positions in offices—even if the telephone offices were initially located under the roof, where the telephone cables entered the phone company's building, exposing the operators to heat in the summer and cold in the winter. Telephone operators received minimal wages, usually requiring the women to continue living with their families and saving on their lunch costs, and they had few chances of career development. Many women worked as telephone operators only for a few years, between their high school graduation and marriage, and then left the workforce. The few women who remained in the industry typically stayed single, and they could advance into lower managerial positions, but not further. In other words, telephone operator turned into a low-paid, dead-end, yet vital, job. Notably, the Bell company delayed automating its telephone switchboards because executives felt that the personal service provided by the women phone operators were an important part of the company's connection to its clients (Green 1995; Lipartito 1994). Some companies already introduced automatic switchboards—which would require customers to dial a number rather than pick up the receiver and request a connection from an operator—in the late 1890s; but the Bell company waited until 1919. Women telephone operators had become an important part of the telephone system, although they themselves reaped few benefits from their occupation.

As telephone operations became gendered feminine, so did telephone technology (Martin 1991; Rakow 1992). Once the Bell patents expired in 1894,

---

[15] For telephone operators, see also Green (2001).

farmers and other Americans eagerly adopted telephone technology, sometimes running barbed wires along fields to hook up party lines to local phone companies. Bell competitor companies decreased prices, making the technology available to larger parts of the population, and the telephone became a social communication medium as much as a business device. Farmers listened to the news or the weather forecast on the phone, they called in for church sermons or concerts and they 'visited' on the phone on Sundays when the weather or other circumstances forbade face-to-face visits. The telephone allowed staying in touch with family, neighbours and friends—a form of social work often conducted by women. Soon, phone companies branded women chatting on party lines as problems to be controlled (Kline 2000).

With the emergence of radio technology, the gendering shifted again. While knowledge about electromagnetic waves had grown during the late nineteenth century, at the turn of the twentieth, entrepreneurial inventors like Guglielmo Marconi, Reginald Fessenden, Lee de Forest and John S. Stone developed practical radio transmitter and receiver devices, commercialized the technology and improved it over the next two decades to allow for voice transmission and higher quality in transmission and reception (Hong 2001). During this time, countless hobbyists took to the airwaves. Mostly young, white, middle-class men devised crystal radio receiver sets from cheaply available materials, such as copper wires wound around a Quaker oatmeal box and a telephone receiver lifted from a local phone booth; only the crystal needed to be purchased for a few dollars. Listening to the airwaves from their attics, the hobbyists sought to reel in far-away stations, formed clubs that met on the air to practice their transmission of Morse code—'wireless telegraphy'—formed relay leagues to transmit messages from coast to coast, and excitedly partook in current events, for instance listening to radio messages related to the Titanic disaster.

Journalists hailed radio hobbyists as innovative tinkerers whose technical expertise opened great careers. One such example was Walther Willembourg, a student at Stevens Institute of Technology outside of New York City, who had built himself a powerful radio transmitter. Willembourg was not above the kinds of pranks that telegraph operators had also committed, interfering with other messages and overpowering a military officer with his impressive equipment. Despite such behaviour, Willembourg was hailed as a model for the future, with the mastery of radio technology preparing Willembourg and his likes for successful careers. Along the way, they also became heroes for their selfless use of radio technology in times of disaster—for example, when ship radio operators sent distress signals without a pause to call for help, or when radio relay league operators allowed for communication into a disaster zone after a storm or earthquake (Douglas 1987). These accolades for technical mastery fell into a time of shifting social ideas about masculinity. Up to the late nineteenth century, the United States had been a pioneer farming country that required rugged individualism and muscular strength. With the shift from an agricultural to an industrial society, and the growth of large corporations with managerial hierarchies, professional success would demand different qualities from young men. Rugged individualists would be replaced by team players in the corporate hierarchy, and muscular brawn would be replaced by technological mastery. In this new atmosphere, radio hobbyists like Willembourg seemed to be poised for success.

The gendering of information and communication technologies in the nineteenth and early twentieth century thus varied. Some technologies, such as telegraphy and the radio, became gendered masculine. Operators exhibited competitive traits, committed more than the occasional prank, and at the same time had chances for professional advancement and successful careers. Other technologies, like the telephone, became gendered feminine. Operators were expected to provide personal service to customers and comport themselves appropriately although they received poor payment and had few career opportunities. Women also worked in other office positions that developed from the need to process increasing amounts of information during the rapid corporate expansion of the late nineteenth century, for example as typists, stenographers and filing clerks. In these positions, they faced similar gendering and working conditions as the telephone operators. Less is known about the gendering of another corporate information technology emerging at the turn of the twentieth century, punchcard machines, which are sometimes considered predecessors to electronic computers. While the historical record is mixed, in the United States women seem to have mostly entered data onto punch cards, and men operated the sorting and tabulating machines that processed the cards.[16] The examples discussed here demonstrate that information and communication technologies typically became gendered, with immediate social implications for the technology and its operators and users; yet, whether a given technology would be gendered masculine or feminine was not predetermined.

Electronic Computing: Social Shaping of a Technical Field

Electronic digital computers were first developed during World War II. As the statistical comparison at the beginning of this chapter demonstrated, in the United States, the field initially welcomed women practitioners, with one in two graduates in computing being women by the early 1980s. Open questions included whether computing was a science or an art; what a typical career in computing looked like; and whether the field would be gendered masculine or feminine. Historians have shown that women carved out niches in computer programming, from programming the initial electronic computers to founding women-only software services companies in the 1960s, and later women-only listservs, conferences and associations. But women also faced stereotyping of their work as routine, and they sometimes concealed their identities as women

---

[16] In Great Britain, women also appear to have operated tabulating and sorting machines.

programmers. In the wake of the software crisis of the late 1960s, software engineering methodology helped professionalize the field and raise its status; also, the number of women —and other minorities—plummeted. A closer look allows us to recognize that the social forces shaping the field as computer programming became a technical and engineering field, not an art; professional careers were to be built on degree qualifications, not interest and aptitude; and computing became masculine.

Women have worked as programmers from the early days of electronic computing. In the United States, women programmed the ENIAC, an electronic digital computer built at the University of Pennsylvania's Moore School for Engineering in Philadelphia during World War II. Originally devised for the calculation of ballistic trajectories for the US military, the machine was completed only after fighting had ended, and the first—highly classified—program to run on it supposedly was a Monte Carlo calculation for the detonation of the hydrogen bomb.[17] A group of six women—Frances Elizabeth Snyder Holberton, Kathleen 'Kay' McNulty Mauchly Antonelli, Frances Elizabeth Snyder Holberton, Marlyn Wescoff Meltzer, Frances Bilas Spence and Ruth Lichterman Teitelbaum—'programmed' the computer by changing its wire connections.[18] The women had begun as human 'computers' trained at the Moore School who manually performed ballistics calculations; advanced to operating an analogue calculator, the Differential Analyzer, for the same purpose; and eventually were selected for work on the ENIAC. Without any instructions or models to refer to, the women studied technical drawings for the ENIAC to understand the machines' logic, and how to program it; they performed highly innovative and original work. Like their predecessors in telephone operations, the women relished their work and that it opened opportunities to them beyond the typical teaching jobs for mathematically trained women (Light 1999; Abbate 2012).[19]

Yet, the women were not recognized for their pioneering work at the time, even though one of them, Frances Elizabeth Snyder Holberton, saved the day of the public ENIAC inauguration in February 1946 when she fixed the demonstration program at the last minute. None of them was invited to the celebratory dinner, and the official press release even used a photo that was manipulated so that it showed the computer with a male engineer, and

---

[17] For the construction and use of ENIAC, see Haigh et al. (2016).

[18] Technically, the ENIAC was not yet a programmable computer because it didn't have a stored program; instead, the program was expressed in wired connections. To avoid confusion the women programmers here are listed with their complete names, including birth and married names. In Great Britain, similarly, the wartime decryption efforts at Bletchley Park relied heavily on women to program and operate the so-called Colossus computers. See Abbate (2012, pp. 11–38).

[19] By now, two documentaries tell the story of the ENIAC programmers: *Top Secret Rosies: The Female 'Computers' of WWII* by LeAnn Erickson, (2010) which uses interview footage by ENIAC programmers as well as by women involved in the manual and mechanical ballistics calculations at the Moore School, and *The Computers* by Kathy Kleiman, (2015) which presents Light's arguments without explicit references.

cropped out a woman programmer in the background (Light 1999). Three of the six women led distinguished careers in programming, accompanying the ENIAC on its delivery to the Aberdeen Proving Ground and later programming the UNIVAC computer, the first commercially available computer that the ENIAC lead-engineers, John Mauchly and John Presper Eckert, built in the early 1950s. The others dropped out of the field. They may have followed the trend of hundreds of young women leaving their wartime occupations as young men returned from the front; or, they may have fallen victim to the gendering of programming envisioned by lead consultants and collaborators to the ENIAC project, such as John von Neumann and Herman Goldstine. Both saw coding as a two-step process consisting of the intellectual work of devising the program logic and flow-chart (thought of as a man's and scientist's job), and the routine task of coding the flow-chart in machine language (to be done by menial labour, including women) (Goldstine and von Neumann 1947).[20]

As increasing numbers of academic institutions and large corporations acquired computers, programming proved to be an essential bottleneck. While scientific computing installations typically required a degree in mathematics, business computing proved different. Managers observed that some programmers were significantly more productive than others, and they wondered whether programming abilities built on innate talent that could be identified, like in an artist, or whether programming abilities could be taught through systematic instruction, like a science. By the late 1960s, 80% of companies used the IBM Programmer Aptitude Test, and more than 700,000 persons took the test in 1967 alone. The test could be administered cheaply and quickly to large numbers of persons from secretaries to managers; it also relied on the assumption that productive programmers had some innate qualities that could be detected. Some computer historians have argued that personality tests privileged personality traits such as 'disinterest in people' and a focus on mathematical trivia, logic puzzles and work games, thus establishing and enforcing the stereotype of programmers as introvert, object-oriented geeks—qualities that were more socially appropriate for men than for women (Ensmenger 2010a). Other historians have pointed out that access to personality tests was easy and open to women—unlike higher education at the time, particularly in technical fields like engineering—and that the tests therefore opened a path to programming for women (Abbate 2012). Regardless of how one interprets the implications of aptitude tests, by the mid-twentieth century, the distinction between the two cultures of the natural sciences and humanities was clearly established, rhetorically and institutionally (Snow 1959). Unlike in Victorian Britain, where Lovelace had mixed technical and artistic approaches in her thinking about flying machines and mechanical calculators, programming now could only be either a scientific or humanities field, not both. While programming skills initially appeared to be based on innate talent, like an

[20] These assumptions soon were proven to be out of touch with reality. See also Ensmenger (2010a).

art, it eventually became a science, with the introduction of academic degree programs from the late 1960s onwards.

In addition to the question of whether programming was a science or an art, it also was an open question of what a career in computing was to look like. Some women used the initial fluidity of the computing field to carve out niches for themselves, and they loved their work. By the 1960s, for example, two women entrepreneurs independently ran software services companies: Elsie Shutt had founded Computations Inc. in Harvard, Massachusetts, in 1957, and Stephanie Shirley started Freelance Programmers Ltd in Hemel Hempstead near London in 1962. Both followed similar corporate strategies: they hired primarily women programmers who worked from home, on a part-time basis, while taking care of their families. For these women programmers, the employment opened an opportunity to stay current in the field, and to remain professionally engaged during their family phase when many companies expected them to leave their jobs. Both companies conceptualized programs as modular units that could be divided between different programmers, and they developed excellent procedures and communication manuals for coordination; thus, they solved some of the problems that software engineering was to address later on. They also benefited from a stable workforce and low overhead costs since the women worked from home, without benefits. Although the companies primarily operated for the social mission of providing work for women programmers, with profit being secondary, they offered high-quality services for low prices (Abbate 2012). Their example showed that successful programmers could work part-time, from home.

Yet, both entrepreneurs occasionally obscured their identity as women as well as the identity of their workforce and work conditions. They typically offered fixed price contracts which required good estimates of the work to be done—and concealed the nature of the workforce. Also, Shirley went by 'Steve' rather than 'Stephanie,' and the women employees typically wore name tags so that they could recognize their colleagues during visits at client sites and avoid awkward introductions if they had never met in person. When the women made phone calls to clients, they sometimes played in the background tapes with office noises, suggesting a professional work environment that belied their domestic setting. Thus, on the one side, they carved out a niche that allowed them to pursue the work they enjoyed; on the other, they hid the nature of their work supposedly out of fear that their identity and domestic setting would diminish their professional recognition. Different from Lovelace's time, the assumption was that programming careers—as most other professional careers—required full-time work, in an office. While the nature of computer programming did not require these work conditions, social expectations about professional careers shaped the assumptions.

By the late 1960s, computer programming faced the so-called software crisis, officially named at a 1968 NATO conference in Germany. Programming appeared to lag behind the development of computer hardware, with computer programs being completed behind schedule and over budget, and often of poor quality, with many remaining bugs. Also, the field appeared to face a severe workforce shortage. Yet, just simply recruiting more programmers to the job was not a solution, as contemporaries quickly and painfully discovered, because rising communication problems and complexity delayed programs even further; there was no silver bullet (Brooks 1957). Multiple solutions were proposed for the software crisis. One of them was the development of easily legible higher-level programming languages, such as IBM's PL/1. Comprehensible to non-specialists, PL/1 promised corporate managers easier control over the arcane work of computer programmers, and could have opened programming to a larger pool of employees without previous programming experience, including 'Susie Meyers,' a petite blonde featured in a PL/1 advertisement which, playing on gender stereotypes, envisioned programmers as young women, presumably easily replaceable and low-paid (Ensmenger 2010a, pp. 133–134). The solution that eventually won out was software engineering.

Software engineering called for the systematic application of scientific and technical knowledge to the design, creation and maintenance of software, and it included procedures to define the goal and scope of a program, divide and document the work, and communicate between programmers. This allowed programmers to claim authority over their work from managerial control and increased the role of academic computer scientists in accredited degree programs in training future programmers—over, for example, psychologists administering personality tests. Software engineering, it has been argued, professionalized computer programming and raised the status of the field (Ensmenger 2010b). It also made it more masculine. Computer science departments often cemented existing stereotypes of programmers as hackers—unkempt geeks working late at night, gulping down pizza and Coke. Requiring an academic degree for entry into the field—rather than interest, experience or other training—software engineering disadvantaged women and other minorities who faced obstacles to entering higher education in a technical field, and who more often dropped out than their white male peers.

In addition, workplace segregation and discrimination carried over from office work and data processing into electronic computing, curbing women's careers. In the United States, women had long operated office technologies such as typewriters and filing cabinets—often in routine low-paid positions without perspective for career advancement (Davies 1982; Strom 1992; Schlombs 2010). In data processing, it appears that women entered data by punching it on the cards, while mostly men operated the tabulating machines. This gendered division of labour carried over into electronic computing, with women becoming responsible for data entry and men operating and programming the new machines (Haigh 2010). In Britain, by contrast, women had operated tabulating machines; for example, they had dominated the so-called 'machine ranks' and worked as operators of punchcard machines in public service, for operations such as social security calculations. Women operated the machines and programmed them. Yet, job classifications ensured that women were relegated to lower-paying grades, and although women rose to supervisory positions, they were not allowed any career perspective. On the contrary, when electronic computers began to replace punchcard machines in the British public service, women were forced out of their jobs as operators and programmers, and men took their positions. In some cases, women even were made to train their male successors (Hicks 2017).

Although computer programming became more masculine over time, women remained in the field and continued to carve out their niches—even if in lower numbers. By the 1980s, women began to associate and create environments in which they could thrive. For example, the computer scientist Anita Borg formed a mailing list, called Systers, for women computer programmers in 1987. Borg, who herself had learned to program while working at an insurance company and then returned to college for a graduate degree in computer science, worked passionately towards increasing the number of women in technical fields. The mailing list was restricted to highly trained women and strictly devoted to technical issues; yet, it avoided the often competitive and acerbic tone on many listservs and thus created a place where women could ask questions without the fear of being put down, engage in difficult questions, and advance each other and their field. Borg later co-founded the Grace Hopper conferences, a series of computer conferences for women named after the pioneer programmer who led the development of the first computer language compiler. These conferences provided a space for women to meet, network, and inspire each other, and led to the founding of the Institute for Women and Technology—now renamed the Anita Borg Institute. Today, major funding organizations, employers and institutions of higher education have created outreach programs for school-age girls to gain experience themselves as programmers, such as through the Obama administration's *Hour of Code*, and associations such as *Women in Computing* allow women undergraduate students to network with each other and develop their leadership skills (Abbate 2012, 2018; Bix 2017).

Many factors contributed to turning computing from a more open to a more masculine field: the understanding of the field as a science rather than an art, requiring an academic degree in a technical field that male students were more likely to obtain; the expectation of working full time, over a lifelong career that was more typical for men than for women; computer engineering and professionalization strengthening these factors; gendered stereotypes about professionals that did not appeal to young girls and women; and outright discrimination disadvantaging women's success in the field. It remains an open question how far these factors account for the drop of women completing computing degrees in the mid-1980s and again in the early 2000s; other factors, such as curricular changes, the shift to personal computing, or economic and political events may have contributed to those drops as well.[21]

---

[21] Early journalistic treatments suggest personal computing culture may have contributed to the masculinization of computing; see Thompson (2019).

While much needs to be done, awareness of the importance of women working in computing and creating computing devices and programs continues to grow, and the women who elect to work in the field enjoy doing so, using and advancing the technology.

Conclusions

This chapter demonstrates that after technical disciplines had formed in the nineteenth century, led by pioneering women such as Lovelace and Somerville, information and communication technologies became gendered. If gendered masculine, like the telegraph and radio, a competitive work culture typically developed, and operators enjoyed good pay, attractive career opportunities and high social recognition. If gendered feminine, like telephony, operations were typically considered routine work, and operators received lower pay and limited career opportunities—if they existed at all—as well as less social recognition. The case of electronic computing, finally, indicated that the gendering of a field could change over time, in the United States from one open to women to one dominated by men, and that the gendering of the field varied between countries and geography regions. Already in the early days of computing, women like the ENIAC programmers did not receive recognition for their original work; they often found themselves relegated to routine work, such as coding; and they sometimes elected to hide their identity as women. Social expectations surrounding technologies, operators and users gendered these technologies; there was nothing inherently masculine or feminine about computing or other information and communication technologies.

Yet, this history also indicates that gender was not the only identity dimension at work. For example, the radio operators not only were men; they also were young, white and middle class. Likewise, with the introduction of software engineering, the need for higher education to enter the field advantaged not only men, but white middle-class men over other minorities that had traditionally been disadvantaged in higher education: white women as well as people of colour and those with working-class backgrounds. These observations raise questions of intersectionality. While a growing number of histories of computing have begun to investigate questions of gender, more work needs to be done on how gender intersected with other dimensions of diversity, such as race, class, age and ability.[22] Exploring the implications of intersectionality in computing remains a subject in need of further research to show the diversity of the field, as well as its patterns of exclusion, and to more fully

---

[22] While Venus Green (2001) was one of the first authors to draw attention to issues of intersectionality in her study of race and gender in telephone operations, public awareness appears to have outpaced the scholarly discourse with the blockbuster success of Shetterly's *Hidden Figures* (2016).

understand the implications of these patterns for developers, operators, users, technologies and society at large (Nelsen 2017).[23]

In addition, the global differences in the participation of women in computing in the United States, Europe, Southeast Asian and Arab countries point to the importance of a comparative approach. The historiographical debate about the masculinization of computing has mostly focused on the United States and, to a lesser degree, Europe, and the temptation is large to overgeneralize from one or a few countries to others. Yet, even the existing research already points to international differences. In Britain, for example, the public service sector played a larger role in excluding women from computing, and this exclusion process happened earlier than in the United States. These differences underscore that social and cultural conditions shaped the gendering of computing.

To develop a fuller picture, research will have to consider larger social and political factors. We have seen how assumptions about computing shaped the workforce. For example, if computing was seen as an art for which practitioners brought an inherent ability, recruitment could cast a wide net that captured women as well as men. However, if an academic degree was required for entry into the field, those disadvantaged in higher education, such as women and minorities, faced higher burden of entry. And if a professional career demanded full-time work without interruptions, such assumptions often conflicted with women's expected family roles. In addition, larger external factors, such as family policy, women's rights and expectations of women's work shaped the field. To understand the history of gender and computing in countries as disparate as the United States, Great Britain, Germany, Malaysia and the United Arab Emirates, historians need to understand why women were encouraged or expected to take on salaried work, or urged to refrain from doing so; which rights women had, for example to own property, earn money and make decisions about work; and which expectations were placed on women with regard to their roles as housewives and mothers. This may require new measures to assess the field. For example, while data on undergraduate degrees is a convenient measure, it assumes the 'pipeline' image that most practitioners enter the field with a formal degree, which is not necessarily the case for women and minorities. It certainly means that change will not come from merely encouraging women and other minorities to fit into the dominant culture in computing; only changes in larger social and political factors will enable the field to become more inclusive and diverse.

Acknowledgements I would like to acknowledge help by Caroline Hayes on the data for Table 15.1, comments by Mar Hicks on this chapter, and many discussions with my student Dominique Molee in the preparation of this chapter.

---

[23] To do so will require unlocking new historical sources beyond the traditional technical archives that often reinforce the historiographical dominance of white, male, middle-class professionals.

REFERENCES

Abbate, Janet. 2012. *Recoding Gender: Women's Changing Participation in Computing*. Cambridge, MA: MIT Press.

Abbate, Janet. 2018. "Code Switch: Alternative Visions of Computer Expertise as Empowerment from the 1960s to the 2010s." *Technology and Culture* 59:4: S134–S159.

Aspray, William. 2016. *Women and Underrrepresented Minorities in Computing: A Historical and Social Study*. Cham, Switzerland: Springer.

BMBF (Federal Ministry of Education and Research, Germany). 2020. "University Graduates by Subject Group, Examination Group and Gender." https://www.datenportal.bmbf.de/portal/de/Tabelle-2.5.46.html#A1 Accessed on 29 March 2021.

Bodden, Valerie. 2017. *Programming Pioneer Ada Lovelace*. Minneapolis: Lerner Publications.

Brooks, Frederick P. 1957. *The Mythical Man-Month: Essays on Software Engineering*. Reading, MA: Addison-Wesley.

Bix, Amy S. 2017. "Organized Advocacy for Professional Women in Computing: Comparing Histories of the AWC and ACM-W." In *Communities of Computing: Computer Science and Society in the ACM*, edited by Thomas J. Jisa, 142–172. New York: Association of Computing Machinery.

Bruton, Elizabeth, and Mar Hicks. 2020. "A History of Women in British Telecommunications: Introducing a Special Issue." *Information and Culture* 55:1:3.

Catalyst. 2018. "Women's Share of Bachelor's Degrees in STEM Fields, 2018". https://www.catalyst.org/research/women-in-science-technology-engineering-and-mathematics-stem/ Accessed on 29 March 2021.

Davies, Margery W. 1982. *A Woman's Place Is at the Typewriter: Office Work and Office Workers, 1870–1930*. Philadelphia: Temple University Press.

Doudna, Kelly. 2018. *Ada Lovelace: Pioneering Computer Programming*. Minneapolis: Abdo Publishing.

Douglas, Susan J. 1987. *Inventing American Broadcasting, 1899–1922*, 200–202. Baltimore, Maryland: Johns Hopkins University Press.

Ensmenger, Nathan L. 2010a. "Making Programming Masculine." In *Gender Codes: Why Women Are Leaving Computing*, edited by Thomas J. Misa, 151–141. Hoboken, NJ: IEEE Computer Society.

Ensmenger, Nathan L. 2010b. *The Computer Boys Take Over: Computers, Programmers, and the Politics of Technical Expertise*. Cambridge, Mass.: MIT Press.

Ensmenger, Nathan L. 2015. 'Beards, Sandals, and Other Signs of Rugged Individualism': Masculine Culture Within the Computing Professions. *OSIRIS* 30:1:38–65.

Erickson, LeAnn. 2010. "Top Secret Rosies: The Female 'Computers' of WWII". Public Broadcasting Service. Film.

Essinger, James. 2014. *Ada's Algorithm: How Lord Byron's Daughter Ada Lovelace Launched the Digital Age*. Brooklyn: Melville House.

Fuegi, John and Jo Franci. 2003. Lovelace & Babbage and the Creation of the 1843 'Notes'. *IEEE Annals of the History of Computing* 25:4:16–26.

Gabler, Edwin. 1988. *The American Telegrapher: A Social History, 1860–1900*. New Brunswick: Rutgers University Press.

Green, Venus. 1995. Goodbye Central: Automation and the Decline of 'Personal Service' in the Bell System, 1878–1921. *Technology and Culture* 36:4:912–949.

Green, Venus. 2001. *Race on the Line: Gender, Labor, and Technology in the Bell System, 1880–1980*. Durham, North Carolina: Duke University Press.

Goldstine, Herman H., and John von Neumann. 1947. *Planning and Coding of Problems for an Electronic Computing Instrument*. Princeton, New Jersey: Institute for Advanced Studies.

Haigh, Thomas. 2010. "Masculinity and the Machine Man: Gender in the History of Data Processing." In *Gender Codes: Why Women Are Leaving Computing*, edited by Thomas J. Misa, 51-71. Hoboken, NJ: IEEE Computer Society.

Haigh, Thomas, Mark Priestley, and Crispin Rope. 2016. *ENIAC in Action: Making and Remaking the Modern Computer*. Cambridge, MA: MIT Press.

Hartmann, Doreen. 2015. "Zwischen Mathematik und Poesie. Leben und Werk von Ada Lovelace." In *Ada Lovelace. Die Pionierung der Computertechnik und ihre Nachfolgerirnnen*, edited by Sybille Krämer, 25–26. Paderborn: Wilhelm Fink.

Hayes, Amy. 2018. *Ada Lovelace: First Computer Programmer*. New York: PowerKIDS: 2017.

Hayes, Caroline. 2010. "Computer Science: The Incredible Shrinking Woman." In *Gender Codes: Why Women Are Leaving Computing*, edited by Thomas J. Misa, 25–50. Hoboken, NJ: IEEE Computer Society.

Hicks, Mar. 2017. *Programmed Inequality: How Britain Discarded Women Technologists and Lost Its Edge in Computing*. Cambridge, MA: MIT Press.

Hollings, Christopher, Ursula Martin, and Adrian Rice. 2018. *Ada Lovelace: the Making of a Computer Scientist*. Oxford: Bodleian Library, 2018.

Hong, Sungook. 2001. *Wireless: From Marconi's Black-Box to the Audion*. Cambridge, MA: MIT Press.

Huskey, Velma R. and Harry D. Huskey. 1980. Lady Lovelace and Charles Babbage. *Annals of the History of Computing* 2:4:299–329.

Jepsen, Thomas C. 2000. *My sisters telegraphic: Women in the Telegraph Office, 1846–1950*. Athens: Ohio University Press.

John, Richard R. 2010. *Network Nation: Inventing American Telecommunications*. Cambridge, MA: Belknap.

Kleiman, Kathy. 2015. "Great Unsung Women of Computing: The Computers, The Coders and The Future Makers". Seattle: Seattle International Film Festival. Film.

Kline, Ronald R. 2000. *Consumers in the Country: Technology and Social Change in Rural America*. Baltimore, Maryland: Johns Hopkins University Press, 2000.

Krämer, Sybille. 2015. "Im Mittelpunkt steht Ada. Zur Einleitung in diesen Band." In *Ada Lovelace. Die Pionierung der Computertechnik und ihre Nachfolgerirnnen*, edited by Sybille Krämer, 7–8. Paderborn: Wilhelm Fink.

Labrecque, Ellen. 2017. *Ada Lovelace and Computer Algorithms*. Ann Arbor, MI: Cherry Lake Publishing.

Lagesen, Vivian Anette. 2008. A Cyberfeminist Utopia? Perceptions of Gender and Computer Science among Malaysian Women Computer Science Students and Faculty. *Science, Technology, & Human Values* 33:1:5–27.

Lew, Kristi. 2018. *Ada Lovelace: Mathematician and First Programmer*. New York: Britannica Educational Publishing.

Light, Jennifer S. 1999. When Computers Were Women. *Technology and Culture* 40:3:455–483.

Lipartito, Kenneth. 1994. When Women Were Switches: Technology, Work, and Gender in the Telephone Industry, 1890–1920. *American Historical Review* 99:4:1075–1111.

Loh-Hagen, Virginia and Jeff Bane. 2018. *Ada Lovelace*. Ann Arbor, MI: Cherry Lake Publishing, 2018.

Margolis, Jane, and Allan Fisher. 2002. *Unlocking the Clubhouse: Women in Computing*. Cambridge, MA: MIT Press.

Markus, Julia. 2015. *Lady Byron and her Daughters*. New York: W. W. Norton.

Martin, Michèle. 1991. *"Hello, Central?" Gender, Technology, and Culture in the Formation of the Telephone System*. Montreal: McGill-Queen's University Press.

McGrath Cohoon, J., and William Aspray, eds. 2006. *Women and Information Technology: Research on Underrepresentation*. Cambridge, MA: MIT Press, 2006.

Mellström, Ulf. 2009. The Intersection of Gender, Race and Cultural Boundaries, or Why is Computer Science in Malaysia Dominated by Women? *Social Studies of Science* 39:6:885–907.

Menabrea, L. F. 1843. "Sketch of the Analytical Engine Invented by Charles Babbage," with notes by the translator Ada Augusta, Countess of Lovelace. *Scientific Memoirs* 3:66–731.

Miller, Claire C. 2018. "Ada Lovelace: A gifted Mathematician Who is Now Recognized as the First Computer Programmer." *New York Times*, 19 March.

Misa, Thomas J., ed. 2010. *Gender Codes: Why Women Are Leaving Computing*. Hoboken, NJ: IEEE Computer Society.

Misa, Thomas. 2016. "Charles Babbage, Ada Lovelace, and the Bernoulli Numbers." In *Ada's Legacy: Cultures of Computing from the Victorian to the Digital Age*, edited by Robin Hammerman and Andrew L. Russell, 11–31. New York: Association for Computing Machinery Books.

Moore, Doris Langley. 1977. *Ada Countess of Lovelace: Byron's Legitimate Daughter*. New York: Harper & Row.

Nelsen, R. Arvid. 2017. Race and Computing: The Problem of Sources, the Potential of Prosopography, and the Lesson of Ebony Magazine. *IEEE Annals of the History of Computing* 39:1:29–51.

Noble, Safiya Umoja. 2018. *Algorithms of Oppression. How Search Engines Reinforce Racism*. New York: New York University Press.

Pohlke, Annette. 2015. "'Princess of Parallelograms' meets 'Queen of Science'. Mary Somerville als Lehrerin, Freundin, Vorbild." In *Ada Lovelace. Die Pionierin der Computertechnik und ihre Nachfolgerirnnen*, edited by Sybille Krämer. Paderborn: Wilhelm Fink.

Rakow, Lana F. 1992. *Gender on the Line: Women, the Telephone, and Community Life*. Urbana: University of Illinois Press.

Robinson, Fiona. 2016. *Ada's Ideas: The Story of Ada Lovelace, the World's First Computer Programmer*. New York: Abrams Books.

Sanchez Vegara, Isabel. 2017. *Ada Lovelace*. Minneapolis: Lincoln Children's Books.

Schaffer, Simon. 1994. Babbage's Intelligence: Calculating Engines and the Factory System. *Critical Inquiry* 21:1:203–227.

Schinzel, Britta, ed. 2013. Weltbilder in der Informatik: Sichtweisen auf Profession, Studium, Genderaspekte und Verantwortung. *Informatik Spektrum* 36:3.

Schlombs, Corinna. 2010. "A Gendered Job Carousel: Employment Effects of Computer Automation." In *Gender Codes: Why Women Are Leaving Computing*, edited by Thomas J. Misa, 75–94. Hoboken, NJ: IEEE Computer Society.

Schmitt, Katherine M. 1930. I Was Your Old Hello Girl. *Saturday Evening Post*, July 12.

Shapin, Steven, and Simon Schaffer. 1985. *Leviathan and the Air-pump: Hobbes, Boyle and the Experimental Life*. Princeton: Princeton University Press.

Shetterly, Margot Lee. 2016. *Hidden Figures: The Untold True Story of Four African-American Women who Helped Launch Our Nation Into Space*. New York: William Morrow.

Snow, Charles P. 1959. *The Two Cultures and the Scientific Revolution: The Rede Lecture*. Cambridge.

Stanley, Diane. 2016. *Ada Lovelace: Poet of Science, the First Computer Programmer*. New York: Simon & Schuster.

Stein, Dorothy. 1985. *Ada: A Life and a Legacy*. Cambridge, Mass.: MIT Press.

Stone, Tanya Lee and Marjorie Priceman. 2018. *Who Says Women Can't be Computer Programmers?: The Story of Ada Lovelace*. New York: Henry Holt.

Strom, Sharon Hartman. 1992. *Beyond the Typewriter: Gender, Class, and the Origins of Modern American Office Work, 1900–1930*. Urbana: University of Illinois Press.

Swade, Doron. 2000. *The Difference Engine: Charles Babbage and the Quest to Build the First Computer*. New York: Penguin Books.

Thompson, Clive. 2019. *Coders. The Making of a New Tribe and the Remaking of the World*. New York: Penguin, 2019.

Toole, Alexandra. 1998. *Ada Enchantress of Numbers: Prophet of the Computer Age*. Mill Valley, California: Strawberry Press.

Unesco Institute for Statistics. 2021. "Percentage of Graduates from Information and Communication Technologies Programmes in Tertiary Education who are Female (%)". The World Bank.

Wallmark, Laurie. 2016. *Ada Byron Lovelace and the Thinking Machine*. Minneapolis: Creston Books.

Warrick, Patricia S. 2007. *Charles Babbage and the Countess*. Bloomington, Indiana: AuthorHouse.

Woolley, Benjamin. 1999. *The Bride of Science: Romance, Reason, and Byron's Daughter*. New York: McGraw-Hill.

CHAPTER 16

The Cultural Context of Gendered Science: India

Carol C. Mukhopadhyay

Scholars, development experts, and nations have long recognized the problem of gender disparities in educational attainment, including in STEM. Yet despite remarkable gains in girls' overall education, globally, the STEM gender gap persists. It is substantial in undergraduate college enrolments, increases at graduate levels, and is most significant in the science research community. Currently just 30% of the world's researchers are women, with many countries, such as India (at 16.6%) having significantly lower rates (UNESCO 2020). Systematic cross-national STEM data is difficult to find, especially on non-Western countries and disaggregated by field. Over the past decade, however, international organizations and countries, encouraged by United Nations 2030 Sustainable Development Goals, have devoted resources for improving and standardizing cross-national data and identifying underlying dynamics. As these efforts move forward and comparative data is used for policy purposes, it is crucial to ask whether the issues surrounding women, gender, and science are the same everywhere. Does the scientific gender gap emerge through the same processes in all societies? Does gendered science in India, with a statistical footprint similar to the United States, reflect the same causal factors? Or are there different cultural, socio-economic, historical, and

---

C. C. Mukhopadhyay (✉)
Department of Anthropology, San José State University, San José, CA, USA
e-mail: carol.mukhopadhyay@sjsu.edu

science-related academic choice processes at work? Without answering such questions, we cannot develop culture-sensitive, appropriate policies.

This chapter addresses these issues using ethnographic, questionnaire, and institutional data collected in India from 1988 to 1990, with brief returns in 1996, 2004, and 2008. Although the acronym STEM encompasses many fields, the focus was engineering, a discipline with the largest gender gap in both the United States and India. The Indian research settings were primarily English-medium, higher educational, science and technology institutions with largely urban, English-speaking, educated élites; this reflects the STEM population then and today, especially in prestigious fields and educational institutions. However, non-élite private and government colleges and schools were also visited. A significant 'information-sharing and scholarly exchange component' included trips to eight Indian cities and thirty education-oriented institutions, two all-India conferences, and discussions with over sixty 'expert consultants'.[1] This produced a wealth of unpublished data, insights into the complex Indian educational system, and an introduction to gender-education-family linkages across varied regions, rural–urban settings, classes and 'communities' ('caste'/religion).

The ethnographic phase of research involved in-depth, student-centred 'fieldwork' at engineering-science oriented colleges, with participation-observation, focus groups, structured interviews, and collection of academic biographies. This included two months' on-campus residence at the Indian Institute of Technology (IIT), Madras, with shorter stays at the Indian Institute of Science, Bangalore (IISc) and Cochin University of Science and Technology (CUSAT) in Kerala.Cognitive anthropology approaches, especially cultural models and ethnographic decision-modelling, were used to understand the processes underlying students' science-related academic decisions (Mukhopadhyay 2004). The initial study expanded in 1989–90 to pre-colleges, incorporating non-science students and students from more diverse socioeconomic backgrounds, school types, and regions. A culturally relevant Student Academic Decision Process Questionnaire (SAQ) was created to obtain information on student academic choices, achievement, family backgrounds, and other educationally significant social-cultural variables. A narrative segment elicited 'folk explanations' for gendered activities and 'mental images' of scientists. Four Western-based Math-Science attitudes surveys were also used.[2] Questionnaires were administered to sixth, ninth, and eleventh grade students at twelve linguistically diverse schools in four major Indian cities.[3] The resulting database contains nearly 5000 questionnaires

---

[1] I use 'expert consultants' to differentiate informants with professional expertise from more typical anthropological informants.

[2] For Mathematics: Fennema-Sherman (1976); for Science: Kelly (1985) and Smail and Kelly (1984).

[3] The cities were Madras-Chennai, Delhi, Bangalore, and Hyderabad. Grade specific SAQs were created in English, Hindi, and Kannada.

from over 1600 students. A subsequent phase used SAQ and questionnaire data to formally test the ethnographic-research-based theory of the Indian scientific gender gap.[4]

The Indian case study presented here illustrates the profound role that cultural models, specifically Indian cultural models of family, gender, and schooling, play in science-related educational and career choices. The Indian data also reveals how the cultural and social context in which science is selected, learned, and practiced contributes to the gendering of science.

India's STEM Gender Gap

India government statistics show that women have always been significantly underrepresented in science-related fields.[5] In 1950–1951 girls were only 7.1% of students enrolled in university science degree programmes.[6] By the late 1980s, this had risen to 30%, with just under 40% in 2000. Only recently have girls reached parity. The gender gap is most dramatic in engineering and technology. In 1950–1951, women were one-tenth of 1% of students in degree programmes, only 1% in 1970–1971, and barely 6% in the late 1980s. Even today, females are less than a third of undergraduates in engineering and technology. The gender gap becomes a chasm at premier engineering institutions, like the Government of India funded Indian Institutes of Technology (IITs). In 1988 and 1996 women constituted less than 6% of IIT Engineering (B.Tech) students. Women were still well under 10% as recently as 2017 (Verma 2018).

Historically, one major source of the scientific gender gap has been overall gender disparities in education. In 1950–1951, only 20% of girls aged six to

---

[4] Decision-modelling and conventional statistical testing procedures were employed. A composite patrifocal family variable with four sub-components was created and used, along with academic and economic variables, in the logistic regression analysis. SAQ narrative vignettes and math-science questionnaire responses allowed comparison of Indian and Western cultural models of causality, gender, science, mathematics, and personhood. Results strongly support the ethnographically based analysis presented here. For details on research phases, pre-college sample and the full analysis, see Mukhopadhyay (2001).

[5] Indian educational statistics are complex. Enrolment data is generally given but actual attendance is lower, and exam pass rates even lower. Data by gender is often simply raw numbers; other methods are females per 100 males, ratio females/males, percentage of females in relation to total students. Gender comparisons can be misleading unless student population sex ratios are equal, something generally not the case. One solution used in India is the Gender Parity Index (GPI) where 1.0 equals parity; see AISHE (2019).

[6] Educational statistics for the 1950s through early 2000s come from data summary tables in Mukhopadhyay and Seymour (1994), Mukhopadhyay (1994) and from unpublished data I collected in India and at a 2004 National Institute for Science Technology and Development (NISTAD) conference (see Kumar 2009). Current statistics, unless otherwise indicated, come from the comprehensive Government of India (GOI) annual report, All India Survey on Higher Education 2018–2019 (AISHE), available on-line. Acronyms for GOI reports, for example AISHE, and ESAG for 'Educational Statistics at a Glance', will be used in the text. For IIT see Council of Indian Institute of Technology (2021).

eleven years were in school compared to 55% of boys. In the late 1980s, virtually all boys were enrolled but only around 77% of girls. Historically, attrition—for both sexes—has been significant at middle grades but greater for girls; these gender gaps increase at secondary school as puberty increases pressures for girls to marry. In 1950–1951, nearly ten times more boys than girls, age fourteen to seventeen years, were in school; in 1970–1971, almost three times more boys; and in the late 1980s, twice as many boys as girls were enrolled. Finally, in 2014, girls reached near parity with boys (WMI 2019). The Indian STEM gender gap, then, is partially an educational attrition pipeline issue. By the time one reaches college and can pursue STEM degrees, the proportion of females, historically, is significantly lower than males.[7]

Patrifocality and Educational Decisions

In India educational decisions are family decisions, involving substantial resources and with significant, long-term family social and economic, impacts. Families are guided by a 'patrifocal' family cultural model (Mukhopadhyay and Seymour 1994). Among its characteristics are an emphasis on collective family goals and welfare; patriarchal structural features (patrilineal descent, patrilocal residence) which reinforce the centrality of sons and the peripheral status of daughters[8]; gender-differentiated responsibilities; regulation of female sexuality (to maintain the purity of the patriline) through arranged marriages and restricted male–female interactions; and female standards which emphasize obedience, self-sacrifice, adaptability, restraint, and other traits conducive to (joint) family harmony. While other cultural models of family exist (Kolenda 1987; Pai 2002), this is a prominent one to which most Indians have been exposed. As such, it provides a significant culturally rooted framework for thinking about and making educational decisions.

Education in India was heavily promoted after Independence. Literacy, formal education, and science and technology were seen as keys to Indian national development, a theme that continues today. Being a 'matriculate' (completing secondary school) opened-up economic opportunities. College degrees offered even greater prestige and more lucrative careers. This fueled an expansion of the Indian educational system, producing an academic hierarchy of subjects, degrees, and institutions, with intense competition for 'seats' in high-ranked fields—science and engineering—at high-ranked schools. Such degrees brought jobs 'with scope'—that is, careers with financial and career

---

[7] As noted earlier, the percentage of females in STEM is misleading unless overall school enrolments sex ratios are equal. This is important in cross-national comparisons.

[8] Patrilineality refers to the tracing of kinship lineage affiliation (and property and other rights) through males rather than females, from father to sons (and daughters) and from sons to their offspring. Patrilocality refers to post-marriage household arrangements, with sons remaining with and responsible for their natal family, bringing in wives to help. Daughters, after marriage, go to their husband's family. The result is a multi-generational 'joint' family centered around related same-lineage males.

advancement potential—as well as opportunities to study abroad. Education, especially higher education, became a vehicle for expressing and improving family economic and social status. Families, particularly upper-middle-class families, increasingly preferred educated grooms, stimulating the growth of dowry/groom price, despite its prohibition. This broader context influenced how families approached educational decisions for both sons and daughters. Consistent with patrifocal cultural models, families frame educational decisions in terms of their projected impact on collective family welfare. These decisions involve family resources, status, and marriage considerations, and (like marriage) are 'too important' to leave in the hands of young people. Indeed, decisions about students' education are viewed as major family obligations, demanding careful research and broad consultation with family members.

Despite the perceived importance of education, investing in boys' education is viewed differently than investing in the education of girls. Sons, in the traditional patrifocal model, have long-term obligations to care for their natal families, financially and in other ways, and they remain in the household after marriage. Thus, investment in a son's education benefits the family directly. In contrast, daughters will marry, 'leave' the family and acquire obligations towards their husband's family. A daughter's education, after marriage, mainly benefits her in-laws rather than her natal family. The patrifocal family's primary obligation to daughters is to see that they 'marry well' (are 'well-settled') and uphold family honour. From this perspective, girls' education has always had both benefits and costs, and education can enhance a girl's marriage prospects. Historical arguments for girls' education emphasized how schooling cultivated attributes consistent with patrifocality. Educated wives could better supervise children's education and interact in their spouse's world. Educated husbands demanded more educated wives, motivating some rural families to provide schooling for their daughters. These themes persist today (Gold et al. 2019; Ullrich 2019).

But education could also pose social dangers to unmarried girls, exposing them to situations that could harm their 'reputation' as a potential bride, wife, and daughter-in-law. Schooling requires going 'outside the family' into the male world of public spaces, whether travelling to or studying in school. Education could 'spoil a girl's character', cultivating traits, such as independence and outspokenness, that could undermine patrifocal family values and make it difficult for a girl to adjust to a husband's family. Boys' education, on the other hand, brings both social and economic benefits. Not only does education increase occupational opportunities and prestige which benefit the entire family, it also enhances a boy's marriageability and his dowry, offsetting the expenses of his education.

However, investing in education beyond primary school is costly, especially for poorer or rural families who depend on children's labour to augment family income or for agricultural or household work. Only a fraction of families can afford college. In 1981, barely 2% of males and less than 1% of females had university degrees. In the early 2000s, despite the growth of universities, only

6% of the population had access to higher education. Recent educational data shows dramatic improvement, partially due to government policies designed to eliminate traditional disparities based on class, caste, and gender. Higher education enrolments are now estimated at over thirty-seven million.[9] Yet some argue that formal education and the proliferation of private schools and colleges has exacerbated, rather than reduced traditional inequalities (Kamat 2014; Wadhwa 2018). Even today, barely 26% of students age eighteen to twenty-three are enrolled in higher education and class/caste/regional disparities remain, affecting girls more than boys. Girls' drop out of secondary school for financial reasons (as do boys) but also because of domestic work and the lack of female teachers and girls' toilets (ESAG 2018, p. vi). Statistics showing gender parity mask more subtle disparities. More girls than boys attend less expensive, lower-achieving government schools, especially in rural areas. If families have limited resources for private schools, they tend to send sons rather than daughters regardless of achievement (ESAG 2018, p. ii).[10]

Patrifocality and STEM

Overall gender educational disparities, however, do not fully explain the scientific gender gap, especially in engineering and technology. Some families have always had sufficient resources and motivation to send both sons and daughters to college. Yet these girls pursued arts more than science or, for professional degrees, teaching and medicine rather than engineering and technology. In the 1950s, girls were only 11% of university enrolments but 16% of those in arts, 32% in education, and 16% in medicine, while representing only 7% of science enrolments and well under 1% in engineering and technology. By the late 1980s, girls were 30% of undergraduates but still disproportionately in arts and less than 4% in engineering and technology. These patterns were similar in 2001. Since 2000, however, girls have made remarkable educational progress. Government policies addressed social issues, such as the need for sex-segregated education, all-girls hostels, women teachers, safe transportation, and financial incentives.[11] There are now sixteen All-Women Universities and 11% of all colleges are for women. Girls are near parity in college enrolments, nearly half of undergraduates (49%), over half of post-graduates, and 44% of Ph.D. students. Clearly education plusses now outweigh minuses for most families with the means to send children to college (Ullrich 2019). And the pipeline component of the STEM gender gap has virtually disappeared.

---

[9] Higher education in India refers to a complex array of degree-granting and non-degree granting institutions. For a comprehensive overview see AISHE (2019).

[10] In SAQ pre-college data, girls with better grades, and from similar families, socio-economically, were more likely than boys to attend less expensive government municipal schools rather than private schools.

[11] There are now almost equal numbers of girls' and boys' hostels and residents (AISHE 2019, Table 32).

Yet the most current higher education enrolments show some familiar patterns (AISHE 2019). Girls are still more apt than boys to pursue arts and education (53% and 66% respectively of all students), but they are now a majority of medical science students (66%) and over half (51%) of science undergraduates. When it comes to engineering, however, the gender gap persists. Females are less than one-third of BEng/BTech students (28.9%) and less than one-fifth (18%) of polytechnic students, concentrated in nursing (86%) and teacher training (63%). Equally telling, females are most under-represented in Government of India designated 'Institutions of National Importance' (INI).[12] This STEM-oriented group includes all the IITs, the newer Indian Institutes of Information Science (IIITs), the National Institutes of Engineering, Indian Institute of Science, and All-Indian Institutes of Management. It also includes top-ranked medical institutions. Yet women are only 23.9% of all INI students.[13] As for IITs, still the premier engineering and technology institutions, women's BEng/BTech enrolments averaged 8% in 2017, with far less at some campuses. Thus, while more women pursue engineering and technology, they remain barely visible in the IIT world. Once again, Indian cultural models of family, gender, and schooling are at work. The same processes responsible for past gender educational disparities continue to impact STEM. Studying STEM compared to other fields like arts, exacerbates the traditional problems of girls' education, increasing financial investments and social-marriageability risks. These are far greater for engineering degrees, and greatest at prestigious institutions.

STEM: Costly Alternatives

STEM degrees have long been more desirable, difficult, and costly than non-science degrees such as arts or commerce. To pursue college science, it is mandatory to take science streams in secondary school and achieve sufficient marks in key subjects like mathematics, physics, chemistry, and biology. Urban areas often rank secondary schools, public and private, with competition for science 'seats' at the 'best' schools. Private schools require substantial school fees or financial 'contributions', in lieu of or along with sufficient marks. Studying pre-college science, then, requires greater investment of family resources than other disciplines. Rural students may have to travel far to reach any secondary school, much less one offering science subjects, or enrolling girls—and staying in a school hostel may be necessary, increasing costs.[14] Gaining admission at the college level requires academic success in

---

[12] For information on INI and other Indian institutions see Ministry of Education (2021).

[13] Among non-INI institutions, females are predictably overrepresented in State Public Universities as opposed to private ones, and in colleges rather than universities (AISHE 2019, Table 26).

[14] Currently, 12% of rural households do not have a secondary school within five kilometers, vs. 1% in urban areas (ESAG, 2018: iii). Gold et al. (2019) describe a rural family

science prerequisites and entry exams, and this is facilitated by attending 'good' secondary schools, often private and expensive.[15] The increased value of boys' STEM education focuses family resources even more on sons, leaving less for daughters. If girls take science streams, it may be at less expensive municipal schools, without extra tutorials or release from household tasks.

STEM: Greater Social Dangers and Marriageability Risks

Expert consultants, parents, and students viewed the male-dominated social context of STEM as a major barrier to women's participation. Same-sex girls' schools are still preferred by many families, especially conservative religious ones. Yet science streams are often lacking, especially in rural areas where funds are scarce. Girls must either go to a girls' school in another locale or risk the social dangers of attending nearby co-educational schools. In urban areas where girls can live at home, incursions into predominantly male public spaces remain socially problematic. Female students cited family concerns about the 'dangers' of daughters travelling across town, especially 'alone' and at 'odd' hours (after dark). Girls described 'comments', 'pinching and that sort of thing, common in buses, streets, and marketplaces', which, 'of course, prohibits us from going into crowded places'. The primary 'dangers' are social, centred around issues of male–female sexuality—as interpreted through a patri-focal and patriarchal lens. On crowded buses, for example, female students were afraid of 'creating a scene' should they resist an attempted 'pinch'; passengers might say it was the girl who 'tempted that person'. If something 'serious' happens, 'your parents will ask you to keep quiet because it'll not be good for your future [...] if you're not married'. Unless a girl's family can afford to transport her to school, she must confront reputational risks of travelling 'outside' or attend a closer coeducational school.[16] Or, she may just 'go for' arts at a nearby all-girls school.

Pursuing college-level science, particularly engineering and technology, exacerbates these risks. Residence in a student hostel may be mandatory. Until recently, girls could study 'pure' science at all-women's colleges but not engineering. Now hundreds of women's colleges offering engineering

---

whose daughters had to live with relatives in another town just to find an all-girls secondary school.

[15] Relatively well-off expert consultants lamented the growth of costly, highly competitive, academically oriented schools, starting with elementary schools that required entrance exams. Nevertheless, they felt compelled to give their children every advantage in the race for academic success. They also freed children from family chores and provided academic help, tutors, and tutorial courses for crucial exams. Many IIT-CUSAT female students had attended private, academically rigorous, English-language secondary schools.

[16] Schools are starting to provide transportation for females although Gold et al. (2019) report that even in 2017, there was none for their rural village-dwelling informants.

and technology exist, of varying types (public–private), affiliations, reputation, and location. Indian online journals provide rankings, descriptions, and website links. These offer a socially safer alternative but are still, at under 3%, only a fraction of all engineering colleges (AICTE 2021). Most engineering and technology educational institutions, then, require immersion in a virtually all-male environment, unsupervised, close contact with unrelated males, and on-campus residence. Similar conditions confront women at polytechnics and in workplaces. Some employers refuse to hire females on these grounds, further discouraging women from studying STEM.

The issue of 'suitable housing' (all-girls hostels) was a major concern of families and remains a significant constraint on women studying in predominantly male spaces. In 2014, the Indian government allocated funds to assist polytechnics construct women's hostels. New IIIT campus plans include women's hostels as 20–25% of hostel accommodations. Yet while hostel capacity expanded at state public universities and colleges, girls' hostels remain relatively scarce at Institutes of National Importance (AISHE 2019, Table 32). At CUSAT, the girls' hostel was double the intended occupancy.[17] But it was a socially 'safe' space, near the administration building, at the opposite end of campus from boys' hostels. Girls had to be back by 7 p.m. unless at the laboratory, for which they could stay out 'till ten'. Barring special student programmes, after 7 p.m. the campus was virtually an all-male world. At IIT Madras, the girls' hostel was near the campus library, far from boys' hostels, separated from the main road by a long, tree-shaded path. It was overcrowded and another hostel was under construction next door, preserving the all-girls residential compound. IIT girls would 'wander about' at night, going to labs, library, and canteen. The campus was a residential community (staff, faculty, students) with post office, bank and walled, guarded campus entrances. Nevertheless, women students reported social discomfort when going alone for a 'coffee' in the canteen. And they were uncomfortable using the campus swimming pool despite its women-only hours.[18] Girls were a distinct minority on-campus and aware of social dangers, including rumours of social impropriety. Several felt it was inappropriate to initiate conversations with male classmates. Some males reported similar concerns. One freshman, the only girl in her Mechanical Engineering class, said not a single male spoke with her in class the entire semester. Yet girls overcame their discomfort and seemed empowered by their experiences. However, some stayed away from 'industrial tours' because it required travelling in virtually all-male company and there were no 'suitable accommodations' (i.e., for an unmarried girl).

[17] During my research, there were 168 girls in eighty single-occupancy rooms. In contrast, 350 boys were accommodated in 308 rooms.

[18] An issue is swimming attire, at least body-revealing Western style bathing suits. Cultural norms about female bodies, in the presence of males, inhibit girls from participating in many sports in India and elsewhere. Sexual segregation—and related cultural models about gender and sexuality—impacts many spheres of life. For an overview, see Mukhopadhyay and Blumenfield (2020).

For families, the social dangers (and financial costs) of girls' engineering education are lessened by sending them to local colleges, even if eligible for more distant, higher-ranked institutions. This was a recurring theme in expert consultant interviews and girls' academic biographies. Some engineering fields are particularly (socially) 'strenuous' and 'dangerous'. Civil engineering careers connoted 'fieldwork,' a rough 'camp,' full of 'rowdy' male labourers from all castes and classes. Problems included 'suitable accommodations' and having to eat and travel with 'all sorts' of people 'at all hours'. Female engineering students tended to avoid such 'branches'. Electrical and computer engineering offered less socially problematic work contexts and social interactions, and hence were more desirable options. But even engineering 'desk jobs' were associated with late night shifts, 'posting' throughout India, and immersion in mostly male, often working-class social contexts—formidable dangers for marriage-conscious, middle-class girls' families. These concerns remain salient today, even in the modern, globalizing Indian IT world (Fleming 2016). Some companies are providing all-female hostels for employees. And women engineering college websites highlight 'safe and secure' on-campus residential women's hostels.

Other risks relate to the 'education spoils a girl's character' theme. Verbal assertiveness, independent thinking, and leadership were encouraged in CUSAT's graduate management studies programme but some girls seemed shocked at these behavioural expectations. 'It's totally different in a girls' college [...] But here, everybody's so aggressive. And unless you are ready to fight it out, nobody's going to stand back and let you go and give you a chance'.[19] Some IIT girls mentioned discomfort at extra-curricular activities that entailed exerting authority over male peers. Yet students adjusted remarkably well and welcomed, even celebrated, these educational side-effects. From a family perspective, however, cultivating such traits could be a marriageability risk given traditional expectations for Indian wives. Educational 'hypergamy' is another issue: girls are expected to marry up, perhaps reflecting traditions that husbands should rank higher and have authority over wives. The higher the academic qualifications of brides, the higher those needed by grooms. Science generally ranks above arts, so girls with arts degrees are at less risk of out-ranking grooms. Science degrees are more problematic. But rank differences can be slight. Applied science degrees, because of career potential and competitiveness, generally rank higher than 'pure' sciences. One female Ph.D. chemistry student's family arranged her marriage to an 'M-Tech' (Master's in Technology). They considered his degree appropriate, rank-wise, because it was 'applied' while hers was 'pure science'.

---

[19] In CUSAT graduate management studies classes, the six girls (of forty students), seated together, appeared uncomfortable asserting themselves into the lively, communicatively aggressive discussions, especially interrupting males. Similar discomfort occurred when girls were asked to participate, on stage, in male-organized 'skits' involving dating scenarios.

Patrifocal family models, then, have different educational impacts on daughters than sons, but affect girls most when substantial family investments are required and social dangers highest. For girls, pursuing science, especially engineering, especially at premier institutions, has exceptionally high economic and social costs. For boys, STEM degrees facilitate a 'good marriage' (and higher dowry). For girls, finding a suitable spouse can be difficult and expensive. The supply of academically higher-ranked males is smaller while groom price/dowry costs are higher. And issues of being 'too old' arise, especially for girls pursuing postgraduate STEM degrees. Girls should marry young, to boys their age or older.[20]

Increasing Benefits of STEM

Despite these constraints, girls' participation in STEM has grown significantly, especially in pure science and medicine but also in engineering-technology. Even in 1989–1990, pre-college girls planned to pursue STEM at higher levels than previously. There have always been countervailing pressures for girls entering STEM, especially for some families, like those of IIT informants. Clearly the plusses are growing, especially in relatively socially safe fields, institutions, and occupations. As discussed above, education can enhance a girl's marriage prospects and contribute to patrifocal family values. These arguments carry greater weight among élite (class, caste) urban, education-oriented families best positioned to pursue STEM.[21] As more boys acquire STEM degrees, it becomes desirable and possible for girls to do the same. From a marriageability perspective, science degree holders are considered academically successful, implying other positive traits, such as intelligence and perseverance. As mothers/wives they are well-prepared to produce academically successful children and navigate the modern world of ATMs, computers, and IT innovations. A science degree need not threaten educational hypergamy (girls marrying up), given multiple ranking components, such as educational institution, degree type and level. Several informants planned to establish small electronics or IT firms with (future, not yet selected) spouses, suggesting a growing tendency towards educational endogamy (marrying within the same field). Co-educational colleges offer opportunities for girls (and boys) to find personally and socially-economically suitable spouses, making some families

---

[20] Seeta Pai's research (2002) found that in Kerala, age twenty to twenty-four is considered the best age for girls to marry. Some families arrange their daughter's marriage to coincide with graduation, even during final exams. Others arrange them while daughters are still in school, with agreement to continue studies after marriage. According to Fleming (2016) age issues also concern families of unmarried women in the IT industry.

[21] In SAQ data, science-choosers of both genders come from élite families, have higher grades, and attend higher-ranked schools than the sample as a whole. Class, then, somewhat compensates for gender-related barriers. But female science-choosers are even more socio-economically and academically élite than males, and the gap between science-choosers and non-science choosers is greater for females than males (Mukhopadhyay 2001).

more accepting, even encouraging, of 'love marriages'.[22] With family living costs rising, an 'earning' daughter-in-law can be an asset, a high earning one even more so. STEM degrees, in highly ranked fields, from reputable institutions, imply 'good jobs' with 'scope' and opportunities abroad for the girl, her spouse, and in-laws. Virtually all female informants expected to be employed after marriage, although several mentioned in-laws would have to agree. In the pre-college sample, female science choosers were more likely than non-science choosers to anticipate contributing to post-marriage household income.

STEM is not only an asset in the marriage market. Female students and their families emphasized the advantages of being financially independent. Among other things, it reduced pressure to marry, allowing girls and their family more time to find a 'suitable' spouse. Some students in their mid-twenties, with research stipends and good job possibilities, seemed comfortable never marrying, partially because they would not be economic burdens on their families. Their parents, however, felt a moral obligation to see them 'well-settled'—i.e., married. These themes remain today (Fleming 2016; Ullrich 2019). STEM degrees have always been linked to good job opportunities, but these expanded with 1990s liberalization policies, growth of the IT industry, and India's emergence as a global outsourcing site for multinational corporations. Technology parks and Indian-owned IT businesses are springing up all-over urban India, some established by Indian engineers who received graduate degrees abroad. Corporations, like IBM, Microsoft, HP, Amazon, and Intel, are major players in the Indian IT scene, with Indian campuses, large, global—and globalizing—workforces, and a powerful local presence (Fleming 2016; Radhakrishnan 2014). The importance of science and technology in India's economy further fuels the demand for engineering and technology education, both public and private. Multinational companies have entered into 'private–public partnerships' to create new IT institutes, like the Indian Institutes of Information Technology (IIITs), or partnered with established institutions, like IIT Madras, setting up 'incubators', funding applied research, and conducting patent-producing joint-projects. Collaborative business-civic networks, like Global Compact Network,[23] are shaping government technology-oriented initiatives, such as the 2015 Policy on

---

[22] Given the élite student body at many engineering colleges, finding a socially 'suitable' spouse is not difficult although religion and language-region can be problems. Nevertheless, family approval for a 'love marriage' is usually required. The same applies to the IT world (Fleming 2016) and perhaps other professional occupational settings.

[23] Global Compact Network India (GCNI) functions as the Indian Local Network of the UN Global Compact (UNGC), New York, and works towards UN Sustainable Development Goals, including women's rights. Recent reports identify social-cultural constraints on women's workforce participation, especially in IT, similar to those described in this research.

Skill Development and Entrepreneurship, with its gender-inclusivity components.[24] These developments expand women's job opportunities in traditionally 'respectable' settings like schools and colleges. But the private, corporate sector offers new opportunities, with lucrative salaries and global connections. Girls' earnings, or stipends and internships, can compensate families for science education costs and support the education of other family members. One female IIT Ph.D. student's father initially refused to allow her to attend graduate school, even though she was an outstanding student. Her family had limited financial resources and lived in a rural area, so hostel residence was required. She convinced him her degree would improve both her job and marriage prospects. Most importantly, her IIT graduate student stipend could finance her brother's engineering education.

Girls' employment prospects can serve to reduce or offset dowry demands. Dowry in India remains alive although illegal (Pai 2002). Dowry was a subject of heated discussion among CUSAT students with girls uniformly opposed. Boys were ambivalent, partially because of family pressures and concern for sisters' dowries, which their dowries could subsidize. Females expected their future earning power to reduce or substitute for dowry. In a focus group, women computer science students insisted their families would 'absolutely refuse' dowry demands—although whether boys' families would agree is less clear (Seymour 1999). Daughters' earnings can provide 'old-age' insurance for parents and many female science-choosers anticipated some financial responsibility for parents. CUSAT girls, even with brothers, planned to contribute to natal family finances after marriage. As Indian families become smaller, more will end up with only daughters and decisions about daughters' education may involve similar considerations as for sons.

There are other STEM degree benefits. Social risks may be balanced by the prestige it brings to the girl and her family. Educational achievement is celebrated publicly in India. The *Times of India*, online, has an education section with information about admissions processes, upcoming exams, and exam results, by state, school type, and gender. 'Toppers' on secondary school exams are celebrities, with family pictures and biographies. But STEM bias exists, emphasizing science-engineering prizes, guides to 'best' engineering fields, colleges, and test-preparation tutorials. In this context, being exceptionally 'brilliant', interested, and motivated allows some girls,[25] especially from

---

[24] The Government of India in 2015 introduced a comprehensive programme of 'Skill Development and Entrepreneurship'. Aimed at educating the Indian workforce for twenty-first century jobs, it emphasizes non-degree rather than just degree-based technical education. A major goal is increasing women's participation. Policy proposals explicitly address issues of patrifocality identified in this chapter (Government of India 2015).

[25] SAQ female science-choosers have a strong interest in and aptitude for mathematics and science—in some cases more than male counterparts. Some families allow academically successful daughters to delay marriage or continue studies after marriage (Gold et al. 2019). Families without sons, or without academically competent sons, seem more likely to encourage daughters to pursue science.

educationally oriented families, to pursue STEM in highly ranked fields and institutions.[26] The prestige (and other benefits) associated with new opportunities can offset traditional social dangers of females travelling alone or living away from home. This includes going abroad, as with an informant whose family allowed her to accept a prestigious post-doctoral award in the United States. A girl's educational accomplishment can have ancillary family benefits, such as arranging marriages to high-status, education-oriented families, and to Indian professionals living abroad. Indian-born female STEM degree holders are going to the United States (and elsewhere) to work, attend graduate school, or marry Indians working in IT. Within India, as noted earlier, 'respectable' science-related education jobs have expanded. An advanced science degree can pave the way to a socially safe career as a college professor. And the rapid growth of women's colleges offering STEM has resulted in a scarcity of qualified faculty. The demand for women pre-college science teachers has also increased and science degrees can lead to jobs at élite private secondary schools. For postgraduates, there are socially riskier but prestigious research positions at national laboratories (Subrahmanyan 1998). And there is the huge world of IT, including non-IT multinationals, especially as STEM degrees connote accomplishment and can open-up non-science related opportunities.

Among applied sciences, medical degrees have always offered 'safe' options for women. Ironically, patrifocal models of gender segregation facilitated women entering medicine, especially gynecology and obstetrics, since women's families preferred female physicians. Women doctors can establish medical practices at home, or work at established institutions, immersed in virtually all-female occupational settings. Or they can explore other opportunities in medical research, at multinationals, or abroad. Even in the late 1980s, some students took subject prerequisites for both medicine and engineering. Since then, the emergence of medicine-technology collaborations, such as biotechnology and medical engineering, have created new job opportunities, often in socially safe spaces. Mechanical engineering, formerly semi-taboo (for women) has expanded into 'mechanical engineering and automation'. Not all degree holders get jobs and unemployment is a serious problem in India. A look at recent placement statistics for STEM graduates is daunting, often less than 50%, even lower at girls' colleges (AICTE 2021). Women's engineering college websites highlight their 'Placement' services and links to

---

[26] Girls overall have higher school performance outcomes than boys, as seen in ESAG 'pass percentage' for girls on Class X and XII exams. Class X 2015 females equalled or exceeded males in English, mathematics, science, and social science, despite fewer girls attending higher performing private schools (ESAG 2018: 13–16). In college, girls also seem to outperform males, across disciplines, from arts to computer applications and engineering, judging by calculations of raw data on undergraduate 'passed out' rates (AISHE 2019, Table 35).

potential employers while newspapers bemoan the over-supply in STEM. But employment problems are usually worse for non-science degree holders.[27]

For many families, then, the benefits of girls' science degrees outweigh the social costs, especially if one avoids highly ranked fields in male-dominated and socially dangerous educational and work contexts.[28] Not surprisingly, women's engineering colleges offer primarily electronics, computer, and electrical engineering, information technology, and computer applications. Some include mechanical engineering, in particular the version with automation (i.e., robotics), and Ph.D. students work on topics, such as prosthetic design, which combine biology and engineering. Women's engineering colleges may be siphoning off female students from co-educational institutions; women's colleges are only 3% of technical education institutions but enrol approximately 10% of female technical education students (AICTE 2021).[29]

Current female undergraduate enrolments reflect the educational processes described in this chapter.[30] Arts is the most popular degree for both genders, though slightly more so for females (53% of those enrolled). Science is second, with females now 51% of students. Engineering and technology has the third highest enrolments but only 28.9% are female. However, the picture varies for different fields. As expected, women are highly represented in branches most compatible with patrifocality-related concerns: they are over 40% in the highly enrolled branches of computer engineering, electronics engineering, information technology, and architecture; and in computer applications and IT. It is in socially risky fields, like mechanical engineering, that women lag most, a mere 5.1% of the total, as well as in mining and marine engineering at 2.9% and 6% respectively. In all other branches, women are 20–35% of students, including civil engineering. In medical science, females are 61% of all students. While this includes nursing, heavily female at 78%, it is only one of many branches, and only 23% of all medical science students. Today's women have moved far beyond pediatrics and gynecology. They predominate in all but one of the fifteen branches, including pathology, surgery, dentistry, ophthalmology, and administration. The one exception, pharmacy, still has 44.1% female students.

Postgraduate STEM enrolments are also revealing. In the United States, female STEM representation declines from undergraduate to postgraduate

---

[27] Gold et al. (2019) describe repeated attempts of a female informant to pass exams for highly competitive teaching positions, and to pursue additional degrees in hopes of finding a good job.

[28] SAQ sample girls were already expressing academic preferences similar to male counterparts, selecting science over arts-commerce, applied over pure science, and engineering-computer science or bioengineering along with medicine.

[29] Dr. Jayanti Sivasawamy, IIIT Hyderabad, attributes recent dips in female enrolments at her own co-educational institution partially to the rise of women's colleges.

[30] The percentages in this section come from AISHE (2019), mainly my calculations using detailed raw data on specific disciplines and their branches. For undergraduate enrolments, see Table 12. For post-graduate enrolments (Ph.D., M.Phil., Postgraduate Degrees) see Table 13.

levels. The opposite holds for India. This partially reflects patrifocality influences on girls' STEM choices and occupational pursuits, many requiring or facilitated by advanced degrees. AISHE data shows enrolment percentages increasing from undergraduate to postgraduate levels in science, engineering and technology, and IT & Computers. In science, females are 51% of undergraduates but 63% of postgraduates (non-Ph.D.) enrolments, with high percentages in branches like mathematics (66%), biotechnology (71%), and physics (58%). There is some decline at the Ph.D. level (48%) but not in bioscience (59%) or biotechnology (59%). In engineering and technology, females jump from 29% of undergraduates to 37% and 31% of postgraduate PG and Ph.D. students. For the separate IT & Computers degree, females are 41% of undergraduates but over half of postgraduate PG and Ph.D. students (50% and 51%).[31]

Institutes of National Importance (IITs)

Clearly more females are entering STEM. Why, then, do they remain so underrepresented at Institutes of National Importance, especially IITs? Given the previous discussion, it is not surprising. The IITs are an extremely competitive, hence financially costly, educational option in a particularly socially risky, male-dominated environment. Should a girl pass the hurdles for admission—high secondary school marks, sufficient scores on Main and Advanced Joint Entrance Exams (JEE)—she faces patrifocality related restrictions on some branches of engineering and some IIT campuses. Unless she has a high admissions rank, giving her early choices of campus/disciplines, she could be left with socially unacceptable alternatives, such as mechanical engineering or campuses viewed as particularly 'unsafe' for girls. This partially explains why fewer girls than boys with preliminary IIT admission end up actually attending IIT. One study found that on IIT candidate choice lists, girls filled in 'fewer choices than boys, fewer IITs, only a few branches, or a particular geographical area' (Verma 2018).

IIT campus officials, prodded by the Government of India, are addressing low female enrolments and setting ambitious goals for admissions and adding 'supernumerary' (additional) seats or places for women.[32] Yet, adding seats may not be enough. One top official in 2018 blamed families who, for social reasons, send boys and not girls for coaching, which affects their ability to qualify for entry. Other officials recognize the problems presented by predominately male environments and are organizing 'counselling' sessions for girls and their families, to alleviate 'safety' concerns (Verma 2018). These and other

---

[31] These patterns are not evident in education or medical science. Rather, the proportion of females (to males) declines at PG/Ph.D. levels. Social sciences and commerce show Postgraduate/MPhil increase but declines at the Ph.D. level, with branch variations.

[32] These seats add to the total number of IIT seats. It may even benefit males since girls admitted to IIT presumably occupy the supernumerary seats.

attempts to address what one official called the 'societal and parental restrictions' that affect girls, but not boys, plus the addition of 800 supernumerary IIT seats in 2018, have apparently had a significant impact. In 2019 females were 18% of all IIT undergraduate BTech students (Kunja 2020).

FEMALE SCIENTISTS: THE OCCUPATIONAL SPHERE

Why are so few Indian research scientists women? Available evidence shows overall attrition by female degree holders from college to the workplace, regardless of field, and especially in socially unconventional work settings. While women are around 40% of higher education faculty and one-third of higher education staff, their numbers are lower elsewhere. The Government of India's Department of Science and Technology found that women constituted only 14% of the 2017–2018 research and development workforce (Sharma 2020). Radhakrishnan estimated women at 25% of software and 30% of outsourcing workers.[33] Patrifocal family models apparently continue to shape women's occupational paths, even within the globalized, élite world of urban IT. Some argue the workplace may be a harder 'nut to crack' than the educational sphere, especially as more graduates compete for 'good' jobs (Kumar 2009). Current neoliberal ideologies of choice and individual responsibility also produce stresses on women, especially married employed women, to fulfil traditional family obligations, even though neither women's 'double-burden' nor the need for adequate childcare, is adequately addressed (Fleming 2016; Kamat 2014). Instead, the new model of Indianness created in the global IT Indian world draws upon and reinforces, rather than challenges, patriarchal and patrifocal aspects of the traditional Indian family (Radhakrishnan 2014).

Recent government efforts to increase women's participation in workforce-related technology provide culture-sensitive incentives but do not address patriarchal models that also impact female labour participation. These include the need for 'safe' housing, problems of sexual harassment, social constraints on travelling alone, and the demands of traditional in-laws and childcare. It is not surprising that women, once married, at least those with a choice, prefer not to assume a 'second shift'—and provoke in-law and spousal displeasure—by working outside the home, especially after they have children.

CONCLUSIONS

An important lesson of the Indian case, especially from comparative and policymaking perspectives, is the extent to which local social-cultural-historical factors and social processes underlie the Indian scientific gender gap. Indian academic choices are not primarily individual matters involving the pursuit

---

[33] Female labour force participation rates overall are low and declined from 36.7% in 2005 to 26% in 2018 (Global Compact Network India 2019). However, the female work force is largely rural and not highly educated (WMI 2019).

of individual goals; they are embedded in a family context, guided by family goals, circumstances, and long-term obligations.[34] Economic considerations, unlike in wealthy countries, play a major role, and more so for girls. And girls' individual abilities and interests, whether intrinsic or acquired, are not primary factors in their 'leakage' from the STEM pipeline.

Indian explanations for gendered science, whether by expert consultants, Indian students, or ordinary families, in contrast to the United States, consistently focus on social causation rather than locating causality in internal psychological states, character attributes, or intrinsic biological attributes. This reflects a broader explanatory paradigm applied to gendered activities. Indian explanations also emphasize the social attributes of activities. What makes some engineering jobs 'tough, heavy, strenuous, and arduous' for girls is the socially inappropriate location, living facilities, and behaviours. In addition, accounts attribute the scientific gender gap to other social–historical factors: the 'social oppression' of women in Indian society, 'male chauvinism', the preferential treatment of sons over daughters, and fears of educated daughters-in-law.

Perhaps most striking is the absence of essentialist beliefs about intrinsic intellectual differences or psychologically oriented theories of deeply internalized, socialization-rooted barriers to STEM. Indian consultants could not understand how anyone, male or female, would 'fear' academic success, an important source of individual and family pride. Some high-achieving girls were teased and called 'mugpots' by male peers. But their discomfort was social, not anxiety over academic success. And there was no evidence that girls felt pressure to 'play dumb'. Issues of girls' mathematical ability and mathematics as a 'masculine domain' generated surprise, laughter, and bewilderment among Indian expert consultants. How could anyone suggest women were 'naturally' less capable than men at mathematics! They cited famous Indian female mathematicians and pointed out that girls excel in mathematics, are 'toppers' on state-wide exams, and dominate post-graduate mathematics, physics, and chemistry programmes. The Western gender-differentiated brain theory was startling to them; so was the idea that females successful in mathematics experience 'gender role identity conflicts'. They were puzzled by statements appearing on Western Math questionnaires such as, 'Women certainly are logical enough to do well in mathematics' or 'girls who enjoy

---

[34] The patrifocal cultural model described here is itself a 'model'—an attempt to conceptualize a multi-component, flexible, adaptable, context sensitive, and dynamic framing process that families (and anthropologists) can use to think about and make (or comprehend) educational decisions. Patrifocality is not an oppressive family structure in which girls are powerless victims. While patrifocality can limit girls' educational opportunities, it can also facilitate their pursuits before and after marriage. The decline of the educational gender gap, at all levels, reflects this.

studying math are a bit peculiar.' Indian respondents simply did not possess the American cultural models of gender from which such statements are derived.[35]

Educational research in the USA tends to view motivation as an internal, individual process. Indian respondents linked motivation to social processes and social situations. Patrifocal family systems can depress female academic achievement, reducing the motivation to succeed. Some attributed girls' low achievement in mathematics or choice of arts to conservative families who feel too much education lessens opportunities for marriage. 'It's not that they are not bright. They just don't want to work hard. They are not motivated. They don't see any need to work hard'. The social reasons for pursuing particular subjects, and the social implications of academic success, have motivational consequences. They impact academic performance but also shape academic interests. As a group of sisters put it, 'If we cannot get the jobs, we will have no interest [in the subjects one must study to get such jobs]'.

Science and scientists evoke more socially based, less negative images in India than in the West. Findings from the research discussed in this chapter reveal overwhelmingly positive attitudes to science and scientists, reflecting Indian cultural models of science and social personhood as well as post-Independence India's emphasis on science for national development. Thematically, students focus on what scientists *do* rather than who they *are* in a psychological, personality sense. Negative images, when they appear, reflect social concerns, such as corrupt engineering practices or greedy doctors lacking compassion for patients. Absent are personality traits, appearance, and other unappealing personal attributes found in North American and European stereotypes of scientists. Exceptions reflect student exposure to Western images of the 'mad' (male, bearded, disheveled, socially inept) scientist. Even these are reconfigured to conform to Indian cultural models or simply rejected. Finally, most descriptions are not gender-specific, focusing on what scientists do, that is their actions, rather than on their personalities and appearance.

In summary, the most significant barriers to Indian women pursuing STEM are cultural, social, and economic. Among families economically able to send both sons and daughters to college, patrifocal models of family and gender can make science, especially engineering and technology, socially—rather than intellectually or psychologically—'unsuitable' for females. And it is the social attributes of STEM, especially the socially risky context in which engineering-IT is studied and practiced, and the socially strenuous context and behaviours required for some engineering-IT jobs, rather than perceived individual cognitive, physical, or psychological traits or deficits, that have kept many women from entering these fields.

---

[35] US models are still mired in deeply held essentialist views of the 'opposite sex' (biologically or, for socialization-oriented researchers, deeply internalized psychological identities). Indian cultural models of gender and personhood are quite different. See Mukhopadhyay (1982 and 2004).

Acknowledgements Financial support for this research came from American Institute of Indian Studies and Fulbright CIES fellowships and the College of Social Science, California State University, Chico. Support for analysis of the pre-college database came from National Science Foundation (Award #9511725) and San Jose State University, College of Social Science and Department of Anthropology. Thanks also to volume editors for their helpful guidance on the chapter.

References

AICTE (All India Council for Technical Education). 2021. *Government of India*. https://www.facilities.aicte-india.org/dashboard/pages/dashboardaicte.php Accessed on 29 January 2021.

AISHE (All India Survey on Higher Education). 2018–2019. *Government of India*. https://www.education.gov.in/sites/upload_files/mhrd/files/statistics-new/AISHE%20Final%20Report%202018-19.pdf Accessed on 19 January 2021.

Council of Indian Institutes of Technology. 2021. https://www.iitsystem.ac.in/ Accessed on 27 February 2021.

ESAG (Educational Statistics at a Glance). 2018. *Government of India*. https://www.education.gov.in/sites/upload_files/mhrd/files/statistics-new/ESAG-2018.pdf Accessed on 28 February 2021.

Fennema, Elizabeth and Sherman, Julia A. 1976. Mathematics Attitudes Scales: Instruments Designed to Measure Attitudes Toward the Learning of Mathematics by Females and Males. *Journal for Research in Mathematics Education* 7: 324–326.

Fleming, Rachel C. 2016. *Working for a Happy Life in Bangalore: Gender, Generation, and Temporal Liminality in India's Tech City*. Doctoral Thesis: University of Colorado.

Global Compact Network India. 2019. *Opportunity or Challenge? Empowering women and girls in India for the Fourth Revolution*. Delhi: GCNI https://www.globalcompact.in/uploads/knowledge-center/1553517403deloitte-gcni-thought-paper-2019.pdf Accessed on 28 February 2021.

Gold, Ann G., Gujar, Chinu, Gujar, Chumar and Gujar, Madhu. 2019. "Rural Women's Education: Process and Promise." In *The Impact of Education in South Asia*, edited by Helen E. Ullrich, 83–110. Basingstoke: Palgrave Macmillan.

Government of India. 2015. Policies on Skill Development and Entrepreneurship. *Ministry of Skill Development and Entrepreneurship*. https://msde.gov.in/reports-documents/policies/national-policy-skill-development-and-entrepreneurship-2015 Accessed on 15 January 2021.

Kamat, Sangeeta. 2014. "Gender and Education in South Asia." In *Routledge Handbook of Gender in South Asia*, edited by Leela Fernandes, 277–290. London and New York: Routledge.

Kelly, Allison. 1985. The Construction of Masculine Science, *British Journal of Sociology of Education* 6, no. 2: 133–154.

Kolenda, Pauline. 1987. *Regional Differences in Family Structure in India*. Jaipur, India: Rawat Publications.

Kumar, Neelam. ed. 2009. *Women and Science in India: A Reader*. Delhi: Oxford University Press.

Kunju, Shihabuddeen, 02 March 2020. Number of Girls in IITs Grown To 18% For BTech Programmes. *NDTV/Education*, https://www.ndtv.com/education/

number-of-girls-in-iits-grown-to-18-for-btech-programmes-hrd-minister-2188621 Accessed on 15 January 2021.

Ministry of Education. 2021. *Government of India*. https://www.education.gov.in/en/apex-level-bodies Accessed on 29 January 2021.

Mukhopadhyay, Carol. 1982. "Sati or Shakti: Women, Culture and Politics in India." In *Perspectives on Power: Women in Asia, Africa and Latin America*, edited by Jean O'Barr, 11–26. Durham, North Carolina: Center for International Studies, Duke University.

Mukhopadhyay, Carol. 1994. "Family Structure and Indian Women's Participation in Science and Engineering". In *Women, Education and Family Structure in India*, edited by Carol Mukhopadhyay and Susan Seymour, 103–132. Boulder: Westview Press.

Mukhopadhyay, Carol. C. 2001. *The Cultural Context of Gendered Science: The Case of India. NSF Final Report*. https://www.sjsu.edu/people/carol.mukhopadhyay. Accessed on 23 December 2020.

Mukhopadhyay, Carol. C. 2004. A Feminist Cognitive Anthropology: The Case of Women and Mathematics. *Ethos* 32, no. 4: 458–492.

Mukhopadhyay, Carol and Tami Blumenfield. 2020. "Gender and Sexuality". In *Perspectives: An Open Invitation to Cultural Anthropology*, 2nd edition, edited by Nina Brown, Thomas McIlwraith and Laura Tubelle de González. Society for Anthropology in Community Colleges (SACC). https://perspectives.pressbooks.com/chapter/gender-and-sexuality/ Accessed 20 January 2021.

Mukhopadhyay, Carol and Seymour, Susan. 1994. "Introduction and Theoretical Overview." In *Women, Education and Family Structure in India*, edited by Carol C. Mukhopadhyay and Susan Seymour, 1–33. Boulder, Colorado: Westview Press.

Pai, Seeta. 2002. *Family, Childbirth, Marriage, and Schooling Among Nair Women in Kerala, India: Portraits in Cultural Change*. Doctoral Thesis: Harvard Graduate School of Education.

Radhakrishnan, Smitha. 2014. "Gendered Opportunity and Constraint in India's IT Industry: The Problem of Too Much 'headweight'." In *Routledge Handbook of Gender in South Asia*, edited by Leela Fernandes, 234–236. London: Routledge.

Seymour, Susan. 1999. *Women, Family, and Child Care in India: a World in Transition*. Cambridge: Cambridge University Press.

Sharma, Kritika. 12 March 2020. IITs take Women's Quota to 20% as Govt Pushes Gender Balance in Tech Education. https://theprint.in/india/education/iits-take-womens-quota-to-20-as-govt-pushes-gender-balance-in-tech-education/379395/ Accessed on 19 January 2021.

Smail, Barbara and Kelly, Allison. 1984. Sex Differences in Science and Technology Among 11-Year-Old Schoolchildren. *Research in Science and Technological Education. I: Cognitive*. 2, no. 1: 61–76 and *Research in Science and Technological Education II: Affective*. 2, no. 2: 87–106.

Subrahmanyan, Lalita. 1998. *Women Scientists in the Third World: The Indian Experience*. New Delhi: Sage Publications.

Ullrich, Helen. E. ed. 2019. *The Impact of Education in South Asia: Perspectives from Sri Lanka to Nepal*. Basingstoke: Palgrave Macmillan.

UNESCO. 2020. Women in Science. *Fact Sheet No.60*. http://uis.unesco.org/sites/default/files/documents/fs60-women-in-science-2020-en.pdf Accessed on 12 January 2021.

Verma, Prachi. 2018. IITs Creating Extra Seats for Women to Lift Gender Ratio. 18 January 2018. *Economic Times of India E-Paper* https://economictimes.indiatimes.com/industry/services/education/page-1-iits-creating-seats-to-lift-gender-ratio-_-industry-welcomes-move-to-lift-gender-ratio/articleshow/62547957.cms Accessed on 15 January 2021.

Wadhwa, Wilima. 2018. Equity in Learning? *ASER*, 17–19. https://www.asercentre.org/Keywords/p/346.html Accessed on 15 January 2021

WMI (Women and Men in India). 2019. *Government of India*. http://mospi.nic.in/publication/women-and-men-india-2019 Accessed on 15 January 2021.

CHAPTER 17

A Seat at the Table: Women and the Periodic System

Brigitte Van Tiggelen and Annette Lykknes

The periodic system of the chemical elements is one of the most iconic symbols in science.[1] It represents order in the chaos of chemical elements, it summarizes basic chemistry for chemists and students, it may provide a geological map for mineral constituents, and it unites the scientific knowledge of chemistry and physics on both macroscopic and microscopic levels. The periodic system can be considered a representation of the basic substances that constitute all matter—from the spontaneous formation of elemental atoms through a coalition of subatomic particles during the 'Big Bang' to the era of the manufacture of atomic nuclei, which continues today. It has even been regarded a typology in science, akin to ideal specimens of species in natural history collections, albeit on a wider reaching and more symbolic level (Meinel 2009). Indeed, one might think of the periodic system of elements as one of the most important syntheses in the natural sciences of the mid-nineteenth century, along with the theory of evolution and the law of conservation of energy (Knight 2009; Bowler and Morus 2005). Thanks to its unique place in chemistry and chemistry teaching the periodic system is an excellent lens

---

B. Van Tiggelen
Science History Institute, Philadelphia, PA, USA
e-mail: bvantiggelen@sciencehistory.org

A. Lykknes (✉)
Department of Teacher Education, NTNU-Norwegian University of Science and Technology, Trondheim, Norway
e-mail: annette.lykknes@ntnu.no

© The Author(s), under exclusive license to Springer Nature Switzerland AG 2022
C. G. Jones et al. (eds.), *The Palgrave Handbook of Women and Science since 1660*, https://doi.org/10.1007/978-3-030-78973-2_17

through which women's contributions in chemistry can be viewed, as well as the question of how or why historical accounts have not mentioned them.

Traditional accounts of the history of the periodic system often present one man's accomplishments at a particular point in time and space, i.e. the Russian chemist Dmitri Ivanovich Mendeleev's presentation of his periodic system in March 1869. More nuanced narratives also mention other developers of systematizations of the chemical elements (Gordin 2012; Rocke 2019), but histories place emphasis on the creation of a *system* by sorting out similarities between these elements. In our view, this traditional approach to the history of such an important scientific icon does not capture the many layers of knowledge and skill involved in the development of the periodic system, nor the sustained effort needed for it (Lykknes and Van Tiggelen 2019a, b). Indeed, the history of the periodic system started with the different concepts of elements or building blocks of nature long before the nineteenth century, and continued long beyond the 1860s with the continued discovery of new elements and the emergence of further knowledge about them which was critical to their accommodation in the periodic system. Further, the development of an understanding of the atom and its structure contributed to the eventual establishment of the periodic system as a significant theoretical, rather than a mere pedagogical, tool. Within these complexities many women's contributions surface, and their achievements are brought to light, and this also allows an investigation of the question why they were often not acknowledged, and why their research trajectories were rather short. Although it is difficult to discuss the history of chemical elements and the periodic system without mentioning famous female scientists such as Marie Curie, Lise Meitner and Irène Joliot-Curie, the emphasis in this piece will be on lesser-known female figures.

The Periodic System and the Art of Analytical Work: Building on Women's Roles in Kitchen and Laboratory

Isolation and identification of chemical elements involved so-called 'wet-chemical analyses', which constituted an important part of everyday chemical work throughout the nineteenth and well into the twentieth century. This traditionally included meticulous bench work procedures such as repeated dissolution, filtering, and precipitation, and are referred to as 'wet' because they normally involved substances in their liquid phase.[2] During the nineteenth century analytical chemical work was often carried out in professors' homes, sometimes even in the kitchen, and often women both cooked and participated in chemical work. This chemical work borrowed equipment from the kitchen, for example, mortar and pestle for grinding materials, or ovens for sand baths (Lundgren 2019; Guerrini 2016; Brock 2016). The general assumption that women work in the kitchen therefore also made it acceptable for women to take part in analytical chemical work. As Anders Lundgren points out, the skills required in the laboratory were identical to those needed in the kitchen, and more generally, for running a household (2019). Anna Sundström is one example: she took care of the famous Swedish chemist J. Jac. Berzelius' household, and this included his laboratory equipment. She was familiar with all his glassware, was believed to be competent in the execution of chemical operations such as the distillation of hydrochloric acid, and worked in Berzelius' laboratory when he isolated lithium, selenium and vanadium. Indeed, she is today known as the 'first woman chemist in Sweden' (Lundgren 2019, pp. 126–27). Astrid Cleve, the daughter of the Swedish chemistry professor Per Cleve, also participated in her father's analytical chemical work at Kemikum, his office, laboratory and family home (Espmark and Nordlund 2019; Von Euler 1906). In 1898 Astrid Cleve was awarded a doctorate in botany, and thus became the first woman with a doctorate in the natural sciences in Sweden.

The mastery of analytical chemical work, and particularly the determination of atomic weights, was central to the identification and positioning of all elements in the periodic system. After John Dalton published his atomic theory in 1808 and established a correspondence between chemical elements and atoms, atoms gradually became part of the chemists' vocabulary and practice (although many of them treated atoms as hypothetical, not actual entities). Dalton's atomic theory stated that an element consists of only one type of atom with a unique atomic weight. Determining atomic weights as accurately as possible was thus an important task for chemists throughout the nineteenth century. Dalton's work was continued by Berzelius who is best known, among other things, for his comprehensive and accurate determinations of atomic weights (Bensaude-Vincent and Stengers 1996). Atomic weight determinations consisted of meticulous wet-chemical analyses which required the specific skill for controlling the experimental factors necessary for an accurate result. Atomic weight values were, thus, constantly revised. By the early twentieth century atomic weight determinations had become an art that could only be mastered by those trained by one of the very few specialists in the world. Two women who will feature in this chapter, Ellen Gleditsch and Stefanie Horovitz, were trained by the main experts in atomic weight determination in the USA and in Europe respectively, namely by Theodore William Richards at Harvard and by Richard's pupil Otto Hönigschmid at Vienna and Munich (Söderbaum 1915; Forbes 1932).

It was, indeed, the determination of atomic weights that prompted the quest for a system of organization for the elements. Before 1860 there was no consensus on the correct atomic weight value for each element. Experimental difficulties were the underlying problem, but there were also complications due to different assumptions on the chemical formula of composed substances. Depending on the assumed stoichiometry of a chemical compound chemists indeed sometimes arrived at values that were twice as high as those determined by others (Brock 1992). Following the first international congress on chemistry, which took place in Karlsruhe in 1860, chemists gradually adopted the system recommended by the Italian chemist Stanislao Cannizzaro, whose informative pamphlet had been circulated at the congress and had been read by the participants on their home journey (Cannizzaro 1858; Hartley 1966). Cannizzaro's distinction between atoms and molecules as we now understand them enabled chemists to work in a common framework and to adopt a common scale for atomic weights. Dmitri Mendeleev and Lothar Meyer, who had attended the meeting (as had William Odling) and who were developing periodic systems, later stated that the consensus for atomic weight values was pivotal to their efforts of organizing the chemical element into one coherent system (Nye 1984; Rocke 2019). Indeed, all attempts to systematize (and not only classify) the elements in the nineteenth century had one and in fact the main criterion, in common, which was the atomic weight.

Separating and Positioning the Elements

But what was one to do when the atomic weights of two or more elements were close, or even equal, in value, and the chemical properties also overlapped? This was the case for the platinum metals, and later for many rare earth elements.

The metal platinum has been known since the eighteenth century, but the other members of the platinum group, palladium, rhodium, osmium, iridium and ruthenium, were all discovered in the nineteenth century. They occur together in the same mineral deposits and are usually referred to as the platinum metals. However, since they have similar chemical properties, they are hard to distinguish from one another. And ever since chemists tried to sort the platinum metals into the periodic system, there has been some doubt about the accuracy of their atomic weights. Meyer, for example, attempted classifications of these elements as early as 1864 and 1868, and in 1870 added a question mark to the atomic weight of osmium to indicate his uncertainty (Boeck 2019). Mendeleev also experimented with the order of the elements, and also literally added question marks to some atomic weight values. However, both Meyer and Mendeleev considered the methods of atomic weight determinations problematic, and scientists from different laboratories made attempts to improve the atomic weights of the platinum metals. But for accurate values, pure substances were needed. Letters kept in Mendeelev's archive in St. Petersburg indicate that Mendeleev discussed this problem with a Russian chemist, Julia Lermontova (Boeck 2019).

Lermontova was the second woman worldwide to be awarded a doctorate in chemistry, and she received it at Göttingen in 1874. In the following year she was elected a member of the Russian Chemical Society. Prior to this she had attended the lectures of Robert Bunsen in Heidelberg, where she learned about the platinum metals and the separation of platinum from the other platinum metals (Boeck 2019). Here Lermontova probably met Mendeleev, since he was working in Heidelberg in 1860/1861, and later often returned. As mentioned above, like kitchen work, the art of chemical analysis was considered suitable for women; and enlistening Lermontova therefore to separate the platinum metals probably came naturally to Mendeleev, as he knew that she had mastered the current experimental techniques which Bunsen had taught her. Her written report on the separations she achieved forms part of Mendeleev's archive, but was never published (Musabekov 1967). As part of the invisible women's work this shows the importance of a close analysis of unknown records in any investigation of the work behind a scientific discovery. The separation process was certainly complicated, as patents for improved separation processes were issued even in the late twentieth century (Boeck 2019). Lermontova's work was part of the meticulous process, from the acquisition of mineral samples to the determination of the atomic weights of the individual elements, which once more contributed to the task of positioning them in the periodic system.

Another group of elements that were known to be difficult to position in the periodic system were the rare earth elements. Lined up at the bottom of periodic tables today, the elements in this group, even more so than the platinum metals, have very similar properties. And indeed, as the analysis of minerals developed from the very end of the eighteenth century onwards, some substances originally considered to be elemental rare earths turned out to be a mixture of two or more elements of the group. Many aspects regarding these elements, especially their atomic weights, were uncertain, as indicated by the question marks placed before Ce, La, Yt, Di and Er in Fig. 17.1. Further, the discovery and isolation of many new rare earth elements in the 1870s and

	Gruppe I. R^1O	Gruppe II. RO	Gruppe III. R^1O^3	Gruppe IV. RH^4 RO^7	Gruppe V. RH^3 R^2O^5	Gruppe VI. RH^2 RO^3	Gruppe VII. RH R^2O^7	Gruppe VIII. — RO^4
1	H = 1							
2	Li = 7	Be = 9.4	B = 11	C = 12	N = 14	O = 16	F = 19	
3	N = 23	Mg = 24	Al = 27.3	Si = 28	P = 31	S = 32	Cl = 35.5	
4	K = 39	Ca = 40	— = 44	Ti = 48	V = 51	Cr = 52	Mn = 55	Fe = 56 Co = 59 Ni = 60, Cu = 63.
5	(Cu = 63)	Zn = 65	— = 68	— = 72	As = 75	Se = 78	Br = 80	
6	Rb = 85	Sr = 87	?Yt = 88	Zr = 90	Nb = 94	Mo = 56	— = 100	Ru = 104, Rh = 104, Pd = 106, Ag = 104.
7	(Ag = 104)	Cd = 112	In = 113	Sn = 118	Sb = 122	Te = 125	J = 127	
8	Cs = 133	Ba = 137	?Di = 138	?Ce = 140	—	—	—	— — —
9	(—)							
10	—	—	?Er = 178	?La = 180	Ta = 182	W - 184	—	Os = 195, Ir = 197, Pt = 198, Au = 199.
11	(Au = 199)	Hg = 200	Tl = 204	Pb = 207	Bi = 208	—	—	
12	—	—	—	Th = 231	—	U = 240	—	— — — —

Mendeleev's Periodic Table of 1871, redrawn by Jeff Moran, 2013

Fig. 17.1 Mendeleev's Periodic Table of 1871. The platinum metals are placed in group VIII towards the right of the table. Note that they have very similar atomic weight values (Reproduced by kind permission of Jeff O. Moran)

1880s added to these uncertainties, and Mendeleev could not accommodate these new elements in his periodic system (Thyssen and Binnemans 2015).

In 1874–1875 Ellen Swallow, the first female graduate in chemistry at the Massachusetts Institute of Technology, investigated a small sample of the mineral samarskite. The sample originated from Mitchell County in North Carolina, where it had recently been discovered. The sample was given to Swallow by the president of the Boston Society of Natural History, who knew of the 'thoroughness with which she performed the [previous] analysis' (Charbonneau and Rice 2019, p. 149). She had previously analysed samples given to her by the professor of mineralogy Robert Richards, whom she married in 1875. Swallow did two independent analyses of samarskite and detected the same metals in both. At the time six rare earth metals were known, and one of them was cerium. Swallow found an 'insoluble residue' precipitated from the oxalate of cerium, and she reported that this residue might contain as yet unknown elements. As many as ten elements were suggested in the decade following Swallow's investigation, but in the end few were recognized as elements. In 1886, with the help of spectroscopy, one of them, 'mosandrum', was found to contain at least four rare earths, including samarium and gadolinium, which were also present in Swallow's residue (Charbonneau and Rice 2019). Since she never explicitly suggested new elements, Swallow has never been credited with the discovery of the two elements in any way. However, her involvement demonstrates the many steps involved in a scientific discovery, including the work of many who are invisible.

Even more so than platinum metals, rare earth metals were thus difficult to separate from each other. Per Cleve in Sweden was interested in the rare earth elements, their chemistry and position in the periodic system; he had already contributed to the discovery of the elements holmium and thulium, which were found in the mineral sources of the Ytterby mine in the Stockholm archipelago (Espmark and Nordlund 2019). In the mid-1880s, Cleve delegated the collection of the minerals to his students, among them his daughter, Astrid. Astrid Cleve decided to investigate the properties of ytterbium, one of the other rare earth metals which had been found in Ytterby. Additional insights into the individual elements, she thought, would clarify the relationship between the rare earth elements. Like Lermontova and Swallow, Cleve worked on separation processes. Taking up the challenge, she eventually managed to produce what she believed to be a pure sample of the element ytterbium. From this sample, she was able to study ytterbium compounds as well as some of their properties and publish the results in one of the journals of the Royal Swedish Academy of Sciences (Espmark and Nordlund 2019). Her publication was translated into German and was thus made available to an international audience (Cleve 1902). Her work was even acknowledged with an award from the Swedish Academy of Sciences for 'new and important discoveries in the chemical or physical sciences' (Espmark 2012 , p. 70). Cleve was only 27 years old, and her future looked promising.

Her success was not marred by the fact that a few years later (in ca. 1907) her sample proved to consist of the two rare earth elements lutetium and neoytterbium (later renamed 'ytterbium'); improved analytical methods and procedures often resulted in the splitting of rare earths, and the difficulties in obtaining a pure sample almost 20 years later in fact testify further to her achievement (Kragh 1996).

Prior to the 1920s, when atomic numbers were accepted to be the unique characteristic of an element, atomic weights were reckoned to be the defining property. When physical methods such as spectroscopy became commonly used to identify new elements, physical samples of elements nevertheless continued to be important for the determination of their atomic weights. Once a unique atomic weight was established and verified by others, the scientific community was able to acknowledge the discovery. Women (and men) who were well-versed in conducting wet-chemical procedures contributed to the isolation of elements and their identification by means of determining their atomic weight. The Polish-born physicist and chemist Marie Curie and the Austrian-born physicist Lise Meitner are two examples of (well-known) women who co-discovered elements and mastered the necessary analytical chemical procedures. They both spent several years separating new elements or fractions containing these, from other elements or fractions in their mineral samples: Curie formed a team with her husband, the physicist Pierre Curie, and collaborators such as Gustave Bémont and André-Louis Debierne in order to obtain a sample of radium salt from pitchblende (Roqué 2019a). They succeeded in 1902, almost four years and thousands of crystallizations later. Between 1913 and 1918, Lise Meitner, together with the chemist Otto Hahn in Berlin, undertook the difficult task of concentrating a fraction of the mineral that contained the substance that would produce actinium from radioactive decay. They also tracked the substance down further and identified it, analysed its radioactivity and the traces of the actinium residue in pitchblende. The new element they had co-discovered was named 'protactinium' since it was the 'mother substance' of actinium (Roqué 2019b). At the beginning of the twentieth century Marie and Pierre Curie named the ray phenomena 'radioactivity' and became leading scientists in this field. Meitner, too, worked in the radioactivity research, and later co-discovered nuclear fission. She therefore contributed to the birth of the era of nuclear research.

New Opportunities for Women in Radioactivity Research: Understanding Radioactivity and Isotopy

Analytical chemistry continued to offer opportunities for women well into the twentieth century. The emerging field of radioactivity research would even offer female researchers a community: alongside the well-known figures of Marie Curie and Lise Meitner large numbers of less well-known women contributed to the field. Between 1906 and 1934 women constituted between 8.7 and 40 per cent of the researchers at Curie's Paris laboratory, and between

1919 and 1934 around 38% of the staff members at the Institute for Radium Research in Vienna (founded in 1910) were women (Pigeard Micault 2013; Rentetzi 2007; Rayner-Canham and Rayner-Canham 1997). Women also joined Meitner and Hahn's laboratory in Berlin and Ernest Rutherford's laboratories in Canada and Manchester. The new field of radioactivity research offered new opportunities for women in an arena where male hierarchies had not yet been established, and in laboratories and with mentors who were women-friendly. The prospect of being part of cutting edge research probably also encouraged ambitious women willing to take a professional risk to work in these laboratories that pursued a fascinating topic not yet acknowledged as main stream science. There were local differences as well, and for some women routine jobs may have been easier to enter in this field (Lykknes and Van Tiggelen 2019b).

It has been argued that many of the women who worked in radioactivity research in the first quarter of the twentieth century established networks with the other women in the field (Rayner-Canham and Rayner-Canham 1997). Some of them moved between a number of laboratories, some met informally, and they corresponded extensively with and supported each other. Some women, like the Norwegian radiochemist Ellen Gleditsch and the Austrian physicist Berta Karlik, also established connections within the International Federation of University Women (von Oertzen 2014).Gleditsch started her career in Marie Curie's laboratory and continued to visit almost every year after her return to Norway, but she also spent some months in Vienna. She also kept in close contact with Meitner, and when some of the Jewish colleagues she had met in Vienna needed to flee during World War II and continue their research elsewhere, Gleditsch opened up her laboratory to them (Lykknes et al. 2005).Gleditsch herself received the support of her mentor, Marie Curie—whose authority was unsurpassed—in her application for a professorship at her *alma mater* in Oslo (Lykknes et al. 2004).

When the phenomenon of radioactivity was first discovered, it was not well understood. Henri Becquerel had discovered rays that were emitted spontaneously, and Marie and Pierre Curie had established that only very few elements possessed this property. Together these three scientists shared the Nobel Prize in physics in 1903. But what were these rays, and what process caused their emission? As Marie Curie stated in her Nobel Prize lecture of 1911, when she was awarded her second Nobel Prize—this time in chemistry—scientists understood that radioactivity was an 'entirely separate kind of chemistry'—'the chemistry of the imponderable', since both its subject and detection method were undetectable to the human eye (Curie 1911). In 1902 the New Zealand-born physicist Ernest Rutherford and his British collaborator, the chemist Frederick Soddy, both working in Montreal, Canada at the time, proposed that atoms spontaneously break down into new atoms during radioactive decay—a kind of modern alchemy. This theory was based on experiments on what was called 'emanation', i.e., something that was emitted from both radium and thorium. It had been difficult to establish the nature of this emanation. Rutherford initiated a systematic study of the emanation phenomenon, which had been reported by other scientists as well, and in 1901 Rutherford and his graduate student Harriet Brooks published an article in which they described the emanation from radium as a gas of heavy molecular weight which could not be a vapour of radium (Rayner-Canham and Rayner-Canham 2019a). In the following year Rutherford published these results in *Nature*, this time under his own name, but acknowledging Brooks' contribution. Indirectly, Rutherford and Brooks proposed that the emanation was a new element, and the title of their joint paper was, indeed, 'A New Gas from Radium' (Rutherford and Brooks 1901). In 1910, this gas was identified as radon, an element in the noble gas family of the periodic system.

Radioactivity research also led to the discovery of a wealth of new substances, since every product of radioactive decay with a unique atomic weight was considered to be a new element. The problem was that there were not enough blank spaces in the periodic system to accommodate all of them. At a dinner party in 1913 Soddy, who had moved to Manchester, discussed the idea that one element might consist of atoms with different atomic weights. This idea became important for chemistry as it created order in the chaos of radioactive substances; indeed, most of the newly discovered radioactive substances turned out to be isotopes of known elements rather than new elements. Although it was Soddy who developed the concept of isotopes, the term 'isotope' was first proposed by Margaret Todd, a physician who participated in the discussion at the above-mentioned dinner party in Glasgow (Hudson 2019). Her proposal was both philologically and scientifically informed, as the word derives from the Greek words for 'same' and 'place': isotopes occupy the same place in the periodic system. They have the same chemical properties, only their atomic weights differ.

Women also contributed to the corroboration of the concept of isotopes in the scientific community. One potential experimental way of finding evidence for the existence of isotopes was the study of lead, the end product of radioactive decay. Lead from uranium would have a different atomic weight than ordinary lead. The Polish-Jewish chemist Stefanie Horovitz, who was working at the Radium Institute in Vienna, was assigned this research in 1913 or early 1914 by her supervisor Otto Hönigschmid, one of the leading atomic weight determination experts in Europe. Horovitz separated lead from uranium samples and determined the atomic weights of lead in the different samples. The atomic weights she found differed beyond experimental error, which made it the first definite evidence that isotopes exist (Rayner-Canham and Rayner-Canham 2019b). Ellen Gleditsch, after working for many years with Marie Curie in Paris, learned atomic weight determination from Theodore Richards at Harvard. In 1922–1923 she conducted a comparative investigation of the atomic weight of chlorine in samples of different origin. This led her to the conclusion that the atomic weights of isotopes were constant and independent of their origin; the only exception was isotopes resulting from radioactive decay (Lykknes 2019). In those cases, as Horovitz demonstrated, the atomic weight of the lead sample depended on the amount of radioactive material present in the starting material.

Apart from representing some of the many questions that women investigated in radioactivity research in the early decades of the periodic system, these brief case studies illustrate the important roles that male professors or male principal investigators played by supporting, or not supporting, those women who took part in discussions or joint research. Rutherford was known to welcome women chemists in his laboratory and support their research (Rayner-Canham and Rayner-Canham 1997). This was also the case for Brooks. Following her role as Rutherford's first graduate student researcher she took up a position at the women-only Bryn Mawr College and often wrote to Rutherford, confiding in him and telling him about her low self-esteem and the need for his encouragement and support (Rayner-Canham and Rayner-Canham 2019a). Rutherford even helped her secure a place in J.J. Thomson's laboratory in Cambridge, and after one year in England she returned to Rutherford's laboratory instead of returning to Bryn Mawr to complete a Ph.D. Horovitz and Gleditsch also benefitted from training with male experts of international standing to master atomic weight determinations. Naturally, Marie Curie occupied a similar role in many women's careers, but most laboratory leaders and international experts were, indeed, male. New, unestablished fields without male hierarchies such as radioactivity made it possible for women to pursue research.

Filling in Gaps: Opportunities and Limitations for Women Through Marriage

Another common opportunity for women of the late nineteenth and early twentieth centuries to research as well as being supported by male scientists was marriage. For Ida Noddack, née Tacke, meeting a partner who was enthusiastic about the periodic system led her to new research areas, and to both success and failure.

By the early 1920s some predicted, but unidentified elements of the periodic system were yet to be discovered. These undetected elements had been known to exist since 1913, when the British physicist Henry Moseley discovered a link between an element's atomic number and its X-ray spectrum. This discovery confirmed that all of the blank spaces left by Mendeleev and others (see the spaces with lines in Fig. 17.1) were expected to be filled. And while many of these unknown elements had been identified in the late nineteenth century, a few were still undiscovered in the 1920s. Two of these were the so-called 'eka-manganeses', elements 43 and 75, which occupied the spaces beneath manganese in the periodic system. As early as 1914, the German chemist Walter Noddack had been looking for the eka-manganeses in platinum ores, but he did not succeed and had to abandon the search. In around 1920 he met the like-minded Ida Tacke, who was soon to earn her Ph.D. in chemistry, and they decided to search for the eka-manganeses together (Van Tiggelen and Lykknes 2012).

Tacke had graduated in chemical engineering at the Technische Hochschule Berlin in 1919 and won a prize in chemistry and metallurgy, whereupon she decided to pursue a Ph.D. After she received her 'Doktor Ingenieur' in 1921, she secured a job at the Allgemeine Elektricitäts-Gesellschaft (AEG). However, when Walter and Ida, who would marry in 1926, realized the extent of the work needed to prepare for the search for missing elements, Ida decided to quit her job and work with Walter on a voluntarily basis. She first spent nine months sifting through one hundred years of literature in inorganic chemistry to get an overview of the chemistry of the elements near manganese in the periodic system, and also to acquaint herself with all previous attempts to detect elements 43 and 75. After this thorough research, the couple was ready to search for the elements in minerals. They followed the tradition initiated by Mendeleev himself of predicting chemical properties based on the elements' positions in the periodic system. They took this approach even further by using the system as a geochemical navigation tool in their search of the elements. They assumed that the elements would form minerals with neighbouring elements and that they would be very scarce, otherwise, they argued, other scientists would have detected them earlier (Van Tiggelen 2001). At first, they searched in manganese ores, but this search was soon broadened to include ores in which platinum elements would form minerals, and even beyond. In 1925, Ida Tacke and Walter Noddack, together with Otto Berg from Siemens und Halske, announced that they had, indeed, detected elements 43 and 75. The collaboration with Berg was short-lived and related to the cooperation established with Siemens und Halske for the part of the investigation that involved instruments which were not available elsewhere. However, it took four more years for Ida and Walter Noddack to produce the first gram of element 75, which they named rhenium. For this discovery the couple received recognition. The existence of element 43, which they had named masurium, was never confirmed.

When the search for the minerals started, Ida had been accepted as an unpaid guest researcher at the Physikalisch-Technische Reichsanstalt in Berlin, where Walter was head of the chemistry department. There, the couple carried out the analytical chemical investigations of the minerals they collected in the search for the eka-manganeses (Fig. 17.2). Ida soon also accepted the position of unpaid guest researcher at Siemens und Halske, the company with which the couple would collaborate to identify the missing elements by means of X-ray spectroscopy (Van Tiggelen 2001). For most of her career Ida Noddack was an unpaid guest researcher at the institutions employing her husband in Berlin, Freiburg and Bamberg, with the exception of Strasbourg, where she occupied a paid position during World War II. As a married woman, she was not encouraged to work; in Germany as in many countries suffering difficult economic conditions, it was tacitly understood that women should not fill positions needed by men to support their families. In 1932 a law was

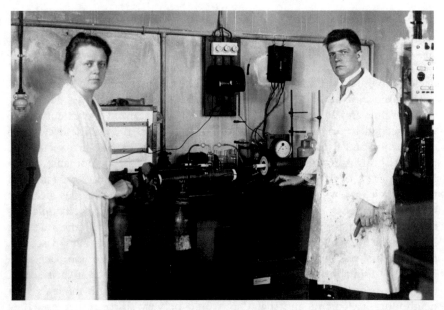

Fig. 17.2 Ida and Walter Noddack in the laboratory, Berlin 1931 (Stadtarchiv Wesel O1a, 5–14-5_02) (Reproduced by kind permission of Stadtarchiv Wesel, Wesel, Germany)

even passed forcing married women to resign (Frevert 1989). But marriage was not all that bad a deal, like many wives who collaborated with their husbands, Ida had access to a laboratory and co-workers, and was able to follow the latest developments in her branch of chemistry. Thanks to her marriage, outside of a formal career, Ida Noddack enjoyed recognition for her work on rhenium as well as for her broad knowledge on its neighbouring elements. This was very useful in 1934, when the physicist Enrico Fermi and his team in Rome announced that they were the first to discover transuranium elements, nos 93 and 94, with the help of induced nuclear reactions. But since some periodic systems considered element 93 to be one of the ekamanganeses, Ida used her knowledge of the manganese family to argue that the Fermi group had not detected elements 93 and 94. Instead, she suggested, they had produced lighter elements that were already known, and that, in fact, the uranium nucleus that they had bombarded with neutron particles had split. This was a very bold claim, without experimental substantiation, and Ida Noddack was first ignored and then ridiculed for her theory (Sime 1996). The idea that the nucleus can actually split was confirmed four years later in publications authored by Lise Meitner, Otto Hahn, Fritz Strassman and Otto Frisch, announcing the discovery of nuclear fission. However, Hahn alone was awarded the Nobel Prize for this work, despite his long-term productive collaboration with Meitner which had begun with their joint discovery of protactinium.

At the time, Meitner was a junior researcher, and forming a team with Hahn was an excellent strategy. But more women followed Ida's 'career by proxy' path of marriage, for better and for worse (Lykknes et al. 2012). Ellen Swallow, for instance, was supported by her husband, the professor, who gave her new challenges on a regular basis, and had good access to literature, bench and publication prospects (Charbonneau and Rice 2019). Astrid Cleve likely hoped for a partnership like Pierre and Marie Curie's when she married Hans-Euler von Chelpin, and Cleve's divorce demonstrated the social value of marriage for a woman's career: the end of her marriage marked the end of her original work in chemistry (Espmark and Nordlund 2012). To women who were already employed, marriage could be a threat and often meant the end of an independent career. This was the case for Harriet Brooks, who was asked to withdraw from her position at Barnard College, the women's college of Columbia University, New York, when she announced that she was considering getting married. She broke up with both her college and her fiancé, but years later her marriage to the former McGill instructor and lecturer Frank Pitcher did indeed put an end to her promising career (Rayner-Canham and Rayner-Canham 2019a).

The history of women in science, and women behind the periodic system in particular, however also includes many women who entered or stayed in science through marriage. Both Marie Curie and Ida Noddack worked with their husbands, but they were never their 'assistants'; both took leading roles in designated areas of the joint research. We have argued elsewhere that Ida and Walter Noddack established a symbiotic work unit, an *Arbeitsgemeinschaft*, always thinking of the best output of the work unit, sometimes at the expense of individual recognition (Van Tiggelen and Lykknes 2012). More often than not, this was at the expense of the wife's recognition.

BIG SCIENCE: INVISIBLE TECHNICIANS AND THE RISE OF FEMALE LEADERSHIP

As scientific laboratories expanded, and with them the number of people working in them, more specialization and a higher level of division of labour ensued. Technical support and research assistants were always part of the laboratories, and they were often forgotten or barely recorded in history, yet essential to the day-to-day work in ensuring the smooth incorporation of new members into the laboratory while maintaining the technical standards in devices and practical experimental processes. As we have mentioned before, women were more often part of this staff than leaders, but in the second half of the twentieth century many more female scientists were able to access to leadership positions, and followed, much later, in the footsteps of Marie Curie.

The shift from the extraction of new elements to their manufacture with nuclear physical means is an excellent example of what is often referred to as 'Big Science'. This generally involves large-scale research plants, large teams working together, enormous budgets, and international collaboration (Nye 1996; Kragh 1999). Nobel laureates Irène and Frédéric Joliot-Curie contributed to the new era of nuclear research when they discovered that radioactivity could be induced by bombarding nuclei with particles, which they called artificial radioactivity. To some extent Marie Curie's laboratory was already a blueprint for Big Science laboratories. In fact, her laboratory was organized to a very high degree, with designated roles and tasks assigned to a specific workforce for which the recruitment procedures were quite different from one role to another. Many women arrived as visiting scientists or guests, others entered the laboratory as students at every level, and then there were also technicians (Pigeard Micault 2013). These were also selected carefully, as Marie Curie always made sure to secure a highly skilled personnel of technicians. For that purpose, she referred to the École d'Enseignement Technique Feminin which trained chemical technicians, and in 1929, she hired the nineteen-year-old Marguerite Perey, whom she trained in the manipulation of radioactive elements. Soon Perey's specialty was the preparation of actinium samples, as pure as possible, for research on the actinium decay series (Rayner-Canham and Rayner-Canham 2019c). One aspect of her bench work was analytical chemistry, entailing the intricate and repeated processes in fractional crystallizations, but since she was working with radioactive material she also needed to separate isotopes formed during decay in order to obtain pure material, and to ensure that she was passing on reactants of the best quality possible to her superior. As she learned to master these techniques, Perey also mastered her element, and was soon in charge of performing basic measurements on actinium isotopes. For instance, one of the tests for the purity of the produced material involved looking for any radioactivity that would indicate the presence of unwanted contaminants—or verifying the radioactive emission to ascertain the identity and the quality of the substance (Fig. 17.3).

Following Marie Curie's death in 1934 the laboratory, and especially Perey, worked under the leadership of both André-Louis Debierne (discoverer of the element actinium) and Irène Joliot-Curie. Both asked her to do the same task—without consulting each other—i.e. to determine a precise value for the half-life of actinium-227 (Adloff and Kauffman 2005). But in the meantime, Perey had acquired enough expertise and autonomy to be able to pay attention to unexpected phenomena rather than conduct mere preparative chemistry or routine measurements. She discovered that some of the radiation was not accounted for in the radioactive decay series and concentrated her experimental work on this short-lived radiation, which she conjectured would be an isotope that had not been previously observed. Soon the chemical behaviour of her sample left no doubt: this was an alkali metal-like substance, heavy and unstable, the missing eka-caesium (element 87). When she reported her findings to her two superiors, all that was left for them was to acknowledge

Fig. 17.3 Analytical work in the chemistry room of the Curie Laboratory at the Institut du Radium, Paris, July 1930. From left to right: Sonia Cotelle, Marguerite Perey, Alexis Jakimach, Tchang da Tcheng (Musée Curie, MCP 1938) (Reproduced by kind permission of the Musée Curie)

the discovery, and since it was unclear for whom exactly Perey was working at the time of her discovery, the entire credit was given to her; she was the sole author of the article announcing the discovery of a new element in 1939 (Perey 1939). Perey was then encouraged to pursue a university degree, passed her Ph.D. after World War II, and then followed an academic career. Her career emphasizes the importance of technical skills. Of course, both happenstance and professional honesty from her superiors allowed Perey to shine in the spotlight of work that would otherwise simply have appeared under the name of the laboratory leader for whom the technician was working. Interestingly, however, the naming process for the element involved both laboratory leaders, and was also inspired by the Curie tradition. Perey's final choice for the patriotic sounding francium was eventually influenced by Irène and Frédéric Joliot-Curie (Adloff and Kauffman 2005), whereas she and Debierne had initially favoured 'catium' which reflected a chemical property rather than a nationalistic statement.

The career of Toshiko Kuki Mayeda provides another story of technical expertise that would go unnoticed were it not for the laboratory head's willingness to include his technical assistant as co-author. Mayeda's story also illustrates how a permanently employed female technician would find a role in the laboratory, displaying some of the household virtues already mentioned above. A young Japanese-American, Mayeda had endured four years of internment after the Japanese air force had attacked Pearl Harbor before she began to study chemistry at Wilbur Wright College, a city college of Chicago. While still an undergraduate she was hired by Harold C. Urey to clean glassware (Shindell 2019, 2020). In 1934 Urey had won the Nobel Prize for his discovery of the isotope deuterium ('heavy hydrogen', D) two years previously. Urey's discovery resulted from his expertise in and knowledge about the separation of isotopes, and in his Chicago laboratory, his research programme focussed on analysing the proportion of isotopes in nature as a means to investigate the history of the climate in the distant past. His idea was to examine marine shells from molluscs, because they capture oxygen while they are alive by fixing the oxygen in the production of calcium carbonate for their shells. He wanted to use the ratio of two oxygen isotopes (the heavier oxygen-18 to the more abundant oxygen-16) in these shells as an 'oxygen thermometer' to deduce the temperature in oceans. All Urey needed was a specific instrument that would allow for the high-precision measurement of the ratio of the two isotopes of oxygen, i.e. a mass spectrometer specifically built for this measurement. He commissioned a custom-built mass spectrometer in his laboratory, and Mayeda, who was moving on from the cleaning of glassware, became involved in its use. Since this instrument was one of a kind, Mayeda's knowledge of the instrument was critical in its calibration, the preparation of samples and the interpretation of the measurements.

At first Mayeda was credited in the acknowledgements of papers written by members of the laboratory, but later her crucial role was acknowledged to the point of co-authorship. Inside the microcosm of the laboratory, despite her comparatively low level of chemical education, she also became a mentor to new students who learned the specifics of mass spectrometry using this homemade device and appreciated her advice beyond the mere technical content. Another piece of evidence for Mayeda's importance in the laboratory was the fact that Urey's successor, Robert Clayton, insisted that she stay when Urey left in 1958, and that he also considered her as an equal collaborator in operating the laboratory (Shindell 2019). Although Clayton specialized in cosmochemistry rather than the geochemistry pursued by Urey he continued the measuring of oxygen isotopes; he had the advantage of having both a unique instrument and the person who could use it at his disposal, and if necessary it would be possible transform their routines for new productive areas. Clayton systematically acknowledged Mayeda as a co-author; nevertheless, the model of oxygen isotope abundances in the solar system that they developed together is not usually referred to as the 'Clayton-Mayeda model', but rather often described as the 'Clayton model' (Shindell 2019, 2020). In 2002, two years before she died, Mayeda was awarded a medal for her contributions to cosmochemistry by the Geochemical Society of Japan, but this was the only public recognition of her work. To some extent her caring mentorship and her domestic skills in the managing of the laboratory and its machines can be perceived as typically female in the gendered space of the laboratory, roles which she gladly embraced and did not seem to wish to depart from. The fact that her supervisor granted her visibility in authorship is rather unusual compared with many laboratory technicians and research assistants, male and female alike, whose task was to ensure the day-to-day business, and who were thereafter left out of the limelight. Mayeda's example also clearly demonstrates that the gendering of the workplace results from a negotiated balance between the expectations and perceived limitations of all involved.

From Naturally Occurring to Artificially Produced Isotopes

Returning to the quest for new elements, and moving back in time, Perey's discovery of francium marked the end of an era characterized by the isolation of new elements from substances found in nature and by the investigation of naturally radioactive substances. Research on naturally occurring isotopes continued, and still continues to this day, but the discovery of artificial radioactivity and nuclear fission opened up new ways of working with missing elements: manufacture instead of separation.

An excellent example of this transition is the story of the discovery of astatine isotopes by the Austrian co-workers Berta Karlik and her assistant Traude Bernert (Fig. 17.4), who announced in 1943 and in 1944 that they had discovered isotopes 215, 216 and 218 of element 85 (Forstner 2019). Unbeknownst to them, element 85 had meanwhile been produced at Berkeley by the physicists Emilio Segrè, Dale Corson and Kenneth MacKenzie, who had bombarded bismuth-209 with α-particles in 1940. In fact, the American team produced one specific isotope of element 85 with atomic weight 211. The acceptance and naming of the new element were completed much later. Following Fritz Paneth's rule of 1947, whoever first produced an isotope of a new element gained the right to name the element (Paneth 1947). Segrè, Corson and MacKenzie were thus acknowledged as the discoverers of the element 85 which they called astatine.

Karlik and Bernert strike as an all-female team, at a time where all-male teams were the norm. Both Karlik and Bernert belonged to a cluster of women working in radioactivity who had been trained at the Radium Institute in Vienna, with Stefan Meyer as their mentor. Karlik also benefitted from another female network thanks to a fellowship from the International Federation of University Women that allowed her to work with the British physicist William Bragg. She was appointed as a paid lecturer at the University of Vienna in 1942 but the research fellowships she had previously received were denied by the university after the annexation of Austria by the Nazi regime with the justification that women had no future in academia. It was only due to the intervention of the director of the Radium Institute Gustav Ortner that she was able to support herself during the wartime years. In a reversal of fate, from 1947 until 1974, she would be the director of the very Institute

Fig. 17.4 An all-female team: Berta Karlik (right) and Traude Bernert (left) with their experimental setup in 1943–1944 (bpk/Liselotte Orgel-Köhne, image no. 70113898, reproduced with permission of bpk Bildagentur)

that had almost denied her right to work. In 1956 Karlik also became the first female full professor in Austria, with a chair specifically created for her (Forstner 2019).

There were good reasons why the discovery of elements immediately before and during World War II had not been circulated openly. Segrè and his collaborators at Berkeley, like others before them, had used a cyclotron to accelerate particles before they hit a chosen target, utilizing induced nuclear processes instead of working on the natural radioactive series. During World War II nuclear research intensified dramatically. The so-called Manhattan Project of producing nuclear weapons indeed relied heavily on new developments in technology and the nuclear sciences, and these could not be shared with the enemy. But this huge undertaking, the combination of science and war, also influenced the business of science in the long term. With its safety measures of an organized hierarchy of tasks and the confinement of expertise a new kind of laboratory emerged in which the large-scale project was divided into smaller assignments, with specific deliverables and a highly specialized workforce. This workforce was naturally sourced from US academia and industry, and many women were hired, among them the American chemist Isabella Karle (Robinson 2019a); but it was also necessary to hire foreign expertise, and it was further provided by immigrants to the USA who had fled Europe, like the Hungarian radiochemist Elisabeth Róna, who had specialized in the preparation of polonium (Rentetzi 2019). After the war ended, the American chemist Glenn Seaborg, who shared the 1951 Nobel Prize in chemistry with Edwin MacMillan for their discoveries of the transuranium elements and their chemistry, was finally able to disclose the discoveries made during the war in scientific publications and conferences, well beyond internal reports to the project's leaders. Secrecy that had been an important part of the Manhattan project was abandoned, but many nuclear chemistry laboratories like those in Los Alamos retained the 'Big Science' organization that had been inspired by the wartime research programme.

While these laboratories accepted many women on their staff, all team leaders were men until the American nuclear chemist Darleane Hoffman was given such an opportunity at the end of the 1970s. In 1974 the team in which she worked was credited with the discovery of element 106 which, after many discussions, was named 'seaborgium' (Ghiorso et al. 2000; Chapman 2019; Robinson 2019b). In 1979 she became Division leader in the Los Alamos Isotope and Nuclear Chemistry Division. Five years later, she was appointed as professor at the University of California Berkeley and succeeded Glenn Seaborg as the leader of the Heavy Element Nuclear and Radiochemistry group of the Lawrence Berkeley National Laboratory (Murray and Wade 2019; Ghiorso et al. 2000; Chapman 2019). Hoffman fostered an entire generation of nuclear scientists, many of them women who followed in her footsteps and became team leaders, such as Dawn Shaugnessy. Shaugnessy and Jacklyn Gates established career paths and ladders that are very similar to those that their male colleagues aspired to.

Conclusions

Kitchen, Marriage, Radioactivity and Big Science. The history of the periodic system is more multifaceted than conveyed in the traditional and popular accounts. Stories like those presented here demonstrate that the history of chemistry cannot be written without considering the many roles women occupied in the joint endeavour. In this chapter we have particularly emphasized four primary opportunities which enabled women to take part in chemical work on the elements and atoms: pursuing analytical chemical work related to work in the kitchen, entering emerging fields with no history of internal hiring or male-dominated structures, teaming up with established scientist-partners, and taking up 'invisible' jobs that enabled (some) women to grow with the job.

In 1993 Margaret Rossiter introduced the term 'Matilda effect', derived from the established concept of the 'Matthew effect'; the reference is to the gospel of Matthew, which states that those who live in abundance will be given more, while those who have almost nothing will have the very little that they do have taken away from them. In academia this translates into the bias that credits an eminent scientist more than an unknown or lesser-known scholar, even if their work is shared or equal in parts. The Matilda effect is the female corollary of this effect, according to which the achievements of women scientists are attributed to their male colleagues, either during their lifetime or afterwards (Rossiter 1993). Ever since this concept was first introduced historians have been looking for examples of such an effect, and while this carries the risk of over-simplifying and over-rehabilitating women from the past, the stories reported here clearly support the lack of credit for women in different scenarios.[3] Harriet Brooks, for example, worked with a mentor who seems to have supported her in many respects, and over many years. However, the fact that Rutherford did not include her as a co-author in the publication of their joint research in *Nature* may, one might argue, testify to the contrary, especially since this particular article is frequently cited and probably one of the reasons why Rutherford is often acknowledged as one of many co-discoverers of radon. Although Rutherford never put forward such claims for himself, this example demonstrates one of the processes by which the historic narrative can make women invisible, especially when these women are no longer in positions to make further contributions in science.

Beyond the Matilda effect, which focusses on accumulation of credit, the life and career trajectories of women tend to lead disproportionally often to niches of invisibility, not only due to the nature of their work but also due to the way that the history of science has been told, placing emphasis on thinkers and supervisors, leaders of laboratories, and winners of awards, while ignoring the crucial importance of the infrastructure of science (teaching, literature search, data gathering and organizing, computing, lab maintenance, routine procedure, calibration) and of those who filled these roles. Niches of invisibility are, of course, not specific to women; this also applies to discriminated minorities (ethnic, religious, ideological) who also occupied such essentially hidden roles. The gendering of the scientific work both in the making of science and in the writing of its history, however, adds new layers to history and needs to be further emphasized, via new stories and also new narratives within history.

NOTES

1. The title for this chapter is borrowed from Sobel 2020, a review of the book "Women in their Element" called "A seat at the table".
2. The preparation, decomposition and isolation of these substances typically involved the use of glassware (flasks, beakers, pipets), while nowadays they are mainly performed through automated systems.
3. A Google Scholar search for 'Matilda effect in science' generates page after page of relevant hits.

References

Adloff, Jean-Pierre and George Bernard Kauffman. 2005. "Francium (Atomic Number 87), the Last Discovered Natural Element." *The Chemical Educator* 10: 387–394.
Bensaude-Vincent, Bernadette and Isabelle Stengers. 1996. *A History of Chemistry*. Cambridge, MA: Harvard University Press.
Boeck, Gisela. 2019. "Ordering the Platinum Metals: The Contribution of Julia V. Lermontova (1846/47–1919)." In *Women in Their Element: Selected Women's Contributions to the Periodic System*, edited by Annette Lykknes and Brigitte Van Tiggelen, 112–23. Singapore: World Scientific.
Bowler, Peter J. and Iwan Rhys Morus. 2005. *Making Modern Science: A Historical Survey*. Chicago, IL and London: The University of Chicago Press.
Brock, William H. 1992. *The Fontana History of Chemistry*. London: Fontana Press.
Brock, William H. 2016. *The History of Chemistry: A Very Short Introduction*. Oxford: Oxford University Press.
Cannizzaro, Stanislao. 1858. *Sketch of a Course of Chemical Philosophy*. Edinburgh: Alembic Club Reprints, 1910.
Chapman, Kit. 2019. *Superheavy: Making and Breaking the Periodic Table*. London: Bloomsbury Sigma.
Charbonneau, Joanne A. and Richard E. Rice. 2019. "From Miss Swallow's 'Insoluble Residue' to the Discovery of Samarium and Gadolinium." In *Women in Their Element: Selected Women's Contributions to the Periodic System*, edited by Annette Lykknes and Brigitte Van Tiggelen, 145–57. Singapore: World Scientific.
Cleve, Astrid. 1902. "Beiträge zur Kenntnis des Ytterbiums." *Zeitschrift für Anorganische Chemie* 32:129–69.
Curie, Marie. 1911. Marie Curie: Nobel Lecture. https://www.nobelprize.org/prizes/chemistry/1911/marie-curie/lecture/. Accessed on 30 June 2020.
Espmark, Kristina. 2012. *Utanför gränserna: En vetenskapshistprisk biografi om Astrid Cleve von Euler* (PhD thesis, Umeå Universitet). https://www.diva-portal.org/smash/get/diva2:564736/FULLTEXT01.pdf&sa=U&ei=QBdiU6mOJIeG2wW734DYDg&ved=0CEIQFjAH&usg=AFQjCNFrNSdzLHfMfAecCEm6_wDhVaflMA Accessed on 10 July 2020.
Espmark, Kristina and Christer Nordlund. 2012. "Married for Science, Divorced for Love: Success and Failure in the Collaboration between Astrid Cleve and Hans von Euler-Chelpin." In *For Better or for Worse? Collaborative Couples in the Sciences*, edited by Annette Lykknes, Don L. Opitz and Brigitte Van Tiggelen, 81–102. Science Networks, Historical Studies, 44. Heidelberg: Springer Birkhäuser.
Espmark, Kristina and Christer Nordlund. 2019. "Astrid Cleve Von Euler on ytterbium and selenium." In *Women in their Element: Selected Women's Contributions to the Periodic System*, edited by Annette Lykknes and Brigitte Van Tiggelen, 134–44. Singapore: World Scientific.
Forbes, George Shannon. 1932. "Investigations of Atomic Weights by Theodore William Richards." *Journal of Chemical Education* 9, no. 3: 453–58.
Forstner, Christian. 2019. "Berta Karlik and Traude Bernert: The Natural Occurring Astatine Isotopes 215, 216, and 218." In *Women in Their Element: Selected Women's Contributions to the Periodic System*, edited by Annette Lykknes and Brigitte Van Tiggelen, 350–57. Singapore: World Scientific.
Frevert, Ute. 1989. *Women in German History. From Bourgeois Emancipation to Sexual Liberation*. Oxford: Berg.

Ghiorso, Albert, Darleane C. Hoffman and Glenn T. Seaborg. 2000. *Transuranium People: The Inside Story*. London: Imperial College Press.

Gordin, Michael D. 2012. "The Textbook Case of a Priority Dispute: D.I. Mendeleev, Lothar Meyer, and the Periodic System." In *Nature Engaged*, edited by Mario Biagioli and Jessica Riskin, 55–82. New York: Palgrave.

Guerrini, Anita. 2016. "The Ghastly Kitchen." *History of Science* 54, no. 1: 71–97.

Hartley, Harold. 1966. "Stanislao Cannizzaro, F.R.S. (1826–1910) and the First International Chemical Conference at Karlsruhe." *Notes and Records of the Royal Society of London* 21, no. 1: 56–63.

Hudson, John A. 2019. "Dr Margaret Todd and the Introduction of the Term 'Isotope'." In *Women in Their Element: Selected Women's Contributions to the Periodic System*, edited by Annette Lykknes and Brigitte Van Tiggelen, 280–9. Singapore: World Scientific.

Knight, David. 2009. *The Making of Modern Science: Science, Technology, Medicine and Modernity, 1789–1914*. Cambridge and Malden, MA: Polity.

Kragh, Helge. 1996. "Elements no. 70, 71 and 72: Discoveries and Controversies." In *Episodes from the History of the Rare Earth Elements* edited by Christopher H. Evans, 67–89. *Chemists and Chemistry* 15. Dordrecht: Springer.

Kragh, Helge. 1999. *Quantum Generations: A History of Physics in the Twentieth Century*. Princeton, NJ: Princeton University Press.

Lundgren, Anders. 2019. "Women and Analytical Chemistry: Reflections on the Chemical Skill Needed for Investigating the Elements." In *Women in Their Element: Selected Women's Contributions to the Periodic System*, edited by Annette Lykknes and Brigitte Van Tiggelen, 124–33. Singapore: World Scientific.

Lykknes, Annette. 2019. "Ellen Gleditsch and Research on Radium, Chlorine and Potassium." In *Women in Their Element: Selected Women's Contributions to the Periodic System*, edited by Annette Lykknes and Brigitte Van Tiggelen, 301–12. Singapore: World Scientific.

Lykknes, Annette and Brigitte Van Tiggelen. 2019a. "The Periodic System: The (Multiple) Values of an Icon." *Centaurus* 61, no. 4: 287–98.

Lykknes, Annette and Brigitte Van Tiggelen. 2019b. "Introduction." In *Women in Their Element: Selected Women's Contributions to the Periodic System*, edited by Annette Lykknes and Brigitte Van Tiggelen, 1–54. Singapore: World Scientific.

Lykknes, Annette, Don L. Opitz and Brigitte Van Tiggelen (eds.) 2012. "For Better or for Worse? Collaborative Couples in the Sciences. Science Networks." *Historical Studies* 44. Heidelberg: Springer Birkhäuser.

Lykknes, Annette, Lise Kvittingen and Anne Kristine Børresen. 2004. "Appreciated Abroad, Depreciated at Home: the Career of a Radiochemist in Norway: Ellen Gleditsch (1879–1968)." *Isis* 95, no. 4: 576–609.

Lykknes, Annette, Lise Kvittingen and Anne Kristine Børresen. 2005. "Ellen Gleditsch: Duty and Responsibility in a Research and Teaching Career, 1916–1946." *Historical Studies in the Physical Sciences* 36, no. 1: 131–88.

Meinel, Christoph. 2009. "Chemical Collections." In *Spaces and Collections in the History of Science*, edited by Marta C. Lourenço & Ana Carneiro, 137–47. Lisbon: Museum of Science of the University of Lisbon.

Murray, Claire A. and Jessica A. Wade. 2019. "The Unsung Heroines of the Superheavy Elements." In *Women in their Element: Selected Women's Contributions to the Periodic System*, edited by Annette Lykknes and Brigitte Van Tiggelen, 390–402. Singapore: World Scientific.

Musabekov, Yusuf Suleimanovich. 1967. *Julia Vsevolodovna Lermontova 1846–1919* (In Russian: Юлия Всеволодовна Лермонтова, 1846-1919). Moskow: Nauka.

Nye, Mary Jo. 1984. "The Question of the Atom: From the Karlsruhe Congress to the First Solvay Conference, 1860–1911. A Compilation of Primary Sources." *History of Modern Physics, 1800–1950*, 4. Los Angeles, CA: Tomash Publishers.

Nye, Mary Jo. 1996. *Before Big Science: The Pursuit of Modern Chemistry and Physics, 1800–1940*. Cambridge, MA: Harvard University Press.

Paneth, Friedrich A. 1947. "The Making of the Missing Chemical Elements." *Nature* 159: 8–10.

Perey, Marguerite. 1939. "Sur un élément 87, dérivé de l'actinium." *Comptes rendus hebdomadaires des séances de l'Académie des sciences* 208: 97–99.

Pigeard Micault, Natalie. 2013. "Le laboratoire Curie et ses femmes (1906–1934)." *Annals of Science* 70: 71–100.

Rayner-Canham, Marelene F. and Geoffrey W. Rayner-Canham. 1997. "Pioneer Women of Radioactivity." In *A Devotion to Their Science: Pioneer Women of Radioactivity*, edited by Marelene F. Rayner-Canham and Geoffrey W. Rayner-Canham, 12–28. Montreal and Philadelphia, PA: McGill-Queen's University Press and Chemical Heritage Foundation.

Rayner-Canham, Marelene F. and Geoffrey W. Rayner-Canham. 2019a. "Harriet Brooks: Radon, a 'New Gas' from Radium." In *Women in Their Element: Selected Women's Contributions to the Periodic System*, edited by Annette Lykknes and Brigitte Van Tiggelen, 269-79. Singapore: World Scientific.

Rayner-Canham, Marelene F. and Geoffrey W. Rayner-Canham. 2019b. "Stefanie Horovitz: A Crucial Role in the Discovery of Isotopes" In *Women in Their Element: Selected Women's Contributions to the Periodic System*, edited by Annette Lykknes and Brigitte Van Tiggelen, 290–300. Singapore: World Scientific.

Rayner-Canham, Marelene F. and Geoffrey W. Rayner-Canham. 2019c. "Marguerite Perey: The Discoverer of Francium." In *Women in their Element: Selected Women's Contributions to the Periodic System*, edited by Annette Lykknes and Brigitte Van Tiggelen, 341-48. Singapore: World Scientific.

Rentetzi, Maria. 2007. *Trafficking Materials and Gendered Experimental Practices: Radium Research in Early Twentieth Century Vienna*. New York, NY: Columbia University Press, online edition, http://www.gutenberg-e.org/rentetzi/. Accessed on 30 June 2020.

Rentetzi, Maria. 2019. "She Is in the Next Room: Elizabeth Róna and Polonium." In *Women in Their Element: Selected Women's Contributions to the Periodic System*, edited by Annette Lykknes and Brigitte Van Tiggelen, 330–40. Singapore: World Scientific.

Robinson, Ann. 2019a. "Isabella L. Karle and the Synthesis of Plutonium Chloride." In *Women in Their Element: Selected Women's Contributions to the Periodic System*, edited by Annette Lykknes and Brigitte Van Tiggelen, 374–81. Singapore: World Scientific.

Robinson, Ann. 2019b. "IUPAC and the Naming of Elements." *Chemistry International* 41, no. 3. Online publication. https://doi.org/10.1515/ci-2019-0314. Accessed on 11 February 2021.

Rocke, Alan. 2019. "Lothar Meyer's Pathway to Periodicity." *Ambix* 66, no. 4: 265–302.

Roqué, Xavier. 2019a. "Marie Skłodowska Curie – Polonium and Radium." In *Women in Their Element: Selected Women's Contributions to the Periodic System*, edited by Annette Lykknes and Brigitte Van Tiggelen, 259–68. Singapore: World Scientific.

Roqué, Xavier. 2019b. "Lise Meitner and Protactinium." In *Women in Their Element: Selected Women's Contributions to the Periodic System*, edited by Annette Lykknes and Brigitte Van Tiggelen, 324–32. Singapore: World Scientific.

Rossiter, Margaret W. 1993. "The Matthew Matilda Effect in Science." *Social Studies of Science* 23, no. 2: 325–341.

Rutherford, Ernest and Harriet T. Brooks. 1901. "The New Gas from Radium." *Royal Society of Canada, Transactions* 3: 21–25.

Shindell, Matthew. 2019. "Toshiko Mayeda and the Isotopes of Oxygen." In *Women in Their Element: Selected Women's Contributions to the Periodic System*, edited by Annette Lykknes and Brigitte Van Tiggelen, 415–21. Singapore: World Scientific.

Shindell, Matthew. 2020. "The Indomitable Toshiko Mayeda." *Chemistry World* (2 March) https://www.chemistryworld.com/culture/the-indomitable-toshiko-mayeda/4011147.article?adredir=1/. Accessed on 30 June 2020.

Sime, Ruth. 1996. *Lise Meitner: A Life in Physics*. Berkeley, CA and London: University of California Press.

Sobel, Dava. 2020. "A Seat at the Table." *Distillations* (16 June) https://www.sciencehistory.org/distillations/a-seat-at-the-table. Accessed on 30 June 2020.

Söderbaum, Henrik Gustaf. 1915. Presentation, Theodore W. Richards, https://www.nobelprize.org/prizes/chemistry/1914/press-release/. Accessed on 30 June 2020.

Thyssen, Pieter and Koen Binnemans. 2015. "Mendeleev and the Rare-Earth Crisis." In: *Philosophy of Chemistry: Growth of a New Discipline*, edited by Eric Scerri and Lee C. McIntyre, 155–82. Dordrecht: Springer.

Van Tiggelen, Brigitte. 2001. "The Discovery of New Elements and the Boundary between Physics and Chemistry in the 1920s and 1930s: The Case of Elements 43 and 75." In *Chemical Science in the 20th Century: Bridging Boundaries*, edited by Carsten Reinhardt, 131–44. Weinheim: Wiley-VCH.

Van Tiggelen, Brigitte and Annette Lykknes. 2012. "Ida and Walter Noddack through Better and Worse: An *Arbeitsgemeinschaft* in Chemistry." In *For Better or for Worse? Collaborative Couples in the Sciences*, edited by Annette Lykknes, Don L. Opitz and Brigitte Van Tiggelen, 103–47. Science Networks, Historical Studies, 44. Heidelberg: Springer Birkhäuser.

Von Euler-Chelpin, Hans. 1906. "Per Teodor Cleve." *Kungliga Svenska Vetenskapsakademiens Årsbok* 4: 187–217.

Von Oertzen, Christine. 2014. *Science, Gender, and Internationalism: Women's Academic Networks, 1917–1955*. New York, NY: Palgrave Macmillan.

PART V

Science Communication

CHAPTER 18

Mediating Knowledge: Women Translating Science

Alison E. Martin

Historical research on women's involvement in science has tended to focus on their role as laboratory helpmeets and amanuenses in the hands-on 'making' of scientific knowledge. Only more recently have studies emerged which highlight the contribution of women to the publication of scientific knowledge, and to its onward circulation in other scientific cultures and languages through translation. These developments are closely linked to the 'linguistic turn' in the history of science, which has kindled greater interest in the diversification of communication practices in the modern period and the reasons underpinning why, where, and through whom, knowledge comes to circulate. James Secord's seminal 2004 article on science as 'Knowledge in Transit' argued that it was time to take more seriously the 'social nature of knowledge', namely how it ceases to be 'the exclusive property of a single individual or group', but rather contributes to the knowledge shared by larger constellations of people (Secord 2004, p. 655). This line of enquiry derives in part from scholarship in the 1980s and 1990s, which explored from two main angles the relationship between language and science. The first examined the confluences of scientific and literary writing, and identified how their shared forms of style, argumentation and rhetoric were used to create persuasive prose, which affirmed the text's credibility and the (scientific) integrity of its author. Scientists in the mid-nineteenth century still 'shared a common language with other educated

A. E. Martin (✉)
Johannes Gutenberg University of Mainz, Germersheim, Germany
e-mail: amarti01@uni-mainz.de

readers and writers of their time' and their texts 'could be read very much as literary texts' (Beer 1983, p. 4). The second approach used insights from linguistics—particularly rhetorical stylistics—to analyse argument and persuasion in non-literary texts and to investigate how language was employed in a functional and audience-based manner (see for example Fahnestock 1999).

These historically informed understandings of scientific writing as a stylistically complex practice are not fully reflected in how we think of science in translation today. As Scott Montgomery rightly observes, 'the common notion that translating science is a linguistically unsophisticated process, based on word-for-word rendering, is false', yet the translator is 'not considered important enough as a creative, producing agent' (Montgomery 2010, p. 303). In her investigation of Western texts imported into modern Arabic, Marwa S. Elshakry emphasises that scientific texts are rarely neutral purveyors of universal truths: rather, they contribute to ongoing processes of socio-political transformation and mediation (Elshakry 2008, p. 701). Moreover, as they are embedded into new contexts, they also gather different 'literary accretions (prefaces, marginalia, footnotes and commentary)' acquired at the service of different 'linguistic strategies (paraphrases, syntactic and lexical additions or substitution)', through which meanings and frames of reference change (Elshakry 2008, p. 704). As cultural 'go-betweens', translators have long been considered marginal social actors who merely produce 'copies' of an original. This model of authoritative original versus derivative copy sets up series of asymmetrical power relationships between dominant, highly visible authors and their subservient, often largely invisible, translators. Yet, as Kapil Raj has shown for colonial scientific knowledge exchange, 'far from being passive informants', numerous apparently subaltern figures operated as illustrators, translators or bookbinders, all of whom were key to the collective enterprise of publication (Raj 2013, p. 344). Indeed, translators and other 'brokers of knowledge' occupied vital positions in the global economies of knowledge emerging by the end of the Enlightenment (Schaffer et al. 2009, p. xxx).

The metaphors of servitude frequently associated with translation echo patriarchal language structures which feminist translation scholars would argue have shaped women's cultural conditioning down the centuries. These determine whether and how women present themselves in the public sphere as linguistically talented individuals. Although translation and gender has only recently begun to interest those working in the history of scientific knowledge-making, second-wave feminism was the catalyst for a number of studies in the 1990s that militated against the notion that women's self-positioning in translation was always one of erasure (see Simon 1996; Flotow 1997). In an era before women had the same rights and possibilities as men, the publication of a translation by a woman in itself embodied a paradox about women's engagement with the public sphere. As Sherry Simon neatly puts it, this begged the question of 'whether translation condemned women to the margins of discourse or, on the contrary, rescued them from imposed silence' (Simon 1996, p. 146). The title of Michèle Healy's 2004 dissertation, *The Cachet* of the *'Invisible' Translator: Englishwomen Translating Science (1650–1850)*, speaks volumes about why respected and prodigious female scientific translators such as Elizabeth Sabine, Ada Lovelace and Mary Somerville long went unnoticed. While most female translators of science have, over the course of history, tended to be modest, self-effacing individuals, some chose to be more prominent participants in the creation of knowledge. They engaged in what Barbara Godard has termed the 'womanhandling' of a text, were 'aware of process, giving self-reflexive attention to practices', and were prepared to display their name immodestly on a title-page, and append footnotes, or even a preface (Godard 1989, p. 50). The interventionist possibilities of translation, through which a particular agenda can be articulated, subsequently led to numerous case studies that began 'recovering' women translators, active from the Renaissance through to the nineteenth century, who had largely been lost in the course of history (see Flotow 1997; Delisle 2002). Luise von Flotow has subsequently proposed that translation scholars should revisit ideas about the discursive and performative nature of gender, enabling them to explore more productively how women actively mobilised translation for their own ends (Flotow 2011, p. 7).

The languages in which women were educated have, by definition, always determined how they could be involved in scientific translation. In Western Europe, vernaculars were already beginning to assert themselves by the late seventeenth century as the vehicles of cosmopolitan scientific exchange. Latin was still used widely by the members of scientific institutions such as the Royal Society. However, other languages were now being recognised as an appropriate medium for knowledge exchange, and communications by renowned men of science would be received not just in English, French or German but also in languages spoken by much smaller communities, including Dutch (Fransen 2017). Within a century, French had started to gain in importance as a *lingua franca* in the European republic of letters, subsequently followed by English, which 'blossom[ed] into continental recognition suddenly and dramatically around 1750' (Oz-Salzberger 2006, pp. 387–88). Cultural traffic between English and German gained markedly by the start of the nineteenth century, making translation a 'tool of both cultural affinity and cultural particularism' (Oz-Salzberger 2006, p. 385). While male scientists were schooled rigorously in Latin and Greek, they rarely knew many modern tongues, with the exception of French, while their wives and sisters had little familiarity with the classical languages but were more likely to know French, German or Italian, depending on where they had completed their education. As Mary Orr has emphasised through her work on the English taxidermist and translator Sarah Bowdich Lee, the 'exchange, circulation and advancement of new ideas [...] always depended on multi-lingual, and intra-lingual, re-articulation' (Orr 2015, p. 28): women's multilingual proficiency, primarily in modern languages, enabled them to make a significant independent contribution to the transmission of scientific findings.

Over the past decade, histories of translation have recognised the urgent need to reinstate women's involvement in textual transfer. Numerous studies devoted to the achievements of female translators amply demonstrate that women valued the intellectual stimulation and creative possibilities of translation. It could be a valuable means for women to enter the public sphere, while not overstepping the bounds of female modesty by being too ostentatious. It was precisely the ambivalence of translation, its 'simultaneous derivativeness and originality' and its 'double-sidedness' which enabled women to gain a foothold in fields from which they might otherwise have been precluded from entering (Stark 2006, p. 126). While translation was a stimulating (literary) pastime for some, crippling financial anxiety forced others to turn their hand to the translation of textual types lying outside the 'female' domain of fictional writing. Although the notion of 'scientific prose' has never been monolithic, more 'popular' genres aimed at non-specialist audiences began to emerge from the late eighteenth century onwards, as publishers recognised the reading needs of a rapidly expanding market. These invited women's involvement in scientific translation precisely because such texts were not written solely for the exchange of factual information between men of science, but targeted a wider readership that sought to 'capture the audience's interest and render its subject relevant and aesthetically appealing by virtue of a good, even gripping, narrative' (Freddi et al. 2013, p. 222).

No systematic statistical information exists about the relative contribution made by women to the field of scientific translation. Even library catalogues like the English Short Title Catalogue (ESTC), which holds over 480,000 items published in the English language between 1473 and 1800, is not helpful in giving much indication of the percentage of female translators that made up the totality of individuals involved in intralingual textual transfer. This is partly because works are infrequently marked in the catalogue as translations, and partly because many translations cannot easily be attributed, since there is no indication of the translator's identity or it is at best hidden behind initials. However, recent case studies of specific translations have identified a handful of women as translators of significant scientific texts, key among them Aphra Behn, Émilie du Châtelet, Helen Maria Williams, Elizabeth Sabine, Clémence Royer, Mary Somerville and Sarah Bowdich, whose translation output serves as the basis for some of the discussions that follow.

That the majority of these women were working in the Enlightenment and Victorian period is indicative of interest shown recently both by historians of science and historians of print culture in uncovering the situations in which translations are produced and received. The 'sociological turn' in translation studies, informed by Pierre Bourdieu's notions of habitus and field, has focused attention on rethinking translation as a social practice carried out by translators who are themselves 'agents' in the translation process (see for example Wolf and Fukari 2007; Milton and Bandia 2009). By contextualising the working environment of translators and their role and status in society, we can investigate more fully how these realities influence the translation process and the implications they have for transfer situations. Although a focus on those involved in the technologies of publishing and information dissemination—Jonathan Topham's neatly termed 'technicians of print' (2004)—has unearthed numerous individual figures, there is still much to be done if we are to gain a broader understanding of the role played by translators and translation. In her discussion of methodological parallels shared by developments in the history of science and of translation, Maeve Olohan proposes that we should move away from focusing on canonical works in science towards an understanding of the 'scenes of inquiry' in which scientific questions emerged, evolved and disappeared (Olohan 2014, p. 17). Adopting this perspective would revise the 'great men' account of the history of science, thereby enabling not just the category of 'women' but that of other subaltern groups to come to the fore, who were likewise not publicly acknowledged scientific practitioners or card-carrying members of scientific institutions.

Women, Science and Translation

Pivotal work in the 1990s by Patricia Phillips (1990), Ann B. Shteir (1996; also 2003), Margaret and Thomas Creese (1998) and Barbara Gates (1998), made important inroads in identifying and teasing out the contribution by women in Europe and North America to eighteenth- and nineteenth-century science. By uncovering the 'uneasy careers' and 'intimate lives' of such figures, Pnina Abir-Am and Dorinda Outram used their edited volume to tackle a number of lacunae at that time in women's studies and the history of science, notably the lack of in-depth studies of individual women's contributions and of accounts about the historical diversity of women's scientific experience (Abir-Am and Outram 1987, p. 3). Through their investigation of the lives of 'pioneering and outstanding individual women', Abir-Am and Outram sought to provide interpretative categories which demonstrated how 'family situations imposed problems and induced strategies and approaches to scientific work that were specific to women' (Abir-Am and Outram 1987, p. 2). However, such models tend to favour a binary opposition between the domestic and the public sphere in their investigation of 'the process by which science became lifted out of the domestic, amateur context and became transformed into an activity at odds with the domestic, intimate family lives of all its practitioners' (Abir-Am and Outram 1987, p. 3). The argument that the exclusion of the domestic realm from science meant the concomitant exclusion of women from scientific debate certainly holds true for more modern 'professional' scientific investigation and experimentation that is laboratory centred. Scientific translation, though, has invariably been an activity that could be undertaken at home. Indeed, its freelance nature still enables translators today to pursue a career alongside family commitments. Translation therefore represents a notable exception to the models hitherto put forward with regard to women's engagement with science, which invite other paradigms to be explored.

We now have a much clearer idea than twenty years ago about the domains in which women were involved as practitioners, readers and translators of science, and when this involvement began. Douglas Robinson (1995) suggests that the Renaissance generally signalled the start of women using translation as a means to give themselves a public voice—what he terms the 'feminisation' of translation—a theory also valid for the relationship between women and scientific translation. Astronomy had swiftly become a recognised discipline in the seventeenth century, as the heliocentric theory of the universe proposed by Copernicus had revolutionised thinking about the place of the earth and humankind in the cosmos and fuelled debate about further avenues of inquiry. This field of study was eminently suited to educated women from prosperous families, since they primarily needed good powers of observation, access to the most powerful telescopes of the day, and sufficient time to peruse the heavens. With the appearance of Isaac Newton's *Philosophiae Naturalis Principia Mathematica* [*Mathematical Principles of Natural Philosophy*] (London, 1687; Cambridge, 1713; London, 1726), a new theory of gravity developed that potentially transformed orbital astronomy by proposing different approaches for the calculation of planetary orbits. A three-volume work, it required an excellent understanding of both mathematics and astronomy, since it aimed to derive from celestial phenomena the gravitational forces by which bodies tended towards the sun or towards individual planets. Although the mathematician Andrew Motte produced an English translation which appeared in London in 1729, the cultural prestige of French in this period meant that relatively few continental Europeans knew English (Zinsser 2001, p. 228) and Motte's translation would have been of little use to them. For women on the Continent, translation from English was therefore 'a route to making a reputation' that offered 'real advantages' (Mason 2006, p. 131).

This challenge was taken up by Gabrielle-Emilie de Breteuil, the Marquise du Châtelet, who produced the first French translation, the *Principes mathématiques de la philosophie naturelle* (1759). It was published in Paris a decade after her death, with a commentary written by its translator and the French mathematician and geophysicist Alexis-Claude Clairaut who had edited her work. Du Châtelet's aptitude for mathematics developed early and she was educated alongside her brothers, gathering the knowledge she would later require to understand Newton's ideas. Fluent in Latin, and able to read English and Italian, she was linguistically well placed to take on this ambitious project. Her association with Voltaire and her election to the Bologna Academy in 1746 would have given her influential contacts among the European intellectual élite. These associations also conferred on her the intellectual capital she needed to ensure she was accepted by male contemporaries as their equal (Zinsser 2001, p. 228). Du Châtelet thus challenged male hegemony in scientific debate through her own learning and her pursuit in 'high-level public forums' of her own scientific interests (Pieretti 2002, p. 478).

Newton's *Principia* was a prominent text that might easily have overshadowed the achievements of its translator. However, Du Châtelet's translation included a preface and commentary which emphasised her pedigree as a female intellectual of the Enlightenment and highlighted the double contribution of her work to posterity, which had involved putting this work into French and clarifying its lines of argument. Her preface also described her working methods, which involved checking the mathematical validity of the French translation she was producing, in line with the motto of the Royal Society '*nullius in verba*', 'take no-one's word for it' (Newton 1759, p. vi). The commentary was formulated in a language 'free of mathematical or scientific jargon', and gave definitions of even the most basic terms, such as 'orbit' or 'ellipse', to ensure that laypersons could gain a grasp of the work's content (Zinsser 2001, pp. 232, 233). Du Châtelet therefore understood translation as a means by which to share her learning with her contemporaries. Her work highlighted the 'pedagogical value of translation' as much as it challenged the 'silence imposed on women by the terrible state of ignorance from which they suffered' (Pieretti 2002, p. 476). In rethinking (and rewriting) the *Principia* for a different audience, Du Châtelet made of translation a process by which she 'decoded and recoded her source texts', allowing her to 'transform and appropriate them freely as her intellectual property' (Mason 2006, p. 125).

In the course of the eighteenth century, another scientific domain emerged which was deemed appropriate for women: natural history. Towards the end of the Enlightenment, a host of introductory works on botany, entomology, geology and chemistry began to flood the market which stressed the practical application of the knowledge they contained. They championed rational recreation, 'which could be practised with a minimum of expensive equipment or education' as a useful way for children, and their educators, to spend their time (Fyfe 2000, p. 283). Botany in particular became a pursuit in which women were both producers and consumers of plant knowledge, because it was a scientific activity sanctioned by women of rank and status like Queen Charlotte, the 'scientific wife' of George III. Geared towards a non-specialist audience, works such as Sarah Trimmer's *Easy Introduction to the Knowledge of Nature* (1782) or the Quaker writer Priscilla Wakefield's step-by-step guide to understanding Linnaean nomenclature, *Introduction to Botany* (1796) demonstrated how scientific study could be an inclusive undertaking. These books were immensely popular in Britain, but also appeared swiftly in French. Similarly, botanical works penned on the continent spurred on British women to try their hand at translation. Poetry appears to have carried particular appeal. An English version by Maria Henrietta Montolieu née Heywood (Hayward?) of Jacques Delille's best-selling *Les Jardins, ou l'art d'embellir* (1789/1801) appeared as *The Gardens, A Poem* (1798/1805). The first English rendering of the Swiss botanist Albrecht von Haller's highly successful poem on the flora and fauna of the Alps, *Die Alpen*, was produced by 'Mrs J. Howorth' for her translation collection *Poems of Baron Haller* published with Bell in London in 1794.

While women were therefore given the opportunity to translate 'popular' works of science that used literary formats, translating core theoretical writings underpinning the development of a particular scientific field firmly remained within the male domain. It is therefore somewhat surprising that Johann Wolfgang von Goethe's seminal essay on plant evolution, *Die Metamorphose der Pflanzen* (1798), should not appear in English until 1863 as the *Essay on the Metamorphosis of Plants*, and that this translation should be the work of one 'Emily M. Cox'. However, although she was named as the translator of this piece, when it was published in the very first issue of the *Journal of Botany*, her achievement was essentially overshadowed by the textual presence of the English physician and taxonomist, 'Maxwell T. Masters, F. L. S.' who provided the footnote apparatus to the piece. Institutional recognition—his Fellowship of the Linnean Society—automatically gave him a cultural prestige to which Cox could not aspire. In addition to this, he appended lengthy and copious notes to the translation, in which he situated Goethe's ideas (now over half a century old) in the context of late Enlightenment thought, and updated them in the light of developments in plant systematics.

Translation, Visibility and Paratext

Women's visibility in translated texts—fundamental in trying to reconstruct a history of women as translators of scientific literature—largely depended on their paratextual presence, whether in the form of a preface, commentary or footnotes. We know, for example, that Margarete (Meta) Forkel played a seminal role in the 'translation factory' established by the scientific explorer Georg Forster, who had accompanied James Cook on his second circumnavigation. However, it is only by carefully combing Forster's correspondence that we can gauge her significance in contributing to his prodigious translation activities (Roche 1994, p. 104). Similarly, the British translator Elizabeth Carter's engagement with Newtonian science was a rather different affair than that of Du Châtelet. In the English rendering of Francesco Algarotti's *Il Newtonianismo per le dame: ovvero Dialoghi sopra la luce e il colore* (1737), translated as *Sir Isaac Newton's Philosophy Explained for the Use of Ladies in Six Dialogues on Light and Colours* (1739), Carter was a deeply 'invisible' figure. Although her translation likewise explained Newton's works in appealing ways (Miller 2013, p. 194), and its conversational format moulded information to make it appropriate to female readers, the name of the translator was not printed on the frontispiece, and no translator's preface appeared with the work. This self-effacement well reflected the role into which many female authors now felt themselves corralled by a need to conform to traditional values and to shy away from public recognition. A subsequent 1742 English edition of Newton's *Principia* had few qualms about freely pirating Carter's work and combining her text with that of an unattributed 1738 translation of the first French version (Miller 2013, p. 193).

How did female translators use translation as a form of self-assertion and overt self-positioning in science? Already by the early modern period, women had recognised that translation could be exploited to give themselves a public voice in intellectual debate. Although the majority of these women are now known primarily for their achievements as novelists, playwrights or poets, a handful were also involved in translating texts from the field of natural philosophy. Aphra Behn, today considered one of the first British women to live by her pen, was a prolific writer of romantic tragicomedies and prose fiction, notably the best-selling *Oronooko* (1688). The same year that this appeared, she also produced an English version of the French Enlightenment philosopher Bernard de Fontenelle's essay on astronomy, *Les Entretiens sur la Pluralité des Mondes* [*Conversations on the Plurality of Worlds*] (1686). This was constructed in the form of a dialogue between an ignorant but eager Marquise, desirous of learning more about the heavens from her overtly knowledgeable male instructor. Behn's translation, *A Discovery of New Worlds* (1688), was bold on several levels. For one thing, it saw a female translator enter into discussion about the astronomical models of the time and defend the Copernican system against scriptural understandings of planetary organisation. For another, it appeared in direct competition with two other versions by male translators. Rarely averse to publicity, Behn used her translator's preface to criticise what she considered Fontenelle's clumsy handling of female intellectual curiosity in this work. While Behn initially 'downplays her own accomplishment and seems to confirm the belief in women's inferiority' by casting herself as a mere beginner in the science of astronomy (Uman 2012, p. 29), she also uses this paratextual space to voice her disagreement with Fontenelle about the construction of his central character, the Marquise. Fontenelle seemed to be working on the stereotypical assumption that women were not able to cultivate their own minds without the help of men and were intellectually inferior to them. In line with the thinking of the time, women were metaphorically equated with nature, and men with the mind. Women's involvement in scientific knowledge transmission that went beyond passive reception and consumption therefore pressed on the conventions of female propriety—a limitation that Behn would question more vocally than Carter half a century later (Agorni 1998, p. 182).

Behn's translation of the *Entretiens* bears interesting comparison with the two other translations which appeared on the market just before or contemporaneously with her own: Sir William Domvile's *A Discourse on the Plurality of Worlds* (1687) and John Glanvill's *A Plurality of Worlds* (1688). Domvile, Attorney General to Ireland, had probably been prompted to translate the work by his son-in-law William Molyneux, who was an Irish natural philosopher and Fellow of the Royal Society. Domvile asserted that he had considered addressing his translation to the fair sex, but considering that women were the 'Planets that influence and adorn our Globe' and that it was the business of men to 'discover their Vortices', the work was more properly one to be consumed by men (Fontenelle 1687, n.p.). Molyneux, by way of a 'Note to the Bookseller', revised this sexually allusive statement by asserting that this was a work of 'Ingenuity and pleasant Fancy' that would be evident to all readers, including women (Fontenelle 1687, n.p.). Glanvill's translation included no translator's preface but did domesticate Fontenelle's discussion of distance, which had originally drawn on the topography of Paris, by placing it in London (Fontenelle 1687, p. 36). Behn returned this discussion to its Paris setting in her translation, since, as she pointed out in her preface, it better conveyed Fontenelle's argument (Behn 1993, p. 86).

Like Behn, the Scottish mathematician and astronomer Mary Somerville, working one-and-a-half centuries later on Pierre Simon Laplace's *Traité de mécanique céleste* [*Treatise of Celestial Mechanics*] (1798–1825; trans. 1831), did more than simply put the foreign text into English. She similarly considered that the translation of a text embodying relatively complex mathematical ideas had to be underpinned by an understanding of the cosmological principles governing the workings of the universe. Somerville approached the translation of this text in different ways, sometimes giving a direct rendering of Laplace's account in those passages she deemed straightforward, sometimes guiding the reader by clarifying apparent contradictions or explaining the strategies which Laplace had adopted (Neeley 2001, p. 97).

Conduits of Communication: Translation and the Nineteenth-Century Popularisation of Science

Scientific travel writing was one of the most important genres to which women contributed as translators. In the nineteenth century, travellers from Europe and the United States voyaged in the interest of commercial and scientific enterprises and their accounts of distant lands, peoples and practices revealed new worlds to armchair readers back home. The reports produced by explorers and naturalists were rarely purely factual in nature. Frequently written in the spirit of adventure, colonial conquest and intellectual curiosity, they incorporated imaginative modes of writing into their prose which made this a literature which became 'the talk of the street, the salon, and the Victorian parlour' (Dassow Walls 2006, p. 499). It was a non-fictional genre particularly accessible to women writers and translators, some of whom—notably the Norwich translator Anne Plumptre, who put François Pouqueville's *Voyage en Morée, à Constantinople, en Albanie et dans plusieurs autres parties de l'empire Othoman* [*Travels through the Morea, Albania, and Several Other Parts of the Ottoman Empire*] (1805; trans. 1813) into English—commanded high social capital and were seminal figures in nineteenth-century intellectual networks (Pickford 2012). Other women, such as Jane Percy Sinnett and Sarah Austin, also used the translation of non-fictional travel writing to make important contributions to contemporary debates about political reform, nationhood and identity in Britain (Johnston 2013, p. 8).

The English translations of works by the Prussian scientist and explorer Alexander von Humboldt exemplify British women's intense engagement with science through translation in the nineteenth century and highlight the range of skills demanded of his female translators (Martin 2018). Humboldt's major works first appeared in English between 1811 and 1858, a period in which science 'transformed European understanding of the natural and human worlds and simultaneously transformed European modes of acting in those worlds' (Fulford et al. 2004, p. 5). The *Relation historique du voyage aux régions équinoxiales du nouveau continent* (1814–1825), intended to be the comprehensive account of Humboldt's five-year journey to the Americas with the French botanist Aimé Bonpland between 1799 and 1804, began to appear in 1814 as what would become the seven-volume *Personal Narrative of Travels to the Equinoctial Regions of the New Continent* (1814–29). This translation was by the English poet and novelist, Helena Maria Williams, who had already won literary acclaim for her sentimental fiction in the late 1780s. Williams's translation of the *Personal Narrative* was essentially a collaboration with Humboldt (Leask 2001, p. 235), since she worked directly from his French print proofs, rather than from the published volumes, to produce her first draft. This was then corrected meticulously by Humboldt, before he returned it to her for further reworking. Humboldt's queries and corrections point up the discursive process-oriented nature of translation and its potential to allow exchange between author and translator. Humboldt found numerous problems with the translation of nautical, geographical and astronomical terminology, which, as he impressed upon his translator, needed urgent correction to ensure that he did not lose his reputation in the international scientific community (Martin 2018, pp. 93–101). Yet Humboldt was far more lenient in those passages where Williams, applying her talents as a writer of exuberant fictional prose, reinforced his own exotic images of nature. Her translation mediated the original author's style and ideas, 'strategically altering their presentation to suit her voice and vision, as well as her personal and political ends' (Bailes 2017, p. 122). The translation thus became a hybrid, multivocal piece, particularly in those passages where Humboldt actively encouraged his translator to employ her expertise as a literary writer to create vibrant, engaging scenes of the tropics. Humboldt's *Personal Narrative* was therefore integrally related to Williams's own oeuvre, and she left a stylistic imprint on it that would contribute significantly to the imaginative appeal of this work throughout the nineteenth century and beyond.

Some thirty years later, the journalist and translator Thomasina Ross revived and repackaged Williams's translation of Humboldt's travel account in a three-volume edition (1852–1853) for the publisher Henry Bohn, to be included in his recently launched 'Scientific Library' series. The extensive abridgement required to slim the *Relation historique* down to three volumes—the same format as the immensely popular nineteenth-century 'three-decker novel'— was considered a great improvement by critics of the day. Ross's judicious editing cut Humboldt's references to his own place in the scientific community and his warm testimony of fellow scientists' achievements. Her rendering of the *Relation historique* thus drew attention away from the collaborative and 'polite' nature of scientific knowledge generation. It also removed references to the importance of observation and (failed) experimentation, and in so doing made the establishment of scientific fact seem a more self-evident and straightforward undertaking (Martin 2018, pp. 131–48). Rhetorical devices that articulated modesty and uncertainty, which were essential in shaping Humboldt's speculative, evaluative approach, were also removed to foreground his actual findings over his thought processes. Ross opted for a more linear, dispassionate form of communication which arguably better aligned his writing with scientific styles of reporting at mid-century, and which certainly brought Humboldt's work to a wider audience in Britain through this cheaper, more easily accessible edition.

Meanwhile, Humboldt's short essay collection, the *Ansichten der Natur*, initially published in 1808, revised by Humboldt in 1826 and, finally, modernised in 1849, appeared in English in two different editions that openly competed with each other. The first was by Elizabeth Sabine, wife of the Irish explorer and geophysicist Colonel Edward Sabine, and appeared as *Aspects of Nature* (1849) with Longman and Murray. The second was by the linguist and marine biologist Elise C. Otté and came out a year later as *Views of Nature* (1850) with Bohn. Elizabeth Sabine, who also translated prestigious texts for her husband's French and German colleagues, was used to working at the forefront of the geological and physical sciences. Unusually, this translation was published in her own name, rather than 'under the superintendence' of her husband, as was the case with her other translations of Humboldt's works. Indeed, Longman and Murray incorporated into their advertising strategy the fact that this was a work which had appeared under Humboldt's sanction and cooperation, and that he had expressly asked Elizabeth Sabine to undertake it (Martin 2018, p. 165). Otté, whose rendering appeared on the market slightly later, had the advantage that Sabine had already negotiated problems related to the translation of terminology or of complex scientific ideas. However, her work was complicated by the fact that her rendering needed to be sufficiently different from the version immediately preceding hers, in order to be recognisable as a new work in its own right. Style, sentence structure and idiom all played a subtle, but important, role in differentiating the later translation from Elizabeth Sabine's version. The different footnoting and endnoting strategies used by the competing publishers also helped to give the two translations an individual paratextual character.

Fierce competition between Bohn and his rivals Longman and Murray also overshadowed the translation history of Humboldt's final, magisterial work, *Kosmos* (1845–1862). The ambitious scope of this holistic project, which covered everything known about the universe and the earth, made it a challenging work to write. It was just as challenging to translate. Again, Sabine and Otté vied for public recognition as Humboldt's translators as they produced their renderings for Longman and Murray (1846–1858) and Bohn (1849–1858) respectively. This time, though, they were competing not only against each other but also against the Bristol eye-surgeon, Augustin Prichard, who only managed to translate the first two volumes (1845–1848) for the London publisher Hippolyte Baillière, before giving up. Once again, both Sabine and Otté were committed to conveying the original as faithfully as possible, but they were disadvantaged by Prichard's rendering of the first two volumes of *Kosmos*, from which they had to distance themselves stylistically as well as commercially, if their translations were to be successful. Paratextual additions helped to give these translations rather different trajectories, as Edward Sabine added a series of 'Editor's Notes', endorsed by Humboldt, which emphasised the importance of British institutions in the advancement of science (Martin 2018, pp. 224, 225).

By the middle of the nineteenth century, scientists were therefore becoming increasingly aware of the transformative possibilities, but also disempowering potential, of translation. The collaborative translation project undertaken with Williams taught Humboldt the dangers of putting a text into the hands of a non-specialist, who might have a flair for stylish literary prose but knew little of the scientific concepts and terms being translated. Around the time that Humboldt's translators were working on *Kosmos*, Darwin was thinking about having his works translated into other languages. But he was arguably rather hasty in recruiting translators, and placed a rather 'naïve trust in individuals who appeared to be receptive to evolution' (Lightman 2015, p. 403). He struggled in particular with the French translator of his *Origin of Species* (1859), Clémence Royer, a teacher, who was an autodidact in scientific matters (Harvey 2008, p. 359), and had no qualms about producing a rather free, and certainly inaccurate, translation of his work. Although he did not reject her translation outright, he did use the appearance of a new edition to correct some of her terminology, correct errors in her scientific notes, and even omit some of them. While he was irritated by her failure to update her translation at this opportunity, this 'never obscured his admiration for the vigorousness of her style' (Harvey 2008, p. 360). Scientists were therefore aware that putting their work into another language required a rare combination of scientific expertise and an up-to-date knowledge of what was often already a quite specialised field, coupled with an ability to fashion clear, elegant prose in the target language, and full mastery of the source language from which the work was being translated. This extremely complex skill set is fundamental to scientific translation today, yet is still under-appreciated—a situation that will only alter with shifts in the self-perception of its practitioners and the increasing professionalisation of the field.

Conclusions

Historical research into women as translators of science has tended to concentrate mainly on case studies of individuals or specific works in translation, reflecting a lack of larger amounts of data for any given language area. We are gradually compiling the pieces that give us a clearer picture of just how important translation was to scientific endeavour, and transnational approaches will help to reveal further the networks and connections through which knowledge and texts flowed. The near invisibility of women like Cox, about whom we still know painfully little, is the norm rather than the exception. While the discovery of new instances of women's involvement in scientific translation adds to this mosaic, it is important not to mis-represent these figures as anomalous figures, as proto-feminists blazing a trail through the realm of 'male' science. Patricia Fara eloquently warns of the distorting nature of gendered histories and biographies which over-enthusiastically try to reconstruct women's roles in science. She rightly voices irritation at biographies of female scientists which 'emphasise the singularity of their subjects', in order to boost the appeal of their research, or to make 'sweeping generalizations that female scientists are a race apart' (Fara 2013, pp. 43–44). Translation was certainly one of the easier ways by which women could participate seriously in science, and translations of key scientific treatises were often badly needed by male scientists who did not have the linguistic skills, time or inclination to undertake such work. Yet, as Jean Delisle observes, it would present a false picture of history to suggest that all female translators used translation to advance the cause of women (Delisle 2002, p. 9). Moreover, it was also not necessarily the case that work produced by female translators was scientifically sound and accurate. While the 'recuperating stories' now being written are important in demonstrating how some 'ingenious' women may indeed have been at the forefront of facilitating knowledge exchange, we should also be cautious of using these narratives to over-weight female contributions to science.

The current dominance of Global English presents a distortedly monolingual picture of the languages of scientific translation (Dupré 2018, p. 302). By tracing just a few of the key moments of translation activity in Western Europe, we have seen how specific languages have fluctuated in and out of fashion as vehicles conveying scientific knowledge. Similar studies for the Asian languages (see Sarukkai 2016) have begun to make important inroads in understanding the relationship between translation and transmission elsewhere in the world. More connected histories of science need to be written in order to tell a less fragmentary narrative about the role that (female) translators have played in enabling scientific knowledge to circulate between different cultures of scientific knowledge. Indeed, understanding the relevance of language in the emergence of national scientific traditions (see Lightman 2015; Gordin and Tampakis 2015) will go some way to understanding how particular modes of scientific enquiry and different narrative styles emerged that encouraged, or perhaps hindered, women's involvement in science through translation. Thinking about translation less in terms of the ease of communication flow, and more about its moments of disunity, asymmetry and difference, may indeed help us to chart better the ebb and flow of scientific texts, concepts and ideas between languages, and the contribution of women to these patterns of knowledge exchange.

REFERENCES

Abir-Am, Pnina G., and Dorinda Outram, eds. 1987. *Uneasy Careers and Intimate Lives: Women in Science, 1789–1979*. New Brunswick, NJ: Rutgers University Press.
Agorni, Mirella. 1998. "The Voice of the 'Translatress': From Aphra Behn to Elizabeth Carter." *The Yearbook of English Studies* 28: 181–95.
Bailes, Melissa. 2017. *Questioning Nature: British Women's Scientific Writing and Literary Originality, 1750–1830*. Charlottesville, VA: University of Virginia Press.
Beer, Gillian. 1983. *Darwin's Plots: Evolutionary Narrative in Darwin, George Eliot and Nineteenth-Century Fiction*. London: Routledge and Kegan Paul.
Behn, Aphra. 1993. *Works: Volume 4—Seneca Unmasqued and Other Prose Translations*, ed. Janet Todd. London: Pickering.
Creese, Mary R. S. with contributions by Thomas M. Creese. 1998. *Ladies in the Laboratory? American and British Women in Science, 1800–1900. A Survey of Their Contributions to Research*. Lanham, MD/London: Scarecrow Press.
Dassow Walls, Laura. 2006. "Exploring the World." In *The Oxford History of Literary Translation in English. Vol. 4. 1790–1900*, ed. Peter France and Kenneth Haynes, 498–504. Oxford: Oxford University Press.
Delisle, Jean. 2002. "Présentation." In *Portraits de traductrices,* ed. Jean Delisle, 1–11. Ottawa: Presses de l'Université d'Ottawa.
Dupré, Sven. 2018. "Introduction: Science and Practices of Translation." *Isis* 109, no. 2: 302–7.
Elshakry, Marwa. S. 2008. "Knowledge in Motion: The Cultural Politics of Modern Science Translations in Arabic." *Isis* 99, no. 4: 7041–730.
Fahnestock, Jeanne. 1999. *Rhetorical Figures in Science*. Oxford: Oxford University Press.
Fara, Patricia. 2013. "Weird Sisters?" *Nature* 495: 43–44.
Flotow, Luise von. 1997. *Translation and Gender. Translating in the 'Era of Feminism'*. Manchester: St. Jerome.
Flotow, Luise von. 2011. Preface to *Translating Women*, ed. Luise von Flotow, 1–10. Ottawa: University of Ottawa Press.
Fontenelle, Bernard le Bovier de. 1687. *A Discourse on the Plurality of Worlds*. Translated by Sir W. D. Knight. Dublin: A. Crook and Son.
Fontenelle, Bernard le Bovier de. 1688. *A Plurality of Worlds*. Translated by John Glanvill. London: Bentley and Magnes.
Fransen, Sietske. 2017. "Anglo-Dutch Translations of Medical and Scientific Texts." *Literature Compass*. https://onlinelibrary.wiley.com/doi/full/10.1111/lic3.12385. Accessed 21 September 2020.
Freddi, Maria, Barbara Korte, and Josef Schmied. 2013. "Developments and Trends in the Rhetoric of Science." *European Journal of English Studies* 17, no. 3: 221–34.
Fulford, Tim, Debbie Lee, and Peter J. Kitson. 2004. *Literature, Science and Exploration in the Romantic Era: Bodies of Knowledge*. Cambridge: Cambridge University Press.
Fyfe, Aileen. 2000. "Young Readers and the Sciences." In *Books and the Sciences in History*, ed. Marina Frasca-Spada and Nick Jardine, 276–90. Cambridge: Cambridge University Press.
Gates, Barbara T. 1998. *Kindred Nature: Victorian and Edwardian Women Embrace the Living World*. Chicago, IL: University of Chicago Press.

Godard, Barbara. 1989. "Theorizing Feminist Discourse/Translation." *Tessera* 6: 42–53.
Gordin, Michael D., and Kostas Tampakis. 2015. "Introduction: The Languages of Scientists." *History of Science* 53, no. 4: 365–77.
Harvey, Joy. 2008. "Darwin in a French Dress: Translating, Publishing and Supporting Darwin in Nineteenth-Century France." In *The Reception of Charles Darwin in Europe, vol. II*, ed. Eve-Marie Engels and Thomas F. Glick, 334–74. London: Continuum.
Healy, Michèle. 2004. "The Cachet of the 'Invisible' Translator: Englishwomen Translating Science (1650–1850)." Unpublished Thesis, University of Ottawa.
Johnston, Judith. 2013. *Victorian Women and the Economies of Travel, Translation and Culture, 1830–1870*. Abingdon: Routledge.
Leask, Nigel. 2001. "Salons, Alps and Cordilleras: Helen Maria Williams, Alexander von Humboldt, and the Discourse of Romantic Travel". In *Women, Writing and the Public Sphere, 1700–1830*, ed. Elizabeth Eger, Charlotte Grant, Clíona Ó Gallchoir, and Penny Warburton, 217–35. Cambridge: Cambridge University Press.
Lightman, Bernard. 2015. "Scientific Naturalists and Their Language Games." *History of Science* 53, no. 4: 395–416.
Martin, Alison E. 2018. *Nature Translated: Alexander von Humboldt's Works in Nineteenth-Century Britain*. Edinburgh: Edinburgh University Press.
Mason, Adrienne. 2006. "'L'air du climat et le goût du terroir': Translation as Cultural Capital in the Writings of Mme Du Châtelet." In *Emilie Du Châtelet: Rewriting Enlightenment Philosophy and Science*, ed. Judith P. Zinsser and Julie Candler Hayes, 124–41. Oxford: Voltaire Foundation.
Miller, Laura. 2013. "Publishers and Gendered Readership in English-Language Editions of *Il Newtonianismo per le Dame*." *Studies in Eighteenth-Century Culture* 42: 191–214.
Milton, John, and Paul Fadio Bandia, eds. 2009. *Agents of Translation*. Amsterdam: Benjamins.
Montgomery, Scott L. 2010. "Scientific Translation." In *Handbook of Translation Studies. Vol. 1*, ed. Yves Gambier and Luc van Doorslaer, 299–305. Amsterdam/Philadelphia: Benjamins.
Neeley, Kathryn. A. 2001. *Mary Somerville: Science, Illumination, and the Female Mind*. Cambridge: Cambridge University Press.
Newton, Isaac. 1759. *Principes mathématiques de la philosophie naturelle, Par feue Madame la Marquise du Chastellet*, vol. 1. Paris: Desaint & Saillant/Lambert.
Olohan, Maeve. 2014. "History of Science and History of Translation: Disciplinary Commensurability?" *The Translator* 20, no. 1: 9–25.
Orr, Mary. 2015. "The Stuff of Translation and Independent Female Scientific Authorship: The Case of Taxidermy . . . anon. (1820)." *Journal of Literature and Science* 8, no. 1: 27–47.
Oz-Salzberger, Fania. 2006. "The Enlightenment in Translation: Regional and European Aspects." *European Review of History* 13, no. 3: 385–409.
Phillips, Patricia. 1990. *The Scientific Lady: A Social History of Women's Scientific Interests 1520–1918*. New York: St Martin's Press.
Pickford, Susan. 2012. "Writing with "Manly Vigour": Translatorial Agency in Two Early Nineteenth-Century English Translations of François Pouqueville's *Voyage en Morée, à Constantinople et en Albanie* (1805)." In *Travel Narratives in Translation,*

1750–1830: Nationalism, Ideology, Gender, ed. Alison E. Martin and Susan Pickford, 197–217. London: Routledge.

Pieretti, Marie-Pascale. 2002. "Women Writers and Translation in Eighteenth-Century France." *The French Review* 75, no. 3: 474–88.

Raj, Kapil. 2013. "Beyond Postcolonialism... and Postpositivism: Circulation and the Global History of Science." *Isis* 104: 337–47.

Robinson, Douglas. 1995. "Theorising Translation in a Woman's Voice: Subverting the Rhetoric of Patronage, Courtly Love and Morality." *The Translator* 1: 153–75.

Roche, Geneviève. 1994. "'Völlig nach Fabrikenart': Handwerk und Kunst der Übersetzung bei Georg Forster." In *Weltbürger—Europäer—Deutscher—Franke. Georg Forster zum 200. Todestag*, ed. Rolf Reichardt and Geneviève Roche, 101–19. Mainz: Universitätsbibliothek Mainz.

Sarukkai, Sundar. 2015. "Translation as Method: Implications for History of Science." *Indian Journal of History of Science* 51, no. 1: 105--17.

Schaffer, Simon, Lissa Roberts, Kapil Raj, and James Delbourgo. 2009. Introduction to *The Brokered World: Go-Betweens and Global Intelligence, 1770–1820*, ed. Simon Schaffer, Lissa Roberts, Kapil Raj, and James Delbourgo, ix–xxxviii. Sagamore Beach, MA: Watson Publishing International.

Secord, James A. 2004. "Knowledge in Transit." *Isis* 95, no. 4: 654–72.

Shteir, Ann B. 1996. *Cultivating Women, Cultivating Science: Flora's Daughters and Botany, 1760–1860*. Baltimore, MA: Johns Hopkins University Press.

Shteir, Ann B. 2003. "Finding Phebe: A Literary History of Women's Science Writing." In *Women and Literary History: 'For There She Was'*, ed. Katherine Binhammer and Jeanne Wood, 152–66. Newark, DE: University of Delaware Press.

Simon, Sherry. 1996. *Gender in Translation. Culture and Identity and the Politics of Transmission*. London: Routledge.

Stark, Susanne. 2006. "Women." In *The Oxford History of Literary Translation in English. Vol. 4. 1790–1900*, ed. Peter France and Kenneth Haynes, 125–31. Oxford: Oxford University Press.

Topham, Jonathan R. 2004. "Technicians of Print and the Making of Natural Knowledge." *Studies in History and Philosophy of Science* 35, no. 2: 391–400.

Uman, Deborah. 2012. *Women as Translators in Early Modern England*. Newark, DE: University of Delaware Press.

Wolf, Michaela, and Alexandra Fukari, eds. 2007. *Constructing a Sociology of Translation*. Amsterdam/ Philadelphia: Benjamins.

Zinsser, Judith P. 2001. "Translating Newton's 'Principia': The Marquise du Châtelet's Revisions and Additions for a French Audience." *Notes and Records of the Royal Society of London* 55, no. 2: 227–45.

CHAPTER 19

Queen Lovisa Ulrika of Sweden (1720–1782): *Philosophe* and Collector

Anne E. Harbers and Andrea M. Gáldy

Queen Lovisa Ulrika[1] of Sweden was an exception to most women of her age and position in both her educational and collecting ambitions. While aristocratic women by and large received a much better education than their bourgeois or lower-class contemporaries, they too were seen as having distinct 'womanly' goals in life such as getting married and producing legitimate aristocratic children. Their education was dependent on family and social attitudes and on their own aspirations and energy to pursue self-improvement and intellectual endeavours. From early modern times their occupation with plants and herbal remedies may have been regarded as appropriate across Europe (e.g. in Denmark and Saxony), even though collecting and display of most categories of collecting pieces was not a pastime easily undertaken by many women, including noble ladies. Queen Lovisa Ulrika of Sweden challenged these ideas and amassed a notable collection of plants and invertebrates, minerals, ores and corals, all of which was donated by her grandson, King Gustav IV Adolf to Uppsala University in 1803. These eighteenth-century Royal Collections, still housed in the natural history collections of Uppsala University, including the Museum of Evolution founded in 1999, remain valuable to modern-day scientific research (Hämäläinen and Orr 2017; Lindkvist 2016). The specimens themselves as well as the legacy of the classification and detailed catalogue work

---

A. E. Harbers
Radboud University, Nijmegen, The Netherlands

A. M. Gáldy (✉)
Collecting & Display, London, UK

© The Author(s), under exclusive license to Springer Nature Switzerland AG 2022
C. G. Jones et al. (eds.), *The Palgrave Handbook of Women and Science since 1660*, https://doi.org/10.1007/978-3-030-78973-2_19

of Carl von Linné or Carl Linnaeus (1707–1778) are regarded as important to this day. This chapter will explore how and why the Royal Collections of Queen Lovisa Ulrika came to influence modern science as well as investigate the role they played in significant publications that are regarded as seminal to the development of taxonomy.

Lovisa Ulrika—Early Life and Education

When Lovisa Ulrika died in 1782, a contemporary diarist wrote that she had introduced to Sweden 'foreign taste and luxury' and had 'encouraged arts and sciences to a poor land [that] could do better without' (Barton 1972, p. 11). Why were her attitudes and interests in the arts and sciences regarded in this way, nearly four decades after she had arrived in Sweden as the new Crown Princess from Prussia?

Lovisa Ulrika was born on 24 July 1720 at the Hohenzollern city palace of Berlin (Fig. 19.1). She was the youngest of fourteen children born to her royal

Fig. 19.1 Queen Lovisa Ulrika of Sweden (1720–1782). Unknown artist, Lovisa Ulrika, drottning av Sverige, Nationalmuseum: Public Domain

parents, Friedrich Wilhelm I and Sophia Dorothea of Hannover. Lovisa Ulrika was thus part of a line of Hohenzollern women who had played a considerable role in the rise of Prussia, through their involvement with and promotion of cultural politics and through a reinforcement of the status and prestige of the dynasty. In addition, she also came from a line of women who posed and discussed issues related to the universe and to God and Nature with the philosophers, theologians and scientists (called 'natural philosophers') of their day.

Lovisa Ulrika's early years were spent with her siblings and her parents at the royal court and residences in Berlin. In her younger years, she was in the care of Huguenot governesses who had also instructed her brothers, including Crown Prince Frederick, the future Frederick the Great of Prussia. As Lovisa Ulrika grew up, she was educated by the French scholar Maturin Veyssière de la Croze (1661–1739) who had previously been a tutor to her siblings. He was appointed to the position of Royal Librarian in 1697, promoted in due course to the Academy of Sciences in Berlin as well as becoming Professor of Philosophy at the French Collegium in Berlin from 1720. Previously a Benedictine scholar in Paris, he fell out with his Prior due to his studies in areas outside the Catholic doctrine and, in 1696, fled Paris for Basel where he converted to Protestantism. In addition to European languages and Latin, he spoke Armenian, Slavic and Semitic languages and held a position in Armenian Studies at the University of Berlin. On his death, he left unpublished dictionaries in Armenian, Slavic, Coptic and Syriac.

In her studies of Lovisa Ulrika, Elise Dermineur convincingly argues that De la Croze's tuition developed in Lovisa Ulrika a taste for books and for the natural sciences and that within the court at the Berlin Schloss 'she rubbed shoulders with her relatives, courtiers, army officers, ministers, invited artists and thinkers, and scientists of the Academy' (2017, p. 24). The presence of the scientists of the Academy at court might have been related to the interests of Lovisa Ulrika's grandmother, Sophie Charlotte von Hannover (1688–1705), who had used current ideals of cultural politics to promote the dynastic ambitions of the new Prussian monarchy. As the only daughter of Elector Ernst August of Hannover (1629–1698), she had married Friedrich (1657–1713) at sixteen. Friedrich eventually became Elector of Brandenburg in 1688 and, in 1701, declared himself first King of Prussia. Sophie Charlotte thus rose to the title of Electress of Brandenburg (from 1688) and Queen Consort of Prussia from 1701.[2]

In 1697, Sophie Charlotte proposed the building of an astronomical observatory in Berlin. The mathematician and philosopher Gottfried Wilhelm Leibniz (1646–1716) who had been attached to her father's court at Hannover responded to the invitation. Today he is remembered as the first scientist to publish on calculus. Leibniz wrote to the Electress suggesting that the observatory should be extended to encompass an Academy of Sciences, such as had previously been established in Paris and London. As the result of their correspondence and his visits, Sophie Charlotte invited Leibniz to be her friend and teacher, as he had been to her mother at the Hanoverian court (Strickland 2011; Brown n.d.). In 1700, the Elector approved the establishment of Sophie Charlotte's proposed astronomical observatory in Berlin and also of a Society of Sciences named *Kurfürstlich Brandenburgische Societät der Wissenschaften* (i.e. the Electoral Brandenburg Society of Sciences); Leibniz was named president of the society. In 1701, when the Elector became King of Prussia, the society was renamed *Königlich Preußische Sozietät der Wissenschaften* (Royal Prussian Society of Sciences) and was one of the first such societies to teach both natural sciences and humanities.

Leibniz and Sophie Charlotte continued their correspondence on many philosophical topics until her unexpected death from pneumonia on 1 February 1705, at the age of thirty-seven. In a note written during her last illness, the Prussian queen remembered both Leibniz and her husband: 'Don't grieve for me, for I am about to satisfy my curiosity about things that even Leibniz was never able to explain—space, the infinite, being, and nothingness—and for my husband, the king, I am about to provide a funeral-spectacle that will give him a new opportunity to display his pomposity and splendour!' (quoted in Mates 1986, pp. 27, 28). Throughout her life, Lovisa Ulrika maintained a strong correspondence with her Prussian family, by means of many letters written in French to her mother, sister and brother Frederick. Beyond his military successes, her brother King Frederick II of Prussia (r. 1740–1786) was also fond of philosophy and Enlightenment ideas and invited Voltaire to Potsdam in 1750. Described as a philosopher-king, a latter-day renaissance man, he believed princes were duty-bound to implement the rational teachings of philosophers (Blanning 2015). Although Frederick seems to have bullied his brothers, despite their being important to his military campaigns and to the monarchy, he behaved more kindly towards his sisters. Indeed his respect for female intellect is illustrated in his writing about his grandmother, Sophie Charlotte: 'This princess had the genius of a great man and the knowledge of a savant; she did not deem it unworthy of a queen to admire a philosopher; the philosopher was Leibniz, and she bestowed her friendship on him with the thought that those to whom Heaven has given noble minds are the equivalent of kings' (Mates 1986, p. 26).

Frederick of Prussia continued his grandmother's long-standing interests in the natural sciences, as manifested by her contact with Leibniz and with the Royal Prussian Society of Sciences. During his reign, in 1744, the *Nouvelle Société Littéraire* and the Society of Sciences were merged to form the *Königliche Akademie der Wissenschaften* (Royal Academy of Sciences). Under the new statutes, the focus rested on research to settle unsolved scientific questions, including offers for financial rewards. The Academy established many research entities, now part of the University of Berlin, for example, an observatory in 1709; an anatomical theatre in 1717; a *Collegium Medico-Chirurgicum* in 1723; a botanical garden in 1718 and a laboratory in 1753.

As Lovisa Ulrika reached her late teenage years, a number of dynastic unions with the marriageable princes of Europe were negotiated by her father;

however, he died in 1740. When her brother Frederick succeeded to the throne, his attention by necessity turned to military matters, rather than to her marriage plans. In December 1743, when Lovisa Ulrika was twenty-three, he appointed her as *coadjutrix* and princess-abbess to the Abbey of Quedlinburg. This was a position of some prestige with a significant income, since the Abbey gave a home to the unmarried daughters of noble and royal families (Dermineur 2017).

Even though Lovisa Ulrika took up the position at Quedlinburg Abbey, circumstances quickly changed when the Swedish parliament invited Adolf Frederik of Holstein-Gottorp to become Crown Prince of Sweden (1710–1771). The Swedish parliament and aristocracy regarded Adolf Frederick as a mere figurehead, since the real power was held by the *Riksdag* of Sweden made up of the four estates of the Nobility, Clergy, Burghers and Peasants. Nonetheless, this invitation provided the opportunity for a dynastic marriage alliance between Sweden and Prussia and Lovisa Ulrika was married by proxy in Berlin in July 1744; she arrived in Stockholm in late August 1744. Adolf Frederick was distantly related to the Swedish Vasa dynasty but his immediate family had distinguished itself as important collectors at Gottorp castle at Schleswig. Situated in the borderlands between Schleswig–Holstein-Gottorp and Denmark, they repeatedly changed allegiance. In 1713, the Danish King Frederick IV conquered Gottorp and the collections became the property of the Danish Crown. By 1743, the collections were inventoried; they were packed up and transported to Denmark in 1750. There they were taken to the Royal *Kunstkammer* in Christiansborg but later divided up and redistributed with the mathematical instruments, coins and medals and *naturalia* being taken to Rosenborg Castle (Bencard et al. 1997). Thus Adolf Frederick was used to a family tradition of collecting and possibly tried to emulate his ancestors by making a show of strength via rebuilding a new collection as King of Sweden.

As they met in their respective roles as Crown Prince and Crown Princess of Sweden, Lovisa Ulrika brought the tradition of welcoming philosophers and thinkers, mathematicians and theologians from the Prussian Court, considered to be an integral part of the promotion of status and prestige as the result of intellectual and higher levels of thinking that reflected ideals of the French Enlightenment. Just prior to her arrival in Sweden, in 1740 and 1743, she had had direct contact with Voltaire at the Prussian Court, to which he was invited by her brother Frederick.

Hence, Lovisa Ulrika arrived in Sweden with a number of ambitions which aligned with Frederick of Prussia's ideals of the French Enlightenment. She aimed to promote the interests of her brother (Dermineur 2017)[3] and maintained a belief in absolute monarchy; therefore, she was at odds with the Swedish political system. Her ambitions contributed to her self-image, her self-education and as will be shown, were manifested in her Royal Collection.

Lovisa Ulrika—Collector

Lovisa Ulrika has been described as having 'an insatiable intellectual curiosity which extended to astronomy, mathematics, the sciences and the applied sciences', while also having 'a masterful temper, unflagging interest in politics and abilities superior' to her husband, as well as being a connoisseur of the arts and literature (Roberts 1986, p. 178). She was also described as 'ambitious, fiercely proud of her Hohenzollern birth, devious, incorrigibly intriguing, a bad judge of what was practicable, prone to mistakes arising out of wishful thinking, but steel-hard and courageous'. By many, such as the Swedish diplomat Count Carl Gustaf Tessin who was involved in the political negotiation of her marriage alliance and later with the signing of the 1747 treaty with Prussia, she was considered fascinating in her charm and intelligence (Roberts 1986). Her correspondence with her family in Prussia indicates a close relationship with her brothers and sisters, including her brother Frederick, but also reveals her as a complex personality, with a lively mind and an insatiable thirst for intellectual and social stimulation (Rivière 2004).

On her arrival in Sweden in 1744, the then Swedish King Fredrik I transferred the fiefdom and Baroque palace of Drottningholm to the new Swedish Princess Lovisa Ulrika. The palace of Drottningholm, with its lavish interiors and generous outdoor spaces provided an environment in which Lovisa Ulrika and her husband Adolf Fredrik would lead a successful marriage over the next thirty years or so. It was a place in which to share interests in collecting, although they had differing collecting aims and their respective Royal Collections were kept in separate physical locations. There had been a rudimentary Swedish Royal Collection before their arrivals as prince and princess to Sweden but the new royal couple would largely build their respective collections through astute acquisitions of established collections, mostly purchased in Holland—'the land of Museums [...] where the wealthy and educated of all classes vied with each other for possessing the choices of rarities' (Lovén 1887, p. 4).[4] Large amounts were spent between 1750 and 1752 in Amsterdam and Utrecht, including 4000 guilders for parts of the collection of Albert Seba, auctioned on 14 April 1752 to a total collection estimate of 50,000 guilders. Purchased Seba items were distributed between the collections of king and queen.

In her studies on Lovisa Ulrika as crown princess and queen, Merit Laine (1998a) proposes that she used Drottningholm as a visual communication tool of the status and taste of the royal couple with the aim to increase the power of the Swedish monarchy.[5] Lovisa Ulrika's collecting interests were extensive, ranging from paintings commissioned on her behalf by Swedish diplomats in Paris to coins, medals, antiquities, books and manuscripts. In addition, she owned a collection of *naturalia* comprising dried invertebrate and plant materials, all housed at Drottningholm. Adolf Fredrik collected vertebrates preserved in alcohol held in his cabinets at Ulriksdal palace to the northeast of Stockholm. At Drottningholm, Lovisa Ulrika commissioned her architects to add a new storey to the existing wings, with five new rooms surrounding the courtyards; work commenced in 1747. These display rooms were located between the king's and queen's apartments and included a library, a study room, the cabinet of coins and medals, the cabinet of minerals and the cabinet of natural history (Fig. 19.2). They were decorated in the style of Louis XVI and included Baroque and Rococo decorative elements. With the exception of the cabinet of coins and medals, Lovisa Ulrika had all the rooms redecorated in 1753, following Adolf Fredrik's succession to the throne in 1751.

Fig. 19.2 The Cabinet of Natural History at Drottningholm Palace, Sweden. Reproduced by kind permission of Alexis Daflos (photographer) and The Royal Court, Sweden

Her timing suggests an intention to emphasise the role of the new Swedish monarchs as patrons of literature, arts and sciences and, as a result, to enhance their status (Laine 1998a). At this point, she also commissioned or acquired a collection of Dutch, French, English and Swedish paintings. Her antiquities collection was distributed between the cabinet of coins and medals, the library and the study, since these items were considered objects of study. In her correspondence, Lovisa Ulrika regularly referred to her collections, to the new acquisitions as well as to the architectural modifications (Laine 1998b).[6]

When redecorating the library, a colour scheme of white and gold was adopted with Corinthian pilasters to give a sense of contemporary style and decorum. Six reliefs were included, featuring one of Lovisa Ulrika in the guise of Minerva and thus in the role of a patron of the arts, sciences and literature. The other five reliefs included depictions of the Muses Polyhymnia, Erato, Calliope, Clio and Urania. Science was represented by Urania, as the muse of astronomy, wisdom and virtue. In 1753, Lovisa Ulrika founded the Swedish Royal Academy of Letters to encourage eloquence, historical research and the reform of the Swedish language. Her decoration of the library at Drottningholm was a consequence of this interest in linguistics and displayed quotations from Virgil and Ovid; it also underlined her interest in Enlightenment concepts and French Classicism.

The cabinets of minerals and natural history were of smaller proportions but kept in the same colour scheme of white and gold as the library. The cabinet of natural history displayed six relief portraits in the classical style of eighteenth-century Swedish scientists, each of whom had associations with Uppsala University. Uppsala University, founded in 1477, is the oldest university in Sweden with a strong focus on studies in the natural sciences to this day. The six scientists depicted are Torbern Bergman (1735–1784), Anton von Swab (1702–1768), Nils Rosen von Rosenstein (1706–1773), Samuel Klingenstierna (1698–1765), Johan Gottschalk Wallerius (1709–1785) and Carl Linnaeus (1707–1778). Bergman was a chemist and mineralogist and known for publishing a *Dissertation on Elective Attractions* on chemical affinity in 1775. He became Professor of Chemistry at Uppsala University and was elected to the Royal Swedish Academy of Sciences in 1764. Anton von Swab (1702–1768) also was a mineralogist and chemist. He had received his education at Uppsala University and his work focussed on the advancement of mining in Sweden; he would be elected to the Royal Academy of Sciences in 1742.

Nils Rosen von Rosenstein (1706–1773) was a physician and is regarded as the founder of modern paediatrics for his publication on *The Diseases of Children and their Remedies*. He was a professor at Uppsala University and also worked as physician, first for King Fredrik I and then for Adolf Fredrik. He was present at all deliveries of Lovisa Ulrika's four children and she believed he had saved the life of her first-born, Crown Prince Gustav in 1746. Samuel Klingenstierna (1698–1765) was a mathematician and scientist interested in natural philosophy, i.e. the study of the philosophy of nature and regarded as a pre-runner to modern science. As a student, he lectured at Uppsala University on the mathematical discoveries of Newton and Leibniz and, in 1728, was appointed as Professor of Geometry at Uppsala University. After his retirement in 1756, he became tutor to the young crown prince, the future King Gustav III (r. 1771).

Johan Gottschalk Wallerius (1709–1785) was a chemist and mineralogist. He was the younger brother to the Swedish physicist, philosopher and theologian Nils Wallerius. In 1750, he took the new Chair of Chemistry, Medicine and Pharmacy at Uppsala University and in the same year was elected to the Royal Academy of Sciences. He is regarded as a founder of agricultural chemistry. He retired from the Chair of Chemistry in 1767, when his student Torbern Bergman took up the post. Finally featured among these portraits was Carl Linnaeus, later Carl von Linné. As a botanist, physician, zoologist and professor at Uppsala University, Linnaeus was the founder of binominal nomenclature taxonomy (the science of identifying, naming and classifying organisms such as plants and animals). Of these natural philosophers, the closest and most mutually beneficial relationship that Lovisa Ulrika developed was with Carl Linnaeus as will be explored later in this essay.

Lovisa Ulrika's collection defined her own particular focus by her display rooms being turned into specific cabinets of minerals and of natural history. The scientists whose portraits she included had interests in mineralogy and chemistry as well as in natural history (shells, insects, plants, corals). Her interest in mineralogy reflected the economic importance of ores and minerals to Sweden (continued to this day). Sweden is still among the largest European producers of iron, silver, lead, gold, copper and zinc (*Treasures from Inside the Earth*, 2001–2002, p. 15). The paintings at Drottningholm were not given their own gallery but were distributed over several rooms, including Lovisa Ulrika's own apartments. They were hung in accordance with national schools and made available to young painters as part of their training (Laine 1998b, p. 499).[7] Correspondence by Lovisa Ulrika shows that she sympathised with contemporary views that valued the notion of princely collections being available for a wider audience (Laine 1998b, p. 499).[8]

Originally, princely collections of early modern Europe used to be displayed in specialised display rooms, for example, small study rooms, to which access was mostly reserved for the collectors' families and to especially privileged guests. The collection and its display thus played a form of hide and seek as something highly interesting and advertised in writing and by word of mouth but not open to all those who wished to visit. This form of display would change in the later sixteenth century when collecting items tended to be of larger dimensions so that their display was more appropriate for galleries or gardens. In addition, plaster casts of such exhibits developed into a new and useful medium adopted in an academic environment (Vasari 1881, p. 517).

From the eighteenth century, galleries, e.g. the Uffizi in Florence, the Electoral galleries in Dresden, the Fridericianum at Kassel and later the National Gallery in London, started to welcome a more diverse audience. Men and women were among the visitors. Attendees needed to be properly dressed (Taylor 2012, p. 265; Linnebach 2016, pp. 207, 208), often had to sign a Visitors Book and they also sometimes had to pay an entrance fee or give a gratuity to the personnel in charge (Findlen 2012, pp. 83, 100, 104; Linnebach 2016, pp. 193–97). While at first it was necessary to book ahead and tour the exhibition guided by the gallery's custodian, eventually it became possible to turn up during quite generous opening hours. The Fridericianum seems to have been among the earliest institutions to offer such a service, long before the British Museum in London (founded in 1753) (Linnebach 2016, pp. 195, 196). Formerly private princely collections thereby became an increasingly public display with a multiple function: to attest to the princes' taste, wealth and scientific and/or artistic knowledge, to educate the rising middle class and to provide the necessary 'raw material' for the training of artists and scientists employed at the court (Bracken et al., forthcoming 2022).[9] After all, well into the eighteenth century, such galleries combined the display of works of art with an armoury, stuffed exotic animals or scientific instruments, all of which were admired by male and female visitors. Even though the female audience was much less numerous, amounting to about 10%, and included mostly aristocratic women with much better access to education at home and abroad, such visits to museums and galleries opened new possibilities to gain information and knowledge (Linnebach 2016, p. 207). The increased opportunity to travel and observe art and antiquities, as well as natural science collections, was greatly appreciated and even formed the foundation of publications, written by female authors on the basis of their observations and travel diaries (Müller 2016). One should also not forget that in the case of the Uffizi, it was the sister of the last Medici Grand Duke, Anna Maria Louisa (1667–1743), who understood the galleries' importance and ensured by means of the Medici-Lorraine Family Pact (1737) that these collections would not be sold or otherwise dispersed after her death (Whitehead 1983; Findlen 2012; Reid 2021).

Lovisa Ulrika's use of her collections followed these developments as part of Enlightenment court culture. Her collection was displayed next to her apartments and library, similar to the early modern examples in sixteenth-century palaces. A significant difference, however, at Drottningholm, was the importance accorded to her host of scientists at court by displaying their portraits in conjunction to the actual collection. Lovisa Ulrika's interest in the sciences was a particular one and extended beyond an interest to promote her image as Swedish monarch in the personification of Minerva, as patron of the arts, science and literature. In her library she made reference to the ancient Muses, not to Swedish writers or poets despite her founding the Swedish Royal Academy of Letters. In her natural history cabinet she chose to feature contemporary Swedish scientists whose areas of expertise reflected her collecting focus. This focus within her collecting practice was an extension of her self-image of her learning and how she personified her Enlightenment ideals in the name of the monarchy.

Lovisa Ulrika—Philosophe

Prior to her arrival in Sweden, Lovisa Ulrika had embraced ideas of the French Enlightenment developed through personal contact with Voltaire and reflected through books within her library at Drottningholm. During a visit by Voltaire to Berlin and to the Hohenzollern court from 30 August to 12 October 1743, the young princess Lovisa Ulrika seems to have become 'infatuated' with the French philosopher. As Marc Serge Rivière's research on Lovisa Ulrika and Voltaire has shown, the philosopher had visited the court back in 1740, but this longer visit in 1743 involved a closer relationship with the royal family (Rivière 2003). A contemporaneous writer describes Voltaire as charming the Queen Mother, as well as Princess Amalia and Lovisa Ulrika, during private dinners and readings of his plays; Voltaire went on to write a madrigal to Lovisa Ulrika. Did Voltaire overreach himself or was this flattery a *déclaration galante*? Rivière, in discussing the life-long relationship between Voltaire and Lovisa Ulrika, draws parallels with other influential women, such as Mme de Pompadour and Catherine II of Russia, as well as with many German and Prussian queens and princesses, with whom Voltaire's objective was to seek patronage to promote his work throughout Europe. His relationship with Lovisa Ulrika however changed in tone in 1744, when she became married, and again in 1751, when she became Queen of Sweden. Catalogues of the library at Drottningholm, listing volumes of books owned by Lovisa Ulrika include thirty-five works by Voltaire, for example, one inscribed by him to 'Queen Lovisa Ulrika', indicating it was given after 1751, and which includes many corrections in his own hand (Rivière 2003). Therefore, we can conclude that Lovisa Ulrika followed her family's tradition and embraced the ideas of the French Enlightenment. Indeed, her patronage and promotion of the arts and sciences in Sweden seems to reflect this attitude.

Lovisa Ulrika's strong interest in what is now regarded as the 'natural sciences' may have been prompted by an awareness of her grandmother's patronage of Leibniz in Prussia.[10] Or was it encouraged by her own interest in natural philosophy, which in turn fed into her engagement with natural history? Lovisa Ulrika referred to herself as a *philosophe* in letters to her mother in 1746, to her sister Amalie in 1748 and to her brother Wilhelm in 1751 (Rea Radisich 2003).[11] A contemporary (1765) *Encyclopédie* defines a person's *esprit philosophique* as being the result of marvellous intelligence, of forceful reasoning and of sure and contemplative taste (Rea Radisich 2003). It communicates an ideal of learning and education, expressed by Lovisa Ulrika in a letter dated 25 November 1746, in which she refers to the pleasure she gained from re-reading the six-volume work *Traité de L'Opinion VI: Ou Memoires Pour Servir A L'Histoire de L'Esprit Humain* (1733) by Gilbert Charles Le Gendre, a publication that included information on philosophy, metaphysics, mathematics and natural sciences. In this letter she writes of her reading material: *Il est écrit avec tout l'esprit imaginable et tout propre à donner une idée superficielle des sciences, ce qui est tout qu'une femme doit savoir...* [It is written with all the imaginable mind and everything to give a broad idea of the sciences, which is what a woman must know ....].[12]

Lovisa Ulrika also showed an interest in the link between children's education and play in her letter to her mother (1749), in which she describes the *Bureau Typographique*, an invention to teach her son Gustav to read 'before he was four' (Rea Radisich 2003). Earlier, on 31 May 1746, while pregnant with Gustav, she had written to her mother: *Je fais batir aussi deux galeries, qui seront ornées de tableaux que j'ai fait venir de Paris. C'est Boucher et Chardin qui en sont les maitres. J'ai donné pour sujet au premier 'Les quatre heures du jour' et à l'autre l'éducation sévère et l'éducation douce et insinuant. Ils doivent arriver incessament.* [I will have two galleries built that will be decorated with paintings I have ordered from Paris. Boucher and Chardin are the masters thereof. I have given as a subject to the former "The Four Hours of the Day" and to the other "The Harsh Education and The Gentle Insinuating Education." The paintings are due to arrive at any moment'].[13]

The paintings commissioned by Lovisa Ulrika as well as those acquired through Count Tessin's collection indicate her subjects of interest. For example, in Jean-Baptiste-Siméon Chardin's *The Morning Toilette* (now at the Nationalmuseum, Stockholm), the theme of education is referred to by means of the little girl's glance at the mirror as a forewarning of vanity that a good mother makes her child avoid (Grate 1994).[14] In the letter to her mother quoted above, Lovisa Ulrika perceived her role as a partner in the creative production of the works from Chardin (1699–1779). The actual works delivered from Chardin and received by Lovisa Ulrika in June 1747 were however not of the subjects requested and detailed in her letter, but represented *Domestic Pleasures* and *The Housekeeper* (both now at the Nationalmuseum, Stockholm).

The Swedish scientist Carl Linnaeus who had featured in Lovisa Ulrika's significant renovations of 1753 played an important and synergistic role in the Royal couple's collecting activities. Linnaeus had been born in Småland in 1707 and, after studying medicine at Lund University and then Uppsala University, he had travelled to Holland in 1735 to gain his doctorate at Leiden University. He returned to Sweden in 1738. At Leiden, he would have seen the University garden, *Hortus Botanicus*, a leading example of a medical garden, established for the cultivation of rare and exotic plants newly imported to Europe and for the study of plants as pharmaceutical and medical material. The garden had been established in the 1590s by the botanist Carolus Clusius as a *hortus academicus*, with the extensive trading routes of the Dutch East India Company (VOC) providing many exotic plants to The Netherlands. The sixteenth-century traditions of cabinets of curiosity or *Wunderkammern* were continued and developed by the seventeenth- and eighteenth-century interest in the natural world through natural philosophy and natural history. It resulted in collecting for a medical or scholarly purpose and paralleled aristocratic or princely collecting (Swan 2007).

In 1741, Carl Linnaeus was appointed to the Chair of Medicine and Botany at Uppsala University, a position he held until his death in 1778 (Harbers 2014). When he arrived at this post, the academic garden had fallen into disrepair and he set about bringing it back to life. In 1744, Crown Prince Adolph Fredrik and his newly-arrived bride, Crown Princess Lovisa Ulrika, visited Linnaeus in Uppsala. Richard Pulteney (1730–1801), an English botanist and physician and advocate of Linnaean taxonomy became the first English-language biographer of Linnaeus in 1805 and noted the special attention given to Linnaeus over others when he went to congratulate Lovisa Ulrika on her arrival to Sweden:

> In the year 1744, Linnaeus improved botany very much, and worked on the necessary books, without which the Professorship would not have been of so much use as it ought to be. He also laid out the garden agreeably to his system.[...].When his Royal Highness Prince Adolph Fredrik viewed the university, and the Professors were presented to him by the Chancellor Count Gyllenborg, Professor Andreas Celsius and Carl Linnaeus were denominated *Lumina Academica*, on account of their knowledge, which was celebrated as well within as without the kingdom; and the same year, when the Rector and four of the Professors (of whom Linnaeus was one) waited on her Royal Highness [...] to congratulate her on her delivery [i.e. arrival], Linnaeus was the only one who was ordered to proceed to Ekholmsund, and he had there a special audience of her Royal Highness. (Pulteney 1805, p. 539)

Around this time, Linnaeus was writing a publication *Fauna Svecica* which was a synopsis of the 1357 species of animals known to him in Sweden and which he published in 1746. He was still developing his system of binominal nomenclature, at this time using a system of Latin words following the genus name. In this publication he names two dragonflies (*Libellula*), using a special name (the *Vulgo*) intended for everyday use and, for the first time, the name of a particular person in his system. He named species 757 (now known as the male of *Calopteryx virgo*) as the 'Lovisa' and species 758 (now known as the female of *Calopteryx virgo*) as the 'Ulrica'. As noted by Hämäläinen and Orr (2017), this dedication of what could be considered the most gorgeously arrayed and charming demoiselle damselfly to occur in northern Europe to Princess Lovisa Ulrica indicates a flattering reflection of the qualities he admired in Her Royal Highness (Fig. 19.3).

In 1751, Lovisa Ulrika asked Linnaeus to study the collections at both Ulriksdal and Drottningholm and work with appointed artists and engravers to publish the king's and queen's collections in two separate works. The king's vertebrates from his natural history collection were published in 1754 under the title of *Museum S.R.M. Adolphi Friderici Regis*, in folio and paid for by the king. It was the first zoological work, in which the binominal nomenclature was used. The 1754 published material from the king's collection was to be incorporated in the tenth edition of Linnaeus' *Systema Naturae*—published in two volumes in 1758 and 1759. The tenth edition marked the starting

Fig. 19.3 A male *Calopteryx virgo*, the species to which Linnaeus (1746) gave the names 'Lovisa' and 'Ulrica' after the Princess. Photographed by Holger Gröschl, 2003: Creative Commons license

point for zoological nomenclature, introducing bionomical nomenclature for animals which he had already adopted for plants in *Species Plantarum* in 1753. Linnaeus would publish the queen's collection, including many plants and shells under the title of *Museum S.R.M. Ludovicae Ulricae* in 1764. He also included the information of both the king's and queen's respective catalogues in the twelfth edition in 1766. By this time, Linnaeus had been ennobled by the king and thus became Carl von Linné in 1757.

THE SIGNIFICANCE OF LOVISA ULRIKA'S COLLECTION TO SCIENCE

When lecturing in 1752, Linnaeus commented that 'Her Majesty possesses a cabinet of shells that surpasses every other in Europe' (Lovén 1887, p. 5). The significant amounts of money that Lovisa Ulrika spent on her collections of insects and shells particularly had built a superior collection and provided Linnaeus with the opportunity of observing many fine and very costly objects which he otherwise would not have had the opportunity of describing.

Queen Lovisa Ulrika's patronage of Linnaeus extended beyond granting access to her collections and discussions on natural history at court. Linnaeus had the ambition of describing the world's entire flora and fauna (Nordenstam 2009).[15] To achieve this aspiration, access to the Royal Collection was important as well as the creation of his own collections, owned by him as a professor rather than by Uppsala University, even though these were of less value in terms of financial expenditure. He was active in an international network of over 600 correspondents and encouraged his students to be collectors and explorers, to venture out far beyond Sweden and to bring back specimen samples. He called these young male students who travelled outside Europe his 'apostles'. Queen Lovisa Ulrika financially contributed to some of the expeditions undertaken by Linnaeus' apostles, receiving plant specimens for her Royal Collection on their return or purchasing natural history objects directly from them.

These journeys also had the objective of improving the economic position of Sweden by returning with plants that had commercial potential (Koerner 1999). An example is the 1746 journey to China, accomplished by the first apostle Christopher Tärnström (1711–1746). His instructions from Linnaeus were to return to Sweden with a living tea plant, seeds of mulberry trees and live goldfish for Lovisa Ulrika (Nordenstam 2009). Linnaeus also had an interest in the potential of plants from North America and the student he sent, Pehr Kalm (1716–1779), departed in 1747 and collected seeds in Pennsylvania and New Jersey as well as in Canada. On his return in 1751, his collection of plants and seeds was divided between Linnaeus, the KVA (The Royal Swedish Academy of Science) and, in 1754, Queen Lovisa Ulrika. The plant material from Queen Lovisa Ulrika's collection was later donated to Uppsala University and is today part of Queen Lovisa Ulrika's Herbarium at the University. These specimens duplicate material in the Linnaean collections at the Linnaean Society of London.

Queen Lovisa Ulrika's Herbarium also contains plant material collected by 'apostle' Fredrik Hasselquist (1722–1752) who travelled to Egypt and the Holy Land in 1749. He died near Smyrna (Izmir, Turkey) on 9 February 1752, 'fatigued from his travels': all his manuscripts and collections were sequestered at Smyrna and would only be released against payment of 14,000 silver dollars (Pulteney 1805, p. 545). Queen Lovisa Ulrika paid the debt and the materials were handed over a year or more after Hasselquist's death to Linnaeus who ensured that some of the valuable plant material was given to the Queen's collection. Linnaeus also posthumously published Fredrik Hasselquist's travelogue in 1757 under the title of *Iter Palaestinum Eller Resa Til Heliga Landet, förrättad ifrån År 1749 Til 1752* (Pulteney 1805; Nordenstam 2009; Gentry 2008).

In addition to her interactions with Linnaeus in relation to her interests in natural history, Queen Lovisa Ulrika also made donations to the Royal Swedish Academy of Sciences, as part of which an observatory was inaugurated in 1753. At the time of the foundation of the observatory, Queen Lovisa Ulrika gave the academy an unusually large wasps' nest and as a special gift for the inaugurations she presented silk made by silkworms in Sweden. The observatory was mainly designed as an astronomical observatory, but the building also housed a workshop for a scientific instrument-maker, archives, a library and a cabinet of *naturalia*. Research by Söderlund (2008)[16] confirms that it was specified at least from the 1780s (previous information is scant) that the museum should be open on either Wednesdays or Saturdays from 10 a.m. to

1 p.m. and that anyone should be allowed to enter free of charge. Unfortunately, the Visitor's Book has not been located and therefore no further details as to the classes and gender of visitors are known. The demonstrator/curator was not allowed to leave the city for more than two weeks and, whenever he had to be absent, he had to put a sign on the door with information about his return.

It has also been shown that Lovisa Ulrika was responsible for the employment of a French scholar, Philibert de Commerçon (1727–1773), who had studied medicine and botany at Montpellier. He travelled widely in Europe, in France and also to Switzerland and, in around 1757, was employed by the Queen, at the request of Linnaeus, to work on fish from the Mediterranean for the tenth edition of Linnaeus' *Systema Natura*. Later, de Commerçon accompanied the first French voyage of circumnavigation of the globe (1766–1769), led by Louis-Antoine Comte de Bougainville (1729–1811) (Jansen 2015).

Where Are Lovisa Ulrika's Collections Today?

After King Adolf Fredrik's death in 1771, his collections were moved from Ulriksdal to Drottningholm in 1773, while in 1777 Drottningholm and its collections were purchased by the Swedish state. In 1801, the specimens preserved in alcohol were handed to The Royal Swedish Academy of Science but the mammals and birds went to the Natural History Museum in 1828. A number of objects remain on exhibition there today.[17]

In 1803, King Gustav IV Adolf made a donation to Uppsala University that included the zoological material from the Royal Collection of his paternal grandmother Queen Lovisa Ulrika. This gift consisted of her collection of *conchilia* (seashells and mussel shells), insects, but also plants, fossils and minerals. Linnaeus had used this material in his publications. The plant material came from the North American excursion of Pehr Kalm and the plants from Palestine were collected by Fredrik Hasselquist (Moberg 2008). These plants are today part of Queen Lovisa Ulrika's University Herbarium. The minerals, now kept at the Museum of Evolution at Uppsala University, Linnaeus had meant to use in a future project, never completed, since he refers to them as: 'The other part of Your Majesty's rich collection, namely, the magnificent Corals, the clear Crystals, and the rich Ores, I have left to be the work of another day' (Moberg 2008, p. 143).

In a Communication to the Royal Swedish Academy of Sciences, Sven Lovén stated in 1887:

> To him, who had at his own command but a small supply of specimens out of lower classes of the Animal Kingdom, the Museum of the Queen was the principle source of information with regard to exotic Insects, Vermes and Zoophta. Without it he never could have effected as successfully as he did, his great scheme of the *Systema Naturae* of 1758, and it may be allowable to suppose that his enlightened friend Count Tessin, when in later years he was blamed for

having squandered the Queen's money on the whim of a Museum, found some notice in recollecting that he had done so, seemingly to follow the fashion of the day but in reality just as much for the sake of science and Linnaeus. (Lovén 1887, p. 6)

It is the aim of this essay to show that Lovisa Ulrika's role as a *philosophe* and collector as well as her patronage and support of the scientific work of Carl Linnaeus went well beyond her ambitions to achieve publication and thus exposure for her natural history collections. Rather, it was a genuine personal intellectual pursuit until political issues took over the royal couple's attention. The active years of the relationship between the Royal couple and Carl Linnaeus extend from 1746, when Linnaeus ordered his student Christopher Tärnström to undertake a journey to China to obtain a live goldfish for her, until around 1756, when she and her husband were embroiled in a *coup d'état* that politically sidelined the monarchy and brought into question their spending, as referred to by Lovén in 1887.

Current Scientific References to the Collections of Queen Lovisa Ulrika

Despite the dispersal of the Royal Collections nearly 250 years ago, the natural history collections of Queen Lovisa Ulrika continue to contribute to scientific research and scholarship. Due to the publications by Linnaeus of the separate branches of the Royal Collections of king and queen, we can trace the zoological and vertebrate material from the collection of Adolf Fredrik and the invertebrate marine, plant and insect collection of Lovisa Ulrika, whenever the collections are consulted for continued reference.

A survey of scientific research publications over the last fifty years includes papers that reference the natural history collections of Queen Lovisa Ulrika and include such broad subjects of study as: the Echinoids of the Easter Islands (Fell 1974), shells (Way 2007; Kohn 2014), birds (Svanberg 2007, 2016) orchids (Jarvis and Cribb 2009), butterflies (Vane-Wright 2010), herpetological studies which include the branch of zoology that studies amphibians, including frogs, toads, salamanders, newts, and reptiles such as snakes, lizards, turtles, terrapins, tortoises, crocodilians and amphisbaena (Bell 2012), and dragonflies (Hämäläinen and Orr 2017) to name but a few. Publications on the topic of taxonomy also continue to reference to her collections.

In 2016, the Museum of Evolution at Uppsala University listed the collection of natural curiosities once in the possessions of Queen Lovisa Ulrika as especially valuable because of 'its accurate documentation of Linné and the specimen's type status.' Further to this, the important mineral collections that they hold include those of Queen Lovisa Ulrika and King Adolf Fredrik among others (Lindkvist 2016).

Conclusions

The role that Queen Lovisa Ulrika of Sweden adopted as a self-styled *philosophe* through her focus on self-education and French Enlightenment ideals contributed to her interest in natural history and her collecting in both the arts and sciences. Her collection, and her invitation to Swedish scientist Carl Linnaeus to catalogue and publish the Royal Collections, prompted and supported new scientific developments in taxonomy that have continued to bear fruit to this day. Her support of Linnaeus extended beyond her own collecting to her active patronage in terms of making specific requests and financial support to his students travelling outside Sweden and also contributed to his work and publications. Her role as a woman involved in scientific interests matched the work of other women in pursuit of enlightened learning in both scholarly and artistic collecting and attainment of knowledge to support a cultural environment.

Eminent female collectors and scientists were active from at least the seventeenth century onwards. The well-known author of the *Metamorphosis insectorum Surinamensium* [*The Metamorphosis of the Insects of Suriname*], Maria Sibylla Merian was a naturalist, entomologist and artist who specialised in artistic renderings of closely observed exemplars of flora and fauna. In 1699, at the age of 52, she accepted an invitation to join a five-year expedition to Surinam in the north of South America. Before leaving for Surinam in 1699, she sold the contents of her study in Amsterdam, but upon her return she 'arranged her collections in her house, pressed and well-displayed in boxes where they can be seen by all' (Heard 2016). In 1740, the French philosopher, mathematician and physicist Émilie du Châtelet published her textbook on physics, *Institutions de physique* [title translated into English as *The Foundations of Physics*] (Detlefsen 2018). From 1745 until her death in 1749, she worked on a French translation of Isaac Newton's *Principia Mathematica* which was published posthumously by Voltaire, in part in 1756 and in full in 1759. For many years, this translation would remain the main French translation of the English mathematician's work. While these are examples of single, particularly gifted women with an interest in Botanics and Mathematics, in 1785 a group of women formed the *Natuurkundig Genootschap der Dames* [*Ladies' Society for Physical Sciences*] in Middelburg in the Dutch Republic that continued to exist for nearly one hundred years, well into the nineteenth century. This Society met regularly to have lectures and learn about the physical sciences and exercise their pursuit of knowledge in this formal way. The library of the Society included a copy of one of Linnaeus' publications translated into Dutch, showing the breadth of their interests beyond the physical sciences to include natural sciences too (Harbers and Gáldy 2020). Although from the upper levels of society, with the necessary leisure time to champion women's education, this group of women were also determined to increase their own education by means of lectures and self-study.

Due to her superior education, active pursuit of knowledge, and interactions with the Swedish academic world, more than 230 years after her death, Lovisa Ulrika's collections are still used and referenced in scientific publications and remain a testament to her. Nonetheless, Lovisa Ulrika of Sweden, whose name is often associated with a period of political intrigue, still needs to be recognised for the contributions made by an enlightened mind and for her far-reaching global scientific legacy.

Acknowledgements Our thanks go to Kerstin Hagsgård of The Royal Court, Sweden, Isabelle Charmantier of the Linnean Society of London, Maria Asp of The Royal Swedish Academy of Sciences archives, and Jan Over for translations from the French.

Notes

1. There are many different names/spellings for Lovisa Ulrika, depending on her country of residence and her position. Throughout the present essay we shall use the spelling of Lovisa Ulrika that seems to be the most widely accepted form in scholarly literature.
2. For the relationship between Sophie Charlotte and the mathematician & philosopher Leibniz, see Strickland (2011); Brown (n.d.).
3. Dermineur (2017) discusses the wish of her brother, Frederick the Great, to use Lovisa Ulrika's position as the wife of the king of Sweden as an unofficial Prussian lobbyist and diplomat.
4. For a discussion of the early purchases by the King and Queen for the Royal Collections, an account is given in Lovén (1887).
5. In her 1998 Ph.D. thesis Laine (1998a) examines the role of the collections of Lovisa Ulrika at Drottningholm Palace, Stockholm.
6. Laine (1998b) gives detailed descriptions of the collecting rooms. These were visited by Anne Harbers in June 2019 with the kind permission and under the guidance of the Royal Collections, Sweden.
7. Laine (1998b) describes the placement of paintings and access offered to young painters.
8. Laine (1998b) identifies correspondence in which Lovisa Ulrika expresses her opinion on the accessibility of princely collections to a wider audience.
9. See for example the papers of the recent conference 'A Matter of Access' organised by the international forum *Collecting & Display* and held in Munich and London, 22 and 24 June 2019. To be published as Bracken et al. (2022).
10. In the present essay, we understand 'patronage' as the support given to a person by another person in kind or by financial means or through introductions and connections.

11. Rea Radisich (2003) considers the books that are detailed in her letters and paintings that Lovisa Ulrika commissioned from Chardin & Boucher in Paris to reflect her view of herself as a *philosophe*.
12. Letter in French quoted from Rea Radisich (2003, p. 48): *Il est écrit avec tout l'esprit imaginable et tout propre à donner une idée superficielle des sciences, ce qui est tout qu'une femme doit savoir....*
13. Letter in French quoted from Rea Radisich (2003, p. 55): *Je fais batir aussi deux galeries, qui seront ornées de tableaux que j'ai fait venir de Paris. C'est Boucher et Chardin qui en sont les maitres. J'ai donné pour sujet au premier "Les quatre heures du jour" et à l'autre l'éducation sévère et l'éducation douce et insinuant. Ils doivent arriver incessament.*
14. This painting is discussed by Grate (1994). He also relates that it was noted as "en Gouvernante som satter til ratta hufwan pa en liten flicka." [A governess arranging the hood of a little girl] when in the collection of Count Tessin in Paris. The work was to enter Lovisa Ulrika's collection from that of Count Tessin.
15. Nordenstam (2009) has written about Linnaeus' ambitions to describe the known flora and fauna of the eighteenth-century world and the role his students/ apostles play in that ambition.
16. Söderlund (2008) describes the contents and the accessibility and display of the *naturalia* cabinet of the Royal Academy of Sciences at the end of the eighteenth century.
17. Personal visit by Anne Harbers in June 2019.

References

Barton, H. Arnold. 1972. "Gustav III of Sweden and the Enlightenment." *Eighteenth-Century Studies* 6, no.1: 1–34. https://doi.org/10.2307/3031560.

Bell, Christopher. J., ed. 2012. "The Herpetological Legacy of Linnaeus: A Celebration of the Linnaean Tercentenary." *Bibliotheca Herpetologica* 9, no. 1–2: 98–103.

Bencard, Mogens, Jørgen Hein, Bente Gundestrup, and Jan Drees, eds. 1997. *Gottorf im Glanz des Barock. Kunst und Kultur am Schleswiger Hof 1544–1713, Vol. II*. Schleswig: Schleswig-Holsteinisches Landesmusem.

Blanning, Tim. 2015. *Frederick the Great: King of Prussia*. London: Allen Lane Publications.

Bracken, Susan, Andrea M. Gáldy and Adriana Turpin, eds. 2022 *Collecting and Access*. Newcastle: Cambridge Scholars Publishing.

Brown, Gregory. n.d. Dept of Philosophy, University of Houston, Website on G. W. Leibniz (1646–1716). http://www.gwleibniz.com/sophie_charlotte/sophie_charlotte.html. Access 28 November 2020.

Dermineur, Elise M. 2017. *Gender and Politics in Eighteenth-Century Sweden. Queen Louisa Ulrika (1720–1782)*. New York: Routledge.

Detlefsen, Karen. 2018. "Émilie du Châtelet." In *The Stanford Encyclopedia of Philosophy*, ed. Edward N. Zalta. Stanford: Stanford University Press.

Fell, F. Julian. 1974. "The Echinoids of Easter Island (Rapa Nui)." *Pacific Science* 28, no. 2: 147–58.
Findlen, Paula. 2012. "Uffizi Gallery, Florence: The Rebirth of a Museum in the Eighteenth Century." In *The First Modern Museums of Art: The Birth of an Institution in 18th- and Early 19th-Century Europe*, ed. Carole Paul, 73–111. Los Angeles: J. Paul Getty Museum.
Gentry, Anthea. 2008. "Linnaeus' Specimens of Mammals and Birds." In *The Linnaean Legacy: Three Centuries after His Birth, The Linnaean Special Issue No. 8*, ed. Mary J. Morris and Leonie Berwick, 145–52. London: The Linnaean Society of London.
Grate, Pontus. 1994. *French Paintings II: Eighteenth Century*. Stockholm: Swedish National Art Museums.
Hämäläinen, Matti, and Albert G. Orr. 2017. "From Princess Lovisa Ulrika to the Gyalsey, Dragon Prince of Bhutan—Royalty in dragonfly names from 1746 to 2017." *Agrion* 21, no. 2: 61–71.
Harbers, Anne. 2014. "Carl Linnaeus and the Natural History Collections of Lovisa Ulrika of Sweden at Drottningholm Palace." In *Collecting Nature*, ed. Andrea M. Galdy and Sylvia Heudecker, 137–50. Newcastle upon Tyne: Cambridge Scholars Publishing.
Harbers, Anne, and Andrea Gáldy. 2021. "Science, Gender and Collecting: The Dutch Eighteenth-Century Ladies' Society for Physical Sciences of Middelburg." In *Women and the Art and Science of Collecting in Eighteenth Century Europe*, ed. Arlene Leis and Kacie L. Willis, 21–38. New York: Routledge.
Heard, Kate. 2016. *Maria Merian's Butterflies*. London: Royal Collection Trust.
Jansen, Justin. 2015. "The Bird Collection of the Muséum national d'Histoire naturelle, Paris, France: The First Years (1793–1825)." *Journal of the National Museum (Prague), Natural History Series* 184, no. 5: 81–111.
Jarvis, Charlie, and Phillip Cribb. 2009. "Linnaean Sources and Concepts of Orchids." *Annals of Botany* 104, no. 3: 365–76. https://doi.org/10.1093/aob/mcp005. Accessed 14 October 2018
Koerner, Lisbet. 1999. *Linnaeus: Nature and Nation*. Cambridge, MA: Harvard University Press.
Kohn, Alan J. 2014. *Conus of the Southeastern United States and Caribbean*. Princeton and Oxford: Princeton University Press.
Laine, Merit. 1998a. "En Minerva för vår Nord: Lovisa Ulrika som samlare, uppdragsgivare och byggherre: [Queen Lovisa Ulrika, Her Collections and Commissions of Art and Architecture]." PhD diss., Uppsala University, Sweden.
Laine, Merit. 1998b. "An Eighteenth-Century Minerva: Lovisa Ulrika and Her Collections at Drottningholm Palace 1744–1777." *Eighteenth-Century Studies* 31, no. 4: 493–503.
Lindkvist, Maria. 2016. "The Importance of Curation: A Case-Study of the Subfossil Lemur Collection in the Museum of Evolution, Uppsala University." PhD diss., Uppsala University, Sweden.
Linnebach, Andrea. 2016. "Das Publikum der Antiken: Kunsthaus und Museum Fridericianum in Kassel als Ziel von Bildungs- und Forschungsreisen der europäischen Aufklärung." In *Auf dem Weg zum Museum. Sammlung und Präsentation antiker Kunst an Deutschen Fürstenhöfen des 18. Jahrhunderts*, ed. Alexis Joachimides, Charlotte Schreiter, and Rüdiger Splitter, 191–210. Kassel: Kassel University Press.

Lovén, Sven L. 1887. *On the Species of Echinoidea Described by Linnaeus in His Work Museum Ludovicae Ulricae*, in a Communication to the Royal Swedish Academy of Sciences, 11 May 1887. Re-printed on demand by Bibliolife, June 2019.
Mates, Benson. 1986. *The Philosophy of Leibniz: Metaphysics and Language*. New York: Oxford University Press.
Moberg, Roland. 2008. "The Linnaean Collections at Uppsala University." In *The Linnaean Legacy: Three Centuries after his Birth, The Linnean Special Issue No 8*, ed. Mary J. Morris and Leonie Berwick, 141–44. London: The Linnaean Society of London.
Müller, Adelheid. 2016. "Strategie und Leidenschaft: Weibliche Wege zur Antikensammlung." In *Auf dem Weg zum Museum. Sammlung und Präsentation antiker Kunst an Deutschen Fürstenhöfen des 18. Jahrhunderts*, ed. Alexis Joachimides, Charlotte Schreiter, and Rüdiger Splitter, 211–41. Kassel: Kassel University Press.
Nordenstam, Bertil. 2009. "Linnaeus's Global Project – The Exploration of the World's Flora." *Rheedea* 19, no. 1–2: 1–22.
Pulteney, Richard. 1805. *A General View of the Writings of Linneaus*, ed. William George Maton. London: Royal College of Physicians of London.
Rea Radisich, Paula. 2003. "Lovisa Ulrika of Sweden, Chardin and Enlightened Despotism." In *Women, Art and the Politics of Identity in Eighteenth-Century Europe*, ed. Melissa Lee Hyde and Jennifer Dawn Milam, 46–63. Aldershot: Ashgate Publishing.
Reid, Callum. 2021. *Collecting and Display in the Uffizi Gallery: Art in the Age of the Grand Dukes*. Routledge.
Rivière, Marc Serge. 2003. "'Divine Ulrique': Voltaire and Louisa Ulrica, Princess of Prussia and Queen of Sweden (1751–1771)." *Irish Journal of French Studies* 3: 41–62. https://doi.org/10.7173/164913303818644432. Accessed 5 August 2018
Rivière, Marc Serge. 2004. "The Pallas of Stockholm, Louisa Ulrika of Prussia and the Swedish Crown." In *Queenship in Europe 1660–1815: the Role of the Consort*, ed. Clarissa Campbell Orr, 322–43. Cambridge, UK: Cambridge University Press.
Roberts, Michael. 1986. *The Age of Liberty Sweden 1719–1772*. Cambridge, UK: Cambridge University Press.
Söderlund, Inga Elmqvist. 2008. "The Cabinet of Naturalia of the Royal Swedish Academy of Sciences at the End of the 18th Century." *NaMu IV*, http://www.ep.liu.se/ecp/030/015/ecp0830015.pdf. Accessed 13 October 2018
Strickland, Lloyd, ed. 2011. *Leibniz and the Two Sophies: The Philosophical Correspondence*. Toronto, Ont.: Victoria University.
Svanberg, Ingvar. 2007. "Golden Pheasant (Chrysolophus pictus) in Sweden in the 1740s." *Der Zoologische Garten* 77, no. 2: 24–28. https://doi.org/10.1016/j.zoolgart.2007.05.003. Accessed 14 October 2018.
Svanberg, Ingvar. 2016. "Carl Linnaeus as an Aviculturalist." *Avicultural Magazine* 122: 53–58.
Swan, Claudia. 2007. "Collecting Naturalia in the Shadow of Early Modern Dutch Trade." In *Colonial Botany: Science, Commerce, and Politics in the Early Modern World*, ed. Londa Schiebinger and Claudia Swan, 223–36 Philadelphia: University of Pennsylvania Press.
Taylor, Brandon. 2012. "National Gallery, London: For 'All Ranks and degrees of men.'" In *The First Modern Museums of Art: the Birth of an Institution in 18th- and Early 19th-Century Europe*, ed. Carole Paul, 282–303. Los Angeles: J. Paul Getty Museum.

Treasures from Inside the Earth. Exhibition Catalogue (2001–2002). The Swedish Museum of Natural History.

Vane-Wright, Dick. 2010. "Papilio enceladus Linnaeus, 1758." *The Linnean* 26, no. 3: 12–18.

Vasari, Giorgio. 1881. *Le Opere con nuove annotazioni e commenti*, vol. VI, ed. Gaetano Milanesi. Sansoni Editore.

Way, Kathie. 2007. "The Linnaean Shell Collection at Burlington House." In *The Linnaean Special Issue No 7*, ed. Brian Gardiner and Mary J. Morris, 37–46. London: The Linnaean Society of London.

Whitehead, Jane S. 1983. "'The noblest Collection of Curiositys': British Visitors to the Uffizi, 1650–1789." In *Gli Uffizi, quattro secoli di una galleria: atti del convegno internazionale di studi, Firenze 20–24 settembre 1982*, vol. I, ed. Paola Barocchi, Giovanna Ragionieri, 287–307. Florence: L.S. Olschki.

CHAPTER 20

Marianne North and Scientific Illustration

Philip Kerrigan

In her autobiography, memorably entitled *Recollections of a Happy Life*, the Victorian botanist, traveller and painter Marianne North (1830–1890) relates a conversation she overheard between two men who were visiting a private exhibition of her works which she organised in the Kensington Museum in 1877. One of the men, 'Mr. Thompson', had remarked to his companion on arrival: 'We must get out of this civilly somehow. I know what all these amateur things always are!' and then on the way back: 'We must have these things at any price' (North 1993, p. 321). The perspectives and prejudices which underlie this exchange are illustrative of some of the challenges that North faced as a Victorian upper-middle class woman seeking to pursue and secure recognition for her work. They also form the impetus behind this chapter.

This chapter will first situate North and her work within the relevant social, artistic and scientific contexts of her time, particularly in relation to her position as an upper-middle class woman. I shall then demonstrate how North's paintings express ideas about the physiology and behaviour of plants that she came across through the work of Charles Darwin. Finally, I shall consider the extent to which an autobiographical dimension spills over from her writings into her paintings, and in more than just the literal sense that the paintings are records of the plants and places that she encountered.

P. Kerrigan (✉)
University of York, York, UK
e-mail: philip.kerrigan@york.ac.uk

© The Author(s), under exclusive license to Springer Nature Switzerland AG 2022
C. G. Jones et al. (eds.), *The Palgrave Handbook of Women and Science since 1660*, https://doi.org/10.1007/978-3-030-78973-2_20

Marianne North was born in 1830, the middle of three daughters, into a very well-connected and well-to-do family. Her father, Frederick North, was the Liberal MP for Hastings in Sussex and was a keen amateur naturalist and botanist, travelling widely throughout Europe and the Near East with his family during the long Parliamentary recesses and periods when he lost his seat in Parliament (North 1993, pp. 1–38). Through his interests Frederick North knew many important scientific figures, including the President of the Royal Society, Sir Davie Gilbert, the writer and pioneer of eugenics, Francis Galton, and the first director of Kew Botanic Gardens, Sir William Hooker. These were connections that his daughter inherited. In July 1871, one and a half years after the death of her father, North embarked on her first independent voyage to the United States and Canada and from there straight on to her first tropical destination, Jamaica. It is unlikely that, at this point, North had a clear plan to initiate a painterly survey of as much of the world's flora as possible. During these initial solo trips, she was fairly cosmopolitan in her choice of subject matter. In her account of her expedition to the US and Canada there is comparatively little about the plants, in comparison to what would come later, but plenty about the people she met. The majority of paintings she produced were landscape views, especially of the Niagara Falls (a web-link to view one of these paintings is given in Table 20.1 at the end of the chapter under Fig. 1), and there are relatively few close-up studies of the indigenous flora, although she painted an example of what was to become one of her favourite subjects, a tree apparently growing out of bare rock and in the most harsh and exposed position—an old red cedar near West Manchester, Massachusetts (Table 20.1, Fig. 2) (North 1993, p. 48).[1]

It seems to have been her first experience of the Tropics during her stay in Jamaica, which directly followed on from her North American tour, which precipitated her enduring interest in plants. She had been prompted to visit that island by Charles Kingsley's description of the West Indies in his book *At Last* of 1870 (North 1993, p. 39).[2] Her high expectations were not disappointed—she wrote in a letter to her great friend and fellow adventurer, Amelia Edwards,[3] 'the first sight of the Tropics nearly drove me wild', and professed that 'from this time [i.e. her stay in Jamaica] I began making the collection of oil paintings now at Kew'.[4] Her next stop was Brazil, a Mecca for many famous contemporary naturalists including Alfred Russel Wallace (the co-founder of natural selection with Charles Darwin), the celebrated American comparative anatomist Charles Agassiz, whom she had met on her trip to the United States, and the butterfly enthusiast and proponent of Darwin, Henry Bates (Bates 1863; Russell Wallace 1853; Agassiz 1868). It was here that North really started to hunt out the characteristic indigenous plant species which are the organising principle behind the collection at Kew. Her travels continued, on and off, until the beginning of 1885 during which time she visited every continent apart from Antarctica, though she focussed on those areas lying within the tropics. This fourteen-year odyssey lives on both in the paintings she produced and in a two-volume autobiography she wrote relating all of its details, botanical and otherwise. These volumes were edited by North's sister, Catherine Addington Symonds and were published in London in 1892, two years after her death. They constitute a wonderfully rich source, alongside the many archived letters she left, for interpreting the images.

Without the creation of the Marianne North Gallery in Kew Royal Botanic Gardens, it is possible however that her prodigious artistic output might never have achieved the recognition it did. The Gallery was designed and built at her own expense by her architect friend James Fergusson. According to a letter she wrote to Joseph Hooker in August 1879, the idea of finding a permanent home for her work at Kew did not occur for nearly a decade after her first solo expedition:

> It came into my head the other day after reading that bit in the *Pall Mall Gazette* about my flower paintings, that I should like very much to place them near their live neighbours and if a piece of spare ground could be found in or close to the pleasure grounds at Kew I would build a suitable gallery for them [...] & it would be a great happiness to know my life has not been spent in vain, & that I can leave something behind which will add to the pleasure of others & not discredit my father's old name.[5]

North had already organised two exhibitions of her work in London, including the one at the Kensington Museum in 1877 and a second two years later at a room she hired in Conduit Street. It is reasonable to assume that had it not been for the favourable reception of her paintings at her two previous exhibitions, it is unlikely that she would have ever made such a bold proposal.

Although they have always been popular with visitors to Kew, North's paintings have been left out of many textbooks on the art of the period, including those focussing on nineteenth-century landscape and representations of nature. It is only in volumes dedicated to botanical illustration that North's work is discussed and, even then, she is afforded little space. She is mentioned only in passing in Martyn Rix's *The Art of The Botanist* and fares little better in Wilfrid Blunt's *The Art of Botanical Illustration* where her work is described in rather negative terms. From the perspective of an art historian like Wilfrid Blunt, who is primarily interested in constructing a neat linear narrative around the development of compositional practice and style in botanical drawing and painting, her work is dismissed as a curious but insignificant anomaly.

Blunt's comment that 'Botanists consider her primarily as a painter; but artists will hardly agree' (Rix 1981, p. 218; Blunt 1951, p. 237), while unfairly disparaging does highlight the complex hybrid status of her output between botanical illustration, landscape and still-life painting. Such a style of work is not, however, without precedent. Robert Thornton's *The Temple of Flora*, a collection of 'picturesque botanical plates' intended to accompany his *A New Illustration of the Sexual System of Carolus von Linnaeus*—an important, though financially disastrous, production from the turn of the nineteenth century—equally attempted to represent many of the plants in their natural surroundings (King 1981; Bush 1974)[6] (Table 20.1, Fig. 3). Although the backgrounds are not generally as faithfully realised or as detailed as in North's work, the artists working on them had not had the advantage of personally having visited the localities where the plants grew but instead had to rely on other accounts and images. Closer in time to North, Joseph Hooker, the celebrated botanist and soon-to-be Director of the Royal Botanic Gardens Kew, had commissioned Walter Fitch to work up a partly coloured field sketch he had made of the *Rhododendron dalhousiae* growing epiphytically on a tree as a frontispiece for his *The Rhodendrons of the Sikkim Himalaya* of 1849–51 (Table 20.1, Fig. 4).

Equally, a number of artists before North shared her interest in representing the interactions between one plant and another and between plants and the insects and bird life which they supported. One of the most famous of these was another female painter of the late seventeenth and early eighteenth century, Maria Sibylla Merian, who was an entomologist and the artist-author of the celebrated *Metamorphosibus de Insectorum Surinamensium* of 1714 (Table 20.1, Fig. 5). A plate from the latter showing beetles and beetle larvae on a prickly poppy has a very clear echo in an image by North of 'A Chilean Stinging Nettle and Male and Female Beetles' from the 1880s (Table 20.1, Fig. 6).

The majority of the scholarly literature on North has tended to focus more on her written output than on her paintings, although Suzanne Le-May Sheffield and Antonia Losano offer some interesting close-readings of the images (Le-May Sheffield 2001; Losano 1997). The majority of the analyses of North's work have been from a largely post-colonial angle, including Losano's article (Guelke and Morin 2001; Morgan 1993, 1996).[7] These accounts consider how North, in her capacity as an upper-class woman, contributed to British imperial authority through the intellectual mapping of one aspect of an area's natural production and by identifying species of horticultural—and perhaps even agricultural and manufacturing interest—for the home market. Guelke and Morin argue that North's complicity with such an agenda is evident from the tone and content of her autobiography which frequently emphasises Britain's racial and cultural superiority and thus 'places her firmly within European traditions of imperialism' and 'ethnocentrism' (Guelke and Morin 2001, p. 321). Guelke and Morin further argue, as does Susan Morgan, that North's project was governed by an external and semi-official mandate, that of the Royal Botanic Gardens at Kew (Guelke and Morin 2001, p. 314). This ignores, however, that it was North who approached Sir Joseph about the creation of a gallery rather than the other way round as Hooker himself made clear.[8]

Le-May Sheffield has been one of the first to identify a more positive dimension to North's work in terms of its scientific role and significance. She argues that North's departure from conventional botanical illustration in figuring her plants into their natural environment reflects North's interest in the 'interconnectedness of the natural world' (Le-May Sheffield 2001, p. 119)—she wished to 'comment on the relationship between human beings and the natural world' and 'to plead for the preservation of the fast-disappearing wilderness the world over' (Le-May Sheffield 2001, p. 117). The review in the *Times* on the occasion of the opening of the Gallery in 1882 certainly perceived such significance to her work: 'Such scenes can never be renewed by nature, nor when once effaced can they be pictured to the mind's eye, except by means of such records as this lady has presented to us and posterity' ("Miss North's Paintings" 1882, p. 4). It is undoubtedly true that North, wherever she saw it occurring, deeply regretted the destruction of the natural environment, for instance, she wrote of the cutting down of the California redwood trees, that 'it broke one's heart to think of man, the civiliser, wasting treasures in a few years to which savages and animals had done no harm for centuries' (North 1993, p. 212).

LIFE AND CONTEXT

At a time when women were expected to devote themselves to their husbands and children or, failing that, to engage in some nurturing role or philanthropy, North's decision to go off exploring on her own was a relatively unusual one that even risked accusations of selfishness. This was in spite of her having spent almost twenty years of her life fulfilling such feminine duties of care as a companion and housekeeper, and latterly nurse, for her father.

Although North's practice of travelling alone might have been frowned upon by some, she was generally treated with respect and indeed welcomed and granted privileges by officials, professionals and other significant people she met and stayed with while travelling. Alison Blunt makes the point that upper-middle class women like North were granted a 'temporary licence to behave in ways constructed as masculine while travelling' (Blunt 1994, p. 58) as racial and class hierarchies tended to take precedence over sex and gender (Blunt 1994, p. 60). The name, social position and connections North had inherited from her father meant more often than not that she could rely on the assistance of local dignitaries and government officials wherever she went.[9] Her association with Kew gardens through her friendship with Sir Joseph and Lady Hooker, as Susan Morgan observes, gave her privileged access to the facilities and expertise of Kew's many impressive garden outposts abroad (Morgan 1996, p. 122). North further enjoyed free railway passes in Australia, South Africa and Chile courtesy of their governments (North 1892, p. 325).

Justifying the pursuit of an independent life, however, was not the only or the least of the struggles North faced as a woman. It was a real challenge at that time for a woman to achieve respect or prominence in either the scientific or the artistic fields because of the negative and restrictive attitudes and beliefs concerning the abilities and role of women which limited the possibilities for formal training. Barbara Gates writes that 'science for women was viewed mainly as a female "accomplishment" rather than as a serious contribution to the field of natural history' (Gates 1998, p. 67). Gates also relates the difficulties of women entering professional organisations (Gates 1998, p. 66).

Similarly Deborah Cherry writes that 'Women artists were located in asymmetrical and unequal relations to art education, art administration and professional status' (Cherry 1993, p. 53; Casteras and Peterson 1994; Yeldham 1984; Prettejohn 2000, ch. 2).

Although North inherited close contacts with men in the world of science from her father, including William Hooker and his son and successor in the post of Director of Kew, Joseph Hooker, she was almost entirely self-taught in botany. She was also largely self-taught as an artist, although she describes herself as having had some lessons in watercolour from a Dutch flower-painter, one Miss van Fowinkel and later on in oils from the Australian artist Dowling, who spent the Christmas and New Year of 1867–1868 staying with her family (North 1993, p. 26).

David Noble observes that the world of science had always traditionally been a masculine environment, this being in some ways a legacy of its strong links to monastic communities in the past; but even when in the second half of the nineteenth-century the field started to become increasingly secular and professionalised, it remained still predominantly the preserve of men (Noble 1992, p. 276). The first professor of botany at the newly formed University of London, John Lindley, was especially hostile to women's involvement in the professional sphere. He regarded Botany's traditional association with women as having undermined its scientific reputation, and fought hard to raise the discipline's status through the professional work of men (Shteir 1997, pp. 32, 33). Women were also excluded from membership of most professional, and certainly the more prestigious national, societies (Guelke and Morin 2001, p. 312).

These circumstances reflected broader attitudes at the time regarding the 'proper' role of women. These views are summed up in a verse that the magazine *Punch* published in the context of an ongoing argument in the Royal Geographic Society about the admission of women fellows:

> A lady an explorer? A traveller in skirts?
> The notion's just a trifle too seraphic:
> Let them stay and mind the babies, or hem our ragged shirts;
> But they mustn't, can't, and shan't be geographic
> (quoted in Blunt 1978, p. 181)

In the amateur domain, the story was rather different and, especially in the field of botany, there were opportunities for women providing, like North, they had sufficient enthusiasm and wealth (Guelke and Morin 2001, p. 311).[10] At the time North was working there were several prominent and well-respected women botanists and naturalists, including Lydia Becker (1827–1890), the author of *Botany for Novices* of 1864 (Shteir 1996, pp. 227–31).[11]

Popularising science and making natural history accessible was one of the chief areas of competence claimed for and by women and this certainly counted amongst North's aspirations: 'I found people in general woefully ignorant of natural history, nine out of ten of the people to whom I showed my drawings thinking that cocoa was made from the coconut' (North 1993, p. 321). As well as making the natural sciences accessible to a wider audience, the accurate representation of plants was regarded as one of the special accomplishments of a lady, as testified to in works including George Brookshaw's *New Treatise on Flower Painting; or, Every Lady Her Own Drawing Master* of 1816. Watercolour however was the stipulated medium, working in oils was not considered suitable—*The Lady's Magazine* for instance dismissed the practice as involving a 'sort of squalid discomfort in its pursuit that makes it exceedingly repugnant to elegant women' ("The Society of British Artists" 1834, p. 298).[12] The term 'accomplishment' was also significant for the practice of botanical drawing and painting was aimed at female self-improvement and certainly was not intended as training to compete with men in the professional sphere (Gates 1998, p. 67).

It seems that North was keen that her work would be recognised as having a value far above the amateur. For instance, she expressed the hope that her gallery at Kew, would, being positioned 'far off from the usual entrance gates' be reached only 'by those who cared sufficiently for plants' and escape the notice of those dilettantes who 'merely cared for promenading' (North 1892, pp. 86, 87). She certainly seems to have taken very seriously the scientific authority and reputation of her work. She commissioned her botanist friend William Botting Hemsley to complete her own observations on the plants figured into an accompanying catalogue (North 1892, p. 211), and when a notice was published in *The Garden*, following her return from South Africa, to the effect that a 'supplement' would soon be added to the catalogue to describe the newly added paintings of flora from that continent, she wrote to Hemsley stating that she 'quite disapprove[d] of that – the whole must be reprinted and renumbered so as to be complete in itself [...] I do not wish my catalogue to be a patched-up affair but complete'.[13]

North was equally concerned that the accuracy of her representations should be recognised. A review in the *Times* related how North painted 'on the spot and literally 'with her eye on the subject [...] steadily on for hours and hours until the sketch was complete, lest any change or accident should interfere with its perfect accuracy' ("Miss North's Paintings" 1882, p. 4). In a letter to Hooker, North relates her anxiety about achieving an accurate representation of the Capucin tree: 'Mr Estridge [...] sent me the dried flowers; he is no botanist and quick to jump to conclusions, so that I have never felt sure of the correctness of my painting'.[14]

As well as having a concern for the accuracy of the individual species, North was keen that the overall collection of her paintings would be a representative, if not comprehensive, record of the world's flora. She wished if possible to include representatives of all the known orders of plants, writing 'the omissions amongst the orders are most interesting to me and a few could easily be added here and there in groups and make my list of orders look much grander'.[15] She also stated in another of her letters that 'till I have been to the fourth quarter of the globe the set is not complete', and in the *Official Guide* to the Gallery of 1892 the editor relates how 'At the suggestion of the late Mr. Darwin, and in order to render the collection more nearly representative of the Flora of the world, Miss North next proceeded to Australia, Tasmania and New Zealand [...] very fully illustrating the most striking features of the marvellous Australasian flora' (Hemsley 1892, p. vi).

North aspired however not just for scientific but also artistic recognition, writing to her friend, the botanical scholar Dr. Coke Burnell, in January 1880, that she had 'just finished a large picture of Kinchinjinga to try its fate at the RA and hope[d] to paint a companion picture of a south country swamp to match it before April',[16] and she announced proudly in her autobiography how Frederic Church, then one of the most celebrated living American landscape painters, 'looked through all my paintings with real interest' (North 1892, p. 208).[17]

North's departure from one of the usual key conventions of botanical illustration—isolating specimens against the white background of the paper—allowed her to incorporate greater formal interest and variety into her work, by for instance incorporating a landscape element or by grouping the plants together to create a still-life. Traditionally botanical illustration was meant to prioritise scientific clarity over and above aesthetic appeal, and inventive composition and expressive emphasis were potentially seen as standing in the way of this clarity. These priorities were articulated as early as 1542 by Fuchs in the Preface to his *De Historia Stirpium*: 'we have purposely [...] avoided the obliteration of the natural form of the plants by shadows and other less necessary things by which the delineators sometimes try to win artistic glory: and we have not allowed the craftsmen to so indulge their whims as to cause the drawing not to correspond accurately to the truth' (Saunders 1995, pp. 20, 21). Several centuries later Joseph Banks, in his Preface to Franz Bauer's *Delineations of Exotick Plants* (1796–1803) wrote: 'each figure is intended to answer itself every question a Botanist can wish to ask, representing the structure of the plant it represents' (quoted in Saunders 1995, p. 8).

North rejected line and watercolour which was the preferred choice of botanical illustrators in favour of using oils and a looser, more painterly technique (De Bray 1989, p. 40). By choosing to work in oils North avoided directly competing with (mostly male) contemporary botanical artists and was at the same time able to distinguish her work from that of female amateur and accomplishment artists for whom watercolour was the accepted medium (Casteras and Peterson 1994, p. 11; Prettejohn 2000, p. 76).

In a number of ways however, North's images continue to respect the traditional requirements of botanical illustration. Her manner of painting directly onto the white of the canvas, of using bright and highly saturated colours and avoiding much modelling in light and shade, echo Fuch's strictures. Such an approach was also common to the work of North's contemporaries, the members of the Pre-Raphaelite Movement who frequently included detailed depictions of plants in their paintings. This group of artists were committed to scientific truthfulness in their work and importantly saw this goal as complementary with, rather than as at odds with, artistic value (Stanford 1973, p. 73). It is likely that North shared this opinion. She was a good friend of the landscape artist and pre-Raphaelite associate, John Brett, and her autobiography and letters clearly demonstrate her sympathy for this new way of making art. North's still-life arrangements also carry strong echoes of a long tradition of flower painting originating in the work of seventeenth-century Dutch artists, including another woman, Rachel Ruysch, and continuing into that of contemporaries like the Frenchman Henri Fantin-Latour. By evoking the work of these artists and artistic traditions, North was no doubt hoping to enhance the appeal and status of her own work.

RECEPTION OF PAINTINGS

It is interesting in the light of North's artistic and scientific aspirations, and the prevailing cultural positions at the time, to consider the reception that her paintings received from the press. It might be expected that this would be coloured, at least to some extent, by negative attitudes towards women as 'serious' artists and scientists; however, both in the popular press and more specialised scientific and artistic journals, there is little if any evidence of such prejudice and the reviews are, almost without exception, overwhelmingly positive and complimentary. The *Daily News* declared, for instance, following the opening of the Gallery, that the collection therein 'will doubtless add greatly to the attraction of the beautiful gardens, in which it finds so appropriate a place' and 'a record of research into natural history such as has never before been made by a single hand, and the like of which we believe exists nowhere else'.[18]

There were, however, differences of opinion as to the respective artistic and scientific value of the works with only the *Daily News* claiming that the paintings succeed equally on both accounts.[19] The review in *The Magazine of Art* declared North to be 'a pertinacious, and – what is better, perhaps – an intelligent botanist', and one who commands 'an able and accurate pencil' (Barnett 1881–1882, p. 430). Although the reviewer conceded that 'several individual pictures have good artistic qualities', otherwise he concluded that 'throughout there has been little, if any, attempt at artistic grouping, but slight effort to achieve effect either of colour or line' and that their 'chief interest is essentially scientific' (Barnett 1881–1882, p. 431). By contrast, the *Times* review came down in favour of the artistic over the scientific merits of North's work, declaring North an 'accomplished artist' and remarking on her 'wonderful power of selection and no less wonderful skill in draughtsmanship', but making clear that she was not a botanist 'in the strict sense of the term' ("Miss North's Paintings" 1882, p. 4).

North's paintings received a complimentary review in the prestigious scientific journal *Nature*, albeit written by North's friend and author of the catalogue, W Botting Hemsley: 'the paintings [...] are so thoroughly naturalistic, that a botanist has little difficulty in determining such as are not known to him by sight' (Hemsley 1882, p. 155). Such statements were no hollow rhetoric: several botanists were moved to retrace North's footsteps in search of plants they had seen depicted in her paintings that were unfamiliar to them. Many of these were indeed confirmed as previously unknown species and five of them were named in honour of North (North 1892, p. 337).

Charles Darwin

North's paintings express ideas about the physiology and tendencies of plants that were prompted by the work and theories of her contemporaries, in particular Charles Darwin. This argument is suggested partly through analogy to North's published writings in which Darwin's name is cited in a number of places in the most complimentary terms. It is also evidenced by the clear familiarity that North shows with his ideas—or at least ideas brought again to prominence by his writings.

North visited Darwin in 1880, shortly before her second trip to Borneo, and described him as 'the greatest man living, the most truthful, as well as the most unselfish and modest' (North 1892, p. 241). North was very likely introduced to Darwin through her father's friendship with Francis Galton, Darwin's cousin, with whom they used to holiday in the Lake District (North 1993, p. 32). Also North was, even before the construction of the gallery, a close friend and correspondent of Joseph Dalton Hooker, who was himself a very close friend and champion of Darwin.

As well as being the author of a whole new theory about the development of life, Darwin wrote many influential botanical volumes, including one on insectivorous plants, another on movement in plants and yet another on modes of fertilisation in orchids (Darwin 1862, 1875; Darwin and Darwin 1880). All of these were concerned with the particular strategies plants adopted in the 'struggle for life'—a favourite expression of Darwin's and one of the key components of his theory of natural selection (Darwin 1859).[20] Although North never refers directly to natural selection in her autobiography or letters, she was certainly aware of how a species' adaptation to its environment was necessary to its survival and cited Darwin in this respect. For instance, she wrote to her friend the botanical scholar, Dr. Coke Burnell, of a plant she found in Northern India that 'as the seeds ripen the leaves grow up and hide them from the eyes of the birds [...] [which is] a fact for Darwin!'.[21] She was also greatly engaged by the phenomenon that alongside adaptation drove forward natural selection—that is competition and its inevitable outcome, conflict.

North's decision to paint plants in their natural environment, rather than isolate them against the white of the pages as per usual botanical practice, allowed her to introduce more formal and compositional variety and expression and hence raise the artistic value of her work. Perhaps an even more compelling reason for her choice is that plant classification for its own sake was of less interest to her than describing plants in relation to their natural environment—or what Darwin repeatedly called their 'conditions of life'. Reflecting this, North sought to convey the type of environment in which a plant grew, the odds against which it had to contend, including the competition and predation of other plants, and the many different strategies plants adopted to further their survival in sometimes extreme situations. This is not just evident in her paintings but also in the narrative contained in her autobiography; here she never spends long speculating as to which genus or family a plant might belong to, or whether it might be a previously un-described species, but often relates at length a particular plant's habit of growth or the nature of the environment in which it is found.

Another way in which North's painting practice differs from conventional botanical illustration is in representing plants exactly as she found them; she excludes any idealisation or generalisation, in contrast to the demand of botanical illustrators to provide a 'typical' or 'exemplary' specimen of a particular species of plant. The reviewer in *The Magazine of Art* recognised this quality in North's work, declaring that each of the pictures had been treated with 'unusual conscientiousness, and with a self-denying faithfulness to nature that is rare indeed' (Barnett 1881–1882, p. 431). In her image of an angel's trumpet and hummingbirds, for instance (Table 20.1, Fig. 7), she included the ragged holes in the plant's leaves, a practice which would have been inconceivable in a conventional botanical drawing of the period. Le-May Sheffield at one level accounts for this practice in terms of North's working methods—painting in front of the live object out in the field rather than copying an isolated and specially selected and prepared specimen in the comfort of the studio (Le-May Sheffield 2001, pp. 115, 116), and also less charitably 'because she did not have the skill or the interest in generalising from the specific' (Le-May Sheffield 2001, p. 113). However, it seems likely that this practice was as much purposeful as it was perhaps easier. Taxonomical clarity was of less importance to North than showing each plant, in accordance with Darwin's outlook, as an individual struggling to survive in its own particular environment.

One type of adaptation North frequently figured was the aerial roots and supports which were one of the features of the tropical forest which especially captured the imagination of contemporary explorers. North produced about a dozen images of such trees including the Indian rubber or banyan tree with their 'long tangled roots creeping over the outside of the ground, and huge supports growing down into it from their heavy branches' (North

1993, p. 309). Although perhaps less obviously demonstrating a specific adaptive strategy, North also painted no less than eight images of trees apparently growing out of bare rock, many of them contorted into fantastic shapes by the wind like the tree growing out of bare rock high in the American Rockies mentioned earlier. Another example is an old red cedar near West Manchester, Massachusetts (Table 20.1, Fig. 2) which she described as 'perfectly shaved at the top [...] by the sea winds, with its branches matted and twisted in the most fantastical way underneath, and clinging to the very edge of the precipice, its roots being tightly wedged into a crack without any apparent earth to nourish it' (North 1993, p. 48).

These adaptations are of a fairly innocuous character; however North also portrayed many plants that had evolved mechanisms and habits which not only furthered their chances of survival in a particular environment but did so at the direct expense of other plants. The idea of conflict in the plant kingdom is a key theme in *The Origin* and, indeed, is the focus of a whole chapter entitled 'The Struggle for Existence' as well as of some of the most celebrated quotations, including the famous tangled bank analogy at the very end. Having said this, the idea of conflict in nature per se was not a particularly new one. It had been identified by Darwin's grandfather Erasmus Darwin in his *The Temple of Nature* of 1803 and had been rearticulated in Charles Lyell's *Principles of Geology* of 1830–1833. Awareness of conflict in the plant kingdom was also emphasised in the accounts of tropical explorations offered by mid-nineteenth-century naturalists including Henry Walter Bates, Alfred Russel Wallace and Charles Kingsley. Bates observed that though 'The competition between organised beings exists everywhere, in every zone, in both the animal and vegetable kingdoms. It is doubtless most severe, on the whole in tropical countries'. Moreover, in 'vegetable forms', this conflict is 'more conspicuously exhibited' (quoted in Taylor 1884, pp. 230, 231).

North's first independent trip abroad to the States and Canada in 1871, coincided with the publication of Darwin's second major work *The Descent of Man*, as well as Russel Wallace's *Contributions to the Theory of Natural Selection* and St George Mivart's *Genesis of Species*. The issue of evolution and the fierce debates surrounding it were therefore very much at the forefront of not only natural history but intellectual discourse as a whole.

Although as stated, awareness of conflict in nature preceded Darwin, what was different and new with Darwin's theory was that the whole principle of the organisation of nature was now founded on conflict; indeed, it was this that upset many of Darwin's contemporaries. A natural system based on conflict is not easy to reconcile with the idea of a benevolent, all-wise Deity. Equally the enjoyment of nature had traditionally been founded on a natural theological belief in a harmonious, benign and providentially-designed natural world which spoke reassuringly of God's concern for man and through which one could come closer to knowing God.

The chief perpetrators of vegetable conflict that North chose to represent were the strangler figs and parasites, plants whose habits had caused the botanical author John Ellor Taylor to group them in a chapter evocatively entitled 'Robbery and Murder' in his popular *The Sagacity and Morality of Plants* of 1884. The stems of these figs slowly envelop the tree which supports them in a rigid and suffocating lattice, eventually quite literally strangling the tree beneath which slowly rots away. North herself described the strangler figs as 'murderers', writing 'It seemed difficult to believe that those delicate velvet leaves and crimson stalks which ornament the tree so kindly at first, should start with the express intention of murdering it and taking its place!' (North 1993, p. 246). The degree of anthropomorphism here is striking. Their behaviour, however, according to Darwin's theory, was not wilful malevolence but rather a particularly effective, if unpleasant, survival strategy whereby the fig exploits the support provided by the tree to allow it to reach much more quickly the sustaining light and air above. North can however perhaps be forgiven for describing their behaviour in such anthropomorphic terms when Darwin equally often personifies nature as something sentient and active in his writings, despite his underlying contention that natural processes are utterly unconscious and impersonal.

Interestingly, in her images of the fig, North seems reluctant to figure the plant in its full malignancy. For instance, she painted a specimen from Borneo slowly enveloping a poison tree and thus at least taking on a worthy foe (Table 20.1, Fig. 8). In another of her several representations of the plant (Table 20.1, Fig. 9), it appears to pose little threat to the very healthy looking palm it surrounds and, moreover, provides a convenient support for the nests of the sociable oriole which hang from it like giant pendant earrings. The sunny, cloudless atmosphere and fresh greens of the foliage further detract from any potential sense of menace. It is tempting to speculate that North felt some lingering, and maybe even unconscious, attachment to an earlier natural theological vision of an essentially benign nature. This was despite her occasional unfavourable comments about evangelical Christian groups she came across on her travels and favourable mention of John Tyndall's controversial address to the Belfast Association in 1874, which she attended, on the relations between science and theology (North 1993, p. 191). Alternatively this disinclination to highlight the sinister nature of the figs' activity may reflect what the historian Gillian Beer describes in relation to Darwin's own reflections on nature as the 'sheer imaginative difficulty of bearing constantly in mind the struggle for life' (Beer 2000, p. 65). One of the most famous passages from Darwin's *Origin* describes how: 'We behold the face of nature bright with gladness, we often see superabundance of food; we do not see, or we forget that the birds which are idly singing round us mostly live on insects or seeds, and thus are constantly destroying life' (Darwin 1996, pp. 52, 53).

North's studies of twisting aerial roots, contorted trees and encircling strangler figs all relate to Darwin in a further way by demonstrating the effects of circumnutation; this is the principle of orbital movement in the leaves, shoots and root-tips of plants which he set out in his *On the Power of Movement in Plants* of 1881. This movement is fundamental to the ability of the plant to search for light and shade, to overcome obstacles in its course and to regulate the exposure of its leaves to cold at night and to solar radiation during the day. Darwin observed that the same principle of movement, when combined with a geotrophic instinct, allows the root radical to penetrate down through the earth, while avoiding or circumnavigating large obstacles such as stones. North describes how this process allows the fig seedling, which germinates from 'a seed dropped in the branches [of a tree] by birds or wind' to extend its roots downwards 'by the most efficient path, to eventually meet the soil' (North 1993, p. 246). She observed a similar process occurring in the mangrove swamps, but here the seed had not been deposited by any bird or mammal but germinated while still attached to the inflorescence (Table 20.1, Fig. 10).

North was also very much intrigued by the habits of parasitic plants, though she did not condemn these plants in her writings in the same way as she did the strangler figs. Perhaps this was because the damage or disadvantage they brought upon their hosts was rather less obvious. Taylor in his *The Sagacity and Morality of Plants* noted of parasitic plants that 'Their flower-stems [...] rise above the ground, looking as if they had been developed in a fair and honest manner' (Taylor 1884, p. 249). North noted how one species from the Chilean Andes grew on one side only of a cactus, which flowered on the opposite side, thus indicating a degree of accommodation as well as competition between the two species (Table 20.1, Fig. 11). Another parasite from the same area she described as being the 'pet' of the Southern beech on which it grew. North may further have overlooked the mercenary character of many parasitic plants because she found them so beautiful, for instance, she described the California snow-flower as 'a gorgeous parasite of the purest crimson and white tints, which grows at the roots of the sequoia, about 5000–6000 feet above the sea' (Table 20.1, Fig. 12).

Many of the remarkable and beautiful plants North came across were not just hostile to each other but had adapted ways of deterring predators. For instance, she describes the South African amaryllid Boophone toxicari which 'the Kafirs poison their arrows with' (North 1892, p. 249). The formidable stinging hairs of the Chilean loasa plant she painted are alluded to by the literally devilish-looking black beetle set alongside (Table 20.1, Fig. 6).

As well as defending themselves against insect and animal predators through poison and painful stings, other plants had evolved, as J. E. Taylor put it, so as "to turn the tables' on their ancient and hereditary foes the insects' (Taylor 1884, p. 257). Insectivorous plants had evolved the habit of feeding on insects in order to allow them to survive in very low nutrient situations such as bogs. The notion of an insectivorous or flesh-eating plant was, however, particularly monstrous and unsettling to those who had held up the plant kingdom as evidence of providential design. Particularly repugnant was the way in which plants lured insects to their death through a sticky, slippery trail of nectar. The popular naturalist M. C. Cooke in his *Freaks and Marvels of Plant Life* of 1882 quoted a poem of 1875 written in response to the publishing of Darwin's *On Insectivorous Plants* which stated:

> Surely the fare
> Of flowers is air,
> Or sunshine sweet;
> They shouldn't eat,
> Or do aught so degrading. (Cooke 1882, p. 148)

Paintings of carnivorous plants (one species of which, the *Nepenthes northiana*, was named after her) figure throughout North's entire painting career, with representatives from each continent. Although they had long attracted the interests of naturalists, it is tempting to think that North's focus on them might in part have been inspired by Darwin's writings. In contrast to the images of the parasitic plants, there is a definite element of menace in many of her representations of plants of this class. They are often arranged in crowded compositions, as in the collection of North American species (Table 20.1, Fig. 13) and another painting where the pitcher plants form a menacing circle around a white orchid (Table 20.1, Fig. 14). Their literal struggle for canvas space evokes a more literal conflict and competition for survival between them. Their proximity to the picture plane and the close-up viewpoint and scale adds to their naturally rather anthropomorphic appearance which in turn enhances the sense of the sinister and of impending conflict. This anthropomorphising also echoes the almost animal-like sentience Darwin attributes to plants in many of his writings including his *On Carnivorous Plants* of 1875.

Autobiography

The influence of contemporary scientific thinkers, especially Charles Darwin, on North and her work, plus her acute awareness of her gender and social and scientific status, can both be identified in her artistic output. This suggestion requires a shift from a relatively literal to a more metaphorical and symbolic level of interpretation: no resort to analogy is required to claim that North's images of strangler figs swallowing up hapless host trees were inspired by the writings of Charles Darwin, whereas to argue that such representations evidence anxiety on North's part about her fate as an artist and scientist in these two male-dominated worlds is to introduce an additional set of assumptions. Did North consciously or unconsciously weave symbolism into her art? Did she so love plants that she started to identify with them and, if so, was there a wider cultural tendency to draw parallels between the experience of plants and of people?

Ascribing symbolic meanings to North's paintings is not without precedent: Antonia Losano gives a gendered twist to her interpretation of North's painting of a group of Bornean pitcher plants surrounding a white orchid, describing them in somewhat anachronistic terms, as 'a quintet of phallic saxophonists jammin' round an innocent white orchid' (Losano 1997, p. 446) (Table 20.1, Fig. 14). A slightly different interpretation to the psycho-sexual one of Losana is that the 'phallic saxophonists' could equally be seen as the scientific and/or artistic establishment, crowding round North or rather her creative enterprise. This image dates from either her first visit to Borneo in 1876 or her second in 1880, and it was when she returned to England in 1877 that North started to make enquiries about the possibility of showing off her works for the first time (North 1993, p. 321).

The prominence of gender in mid-late-nineteenth-century discourse provides circumstantial evidence in favour of such a theme permeating through into North's painting. Rachel Malane in her *Sex in Mind: The Gendered Brain in Nineteenth-Century Literature and Mental Sciences* argues that 'The nineteenth century experienced a widespread, almost obsessive inquiry into gender difference, with special attention to discerning the mental capabilities of the sexes' (Malane 2005, p. 22). Although such attitudes predated the Victorian period, contemporaneous theories about the respective mental capabilities of women and men gave them renewed emphasis. Studies by a succession of phrenologists and craniologists made the case for the inferior morphological properties of a woman's cranium and brain and argued that this resulted in less developed mental powers. Although, as noted by Malane, the scientific authority of these theories was increasingly challenged as the century wore on, the ideas persisted 'in more popularised literature' and also 'in certain scientific contexts' (Malane 2005, p. 11).

Darwin in his *Descent of Man and Selection in Relation to Sex* of 1871, while not specifically referencing the work of the craniologists (in)famously declared that 'the chief distinction in the intellectual power of the two sexes is shown by man's attaining a higher eminence, in whatever he takes up [...] requiring deep though, reason, or imagination' (Darwin 1871, p. 613). Darwin and his supporters explained this difference in part through his theory of evolution through natural selection within which women, were somewhat shielded from selection pressures whereas men were exposed to their full force (Malane 2005, p. 13). In Darwin's view this situation was likely to be ongoing and to 'keep up or even increase their mental powers, and, as a consequence, the present inequality between the sexes' (Darwin 1871, p. 560). Both Darwin in his *Descent of Man* and Francis Galton in his *Hereditary Genius: An Inquiry into its Laws and Consequences* of 1869 argued that men were the superior of women in all occupations demanding intellectual acumen, which included art as well as science, and that this had been built up through inheritance down the male line (quoted in Malane 2005, p. 16; Darwin 1871, p. 565). Indeed, it was strongly argued that this difference in the sexes had not been brought about through lack of opportunity or the oppression of women but was part of the natural order. Darwin's contemporary Herbert Spencer argued that a woman's naturally evolved primary function of bearing children imposed a higher energy cost than men's reproductive functions and left less energy over for mental development (Malane 2005, p. 38). Any attempt by women to compete in intellectual terms with men was therefore futile and, moreover, compromised their reproductive capacity dangerously (quoted in Malane 2005, p. 18; Galton 1869, pp. 328, 329).

Against this ideological background, North's situation as a childless spinster who had forsaken a 'suitable' feminine occupation in favour of energetically pursuing an independent creative endeavour, left her vulnerable to accusations of unnatural and unwomanly conduct. It is apparent from her writings that North was aware of the potential for these attitudes. She relates for instance how a (male) visitor to the newly opened Kew Gallery remarked to his friend: 'It isn't true what they say about all these being painted by a woman, is it?' This anxiety is suggested by a letter she wrote to Joseph Hooker, recommending her friend James Fergusson as architect of the Gallery, so that at least 'people will not then say that I have built a horrid eyesore of brick and mortar and spoilt the Kew Gardens for my own magnification'.[22]

One of the most distinguishing features of North's paintings are the highly saturated colours. Although such a palette may have been inspired by the strong and bright colours of the tropics which we know made a strong impression on her, it is possible that there may also be a connection to Darwin's writings on sexual selection and analogous aesthetic preference in birds and insects, even if only at a subconscious or incidental level. In *The Descent*, Darwin introduces the section devoted to birds by stating that they are the 'most aesthetic of all animals [...] and they have nearly the same taste for the beautiful as we have. This is shewn [...] by our women, both civilised and savage, decking their heads with borrowed plumes, and using gems which are hardly more brilliantly coloured than the naked skin and wattles of certain birds'. (Darwin 1871, pp. 32, 371, 372).

Evelleen Richards in her *Darwin and the Making of Sexual Selection* makes the case that Darwin in his *Descent of Man* of 1871 sought to naturalise contemporary codes of dress and behaviour for women through analogy to the appearance and habits of birds and his theory of sexual selection, albeit these attempts were complicated by the fact that amongst birds it was generally the males who had the greater ornamentation and the females who did the choosing.

Richards further draws an analogy between women's choice of dress and 'image making'—she writes:

> Many contemporary accounts attest to the aesthetic pleasure—even to the sense of empowerment—that women experienced through their dress and decor. These were socially accepted, even encouraged, practices that allowed women to experience themselves as agents, where the results of their creative endeavors accrued aesthetic value within the terms of the moral exhortation to be beautiful and to beautify. (Richards 2017, pp. 239, 240)

Richards observes, however, that while 'Married women might wear richly coloured silks and velvets, and display ornate jewelry or exotic furs and feathers', 'unmarried girls were expected to dress simply and demurely in cottons and muslins' and 'older women wore darker, heavier clothes' (Richards 2017, p. 226).

North as an unmarried woman was required to dress simply and soberly but it is perhaps not going too far to suggest that her 'image-making' and exhibitionistic energies were projected into both art works and the creation of a gallery to display them. The purpose of this display may not have been to attract a mate, as was the purpose ascribed to such decorative display by Darwin, but rather to court, entrance and win over a predominantly male establishment.

As well as accounting for avian ornamentation, Darwin also considered a form of behaviour amongst certain birds that he believed arose from a similar aesthetic impulse wedded to sexual selection—that exhibited by the bower bird (Darwin 1871, pp. 112, 113). Evelleen Richards draws a parallel here to the role of the Victorian woman as a home-maker and the fashion for appropriating natural objects including flowers, fruit, leaves, moss, seaweeds, shells, feathers, birds, butterflies—dried, stuffed or reworked in various media to decorate not only the self but also has the household and in particular the place where guests were received, the parlour (Richards 2017, p. 245). Is it possible to see the North Gallery as a variation on the Victorian parlour, given its equal status as a place for impressing external visitors?

Conclusions

It is clear that North worked within a culture that was at best suspicious of and, at worst, hostile to women operating within the professional sphere, whether botanical or artistic. It is plausible therefore that her interest in themes of, often brutal, competition in the plant kingdom, brought to the fore in the writings of Charles Darwin, stem at least in part from the social situation she found herself and her determination to succeed and carve out a niche for herself within this adverse environment. The idiosyncratic characteristics of her paintings reflect a need to create something exceptional to her peers, and indeed to her own sex, in order to stand out and receive recognition as not just another amateur female botanist or painter but as a credible artist and scientist.

20 MARIANNE NORTH AND SCIENTIFIC ILLUSTRATION 441

Table 20.1 Marianne North and scientific illustration

Figure no	Title	Weblink
1	MN187: Marianne North, *View of Both Falls of Niagara*, oil on board, 1871	https://artuk.org/discover/artworks/view-of-both-falls-of-niagara-88471/search/venue:marianne-north-gallery-5000/page/23
2	MN207: Marianne North, *An Old Red Cedar on the Rocks near West Manchester, Massachusetts*, oil on board, 1871	https://artuk.org/discover/artworks/an-old-red-cedar-on-the-rocks-near-west-manchester-massachusetts-87692/search/venue:marianne-north-gallery-5000/page/24/view_as/grid
3	*The Nodding Renealmia*: engraving in aquatint, stipple and line from a painting by Peter Henderson, Plate XIII of Robert John Thornton, *New illustration of the sexual system of Carolus von Linnaeus: and the temple of Flora, or garden of nature*, 1807. Biodiversity Heritage Library via Wikimedia Commons	https://commons.wikimedia.org/wiki/File:New_illustration_of_the_sexual_system_of_Carolus_von_Linnaeus_BHL307006.jpg
4	Rhododendron dalhousiae: hand-coloured lithograph by W. H. Fitch from a drawing by J. D. Hooker, frontispiece to Hooker's *The Rhododendrons of Sikkim Himalaya* (1849–1851). Biodiversity Heritage Library via Wikimedia Commons	https://commons.wikimedia.org/wiki/File:The_rhododendrons_of_Sikkim-Himalaya_(Tab._I)_(8221577520).jpg
5	An illustration of beetles and larvae on a Mexican or prickly poppy: hand-coloured engraving by P. Sluyter from a drawing by Maria Sibylle Merian, from Merian's *De Metamorphosibus Insectorum Surinamensium* (1714). Biodiversity Heritage Library via Wikimedia Commons	https://commons.wikimedia.org/wiki/File:Metamorphosis_insectorum_surinamensium_(Pl._24)_BHL41398771.jpg
6	MN7: Marianne North, *A Chilean Stinging Nettle and Male and Female Beetles*, oil on board, 1880s	https://artuk.org/discover/artworks/a-chilean-stinging-nettle-and-male-and-female-beetles-87621/view_as/grid/search/keyword:mn7--venue:marianne-north-gallery-5000/page/1
7	MN47: Marianne North, *Flowers of Datura and Humming Birds, Brazil*, oil on board, c. 1873	https://artuk.org/discover/artworks/flowers-of-datura-and-humming-birds-brazil-87862/search/venue:marianne-north-gallery-5000/page/24/view_as/grid

(continued)

Table 20.1 (continued)

Figure no	Title	Weblink
8	MN781: Marianne North, *Poison Tree Strangled by a Fig, Queensland*, oil on board, early 1880s	https://artuk.org/discover/artworks/poison-tree-strangled-by-a-fig-queensland-88203/view_as/grid/search/venue:marianne-north-gallery-5000--identifier:mn781/page/1
9	MN92: Marianne North, *'Scotchman Hugging a Creole', Brazil*, oil on board, c. 1873	https://artuk.org/discover/artworks/scotchman-hugging-a-creole-brazil-88244/view_as/grid/search/venue:marianne-north-gallery-5000--identifier:mn92/page/1
10	MN563: Marianne North, *A Mangrove Swamp in Sarawak, Borneo*, oil on board, 1876	https://artuk.org/discover/artworks/a-mangrove-swamp-in-sarawak-borneo-87639/view_as/grid/search/venue:marianne-north-gallery-5000--identifier:mn563/page/1
11	MN23: Marianne North, *A Chilean Cactus in Flower and Its Leafless Parasite in Fruit*, oil on board, 1880s	https://artuk.org/discover/artworks/a-chilean-cactus-in-flower-and-its-leafless-parasite-in-fruit-87620/view_as/grid/search/venue:marianne-north-gallery-5000--identifier:mn23/page/1
12	MN210: Marianne North, *California Flowers*, oil on board, 1875	https://artuk.org/discover/artworks/californian-flowers-87749/view_as/grid/search/venue:marianne-north-gallery-5000--identifier:mn210/page/1
13	MN212: Marianne North, *North American Carnivorous Plants*, oil on board, 1870s	https://artuk.org/discover/artworks/north-american-carnivorous-plants-88139/view_as/grid/search/venue:marianne-north-gallery-5000--identifier:mn212/page/1
14	MN628: Marianne North, *Wild Flowers of Sarawak, Borneo*, oil on board, 1876	https://artuk.org/discover/artworks/wild-flowers-of-sarawak-borneo-88568/view_as/grid/search/venue:marianne-north-gallery-5000--identifier:mn628/page/1

Notes

1. See North (1993, p. 48) for a description of this subject.
2. Kingsley was an early convert to Darwin's outlook and attempted to square this with a Christian perspective. In his book, *At Last*, this project is clearly apparent with musings on the pervasiveness of natural competition and variation and how God works through these laws.
3. Amelia Ann Blanford Edwards (7 June 1831–15 April 1892), was an English novelist, journalist, traveller and Egyptologist. As well as sharing with North a passion for travel, Edwards was an artist and illustrated some of her own books. Like North too she trained early on as

a singer, but her voice failed. In 1882, Edwards co-founded the Egypt Exploration Fund.
4. Somerville College, Oxford, Amelia Edwards Archives, fol. 258.
5. Royal Botanic Gardens, Kew, MN/1/4, fol. 2.
6. For a contextual analysis of Thorton's *Temple of Flora*, together with reproductions of the plates, see King (1981); see also Bush (1974).
7. See also Susan Morgan's introduction to North (1993) which is reproduced in slightly revised form in Morgan (1996).
8. Royal Botanic Gardens, Kew, MS MN/1/4, fol. 22.
9. She even managed a meeting with the American President in the White House on the grounds of her being descended from the illustrious Roger, the fourth Lord North who had been Attorney-General under James II.
10. Barbara Gates writes that travel accounts were 'one literary form in which women might gain popular credibility for their knowledge of the natural world' Gates (1998, p. 99).
11. For a discussion of Lydia Becker's work see Shteir (1996).
12. Susan Casteras further declares that sketching and watercolour were popularly viewed as the preferable media for women because 'they were more 'feminine' in nature – neater and easier to conceal, transport, or remain in the background': Casteras and Peterson (1994, p. 11).
13. Royal Botanic Gardens, Kew, W. Botting Hemsley, Letters, vol. 2, fol. 68.
14. Royal Botanic Gardens, Kew, Director's Correspondence, vol. 97, fol. 169.
15. Royal Botanic Gardens, Kew, W. Botting Hemsley, Letters, vol. 2, fol. 58.
16. Royal Botanic Gardens, Kew, MN/1/1, fol. 44.
17. North stopped off at Church's home, Olana, on the Hudson River at the end of her second tour across the United States of 1881. She had previously met Church on her first visit to the US ten years earlier in 1871.
18. Cutting in Royal Botanic Gardens, Kew, MN/1/4, fol. 97.
19. Royal Botanic Gardens, Kew, MN/1/4, fol. 97.
20. This theory was first substantively set out before the public in Darwin (1859).
21. Royal Botanic Gardens, Kew, MN/1/1, fol. 25.
22. Royal Botanic Gardens, Kew, MN/1/4, fol. 3.

References

Agassiz, Louis. 1868. *A Journey in Brazil*. Boston: Ticknor and Fields.
Barnett, H. V. 1881–1882. "Miss Marianne North's Paintings at Kew." *The Magazine of Art* 5: 430–31.
Bates, Henry Walter. 1863. *The Naturalist on the River Amazons: A Record of Adventures, Habits of Animals, Sketches of Brazilian and Indian Life, and Aspects of Nature under the Equator during Eleven Years of Travel*. London: John Murray.
Beer, Gillian. 2000. *Darwin's Plots: Evolutionary Narrative in Darwin, George Eliot, and Nineteenth-Century Fiction*. Cambridge, UK: Cambridge University Press.
Blunt, Alison. 1994. "Mapping Authorship and Authority: Reading Mary Kingsley's Landscape Descriptions." In *Writing Women and Space: Colonial and Post-Colonial Geographies*, ed. Alison Blunt and Gilliam Rose, 51–72. New York and London: Guilford Press.
Blunt, Wilfrid. 1951. *The Art of Botanical Illustration*. London: Collins.
Bush, Clive. 1974. "Erasmus Darwin, Robert John Thornton, and Linnaeus' Sexual System." *Eighteenth-Century Studies* 7, no. 3: 295–320.
Casteras, Susan P., and Linda H. Peterson. 1994. *A Struggle for Fame: Victorian Women Artists and Authors*. New Haven: Yale Center for British Art.
Cherry, Deborah. 1993. *Painting Women: Victorian Women Artists*. London and New York: Routledge.
Cooke, Mordecai Cubitt. 1882. *Freaks and Marvels of Plant Life; or, Curiosities of Vegetation*. London and New York: Society for Promoting Christian Knowledge.
Darwin, Charles. 1859. *On the Origin of Species by Natural Selection*. London: John Murray.
Darwin, Charles. 1862. *On the Various Contrivances by Which British and Foreign Orchids Are Fertilised by Insects*. London: John Murray.
Darwin, Charles. 1871. *The Descent of Man and Selection in Relation to Sex. Vol. 1*. London: John Murray.
Darwin, Charles. 1875. *Insectivorous Plants*. London: John Murray.
Darwin, Charles. 1996. *The Origin of Species*, ed. Gillian Beer. Oxford: Oxford University Press.
Darwin, Charles and Sir Francis Darwin. 1880. *The Power of Movement in Plants*. London: John Murray.
De Bray, Lys. 1989. *The Art of Botanical Illustration: The Classic Illustrators and Their Achievements from 1500–1900*. Bromley: Helm.
Galton, Sir Francis. 1869. *Hereditary Genius: An Enquiry into Its Laws and Consequences*. London: Macmillan.
Gates, Barbara T. 1998. *Kindred Nature: Victorian and Edwardian Women Embrace the Living World*. Chicago and London: University of Chicago Press.
Guelke, Jeanne K., and Karen M. Morin. 2001. "Gender, Nature, Empire: Women Naturalists in Nineteenth-Century British Travel Literature." *Transactions of the Institute of British Geographers* 26, no. 3: 306–26.
Hemsley, W. Botting, ed. 1892. *Official Guide to the North Gallery*. London.
King, Roland. 1981. *The Temple of Flora by Robert Thornton*. London: Weidenfeld & Nicolson.
Le-May Sheffield, Suzanne. 2001. *Revealing New Worlds: Three Victorian Women Naturalists*. London: Routledge.

Losano, Antonia. 1997. "A Preference for Vegetables: The Travel Writings and Botanical Art of Marianne North." *Women's Studies* 26, no. 5: 423–48.

Malane, Rachel. 2005. *Sex in Mind: The Gendered Brain in Nineteenth-Century Literature and Mental Sciences.* New York: Peter Lang.

"Miss North's Paintings of Plants." *The Times*, 8 June 1882.

Morgan, Susan. 1993. Introduction to *Recollections of a Happy Life: Being the Autobiography of Marianne North*, by Marianne North, xi–xl. Charlottesville and London: University of Virginia Press.

Morgan, Susan. 1996. *Place Matters: Gendered Geography in Victorian Women's Travel Books about South-East Asia.* New Brunswick: Rutgers University Press.

Noble, David F. 1992. *A World without Women: The Christian Clerical Culture of Western Science.* Oxford and New York: A. A. Knopf.

North, Marianne. 1892. *Recollections of a Happy Life: Being the Autobiography of Marianne North. Vol. 2.* London and New York: Macmillan and Company.

North, Marianne. 1993. *Recollections of a Happy Life: Being the Autobiography of Marianne North*, ed. Susan Morgan. Charlottesville and London: University of Virginia Press.

Prettejohn, Elizabeth. 2000. *The Art of the Pre-Raphaelites.* London: Tate Publishing.

Richards, Evelleen. 2017. *Darwin and the Making of Sexual Selection.* Chicago and London: University of Chicago Press.

Rix, Martyn. 1981. *The Art of the Botanist.* Guildford and London: Lutterworth.

Russell Wallace, Alfred. 1853. *A Narrative of Travels on the Amazon and Rio Negro.* London: Reeve and Company.

Saunders, Gill. 1995. *Picturing Plants: An Analytical History of Botanical Illustration.* Berkeley and Los Angeles: Zwemmer.

Shteir, Ann B. 1996. *Cultivating Women, Cultivating Science: Flora's Daughters and Botany in England, 1760 to 1860.* Baltimore: John Hopkins University Press.

Shteir, Ann B. 1997. "Gender and 'Modern' Botany in England." *Osiris* 2, no. 12: 29–38.

Stanford, Derek, ed. 1973. *Pre-Raphaelite Writing: An Anthology.* London: Dent.

Taylor, John Ellor. 1884. *The Sagacity and Morality of Plants: A Sketch of the Life and Conduct of the Vegetable Kingdom.* London: Chatto and Windus.

"The Society of British Artists." 1834. *The Lady's Magazine and Museum of Belles Lettres.*

Yeldham, Charlotte. 1984. *Women Artists in Nineteenth-Century France and England.* New York and London: Garland.

CHAPTER 21

The Cycle of Credit and Phatic Communication in Science: The Case of Catherine Henley

Jordynn Jack

The most successful of these women found it advisable (to use a biological analogy) to develop certain 'adaptive behavior' in order to survive; they would seek out a strong protector and, on occasion, take on the coloration or attitudes of the dominant group in an attempt to 'pass' as something they were not. (Rossiter 1982, p. 180)

In their now-canonical text, *Laboratory Life*, Bruno Latour and Steve Woolgar outline how scientists are motivated by a cycle of credit. Within this cycle, scientists invest time and money into equipment that generates data. In turn, data lead to arguments, articles, and, if those are successful, to recognition that scientists can 'cash in' in the form of grants they can use to produce still more data. The most successful scientists are those who are able to generate enough credit to keep moving through the system. This often entails a combination of luck and of conscious decisions to study a particular topic, apply to particular labs, or find positions at particular institutions. Those who are less successful, they explain, 'such as some of the technicians' find themselves 'with careers which were inextricably bound up with the material elements of the laboratory' (Latour and Woolgar 1986, p. 188).

J. Jack (✉)
Department of English and Comparative Literature, University of North Carolina, Chapel Hill, NC, USA
e-mail: jjack@email.unc.edu

Latour and Woolgar pay no attention to the gendered dimensions of this cycle of credit, apart from one offhand mention of a scientist who identified a central problem in his discipline and 'grasped the chance opportunity of assistance from a lady whose skill perfectly matched his goal' (Latour and Woolgar 1986, p. 213). In their account, the cycle of credit is neutralized, with technicians and secretaries often relegated to the 'material elements of the laboratory' (Latour and Woolgar 1986, p. 188). Take the depiction of the laboratory as a highly masculinized space in this passage: 'The tension of a battalion headquarters at war, or of an executive room in a period of crisis does not compare with the atmosphere of a laboratory on a normal day! This tension is directed toward the secretaries in efforts to persuade them to type manuscripts in time and towards the technicians to effect the rapid order of animals and supplies and to the careful execution of routine assay work' (Latour and Woolgar 1986, p. 229). While Latour and Woolgar do not describe the gendered labor force in the laboratory they observed, it is apparent that some people working in the lab (many of them female) are positioned in subordinate roles. While not all technicians were women, that role was often one of the only possibilities for women trained in the sciences during the period from 1940 to 1970s (and probably longer than that). If, as Latour and Woolgar describe it, the cycle of credit largely excluded technicians and therefore many women, then those who did work in those positions had to develop their own strategies and resources for getting along (if not getting ahead) in science.

To date, studies of women in science have focused heavily on the smaller percentage of women who were able to make inroads within the traditional cycle of credit—even if, like Rosalind Franklin and many others like her, they still faced significant challenges in getting recognition and credit for their work. Other research has focused on citation practices and other credit-granting mechanisms (such as prizes and awards). At every stage in the 'cycle of credit,' women are disadvantaged: men are more likely than women to get hired, tenured, and promoted; to receive funding; and to be cited (Shen 2013). Women often deal with sexual harassment, intimidation, and hostility in the scientific workplace, which also reduces their ability to gain credit (Settles et al. 2006). And Latour and Woolgar's 'cycle of credit' overlooks the acts of self-promotion involved, such as asking questions at conferences (Hinsley et al. 2017) or seeking patents for their work (Göktepe-Hulten and Mahagaonkar 2010), both of which women are less likely to do.

Despite these barriers, we also know that women have often been the unsung heroes of science, in some cases performing important work that enabled key discoveries. There are many cases where women made core contributions to key discoveries but were passed over for the Nobel Prize in favor of a male scientist (Flint 1997). Historian of science Margaret Rossiter refers to this as the 'Matilda Effect,' a phenomenon that is due to systematic undervaluing of women's contributions to science (Rossiter 1993, p. 334; Lincoln et al. 2012). The cycle of credit, to put it simply, is masculinist and fails to account for the differences in women's careers in science. Yet we know much less, on a fine-grained level, about the women who served in the even more workaday roles of research associates and technicians whose work did not contribute to a Nobel Prize or a major discovery but who nonetheless made contributions to scientific fields. This chapter addresses this by identifying the rhetorical negotiations women performed within the scientific cycle of credit. The chapter first provides background information about the roles women took on during the post-war period as research associates. This is followed by an overview of Catherine Henley's career as a research associate in zoology at the University of North Carolina, Chapel Hill, a large, public research university in the United States southeast. The remainder of the chapter examines three aspects of phatic communication—communication used to create goodwill and maintain relationships—that enabled Henley's participation in zoological research at a time when few women held tenured positions: her ability to establish interpersonal relationships, to differentiate herself from other women, and to perform precision work as a means of facilitating her mentor and collaborator's research. Ultimately, archival documents illuminate aspects of the cycle of credit that are not accounted for in Latour and Woolgar's model but that, nonetheless, play a key role in scientific careers, especially for women who lacked access to resources, power, and prestige during the post-war period.

Background: Women as Research Associates

Latour and Woolgar's narrative ultimately erases the material elements of the laboratory. As they put it, 'Once the data sheet has been taken to the office for discussion, one can forget the several weeks of work by technicians and the hundreds of dollars which have gone into its production' (Latour and Woolgar 1986, p. 69). This erasure represents a paradox, since 'Without the material environment of the laboratory none of the objects could be said to exist, and yet the material environment very rarely receives mention' (Latour and Woolgar 1986, p. 69). This is true for those who worked with those materials themselves. As Rossiter explains, there is little evidence to confirm how often women served as technicians or research associates in labs during the post-war period. One study from 1960 found that 23.2 percent of research associates in the physical and biological sciences were women (Rossiter 1995, p. 151). Data such as this are difficult to come by since, Rossiter explains, these jobs 'were designed to be invisible and to gain the worker no professional recognition' (Rossiter 1995, p. 149).

As staff members, research associates prepared specimens, ran experiments, typed manuscripts, and even edited professional journals, but they were rarely credited as authors for their work. During this time period, a footnote of acknowledgment was often the most that could be expected. As Rossiter explains, research associates could potentially advance to senior research associate, a position that might have entailed a slightly better salary or benefits, but there was no pathway from this track onto the tenure track, in most cases. Often, women were hired into these positions due to a personal relationship, for example, because they were married to a male scientist or had served as his graduate student or research assistant.

For many female scientists, any hope for a career in academia depended upon a mentor. In some fields (such as nutrition) a female mentor might exist with the power to arrange positions for her mentees. In most fields, however, women were dependent upon a male mentor. While not all men would take on female students or advocate for them, in each field a few such men existed who would make an effort either to obtain academic positions for their mentees or, even more rarely, to hire them themselves (Rossiter 1982, p. 185). For these women, participation in the 'cycle of credit' depended largely upon interpersonal skills as well as scientific ones: 'Under such a system of personal patronage, about all the favored women could do was keep up the good work, be loyal to their benefactors, and stay out of "politics"' (Rossiter 1982, p. 186). While such mentorship was often the only option for women scientists, though, dependence on a male faculty member also left them open to charges of favoritism or the assumption that they were merely 'assisting' a professor rather than doing their own scholarly research.

We have little evidence of the communicative or rhetorical work expected of women in these positions. However, the following analysis draws on archival materials from the papers of Catherine Henley (1922–1999), a zoologist who worked as a research associate for her mentor, Donald Paul Costello, for 30 years. Henley likely met Costello first as an undergraduate student in one of his courses (although this is not clear from the archival record). Henley pursued a Master's degree at the University of North Carolina (UNC) in Chapel Hill but ultimately left for Johns Hopkins University. Upon completing her Master's in 1947, she returned to UNC to complete her Ph.D. in 1949 with Costello as her advisor. Costello co-authored 27 articles with Costello during that time period, most of them anatomical microscopy studies of marine creatures such as salamanders, protosmoia, and turbellaria. She also published 16 single-authored articles between 1946 and 1977.

The archival materials, available at the UNC-Chapel Hill's Wilson Library, show how Costello helped Henley by providing her with advice, finding her positions (within his own lab and that of colleagues), and then supporting her on multiple grants until he retired. In turn, the evidence (much of it correspondence between the two when one or both were working away from UNC) describes how Henley performed craft scientific and rhetorical work that supported Costello's career: helping him edit a professional journal, collecting and preparing specimens, conducting experiments, and writing papers. Costello named Henley a co-author on much of the work but was unable to help her secure a more permanent position with tenure. She remained a research associate at UNC until 1976, when, with Costello nearing retirement, she left UNC to work for the National Institutes of Health in the grant department.

Henley entered into a field in which women were both well-represented and under-acknowledged. In 1938, a few years before Henley would have entered UNC as an undergraduate, a survey of female scientists reported 196 employed in zoology—more than any other discipline. Yet, only five women were faculty in zoology in top research institutions. (UNC was not included among these.) As Rossiter puts it, 'Obviously the men in zoology were fiercely resisting even the slightest tendency toward feminization' (Rossiter 1982, p. 181). By the 1950s, women's representation in zoology was not much higher; according to a survey by the National Science Foundation in 1956–1958, women represented 12.33 percent of zoology at that time (this including researchers of all ranks) (Rossiter 1995, p. 101). The proportion who even made it to assistant professor was staggeringly low; in 1960, out of 255 women employed at 20 leading universities, only five were assistant professors (and none were full professors) (Rossiter 1995, p. 129). However, there were plenty of women working in zoology—just not as tenure-track professors. Most served, like Henley, as research associates, teaching assistants, and the like (Rossiter 1995, p. 129). This type of work became 'women's work' in many fields, and it carried little job security: 'even though such a worker's efforts added to the reputation of the professor, the project, the university, and even the funding agency, his or her own status remained precarious, for even after many years or decades the university usually was not obligated to offer the staff member any tenure or long-term security' (Rossiter 1995, p. 149). The conditions for women in zoology (as in most scientific fields) were poor throughout Henley's career: the best she could hope for, it seemed, was a stable position as a research associate. For Henley to have any opportunities at all, she had to ensure that she maintained the support of her mentor and stayed in his good graces.

Establishing Interpersonal Relationships

One key feature of all professional communication is that it is fundamentally geared toward maintaining goodwill, or what is often called 'phatic communication.' This type of communication performs the rhetorical function of 'creating effective communication channels, keeping them open, and establishing ongoing and fruitful relationships' (Porter 2017, p. 175). Traditionally phatic communication has been considered as less significant than communication that conveys meaning or information. However, Porter argues that it is actually a primary purpose of professional communication, since writers must build relationships, ensure cooperation, develop new partnerships, and otherwise enact relational goals through writing (and other forms of communication).

Henley's correspondence provides multiple examples of how she engaged in phatic communication to ensure her position within the scientific field, in general, and at UNC-Chapel Hill, in particular. The first letter from Costello to Henley appears in the archive in 1944. After receiving criticism from a professor in the department, Henley left unceremoniously. Costello wrote to her on 29 May, urging her to reconsider and offering to help her find a job elsewhere, should she decide to leave for good. Here, it is Costello who does much of the phatic communication work. He writes that he has 'perfect faith in your ability, intelligence, and integrity and would be more than glad to recommend you to anyone.' He goes further to explain that the professor in question, a Dr. Beers, was a known misogynist and 'always reacted against women of forceful personality.' Thus, he assures Henley, she should not 'take too much to heart anything he may have said' as it was 'directed against women in general rather than against you in particular.' And, Costello assures Henley that she could still go to Woods Hole (the marine biology lab in Cape Cod where researchers gathered each summer to take courses, gather specimens, and conduct research), especially since it was there that 'I can do most for you.'[1]

On 30 May 1944, Henley writes to Costello asking for his advice about the exam situation with Dr. Beers: 'If you have the time, I should appreciate having your advice on what to do about this matter; knowing that Dr. Beers and Dr. George are friends has made me think twice about taking any positive action in the affair'.[2] This exchange sums up, in a nutshell, how Costello, as a senior male advisor, and Henley, as a younger, female student, began to form a professional relationship. On his part, Costello dispensed advice and arranged job opportunities for Henley; for her part, Henley asked for advice and sought to maintain goodwill.

While Henley did agree to return to Chapel Hill to finish her outstanding exams, she ultimately decided to leave to pursue her Master's degree at Johns Hopkins University. However, even as she prepared to move on to Hopkins, Henley employed phatic communication to help maintain her relationship with Costello. After all, as a woman in a male-dominated field, she would need a mentor. During this time period, prominent scientists participated in a network of contacts (as they still do today, of course) and could often secure opportunities for their students by tapping into that network. It becomes clear from their correspondence that both Henley and Costello were well aware that their relationship partly existed as a means of granting Henley access to that network.

In October 1944, for instance, Henley writes a newsy letter to Costello (and his wife, Helen, herself a former scientist), full of information about her own situation and that of some other former UNC students. 'Helen Strong is back at Smith and writes hilarious tales of her adventures with travels, students and fish-breeding for her thesis,' Henley writes. She also hints that she would like to be considered for a position at Woods Hole the next summer: 'It appears that there is some slight chance of my getting to Woods Hole again next summer; I'll have to talk to Dr. Willier about this more definitely, but am hoping greatly that the answer will be in the affirmative. Has anything more been said about the assistantships in Embryology for next year?'[3] By recounting news, generally expressing gratitude, and then asking questions ('Has anything more been said about the assistantships?') she is able to obtain a position at Woods Hole without asking directly for one. For his part, Costello agrees to arrange something for Henley at Woods Hole, and she ultimately gets an assistantship there due to Costello's string-pulling with a Dr. Hamburger, the researcher in charge of the embryology course.

After finishing her degree at Hopkins, Henley ultimately returned to Chapel Hill to work for Costello in his lab as a doctoral student and research assistant. She earned her Ph.D. in 1949 for her thesis, 'Chromsomal mosacisim and abnormalities of mitosis by experimental temperature shock on the developing embryos of *Triturus torosus*.' Eventually, Henley's relationship with Costello appears to have evolved, and with it the nature of the phatic communication she engaged in. In 1949, she became research associate, and by all measures Costello and Henley became collaborators, publishing work together for the next 20 years. The two often co-authored articles and even a manual, titled *Methods for Obtaining and Handling Marine Eggs and Embryos* (Costello and Henley 1971).

As their relationship changed, Henley increasingly took on phatic communication not simply to maintain her relationship with Costello, but often on behalf of maintaining Costello's relationships with others. In December 1956, for instance, Henley wrote to Costello in an attempt to smooth over his relationship with another assistant named Louise:

> There is a matter which I thought you might want to attend to: Louise came back to work today, and from her conversation, I gather that she was quite seriously hurt at what appears to have been the rather curt instructions you left for her. I'm sure you didn't mean them that way, and also that when she is a little less 'down' with her bug, she won't be quite so upset—but perhaps it would help if you could write her as cordial and communicative a note as possible. She is so fond of you, and so loyal to you that I thought maybe you'd like to know this. The same applies, to a much less marked degree of hurt feelings, to Gene and Jim, too.[4]

Here, we see Henley trying to smooth out Costello's relationships with three other people—presumably fellow research assistants or technicians in Costello's lab.

When Costello became editor of the *Biological Bulletin* in 1951, Henley effectively became the editorial assistant, handling much of the correspondence (as well as editorial work, as explained below) between the journal and its reviewers and authors. This too, meant that Henley was often performing phatic communication on Costello's behalf. On 6 July 1954, for instance, she writes to Costello with a list of updates about submissions to the journal, reviewers assigned, and submissions still needing review. At the end of the letter, she weighs in on a manuscript that had recently been rejected by reviewers, and asks: 'Would you like me to draft a letter of refusal and send it for you to sign? Or would you rather do your own?'[5] By offering to write the letter, Henley was taking on phatic communication work on Costello's behalf. As an editorial assistant, she was in effect taking on the interpersonal work that is required to maintain goodwill among contributors, reviewers, and editors of the journal.

From this example, at least, it is apparent that research associates not only performed scientific work. For Henley, at least, her responsibilities included phatic communication on Costello's behalf that enabled him to maintain relationships with the research assistants and graduate students he worked with, and, especially during the 18 years he served as editor for the *Biological Bulletin*, with other scientists. This kind of work might normally be considered secretarial work—another gendered position. Thus, the female research associate could be, in practice, both secretary and scientist.

Working with Precision

An additional type of phatic communication requires mentioning here, and it is one that may not traditionally be understood as such within scientific communication but that fits well with the idea of scientific-secretarial work and its function within the cycle of credit. If the goal of phatic communication is to establish and maintain communication channels as well as relationships, then part of Henley's role was to do so using both traditional communication forms (letters, reports, etc.) but also by furnishing scientific information that would help Costello to continue to build scientific relationships. In fact, the cycle of credit itself depends heavily on phatic communication. A successful scientist has to establish relationships with others in the field (scientists, grant reviewers, etc.) who may have resources (materials, data, money) that might be useful to them. For scientists, phatic communication may involve sharing news and small talk, but it may also involve sharing and discussing data or, in the case of zoologists, sharing slides, photographs, and specimens. By supplying Costello with a steady stream of carefully collected and tended specimens, well-prepared slides, and precise notes about their research, Henley was providing 'inputs' that could be entered into the cycle of credit.

As a research associate, much of Henley's job entailed performing experimental work on Costello's behalf, which could range from collecting specimens (often while at Woods Hole), managing shipments of specimens, attending to the care and feeding of the various 'beasties' (as they called them), preparing slides, and keeping meticulous records. In a 1946 article about *Leptosynapta* (sea cucumbers), for instance, Costello notes that Henley collected some of the specimens described (along with a Roberta Lovelace), noting that six *Leptosynapta* specimens, in particular, were identified while the researchers were collecting *Nereis*, which they did almost every night from early June until the end of September 1945) (Costello 1946, p. 95). During these collection outings, Henley and Lovelace would have been responsible for locating, capturing, labeling, and then caring for the tiny marine creatures to be studied in the lab. (It is unclear whether they did so alone at night or whether Costello would have accompanied them.) In 1946, when Costello wrote to Henley offering her a job if she wished to return to UNC, he described her work as help with 'my Polychoerus material—staining, sectioning, studying slides, etc.'[6] Throughout his career, Costello depended on Henley (and presumably other female research associates mentioned in their correspondence and articles, such as Lovelace) to perform such work accurately. This work fed into his cycle of credit, typically, generating material for research articles, which could lead to more grants and more recognition.

In a letter dated 14 May 1968, for instance, Henley encloses material she had prepared at UNC and sends it to Costello while his is at Pacific Grove, California, doing research:

> Enclosed are the Ems of negatively stained Polychoreus material. You already know, I'm sure, that the material is teased in phosphotungstic acid, pipetted onto coated grids, dried down and examined without further treatment. The PTA precipitates around structures in a dark mass, and thus shows up the structures as white. Nos. 85-9, 86-3 and 86-8 are of special interest because of the previously unobserved structure at the tip of the sperm. I suspect this may be the head end; there is a suggestion of white nuclear material immediately behind the cap. Note, also, the striking configuration of the nucleus in some of the others, explaining the 'beaded' appearance seen by light.[7]

The correspondence between the pair is rife with such detailed descriptions of the work performed (usually by Henley) or the work to be performed (usually as dictated by Costello). In this case Henley has not only prepared the material meticulously, but she also describes that preparation and then points out for Costello which slides are particularly interesting. In this way, she performs not just mechanical tasks but lays the groundwork for observations (and eventual publications) that Costello might undertake.

While Costello and Henley often co-authored articles, in other cases, Costello published work as a single author to which Henley and others contributed extensively. In a 1973 article, for instance, he includes a footnote that reads: 'Sincere appreciation to Dr. Henley for taking practically all the electron micrographs upon which the supporting observations were based is here recorded' (Costello 1973, p. 307).[8] In today's cycle of credit, such extensive work would likely have placed someone onto the author list for an article. Henley's work for Costello on the *Biological Bulletin* takes a similar form: keeping track of submissions and to which readers they had been assigned, providing updates on journal acceptances, and so on. In one letter, for instance, Henley lists articles submitted to the journal for which a decision was needed and appends notes to help Costello in making determinations. In one instance she notes: 'We've had other papers from him which were carefully done'; in another she adds 'author is a student and is now working in Brazil with Karl Wilbur.'[9] The notes suggest two types of information: one, such as the first comment, about whether the work seems well done, and another about the authors' connections to other researchers. (This shows how, following Latour's suggestions, scientists assess quality of work not only based on its intrinsic qualities but also based on what is known of the author and his connections to other prominent researchers.) By appending these details, Henley arguably again engages in phatic communication—by performing work to ease Costello's job, she is maintaining her own relationship to him and assuring his goodwill.

Later evidence suggests that this type of labor was not entirely enjoyable for Henley. In a letter to 'people' (presumably, her family members), she writes: 'Besides improved living conditions, one of the factors that's made a big difference in my outlook on Woods Hole is not having to cope anymore with the journal which Don and I labored with for 18 years. I'm interested to find that it takes two people full time to do what I squeezed into half time, all by myself!'.[10] While Henley rarely betrayed her sentiments to Costello, this work was clearly onerous. This statement also suggests that Henley single-handedly performed much of the labor that Costello got credit for as editor.

Among the 30 years of correspondence between Henley and Costello, there is only one occasion in which Henley directly expressed her annoyance at Costello's expectations. As he prepared to leave Pacific Grove in spring of 1968, Costello wrote to Henley suggesting that his plan was to go immediately to Woods Hole to take up his summer research work. Henley responded as follows:

> You took me so completely by surprise Saturday evening with your airy statement that you were planning to go directly to Woods Hole that I didn't have the wits to say what I should have: Unless you have some very sound reason for that course of action (such as health), I'm afraid I'll have to refuse point-blank. Things already have me swamped and I simply cannot take on the task of packing all the Bulletin stuff and the Zeiss and the ten million other things needed. Perhaps you do have a good reason; if so, I'd appreciate hearing it. If not, I think you're being very unfair and making completely unreasonable demands, and I won't do it. Flitting from one pleasure dome to another while someone else does the drudgery doesn't become you.[11]

Costello responded with an apology, agreeing to come back to Chapel Hill first to help with the packing (and signing the letter 'dunderheaded but loving', in a note that perhaps indicates something of the relationship between the two).[12] The rarity of this type of exchange suggests that, throughout their work together, Henley was doing relationship management or what is often called nowadays as 'emotional labor'—that is, carefully managing or hiding her own feelings in order to smooth over and maintain goodwill.

Henley's success at this type of labor is apparent despite the fact that she performed labor required to maintain Costello's position (and, thereby, her own) in the cycle of credit. Costello did, to some extent, try to share some of the credit with Henley. She is often listed as co-author on his publications, and in at least some cases, he encouraged Henley to claim single authorship.

In 1968, for instance, he responds to a manuscript Henley sent him while he was at Pacific Grove: 'It is a beautiful piece of work, and you deserve full credit for it … My name need not be on the title page <u>unless you wish it to be</u>. I am flattered, but would like to have you get full credit for this.'[13] This passage indicates some of the vagaries of Henley's position. While the manuscript clearly reflected independent research on her part, her position as a research associate in Costello's lab may have meant that she felt compelled to put his name on the title page. Alternatively, she may have felt Costello's name would carry greater weight, helping her to get the piece published.[14] It is notable that Costello insisted on her publishing the article as sole author because it suggests Costello's investment in helping Henley to establish her own reputation.

Of course, over time the relationship between Henley and Costello apparently became a friendship, not just a work arrangement, and much of this work, to be sure, Henley may have performed out of friendship rather than obligation. Nonetheless, the fact remains that meticulousness was a skill women were expected to portray but were not always rewarded for during this period. As Rossiter put it, being meticulous was a double-edged sword: 'when a woman did good work as an instructor, she was rarely rewarded in these years with promotion to a higher rank. Instead she was told that being good at such menial work meant that she could do nothing else and would therefore have to remain an instructor for the rest of her career' (Rossiter 1995, p. 167). Presumably, the same goes for a research associate.

Differentiating: The 'Other Woman': Phatic Communication and Ethos

Throughout the correspondence between Costello and Henley, it becomes clear that phatic communication not only served interpersonal purposes but also helped to construct ethos for both participants. As Porter puts it, phatic communication is 'fundamental to the rhetor's identity (ethos) as their overall behavior and practice as a rhetor' (Porter 2017, p. 179). By maintaining goodwill, one is also portraying oneself as a *person* of goodwill. For Henley, however, this performance at times took on a particularly gendered aspect. To be a good scientist during this time period—or a good female scientist, in particular—it seems it was necessary to differentiate oneself from other women. Both Henley and Costello reinforced this idea at various times.

Early in their correspondence, in 1945, Costello wrote a response to Henley, who had received news that some of the Woods Hole colleagues had viewed her as having an unfavorable attitude that was 'a disgrace' to the Embryology course for which she was assisting.[15] Costello responded by reassuring Henley that, while she might be considered to have a 'difficult personality' by some, he appreciated her intelligence and the 'frankness that is exhilarating compared to the usual evasive, simpering female'.[16] It became apparent through this exchange and others that followed that Henley was to perform the 'exceptional woman' role, demonstrating her dissimilarity from other 'simpering' women.

Sometimes, it was Henley who initiated such performances in her letters. In November, 1946, while she was away working on her Master's at Hopkins, Henley wrote to Costello of a female scientist who had visited the lab: 'My main impression of her was the reiterated wonder if dowdiness is necessarily a prerequisite for women in science—or, as Hank used to say, "lady biologists"!'[17] Henley also engaged in banter with Costello, teasing him for having admitted to be interviewing 'young females by the score' for a position in his lab.[18] Henley responds by teasing him for this statement: 'Re interviewing "young females": I can just imagine what a chore that was—or weren't any of them blondes?'[19] This behavior seems meant to assure Costello that she is not threatened by these other female candidates. This, presumably, sets her apart from other women who may be (stereotypically) jealous or petty.

Finally, Henley extended this type of distinction to other women as well. When Costello was looking for an assistant for a summer position at Woods Hole, she wrote with a recommendation for him to hire someone she had met while working for a semester at Stanford, calling this candidate 'one of the most thoroughly nice and likable people I've ever encountered—amazing naïve and unspoiled and devoid of the usual female vices'.[20] Though perhaps not distinctive separately, each of these small mentions of female vices or 'simpering' suggests that women scientists (and Henley herself) must not be taken in such a way. In fact, Costello made it clear to Henley that her exhilarating frankness differentiated her from other women, and Henley, in turn, continued to play that role in their correspondence. This might seem like an isolated example, but, in fact, the idea of the 'exceptional woman' in science has a deeper history.

Throughout the early twentieth century (in what would have been Henley's formative years), female scientists like Florence Sabin and Marie Curie appeared regularly in the media in epideictic articles celebrating their talents (Jack 2013). These articles often depicted such women as exceptional in many ways—exceptionally intelligent, motivated, and prolific. The message for women, unfortunately, was that they too could only succeed in science if they possessed this exceptional level of ability. In 1979, Adrienne Rich wrote of academia more generally that 'The exceptional women who have emerged from this system and who hold distinguished positions in it are just that: the required exceptions used by every system to justify and maintain itself' (Rich 1979, p. 127). In other words, the successes of women like Curie or Sabin could be used to show that anyone could succeed. Yet the fact that only such women were celebrated or lauded enabled institutions to gloss over the discrimination and prejudice that actually existed within those fields.

Within Henley's department, for instance, it was quite clear that a tenure-track position was not a possibility for a woman. In fact, Henley felt that the department was decidedly anti-woman, well into the 1960s. In 1969, Henley wrote to a friend, Mimi, that she had decided that at my age, it's high time to settle down in a job with tenure; I'm getting too old to go through the waiting-to-hear-about-the-grant bit every year, particularly since things in that line are getting leaner and leaner .... Since the chairman of our Department at Chapel Hill has come right out in the open and admitted that if he could, he'd get rid of the women he already has on his staff and isn't about to hire any more, no matter how good, that seems to settle that possibility.[21]

While Henley stayed on at UNC for seven more years, this note does suggest that she found her position precarious and that at her institution, at least, there were no exceptions, even for women who managed to conduct research and publish articles (like Henley did).

Ultimately, Henley's tenuous academic position as a research associate and the sexism she faced in her own department made her leave research altogether in the latter years of her life. Once Costello retired in 1976, Henley no longer had access to a lab at UNC and could not maintain the title of research associate, since she was supported on Costello's grant money. Correspondence shows that Henley and Costello discussed the idea of her moving permanently to Woods Hole to do research, given the climate at UNC. Henley also tried applying for her own grant through the NIH, which would have allowed her to continue researching (likely at Woods Hole). The grant was rejected, though, and Henley ultimately found a job reviewing grants through the NIH, which meant leaving Chapel Hill (and Woods Hole) to live in Bethesda, Maryland.

While a single grant proposal does not make a large sample size, it is striking that Costello was quite successful in stringing together enough grants to fund Henley for 30 years. Henley, for her part, contributed to his success and had many publications (both as co-author and single author). Yet, even though she had fed into the cycle of credit she was not able, on her own, to garner enough credit to be able to cash in enough chips to establish her own position, either at UNC or elsewhere. For the rest of her career, Henley worked in positions for the NIH and at the National Eye Institute that dealt with grants review and administration.

Conclusions

It is apparent from their correspondence that Henley performed her fair share of the work she and Costello published together. In the last year or two of his life, Costello was dealing with congestive heart failure, which meant he spent much of his time (according to the correspondence) reading, writing letters, and reflecting back on his life. At several points in their letters, Costello hints at the extent to which Henley was actually central to his career. In January 1977, Costello reflects that 'I am beginning to suspect I frittered away an awful lot of time in my life doing things that weren't necessary or worth doing, and would have made even more of a mess of things if you hadn't been here to suggest worthwhile things to keep my nose closer to the grindstone'.[22] And in the last letter he sent her in December 1977, he acknowledges his goal was 'to give as much as I could (but always less than half) to our joint enterprises and decisions' and also refers to the *Biological Bulletin*, 'which you kept going for 18 years'.[23] These admissions, however, did not shape the public recognition of Henley's work after Costello's death. In fact, Henley was called upon to contribute obituaries and memorial essays about Costello to scientific journals, and she did so without taking credit, in any way, for her part in his work.

Henley is just one of many women in science who never received widespread credit for her work, in part because she (like many others) remained untenured, in the precarious position of a research associate, and tied to a male collaborator. This chapter illustrates how archival materials, when available, can help us to identify some of the work research associates performed—work that was both scientific and rhetorical. In particular, we can view much of this work through the lens of phatic communication. For Henley, at least, both the work she did through language (to establish interpersonal relationships and to develop her ethos) and the work she did with scientific materials (such as carefully collecting specimens and preparing slides) shored up her tenuous position within a hostile scientific department. Her best and only hope at stability within such an environment was to attract and maintain the goodwill of a mentor and collaborator.

More broadly, this example points to the importance of phatic communication for women seeking to make a career in the sciences during this period (and arguably still does today). While all professional communicators must of course maintain goodwill among their colleagues, for those in a marginalized position this requirement takes on greater significance. Examining this type of communication in historical contexts is difficult due to the often-ephemeral nature of such exchanges. In this case, Henley's extensive records of her correspondence with Costello makes it possible to examine how phatic communication between them worked. Future research might identify additional archives that could provide a broader picture of phatic communication in the past. Alternatively, interview methods might allow for contemporary investigations of whether and how phatic communication continues to shape careers for women in science.

Acknowledgements The author wishes to thank Bob Costello, George Costello, Jean DeSaix, and Mary Beth Thomas for sharing their remembrances of Catherine Henley, Donald Costello, and the Department of Zoology at the University of North Carolina during this era.

NOTES

1. Costello to Henley, 29 May 1944, Folder 1, Catherine Henley Papers, Collection #4979, The Southern Historical Collection, Louis Round Wilson Special Collections Library, University of North Carolina,

Chapel Hill. All subsequently cited correspondence is from this collection. Permission granted to quote from Costello's letters from Robert Costello. Due diligence was used to pursue possible executors for Henley's papers, as she never married nor had children.
2. Henley to Costello, 30 May 1944, Folder 1, Box 1, Catherine Henley Papers. Collection #4979, Wilson Library, University of North Carolina, Chapel Hill.
3. Henley to Dr. & Mrs. Costello, 8 October 1944, Folder 1, Box 1, Catherine Henley Papers.
4. Henley to Costello, 5 December 1956, Folder 6, Box 1, Catherine Henley Papers.
5. Henley to Costello, 6 July 1954, Folder 5, Box 1, Catherine Henley Papers.
6. Costello to Henley, 4 November 1946, Folder 2, Box 1, Catherine Henley Papers. Henley and Costello worked for many years with polychoerus and other species of turbellaria or flatworms.
7. Henley to Costello, 14 May 1968, Folder 8, Box 1, Catherine Henley Papers.
8. Costello, "A New Theory," 307.
9. Henley to Costello, 5 May 1968, Folder 8, Box 1, Catherine Henley Papers.
10. Henley to "people," 23 June 1969, Folder 9, Box 1, Catherine Henley Papers.
11. Henley to Costello, 14 May 1968, Folder 8, Box 1, Catherine Henley Papers.
12. Costello to Henry, 4 June 1968, Folder 9, Box 1, Catherine Henley Papers.
13. Costello to Henley, 17 March 1968, Folder 8, Box 1, Catherine Henley Papers.
14. I thank Donald's sons, George and Bob Costello, for this insight.
15. Henley to Costello, 5 October 1945, Folder 1, Box 1, Catherine Henley Papers.
16. Costello to Henley, 12 October 1945, Folder 1, Box 1, Catherine Henley Papers.
17. Henley to Costello, 5 November 1946, Folder 2, Box 1, Catherine Henley Papers.
18. Costello to Henley, 7 January 1948, Folder 3, Box 1, Catherine Henley Papers.
19. Henley to Costello, 10 January 1948, Folder 3, Box 1, Catherine Henley Papers.
20. Henley to Costello, 29 April 1948, Folder 4, Box 1, Catherine Henley Papers.
21. Henley to Mimi, 11 July 1969, Folder 9, Box 1, Catherine Henley Papers. Mimi is mentioned at times in the correspondence between Henley and Costello, so she was presumably a mutual acquaintance and

likely worked in the field of zoology. George and Bob Costello suggest Mimi was Mimi Bennett, a friend of Costello and Henley who taught at Sweet Briar College.
22. Costello to Henley, 5 January 1977, Folder 20, Box 2, Catherine Henley Papers.
23. Costello to Henley, 18 December 1977, Folder 21, Box 2, Catherine Henley Papers.

References

Costello, Donald P. 1946. "The Swimming of Leptosynapta." *Biological Bulletin* 90, no. 2: 93–96.

Costello, Donald P. 1973. "A New Theory on the Mechanics of Ciliary and Flagellar Motility. II. Theoretical Considerations." *Biological Bulletin* 145, no. 2: 292–309.

Costello, Donald P., and Catherine Henley. 1971. *Methods for Obtaining and Handling Marine Eggs and Embryos*. Woods Hole, MA: Marine Biological Laboratory.

Flint, Anthony. 1997. "Behind Nobel, a Struggle for Recognition: Some Scientists Say Colleague of Beverly Researcher Deserved a Share of Medical Prize." *The Boston Globe*, 1–4.

Göktepe-Hulten, Devrim, and Prashanth Mahagaonkar. 2010. "Inventing and Patenting Activities of Scientists: In the Expectation of Money or Reputation?" *Journal of Technology Transfer* 35, no. 4: 401–23. https://doi.org/10.1007/s10961-009-9126-2. Accessed 10 December 2020.

Hinsley, Amy, William J. Sutherland, and Alison Johnston. 2017. "Men Ask More Questions Than Women at a Scientific Conference." *PLoS One* 12, no. 10: 1–15. https://doi.org/10.1371/journal.pone.0185534. Accessed 10 December 2020.

Jack, Jordynn. 2013. "'Exceptional Women': Epideictic Rhetoric and Women Scientists in America, 1918–1940." In *Women and Rhetoric Between the Wars*, ed. Elizabeth M. Weiser, Ann George, and Janet Zepernick, 223–39. Carbondale, IL: Southern Illinois University Press.

Latour, Bruno, and Steve Woolgar. 1986. *Laboratory Life: The Construction of Scientific Facts*. Princeton, NJ: Princeton University Press.

Lincoln, Anne E., Stephanie Pincus, Janet Bandows Koster, and Phoebe S. Leboy. 2012. "The Matilda Effect in Science: Awards and Prizes Is the US, 1990s and 2000s." *Social Studies of Science* 42, no. 2: 307–20.

Porter, James. E. 2017. "Professional Communication as Phatic: From Classical Eunoia to Personal Artificial Intelligence." *Business and Professional Communication Quarterly* 80, no. 2: 174–93. https://doi.org/10.1177/2329490616671708. Accessed 10 December 2020.

Rich, Adrienne. 1979. "Toward a Woman-Centered University." In *On Lies, Secrets, and Silence: Selected Prose, 1966–1978*, 125–55. New York: W. W. Norton.

Rossiter, Margaret W. 1982. *Women Scientists in America: Struggles and Strategies to 1940* (Vol. I). Baltimore: Johns Hopkins University Press.

Rossiter, Margaret W. 1993. "The Matthew Matilda Effect in Science." *Social Studies of Science* 23, no. 2: 325–41. https://doi.org/10.1177/030631293023002004. Accessed 10 December 2020.

Rossiter, Margaret W. 1995. *Women Scientists in America: Before Affirmative Action, 1940–1972* (Vol. II). Baltimore: Johns Hopkins University Press.

Settles, Isis H., Lilia M. Cortina, Janet Malley, and Abigail J. Stewart. 2006. "The Climate for Women in Academic Science: The Good, the Bad, and the Changeable." *Psychology of Women Quarterly* 30: 47–58.

Shen, Helen. 2013. "Mind the Gender Gap: Despite Improvements, Female Scientists Continue to Face Discrimination, Unequal Pay and Funding Disparities." *Nature* 495: 22–24. Retrieved from http://link.galegroup.com/apps/doc/A321682869/SCIC?u=unc_main&sid=SCIC&xid=b8babb0f. Accessed 10 December 2020.

CHAPTER 22

Rachel Carson: Scientist, Public Educator and Environmentalist

Ruth Watts

Best known for her powerful, internationally influential yet very controversial book of 1962, *Silent Spring*, the American scientist Rachel Carson (1907–1964), is often cited as beginning the environmental movement (Lytle 2007, pp. 199–257). Previously she had become famous for beautifully written books, articles and pamphlets on the sea, nature and the natural world. As a pioneer environmentalist and ecologist increasingly convinced that in a time of great scientific, technological and political change it was vital to enlarge public understanding of science, her work as a marine biologist and editor in what became the US Fish and Wildlife Service (FWS), had enabled her to gather together the latest scientific knowledge and research. This she conveyed to the public through extraordinary writings. Emphasising that public discourse must be underpinned by rigorous scientific knowledge, she nevertheless retained a deep sense of wonder at the natural world which she believed vital for all to have, not least so that they should appreciate how humanity was in danger of destroying the very eco-system that grants us life. This chapter will analyse Carson's position and role as a scientist, public educator and environmentalist/ecologist.

R. Watts (✉)
School of Education, University of Birmingham, Birmingham, UK
e-mail: r.e.watts@bham.ac.uk

© The Author(s), under exclusive license to Springer Nature Switzerland AG 2022
C. G. Jones et al. (eds.), *The Palgrave Handbook of Women and Science since 1660*, https://doi.org/10.1007/978-3-030-78973-2_22

Education and Early Career: The Development of a Public Educator in Science

From childhood Carson wanted to be a writer and was interested in nature, both interests fostered keenly by her mother, Maria. Together they went out almost daily on their impoverished rural property in Springdale, Pennsylvania, observing, discovering and then studying living things in woods and fields nearby, in keeping with the advice in Anna Comstock's *Handbook of Nature Study* of 1911. This stimulated in Rachel an abiding love and understanding of nature, particularly birds. Maria also introduced Rachel to the *St. Nicholas Magazine* for the young whose regular feature, 'Nature and Science for Young Folks', was underpinned by ideas of the nature studies movement. The magazine featured both well-known authors and promising ones including Rachel, five of whose essays they published before she was 15. 'My Favourite Recreation' in 1922, revealed her early sharp powers of observation of the natural world. From a young age Rachel read avidly and continued to do so throughout her life, particularly loving the works of nature writers, books about the sea and poetry—Tennyson's 'Locksley Hall' being a favourite (Carson [1922] 1998, [1945] 1998; Lear 1998, pp. 11–22; Freeman 1995; Brooks 1989, pp. 5–7; Souder 2012).

Scholarships enabled Carson's attendance at Pennsylvania College for Women (PCW) where one brilliant teacher, Mary Scott Skinker, inspired her to change from English to biological studies and, after graduation, to spend a summer at Woods Hole Marine Biological Laboratory, Massachusetts, Carson's first real experience of the sea. She loved working and socialising in this stimulating yet relaxed coeducational, scientific research community, with its riches of practical and literary research on marine organisms. From here arose her first networking with expert marine biologists and her lifelong love of exploring rocky sections of shoreline at low tide to examine the marine life teeming in tidal pools. She went on research trips collecting plants and animals from the sea floor, her first realisation of the great variety of the interdependent life of the unseen sea world (Lear 1998, pp. 39–63; Carson [1942] 1998, pp. 53–54; Souder 2012, pp. 43–44).

Carson's subsequent studies in marine biology at Johns Hopkins University were also enabled by scholarship, a rare distinction then for women, generally struggling to enter the higher levels of scientific research and employment (Lear 1998; Rossiter 1984). Underprepared for laboratory work, she nevertheless earned her master's through her thesis on 'The development of the pronephros during the embryonic and early larval life of the catfish (Ictalurus punctatus)'. She also imbibed, particularly from her reading of Charles Elton's *Animal Ecology*, ecological principles concerning the food chain and communities of living things being at the heart of all natural systems and the vital importance of field work as well as laboratory work which were to underpin her subsequent scientific work. Family obligations meant she also needed to work part-time; a challenge met by two successive assistantships in scientific research which, in turn, made her lasting, significant scientific contacts and won her accolades from academic colleagues (Lear 1998, pp. 63–76; Lytle 2007, pp. 201–2).

Carson's hopes, however, of pursuing a Ph.D. at Johns Hopkins were dashed by the need to support her family, a situation exacerbated by the Depression, but one that was to pertain for the rest of her life. Instead, she took civil service exams in parasitology, wildlife and aquatic biology to qualify for government service—the only woman competing. Her brilliant results, together with admiration of her lucid style and scientific accuracy, demonstrated in radio scripts she wrote for the Bureau of Fisheries and in local newspaper articles (Carson [1938] 1998), took her in 1936 into the US federal service as an aquatic biologist and editor, a rare appointment for a woman. Her seemingly inferior post was to give her lasting access to core research, frontline researchers and locations as she analysed the data of marine scientists, wrote their reports and informed the public of their findings. Building up vital networks among scientific professionals, she put the wide and deep scientific knowledge she amassed to good use in her writings. A shy woman but a good listener, she developed to become a firm, scrupulous and efficient editor, enlivening work with her zest and humour. Her responsibilities gradually growing over the years, she rose by 1949–1952 to become Editor-in-Chief of all publications for what by then had become the larger US Fish and Wildlife Service (FWS) (Lytle 2007, pp. 39–38; Lear 1998; Brooks 1989).

As a professional civil servant writing public pamphlets on conservation and natural resources, Carson became known for her lyrical prose. Even the factual conservation bulletins of 50–60 pages she wrote during the War to popularise little-known seafoods, showing their nutritive value and where and when to buy them, were praised for a 'vivid exposition' rare in government handbooks. Her ever growing concerns about the human destruction of wildlife and the resultant damage done to the environment and local economy (Carson [1938] 1998, pp. 15–23, [1956] 1998a, [1959] 1998), strengthened her determination, while earning more money for her family, to inform the public on scientific topics and help general readers understand the hidden processes of nature. Thus, she regularly published popular short articles culled from her own research, her knowledge of developments in government science and a rich scientific and semi-popular literature (Carson [1944] 1998).

A prime example of such public education was seen when, from 1946 to 1950, Carson initiated and edited for the FWS a series of mind-awakening booklets entitled *Conservation in Action*, to supply accurate knowledge on the individual national wildlife refuges of the US. She researched and wrote the first five of these herself, exploring down the eastern coast and north-west to Oregon. She was often accompanied in turn by Shirley Briggs and Kay Howe, both graphic designers, photographers and illustrators for the FWS, making an 'island of femininity in the otherwise heavily masculine agency'

(Souder 2012, p. 112). The series became 'a classic in the ordinarily unimaginative world of public information', exemplifying Carson's work in 'exposing popular audiences to the most authoritative science'. A model of public education in style, illustrations and layout, it simultaneously produced accurate, up-to-date and full knowledge on the national wildlife refuges of the US. Importantly, it introduced an ecological viewpoint, Carson combining meticulous science and sensual delight in her findings with fears concerning the deleterious effects of the human pollution of nature (Souder 2012, pp. 114–24; Lytle 2007, p. 70; Carson [1947] 1998). In the fifth booklet, having listed the 'many chapters of reckless waste and appalling destruction' by humans in the Western hemisphere in their 'relatively short history of [...] exploitation of its natural resources', she highlighted the interdependence of living beings—humans and animals—a significant point underlying all her subsequent work. For all, the preservation, conservation and effective use of the basic resources of the earth—wildlife, water, forests and grasslands—were essential (Brooks 1989, pp. 100–1, quoted from Carson 1948).

Rachel Carson: Writer on Science

In a period when the printed page dominated all media, it was Carson's superb literary ability, marshalling masses of up-to-date and thoroughly researched facts in very readable narratives with coherent structure and compelling arguments that made her name as a scientist (Souder 2012, p. 152; Carson [1949] 1998). Her talents appeared best in her trilogy of best-selling books on the sea—a biography of the ocean that brought her even international fame as a naturalist and science writer for the public. Firstly, a magazine article, 'Undersea', in 1937, written to make the mystery and beauty of underwater accessible to the non-scientific reader, argued what 'became her signature themes'—the ancient, lasting ecology of ocean life and the material immortality of all organisms, even the smallest (Carson [1937] 1998, pp. 5–11, [1942] 1998, pp. 55–62). It became the basis of her first book, *Under the Sea Wind*. This drew from both other research of the time and her own extensive observations, explorations and researches, particularly those at Woods Hole and at Beaufort, North Carolina whose fisheries station had the largest research facility on the East Coast. Without anthropomorphising them, she wanted the narrative to be of sea animals themselves and their relentless struggle for survival, the central character being the sea itself. The resulting scientifically accurate yet sympathetic focus on the hazardous lives of a sanderling, a mackerel and an eel, in turn introduced the reader to the intricate ecosystems of the sea coast, the open ocean and the sea bottom as it gently slopes to the deep sea. Readers could visualise the skill and economy of sea creatures in a world, unknown to most, where humans were just one more predator, whose wasteful procedures at least provided food for sea and beach creatures. Without being bathed in scientific labels or technical jargon, readers could gain vivid understanding of the holism of the food chain and the scale and complexity of ocean ecosystems and of the need to value and preserve both. Much to Carson's delight, the book was acclaimed by other scientists, but, coming out as it did just before the Japanese attack on Pearl Harbour, was not the popular success at first that it became when a new edition was published a decade later (Carson 1996; Lear 1961, pp. ix–xv, passim; Brooks 1989, pp. 30–35, 70; Lytle 2007, pp. 41–56).

Carson's unorthodox writings were full of references to the current scientific and technological advances which were transforming biological and ecological sciences and creating new advances and new dangers alike in agriculture and warfare. She supplemented her own wealth of knowledge with that of experts such as the biologist and natural history writer, William Beebe, inventor of the bathysphere (deep-sea submersible) and the oceanographer Henry Bigelow, both eminent American scientists who encouraged her writings (Brooks 1989, pp. 109–14; Kroll 2008). Her second book, *The Sea Around Us* (1951) was an outstanding example of this, conveying a tangible love of dynamic nature plus delight about new discoveries made during and since World War II. The 1961 edition heralded further subsequent breakthroughs. This very readable synthesis of contemporary marine science popularised the new science of oceanography while, respectively, recognising the food chain from the tiniest organisms up to humans. Rejoicing that '[N]othing is wasted in the sea; every particle of material is used over and over again, first by one creature, then by another', Carson was increasingly disturbed by the irretrievable mistake of depositing harmful, especially radioactive, elements in the sea (Carson 1961, Preface, p. 30, passim). She introduced most readers to a wealth of new knowledge as well as new concepts and words such as 'ecology', 'food chain', 'biosphere' and 'ecosystem' (Zwinger 1989, pp. xxi–xxv). Her own view of nature was biocentric, 'informed by a Darwinian approach to evolution, in which humans were just another species, albeit a dangerous and often destructive one' (Lytle 2007, p. 87), Carson reluctantly correcting her earlier optimistic belief that humans could not damage the ocean (Freeman 1995, pp. 248–49). For example, she echoed her prize-winning essay 'The Birth of an Island' of 1950, by deploring the devastation humans caused in the unique oceanic islands, 'natural museums filled with beautiful and curious works of creation, valuable beyond price because nowhere in the world are they duplicated' (Carson 1961, p. 96; Lear 1998, pp. 183, 189).

The Sea around Us became a runaway bestseller, winning her prestigious national awards, including from a range of both literary and scientific societies, translated into 32 languages and allowing her to retire from her government post to concentrate on her research and writing. The general public, literary critics and, importantly, well-known scientists, were amazed by her ability to present such comprehensive, clear, balanced, scientific accounts in such poetic language (Lytle 2007, pp. 70–86; Lear 1998, pp. 198–208, 215–17). In such ways Carson developed public awareness, her fan mail revealing 'an immense and unsatisfied thirst for understanding of the world around us' (Carson [1952] 1998a, p. 96).

Carson's Guggenheim Fellowship in 1951, a rare award for a woman, enabled her to research for what became *The Edge of the Sea*. From her long investigations along the Atlantic coast of the USA, she realised how crucial local environmental conditions were. Alive to the importance of context and the need to comprehend the whole life of a creature or plant, she, unusually, grouped marine subjects not by taxonomy but by the communities they shared and the habitats where they were found. This new type of nature book was based on the study of living marine communities in three main types of coastline: rugged, rocky sea shores adapted to the tides (particularly the Maine coastline that she loved so much and where she had recently bought land and built a cottage to live in every summer); sand beaches from there southward, dominated by the waves; and the coral reefs and mangrove coasts of the far south, governed largely by the ocean currents (Carson 1955, Preface, passim; Brooks 1989, pp. 151–63). With many references to her own extensive and meticulous research and to that of other scientists, and the use of modern technology, she built up an understanding of the beauty, wonders yet harsh realities of the sea, seashore and its creatures and plants. In this way she made clear their evolving adaptation to circumstances, the significance of location and climate to all living things, and the interdependence of all life. Informatively illustrated throughout by Bob Hines and with a comprehensive Appendix on the plants and animals referred to, this became her third highly popular science bestseller (Lytle 2007, pp. 103–12).

In the only purely scientific paper she ever gave to a professional academic organisation, Carson said that her recent work on 'the ecology of the seashore' had made her aware 'of the complex pattern of life'; everything was related 'No thread is found to be complete in itself, nor does it have meaning alone'. Thus, biologists had to be aware of related sciences if they were to understand marine creatures. Her work led her to many questions, including the prescient one of the potentially dangerous effects of climate warming (Carson [1953] 1998a).

Carson's books on the sea, despite being densely packed with up-to-date scientific information, gracefully evoked the wonders and marvels of the earth, sky and sea as well as the creatures and plants which lived on or in them, however small. Yet she was careful not to express in print her almost mystical love of nature, deeply aware as she was of her need for scientific credibility, especially from scientists eager to spot sentimentality or nebulousness. For all her books she researched long, deeply and thoroughly, consulted leading relevant scientists and had them check her chapters. She habitually built up widening networks of scientists and correspondents to support her work, including ones from the Audubon Society, both national and regional, of which she was long an enthusiastic active member (Lytle 2007) and publicly argued against the replacement of professional scientists by political appointees with the change of Government in 1952, fearing commercial aims would threaten conservation of the nation's wildlife resources (Carson [1953] 1998b). Many scientists were excited by the amount of material assembled, the new facts she revealed and her ability to 'translate hard science into digestible prose' (Souder 2012, p. 131). Others were less enthusiastic, seemingly prejudiced against her status as a scientist, especially as she was a writer for the public not academia, and, to boot, a woman. Indeed, they were perplexed both by her poetic sensibility and how a woman could produce work of such scope and complexity (Lytle 2007, pp. 13, 56; Lear 1998, pp. 203, 206, 214). Naturally shy, Carson gradually learned to be a very effective speaker and in some of her increasing number of public speeches she wryly addressed such concerns, amused, for example, both by the man who insisted on addressing her as 'Dear Sir' and by those who even if they accepted her sex, were surprised to find that she was not a tall, oversize, Amazon-type female (Carson [1951] 1998, p. 77). She also explained that, although her years at the FWS let her enter 'certain places where few other women have been', there had been limits to this. However, she stressed how imperative it was for women to realise the threats to the beauty of nature and thus to the means to physical, mental and spiritual health (Carson [1954] 1998).

To those who criticised her scientific writings as being too poetic, Carson replied that she simply wrote 'as the subject demanded': any poetry in her writing was there because it was there in nature. Privately she confessed how important she thought awakening an emotional response to nature was (Carson [1952] 1998b, p. 91; Freeman 1995, p. 231). Equally, in contrast to those scientists suspicious of her popular audience, Carson challenged the remoteness of science and the idea that in a scientific age only a few human beings 'isolated and priestlike in their laboratories' should have scientific knowledge. Science, like biography, history or fiction, aimed to 'discover and illuminate truth', so its literature should not be separate (Carson [1952] 1998b, p. 91). She chivvied nature writers and publishers to realise the 'immense and unsatisfied thirst for understanding the world about us' to which her correspondence from a hugely varied public testified (Carson [1952] 1998a, pp. 94–96).

Carson used a variety of media to educate the public, including brief but successful forays into radio and television, although she hated the documentary film of 1953, supposedly based on *The Sea around Us*, for its scientific inaccuracies and sensationalism (although it won an Academy Award) (Lear 1998, pp. 238–40). In 1957 she wrote a television script on clouds for *Omnibus*, the research for which renewed her interest in global climate change (Carson [1957] 1998). Carson typically used newspapers, nature and women's magazines to spread her ideas on conservation and ecology, as exemplified in her most successful magazine article, 'Our Ever-changing Shore' in 1958, which ended with an eloquent plea to preserve at least some of the nation's seashore for people to explore (Carson [1958] 1998). She was also involved in local conservation projects in Maine (Brooks 1989, pp. 209–12). For the National Council of Teachers of English she wrote an essay on 'Biological sciences' in which she emphasised the new science of ecology, explaining that no living being 'may be studied or comprehended apart from the world in which he [sic] lives'. Any student's first conscious acquaintance with biology, however, should come through nature itself before any research in the laboratory, as had been her own case and that of many other biologists and memorable nature writers (Carson [1956] 1998b; Freeman 1995, p. 7).

Carson emphasised this point in an article in the *Women's Home Companion* in 1956, entitled 'Help your child to wonder' (published posthumously in book form as *The Sense of Wonder*). Written to help parents and teachers of children awaken an appreciation of nature in children, she urged that this should be through enjoyment of the beauties of nature through the senses rather than teaching children facts they were too young to assimilate. She vividly exemplified this educative method from her own recent explorations of woods and shoreline near her beloved cottage on Southport Island, Maine, with her little great-nephew, Roger whom she later adopted when he was orphaned. She cautioned adults that outgrowing the capacity for wonder and awe alienated them from their 'sources of strength' (Carson 1965b, pp. 22–23, 59). Carson's credibility relied on her insistence on scientific accuracy, yet she insisted, 'it is not half so important to *know* as to *feel*. [...] It is learning again to use your eyes, nostrils and finger tips, opening up the disused channels of sensory impression. [...] For most [...] knowledge of our world comes largely through sight, yet we look about with such unseeing eyes that we are partially blind' (Carson 1965b, pp. 23, 27–28).

In particular, Carson wanted children to enjoy the beauties of nature as she and little Roger did, exploring the world of small and inconspicuous things in nature, often ignored by adults, although some 'of nature's most exquisite handiwork is on a miniature scale, as anyone knows who has applied a magnifying glass to a snowflake'. Using such inexpensive tools, you could examine with children 'objects you take for granted as commonplace or uninteresting' (Carson 1965b, pp. 4–13, 34). Adults did not need to know detailed facts about nature but be alive to its many 'avenues of delight and discovery', listen and talk to children and help them develop a 'sense of awe and wonder', for those who dwell 'among the beauties and mysteries of earth are never alone or weary of life' (Carson 1965b, pp. 37–49).

This article was unusual for Carson in allowing her to display her deep love of the nature she scientifically analysed. Yet the way she was 'alert for every impression, with keen delight in all manner of small creatures as well as the vast horizons and far reaches' taught her friend and artist collaborator Shirley Briggs, a new 'way of seeing' back in the 1940s and Carson's thrilled wonder lasted throughout her life (Brooks 1989, pp. 84, 159; Freeman 1995).

'Silent Spring' and Its Aftermath

Thus, with many awards and mounting fame, Carson could attempt a very different type of book. At the time of the Cold War, Sputnik and nuclear anxieties, she was becoming increasingly alarmed that rapidly growing industrial developments were destroying the natural environment: 'It is one of the ironies of our time that, while concentrating on the defense [sic] of our country against enemies from without, we should be so heedless of those who would destroy it from within' (Carson [1953] 1998b, p. 100; Freeman 1995, pp. 248–49). Since the 1940s, together with a small but growing number of scientists in both the USA and abroad, she had been disturbed by the effects of new synthetic chemical pesticides sprayed unremittingly in large quantities in US agriculture, especially the seemingly wonder pesticide, dichlorodiphenyltrichloroethane (DDT). These pesticides were wiping out not only many insects but also, as a rising number of case studies illustrated, the creatures which fed on them, yet the surviving insects were emerging stronger and resistant to the insecticides (Carson [1959] 1998; Brooks 1989, pp. 229–46; Souder 2012, pp. 244–96). Carson, initially reluctant to lead opposition to this, ceaselessly sought the advice of many experts: a widening network of scientists, sympathetic government officials and concerned citizens, including those from the Audubon Society, humane societies and garden clubs. Her editor at Houghton Mifflin, Paul Brooks, and her network of close women supporters, particularly Dorothy Freeman and Carson's agent Marie Rodell, were crucial in giving her moral and practical support. Her ecological outlook developing—eschewing some other conservationists' view that humans and nature were opposed—prompted her to investigate the possible effects of pesticides on human body cells, for example, cancer relationships, led her to consult many distinguished physicians (Paull 2013; Brooks 1989, pp. 247–55; Lytle 2007; Lear 1998, pp. 320–46, 357, 362). The ensuing work was replete with scientific evidence yet written as usual in persuasive prose (Carson 1965a).

Silent Spring, first published in 1962, was written to inform people of the new dangers of synthetic pesticides, unnatural 'creations of man's [sic] inventive mind' (Carson 1965a, p. 24). She was not against all use of these but did 'contend that we have put poisonous and biologically potent chemicals into the hands of persons largely or wholly ignorant of their potentials for harm. We have subjected enormous numbers of people to contact with these poisons, without their consent and often without their knowledge' (Carson 1965a, p. 29).

Carson's detailed synopses of practises by chemical companies, agricultural scientists and the government, warned of the devastating effects of the unrestrained use of scientific developments without sufficient testing. 'By their very nature' she argued, 'chemical controls are self-defeating, for they have been devised and applied without taking into account the complex biological systems against which they have been blindly hurled' (Carson 1965a, p. 214).

Tapping into contemporary concerns, she opened with a fable of a desolate, stricken town with no new life, the victim of the community's unthinking contamination of nature. Then, in a telling narrative with chapters headed by phrases such as 'And no birds sing', 'Rivers of death' and 'Indiscriminately from the skies', she built up to 'The human price'. She explained how through the 'web of life' or ecology, the environment was weakened, while humans were storing up toxic materials in their bodies from birth from constantly absorbing minute portions of these poisons, including from daily food, itself in a food chain attacked by such harmful carcinogens in increasing ways. As she depicted human bodies facing considerable health perils, Carson also drew a metaphorical picture of bodies of scientific and technical knowledge running amok in their headlong rush to control nature for the shorter-term gains of humans. Yet, although chemical companies poured money into universities for research on insecticides, little was spent in them on research into biological controls (Carson 1965a, p. 214). Crucially, she questioned who had the right to decide on these matters, to decide on a 'sterile' world without insects or birds, 'during a moment of inattention by millions to whom beauty and the ordered world of nature still have a meaning' (Carson 1965a, p. 121). She ended with a warning against the arrogant, 'Neanderthal' attempt of controlling nature for 'the convenience of man' [sic] and turning 'terrible weapons [...] against the earth' (Carson 1965a, p. 257).

The research difficulties of amassing and checking the evidence from a range of disciplines and current cases, structuring and writing it up in the best way possible were compounded by knowing the results would arouse hostility in much of the chemical, agricultural and medical establishment. Such pressures were exacerbated by both distressing family problems and the fact that during these years, Carson was struggling with spreading cancer, compounded by heart disease, an ulcer, iritis and rheumatoid arthritis. She masked her condition to all but the very closest of friends so that chemical companies would not think that her arguments were affected by her condition, although her notes reveal she was collecting pertinent evidence long before she knew she had cancer. Even so, she insisted to Dorothy Freeman that she 'could never again listen happily to a thrush song if I had not done all I could' (Freeman 1995, pp. 282, 395; Lytle 2007; Brooks 1989, p. 25).

The condensed version of *Silent Spring* published in the prestigious *New Yorker* in three parts in June 1962 was an immediate sensation even before the book appeared in October. It stimulated President Kennedy, whom Carson had actively supported for his push towards conservation and against radioactive pollution, into establishing a special investigation which, in 1963, vindicated her findings. Carson won many awards and in public speeches, including before the US Senate Subcommittee on environmental hazards (where she asked for strict control of spraying of pesticides and reduced to the minimum strength to be effective), she came over as an accomplished scientist, brilliant writer and woman of conscience (Lear 1998, pp. 408,

451–55; Brinkley 2012). Dr. Wilhelm C. Hueper of the National Cancer Institute, for example, praised Carson for her sincerity, 'unusual degree of social responsibility', courage and 'sound and valid' scientific facts and 'reasonable' interpretations (Brooks 1989, p. 255).

Carson built up a 'formidable cadre of allies' of scientists, women in public life and conservationists. Her agent Rodell ensured that Houghton Mifflin inscribed on the jacket version of and publicity materials for *Silent Spring* her degrees and professional scientific work, yet Carson's conclusions were bitterly disputed. Continuous virulent opposition to her arguments arose from entomologists, chemical companies and agribusinesses—powerful, influential groups. DDT, especially, was long to become the icon for those on both sides of the arguments concerning its toxicity. Yet, although Carson had not discussed the benefits of chemical pesticides, she never argued for abandoning them but for their risks to be properly researched, assessed and publicly debated and voted upon (Murphy 2005, p. 44; Maguire 2008). There was an enormous, expensive campaign mounted against her. Chemical companies and related university scientists were furious at her warnings about the various liaisons between industry and university science and alarmed that such a popular writer was tilting at their authority, reputation and sources of funding and possibly stimulating increased regulation. Government departments, especially the Nutrition Foundation, were also wary for their reputation. They too turned to the media where a huge campaign questioned Carson's scientific credibility while asserting its own, simultaneously undermining her as a cat-loving, sentimental, hysterical spinster and attacking her as fanatical or member of some subversive cult, even a 'Communist' (Lytle 2007). Brooks stated that *Silent Spring* was more bitterly attacked than any single book since Darwin's *Origin* arguing that her detractors had realised that Carson was actually questioning the 'basic irresponsibility of an industrial society toward the natural world' and refusing to accept that the 'damage to nature was the inevitable cost of "progress"' (1989, p. 293).

Priscilla Coit Murphy has detailed the way in which both Carson's team—agent, publisher, editors, supportive scientists, conservationists and naturalists, including Audubon societies—and the different strands of the opposition, utilised the media including radio and burgeoning national television. In the latter, Eric Sevareid's *CBS Reports* programme, eagerly viewed by millions, secured the appearance of a very calm, articulate and effective (though really very sick) Carson, faced by a loud-voiced, arrogant and white-coated Dr. Robert White-Stevens plus government officials who appeared to be woolly and confused. A stream of publications issued from scientific and industrial bodies, including widely distributed pamphlets. In particular, *The Desolate Year*, Monsanto's vivid pastiche of *Silent Spring*'s opening fable dramatised 'the horrors of a world without pesticides' (Murphy 2005, passim; Lear 1998, pp. 435–50). Newspapers and magazines varied in their reactions—their locality and the interests they represented usually conditioning their support or opposition—while thousands of related letters varied in their support, a huge engagement in the public domain (Murphy 2005, pp. 133–82).

Carson defended herself in speeches, public hearings and in print, despite illness. She refuted inaccuracies and statements she had never made, scornfully resisting the reviewer who called her giving sources of her information as '*name-dropping*' and her bibliography 'padding', commenting 'Well, times have certainly changed since I received my training in the scientific method at Johns Hopkins!' (Carson 1998b, pp. 201–10; Brooks 1989, pp. 308–16). She won some unexpected allies such as a strong grassroots force of hunters and outdoorsmen who fulminated about the power of large corporations to use science and technology to poison their environment (Hazlett 2004, pp. 711–13). She presented new evidence and, as an 'ecologist', stressed that we all, 'like all other living creatures', live within a continuous, never-ending 'complex, dynamic interplay of physics, chemical, and biological forces' so that no dumping of waste anywhere lacked ecological consequences (Carson [1963] 1998a, [1963] 1998b). At the same time further pollution scandals such as that in Mississippi, traced back to one of her leading opponents—Velsicol—helped cement her science (Lytle 2007, p. 190). By the time she died in April 1964, *Silent Spring* had sold nearly a million copies and had been both rapturously praised and violently excoriated throughout America. In print ever since and translated into 32 languages, it has been termed 'one of those rare books that changed the course of history' (Brooks 1989, p. 227; Lear 1998, pp. 215, 422–72; Murphy 2005, p. 17).

Carson certainly generated a lasting debate with enormous impact, raising public awareness and activity on both sides of the Atlantic. She has been much credited as the inspiration for the establishment of the vitally important US Environmental Protection Agency (EPA), a raft of environmental legislation and the environmental movement. Much of this took place after her death but even before then strong evidence proving her arguments was emerging (Souder 2012, pp. 391–96; US Fish and Wildlife Service). On the other hand, bitter, often virulent, divisions over environmentalism have lasted ever since, owing much in the US certainly, to changing political regimes. Fresh onslaughts on Carson's scientific credibility from the 1990s included assertions that she was responsible for 'literally millions of deaths among the poor people of underdeveloped nations' because of the bans on DDT from 1972, and was a part of a modern 'dark age of anti-science ignorance' (Lytle 2007, pp. 212–20; Edwards 1992; Hecht 1992). Counter arguments indicate that Carson has become a useful anti-heroine to modern neo-liberals who attack environmentalism and an expanded state while using considerable skills to keep the public in two minds about ecology (Hecht 2012).

WAS CARSON A SCIENTIST?

The continued debate about Carson and ecology raises the question of her scientific credibility. Self-described as 'a biologist whose special interests lie in the field of ecology, or the relation between living things and their environment [...]' (Carson [1960] 1998, p. 195) she was, however, attacked for writing for the public not academia, having no doctorate, academic appointment or publications in peer-reviewed journals and being a woman. All this was fuelled by resentment at her confrontation with the scientific establishment and industry, both male-dominated (Lear 1998, p. 430).

It is true that Carson was not an academic research scientist, but, at a time when few women were, she became a respected scientific reporter and editor. She was meticulous in her secondary research, always consulting a wide range of scientists and scientific writings, keeping up-to-date with and carefully checking the latest scientific thinking. She was very aware that knowledge was constantly being transformed (as was illustrated by the first chapter of *The Sea around Us*). She loved networking and drawing together different facets of science, valuing not only new technologically sophisticated research but also direct personal observations and fieldwork. Scientists appreciated her as 'someone who made their work accessible' while she saw giving essential, well-illustrated facts vital to a public ill or uninformed by scientists and industry. She enjoyed the praise of respected scientists and the awards of prestigious scientific societies which continued until her death (Lytle 2007; Souder 2012, p. 149; Freeman 1995; Brooks 1989; Lear 1998, pp. 286–89). As an early scientific communicator she had to write in a non-technical way, although some scientists were appalled that *Silent Spring* began with an allegory (Carson 1998a, p. 197). But Carson delighted in making complicated scientific subjects into works of literature, despairing of 'illiterate scientists' (Freeman 1995, p. 62; Brooks 1995, p. xxviii). Very importantly, she conveyed much new knowledge about the physical environment and ecological concepts to a wide audience, teaching geology, physics and biology gently in beguiling prose (Souder 2012, p. 6).

Furthermore, Carson was significant and pioneering in challenging anthropocentric values with her ecological conception of a biotic community, in her environmental ethics and her awareness of climate change (Cafaro 2008; Souder 2012). A strong Darwinian evolutionist, she saw no conflict between this and religious belief. For her, 'To stand at the edge of the sea, to sense the ebb and flow of the tides, to feel the breath of a mist moving over a great salt marsh, to watch the flight of the shore birds that have swept up and down surf lines of the continents for untold thousands of years, to see the running of the old eels and the young shad to the sea, is to have knowledge of things that are as nearly eternal as any earthly life can be' (Carson 1996, p. 1; Sideris 2008). Above all, she deeply wished everyone to retain such a sense of wonder and pursue the meaning of science so that they would understand the beauties and mysteries of the earth and the necessity to care for, not pollute it. In this she was neither mystical nor sentimental but urging a new way of thinking (Norwood 2008, pp. 251–66; Moore 2008, pp. 267–80; Brooks 1989; Burnside 2002, pp. 1–2).

Yet, after *Silent Spring* especially, Carson was often termed by opponents as 'mystical', 'hysterical' and 'emotional' thus irrational not 'scientific', conforming to gender stereotypes. Although Carson worked well with many male scientists, authors and editors and gradually achieved much in science for a woman of that era, her pioneering holistic views of nature and her fluid, poetic style helped rouse contemporary prejudices. Some critics invoked images of witchcraft in charging her with raising social as well as scientific chaos by opposing 'masculine' technological, scientific and industrial progress. Most of Carson's critics were men who did not want a woman shaking the public's credulity in the ability of science and technology to solve all problems, especially at the time when America's rapidly expanding economic and technological power was challenged by the thalidomide scandal and nuclear fallout. Fundamentally Carson was radical and subversive in questioning human rights to upset the balance of nature at will (Hazlett 2008; Smith 2008). Furthermore, she raised domestic, reproductive and sexual anxieties of continuing relevance by stressing how pesticides made their way into both home and body and possibly had transgenerational effects (Langston 2012).

Carson never fought for women's rights, but she took great satisfaction in winning awards usually reserved for men. Her deepest friendships were with women, for example, her early mentor Mary Skinker, her agent Marie Rodell, artist Shirley Briggs and a few others. Carson's mother, who shared her reverence for nature and literature, lived with her, long doing her housekeeping and typing, until her death in 1959. In particular over the last 12 years of her life, Carson enjoyed a close friendship with Dorothy Freeman who lived near her with her husband each summer. Their intense correspondence shows their deep emotional ties and their shared delights in nature, classical music, literature, pets and their families (Hynes 1989; Freeman 1995).

Conclusions

Rachel Carson was a public educator in science, employing her special literary talents to bring a more holistic understanding of scientific and ecological issues before the public. She developed important networks related to her fields in both science and literature which helped her struggle courageously against powerful vested interests and contemporary awe of science and technology. She raised issues of the political dimensions of knowledge and funding and demonstrated that, as in all public knowledge, thought is needed on where knowledge is coming from and who is speaking. Thus, she combined literary talents and scientific knowledge to alert the public to the burgeoning conservation movement and ecological concerns, to new scientific discoveries, and to some of the unseen wonders of nature around them. This set in motion the concern for environmentalism and ecology that was to become a worldwide movement. Despite renewed virulent disputes on climate change, the destruction of nature, conservation and attacks on the EPA, this movement has grown in strength internationally. In our present times it is championed by ecologists and environmentalists backed by politicians, journalists and environmental activists such as Al Gore, Naomi Klein, Bill McGibben and Tony Juniper (Juniper 2013).

Silent Spring, particularly, never out of print and widely considered the most important environmental book of the twentieth century, has affected many scientists including Sandra Steingraber who learned through Carson how to make visible the 'intercourse between our bodies and the environments [they] inhabit' (Steingraber 2008) and Patricia de Marco who, despite many of Carson's worst fears being realised, calls her, 'one of the most influential thought leaders of the twentieth century [...] a role model for the application of science in public policy' (DeMarco 2017). Paul Brooks recalled that Carson believed it 'would be unrealistic to believe one book could bring a complete change', adding '[B]ut it did' (Brooks 1989, p. 12).

Mark Hamilton Lytle who has detailed his own conversion to and use of environmental history, said Carson's 'dedication as a scientist made her credible; her gifts as a writer made her inspirational [...] [She and her work] breathed life into a fledgling environmental movement' (Lytle 2007, pp. 191, 199–257). He termed Carson a 'subversive' who 'encouraged an entire generation of Americans to rethink fundamental values defining the relationship between human beings and nature [...] [so] they take responsibility for the wanton destruction of nature done by others on their behalf and [thus]reduce the growing threat to life itself' (Lytle 2007, p. 237).

References

Brinkley, Douglas. 2012. "Rachel Carson and JFK, an Environmental Tag Team." *Audubon Magazine*. http://www.audubon.org/magazine/may-june-2012/rachel-carson-and-jfk-environmental-tag-team. Accessed 20 May 2015.

Brooks, Paul. 1989, 1st ed.1972. *The House of Life: Rachel Carson at Work*. Boston, MA: Houghton Mifflin Company.

Brooks, Paul. 1995. Introduction to *Always, Rachel. The Letters of Rachel Carson and Dorothy Freeman, 1952–1964*, by Martha Freeman, xxii. Boston: Beacon Press.

Burnside, John. 2002. "Reluctant Crusader." *Guardian*, May 18.

Cafaro, Philip. 2008. "Rachel Carson's Environmental Ethics." In *Rachel Carson: Legacy and Challenge*, ed. Lisa H. Sideris and Kathleen Dean Moore, 60–78. Albany: State University of New York Press.

Carson, Rachel. [1922] 1998. "My Favourite Recreation." In *Lost Woods: The Discovered Writing of Rachel Carson*, ed. Linda Lear, 12–13. Boston: Beacon Press.

Carson, Rachel. [1937] 1998. "Undersea." *The Atlantic Monthly*, 1937, 160: 322–25. In *Lost Woods: The Discovered Writing of Rachel Carson*, ed. Linda Lear, 3–11. Boston: Beacon Press.

Carson, Rachel. [1938] 1998. "Fight for Wildlife Pushes Ahead/Chesapeake Eels Seek the Sargasso Sea." *Baltimore Sunday Sun*, 9 October 1938. In *Lost Woods: The Discovered Writing of Rachel Carson*, ed. Linda Lear, 14–23. Boston: Beacon Press.

Carson, Rachel. [1942] 1998. "Memo to Mrs. Eales on *Under the Sea-Wind*." *RC Papers*, 1942. In *Lost Woods: The Discovered Writing of Rachel Carson*, ed. Linda Lear, 53–62. Boston: Beacon Press.
Carson, Rachel. [1944] 1998. "Ace of Nature's Aviators." *MSS RC Papers*, 1944. In *Lost Woods: The Discovered Writing of Rachel Carson*, ed. Linda Lear, 24–29. Boston: Beacon Press.
Carson, Rachel. [1945] 1998. "Road of the Hawks." In *Lost Woods: The Discovered Writing of Rachel Carson*, ed. Linda Lear, 30–32. Boston: Beacon Press.
Carson, Rachel. [1947] 1998. "Mattamusket: A National Wildlife Refuge." *Conservation in Action*, 4. U.S. Fish and Wildlife Service, Washington, DC: U.S. Government Printing Office, 1947. In *Lost Woods: The Discovered Writing of Rachel Carson*, ed. Linda Lear, 41–49. Boston: Beacon Press.
Carson, Rachel. 1948. "Guarding Our Wildlife Resources." *Conservation in Action* 5, no. 1. U.S. Fish and Wildlife Service, Department of the Interior, Washington, DC: Government Printing Office.
Carson, Rachel. [1949] 1998. "Lost Worlds: The Challenge of the Islands." *The Wood Thrush*, 1949, 4, no. 5: 179–87. In *Lost Woods: The Discovered Writing of Rachel Carson*, ed. Linda Lear, 63–75. Boston: Beacon Press.
Carson, Rachel. [1951] 1998. "New York Herald-Tribune Book and Author Luncheon Speech." *RC Papers* New York, 16 October 1951. In *Lost Woods: The Discovered Writing of Rachel Carson*, ed. Linda Lear, 76–82. Boston: Beacon Press.
Carson, Rachel. [1952] 1998a. "Design for Nature Writing." *Atlantic Naturalist*, May–August 1952, 232–34. In *Lost Woods: The Discovered Writing of Rachel Carson*, ed. Linda Lear, 93–97. Boston: Beacon Press.
Carson, Rachel. [1952] 1998b. "Remarks at the Acceptance of the National Book Award for Nonfiction." New York, 29 January 1952. In *Lost Woods: The Discovered Writing of Rachel Carson*, ed. Linda Lear, 90–92. Boston: Beacon Press.
Carson, Rachel. [1953] 1998a. "The Edge of the Sea." *RC Papers*, 1953. In *Lost Woods: The Discovered Writing of Rachel Carson*, ed. Linda Lear, 133–46. Boston: Beacon Press.
Carson, Rachel. [1953] 1998b. "Mr Day's Dismissal." *Washington Post*, 22 April 1953, A26. In *Lost Woods: The Discovered Writing of Rachel Carson*, ed. Linda Lear, 98–100. Boston: Beacon Press.
Carson, Rachel. [1954] 1998. "The Real World Around Us." *RC Papers*, 1954. In *Lost Woods: The Discovered Writing of Rachel Carson*, ed. Linda Lear, 147–63. Boston: Beacon Press.
Carson, Rachel. 1955. *The Edge of the Sea*. Boston, USA: Houghton Mifflin Company.
Carson, Rachel. [1956] 1998a. "The Lost Woods: A Letter to Curtis and Nellie Lee Bok." *RC Papers*, December 12, 1956. In *Lost Woods: The Discovered Writing of Rachel Carson*, ed. Linda Lear, 172–74. Boston: Beacon Press.
Carson, Rachel. [1956] 1998b. "Biological Sciences." *Good Reading*, 1956. In *Lost Woods: The Discovered Writing of Rachel Carson*, ed. Linda Lear, 164–67. Boston: Beacon Press.
Carson, Rachel. [1957] 1998. "Clouds." CBS *Omnibus*, 11 March 1957. In *Lost Woods: The Discovered Writing of Rachel Carson*, ed. Linda Lear, 175–86. Boston: Beacon Press.
Carson, Rachel. [1958] 1998. "Our Ever-Changing Shore." *Holiday*, 1958, 24: 71–120. In *Lost Woods: The Discovered Writing of Rachel Carson*, ed. Linda Lear, 113–24. Boston: Beacon Press.

Carson, Rachel. [1959] 1998. "Vanishing Americans." *Washington Post*, 10 April 1959. A26. In *Lost Woods: The Discovered Writing of Rachel Carson*, ed. Linda Lear, 189–91. Boston: Beacon Press.

Carson, Rachel. [1960] 1998. "To Understand Biology/Preface to *Animal Machines*." *Humane Biology Projects*, 1960. In *Lost Woods: The Discovered Writing of Rachel Carson*, ed. Linda Lear, 192–96. Boston: Beacon Press.

Carson, Rachel. 1961, 1st ed. 1951. *The Sea Around Us*. New York: Oxford University Press.

Carson, Rachel. [1963] 1998a. "A New Chapter to *Silent Spring*." *RC Papers*, 1963. In *Lost Woods: The Discovered Writing of Rachel Carson*, ed. Linda Lear, 211–22. Boston: Beacon Press.

Carson, Rachel. [1963] 1998b. "The Pollution of Our Environment." *RC Papers*, 1963. In *Lost Woods: The Discovered Writing of Rachel Carson*, ed. Linda Lear, 227–45. Boston: Beacon Press.

Carson, Rachel. 1965a, 1st ed. 1962. *Silent Spring*. London: Penguin.

Carson, Rachel. Publ. posthumously, 1965b. *The Sense of Wonder*. New York: Harper & Row.

Carson, Rachel. 1996, 1st ed. 1941. *Under the Sea-Wind*. London: Penguin.

Carson, Rachel. 1998a. "Women's National Press Club Speech." In *Lost Woods: The Discovered Writing of Rachel Carson*, ed. Linda Lear, 201–10. Boston: Beacon Press.

Carson, Rachel. 1998b. "A Fable for Tomorrow." In *Lost Woods: The Discovered Writing of Rachel Carson*, ed. Linda Lear, 197–200. Boston: Beacon Press.

DeMarco, Patricia. 2017. "Rachel Carson's Environmental Ethic—A Guide for Global Systems Decision Making." *Journal of Cleaner Production* 140, no. 1: 127–33.

Edwards, J. Gordon. 1992. "The Lies of Rachel Carson." *21st Century Science and Technology Magazine*. http://www.21stcenturysciencetech.com/articles/summ02/Carson.html. Accessed 20 May 2015.

Freeman, Martha, ed. 1995. *Always, Rachel: The Letters of Rachel Carson and Dorothy Freeman, 1952–1964*. Boston: Beacon Press.

Hazlett, Maril. 2004. "'Woman vs. Man vs. Bugs': Gender and Popular Ecology in Early Reactions to *Silent Spring*." *Environmental History* 9, no. 4: 701–29.

Hazlett, Maril. 2008. "Science and Spirit: Struggles of the Early Rachel Carson." In *Rachel Carson: Legacy and Challenge*, ed. Lisa H. Sideris and Kathleen Dean Moore, 149–67. Albany: State University of New York Press.

Hecht, David K. 2012. "How to Make a Villain: Rachel Carson and the Politics of Anti-Environmentalism." *Endeavour* 36, no. 4: 149–55.

Hecht, Marjorie Mazel. 1992. "Bring Back DDT and Science with It!" *21st Century Science and Technology Magazine*, Editorial. http://www.21stcenturysciencetech.com/articles/summ02/DDT.html. Accessed 20 May 2015.

Hynes, H. Patricia. 1989. *The Recurring Silent Spring*. New York: Pergamon Press.

Juniper, Tony. 2013. *What Has Nature Ever Done for Us?* London: Profile Books.

Kroll, Gary. 2008. "Rachel Carson's *The Sea Around Us*, Ocean-Centrism, and a Nascent Ocean Ethic." In *Rachel Carson: Legacy and Challenge*, ed. Lisa H. Sideris and Kathleen Dean Moore, 118–35. Albany, NY: State University of New York Press.

Langston, Nancy. 2012. "Rachel Carson's Legacy: Endocrine Disrupting Chemicals and Gender Concerns." *GAIA* 21, no. 3: 225–29.

Lear, Linda. 1961. Introduction to *Under the Sea-Wind*, by Rachel Carson, ix–xx. London: Penguin.

Lear, Linda. 1998, 1st ed. 1997. *Rachel Carson: The Life of the Author of Silent Spring*. London: Allen Lane, The Penguin Press.

Lytle, Mark Hamilton. 2007. *The Gentle Subversive: Rachel Carson, Silent Spring, and the Rise of the Environmental Movement*. New York and Oxford: Oxford University Press.

Maguire, Steve. 2008. "Contested Icons: Rachel Carson and DDT." In *Rachel Carson: Legacy and Challenge*, ed. Lisa H. Sideris and Kathleen Dean Moore, 194–214. Albany: State University of New York Press.

Moore, Kathleen Dean. 2008. "The Truth of the Barnacles: Rachel Carson and the Moral Significance of Wonder." In *Rachel Carson: Legacy and Challenge*, ed. Lisa H. Sideris and Kathleen Dean Moore, 267–80. Albany: State University of New York Press.

Murphy, Priscilla Coit. 2005. *What a Book Can Do: The Publication and Reception of Silent Spring*. Amherst and Boston: University of Massachusetts Press.

Norwood, Vera. 2008. "How to Value a Flower: Locating Beauty in Toxic Landscapes." In *Rachel Carson: Legacy and Challenge*, ed. Lisa H. Sideris and Kathleen Dean Moore, 251–66. Albany: State University of New York Press.

Paull, John. 2013. "The Rachel Carson Letters and the Making of *Silent Spring*." *Sage Open* 3/3. https://doi.org/10.1177/2158244013494861.

Rossiter, Margaret W. 1984, 1st ed. 1982. *Women Scientists in America: Struggles and Strategies to 1940*. Baltimore, MD: The Johns Hopkins University Press.

Sideris, Lisa H. 2008. "The Secular and Religious Sources of Rachel Carson's Sense of Wonder." In *Rachel Carson: Legacy and Challenge*, ed. Lisa H. Sideris and Kathleen Dean Moore, 232–50. Albany: State University of New York Press.

Smith, Michael. 2008. "'Silence, Miss Carson!': Science, Gender, and the Reception of *Silent Spring*." In *Rachel Carson: Legacy and Challenge*, ed. Lisa H. Sideris and Kathleen Dean Moore, 168–82. Albany: State University of New York Press.

Souder, William. 2012. *On a Farther Shore: The Life and Legacy of Rachel Carson*. New York: Crown.

Steingraber, Sandra. 2008. "Living Downstream of *Silent Spring*." In *Rachel Carson: Legacy and Challenge*, ed. Lisa H. Sideris and Kathleen Dean Moore, 220–29. Albany: State University of New York Press.

US Fish and Wildlife Service. *Remembering Rachel*. www.fws.gov/rachelcarson/resources/FWS_perspectives.pdf. Accessed 31 August 2015.

Zwinger, Ann W. 1989, 1st ed. 1951. Introduction to *The Sea Around Us*, by Rachel Carson, xix–xxvii. New York: Oxford University Press.

CHAPTER 23

Representing Women in STEM in Science-Based Film and Television

Amy C. Chambers

Penny: Oh, wow, a girl scientist.
Leslie: Yep, come for the breasts, stay for the brains.
—'The Hamburger Postulate', *The Big Bang Theory*.

The first woman scientist to appear on the popular science-based sitcom *The Big Bang Theory* (2007–2019) was experimental physicist Leslie Winkle (Sara Gilbert) whose apparently humorous introduction relies on the assumed expectations of both the audience, and their non-scientist mediator Penny (Kaley Cuoco), that a woman scientist is an anomaly. Winkle was not made a series regular and appeared as one of only a small sample of women in the hard sciences, women who were notably defined by their scientific acumen *and* their sexual promiscuity.[1] The physical sciences were left almost entirely up to the men in the show and when women scientist characters were added as regulars, they were both bio-scientists. *The Big Bang Theory* is situated in a history of women scientists' representation which presents them as 'practical, beautiful, anomalous (as women in a male-dominated field), and young' (Szwydky and Pribbernow 2018, p. 307).

Historically, women in STEM (science, technology, engineering, mathematics) on screen have been offered limited representation regardless of

---

A. C. Chambers (✉)
Department of English, Manchester Metropolitan University, Manchester, UK
e-mail: amy.c.chambers@mmu.ac.uk

whether they are biographical or imagined. They are often defined by their male counterparts (mentors, fathers, brothers, lovers) and framed as 'sci-candy' or plain brains rather than fully realized scientists (Attenborough 2011). With broader and more diverse representation of the sciences on screen it may be possible to begin shifting the 'deeply embedded' expectations of what science *is* and scientists *are* (and what they look like), which forms part of 'the institutional fabric of [our] culture' (O'Keeffe 2013, p. 18). Scientists are expected to look like older white men, but fictional media can be utilized to unsettle ideas about 'who should study, practice, and deploy science and technology' (Colatrella 2011, p. 8). Representation of scientists as experts who do not simply reflect the current biases of science (straight, white, male) in both fiction and non-fiction (Chambers and Thompson 2020), can offer role models and examples of women scientists who are not defined by their gender and race but by their scientific proficiency, thus showing that there is space and a place for everyone in the sciences.

Contemporary examples of science-based films with complex woman scientist characters such as *Hidden Figures* (Melfi, 2016), *Black Panther* (Coogler, 2018), and *Annihilation* (Garland, 2018) offer some alternative approaches to the representation of women's scientific expertise and role in science-based narratives. *Black Panther*'s Shuri (Letitia Wright) is Wakanda's chief scientist and technological innovator who, as Carol Azunghi Dralega argues, 'demonstrates through playful and highly cerebral manifestations what a young, black female (who is not adulterated by colonial and socially constructed Western hegemonic femininity and indeed toxic masculinity) can achieve when presented with opportunity, resources and a conducive environment' (2018, p. 463). Notably, Shuri is not an isolated woman with power and ability but rather one of many in a society where women are also dominant in other traditionally male-dominated fields including the military and government. Films like *Black Panther* importantly position women scientists as protagonists within a community of professional agentic women, and as women who can handle and excel in the 'hard' sciences including mathematics, engineering, and physics. Their plentiful success shows that increasing the number of women scientists in fictional film and television shows, and having women-led science-based stories, does not constitute a barrier to critical and financial success.

This chapter focusses on the representation of women scientists in contemporary mainstream fiction film and television and considers the different ways in which we might approach the framing and proliferation of women's representation as scientists. When discussing women scientists, it is not possible to categorize discussions within a single film genre. Eva Flicker uses the label 'fiction film with a scientific theme' to indicate films that 'are set in a scientific milieu' (2008, p. 242). Here, I will use the shorter term 'science-based' to indicate a similar concept. Women scientists can be seen across a broad spectrum of genres including, but not limited to, science fiction (e.g. Ellen Ripley [Sigourney Weaver] in the *Alien* franchise), forensic/police procedurals (e.g.

Clarissa Mullery [Liz Carr] from *Silent Witness* [1996–]), medical dramas (e.g. Jing-Mei Chen [Ming-Na Wen] in *ER* [1994–2009]), and comedies (e.g. Erin Gilbert [Kristen Wiig] in *Ghostbusters* [2016]). In all of these examples and those to follow, scientific concepts, scientific expertise, and representations of scientific culture, or what David A. Kirby terms 'the systems of science' (2008, p. 42), are central to each form of science-based fiction.

1966 was a key moment in the representation of women scientists on screen with the broadcast of the series *Star Trek* (1966–1969) that included the iconic communications officer Uhura (Nichelle Nichols). Uhura is a 'powerful image' of an African-American woman with scientific expertise that had 'a ripple effect that lasted for decades' (O'Keefe 2013, p. 19). The show, the character, and the actress became closely aligned with the real-world science institution NASA with Nichols supporting recruitment (focussed on minority engagement and applications) and the public image of NASA alongside other members of the *Star Trek* cast (Penley 1997, pp. 18–19).

On 12 September 1992 Mae Jemison became the first black woman astronaut. She began each shift on the space shuttle *Endeavor* as a science specialist with Uhura's famous line 'hailing frequencies open' as a hat-tip to her onscreen inspiration (Carrington 2016, pp. 81–84).[2] Despite Uhura's groundbreaking nature, she remains one of only a handful of images of black women scientists to appear on both film and television. The J.J. Abrams reboot of the *Star Trek* franchise introduced a 'new Uhura' (Zoe Saldana) but she has been 'superfluous' to the films' narratives and 'signals a kind of retrograde for the character' (Mafe 2018, pp. 144–45). For Saldana's Uhura science is secondary to her gendered roles, she first appears on screen in *Star Trek* (Abrams, 2009) shot from behind as a desirable body for Kirk (Chris Pine) to admire, and later as a humanizing aspect of Spock's (Zachary Quinto) character as his girlfriend. Despite the missed opportunity of this new Uhura, there have been notable but limited images of women of colour with narrative agency and scientific expertise who are key to the plot rather than simply framed as exceptional or Other.[3]

Although some people may be inspired by and get information about the sciences from actual scientists including their family, friends, and teachers, for the majority the main source of information comes from media representation rather than direct engagement with scientists (Kitzinger et al. 2008). Research indicates that onscreen portrayals of science and scientists can affect the way people perceive the sciences and those who work in STEM fields (Steinke 2010, 2017; Haynes 2017; Haran et al. 2008; Kirby 2008, pp. 41–56; Pansegrau 2008). By having and increasing the number of women scientists on screen, entertainment media can 'reduce stereotyping of science that can lead to misperceptions about the appropriateness of scientific careers for women' (Steinke 1998, p. 143). Where the scientific fields offer a 'dearth' of roles (Steinke 1997, p. 410), entertainment can provide aspirational role models by showing that science is an appropriate career for a woman and one that is not automatically beset by harassment and sexism.

A 2018 study conducted by Jocelyn Steinke and Paola Tavarez identifies how US movies 'fall short in providing positive, inspirational portrayals' of women in STEM (2018, p. 246). Portrayals typically feature comparatively fewer women and tend towards representations that reify existing misogynistic discourses that define women as either lesser scientists or failed women ('bad' wives/mothers). This study uncovered a ratio of almost exactly 2:1 for men and women scientist speaking roles showing that, even in films with women scientists, women are outnumbered and often framed as exceptional 'superscientists' (Elena 1997, p. 270), rather than being normalized as part of the general workforce. This ratio also aligns with broader studies on women's speaking parts in Oscar-winning movies produced from 1977 to 2010 that saw men speaking over women by a ratio of 2.66:1 (Smith et al. 2014). Steinke and Tavarez conclude that their study '[points] to a larger cultural issue related to inclusion and diversity in film portrayals' (2018, p. 259), an issue that is exacerbated at the intersection of science, gender, class, ability, and race/ethnicity.

Studies commissioned and conducted by the Geena Davis Institute on Gender in Media similarly showed that men STEM characters outnumber women at nearly a rate of two to one (62.9–37.1 per cent) with the majority of those being white (71.2 per cent) with fewer represented as black (16.7 per cent), Asian/Asian-American (5.6 per cent), Latinx (3.9 per cent), or Middle Eastern (1.7 per cent) (Davis 2018). Although marginally better than the reality, where only one quarter of scientists in the US are women (Beede et al. 2011), the storyworlds created by the entertainment industry should provide role models that the science industries have yet to manage.

A number of approaches to the study of women of science in entertainment media will be discussed below, beginning with a consideration of how women have been negatively stereotyped or made entirely absent from science-based entertainment media. This leads into a discussion of the idea of historically hidden figures and the ways in which this narrative trope has been utilized. Scholarship concerning the representation of women scientists has often focussed on analysis of dissemination and reception, but there should also be attention paid to the production processes. Who is writing, directing, advising on, and producing these stories? Why must we consider who is seen to 'speak for science' and how this affects audience perceptions about who holds scientific expertise (LaFollette 1988, p. 262)?

The genre context and broadcast platform of the woman scientist will also be examined as there are more examples of women scientists in serial television than on film. All of this must also be underpinned by a recognition of intersectionality, as improving representation has been generally confined to images of white women who are often straight and able-bodied rather than women of colour, the differently abled, queer women, and the intersections of these identities. Even when women are offered space in screen fictions, diversity is side-lined as predominately privileged white women represent their gender in the sciences.

Annihilating/Stereotyping the Woman Scientist

Science is political, embedded, and 'fully embodied' into society and thus attempts to improve the number of women pursuing STEM careers (and the retention of those women by the industry) still needs to move beyond what Susan Harding termed the 'just add women' approach to improving diversity (1995, p. 55). By failing to address the institutional and structural problems that limit women's participation in scientific cultures—both real and imagined—it is impossible to fully diversify and decolonize the representation and perception of the scientist.

Women are faced with additional challenges that are specific to their gendered experiences and societal expectations. As Roslyn Haynes argues, women are shown to be 'torn between professional aspirations and the gender roles expected by [a] society that [locates] them firmly in the domestic sphere' (2017, p. 308). In both fiction and reality, women become 'the "human face" of science while male scientists embody "objective" science and the face of authority and expertise' (Chimba and Kitzinger 2010, p. 617). Women's stories are tempered with gender-specific 'forms of risks' including 'seduction, denigration, criticism of credentials, intimidation, physical violence, manipulation of results, sexist comments, and cultural inequality'—issues and narrative arcs that are rarely if ever part of stories of men of science (Haynes 2017, p. 318). The types of stories that are told about women in the sciences become part of the expectations for women entering the field, an expectation of harassment, bullying, and gaslighting that may result in their contribution being ignored or stolen.

This lack of representation, gendered-focussed representation, and misrepresentation constitutes a form of 'symbolic annihilation'. As Gerbner and Gross argue, 'representation in [a] fictional world signifies social existence' whereas 'absence means symbolic annihilation' (1976, p. 182). Gaye Tuchman specifically notes that in the media, women are largely invisible and when they are visible, they are marginalized or used as a symbolic representation of gender equality (Tuchman 1978). Both of these scholarly interventions are from the 1970s but still alarmingly prescient as women scientists are seen to repeatedly fall into stereotypical roles and narratives. Women scientists, that are predominately white women, are incorporated in science-based narratives as visual symbols of gender equality and onscreen diversity but their storylines often revolve around men and resolve in romantic/domestic success. Even today, the woman scientist cannot avoid the baggage of societally constructed notions of femininity/womanhood.

Despite Uhura's iconic role in *Star Trek*, ethnic-minority women are regularly placed in secondary and background roles if given representation as scientists at all.[4] *Hidden Figures* briefly disrupts this trend by dramatizing the history of black women computers, programmers, engineers, and mathematicians that had previously been lost to stories of great white men and the overarching institutional narratives of NASA. Although the film received criticism for the prevalence of a fictional male figure (Al Harrison [Kevin Costner]) who acts as a 'white saviour' (Page-Kirby 2016; Blay 2017), *Hidden Figures* positively places black women at the centre of a science-based narrative. This film emerges alongside other contemporary examples of black representation including *Black Panther*'s Shuri, *Annihilation*'s astrophysicist Josie Radek (Tessa Thompson), and *Star Trek: Discovery*'s (2018–) science officer Michael Burnham (Sonequa Martin-Green). These black women contribute to this increased visibility and agency, but despite these notable examples it is very 'rare' for black women scientists to be seen in science-based fictions (Meyer 2018).

Eva Flicker offers a typology of women scientists on film that sees them 'subjected to intense simplifications along three basic dimensions: gender, profession, and private life' (2008, p. 246). She argues that 'the woman scientist tends to differ greatly from her male colleagues in her outer appearance: she is remarkably beautiful and compared with her qualifications, unbelievably young. She has a model's body... is dressed provocatively and is sometimes "distorted" by wearing glasses' (2003, p. 316). Professional and gendered stereotypes overlap, and women's representation is often 'oriented on their deficiency' of being either 'not a "real" woman or not a "proper" scientist' (2003, pp. 316–17).

Flicker identifies seven categories of women scientists: the old maid, the gruff woman's libber, the naive expert, the evil plotter, the daughter or assistant, the lonely heroine, and the clever digital beauty (2008). All of these stereotypes consider the relationship of the woman character to male colleagues and alongside her personal relationships (lover, father, brother). These stereotypes are built from a number of tropes that place traits in opposition including masculinity and femininity, naivety and intelligence, and good and evil. They often embody male fantasies and patronizingly frame women as childlike and in need of protection, guidance, and affection.

Flicker updated her stereotypical women scientist taxonomy, first published in 2003, with the addition of 'the clever digital beauty' in 2008 reflecting the growing popularity of the hacker or coder character and showing how images of women scientists can and have evolved. These hackers are young and beautiful and often have stories of loss and betrayal concerning male characters. For example, both Skye/Daisy Johnson (Chloe Bennett) from *Agents of S.H.I.E.L.D.* (2014–2020) and Felicity Smoak (Emily Bett Rickards) from *Arrow* (2012–2019) are former hackers recruited into crime-fighting groups. Their expertise is essential to the plots of their respective shows, but they both have backstories that revolve around absent and manipulative fathers and the development of their abilities are inextricably linked to those relationships. These characters also tend to go through glamorous makeovers when they go undercover, Smoak and more recently Nine Ball/Leslie (Rihanna) in *Ocean's 8* (Ross, 2018) reveal their 'hidden' beauty as their capacity to wearing contact lenses and a figure-hugging evening gown become more useful 'skills' than their computer science capabilities. Women scientist's representation has expanded but men continue to be central to their stories and the construction of their expertise.

Female-coded characters are assumed to be emotional, maternal, and benevolent, and in need of a man either professionally or personally. Their agency is ultimately contained by the promise or acquiescence of a heterosexual coupling. For example, Jane Foster (Natalie Portman) in *Thor* (Brannagh, 2011) is an astrophysicist whose work is commandeered by the US government, she is shown to be young and naïve about her research and quickly distracted from it in a literal collision with Thor (Chris Hemsworth). She is confirmed as an expert by an older male mentor (Stellan Skarsgård) and saved and swept off her feet by the God of Thunder who also returns some of her research notes to her when her own efforts fail. By perpetuating stereotypes about women scientists, media producers continue to normalize ideas about traditional binary gender roles and perpetuate a normative 'cultural imaginary [that] has often struggled to find a place' for the woman scientist (Palmer and Purse 2019, p. 5). They are damaging to all gender identities in their separation of the science entirely from the domestic, and in their promotion of the idea that in order to be a successful scientist you must reject family, home, and hopes of a healthy work–life balance.

Uncovering the Woman Scientist

The phrase 'hidden figure' has received renewed popularity, since the release of *Hidden Figures*, as a term to describe the forgotten, ignored, invisible, supressed, and unrecorded labours and histories of those involved in major historical moments (Rowbotham 1973). Some feminist history of science research has focussed on '[reclaiming] forgotten women scientists and [restoring] their lost voices' as via their very presence these uncovered women scientists can disrupt 'gendered scientific assumptions and practices' that have long been implicit in the history and knowledge of the sciences (Jordanova 1993, p. 474). *Hidden Figures* not only reclaims the contributions of specific women to the Apollo missions but also the input of community of people of colour who worked as researchers and computers at Langley. Attitudes to women and ethnic minorities in the sciences have been formed across decades of mediations of science by scholars and society (including the media) that have made their contributions seemingly 'invisible', thus constituting an 'ellipsis from official recognition' (Oreskes 1996, p. 91).

There have been a number of contemporary films and TV shows that have focussed on historical/contemporary women scientists including *Hidden Figures*, *The Imitation Game* (Tyldum, 2014), *Chernobyl* (2019), and *Timeless* (2016–2018). Even when real women scientists are offered representation, they become illustrative of whole groups of women scientists. *The Imitation Game* represents codebreaking as a male endeavour, with Keira Knightley playing the English mathematician Joan Clarke (1917–1996). In the film

Clarke is seen as exceptional due to her role as the only woman codebreaker and much of her story is aligned to the characterization and progression of Alan Turing (Benedict Cumberbatch). However, Clarke was actually part of a larger group of women—three quarters of the 10,000 strong workforce—who did calculations at Bletchley Park (Dunlop 2015). Similarly, the focus on Katherine Johnson (Taraji P. Henson) in *Hidden Figures* allows for a tightly narrativized version of events, however it also overshadows the contributions of other women and, interestingly, the men of colour who were involved in the Apollo Missions (Shetterly 2015).[5] Hidden figures are representative of the experiences of women—and indeed men—who have previously been sublimated in the histories of great white men of science (Chambers 2017, p. 14).

The television series *Timeless* follows the attempts of white woman history professor (Lucy [Abigail Spencer]), a black male scientist (Rufus [Malcolm Barrett]), and a white male soldier (Wyatt [Matt Lanter]) to stop agents working for Rittenhouse, a mysterious organization founded in 1778 'that wants to control the past, present, and the future', from travelling back in time and changing history for their members' benefit. The series uses the time travel format to explore moments in history, key and often hidden figures, and counter-historical narratives (the team are not always successful). The show also recognizes the issues with maintaining history; the US treatment of black citizens is a recurring theme as the team effectively uphold a system of institutional racism in their pursuit of historical preservation. As Rufus laments: 'we're saving rich white guys' history, a lot of my [black US] history sucks'.

Across the two seasons, *Timeless* includes four women of science as characters: mathematician, Katherine Johnson (Nadine Ellis); actor and inventor, Hedy Lamarr (Alyssa Sutherland); physicist, Marie Curie (Kim Bubbs); and chemist, Irène Curie (Melissa Farman). 'Hollywoodland' (season 2, episode 3) features Hedy Lamarr and explores her invention of the frequency-hopping spread spectrum, a technology that is still used for WIFI and Bluetooth (Barton 2011). *Timeless* uses Lamarr and the way she is often underestimated, and her generally unknown history as an inventor for their story. A similar technique is used with NASA mathematician Katherine Johnson in 'Space Race' (season 1, episode 8). In both cases, it is the woman historian's knowledge of these 'hidden' women's expertise, and their apparent invisibility, that allows them to save the (historical) day.

Praised for its attention to historical detail and accuracy, *Chernobyl* (2019) condenses the work of Soviet scientists into a single fictional nuclear scientist called Ulana Khomyuk (Emily Watson). The Sky/HBO limited series is about the events surrounding the Chernobyl nuclear accident in 1986. Composite characters are a common practice in adaptations and historical dramas, and as with the presentation of science, screenwriters must balance the need for entertainment with accuracy, both scientific and historical (Kirby 2011). But rather than adding another man to the already men-dominated cast the producers actively chose to make this tenacious scientist character a woman.

Chernobyl does not uncover a specific scientist, but it does draw attention to the progressive position of women medics and scientists in the USSR and the involvement of women at senior levels during the Chernobyl disaster (Alexievich 2016). As *Chernobyl*'s writer/producer Craig Mazin remarked when asked about why he thought Khomyuk was an appropriate fictionalized addition: 'one area where the Soviets were actually more progressive than [the West] was in the area of science and medicine' with 'quite a large percentage of female doctors and scientists' (Holloway 2019). Khomyuk is a moral centre for the show, something which perhaps follows the stereotypical notion of the woman as the 'human face' of science but, as Watson notes, she importantly 'represents the many scientists who worked fearlessly and put themselves in a lot of danger to help solve the situation' (Christie 2019). She is a character that may be made more noticeable because many will not expect her to be there. Although an invention for *Chernobyl*, the use of a woman as such an important character should not be diminished; Khomyuk is now a widely disseminated representation of a woman scientist with the mission to save people from further consequences of man-made disaster.

It is not that women scientists have been 'written out' but rather that they 'have never been written into' stories about science (Fara 2008, p. 19). Films and TV shows like *Hidden Figures* and *Timeless* have begun to offer new screen-specific approaches to visualizing these women's 'hidden' histories as central narratives rather than as addendums to stories of men scientists. One of the first Hollywood feature-length scientist biopics was of a woman scientist, but since the release of Marie Curie (1943) there have been four further biopics (1977, 1997, 2014, 2016) made for film and TV of that same woman scientist with 2020 seeing the release of the first woman-written/directed Curie film (*Radioactive*, dir. Marjane Satrapi). These texts contribute to the framing of Curie as *the* exceptional woman of science and an emblematic shorthand for women's contributions to the sciences. Curie is undoubtedly a significant figure, but she is often 'where the conversation begins and ends when it comes to women in science' (Phingbodhipakkiya 2017).[6] Not only do we need more stories about exceptional individual women of science, but also narratives that include women experts as part of a diverse team where gender is not simply a defining narrative or character trope.

Genre and the Woman Scientist

The genre in which the woman scientist appears can also influence how their expertise is received. In some instances, when science is represented within a narrative by a woman it becomes less powerful. Holly Hassel (2008) identifies a woman scientist trope that emerges in the action genre of the 1990s and early 2000s: 'the babe scientist'. The babe scientist is a co-protagonist to the action hero, but she is overly 'invested in the science at the expense of "intuition"' (Hassel 2008, p. 190). She of course aligns to the young, thin, and beautiful stereotype, but needs to be re-educated by the man protagonist 'in the ways of the human spirit' rather than the rational sciences (Hassel 2008, p. 192).

In *I, Robot* (2004) Susan Calvin (Bridget Moynahan) has to be 'converted' from her attachment to the scientific method to follow her emotions. It is only once she rejects science and trusts in the (ultimately proven) paranoia of the protagonist—Del Spooner (Will Smith)—that she is able to help save the world from the rise of the robots that she helped to create. In opposition to the majority representations of women scientists, the action movie woman scientist's emotional detachment is gendered as weak. Her rational adherence to the systems science makes her less powerful in comparison to the man action hero who is driven by his emotion and desire. Intuition and feelings are privileged in the non-scientist action hero, but in other genres that have leading men scientists of their dominance attributed to their emotional detachment.

Men, women, and their gendered expectations are aligned to particular fields. In a focus group study by Rashel Li and Lindy Orthia analysing gender in *The Big Bang Theory*, participants agreed that women and men scientists are presented as having equal scientific expertise in the comedy show. But they expressed frustration at the fact that the women are aligned with the 'soft' biosciences and the men with the 'hard' sciences of engineering and physics (Li and Orthia 2016). *The Big Bang Theory* is unusual in its incorporation of women scientists, but these women are undermined by the 'goals [of the comedy format] to entertain while reinforcing the status quo' (McIntosh 2014, p. 196). The recurring women scientists of the show—Amy Farrah Fowler (Mayim Bialik) and Bernadette Rostenkowski-Wolowitz (Melissa Rauch)—are 'caught between their function as supporting characters for the male leads, and the potential to depict women as working scientists' (Weitekamp 2015, p. 89). They are initially stereotyped and primarily played for laughs as part of the comedy format, with Fowler initially introduced as a punchline to a gag about Sheldon (Jim Parsons) using internet dating. Women in the series are 'always a woman first and only a scientist second' and even when they are shown to be more successful (publications, awards, grants) they must navigate the fragile egos of the men on the show (Archer 2015, p. 48). Despite the radical potential for '[exploring] ways of reshaping science to include women' in the less restrictive comedy genre frame, *The Big Bang Theory* as a central/culturally pervasive example instead resorts to stereotypes and 'jokes' about women to reinforce male dominance (Jowett 2007, p. 32).

The incorporation and speculative discussions of science do not only belong within science fiction but to a broad range of science-based fictions. Although a recent upsurge in fictional women scientists have been aligned to science fiction (specifically astronauts) (Lovell 2019), it would not be possible to discuss a long-running medical drama like *Grey's Anatomy* (2005–) as science fiction, despite the series' engagement with new medical science and technology and their imagined uses and consequences. Although medical reality television has tended to focus more on male physicians, women are better represented in fictional medical dramas (Jain and Slater 2013). They are featured most frequently 'as doctors or medical examiners' (Weitekamp 2015, p. 79), and the ensemble casts that often surround them can act to reduce the '"burden of representation" which could typecast an individual woman as a "representation" of all women' in the sciences (Haran et al. 2008, p. ii). Untethered by the restrictions of reality, large ensemble-cast medical dramas have space for diversity, as television productions can be more responsive to audience tastes, cultural trends, and changes in the visibility and acceptance of marginalized groups.

Television series provide more intersectional spaces for women scientists to exist in, to an extent due to the platform's inherent opportunities for writers/producers to develop, drop, and introduce characters over a long-term serial narrative. Women scientist characters on television series also have the potential to significantly impact viewers, especially given the weekly appearances over the course of a season or more. Forensic anthropologist, Temperance 'Bones' Brennan (Emily Deschanel) is the titular character in the 12-season series *Bones* (2005–2017) and she is featured as part of a diverse team of scientists in terms of gender, race, and sexuality. As Lauren Archer argues 'this repetition creates opportunities for viewers to develop a deeper understanding of and identification' with more diverse scientist characters— an image of science that they may not usually come into contact with (2015, p. 31).

Although there is 'not a plethora of examples', television has allowed 'for more ambiguity regarding who is [and can be] the hero' with more opportunities for women characters to develop (Mafe 2018, pp. 125, 141). Films generally have extremely long production scales and offer only one entity on which producers will be judged (critically and financially). Broadcast television, conversely, has a different development and dissemination process that affects how and where intersectional representations of women may appear and develop. TV producers are commissioned to make pilot episodes before a full series is ordered, which means that potentially 'risky' characters can be tested on audiences prior to a full financial commitment. Internet-distributed content offers 'new models of television, with their appeals to ever more niche and activated users' giving media producers the potential to incorporate more gender and race diversity and queer representation (Goddard and Hogg 2018, p. 472).

Streaming services have also offered space to present images of queer women scientists. For example, Netflix's *Sense8* (2015–2018), created by transgender sisters Lana and Lilly Wachowski with J. Michael Straczynski, includes Nomi Marks (Jamie Clayton) who is a hacker, and a proud lesbian transgender woman. *Star Trek: Discovery*, produced for CBS Access/Netflix incorporates a queer woman engineer into its second season, but Jett Reno's (Tig Notaro) 'queerness is not hidden [or] revealed' it is simply there (Chambers 2020, p. 274). Similarly, *Orphan Black*'s biologist Cosima Niehaus (Tatiana Maslany) is one of two queer clones on the show, and one of several women scientists including her girlfriend (Delphine Cormier [Évelyne

Brochu]). Their sexuality is important for character definition but does not propel their narrative arcs. Netflix has the broadcast rights to *Orphan Black* in the UK and Ireland, and the service also became the international distributor for the women scientist-led movie *Annihilation* after distributors chose not to give the film an international theatrical release (Sims 2018). The film premiered on the service alongside original and distributed films and serial shows that offer representation that is often restricted or denied by traditional dissemination methods.

Gender, Media Production, and Representing Women Scientists

Campaigns intended to encourage and improve women's participation in the sciences are reflected and embedded in the media produced (Merrick 2012, p. 752). So, just as it is vital to consider the institutional and structural problems that limit women's participation in scientific cultures, when analysing women scientists' fictional representation we need to be aware of the issues surrounding women working in the media industries. Women-led (both onscreen and in terms of production) entertainment media is burdened with the expectation 'to disrupt, overcome, and dismantle systemic marginalization in a traditionally masculine cinematic space' (Donoghue 2019, p. 4). The entertainment media industry is a male-dominated field; in 2018 women accounted for only 8 per cent of directors, 16 per cent of writers, and 26 per cent of producers working on the top 250 films released that year (Lauzen 2019, p. 1). A 2017–2018 study of prime-time television showed that women constitute 17 per cent of directors, 25 per cent of writers, and 40 per cent of producers (Lauzen 2018, p. 4). Although the statistics show that women are better represented in television production, they are still a long way off achieving parity.

Feminist and women's science fiction literature emerged visibly in the 1960s onwards, and these flourishing and often utopian imaginings of the future 'coincided with (and were invigorated by) the women's liberation movements, which included significantly increased opportunities for women in science and engineering education and careers' (Merrick 2012, p. 753). As Jane Donawerth observes, science fiction literature written by women tends to feature more women scientists and suggests that women writers are more likely to include a variety of women characters (1997, p. 4). However, much of the science-based fiction that is adapted tends to be based upon men-written texts, and then directed and often written/adapted by men too. A large proportion of the media texts under consideration in this chapter are indeed predominately written and directed by men. Women rarely feature in the processes of production as writers, producers, directors, or even as science consultants.

Scientists can be involved behind the scenes of science-based film and television as science advisors. In this role they can consult with media producers to support the development of science-based narratives and storyworlds. It is an opportunity to take a more 'speculative approach to science' and provide 'extrapolative speculations about scientific phenomena' as well as to advise on the representation of realistic science and scientific practice (Kirby 2011, p. 36). But as with the media and science industries, women science advisors are in the minority. The major 'labcoats in Hollywood' associated with popular science-based films and television series are male (Kirby 2011, p. 36), including Jack Horner for *Jurassic Park* (Spielberg, 1993), David Saltzberg for *The Big Bang Theory*, Kip Thorne for *Interstellar* (Nolan, 2014), and Adam Rutherford for *Annihilation*. There are women working as science advisors—Anne Simon for *The X-Files* (1993–2002, 2016–2018), Jessica Coons for *Arrival* (Villeneuve, 2016), Cady Coleman for *Gravity* (Cuarón, 2013), and Cosima Herter for *Orphan Black*—but they are harder to uncover and confirm. All of these women-advised productions feature central woman scientist characters—Dana Scully (Gillian Anderson), Louise Banks (Amy Adams), Ryan Stone (Sandra Bullock), and Cosima Niehaus—whose scientific expertise is central to the stories and structure of each text. Women scientist's expertise is vital for creating entertainment media that not only pictures but also engages with the complexities of women's experiences in the male-dominated sciences and a world that is designed to accommodate men.

The normalization of women's expertise both on and behind the camera is essential to improving the image of the woman scientist and shifting cultural perceptions about who can be a scientist and what kind of life a woman scientist can have. This, like attempts to improve women's participation in the sciences, cannot just be about adding more women but thinking about why so few women are incorporated into these stories in the first place. A lack of women in media production affects the number of women-led fictions, therefore we must consider who is permitted/invited to write and produce science-based stories and how the perspectives of those not represented in the reality of the sciences might be more actively and accurately incorporated.

Conclusions: From Symbolic Annihilation to Normalization

The notion of the exemplary and exceptional woman scientist is a major issue when discussing their representation. Successful women scientists are seen as phenomena rather than a normal part of the scientific community, and they continue to be side-lined in the fictional (and indeed real) world of men-dominated science. Studies on the representation of women in STEM have shown that the images presented reinforce stereotypes and position women scientists in a state of conflict between their personal and professional lives, with few represented as mothers and/or happily married (Haran et al. 2008; Steinke 2005; Lovell 2019; Jain and Slater 2013).

Normalizing women of STEM is central to changing broader attitudes about what and who a scientist is, attitudes that need to be altered across lines of gender, race, class, sexuality, and ability. By involving a diverse range of women in STEM as advisors and collaborators, the representation of women can move from being token figures and anomalies to being regular and entirely expected leading figures in science-based narratives on either the big or the small screen. By making women more visible in stories about science, both fictional and factual, the inspiring images of science that can and are being produced can be associated with women who are not only represented as smart anomalous individuals but as part of a network of diverse and complex professional women.

The 'explicit' rejection and mockery of women's scientific expertise may be lessening, but the 'subtle ways' of making those who are not white, straight, able-bodied men, 'feel like outsiders' shows that, as Patricia Fara argues, 'modern discrimination is elusive, insidious, and stubbornly hard to eradicate' (2018, p. 285). This is achieved through refusing and failing to incorporate women and ethnic-minority voices and faces into the visual and academic cultures of science including: the deficit of portraits of women in institutions; the absence of women's scholarship in student reading and scholarly reference lists; and the lack of representation on screen as both fictional and factual figures. By not being able to see themselves in the ways in which science is mediated and embedded into contemporary culture, those who are under-/not represented can feel like the STEM industries are not a place for them. Improving, increasing, and diversifying media representation is only one way of affecting change, but it is an important part of the/a long-term project to change the sciences in a way that stops constraining people by their race, sexuality, class, ability, and gender identity.

Notes

1. Alongside Leslie, who is presented as sexually confident and occasionally manipulative, the show only offers one other non-biosciences woman scientist: Elizabeth Plimpton (Judy Greer) is a cosmological physicist but her academic prowess is undermined by her voracious sexual appetite ('The Plimpton Stimulation').
2. Jemison also appeared in the 150th episode of *Star Trek: Next Generation* as Lt. Palmer. She was the first real astronaut to appear in the series.
3. Other examples of fictional black women in STEM include: haematologist Karen Jenson (N'bushe Wright) in *Blade* (1998), medic-in-training Martha Jones (Freema Agyeman) in *Doctor Who* (2007–2010), astronaut Molly Woods (Halle Berry) in *Extant* (2014–2015), inventor/engineer Shuri in *Black Panther*, astrophysicist Josie Radek (Tessa Thompson) in *Annihilation*, astronaut/engineer Ava Hamilton (Gugu Mbatha-Raw) in *The Cloverfield Paradox* (2018), and engineer Naomi Nagata (Dominique Tipper) in *The Expanse* (2015–2022).

4. For an in-depth discussion of issues of race terminology, see Song (2020). Ethnic-minority is used throughout this chapter as opposed to Women/People of Color or BME (black, minority, ethnic).
5. Margot Shetterly, author of *Hidden Figures*, is the daughter of Robert B. Lee III who was a black research scientist at NASA Langley Research Center. The film adaptation focusses on the work of black women at Langley and does not present any images of the black men despite their inclusion in the source history.
6. The *Beyond Curie* design project celebrates brilliant women scientists who have not received the recognition they deserve.

References

Alexievich, Svetlana. 2016. *Chernobyl Prayer: Voices from Chernobyl: A Chronicle of the Future*. Transl. Anna Gunin and Arch Tait. London: Penguin Classics.

Archer, Lauren R. 2015. "Science in Stilettos: Shaping Perceptions of Women in Science." In *The Sexy Science of The Big Bang Theory: Essays on Gender in the Series*, ed. Nadine Farghaly and Eden Leone, 26–50. Jefferson: McFarland.

Attenborough, Frederick Thomas. 2011. "Complicating the Sexualization Thesis: The Media, Gender and 'Sci-Candy'." *Discourse & Society* 22, no. 6: 659–76.

Barton, Ruth. 2011. "Rocket Scientist!: The Posthumous Celebrity of Hedy Lamarr." In *In the Limelight and Under the Microscope: Forms and Functions of Female Celebrity*, ed. Diane Negra and Su Holmes, 82–101. New York: Continuum.

Beede, David, Tiffany Julian, David Langdon, George McKittrick, Beethika Khan, and Mark Doms. 2011. "Women in STEM: A Gender Gap to Innovation." Office of the Chief Economist. U.S. Department of Commerce, Economics and Statistics Administration. *ESA Issue Brief* 4, no. 11: 1–11. https://files.eric.ed.gov/fulltext/ED523766.pdf. Accessed 15 July 2019.

Blay, Zeba. 2017. "*Hidden Figures* and the Diversity Conversation We Aren't Having." *The Huffington Post*, February 23. https://www.huffingtonpost.co.uk/entry/hidden-figures-and-the-diversity-conversation-we-arent-having_n_58adc9bee4b0d0a6ef470492. Accessed 1 August 2019.

Carrington, André M. 2016. *Speculative Blackness: The Future of Race in Science Fiction*. Minneapolis: University of Minnesota Press.

Chambers, Amy C. 2017. "*Hidden Figures*: Screening Hidden Histories." *Viewpoint* 113: 14.

Chambers, Amy C. 2020. "*Star Trek* Discovers Women: Gender, Race, Science, and Michael Burnham." In *Fighting for the Future: Essays on Star Trek: Discovery*, ed. Mareike Spychala and Sabrina Mittermeier, 267–86. Liverpool: Liverpool University Press.

Chambers, Amy C., and Shelley Thompson. 2020. "Women, Science and the Media." In *The International Encyclopedia of Gender, Media, and Communication*, ed. Karen Ross. Hoboken, NJ: Wiley. https://doi.org/10.1002/9781119429128.iegmc304.

Chimba, Mwenya, and Jenny Kitzinger. 2010. "Bimbo or Boffin? Women in Science: An Analysis of Media Representations and How Female Scientists Negotiate Cultural Contradictions." *Public Understanding of Science* 19, no. 5: 609–24.

Christie, Janet. 2019. "Emily Watson on Her New TV Drama, *Chernobyl*." *The Scotsman*, May 4. https://www.scotsman.com/arts-and-culture/film-and-TV/emily-watson-on-her-new-TV-drama-chernobyl-1-4920211. Accessed 31 July 2019.

Colatrella, Carol. 2011. *Toys and Tools in Pink? Cultural Narratives of Gender, Science, and Technology*. Columbus: The Ohio State University Press.

Davis, Geena. 2018. "Portray Her: Representations of Women STEM Characters in Media [2018 Executive Findings]." The Lyda Hill Foundation and the Geena Davis Institute on Gender in Media. https://seejane.org/wp-content/uploads/portray-her-executive-summary.pdf. Accessed 20 July 2019.

Donawerth, Jane. 1997. *Frankenstein's Daughters: Women Writing Science Fiction*. New York: Syracuse University Press.

Donoghue, Courtney Brannon. 2019. "Gendered Expectations for Female-Driven Films: Risk and Rescue Narratives around Warner Bros.' *Wonder Woman*." *Feminist Media Studies*, 1–17. https://doi.org/10.1080/14680777.2019.1636111.

Dralega, Carol Azungi. 2018. "The Symbolic Annihilation of Hegemonic Femininity in *Black Panther*." *International Journal of Gender, Science and Technology* 10, no. 3: 462–65.

Dunlop, Tessa. 2015. *The Bletchley Girls. War, Secrecy, Love and Loss: The Women of Bletchley Park Tell Their Story*. London: Hodder and Stoughton Ltd.

Elena, Alberto. 1997. "Skirts in the Lab: Madame Curie and the Image of the Woman Scientist in the Feature Film." *Public Understanding of Science* 6, no. 3: 269–78.

Fara, Patricia. 2008. *Pandora's Breeches: Women, Science and Power in the Enlightenment*. London: Pimlico.

Fara, Patricia. 2018. *A Lab of One's Own: Science and Suffrage in the First World War*. Oxford: Oxford University Press.

Flicker, Eva. 2003. "Between Brains and Breasts—Women Scientists in Fiction Film: On the Marginalization and Sexualization of Scientific Competence." *Public Understanding of Science* 12, no. 3: 307–18.

Flicker, Eva. 2008. "Women Scientists in Mainstream Films: Social Role Models—A Contribution to the Public Understanding of Science from the Perspective of Film Sociology." In *Science Images and Popular Images of the Science*, ed. Bernd Hüppauf and Peter Weingart, 241–56. New York: Routledge.

Gerbner, George, and Larry Gross. 1976. "Living with Television: The Violence Profile." *Journal of Communication* 26, no. 2: 173–99.

Goddard, Michael N., and Christopher Hogg. 2018. "Introduction: Trans TV as Concept and Intervention into Contemporary Television." *Critical Studies in Television: The International Journal of Television Studies* 13, no. 4: 470–74.

Haran, Joan, Mwenya Chimba, Grace Reid, and Jenny Kitzinger. 2008. "Screening Women in SET: How Women in Science, Engineering and Technology are Represented in Films and on Television." UKRC Research Report Series No. 3. UK Resource Centre for Women in Science, Engineering and Technology. http://orca.cf.ac.uk/17535/1/report_3_haran.pdf. Accessed 28 August 2018.

Harding, Sandra. 1995. "Just Add Women and Stir?" In *Missing Links: Gender Equity in Science and Technology for Development*, ed. Gender Working Group, United Nations Commission on Science and Technology for Development, 295–308. New York: UN Development Fund for Women.

Hassel, Holly. 2008. "The 'Babe Scientist' Phenomenon: The Illusion of Inclusion in 1990s American Action Films." In *Chick Flicks: Contemporary Women at the Movies*, ed. Suzanne Ferris and Mallory Young, 190–203. New York: Routledge.

Haynes, Roslynn D. 2017. *From Madman to Crime Fighter: The Scientist in Western Culture*. Baltimore, MD: John Hopkins University Press.

Holloway, Daniel. 2019. "Chernobyl [interview with Craig Mazin]. Variety TV Take [podcast]." May 3. https://variety.com/2019/TV/news/listen-chernobyl-ep-craig-mazin-on-making-a-non-traditional-disaster-show-1203204493/. Accessed 15 July 2019.

Jain, Parul, and Michael D. Slater. 2013. "Provider Portrayals and Patient–Provider Communication in Drama and Reality Medical Entertainment Television Shows." *Journal of Health Communication* 18, no. 6: 703–22.

Jordanova, Ludmilla. 1993. "Gender and the Historiography of Science." *The British Journal for the History of Science* 26, no. 4: 469–83.

Jowett, Lorna. 2007. "Lab Coats and Lipstick: Smart Women Reshape Science on Television. In *Geek Chic: Smart Women in Popular Culture*, ed. Sherrie A. Inness, 31–48. New York: Palgrave Macmillan.

Kirby, David A. 2008. "Cinematic Science." In *Handbook of Public Communication of Science and Technology*, ed. Massimiano Bucchi and Brian Trench, 67–94. London: Routledge.

Kirby, David A. 2011. *Lab Coats in Hollywood: Science, Scientists, and Cinema*. Cambridge, MA: The MIT Press.

Kitzinger, Jenny, Joan Haran, Mwenya Chimba, and Tammy Boyce. 2008. "Role Models in the Media: An Exploration of the Views and Experiences of Women in SET." UKRC Research Report Series No. 1, UK Resource Centre for Women in Science, Engineering and Technology. http://orca.cf.ac.uk/17534/1/report_1_kitzinger.pdf. Accessed 1 August 2019.

LaFollette, Marcel C. 1988. "Eyes on the Stars: Images of Women Scientists in Popular Magazines." *Science, Technology, & Human Values* 13: 262–75.

Lauzen, Martha M. 2018. "Boxed in 2017–18: Women on Screen and Behind the Scenes in Television." Center for the Study of Women in Television and Film, San Diego State University. https://womeninTVfilm.sdsu.edu/wp-content/uploads/2019/07/2017-18-Boxed-In-Report-1v2.pdf. Accessed 31 July 2019.

Lauzen, Martha M. 2019. "The Celluloid Ceiling: Behind the Scenes Employment of Women on the Top 100, 250, and 500 Films of 2018." Center for the Study of Women in Television and Film, San Diego State University. https://womeninTVfilm.sdsu.edu/wpcontent/uploads/2019/01/2018_Celluloid_Ceiling_Report.pdf. Accessed 31 July 2019.

Li, Rashel, and Lindy A. Orthia. 2016. "Communicating the Nature of Science Through *The Big Bang Theory*: Evidence from a Focus Group Study." *International Journal of Science Education* 6, no. 2: 115–36.

Lovell, Bronwyn. 2019. "Cosmic Careers and Dead Children: Women Working in Space in *Aliens, Gravity, Extant*, and *The Cloverfield Paradox*." *Science Fiction Film and Television* 12, no. 1: 73–102.

Mafe, Diana Adesola. 2018. *Where No Black Woman Has Gone Before: Subversive Portrayals in Speculative Film and TV*. Austin: University of Texas Press.

McIntosh, Heather. 2014. "Representations of Female Scientists in *The Big Bang Theory*." *Journal of Popular Film and Television* 42, no. 4: 195–204.

Merrick, Helen. 2012. "Challenging Implicit Gender Bias in Science: Positive Representations of Female Scientists in Fiction." *Journal of Community Positive Practices* 4: 744–68.

Meyer, Karlyn Ruth. 2018. "Dr. Karen Jenson, Hematologist: How 1998's *Blade* Set the Stage for Black Women Scientists on Screen. *Lady Science*. https://www.ladyscience.com/essays/dr-karen-jenson-hematologist-how-1998s-blade-set-the-stage-for-black-women-scientists-on-screen. Accessed 15 July 2019.

O'Keeffe, Moira. 2013. "Lieutenant Uhura and the Drench Hypothesis: Diversity and the Representation of STEM Careers." *International Journal of Gender, Science and Technology* 5, no. 1: 4–24.

Oreskes, Naomi. 1996. "Objectivity or Heroism? On the Invisibility of Women in Science." *Osiris* 11: 87–113.

Page-Kirby, Kristen. 2016. "When the Women of *Hidden Figures* Needed the Man." *Washington Post*, December 23. http://wapo.st/2hZTAj6?tid=ss_tw&utm_term=.42b3d87785cc. Accessed 31 July 2019.

Palmer, Lorrie, and Lisa Purse. 2019. "When the Astronaut Is a Woman: Beyond the Frontier in Film and Television." *Science Fiction Film and Television* 12, no. 1: 1–7.

Pansegrau, Peter. 2008. "Stereotypes and Images of Scientists in Fiction Films." In *Science Images and Popular Images of the Sciences*, ed. Bernd Hüppauf and Peter Weingart, 257–66. New York: Routledge.

Penley, Constance. 1997. *NASA/Trek: Popular Science and Sex in America*. London: Verso.

Phingbodhipakkiya, Amanda. 2017. "About the Project: Beyond Curie." https://www.beyondcurie.com/about. Accessed 29 August 2019.

Rowbotham, Sheila. 1973. *Hidden from History: Rediscovering Women in History from the 17th Century to the Present*. London: Pluto Press.

Shetterly, Margot Lee. 2015. *Hidden Figures*. New York: HarperCollins.

Sims, David. 2018. "The Problem with *Annihilation*'s Messy Release." *The Atlantic*, January 31. https://www.theatlantic.com/entertainment/archive/2018/01/annihilation-paramount-netflix/551810/. Accessed 2 August 2019.

Smith, Stacey L., Marc Choueiti, and Katherine Pieper. 2014. "Gender Inequality in Popular Films: Examining On-Screen Portrayals and Behind-the-Scenes Employment Patterns in Motion Pictures Released Between 2007–2013." *USC Media, Diversity, & Social Change Initiative*. https://annenberg.usc.edu/sites/default/files/MDSCI_Gender_Inequality_in_600_films.pdf. Accessed 31 July 2019.

Song, Miri. 2020. "Rethinking Minority Status and 'Visibility'". *Comparative Migration Studies* 8, no. 5: 1–17. https://doi.org/10.1186/s40878-019-0162-2.

Steinke, Jocelyn. 1997. "A Portrait of a Woman as a Scientist: Breaking Down Barriers Created by Gender-Role Stereotypes." *Public Understanding of Science* 6, no. 4: 409–28.

Steinke, Jocelyn. 1998. "Connecting Theory and Practice: Women Scientist Role Models in Television Programming." *Journal of Broadcasting and Electronic Media* 42, no. 1: 142–51.

Steinke, Jocelyn. 2005. "Cultural Representations of Gender and Science: Portrayals of Female Scientists and Engineers in Popular Films." *Science Communication* 27, no. 1: 27–63.

Steinke, Jocelyn. 2010. "Gender Representations of Scientists." In *Encyclopedia of Science and Technology Communication*, ed. Susanna Hornig Pries, 322–25. Thousand Oaks, CA: Sage.

Steinke, Jocelyn. 2017. "Adolescent Girls' STEM Identity Formation and Media Images of STEM Professionals: Considering the Influence of Contextual Cues." *Frontiers in Psychology* 8, no. 716: 1–15.

Steinke, Jocelyn, and Paola Maria Paniagua Tavarez. 2018. "Cultural Representations of Gender and STEM: Portrayals of Female STEM Characters in Popular Films 2002–2014." *International Journal of Gender, Science and Technology* 9, no. 3: 244–77.

Szwydky, Lissette Lopez, and Michelle L. Pribbernow. 2018. "Women Scientists in Frankenstein Films, 1945–2015." *Science Fiction Film and Television* 11, no. 2: 303–39.

Tuchman, Gaye. 1978. "The Symbolic Annihilation of Women by the Mass Media." In *Hearth and Home: Images of Women in the Mass Media*, ed. Gaye Tuchman, Arlene Kaplan Daniels, and James Walker Benét, 3–50. New York: Oxford University Press.

Weitekamp, Margaret A. 2015. "'We're Physicists': Gender, Genre and the Image of Scientists in *The Big Bang Theory*." *Journal of Popular Television* 3, no. 1: 75–92.

PART VI

Access, Diversity and Practice

CHAPTER 24

Catalysts, Compilers and Expositors: Rethinking Women's Pivotal Contributions to Nineteenth-Century 'Physical Sciences'

Mary Orr

Across its multicultural history, the accepted exceptionalism of women in science turns on an oft-repeated fact: the alleged coinage in 1833 by William Whewell (1794–1866) of the term 'scientist'.[1] His portmanteau label for a status, authority and hence 'career' in the shaping of knowledge of the physical world then directly inform the increasing professionalization of the sciences since the 1830s. Their concomitantly exclusive—and exclusionary—practices have thus sharpened the foci of women's history of science since the 1980s in its multifarious recuperation and recognition of the many unheralded women in science before and after '1833'. Work by Abir-Am and Outram (1989) and Creese and Creese (1998) has been pivotal to making visible the many 'uneasy careers' of hitherto unacknowledged 'ladies in the laboratory', as disclosed by the 'intimate lives' of the wives, daughters and sisters who were the illustrators, translators and amanuenses of (more famous male) scientists. Influential feminist methodologies and frameworks, such as by Fox Keller (1996) and Harding (2004), have also targeted the perennial gender(ing) of science and its publication, to spearhead more thorough investigation of archives and footnotes for the many occluded women who were highly active in fields such as geology (Burek and Higgs 2007). These invaluable re-feminizing critical frameworks, however, overlook other models, outlets and trajectories for women's scientific work. More detrimentally, they may inadvertently perpetuate the paradigms

M. Orr (✉)
School of Modern Languages, University of St Andrews, St Andrews, UK
e-mail: mmo@st-andrews.ac.uk

for 'the scientist' that collocate with the established professional 'pipeline' model for science which so demonstrably affects the situation of women in STEM(M) today.[2] A major woman scientist is thus belated ('the first woman –ologist'), secondary ('Crick et al.') or worse. Athenas, Hypatias and 'Pandoras in breeches' (Fara 2004) are always potentially 'hybrid' and 'leaky', whether as impostor-monsters and/or as '#DistractinglySexy' (Morrison 2019).

Such caricatures and stereotypes for aberrant (read exceptional) women in science are also revelatory as counter faces of the 'archetypes' (Golinski 1999) and paradigms of science, and hence its paradigm shifts (Kuhn 2012).[3] The object of this chapter is therefore to question the 'scientist' paradigm and its professional 'pipeline' model as the authoritative norm for evaluating, and *evacuating* women's (major, important, significant) contributions to science. Our starting point is a close re-examination of William Whewell's much-cited review essay of 1834, *On the Connexion of the Physical Sciences* to inform our main argument: women's history of science has much to gain from renewed critical scrutiny of the published 'standpoint' theories (Harding 2004) by key *men* of science in a given pivotal period of its development, especially when produced in direct response to *a major woman* in their field. If *her* situation, optics and contributions benchmark the assumed parameters for best science practice of the day, they will also demonstrate *her* negotiation of the major models, outlets and trajectories that are available for women's scientific work. If this can be explained neither by her 'intimate life' with men in her science(s), nor by the gendering of her science, the proven viability and sustainability of her models in science will then emerge. To test such an approach, our chapter examines the publication of science in English by women active at the time of Whewell, but not engaged primarily in botany—the 'feminine' science in the tradition of Rousseau—and by women as seemingly different in background, 'career' and 'intimate' scientific circumstances as Mary Somerville (1780–1872), Maria Graham (1785–1842), Charlotte Murchison (1788–1869), Sarah Bowdich (1791–1856), Margaret Gatty (1809–1873) and Athénaïs Michelet (1826–1899). If such women together prove too many exceptions to the new rules for 'the scientist', they reveal other stakes for the 'professionalizing' of science from the 1830s, including its major resetting of national science agendas and official historiography.

Whewell Revisited: A Riposte that Doth Protest Too Much?

The complete title of the anonymous (Whewell's) Article III in the *Quarterly Review* in 1834 is *On the Connexion of the Physical Sciences by Mary Somerville*. Holmes (2014) highlights the unrivalled 'bestseller' importance of her work and its influence on major contemporary 'scientists' such as Herschel, Babbage and Lyell in Britain, and Arago, Biot, Guy Lussac and others in France. What has not been adduced more crucially from these striking qualities is that Somerville's published work of disseminating, clarifying and hence contributing to the latest (French) sciences in 1834 challenges not only Whewell's expertise in her fields in which he has *not yet* published, but it also directly challenges his many preconceptions and premises about the who and how of scientific expertise itself, because these are integral to her own clear objectives as Whewell reports:

> In her simple and brief dedication to the Queen, she says, "If I have succeeded in my endeavour to make the laws by which the material world is governed, more familiar to my fellow countrywomen, I shall have the gratification of thinking, that the gracious permission to dedicate my work to your Majesty has not been misplaced." And if her "countrywomen" have already become tolerably familiar with the technical terms which the history of progress of human speculations necessarily contains [...] if they have advanced so far in philosophy, they will certainly receive with gratitude Mrs. Somerville's able and *masterly* (if she will excuse this word) exposition of the present state of the leading branches of the physical sciences. For our own parts, however, we beg leave to enter a protest, in the name of that sex to which all critics (so far as we have ever heard) belong, against the appropriation of this volume to the sole use of the author's country*women*. We believe that there are few individuals of that gender which plumes itself upon the exclusive possession of exact science, who may not learn much that is both novel and curious in the recent progress of physics from this little volume. (pp. 55–56, emphasis in the original)

In exemplary form in her dedication, Somerville adroitly uses what feminist literary criticism (Poovey 1984; Eger et al. 2001) calls the 'modesty topos', the rhetorical ploy of women authors of Somerville's period precisely to allay fears and maintain a pretence: here is no 'unfeminine' or worse, 'bluestocking', interloper into the territories in Whewell's words of 'that gender that *plumes itself* upon the exclusive possession of exact science' (my emphasis). By thus averring the shocking truth that science in fact has no sex, Whewell must at the same time circumscribe and equal out Somerville's superior intellectual prowess. With pseudo-chivalry that returns her *work* to its place, he can then underscore and promote his expertise in a championing royal 'we'—'we beg leave to enter a protest'—that must acknowledge a '*masterly* exposition' (with mixed comfortableness in his bracket), to then settle the account: 'this little volume'. Such a diminutive, the belittling qualification, thus gives Whewell leave himself in the next main part of his essay to *appropriate* Somerville's main point of 'this volume' as in fact his own, because she had set out in her dedication 'to make the laws by which the material world is governed, more familiar to [her] fellow countrywomen'. His point scoring is then to take her keyword, 'Connexion', and translate it into his term, 'unity', culminating his argument :

[...] we must recollect that her professed object is to illustrate 'The *Connexion of the Physical Sciences.*' This is a noble object; and to succeed in it would be to render a most important service to science. The tendency of the sciences has long been an increasing proclivity to separation and dismemberment. Formerly, the 'learned' embraced in their wide grasp all the branches of the tree of knowledge; the Scaligers and Vossiuses of former days were mathematicians as well as philologers, physical as well as antiquarian speculators. But these days are past; [...] If a [...] poet, like Goethe, wanders into the fields of experimental science, he is received with contradiction and contempt; and, in truth, he generally makes his incursion with small advantage, for the separation of sympathies and intellectual habits has ended in a destruction, on each side, of that mental discipline that leads to success in the other province. But the disintegration goes on, like that of a great empire falling to pieces; physical science itself is endlessly subdivided, and the subdivisions insulated. We adopt the maxim, 'one science only can one genius fit.' [...] between the mathematician and the chemist is to be interpolated a '*physicien*' (we have no English name for *him*), who studies heat, moisture and the like. And thus science, even mere physical science, loses all traces of unity. (pp. 58–59, emphasis in the original)

For Whewell, Somerville's '*Connexion*' 'nobly' counters the prevailing 'disintegration' and specialization of expertise that requires the necessary interpolation that is contemporary physics. Yet in being a throwback to (the best of) 'former days', Somerville's work cannot also now cast its author as the leading contemporary '*physicien*' ('*physicienne*'). Too terrible to contemplate for Whewell is the potential equation by analogy to '(medical) physician' of a pioneering British 'doctoress' of physical sciences. Because so ardently an advocate of 'unity' himself Whewell then casts his maxim, 'one science only can one genius fit', as appropriation *for the present* of a Somerville model in the past, to out-manoeuvre himself into the space of that '*him*' (as English version of the '*physicien*') that is also 'genius'.[4]

Whewell's impasse pivots on equivalence masked as the lack of equivalents, whether established English traditions or specific terms for modern scientific practice. Establishment figures elsewhere, such as the '*physicien*' in France, usefully confer gender (m.), class (level of education) and cultured identity (French), yet grammar and known circumstance in France also permit the 're-gendering' and hence re-appropriation of physics by women such as Somerville. In her publication she is demonstrably adept in the physical sciences (pl.) as well as their multiple *applied* re-translation in English. The lack of an English term for serious students of science thus represents a troubling vulnerability to the twofold threat that is 'Mary Somerville', as the ensuing argument (above) from 'unity' framing Whewell's famous coinage makes very clear:

A curious illustration [...] may be observed in the want of any name by which we can designate the students of the knowledge of the material world collectively. We are informed that this difficulty was felt very oppressively by the members of the British Association for the Advancement of Science [...] in the last three summers. There was no general term by which these <u>gentlemen</u> could describe themselves with reference to their pursuits. *Philosophers* was felt to be too wide and too lofty a term, and was very properly forbidden them by Mr. Coleridge, both in his capacity of philologer and metaphysician; *savans* was rather assuming, <u>besides being French instead of English</u>; some <u>ingenious gentleman</u> proposed that, by analogy with *artist*, they might form *scientist*, and added that there could be no scruple in making free with this termination when we have such words as *sciolist, economist,* and *atheist*—but this was not generally palatable; <u>others attempted to translate the term</u> by which the members of similar associations in Germany have described themselves, but it was not found easy to discover <u>an English equivalent for</u> *<u>natur-forscher</u>*[sic]. The process of examination which it implies might suggest such <u>undignified compounds</u> as *nature-poker*, or *nature-peeper,* for the *naturae curiosi*; but these were indignantly rejected. [...] It is one object, we believe, of the British Association, to remedy these inconveniences by bringing together the cultivation of different departments. To remove the evil in another way is one object of Mrs. Somerville's book. (pp. 59–60, italics in the original, underlining added)

The overwhelmingly 'oppressive' threat that 'Mrs. Somerville' represents for Whewell's own praxis could not be more apparent here in his direct overwriting of the key terms in her quoted dedication—'[her] endeavour to make the laws by which the material world is governed, more familiar to my fellow countrywomen'—in *his* 'students of the knowledge of the material world collectively'. Whewell's 'students' thus include women in theory, because they are *already* present in practice. The incontrovertible proof is Somerville's prowess (despite girls' education normally barring them from learning mathematical and scientific subjects), and the sessions packed by women at the British Association for the Advancement of Science (BAAS) (Ellis 2017). Indeed, the very 'laws by which the material world is governed' that Somerville displays to her 'countrywomen' alongside those of the physical sciences are of the male advantage and privilege prevailing in science by cultural precedent first and foremost. Such assumptions clearly determine Whewell's '*scientist*', both by analogy—with '*artist*' (Mr. Coleridge)—and by equivalence with '*Philosophers*', '*savans*' (m. sing. and pl. Fr.), 'ingenious gentleman', '*naturforscher*' (m. sing. and pl. Ger.) and '*naturae curiosi*' (m. pl. Lat.). Whewell's 'anxiety of influence' (Bloom 1973) is clearly triggered here by perception of the greater prowess of French and German science practitioners and, because they are men, Somerville's (greater) 'mastery' over them through her taking forward of Linnean and former classical (Latin and Greek) 'gentlemanly' models in their new international European vernacular forms. Of modest rank, Whewell had had no direct encounters in 1834 with 'cutting-edge' European science, *savants* and *Naturphilosophen* in the major institutes and academies of Paris or Berlin, unlike Somerville. Moreover, as the writer of the third Bridgewater Treatise in 1833, *Astronomy and General Physics considered with reference* to *Natural Theology*, Whewell also invents 'scientist' in opposition to its other opposites, 'sciolist, economist, and atheist'.[5]

Mary Somerville and her work are therefore a catalyst of major significance for Whewell's essay of 1834 and its proposed term, 'scientist', on at least two counts. First, she already epitomizes how to do science, both to prevent its 'great empire falling to pieces' (the 'unity' question) and to reassert Anglophone pre-eminence in its future 'connexions'. Second, she already embodies the quintessential (modern European) 'scientist' that provokes Whewell's coinage for such status being so necessary for Britain. The remainder of his essay is then a very stark lesson on why the term 'scientist' will remain so vexed for women practitioners if science has no sex and hence covers 'Somervilles' that are indistinguishable, if not superior, in quality from 'Scaligers and Vossiuses' (and 'Whewells'):

> Our readers cannot have accompanied us so far without repeatedly feeling some admiration rising in their minds, that the work of which we have thus to speak is that of a woman. *There are various prevalent opinions* concerning the grace and fitness of the usual female attempts at proficiency in learning and science; [...] But there is this remarkable circumstance in the case,—that where we find a real and thorough acquaintance with comparative ease, and possessed with unobtrusive simplicity, *all our prejudices* against such female acquirements vanish. Indeed, there can hardly fail, in such cases, to be something peculiar in the kind, as well as the degree, of the intellectual character. Notwithstanding all the dreams of theorists, *there is a sex in minds. One of the characteristics of the female intellect is a clearness of perception, as far as it goes*: with them, action is the result of feeling; thought, of seeing; their practical emotions do not wait for instruction from speculation; their reasoning is undisturbed by the prospect of its practical consequences. [...] *In men, on the other hand, practical instincts and theoretical views are perpetually disturbing and perplexing each other. Action must be conformable to rule; theory must be capable of application to action. The heart and the head are in perpetual negotiation, trying in vain to bring about a treaty of alliance, offensive and defensive. The end of this is [...] inextricable confusion—an endless seesaw of demand and evasion.* [...] He learns to talk of matters of speculation without clear notions; [...] to deal in generalities; to guess at relations and bearings; to try to steer himself by antitheses and assumed maxims. Women never do this: what they understand, they understand clearly. (pp. 64–65, emphasis added)

The very 'matters of speculation without clear notions' in which Whewell indulges concerning intellect in 'women' and 'men' nicely illustrate his own 'inextricable confusion' (because a man's). Whewell has also steered himself here 'by antitheses and assumed maxims' about (male) minds and genius, to determine in the coining of 'scientist' the status that defines work(ers) in serious physical sciences. If the key sentence emphasized above 'In men...' were to begin 'In "scientists"...', their powers of rule and 'theory [...] capable of application to action' will indeed dictate the narrowed 'pipeline' model for future scientists (m) that acknowledges very rare inclusions of a Hypatia and

Agnesi (p. 66), or a Madame de Chastelet 'translating and commenting on the "Principia" of Newton' (ibid.), that is exemplars of 'female intellect [as] a clearness of perception, as far as it goes'. Whewell then avoids the defining name that reproves his new rule for the male 'pipeline' in science leaking out the occasional woman: 'on the same subject, in our own time, the "Mécanique Céleste" of Laplace, has been [translated and commented] by a lady of our country' (ibid.). He can name her only in his concluding remark thereby to include her with a qualification: this Mrs. Somerville 'we are obliged to confess, is Scotch by her birth, though we are very happy to claim her as one of the brightest ornaments of England' (p. 68).

Whewell's famous essay therefore unstitches where it most wants to stitch up the term, 'scientist', for modern English gentlemanly science to which he aspires, to ensure that it connotes a distinct class and provenance for 'all students of the knowledge of the material world'. In the 'connexions' of his essay 'scientist' proves altogether charged to defend its gendered coiner from a present danger. This is the equality, if not superiority, of the many subject-defining women in the field as indicative of the future, and despite their many educational disadvantages.

'ONE OF THE CHARACTERISTICS OF THE FEMALE INTELLECT IS A CLEARNESS OF PERCEPTION' THAT GOES FAR?

By 1833 Mrs. Somerville epitomizes the cluster, and hence visible critical mass, of women publishing prominent works of science in Britain in the 1820s and early 1830s. She is, moreover, often the influencer, support and encourager of scientific pursuit in other women, such as Charlotte Murchison (1788–1869) (Kölbl-Ebert 1997a). If Darwin is still aboard the 'Beagle' (1831–1836), Maria Graham (1785–1842), one of Somerville's Scottish women friends, has returned from major expeditions to India (1812), Chile (1822) and Brazil (1824), albeit couched in the lesser disguise of a *Journal of a Residence in....* Graham's article of 1824 on the earthquake in Chile—the first to be published by a woman in the *Transactions of the Geological Society of London* (Thompson 2012)—catalyzes the particular ire of male insiders to geology such as her interlocutor, George Bellas Greenough (1778–1855), the founding President of the Geological Society of London. Despite the scientific evidence and findings of her work, his vituperative print responses to it demonstrate only his blatant, belittling sexism: What can a woman know? In her knowing perhaps too well resides the same 'Somerville' effect. Greenough was himself 'not an original researcher, but saw his scientific task as a diligent gatherer of information' and manifested a 'strong bias against the "Plutonists" or "Huttonians"' (Kölbl-Ebert 2009). His anxiety is that of the cartoon by Henry T. De la Beche in 1832 of Charlotte Murchison "The Light of Science dispelling the Darkness which covered the World". Her Davy lamp in one hand and in her other hand a miner's/geologist's hammer[6] underscores the perilous state of *husbands* in physical sciences of the time, such as Roderick Murchison, William

Buckland and Humphrey Davy. Their reputations built directly on the *prior* scientific interests and expertise of their future wives, respectively, Charlotte Hugonin (Kölbl-Ebert 1997a) and Mary Morland (1797–1857) (Kölbl-Ebert 1997b), or wealth and social position of Jane née Kerr, *olim* Lady Apreece, Lady Davy (1780–1855) (Golinski 1999). Having himself been asked for separation and divorced by his wife Letitia (née Whyte) in 1826, De la Beche was at work (1832–1835) on the first geological map of Devonshire over which he would soon lock horns—also via cartoons—with *Roderick* Murchison in what became the 'Devonian Controversy' (Bate 2010). The known visibility—even nascent female 'club' or 'society'—of such adept women in the coteries of the physical sciences in London social circles might well explain the 'oppressiveness' (Whewell, ibid.) they represented in 1834. Men in gentlemanly and mercantile classes were vying to make their scientific reputations in (manly) physical sciences. Women of such obvious merit in science as 'gatherers' of information and specimens like Greenough, and as 'masterly [...] exposit[ors] of the present state of the leading branches of the physical sciences' like Whewell could not be other than a threat to the self-fashioning of 'scientist' as an immaculately masculine profession, status and career ('pipeline'). The final words in Whewell's essay therefore reveal the ultimate defence. The brightest women in science are its 'ornaments'. Even the best feminist critics fall in with this flagrant pretence when they continue to assume the value of the man in the 'creative couple' (Pycior et al. 1996) as the more worthy because of his un-decorative contributions to the collective of science. The remainder of this essay will now uncouple such assumptions about serious ('pipeline') versus decorative ('sampler') science by examining indicative expert women catalysts, compilers and expositors in the '1833' period working outside the 'pipeline' in two often intertwining categories. These are women who outlive the spousal scientist model literally—through writing science as pre-married women and as widows—or who do not espouse it because their own independent scientific interests and expertise are not those of the partner.

WIDOWS SHAPING SCIENCE FOR THE FUTURE.
CASE STUDY 1: SARAH BOWDICH (LEE), 1791–1856

In 1833, the first biography of George Cuvier (1769–1832) was published in English in London and New York editions as *The Memoirs of Baron Cuvier* (1833a). The French translation appeared in Paris the same year as the *Mémoires du baron Georges Cuvier* (1833b). Its author, however, was not a member of the Geological Society of London, or the recently established Société Géologique de France of 1830, but a 'Mrs. R. Lee'/'Mistress Lee'.[7] The inside title pages reveal more of her identity as 'formerly Mrs. T. Ed. Bowdich'. Why would publishers such as Longman or Harper commit to this publication, and Fournier additional funds for the translation if this author was not an 'authority' despite being a woman? Yet given the array of qualified established national scientific figures (m. pl.) available to write it, including

Cuvier's disciples, why and how is 'Mrs./Mistress Lee' better placed than they? After all, Cuvier was not only France's leading comparative anatomist and 'Father of Palaeontology' for more than two decades. He was also the celebrated official memoirist for deceased colleagues at the Paris Jardin des Plantes: his *Éloges* ran to several volumes and included encomiums to eminent foreign as well as French scientific figures. However, the bitter 'Querelle des Analogues' [Quarrel of the Analogues] of 1830 with his longstanding colleague, Étienne Geoffroy Saint-Hilaire, left hollow Cuvier's narrow victory in the debate (respectively 'transformisme' versus 'révolutions du globe'). It also left moot Cuvier's longer scientific reputations in 1832 for his 'Éloge' writer-successor. Who better qualified to rise to the challenges of this crisis than the best outsider-insider in his sciences in a woman like Somerville with particular additional translation (and inter-relational) skills? The distant future will name this impossible double position the 'glass cliff' for women leaders who have also smashed the 'glass ceilings' in their field (Bruckmüller and Branscombe 2010).[8] The penultimate paragraph of Mrs. Lee's Introduction sums up her particular authority in Cuvier's science and further qualifications yet adherence to her 'secondary' positions in accordance with the modesty topos:

> Mr. Bowdich had returned from his second, and I from my first, voyage to Africa, in the year 1818, and shortly after Mr. Bowdich proceeded to Paris, where his reputation as the successful African traveller, was already known. The letter of Dr. Leach was scarcely necessary with the Baron Cuvier, who received him with that warmth and encouragement which always marked his conduct towards men of talents younger than himself, that interest which he extended to all who were devoted to science. Struck with the facilities afforded for study in the French capital, Mr. Bowdich determined to remain there some time, in order to qualify himself for the principal object of his ambition, a second travel in Africa. We both accordingly went to Paris in 1819; and from that moment the vast library of the Baron Cuvier, his drawings, his collections, were open to our purposes. We became the intimates of the family, with whom, for nearly four years, we were in daily intercourse. We left France with their blessings; and on returning alone to Europe, I was received even as a daughter. My correspondence with M. Cuvier's daughter-in-law, and other branches of the family, has been uninterrupted since that period; I have paid them repeated visits at their own house; and for fourteen years not a single shadow has passed over the warm affection which has characterised our intimacy. (Lee 1833a, p. 5)

It is only in careful reading of the *Memoirs* that the unprecedented record of a woman in Cuvier's science emerges in parallel to her account of his 'biography' (her own term, Lee 1833a, p. 4) as witnessed in the slippage here from 'Mr. Bowdich', to 'we', to 'my', and finally to 'our' (between the Cuviers and Sarah alone). Sarah's highly unusual positioning, rather than position, as a woman in science therefore derives from her work on West African Fish—Cuvier's 'notes' on them appear in the 1826 French edition of the 1825 *Excursions in Madeira and Porto Santo*—and, when widowed, from her onward close collaboration with Cuvier to provide materials, including drawings, for his new twenty-two volume *Histoire naturelle des poissons* [The Natural History of Fishes] (1828–1848) (Orr 2015). As the first major appraisal of Cuvier's long 'uneasy career' at the Jardin des Plantes/Muséum National d'Histoire naturelle, Mrs. Lee's *Memoirs/Mémoires* appraise with the eye-witness authority as his 'disciple' his many scientific publications, his public roles in science and *his* 'intimate life' (Abir-Am and Outram 1989) in science.

The pivotal crisis in science that Cuvier's death epitomizes in 1832 clearly informs the future of French as well as of international science: his major work on the reclassification of vertebrates (including fossil), and especially fishes, is the new standard for all major national scientific studies. Works appearing in the 1830s in Britain on British fauna are no exception, except that they uniformly fail, unlike Cuvier, to acknowledge groundbreaking work before them where this is clearly by a woman with the latest specialist training, such as Sarah Bowdich. One example makes the case and, as further proof of the 'scientist' crisis, Leonard Jenyns's *Manual of British Vertebrate Animals* (1835). The archive edition of the online *Oxford Dictionary of National Biography* entry for 'Blomefield, Leonard, formerly Leonard Jenyns (1800–1893)' written by Thomas Seccombe clarifies his credentials:

> He was the first resident vicar at Swaffham Bulbeck, but [...] Cambridge was within an easy ride, and he was thus able to maintain an intimacy there with such of his contemporaries as shared his love of natural history. These [...] included such names as Henslow, Whewell, Darwin, Adam Sedgwick, Julius Hare, and Bishop Thirlwall. In 1834–5 (preface dated Swaffham Bulbeck, 24 Oct. 1835) he wrote his useful 'Manual of British Vertebrate Animals,' which was issued by the syndics of the Cambridge University Press. [...] Before he had completed it, at the earnest request of Charles Darwin, he undertook to edit the monograph on the 'Fishes' for the 'Zoology of the Voyage of H.M.S. Beagle,' published in 1840. The post of naturalist to the Beagle had first been offered to Henslow and then to Jenyns, but he hesitated to leave his parochial work, and joined Henslow in recommending Darwin for the place.

Jenyns comes to his expertise on Darwin's 'Fishes' by way of his major recent 'gathering' work on British 'Pisces' (including sharks and rays) in the final section of his 1835 *Manual* (pp. 306–524). Its long justificatory introduction (pp. v–xxii) pays particular attention to their regard in his scientific arrangement 'of the Fish is [...] similar to that in the *second edition* [that is 1829–30] of the "Regne Animal" [sic]', p. xvii), method and nomenclatures—'The "Histoire Naturelle des Poissons" of [Cuvier] has been exclusively resorted to in the Class of Fish', p. xix—and British practice underpinning his endeavour:

> To Mr Yarrell in particular, he begs publically to return his sincere thanks for the able help which he has experienced at his hands, and such as alone has enabled him to complete the work upon the plan first contemplated. This help has been

especially felt upon the subject of the British Fishes. Had it not been from the very liberal manner in which that gentleman offered him the almost unlimited use of his Manuscripts and rich collections, the author has no hesitation in saying that he could never have extended the Manual to that department, or presumed to enter upon a field, to which he was previously almost an entire stranger. (pp. xx–xxi)

For 'Fishes', therefore, Jenyns did not operate from first-hand observation in the field as a fisherman (unlike Yarrell), or in situ at the catch as was Sarah Bowdich's practice for her *The Fresh-Water Fishes of Great Britain* (1828–1838). In 1833 she was mid-way through their annual installments of four fishes *per* number, because she accompanied each description with her own exquisite watercolour drawing taken directly from the life. Jenyns's reliance on Yarrell (who publishes his two-volume *A History of British Fishes* in 1836) therefore directly contradicts or stretches the truth of his opening introductory remarks. There Jenyns underscores the necessity '[…] to take care that the descriptions should as far as possible be obtained from the animals themselves, and nothing inserted upon the credit of other writers which was capable of being verified by personal examination. *The day is for ever gone by in which mere compilations will be thought to be of any service to the science of Zoology*' (pp. v–vi, emphasis added). Jenyns in 1835 wants to distance his work from close rivals such as the '"History of British Animals" by Dr. Fleming […] completed in 1827 and published the year following, since which time period a great variety of species have been added to the Fauna of this county, more particularly in the Class of Fish' (p. vi). Yet in this very gap, Jenyns's introduction deliberately fails to mention Sarah Bowdich's new work—a work that already before his employs Cuvier's new classification—despite recording hers clearly in the 'Alphabetical List of Works Quoted' (p. xxv): '*Bowd. Brit. Fr. wat. Fish.*—Bowdich (Mrs. T. E.). The Fresh-water Fishes of Great Britain; drawn and described. Lond. 1828, &c. 4to. (*In course of publication.*)'. Jenyns uses this same wording in the bracket for '*Yarr*' (p. xxxii) since only volume one had appeared in 1835. More egregiously in his own description *compilations*, Jenyns then also deliberately forgets to include her descriptions of the (fresh-water) fish in question in what was clearly a detailed 'personal examination' of Bowdich's publication. As *per* the Cuverian model, he duly records '*Bowd. Brit. Fr. wat. Fish*' in the list of previous authorities heading each of his descriptions, but refers only to the number of her illustration, 'Draw. no.', in every case with one notable exception for the 'Grig Eel' (pp. 477–478).[9] See Table 24.1. Jenyns's work had no accompanying illustrations, or woodcuts (as in Yarrell). Moreover, since Bowdich's book had only a limited circulation of some fifty copies, Jenyns's referral of his readers to her illustrations alone ensures that few will read and compare her text descriptions from the life that precede his own (and also Yarrell's). Such manœuvering to make himself of 'service to the science of Zoology' and Darwin's work on fish constitutes the same move as Whewell's 'scientist'. Contributors trained in the latest science in France, particularly if they are women such as Somerville and Bowdich, brook no argument as rivals: they can only be 'mere' rather than 'serious' compilers whatever the evidence of their scientific work as demonstrably discipline-leading among scientific peers.

SCIENTIFIC HUSBANDS, FATHERS AND THEIR COUSINS IN SECONDARY ROLES AS 'BOOK-KEEPERS', TRANSLATORS AND EDITORS: CASE STUDY 2—MARGARET GATTY, 1809–1873

For men in British science in the early 1830s the powerful 'Somerville' effect, caused by the prominence of genteel women at BAAS and the 'polymathic' work of women such as Bowdich from mercantile classes, can be explained by their accident in history. These women had come to science through the advantages of (self-)education, (self-)expression in print and circumstance in the transitional period of the late Enlightenment (including Napoleonic period) that also permitted men such as Whewell from outside leisured gentlemanly ranks to advance a scientific 'career'. In Britain, the unbroken lineage of 'clergyman science' across the creedal and class spectrum from a Joseph Priestley (1733–1804) to a Gilbert White (1720–1793) could include William Buckland (1784–1856), Leonard Jenyns, Richard Lowe (1802–1874), Hugh Millar (1802–1856) and William Whewell. For women born around the second decade of the nineteenth century, and therefore coming to their science after 1830 when they were also importantly of marriageable age, the 'Somerville' effect was in negative rebound for them. Theirs are the 'secondary' roles—as wives, daughters and sisters—in their engagements with science precisely because more men are vying for prominence in the physical sciences through its hotbed of competing debates and public 'controversies' including in print, for example the 'Cambrian-Silurian Dispute', augured in France by the 1830 'Querelle des Analogues' and epitomized later by Darwin. More to the point these scientific debating platforms mirror pulpits, courtrooms and hustings, to relegate women in science in all social classes from (exceptional) speaker to invariable listener. Enlightened 'clergyman' science in Britain more broadly conceived, however, provided important conduits for women's independent access to, and pursuits in, science. Mary Horner (Lyell) 1808–1873 benefitted immeasurably from her Scottish geologist father Leonard Horner's principled commitments to factory reform and the education of women (starting with his six daughters). Margaret Scott (Gatty), 1809–1873, exemplifies more literally the double benefits of enlightened 'clergyman' science. The immense collections of books belonging to her father, the Rev. Alexander J. Scott, Royal Navy Chaplain to Nelson, that were hers to explore, and then her clergy house upon marriage in 1839 to Alfred Gatty also filled with books, provided her lifelong needs for the home 'laboratory' that is her published work on British seaweeds. Somerville's Scottish Enlightenment legacies of self-education and public self-expression in print deriving from the privileged company of family books and non-conformist parenting for girls therefore reset for the history of nineteenth-century science the importance of rethinking the *men*—principally enlightened fathers and husbands—who play key secondary support roles that enable women into post-1830 science.

The work that Margaret Scott brought to her marriage and continued alongside her more well-known seaweeds research and publication of 'parables' for children is therefore of primary interest here. *The Book of Sun-dials* (1872) was published the year before her death, to be reedited and enlarged in 1889 by her daughter, Horatia R. F. Eden, and Eleanor Lloyd, Margaret's lifelong friend and correspondent 'scientist' in the project.[10] Margaret's original dedication is so revolutionary in its upending of accepted gender conventions (for acknowledging wives) that it escapes notice:

TO THE DEAR HUSBAND
   TO WHOM I AM INDEBTED FOR THE BEST HAPPINESS OF
   THE HOURS OF EARTHLY LIFE,
   AND WITH WHOM I HOPE TO SHARE THE EXISTENCE
   IN WHICH
   TIME SHALL BE NO MORE,
   I DEDICATE THIS VOLUME,
   IN THE COMPILATION OF WHICH HE HAS TAKEN
   SO GREAT
   A PART AND INTEREST

Margaret had started her work in 1835. Alfred's 'part and interest' is therefore as the secondary supporting participant for a project that is about privileging *women*'s time and its making (long before Kristeva in 1981 corners the philosophical question of women's displacements from historical time) as much as it is about sun-dials. Their long and important history in astronomy and the development of scientific instrumentation coequally permit the successful navigation of land and sea and 'domestic' time at home. Moreover, as a work of horology—the art and science of time measurement—Margaret's collection is no 'mere compilation' (Jenyns). It accounts for the numerous time pieces across civilizations that helped regulate work and observance (religious, astronomical, social community). Because their association is with the 'decorative' (in form and use of mottos), sun-dials and Margaret's authority concerning them—'as one of the brightest ornaments' (Whewell)—have been ignored in histories of scientific instrument-making and its assumed non-participations by women. Margaret's work about a women's time-piece collective—Eleanor's and Horatia's parts in her scientific time team—thus reconsiders 'scientific' time as rather more than mechanical (clocks), teleological (including evolutionary), and even eternal. It lies outside this essay, but Margaret Gatty's meticulous observational work on seaweeds and on sundials is co-extensive with her understandings of longer patterns in organic planetary time. Recent science is only catching up with the importance of seaweeds as important fixers of carbon (and sunlight) throughout geological time (Muraoka 2004).

Margaret Gatty's scientific works on time and seaweeds supremely epitomize the importance of women catalysts, compilers and expositors for the futures of science. Her wealthy second half-cousin, Charles Henry Gatty (1836–1903), clearly found her work on seaweeds very much more than an interest that 'possibly influenced [...] his considerable collections among the Channel Islands and along the south coasts of England and although he did not publish his observations he continually passed his information back to those who took an interest in the same pursuit, in particular the St. Andrews' Fisheries Laboratory in Scotland' (Felbridge and District History Group 2003). His private fortune could turn her direct inspiration, knowledge and place in print that he could not match into his place in science as its benefactor. The multiple gifts Charles Henry Gatty makes to the University of St Andrews in 1892–1895 establish its Gatty Marine Laboratory (now Scottish Oceans Institute).

CASE STUDY 3—ATHÉNAÏS MICHELET (1826–1899)

In 1872, a rather different life's work was published in English by T. Nelson and Sons (London, Edinburgh, and New York), *Nature: The Poetry of Earth and Sea*. 'With Two Hundred Designs by Giacomelli (Illustrator of "The Bird"). From the French of Madame Michelet'. Yet no French publication with accompanying illustrations preceded this English translation by W. H. Davenport Adams. His translator's preface is unequivocal:

> The volume [...] was written expressly for its English Publishers by Madame Michelet, and would have been produced at an earlier date but for the interruption caused by the Franco-Prussian War, and, afterwards, by the illness of M. Michelet. [...] The Illustrations were also designed expressly for the Publishers by M. Giacomelli [...]. The conception and execution have occupied him for upwards of two years; and he wishes to be stated that he [...] is willing to rest upon them his future reputation as an artist. (p. v)

Davenport Adams's reputation also rested on this work as T. Nelson and Sons' translator for all four of the 'nature' books by Jules Michelet (1798–1874); in 1872 of *The Bird* (*L'Oiseau*, 1856) and *The Mountain* (*La Montagne*, 1872), and in 1875 of *The Insect* (*L'Insecte*, 1857) and *The Sea* (*La Mer*, 1861). Yet Jules Michelet's otherwise towering importance as the father of French historiography for Gabriel Monod (editor of the Michelet *Œuvres complètes*) and for Roland Barthes his major twentieth-century popularizer is blighted by a key fact. The nefarious actions of Madame Athénaïs Michelet (1826–1899, née

Mialaret) upon Michelet's death—to control the rights to his manuscripts from the clutches of Alfred Dumesnil (Michelet's son-in-law and also a historian) and her subsequent 'editing' (re-writing) of his last works for posthumous publication—show her to be an unnatural wife, wicked step-mother, harpy and witch. These pejorative terms reflect the titles of Michelet's trilogy of the same period as the 'nature' books, *L' Amour* (1858), *La Femme* (1859) and *La Sorcière* (1862). They inform subsequent critical responses to Athénaïs as the 'abusive widow' (Smith 1992). For Gossman (2001) she 'subsequently succeeded (thanks to her husband's name and reputation?) in carving out a minor literary career for herself [for] her *Mémoires d'une enfant*' (p. 333) [*Memoirs of a Girl Child*] (1866, reedited by Monod in 1888, republished in 2004). In consequence she could never be a (French woman) historiographer and natural history writer in her own right.[11]

Nature: The Poetry of Earth and Sea (1872) puts a very different case that independently corroborates Athénaïs's many *leading* roles in the writing of 'Michelet's' 'nature' books. Although Jules Michelet endorsed these severally in his will (published by Athénaïs in 1875 in *La Tombe de Michelet*) and in prefaces (e.g. *L'Oiseau*)—that she masterminded and 'collaborated' in the joint project of the 'nature books', and thereby restored his intellectual and material fortunes—her primary roles are never believed by her many critics, because she can only be the abusive manipulator of posthumous revelations. These include her further defense in print in 1876 of her full 'collaborations' and hence fifty per cent royalty rights for all four of the 'nature books' (*Ma Collaboration à «L'oiseau», «L'insecte», «La mer», «La montagne», Mes droits à la moitié de leur produit*). Davenport Adams's above preface to *Nature* clearly details Athénaïs's sole authorship and conception of her work in its commissioning and at least two-year production schedule that encompasses the delivery of the two hundred accompanying illustrations and her adverse material circumstances. Although *Nature* appears the same year that 'The Mountain' is first published in France, Jules Michelet clearly cannot be the ghostwriter: he is ill. Nor could Athénaïs have executed it entirely from scratch in only two years in the circumstances of a major war, nursing her ailing husband and collaborating (fifty per cent) on 'The Mountain'. Her long preface to *Nature* clearly outlines the provenance of her work as deriving from her earliest, longstanding, interests in natural history:

> *This Book, which I have attempted unaided, gives, under the form of a natural history, a whole life of impressions, of study, of faithful attachment to her who seized my affections as a child, and, if I may dare say so, cherished and fashioned me by successive teachings, by a slow and gentle initiation.*
>
> *This modest exploration of Nature – at first my mother, and afterwards my companion – is, at the same time, an exploration or examination of myself.* (pp. xiii–xxxi, p. xiii, italics in the original)

Childhood exploration of her garden and surrounding landscapes inspired a responsiveness to nature through self-expression that prepared the only daughter of Yves Mialaret (Secretary to François-Dominique Toussaint L'Ouverture (1743–1804) leader of the Haitian Revolution), to correspond with Michelet directly about matters of history during the 1848 Revolution. She married him in 1849 aged twenty-three, after his revolutionary sympathies cost him his government post. Athénaïs's 1872 introduction in English to *Nature* only confirms the latter's conception and *production* as in fact the delivery of her original project of 1854 for the four 'nature books' as outlined in French in her *Collaboration* of 1876:

> In 1854 I started to help him with his history works. I would correct his proofs. I would make extracts for him and I also undertook the major part of his correspondence.
>
> But the taste for natural history that I had enjoyed in the countryside during my childhood remained so strong that while acting as his secretary, I would also find time for these favourite studies.
>
> My project was to write a series of little books for children: I started *The Bird*.
>
> One morning, when running over my first draft, my husband was completely smitten ['séduit'] and said, 'Let's write it together.' [...] Hence my share in the collaboration. I would undertake the first readings. I would go to various libraries, especially those at the Museum [of Natural History/Jardin des Plantes], and excerpt from the rare books and pamphlets that could not be borrowed the passages that we needed. At the same time I would study the collections. I would spend hours with the living animals which even in captivity were a precious assistance to us.
>
> When I got home, I would put my notes and observations in order and little by little the book would take shape.
>
> We would travel extensively together in the interests of our studies [...] While my husband was working on a volume of his *History of France*, I would prepare one for the *Natural History*. I would catch insects or dredge the seabed. My bedroom soon became a field site for observations.
>
> Equipped with a good microscope that Professor Robin had taught me to use I entered the world of discoveries. (*Ma Collaboration* 1876, p. 6, translation by essay author)

Many of the vignettes in *Nature* then also directly correspond to sections in the 'nature books', particularly aspects of *The Mountain* and *The Sea*, albeit in two different keys. In the former is the exuberant 'poetry' of Athénaïs's voice even in English translation. In the latter, the more sober impersonal viewpoint yet often luridly sexualized fascinations with the fecundity, wildness and cruelties of the natural world richly suggest *Jules* Michelet's inadvertent 'examination of [him]self'. Claude Pujade-Renaud's fictional remake of Athénaïs Michelet's widowhood in *Chers Disparus* (2004), drawing on Michelet's diaries, dares to name what Michelet critics cannot. Here is an abusive husband violating her writing, because he also engaged in predatory voyeurism and regular violations of her body, to stimulate his own. Domestic abuse and its difficult legacies for the victim then provide different explanations for Athénaïs's determined fight, including in print, for the rights to his papers including his private diaries, and for a 'sanitized' version of Michelet's reputation. Her *Nature* is therefore an extraordinary survivor narrative of multiple forms of abuse, redolent of the widespread violations of women's (collaborative) work in science: silencing, editing out, belittling. Athénaïs's refusals to suffer these in 1870–1872 concomitantly make *Nature* the restoration narrative of her original early project for the nature books, to write natural history for children.

Nature and all of ('Jules') Michelet's nature books in English translation form T. Nelson's 'Gift Book' list in 1872. The genre as deemed especially suitable for young women readers (Renier 1964) is not without irony. After Michelet's death in 1874, Athénaïs does more than 'popularize' and 'anthologize' Michelet's histories for school curricula (Creyghton 2016). She almost completes in 1899 another sole-authored work, also begun in the late 1850s, entitled *Cats*, published by Monod posthumously with her preface in 1904. If she there reveals that she was offered ten thousand francs for the rights to *Nature* such being *her* reputation for natural history writing, *Cats* also reveals the comprehensive and specialist natural history work she undertook for it as the same for the ('Jules') Michelet nature books. Athénaïs's many researches in the archives were supplemented exhaustively by her requests for information in correspondence with scientific experts of *her* generation. The letters to Athénaïs that Monod collates in the appendix to the 1904 edition include replies from among others Auguste Mariette (1821–1881) for ancient Egypt, Georges Pouchet (1833–1894) for prehistory and Charles Darwin (1809–1882) for natural selection.

Conclusions

*Un*natural deselection, however, turns out to be the overriding issue for recognition of women in science that this essay investigates and challenges through its different critical reappraisals of '1833'. These included the watersheds of Cuvier's death, and the 'Somerville effect' for men in science in this pivotal period for its modern development and formation of the 'pipeline' model. Close scrutiny and reassessment of the published 'standpoint' theories of key men in science, such as Whewell's, therefore prove essential for renewing future research on women in the history of science on two counts. First, such position statements identify the received ideas of a given period that allegedly explain how and why science by women is (naturally) secondary to that of men, with very occasional exceptions (science has a sex). Second, these statements locate particular, as well as accepted, reasons for *perceiving* women's science and its production as less important. Because the position statements of Whewell and Greenough were primarily responses, indeed clear (over)reactions to the published science of Somerville and Graham, they visibly contradicted and thus reset the normative assumptions about women's secondary places and exceptionalism in science (science has a gender). The essay then tested out the corollaries of inverting the automatically assumed primacy of men in science and in science by women. Following Somerville's best science practice and negotiations of the models of her day, the different indicative case studies investigated the published science by women of the period who innovated in their fields. Their science proved part of the rule in its methods of observation, compilation of multiple evidence and exposition of findings, rather than the exception. The essay could then identify how the developments of science by these women in, and outside, the model of 'creative couples' (Pycior et al. 1996) were positively and negatively impacted by the secondary roles of male relations.

The lessons of Mrs. Somerville, Mrs. Bowdich, Mrs. Gatty and Madame Michelet are clear. Women of independence of mind and expression, scientific talents and the means to pursue them find mainstream, alternative and entrepreneurial ways to undertake and publish their work, because science has no sex as Whewell realized. The problem for women in science (and men outsiders) of the period is then the facilitation, critical reception and evaluation of their work, as Cuvier noted in his important blueprint for assessing the merits of all major contributions to science in 1828 (Pietsch 1995, pp. 1–2). As all the indicative case studies in this essay endorse, further research on nineteenth-century history of women in science needs to target the mixed (non-)reception of their work to counter further omissions of its presence, and to investigate other models for its facilitation (or blockage) that lie outside the paradigms of *women*'s 'uneasy careers and intimate lives' (Abir-Am and Outram 1989). The findings of our case studies have particular resonance for women publishing science today. Malign and abusive 'co-authors' and 'editors' still selectively omit, skew, rearrange and even falsify the evidence of the excellence, variety, entrepreneurialism and range of women's work, and the extent of its contributions.

If science has no sex (Cuvier's position) in the crucial '1833' period for the development of modern science, the question of the *un*natural deselection of women in its work highlights the issue and processes that embedded a dominant gender for the study and hence historiography of the physical sciences. The coining of the term, 'scientist', by Whewell is less important for the increasing professionalization of science than his many reasons for doing so. As elucidated in the comparative approach to the history of women in science of this essay, women with 'overseas' and 'home' expertise are of particular interest for future study, especially if their work is 'polymathic' (a qualifier often used of men including Whewell, but not women in science). Somerville and Bowdich differently exemplify in 1833 what Whewell frames as paradigmatic for the (male) 'scientist' henceforth: international-level expertise, reach and renown. His professional status deriving entirely from the academic (and clerical) orders to which he belongs, but women cannot, then elides modern 'scientists' neatly and unequivocally with gendered institutional frames of reference, legitimation and evaluation for their future work. Its 'pipeline' model will increasingly narrow the parameters for acceptable expertise and scientific productions, to distance them from the work of 'popular' science. The distinctions that Whewell and Jenyns make about 'mere', as opposed to 'expert', compilers is already indicative of the innate gendering of 'pipeline' science and its popularizations, including decorative 'sampler' science. If the history of nineteenth-century science is to advance knowledge of the plethora of women's science in print when this cannot exist institutionally, there is an urgent need to reinstate appropriate terms for 'serious' science work so that these circumnavigate the term 'scientist' as the measure of evaluation. Throughout, this essay operated this approach in opting for the term, 'catalyst', in its chemical and metaphorical meanings. Another, and cognate with 'scientist', is the gender-neutral term 'specialist' also exemplified by the women above.

The snapshot of major women catalysts of the 1833 period in this essay therefore richly exemplifies the central, not exceptional, place of women in nineteenth-century science for its increasing development and diversity of specialist interests. Women's science cannot be pivotal within institutionalized nineteenth-century science, but their wider-ranging work is axiomatic and axiological of it. The number and far-reaching influence of women in science of the period has been celebrated negatively in the stereotypes and caricatures of their work in publications by men in science. This essay now invites further critique of 'informed' science writing by men in nineteenth-century science for particular 'standpoint' theories that are caricatured to be absurd. They will then yield up more prominent women in science like Charlotte Murchison, all busy shining lights into the darkness of its many (pipeline) worlds.

Table 24.1 List of References in Leonard Jenyns, *Manual of British Vertebrate Animals* (1835) to the Drawings in *The Fresh-Water Fishes of Great Britain* (1828-1838) by Mrs T. Edward (Sarah) Bowdich

Common Name of Fish as used by Jenyns	Page Number in Jenyns	Reference to Drawing in Bowdich
Perch	p. 330	*Bowd. Brit. Fr. wat. Fish.* Draw. 5
Bull-head (or Miller's Thumb)	p. 343	*Bowd. Brit. Fr. wat. Fish.* Draw. 24
Three-Spined Stickleback	p. 348	*Bowd. Brit. Fr. wat. Fish.* Draw. 20
Common Carp	p. 401	*Bowd. Brit. Fr. wat. Fish.* Draw. 2
& 'Without barbules' (Crucian Carp)	p. 403	*Bowd. Brit. Fr. wat. Fish.* Draw. 23
Barbel	p. 404	*Bowd. Brit. Fr. wat. Fish.* Draw. 9

(continued)

Table 24.1 (continued)

Common Name of Fish as used by Jenyns	Page Number in Jenyns	Reference to Drawing in Bowdich
Gudgeon	p. 405	*Bowd. Brit. Fr. wat. Fish.* Draw. 15
Tench	p. 406	*Bowd. Brit. Fr. wat. Fish.* Draw. 13
(Yellow) Bream	p. 406	*Bowd. Brit. Fr. wat. Fish.* Draw. 18
Roach	p. 408	*Bowd. Brit. Fr. wat. Fish.* Draw. 3
Dace	p. 410	*Bowd. Brit. Fr. wat. Fish.* Draw. 11
Chub	p. 412	*Bowd. Brit. Fr. wat. Fish.* Draw. 6
Rudd	p. 412	*Bowd. Brit. Fr. wat. Fish.* Draw. 21
Bleak	p. 414	*Bowd. Brit. Fr. wat. Fish.* Draw. 4
Minnow	p. 415	*Bowd. Brit. Fr. wat. Fish.* Draw. 8
(Bearded) Loach	p. 416	*Bowd. Brit. Fr. wat. Fish.* Draw. 12
Pike	p. 417	*Bowd. Brit. Fr. wat. Fish.* Draw. 17
Common Trout	p. 424	*Bowd. Brit. Fr. wat. Fish.* Draw. 14
Thames Shad	p. 437	*Bowd. Brit. Fr. wat. Fish.* Draw. 19
(Allis) Shad	p. 438	*Bowd. Brit. Fr. wat. Fish.* Draw. 27
Burbot (Barbolt)	p. 448	*Bowd. Brit. Fr. wat. Fish.* Draw. 30
(Sharp-Nosed) Eel	p. 475	*Bowd. Brit. Fr. wat. Fish.* Draw. 7
Glut Eel	p. 475	*Bowd. Brit. Fr. wat. Fish.* Draw. 22
Snig Eel 'Being unacquainted with this species, I am unable to point out its distinguishing characters. According to Mrs. Bowdich it is the smallest of the Eel tribe, and is caught plentifully in the Thames, but more especially in Berkshire and Oxfordshire. She thinks that Pennant has confounded it with the Glut Eel. Mr. Yarrell informs me, he considers it as distinct from the last species.'	pp. 477–478	*Bowd. Brit. Fr. wat. Fish.* Draw. 28
Sea (?) Lamprey	p. 520	*Bowd. Brit. Fr. wat. Fish.* Draw. 26
River Lamprey	p. 521	*Bowd. Brit. Fr. wat. Fish.* Draw. 16
Pride	p. 522	*Bowd. Brit. Fr. wat. Fish.* Draw. 32
Jenyns does not refer to the relevant drawing in Bowd. for the Salmon (p. 421), the Gwiniad [sic] p. 431; the Vendace (p. 432) or the Sturgeon (p. 493).		

© Mary Orr

NOTES

1. Although Western science is the focus of this chapter, the multicultural nature of science is also in question (Harding 1998). Asian perspectives on women's exclusions from science pivot on similar gender issues, complicated additionally in Indian for example by what Mukhopadhyay and Seymour (1994) call 'patrifocality' (where the woman's education accrues to the husband's family and hence where his mother wields considerable family influence). See also Subrahmanyan (1998).
2. See for example the EU report 'Women in Science and Technology: Creating Sustainable Careers' (2009).
3. Literary criticism including feminist of *fin de siècle* texts and visual arts depicts women as threatening vampires and seductresses luring men to their peril.
4. Feminist reappraisals of the gendering of 'genius' lie outside this essay. See for example Kristeva (2004).
5. The sciolist, 'A superficial pretender to knowledge; a conceited smatterer' (OED) manifests many attributes of the dilettanti, the fop and the modern understanding of an 'amateur' as a non-expert. By counter-analogy, Whewell's direct reference to Coleridge as artist and philosopher, but also major supporter of the upstart Sir Humphrey Davy, is also highly significant for his coinage of 'scientist' aligning with his own, serious self-fashioning as the future Master of Trinity College, Cambridge. See Golinski (1999), but also the longer aesthetic heritage of 'The Poet as Sciolist' (Fraser 1967).
6. A rather grainy reproduction of the image can be viewed on the 'Trowelblazers' website. https://trowelblazers.com/charlotte-murchison/. Accessed 29 January 2019.
7. The inside front cover of the French edition states more ambiguously that the work is 'publiés en anglais par Mistress Lee, et en français par M. Théodore Lacordaire sur les documents fournis par sa famille'. In France, Lacordaire was assumed to be the author (Morren 1870) as in his French Wikipedia entry. https://fr.wikipedia.org/wiki/Th%C3%A9odore_Lacordaire. Accessed 30 January 2019. See Orr (2020). All references are to the English edition of *The Memoirs of Baron Cuvier* printed in London.
8. Our central argument about Whewell's coinage of 'scientist' as a response to the crisis in science for *men* aligns with the findings of Bruckmüller and Branscombe (2010): 'It might seem somewhat counter-intuitive that a higher selection of women for precarious positions should be at least as much, or even more so, about men and leadership than about women and leadership. In fact, it may not be so important for the glass cliff that women are stereotypically seen as possessing more of the attributes that matter in times of crisis, but rather that men are seen as lacking these attributes and that the

attributes that men stereotypically have do not fit with what is perceived as needed in a leader in times of crisis' (p. 448).
9. The argument draws more finely here the place of and usage by Jenyns of Bowdich's *Fresh-Water Fishes* as set out in the only study to date of the *non*-reception of her contributions to British ichthyology (Orr 2014, especially pp. 227–28). Table 24.1 in this chapter enlarges and corrects the information given in note 58 (p. 228) of this publication.
10. See the excellent online text and supporting resources for *The Book of Sun-Dials* at respectively: https://digital.library.upenn.edu/women/gatty/sundials/sundials.html and https://digital.library.upenn.edu/women/ewing/parables/memorial.html.
11. For an extensive discussion of Athénaïs Michelet's work in anthologizing and popularizing Michelet's *œuvre*, not as a co-writer or collaborator in it, see Creyghton (2016).

References

Abir-Am, Pnina G., and Dorinda Outram. 1989. *Uneasy Careers and Intimate Lives: Women in Science, 1789–1979*. New Brunswick and London: Rutgers University Press.

Bate, David George. 2010. "Sir Henry Thomas De la Beche and the Founding of the British Geological Survey." *Mercian Geologist* 17, no. 3: 149–65.

Bloom, Harold. 1973. *The Anxiety of Influence: A Theory of Poetry*. London, Oxford, and New York: Oxford University Press.

Bruckmüller, Susanne, and Nyla R. Branscombe. 2010. "The Glass Cliff: When and Why Women are selected as Leaders in Crisis Contexts." *British Journal of Social Psychology* 49: 433–51.

Burek, Cynthia V., and Higgs Bettie, eds. 2007. *The Role of Women in the History of Geology. Special Publications 281*. London: Geological Society.

Creese, Mary R. S., and Thomas M. Creese. 1998. *Ladies in the Laboratory? American and British Women in Science 1800–1900: A Survey of Their Contributions to Research*. Lanham, Toronto, and Plymouth, UK: Scarecrow Press.

Creyghton, C. M. H. G. 2016. "La survivance de Michelet: Historiographie et politique en France depuis 1870." PhD diss., 95–133. University of Amsterdam. https://pure.uva.nl/ws/files/6667940/03.pdf. Accessed 6 February 2019.

Eger, Elizabeth, Charlotte Grant, Clíona Ó. Gallchoir, and Penny Warburton, eds. 2001. *Women, Writing and the Public Sphere, 1700–1830*. Cambridge: Cambridge University Press.

Ellis, Heather. 2017. *Masculinity and Science in Britain, 1831–1918*. London: Palgrave Macmillan.

European Commission. 2009. *Women in Science and Technology: Creating Sustainable Careers*. Brussels: European Commission. https://www.genderportal.eu/sites/default/files/resource_pool/wist2_sustainable-careers-report_en.pdf. Accessed 7 February 2019.

Fara, Patricia. 2004. *Pandora's Breeches: Women, Science and Power in the Enlightenment*. London: Pimlico.

Felbridge and District History Group. 2003. "Charles Henry Gatty." *Felbridge & District History Group*. http://www.felbridge.org.uk/index.php/publications/charles-henry-gatty/. Accessed 7 February 2019.

Fox Keller, Evelyn. 1996. *Reflections on the Gender of Science*. New Haven: Yale University Press.

Fraser, Russel. 1967. "The Poet as Sciolist." *The Sewanee Review* 75, no. 3: 444–54.

Golinski, Jan. 1999. "Humphry Davy's Sexual Chemistry." *Configurations* 7, no. 1: 15–41.

Gossman, Lionel. 2001. "Michelet and Natural History: The Alibi of Nature." *Proceedings of the American Philosophical Society* 145, no. 3: 283–333.

Harding, Sandra. 1998. *Is Science Multicultural? Postcolonialisms, Feminisms and Epistemologies*. Bloomington and Indianapolis: Indiana University Press.

Harding, Sandra, ed. 2004. *The Standpoint Theory Reader: Intellectual and Political Controversies*. London and New York: Routledge.

Holmes, Richard. 2014. "In Retrospect: On the Connexion of the Physical Sciences." *Nature* 514: 432–33.

Jenyns, Leonard. 1835. *A Manual of British Vertebrate Animals*. Cambridge: John Smith.

Kölbl-Ebert, Martina. 1997a. "Mary Buckland (née Morland) 1797–1857." *Earth Sciences History* 16, no. 1: 33–38.

Kölbl-Ebert, Martina. 1997b. "Charlotte Murchison (née Hugonin) 1788–1869." *Earth Sciences History* 16, no. 1: 39–43.

Kölbl-Ebert, Martina. 2009. "George Bellas Greenough's 'Theory of the Earth' and Its Impact on the Early Geological Society." In *The Making of the Geological Society of London*, ed. C. L. E. Lewis and S. J. Knell, 115–28. Special Publications, 317. London: The Geological Society Publishing House.

Kristeva, Julia. 2004. "Is There a Feminine Genius?" *Critical Inquiry* 30, no. 3: 493–504.

Kuhn, Thomas S. 2012. *The Structure of Scientific Revolutions*, 4th Edition, 50th Anniversary Edition with an Introductory Essay by Ian Hacking. Chicago: Chicago University Press.

Lee, Mistress R. 1833a. *Memoirs of Baron Cuvier*. London: Longman, Rees, Orme, Brown, Green & Longman.

Lee, Mistress R. 1833b. *Mémoires du baron Georges Cuvier*. Paris: H. Fournier.

Michelet, Adéle-Athénaïs. 1872. *Nature: The Poetry of Earth and Sea*. London, Edinburgh, and New York: T. Nelson and Sons.

Michelet, Adéle-Athénaïs. 1875. *La Tombe de Michelet*. Paris: Simon Raçon.

Michelet, Adéle-Athénaïs. 1876. *Ma Collaboration à «L'oiseau», «L'insecte», «La mer», «La montagne». Mes droits à la moitié de leur produit*. Paris: Chamerot.

Michelet, Adéle-Athénaïs. 2004. *Mémoires d'une enfant*, ed. and intro. P. Enckell. Paris: Mercure de France.

Morren, Charles Jacques Édouard. 1870. *Éloge de Jean-Theodore Lacordaire*. Liège: J. Desoer.

Morrison, Aimée. 2019. "Laughing at Injustice: #DistractinglySexy and #StayMadAbby as Counternarratives." In *Digital Dilemmas: Transforming Gender Identities and Power Relations in Everyday Life*, ed. Diana C. Parry, Corey W. Johnson, and Simone Fullager, 23–52. Springer Online: Palgrave Macmillan. https://link.springer.com/content/pdf/10.1007%2F978-3-319-95300-7.pdf. Accessed 3 February 2019.

Mukhopadhyay, Carol Chapnick, and Susan Seymour, eds. 1994. *Women, Education and Family Structure in India*. Boulder, CO: Westview Press.

Muraoka, Daisuke. 2004. "Seaweed Resources as a Source of Carbon Fixation." *Bulletin of Fisheries Research Agency, Supplement* 1: 59–63.

Orr, Mary. 2014. "Fish with a Different Angle: The Fresh-Water Fishes of Great Britain by Mrs Sarah Bowdich (1791–1856)." *Annals of Science* 71, no. 2: 206–40.

Orr, Mary. 2015. "Women Peers in the Scientific Realm: Sarah Bowdich (Lee)'s Expert Collaborations with Cuvier, 1825–1833." In *Women and Science*, ed. Claire Jones and Sue Hawkins. *Notes and Records* 69, no. 1: 37–52.

Orr, Mary. 2020. "Les *Mémoires du baron Georges Cuvier* (1833) de Mistress Lee: mémoires scientifiques, pacte autobiographique, ou réécriture des savoirs?" In *Littérature française et savoirs biologiques au xixe siècle: traduction, transmission, transposition*, ed. Thomas Klinkert and Gisèle Séginger, 183–200. Berlin and Boston: Walter de Gruyter.

Pietsch, Theodore, ed. 1995. *The Historical Portrait of the Progress of Ichthyology: From Its Origins to Our Own Time by Georges Cuvier*. Trans. A. J. Simpson. Baltimore: Johns Hopkins University Press.

Poovey, Mary. 1984. *The Proper Lady and the Woman Writer: Ideology as Style in the Works of Mary Wollstonecraft, Mary Shelley and Jane Austen*. Chicago: University of Chicago Press.

Pujade-Renaud, Claude. 2004. *Chers Disparus*. Paris: Babel.

Pycior, Helena. M., Nancy G. Slack, and Pnina G. Abir-Am, eds. 1996. *Creative Couples in the Sciences*. New Brunswick, NJ: Rutgers University Press.

Renier, Anne. 1964. *Friendships' Offering: An Essay on the Annuals and Gift Books of the Nineteenth Century*. London: Private Libraries Association.

Seccombe, Thomas. n.d. "Blomefield, Leonard, Formerly Leonard Jenyns (1800–1893)." In *Oxford Dictionary of National Biography*. http://www.oxforddnb.com/view/10.1093/odnb/9780192683120.001.0001/odnb-9780192683120-e-2664. Accessed 31 January 2019.

Smith, Bonnie. 1992. "Historiography, Objectivity and the Case of the Abusive Widow." *History and Theory* 13, no. 4: 15–32.

Subrahmanyan, Lalita. 1998. *Women Scientists in the Third World: The Indian Experience*. New Delhi, Thousand Oaks, and London: Sage.

Thompson, Carl. 2012. "Earthquakes and Petticoats: Maria Graham, Geology, and Early Nineteenth-Century 'Polite Science'." *Journal of Victorian Culture* 17, no. 3: 329–46. https://doi.org/10.1080/13555502.2012.686683.

Whewell, William. 1834. "'On the Connexion of the Physical Sciences'. By Mrs. Somerville." *Quarterly Review* 51, no. 100: 54–68.

CHAPTER 25

'The Question Is One of Extreme Difficulty': The Admission of Women to the British and Irish Medical Profession, C. 1850–1920

Laura Kelly

In July 2019, seven women who studied at the University of Edinburgh medical school, but who were ultimately not allowed to qualify, were posthumously awarded their degrees, 150 years after they began their studies. This group of women, who have received extensive historical attention, was known as the Edinburgh Seven, and included Mary Anderson, Emily Bovell, Matilda Chaplin, Helen Evans, Sophia Jex-Blake, Edith Pechey and Isabel Thorne.[1] These women faced significant resistance in their efforts to study medicine and, while at medical school, faced abuse from male students. In particular, the issue of men and women dissecting together resulted in a famous riot at the institution in 1870 (Bashford 1998, p. 112). The male students at the University of Edinburgh medical school protested that women dissecting alongside men signified a 'systematic infringement of the laws of decency' (Bashford 1998, p. 112). The legal exclusion of women from the medical profession in Britain had begun in the mid-nineteenth century with the re-organisation of the medical profession by the 1858 Medical Act which resulted in the exclusion of outsiders. The act did not specifically exclude women but with the emphasis now placed on standardisation of medical qualifications from universities, to which women had no access, it effectively prevented women from making it onto the Medical Register (Wyman 1984, p. 41). British women who wished to attain a medical education went abroad to certain universities

L. Kelly (✉)
School of Humanities, University of Strathclyde, Glasgow, UK
e-mail: l.e.kelly@strath.ac.uk

© The Author(s), under exclusive license to Springer Nature Switzerland AG 2022
C. G. Jones et al. (eds.), *The Palgrave Handbook of Women and Science since 1660*, https://doi.org/10.1007/978-3-030-78973-2_25

which appear to have been more liberal towards the admission of women than those in Britain, such as the Universities of Zurich, Bern, Geneva and Paris. The University of Zurich, for example, admitted women to its medical classes from 1864 (Usborne 2001, p. 109). Jex-Blake attained her MD degree from the University of Bern in 1877. However, qualifications from these European universities did not allow women to practise in the United Kingdom. Thus, leading pioneer British women doctors such as Jex-Blake began to campaign for the right to register as medical practitioners.

However, the question of women's admission to the medical profession in Britain began in the 1850s. Dr. Elizabeth Blackwell, who had studied at Geneva College, New York, became the first female doctor to have her name placed on the Medical Register in 1858 because of a special clause that allowed those in possession of a foreign degree to register (Witz 1992, p. 80). The second was Elizabeth Garrett Anderson. Garrett Anderson had faced a huge amount of opposition in her quest to study medicine, not only from medical schools which refused to admit her but also from British licensing bodies who would not permit her to take their examinations (Witz 1992, pp. 80–83). As a result of her involvement with the Langham Place Circle, a group of women including Emily Davies, who were interested in the promotion of higher education for women, twenty-three-year-old Garrett met Blackwell, and became inspired to study medicine. Garrett's parents were initially strongly opposed to her studying medicine. In a letter to Emily Davies in 1860, Garrett wrote that her father believed 'the whole idea was so *disgusting* that he could not entertain it for a moment' (Garrett Anderson 2016, p. 46). As will be outlined in more depth later, the notion of women engaging in medical training and practice, in particular anatomy dissections, went against Victorian ideals of womanhood and appropriate behaviour.

Garrett was undeterred, however, by the barriers facing her in her pursuit of medical education. With the eventual support of her father, she became acquainted with politician Russell Gurney and his wife Emelia, who were to become active supporters of the campaign to admit women to the medical profession. Because no British medical schools were open to women at the time, Elizabeth undertook nursing experience on the surgical ward of the Middlesex Hospital in London for six months from August 1860, before she was excluded from undertaking further study there as a result of a memorial written by a number of the male medical students. Despite the discouragement of being refused entry to several medical schools in England and Scotland following this, she went on to undertake private study, before qualifying with a licence from the Society of Apothecaries in London in 1865.

Following the award of Garrett's licence, the Society of Apothecaries then closed its licences to individuals who had received private instruction, meaning that women could no longer gain a British qualification (Witz 1992, p. 83). Garrett Anderson set up practice in Upper Berkeley Street, London, before establishing St. Mary's Dispensary for Women and Children in 1866, and later, the New Hospital for Women, in 1871, carving out a sphere of practice in the field of women's and children's health. She attained an MD in Paris in 1870 and was heavily involved in supporting the campaign by early women doctors like Sophia Jex-Blake (1840–1912) to allow women admission to the medical profession. However, Garrett Anderson and Jex-Blake often disagreed on the best approach to this campaign. In a letter to *The Times* in 1873, Garrett Anderson argued that British women should seek 'abroad that which is at present denied to them in their own country', while Jex-Blake disagreed in her response to the letter arguing that prospective women doctors should 'fight it out on this line' and should not 'be driven out of our own country for education' ("Letters" 1873, p. 4). While initially against the idea of a separate medical school for female students, Garrett Anderson later supported the London School of Medicine for Women (LSMW) set up by Jex-Blake in 1874. She taught at the LSMW and was also instrumental in gaining financial support for the school. At a meeting in support of the School in June 1877, she remarked that there was 'nothing injurious to the health, the morals, or the manners of women in a medical education, and that the results were likely to prove beneficial to the female sex and to the nation'. This was followed by almost a decade of credentialist and legalistic tactics by aspiring British women doctors who wished to gain entry to the medical profession (Witz 1992, pp. 86–99).

Today, women medical students are in the majority of entrants in both the United Kingdom and the Republic of Ireland, with the number of registered female doctors in the United Kingdom growing from 43% in 2011 to 45% in 2015. Given the recent advances women have made in the medical profession, it might be easy to forget the struggles experienced by early British women doctors in trying to attain medical qualifications.[2] This chapter therefore examines the history of women in medicine primarily in Britain and Ireland. It explores contemporary arguments both for and against women's entry to British and Irish medical schools in the nineteenth century, as well as the experiences of early women doctors who gained entry to medical schools in this period. The importance of the First World War in further opening avenues to women who wished to study medicine will also be explored, but the chapter will ultimately demonstrate how Victorian attitudes to women in the medical profession persisted well into the twentieth century.

Arguments in Favour and Against Women in the Medical Profession

The question of women studying medicine provoked significant debate in the late Victorian period and beyond. Arguments against women in the medical profession essentially hinged on several key factors: women's supposed physical, mental and emotional unsuitability for medical study; the fact that the medical profession was already well-supplied with doctors and that the introduction of women would increase competition; concerns about the impact upon the family; and concerns about men and women being educated together.

As Alison Moulds has shown, the 'medical-women' question dominated the medical press in the Victorian period and articles and letters on the topic highlight the 'considerable anxiety about both the role of women and the status of medicine' (Moulds 2019, p. 17). Medical journals actively engendered debates on the 'medical-women' question (Moulds 2019, p. 17). For instance, *The British Medical Journal* in 1870 contended that the medical profession was already saturated and that the introduction of women surgeons might displace an equal number of men surgeons ("The Admission" 1870, p. 475). Medical practitioners, particularly specialists in gynaecology and obstetrics who were beginning to notice competition from female doctors in these areas, were instrumental in the attack on the women's higher education movement in nineteenth-century Britain (Burstyn 1973, p. 81). Similarly, an article in the *Irish Times* in 1876 claimed that the 'wholesale irruption of fascinating lady doctors' would result in 'ruinous competition' for medical practitioners, and, concerning lady doctors with foreign, particularly American, qualifications, the writer stated that 'we should tremble to find ourselves in the hands of lady-physicians coming from those districts of the Union where female suffrage reigns supreme' (*Irish Times* 1876, p. 4). An editorial which appeared in the *Medical Times and Gazette* in 1874 on the opening of the London School of Medicine for women espoused similar views:

> We do not believe that women are fitted for the responsibilities, anxieties and physical wear and tear of medical practice; but if they persist in believing the contrary, we have nothing to say against their making the experiment, beyond what we have so often said elsewhere: if they will study medicine, let them do so, but apart from male students. (*Medical Times and Gazette* 1874, p. 714)

This statement is indicative of the common arguments against women studying medicine. Many members of the medical profession believed that women were not fit for the physical and emotional strain of medical practice. However, the author of the article suggested that if women wanted to go ahead with study, they should do so but separately from the male students. In Britain, this appears to have been a significant issue and led to women doctors founding separate medical schools for female students. For instance, in Britain, women were initially educated at single-sex medical schools such as the London School of Medicine for Women (founded in 1874) and the Edinburgh School of Medicine for Women (founded in 1886). These two institutions were largely responsible for the medical education of British women up until the opening of other British medical schools from the 1890s. However, even once medical schools opened their doors to women students, it remained the case that women were educated separately. For example, at the University of Birmingham, female students were educated separately from the male students when they were admitted in 1900 (Reinarz 2009, p. 161). Historians of women in the medical profession have drawn attention to the sense of separatism that women doctors faced in their educational experiences and later in their professional lives.[3]

We may gain further insights from the discussions held at a meeting of the General Council of Medical Education and Registration on 25 June 1875. At the meeting, the letters of Sophia Jex-Blake and Mr. Arthur Norton, with regard to the medical education of women, were ordered to be acknowledged and a statement was provided by the Council that they had been considered. The debate that occurred on the matter provides important insights into contemporary attitudes in the British medical profession to women's admission. While Dr. Sharpey felt that an amendment should be passed in order to allow women to enter the profession, he felt that he would not recommend women from entering, but that he was 'not prepared to stand in the way of women entering on a career on which they think they can safely and properly enter'. In a lengthy speech, Dr. Andrew Wood, on the other hand, was more representative of the views put forward on the issue. He argued that those arguing for women's admission were in the minority, and emphasised that these individuals 'did not know what toils and dangers, what perplexities and constant anxieties there are in the profession; they do not know all the repulsiveness of the dissecting-room; they do not know the bloody scenes of the operating-room; they do not know a great many of those things which are even so repulsive to men that they refrain from entering the profession' ("Meeting" 1875, p. 56). Woods argued that he 'was never told by a woman that she would employ a female doctor. The almost invariable reply to my question when I asked it has been "We would not send for a female doctor; we cannot confide in them; we would not trust our lives in their hands so long as we can get an accomplished, skilful and proper medical man"' ("Meeting" 1875, p. 57). Woods' key issue however, appears to have been with regard to the question of women dissecting, and in particular, the question of women and men dissecting together:

> I would ask any member of this Council whether he would like to see his sister or daughter embracing the practice of medicine? He would revolt at such an idea; he would not like to go into the dissecting-room, as I have gone, and see five or six ladies dissecting a body, with their hands all covered with filth, and five or six male students dissecting another body two or three yards off them. They would not like to see them on their knees covered with blood. They would not like to see them liable to be called out at all hours of the day and night. We should not like to see the toiling and moiling, and noise and discomfort and disquietude to which they would be subjected. Therefore, why should we encourage the sisters and daughters of others to enter a profession that we would not like our own daughters and sisters to take up? ("Meeting" 1875, p. 57)

As Alison Bashford's work has shown, the issue of anatomy dissections was a complex one. It is clear that there was 'considerable cultural investment in a gendered and sexualised understanding of dissection in which the masculine scientist/dissector penetrated, came to "know" the feminised corpse in a dynamic shot through with all types of desire' (Bashford 1998, p. 114). When women medical students entered the dissecting room, they disrupted and reversed 'the gendered subjectivities and the sexualised dynamics which operated there' (Bashford 1998, p. 114). One of the main issues for opponents of women's entry to the medical profession appears to have been that of men and women dissecting together. For instance, the *British Medical Journal* commented in 1870 that it was 'an indelicate thing for young ladies to mix with other students in the dissecting-room and lecture theatre' ("Lady Surgeons" 1870, p. 338). Through the process of dissection, female students 'came to know' the male body. They penetrated it with scalpels and in this way, the power was handed over to the female dissector (Bashford 1998, p. 114).

Dr. Humphry, in response to Woods' speech, argued that while he would not encourage anyone, male or female, to enter the profession of medicine, he felt it was their choice to do so. He argued that he had never witnessed 'an important operation without a woman being present there', and that therefore 'if women of one class are regularly admitted into our operating rooms, I do not see that there is anything in the thing itself to preclude other women from being present there' ("Meeting" 1875, p. 58). Irish doctor Dominic Corrigan argued that although he was concerned about the issue of muscular strength, he felt that women should be admitted, but for the reason that he believed that if women were given a fair field 'they will soon find that they are out of their place, and the agitation will disappear' ("Meeting" 1875, p. 59). In response, Irish doctor Rawdon Macnamara stated that he could

> produce in the city 10,000 or 15,000 ladies with whom you would be very sorry to try conclusions as to physical strength. It is merely begging the question to speak of physical force. There is not much physical force required to handle the stethoscope, or to handle the urinometer or thermometer, so far as medicine is concerned; and as to surgery I do not know what in that case either very much physical force is required. I would be very sorry to allow any gentleman to pass a catheter into my bladder if he was going to exert much physical force […]. ("Meeting" 1875, p. 59)

In Macnamara's view, the moral side of the question was more pertinent, but again, he felt that he could find evidence to discount claims that the study of medicine would have a 'lowering effect on the nature of women', arguing that 'the most delicate-minded women that I know of are at this moment employed as nurses in hospitals. I refer to the Sisters of Charity in the City of Dublin. It would be a calumny to say that the finer feelings of these ladies have been blunted because they have attended in the most serious and critical surgical cases. There is not a more fine-minded, more tenderly-nurtured, or more admirable class of ladies than these Sisters of Charity' ("Meeting" 1875, p. 59). Dr. Stokes argued that 'the question is one of extreme difficulty', asking

'Is the unsexing of women, which certainly would be threatened by a surgical or medical education, a desirable thing?' Stokes also suggested that although physical prowess was not required for most branches of medicine, 'in obstetrics there are conditions in which the full force of the most muscular man is absolutely necessary' ("Meeting" 1875, pp. 59–60). Finally, Dr. Rolleston argued that there was a need for women doctors in other parts of the empire, in particular India, arguing that 'we are responsible not only for the women who are crying out for it here in England, but for those who would be allowed to avail themselves of the well-instructed doctors of their own sex throughout the great empire of India' ("Meeting" 1875, p. 61). The debate was by no means resolved, and discussions over the question continued over the later decades of the nineteenth century. Indeed, from the 1860s to the 1900s, the British medical press was preoccupied with the question of women's suitability for the medical profession (Moulds 2019). As Moulds has shown, medical journals 'held divergent views from one another but that there was dissent and discord within individual titles', with medical journals actively encouraging a range of clashing views on the suitability of women in medicine (Moulds 2019, pp. 16–17).

Conversely, those who argued in favour of women suggested that women's unique qualities made them suitable for a career in the medical profession. Thomas Haslam, for instance, wrote to the *Freeman's Journal* in 1871, making a similar case for the medical education of women. Haslam, who, along with his wife, Anna, would found the Dublin Women's Suffrage Association in 1876, argued that women were much better qualified than men to treat patients of their own sex ("Letter" 1871, p. 3). Similarly, Sophia Jex-Blake argued in her 1872 essay 'Medicine as a Profession for Women', that 'women are *naturally* inclined and fitted for medical practice' (Jex-Blake 1872, p. 8). Sophia Jex-Blake argued that women's emotional natures in fact made them very suited to the medical profession, stating in an 1874 paper that 'Women have more love of medical work, and are naturally more inclined, and more fitted for it than most men' (Jex-Blake 1874, p. 3). Moreover, she insisted that there was a 'very widespread desire […] among women for the services of doctors of their own sex' (Jex-Blake 1874, p. 4). In Ireland, when the Munster branch of the Irish Association of Women Graduates and Candidate Graduates wrote to the Victoria Hospital for Diseases of Women and Children in Cork, requesting that a female doctor be appointed to the hospital's staff in the first decade of the twentieth century, they argued that often women patients preferred to be treated by female physicians (Munster Branch of the Irish Association of Women Graduates and Candidate Graduates n.d.).

Such ideas were supported by some, but not many members of the medical profession. In 1888, Irish doctor Thomas More Madden, gave an address at the Annual Meeting of the British Medical Association in Glasgow. Dr. More Madden was President of the Obstetrics Section of the British Medical Association at the time and supported the admission of women to the medical profession, particularly their admission to the speciality of obstetrics. More Madden is quoted as having said:

> I cannot agree with those who are opposed to the admission of women into the practice of our department of medico-chirurgical science for which their sex should apparently render them so especially adapted. I can see no valid reason why any well qualified practitioner, male or female, should not be welcomed amongst us. Nor if there are any women who prefer the medical attendance of their own sex, does it seem fair that in this age of free trade they should not be afforded every opportunity. (*Freeman's Journal* 1888, p. 4)

However, others disagreed with this viewpoint. In an annual lecture delivered to students at the LSMW in October 1877, Elizabeth Garret Anderson argued against the conventional idea of advocates of women in the profession, that women would be more sympathetic in understanding women's ailments. In Garrett Anderson's view, 'It is often said for instance, "Women will understand women's ailments so much better than men do." I fancy this is only true in a very partial and limited sense and that it is most undesirable that medical women themselves should place much confidence in it. To understand disease the possession of the special organization in which the disease is found, is of very minor importance as compared with the possession of brains and cultivation [...]' (Garrett Anderson 1877, p. 14).

More Madden also argued that there was a distinctive need for women doctors in India and Oriental countries 'where millions of suffering women and children are fanatically excluded from the possibility of any other skilled professional assistance; and I therefore think that such practitioners are entitled to admission into our ranks in the British Medical Association' (*Freeman's Journal* 1888, p. 4). Early women doctors continued to have to defend their place in the medical profession well into the nineteenth century. Elizabeth Blackwell, speaking in 1895, explained that she believed that the movement to allow women to study medicine 'was only a revival of work in which women had always been engaged; but that it was a revival in an advanced form, suited to the age and to the enlarging capabilities of women' (Blackwell 1895, pp. 198–99). Similarly, in an address given by Dr. Mary Scharlieb, a pioneer British gynaecologist, in 1898, for instance, she pointed to the long history of women working in medicine in ancient and early modern times and the important work being conducted by British women doctors in India (Scharlieb 1898). In Scharlieb's view, the situation for British women in medicine was improving year by year, with plenty of schools now being open to women, and women doctors holding 'a fair share of public and private appointments' (Scharlieb 1898, p. 16).

Although it has been argued that aspiring women doctors in Ireland 'met the same determined opposition and prejudice from the medical establishment' as elsewhere, it is evident that the first generations of women doctors found the Irish medical hierarchy to be peculiarly open-minded with regard to the question of women's admission and with regard to their educational experiences (Meenan 1987, p. 81).[4] The King and Queen's College of Physicians of Ireland (KQCPI), later the Royal College of Physicians of Ireland, was the first institution in the United Kingdom to take advantage of the Enabling Act and admit women who had undertaken their medical studies abroad to their licentiate examinations in 1877. The role of the KQCPI in the registration of early British women doctors has been seriously underplayed by historians. The first women to take advantage of the KQCPI's leniency were British female doctors who had studied abroad. Eliza Louisa Walker Dunbar was the first to qualify with a licence from the KQCPI in January 1877 (Kelly 2012, p. 39). She was later followed by many others, including the English doctor Sophia Jex-Blake, a leading campaigner for women's admission to the medical profession. Jex-Blake had been inspired to study medicine after working at the New England Hospital for Women and Children in Boston in 1862 and being inspired by Lucy Sewell, a pioneer American female physician. Jex-Blake then applied to study medicine at Harvard in 1867 but was rejected. She returned to Britain and was accepted to the medical school of the University of Edinburgh in 1869. However, as has been well-documented by historians, Jex-Blake and her cohort, later called 'The Edinburgh Seven', encountered many obstacles and difficulties while studying at Edinburgh and were told in 1873 that they would not be allowed to qualify with medical degrees from the university (Kelly 2012, p. 20).[5] Jex-Blake and some of her cohort went to the University of Bern. She graduated with an MD degree from there in 1877. However, due to the regulations of the Medical Act, a 'foreign' degree would not enable a doctor to practise medicine in Britain. Meanwhile, two politicians, Russell Gurney and William Francis Cowper-Temple, who supported women's admission to the medical profession, actively campaigned for legislation to allow this. In 1876, Gurney's Enabling Act was passed by the British parliament, 'enabling' all of the nineteen recognised medical examining bodies to accept women bodies but stating that they were not obliged to do so. The KQCPI became the first institution to take advantage of this new legislation, thus offering women doctors a means of qualifying as registered medical practitioners (Kelly 2012, p. 20). Sophia Jex-Blake described this decision as 'the turning point in the whole struggle' (Jex-Blake 1886, p. 204).[6] The decision of the KQCPI was arguably the result of a combination of factors: an atmosphere of liberality with regard to women's higher education which existed in Dublin; the fact that women were already being admitted to Irish hospitals to gain their clinical experiences; and finally, and probably most significantly, the question of the financial gain which the KQCPI could make from the fees charged to women graduates. Following the admission of women to the King and Queen's College of Physicians, Irish medical schools soon followed suit and admitted women to study medicine in their classes. In October 1879, the College Council of Queen's College Galway decided to allow women to enter the university, although no woman studied medicine there until 1902. Queen's College Cork permitted the admission of women from 1883, although its first female medical student enrolled in 1890. Women could attend the Royal College of Surgeons from 1884. The first female medical student at Queen's College Belfast began her studies in 1888. Women could attend the Catholic University medical school at Cecilia Street from 1898 (Kelly 2012, p. 44). Trinity College lagged behind, admitting women to its classes from 1904. Lily Baker became the first medical graduate from Trinity College in 1906 (Kelly 2012, p. 174). By contrast, the majority of British universities did not admit women until the 1890s and 1900s, with the exception of the University of London which permitted women to take medical degrees from 1878. There were no graduates from that university until 1882, however, when Mary Ann Scharlieb and Edith Shove graduated following study at the London School of Medicine for Women which had been established by Sophia Jex-Blake in 1874. Other British universities were slower to open their medical classes to women. The University of Bristol opened its classes to women in 1891, the University of Glasgow in 1892 and the University of Durham in 1893 while Cambridge and Oxford allowed women to be admitted from 1916 and 1917, respectively (Kelly 2013, p. 100).

Experiences and Careers

Given the need to defend their place in the medical profession, advice from women doctors to prospective students often advised modesty combined with diligence. In advice to prospective women medical students in Charles Bell Keetley's 1878 *Student's Guide to the Medical Profession* Garrett Anderson appealed to female medical students 'not to be discouraged either by the magnitude of the work to which they have to put their hands, or by the disapproval of many of their friends and acquaintances', putting forward the importance of 'good temper' as a 'precious weapon to those who are working for a cause still regarded by many with prejudiced disapproval'. Yet, Garrett Anderson also suggested that in relation to their dress: 'The less conspicuous medical women make themselves in their dress the better. Here, as everywhere else, the true art lies in the perfect suitability of the dress for the occasion and, if this be attained, grace and refinement cannot be unwelcome' (Garrett Anderson 1878, p. 46). Diligence and hard work were also important qualities with Garret Anderson noting that 'women can less easily afford to be second-rate, their professional work will be more closely scrutinised; mistakes will ruin them more quickly than they will men' (Garrett Anderson 1878, p. 44). Similarly, speaking in 1898 at the Women's Institute, Dr Mary Scharlieb encouraged modesty if women were to continue to progress in the medical profession: 'We have only to be true to ourselves; to be ready to do what is our work, and to be content with our share of success; not to be grasping after honour, money or position but to be faithful and true, and to hope that the days may come when still more opportunities and chances will be given to us' (Scharlieb 1898, p. 16). Similarly, Elizabeth Blackwell, in an address to students at the LSMW in 1890, advised them that their primary motive for undertaking medical study should be an interest in that study and the medical life, and a belief in the positive influence it could have, rather than monetary concerns (Blackwell 1890, p. 4).

For women who went on to study medicine in the late nineteenth and early twentieth century, experiences were mixed. While separate lectures for men and women was an important element of British medical education, Irish medical schools do not appear to have taken any issue with educating men and women students together, with the exception of anatomy dissections. Monsignor Molloy, in a report on the Catholic University medical school in 1907 reported that he had 'taken pains to ascertain whether any practical inconvenience has arisen from having women in our Medical School and I am informed that there has been none whatsoever', although he mentioned that the school had limited space due to numbers of medical students (Progress of the Catholic University School of Medicine 1907, n.d., p. 25). *The Irish Times* reported in 1922 that at the Trinity College, University College and the Royal College of Surgeons, 'men and women are trained together without the slightest awkwardness' and condemned the decision of the London Hospital to admit women on the grounds that the staff 'had found difficulties' in teaching 'unpleasant, but necessary subjects' to a mixed audience. The *Irish Times* argued that this objection to the mixed teaching of medical students 'is as belated as would be nowadays an objection to women cyclists or women hockey-players' ("Women Medicals" 1922, p. 4). Referring to her experiences of mixed classes at the Royal College of Surgeons in Dublin, Clara Williams, a student there remarked very positively that 'Here the classes have been found productive of nothing but good, and they are helping in a large measure to destroy the prejudice against women studying medicine. The present generation of medical men having been educated with women, regard them exactly as their other fellow-students, and respect them according to their merits and capabilities, which is all any of us desire' (Williams 1896, p. 109). In British medical schools on the other hand, male and female medical students tended to be separated for lectures and hospital classes, resulting in a very different educational experience. As Carole Dyhouse has shown, however much emphasis women doctors placed on compassion, or on their sense of a 'woman's mission' to other women, 'the study and practice of medicine was never easy to reconcile with late Victorian or Edwardian notions of femininity' (Dyhouse 1998a, p. 329).

Women medical students often had to demonstrate appropriate demeanour and keep a low profile amidst the rowdiness of the predominantly male medical school (Dyhouse 1998a, p. 332). Indeed, many rites of passage in medical education became imbued with masculine tropes. Masculine displays and the cultivation of the image of the medical student as a rowdy, boisterous, and predominantly male individual therefore became an important force in segregating men and women students. Prank-playing was an important initiation rite for medical students internationally. Challenges to authority at universities in the twentieth century, although usually quite benign and ritualised, were 'both frequent and frequently tolerated as part of the construction of masculinity, part of the "natural order of things"' (Dyhouse 2005, p. 177). Male medical students were usually singled out in the student press for being the perpetrators of pranks with such representations of boisterous male medical students persisting late into the twentieth century. Women medical students, on the other hand, were generally represented as being better behaved, more studious and hard-working than their male counterparts. Medical professors also upheld such representations. Speaking to students of the London School of Medicine for Women in 1938, Lord Thomas Horder (1871–1955), commented that 'women were more thorough, more industrious, more studious and, as a sex, were more curious than men. This curiosity, he pointed out, had already borne fruit in the hands of those women who had made their names in research' (*Irish Examiner* 1938, p. 7).

Moreover, as several historians have pointed out, sport and athletics, in particular rugby, were a crucial part of medical school life in Britain and Ireland in the early twentieth century (Dyhouse 1998b, p. 125; Garner 1998; Kelly 2017). Sport was advocated by professors as being important for promotion of good health and to counteract bad behaviour. As early as 1868, E. D. Mapother, professor of anatomy and physiology at the Royal College of Surgeons in Dublin, remarked on the importance of training 'the physical as well as the mental faculties, which are closely interdependent' in order to combat the threat of contagious diseases to which medical students were often exposed. He suggested that every medical school should provide facilities for the playing of sport and suggested that students' spare time should be spent in the ball-court, gymnasium or cricket field, rather than 'smoking at the dissecting room fires, or in the taverns to which want of occupation will tempt them' (Mapother 1868, p. 117). Additionally, Heaman has suggested that at St. Mary's Hospital Medical School in London sports clubs and games 'were intended to foster "social and tender feelings" between students' (Heaman 2003, p. 85). In addition, rugby became central to an ideology of manliness in sport, as it promoted unselfishness, fearlessness and self-control (Mangan 1986, p. 23).

It was argued by some that female medical students missed out on the social side of medical school life because of the fact that they did not play rugby, which was often viewed as crucial in helping to develop a well-rounded identity among male medical students. Women medical students had the option of joining hockey and tennis clubs if they wished to take part in sporting activities, but there is little sense from the contemporary sources that they were involved to a great degree in these in the period in question. In fact, one critical editorial in *T.D.*, the student magazine of Trinity College Dublin in 1924 suggested that the 'corporate spirit is not inculcated in girls as it is in boys [...] [and that] esprit de corps is lacking from their vocabulary' ("Editorial" 1923, p. 158). In response, a female student argued that women students had fewer facilities available to them for extra-curricular activities than the men students, while also pointing out that the women's residence was several miles away from the college, unlike the men's on-campus accommodation. Furthermore she remarked that female students were 'firmly excluded from the real life of College [...] [but that when] any chance is given to them they only too willingly rise to meet the occasion' ("Correspondence" 1923, p. 176). Separate social spaces also kept men and women students apart. By the 1900s, special ladies' rooms were established for female medical students at some universities, while women students generally rented rooms in 'digs' together or stayed in special accommodation for women students, such as Riddel Hall in Belfast (Kelly 2012, pp. 99–100). The First World War is said to have represented a distinct watershed for women doctors in both Britain and Ireland. As in other spheres of work, women became recognised as a valuable contribution to the workforce. The effect of wartime work, as Thom has outlined, was to demonstrate that women were capable of undertaking a variety of tasks but it did not demonstrate that they were entitled to do them (Thom 1998, p. 45). Advocates of women doctors claimed that qualified women had the right 'to take an active part in the national emergency as the professional equals of male doctors, both at home and abroad' (Geddes 2009, p. 211). Medical journals such as the *Lancet* claimed that these years represented an exceptional time for women to enter into medicine as a result of the great shortage of doctors as consequence of the war ("The Opportunity" 1915, p. 563).

Suffragists seized the war as an opportunity to make demands for an increased role for women in the professions ("War" 1914, p. 194). A distinct change in attitude became discernible as the woman doctor was 'no longer looked at with suspicion' and women doctors came to be of crucial importance in managing hospitals at the front, as well as at home, during wartime ("Women Doctors" 1915, p. 35). Some suffrage magazines, such as *Votes for Women*, claimed that 'the success of women doctors has long ago put such arguments [against the admission of women] to rout; but the war has completed the rout' ("Women Doctors Wanted" 1915, p. 331). In addition, women came to be offered many appointments that had not been open to them before. In 1915, at a meeting to promote the London School of Medicine for Women, Dr. Florence Willey pointed out that many medical women were working among the French and Belgian soldiers as well as filling gaps at home in hospitals ("Medical Women in War Time" 1915, p. 731). Some suffragettes, such as Millicent Fawcett, claimed that 'the advocates of the principle of equal pay for equal work have an encouraging precedent in the successful stand which women doctors have made from the outset that they would not undersell the men in the profession' (Fawcett 1918, p. 5).

From the first year of the war, a distinct change in attitude took place, with the increase in women medical students at universities being praised in Britain by some as a 'great benefit to the nation' ("Women Doctors" 1915, p. 35). By the time of the war, the rate of remuneration for women doctors was equal to that paid to men ("Careers" 1915, p. 5). The role of women doctors in the war effort came to be praised and some newspapers asserted that the war had justified many of the claims made by women, as well as giving added impetus to the demand on the part of many women for fuller opportunities to gain clinical and surgical experience ("The Medical" 1915). Wendy Alexander has commented that women's war services provided collective as well as individual benefits. In the first stages of the war, a few women proved by their actions that women were well-capable of running large mixed general hospitals. In consequence, this meant that after the war, 'it was no longer possible for opponents of women doctors to claim with any legitimacy that they had been excluded on grounds of competency' (Alexander 1987, p. 59). The contribution of Scottish women doctors through the Scottish Women's Hospitals helped to convince many sceptics that women doctors were just as competent as their male counterparts (Alexander 1987, p. 57). In England, there existed a military hospital at Endell Street, London, staffed entirely by women doctors under the command of Louisa Garrett Anderson and the doors of the teaching hospitals of Charing Cross Hospital, St. Mary's, Paddington and St. George's Hospital were thrown open to women medical students as a war emergency ("Women and the Medical Profession" 1916, p. 186).

Some medical journals held a cynical view of women doctors' work during the war. *The Dublin Medical Press* commented in 1915 on the 'snuffling sentimentality' that had been expended upon the work being done by medical women in the war. Without wishing to detract from the efficiency of the lady doctors who had helped in the war effort, the writer wondered, 'why it is that these ladies, who consistently claim to be the equals of men, should be advertised as having done something superhuman when they have done no more than five or six times their number of men, at possibly much great sacrifice, have done and are doing, unsung and unnoticed, in the service of their country' ("Medical Women and the War" 1915, p. 385). Despite the praise given to women doctors for their work in the war effort, this article seems to be suggesting that they were not doing anything that the men were not, and seems to be advocating a more balanced attitude towards women doctors.

Others, however, argued that the First World War had allowed women doctors to prove themselves as eminently capable of medical work. For instance, an article published in *The Times* in 1915 stated:

> During the last twelvemonth, a change of attitude has been clearly discernible – a change which the war has now abruptly completed. The woman doctor is no longer looked at with suspicion if she controls a hospital or directs a dispensary [...] The consequent impetus to the study of medicine by women has been particularly marked during the past year, and now the war has brought about a special state of affairs wherein the great increase of women medical students will be of great benefit to the nation. ("Women Doctors" 1915, p. 35)

It has been suggested too that in England, the war gave women doctors the opportunity to run large general hospitals in a way they had been unable to do previously (Alexander 1987, p. 59). Likewise, several teaching hospitals in London opened their doors to women medical students during the First World

War. This success was short-lived, however, with many of the London teaching hospitals closing their doors again to women a few years later.[7] As Carol Dyhouse has shown, the opportunities for women doctors wishing to obtain clinical experience in London, in particular, were much bleaker in the late 1920s and 1930s (Dyhouse 1998a, p. 337). For example, London Hospital Medical College first admitted women to its wards in 1918 before closing admission in 1922. St. Mary's Hospital admitted women in 1916, closed admission in 1925. Westminster Medical College admitted women in 1916 before discontinuing admission of women in 1926. St. George's Hospital Medical College admitted women in 1916 but discontinued admission in 1919. Although Irish hospitals continued their tradition of allowing women medical students on their wards, numbers of women medical students matriculating at Irish universities decreased dramatically after the war although they did not return to the low numbers of the 1880s and 1890s. Irish universities, like British universities, no longer had the same space for women students wishing to study medicine. Likewise, there was increased competition for posts in hospitals and public health considering the large number of women graduates from Irish and British medical schools after the war (Kelly 2012, p. 155).

After 1918, it seems that the medical marketplace became completely saturated. By 1922, the Irish medical journal *The Dublin Medical Press* reported that the number of women medical students had greatly increased over the years of the First World War and that competition for resident posts in hospitals, infirmaries, sanatoria and asylums was now much keener than it had been during the war. In addition, the journal questioned whether women would continue to be as successful at securing appointments as they had been during the war and pointed towards the demand for medical women in India as a possibility for those unable to secure careers within Ireland and the United Kingdom ("Medical Appointments" 1922, p. 212). Just two years previously, the same journal had claimed that the prejudice against women doctors had long since died out and public appointments for women doctors, such as posts working as inspectors of schools and factories, were increasing every year ("Medical Women" 1920, p. 65). Thus, it was more likely for women graduates in the post-1918 cohort to enter into general practice than their predecessors as a result of the lack of alternative posts available to them. Despite the fears concerning opportunities for women medical graduates after the war, the drop-out rates for students who matriculated prior to the war starting and those who matriculated after were similar, with 41% of those who matriculated before the war not making it to graduation while 39.5% of those who matriculated after the war did not qualify (Kelly 2012, p. 156).

Notably, women medical graduates from Irish institutions were most likely to obtain hospital appointments in general hospitals, rather than in asylums or children's/maternity hospitals, thus indicating that there were genuine opportunities for them outside their expected spheres of employment. This differed to the situation for women medical graduates in England, as Mary

Ann Elston's work has shown. She has illustrated that, of a sample of English women doctors holding 'house posts' in 1899 and 1907, most were likely to be in women-run hospitals than in other types of institutions (Elston 2001, p. 84). These women-run hospitals had close connections with the female medical schools such as the London School of Medicine for Women and the Edinburgh School of Medicine for Women. In contrast, women medical graduates in Ireland were more likely to work in posts in general hospitals rather than in women-run institutions or hospitals for women and children. It is possible that this was due to the system of co-education in Irish medical institutions (Kelly 2012, p. 120).

Finally, the public health sector was an important area of employment for Irish women graduates, especially those who moved to England. This sector of medicine was claimed to be an appealing area of work for new graduates because it did not require the time and capital necessary to build up a private practice ("For Mothers and Daughters" 1922, p. 732). Women doctors working in public health were commonly employed as schools' medical officers, dispensary doctors, assistant Medical Officers of Health (MOH) (Kelly 2012, p. 122). In Ireland, graduates commonly worked as dispensary medical officers (Barrington 1987, pp. 7–8).[8] In Britain, women doctors found a special niche for themselves within the public health movement, working initially as Assistant MOHs and, after 1907, as Schools Medical Officers.

Conclusion

Responding to a letter to the *Lancet* which fretted about the 'glut of women doctors' by William Robinson in 1925, Dr. Mabel Ramsey wrote:

> I am, of course, quite aware that Mr. Robinson's letter is meant as a deterrent and warning to parents and intending medical women candidates as threatening future employment for medical women. Let me tell them to take no heed of such Partingtons [a term for a person engaged in a futile battle], for the public needs and demands medical women in many and ever-widening spheres. Our movement has too strong a hold to be pushed back, much as the reactionary would like it. ("The Glut" 1925, p. 207)

By the 1920s, women doctors had found themselves firmly established within the medical profession in Britain and Ireland, yet they still frequently found themselves having to defend their place there. Evidently, Victorian debates on the question of women's admission to the medical profession were multifaceted, with arguments centring on women's physical, emotional and mental natures, and how these made them either fit or unfit to be doctors. Following their admission to the medical profession, pioneer women doctors encouraged their younger counterparts to be modest and diligent, suggesting that they were aware of how the continuing existence of prejudice meant that women could not occupy a fully equal position alongside their male counterparts. In addition, considering the often macho and boisterous atmosphere of the medical school environment with its emphasis on pranks and rugby, women medical students found it difficult to become fully integrated and formed their own social networks. While the First World War resulted in new opportunities for some medical women, it is evident that they were not fully equal in terms of the spheres of medicine they could find employment in. It is clear today that women have made significant progress in the British and Irish medical professions. Today female students predominate in medical school applications in Britain and Ireland, a pattern which is mirrored by medical schools internationally. In 2017, for instance, 59% of those accepted to medical school in the United Kingdom were women, while in 2016, women represented 56% of those accepted to medical school in Ireland (Moberly 2018, p. 167; "IMO Position on Women in Medicine" 2017, p. 4.). However, while there has been significant progress, there are still many challenges. Many doctors still face gender-based discrimination in the workplace and women are severely underrepresented as consultants (Moberly 2018).

Notes

1. The story of the 'Edinburgh Seven' has been discussed recently by Crowther and Dupree (2007, pp. 152–75). For a biography of Sophia Jex-Blake, see Roberts (1993).
2. For studies of women's struggles in breaking into the medical profession in Britain and the United States, see Blake (1990), Morantz-Sanchez (2000, 1985), More (1999). Recent studies which have focused on women's experiences of careers in the medical profession include Brock (2017) and Kelly (2012).
3. See, for example, Drachman (1986).
4. Similarly, more recently, Irene Finn has argued that the Irish medical profession, with the exception of the King and Queen's College of Physicians of Ireland (KQCPI), held a hostile view towards women in medicine (see: Finn 2000).
5. For more on the story of the 'Edinburgh Seven' see Crowther and Dupree (2007, pp. 152–75). For Sophia Jex-Blake, see Roberts (1993).
6. For more on this, see Kelly (2013).
7. Medical Women's Federation Archives at Wellcome Library: SA/MWF/C.10.

References

Alexander, Wendy. 1987. *First Ladies of Medicine: The Origins, Education, and Destination of Early Women Medical Graduates of Glasgow University*. Glasgow: Welcome Unit for the History of Medicine, University of Glasgow.

Barrington, Ruth. 1987. *Health, Medicine & Politics in Ireland 1900–1970*. Dublin: Institute of Public Administration.
Bashford, Alison. 1998. *Purity and Pollution: Gender, Embodiment and Victorian Medicine*. London: Macmillan.
Blackwell, Elizabeth. 1890. *The Influence of Women in the Profession of Medicine*. Baltimore: Unknown.
Blackwell, Elizabeth. 1895. *Pioneer Work in Opening the Medical Profession to Women: Autobiographical Sketches*. London: Longmans, Green and Co.
Blake, Caitriona. 1990. *Charge of the Parasols: Women's Entry into the Medical Profession*. London: Women's Press.
Brock, Claire. 2017. *British Women Surgeons and their Patients, 1860–1918*. Cambridge: Cambridge University Press.
Burstyn, Joan N. 1973. "Education and Sex: The Medical Case against Higher Education for Women in England, 1870–1900." *Proceedings of the American Philosophical Society* 117, no. 2: 79–89.
"Careers for Boxes and Girls: The Medical Profession: The Outlook." 1915. *Weekly Irish Times*, 20 February, 5.
"Correspondence." 1923. *Trinity College Dublin* 508 (31 May): 176.
Crowther, Anne, and Marguerite Dupree. 2007. *Medical Lives in the Age of Surgical Revolution*. Cambridge University Press.
Drachman, Virginia G. 1986. "The Limits of Progress: The Professional Lives of Women Doctors, 1881–1926." *Bulletin of the History of Medicine* 60, no. 1: 58–72.
Dyhouse, Carol. 1998a. "Driving Ambitions: Women in Pursuit of a Medical Education, 1890–1939." *Women's History Review* 7, no. 3: 321–43.
Dyhouse, Carol. 1998b. "Women Studies and the London Medical Schools, 1914–39: The Anatomy of a Masculine Culture." *Gender & History* 10, no. 1: 110–32.
Dyhouse, Carol. 2005. *Students: A Gendered History*. Abingdon: Routledge.
"Editorial." 1923. *Trinity College Dublin* 507 (24 May): 158.
Elston, Mary Ann. 2001. "'Run by Women, (Mainly) for Women': Medical Women's Hospitals in Britain, 1866–1948." In *Women and Modern Medicine*, ed. Laurence Conrad and Anne Hardy, 73–107. Amsterdam: Rodopi Clio Medica.
Fawcett, Millicent G. 1918. "Equal Pay for Equal Work." *The Economic Journal* 28, no. 109: 1–6.
Finn, Irene. 2000. "Women in the Medical Profession in Ireland, 1876–1919." In *Women and Paid Work in Ireland, 1500–1930*, ed. Bernadette Whelan, 102–19. Dublin: Four Courts Press.
"For Mothers and Daughters: Professions for Girls." 1922. *Catholic Bulletin and Book Review*, November, 12, no. 2: 732.
Freeman's Journal. 1888. 9 August, 4.
Garner, James. 1998. "The Great Experiment: The Admission of Women Students to St Mary's Hospital Medical School, 1916–1925." *Medical History* 42: 68–88.
Garrett Anderson, Luise. 1877. "Inaugural Address." Paper presented at the London School of Medicine for Women, London, 1 October, 3–25.
Garrett Anderson, Luise. 1878. "A Special Chapter for Ladies who Propose to Study Medicine." In *The Student's Guide to the Medical Profession*, ed. Charles Bell Keetley, 42–48. London: Macmillan and Co.
Garrett Anderson, Luise. [1942] 2016. *Elizabeth Garrett Anderson, 1836–1917*. Cambridge: Cambridge University Press.

Geddes, J. F. 2009. "The Doctor's Dilemma: Medical Women and the British Suffrage Movement." *Women's History Review* 18, no. 2: 203–18.
Heaman, E.A. 2003. *St. Mary's: The History of the London Teaching Hospital*. Montreal: McGill-Queen's Press.
"Inadequate Training Facilities." 1938. *Irish Examiner*, 7 October, 7.
Irish Medical Organisation. 2017. "IMO Position on Women in Medicine." https://www.imo.ie/policy-international-affair/documents/IMO-Position-Paper-on-Women-in-Medicine-Final.pdf. Accessed 27 February 2020.
Irish Times. 1876. 15 February, 4.
Jex-Blake, Sophia. 1872. *Medical Women: Two Essays*. Edinburgh: William Oliphant.
Jex-Blake, Sophia. 1874. "The Medical Education of Women." Paper presented at the Social Science Congress, Norwich, October, 1–16.
Jex-Blake, Sophia. 1886. *Medical Women: A Thesis and a History*. Edinburgh: Oliphant, Anderson & Ferrier.
Kelly, Laura. 2012. *Irish Women in Medicine: Education, Experiences and Careers, c. 1880s-1920s*. Manchester: Manchester University Press.
Kelly, Laura. 2013. "'The Turning Point in the Whole Struggle': The Admission of Women to the King and Queen's College of Physicians in Ireland." *Women's History Review* 22, no. 1: 97–125.
Kelly, Laura. 2017. "Irish Medical Student Culture and the Performance of Masculinity, c. 1880–1930." *History of Education* 46, no. 1: 39–57.
"Lady Surgeons." 1870. *British Medical Journal*, 2 April, 338–39.
"Letter to the Editor." 1871. *Freeman's Journal*, 2 February, 3.
Letters to the Editor. 1873. "The Medical Education of Women." *The Times*, 5 August, 3 and 23 August, 4.
Mangan, J. A. 1986. *The Games Ethic and Imperialism: Aspects of the Diffusion of an Ideal*. London: Viking.
Mapother, E. D. 1868. *The Medical Profession and Its Educational and Licensing Bodies*. Dublin: Fannin.
"Medical Appointments for Women." 1922. *Dublin Medical Press*, 13 September, 212.
Medical Times and Gazette. 1874. 26 December, 714.
"Medical Women." 1920. *Dublin Medical Press*, 28 January, 65.
"Medical Women and the War." 1915. *Dublin Medical Press*, 21 June, 385.
"Medical Women in War Time." 1915. *The Common Cause of Humanity: The Organ of the National Union of Women's Suffrage Societies* 6, no. 307 (26 February): 731.
Meenan, F. O. C. 1987. *Cecilia Street: The Catholic University School of Medicine, 1885–1931*. Dublin: Gill and Macmillan.
"Meeting of the General Medical Council." 1875. *Lancet*, 10 July, 55–63.
Moulds, Alison. 2019. "The 'Medical-Women Question' and the Multivocality of the Victorian Medical Press, 1869–1900." *Media History* 25, no. 1: 6–22.
Moberly, Tom. 2018. "Number of Women Entering Medical School Rises after Decade of Decline." *British Medical Journal* 360167.
Morantz-Sanchez, Regina Markell. 1985. *Sympathy and Science: Women Physicians in American Medicine*. Chapel Hill, NC: University of North Carolina Press.
Morantz-Sanchez, Regina Markell. 2000. *Conduct Unbecoming a Woman: Medicine on Trial in Turn-of-the-Century Brooklyn*. Oxford: Oxford University Press.
More, Ellen S. 1999. *Restoring the Balance: Women Physicians and the Practice of Medicine, 1850–1995*. Cambridge, MA: Harvard University Press.

Munster Branch of the Irish Association of Women Graduates and Candidate Graduates. 1902–13. *Letter to the Board of Management of Victoria Hospital, Cork.* Dublin: University College Dublin.
Progress of the Catholic University School of Medicine, 1907. N.d. Dublin: Humphrey and Armour Printers.
Reinarz, Jonathan. 2009. *Health Care in Birmingham: The Birmingham Teaching Hospitals 1779–1939*. Woodbridge: Boydell Press.
Roberts, Shirley. 1993. *Sophia Jex-Blake: A Woman Pioneer in Nineteenth Century Medical Reform*. London: Routledge.
Scharlieb, Mary. 1898. "Women in the Medical Profession." Paper presented at the Women's Institute, London, 25 January, 3–16.
"The Admission of Ladies to the Profession." 1870. *British Medical Journal*, 7 May, 474–75.
"The Glut of Women Doctors" 1925. *Lancet*, 24 January, 207.
"The Medical Woman's Opportunity." 1915. *The Queen*, 27 February (Royal Free Hospital Archive scrapbook).
"The Opportunity for Women in Medicine." 1915. *Lancet*, 13 March, 563.
Thom, Deborah. 1998. *Nice Girls and Rude Girls: Women Workers in World War I*. London: I.B. Tauris & Co. Ltd.
Usborne, Cornelie. 2001. "Women Doctors and Gender Identity in Weimar Germany (1918–1933)." In *Women and Modern Medicine 61*, ed. Laurence Conrad and Anne Hardy, 109–26. Amsterdam: Rodopi Clio Medica.
"War and Professions." 1914. *The Irish Citizen*, 7 November, 194.
Williams, Clara. 1896. "A Short Account of the School of Medicine for Men and Women, RCSI, by Clara L Williams." In *Magazine of the LSMW and RFH* 3, January 1896, 91–132.
Witz, Anne. 1992. *Professions and Patriarchy*. London: Routledge.
"Women and the Medical Profession." 1916. *Votes for Women: Official Organ of the United Suffragists*, August, 9, no. 418: 186.
"Women Doctors: Enlarged Field of Service: Medical Practice in War Time." 1915. *The Times*, 22 January, 35.
"Women Doctors Wanted." 1915. *Votes for Women: Official Organ of the United Suffragists*, 2 July, 8, no. 383: 331.
"Women Medicals." 1922. *Irish Times*, 3 March, 4.
Wyman, A. L. 1984. "The Surgeons: The Female Practitioner of Surgery, 1400–1800." *Medical History* 28, no. 1: 22–41.

CHAPTER 26

The Work of British Women Mathematicians During the First World War

June Barrow-Green and Tony Royle

There is no doubt that the First World War acted as a catalyst to provide women with unique opportunities to infiltrate the male-dominated bastion of industrial engineering. This was particularly apparent at the establishments intimately involved in the advancement of the fledgling discipline of aeronautics and its related fields. The Royal Aircraft Factory (RAF) lay at the centre of the research and development that was driving innovation in aeronautics in Britain, and the facility was responsible for the design and testing of new aircraft for operational use by the Royal Flying Corps (RFC). The National Physical Laboratory (NPL) focused more on the theoretical aspects of aerodynamics. Its primary weapon in this pursuit was the analysis of data obtained from the observation of the behaviour of scale models of aircraft, or aircraft components, in wind tunnels. In London, a department in the British Admiralty was formed specifically to address issues pertaining to the strength and integrity of aircraft structures. Within all three of these domains there was a great demand for individuals who had a background and expertise in mathematics and, as increasing numbers of men were conscripted to bolster the

---

J. Barrow-Green (✉) · T. Royle
School of Mathematics and Statistics, Open University, Milton, UK
e-mail: june.barrow-green@open.ac.uk

T. Royle
e-mail: tony.royle@open.ac.uk

© The Author(s), under exclusive license to Springer Nature Switzerland AG 2022
C. G. Jones et al. (eds.), *The Palgrave Handbook of Women and Science since 1660*, https://doi.org/10.1007/978-3-030-78973-2_26

front line, this requirement was mitigated by the employment and deployment of suitably qualified women. This was a fact that did not go unnoticed by the press, as evidenced by comments in *The Times* of London.[1]

But as much as the War provided a stimulus for the development of aeronautics, it was also a fillip for the development of ballistics, notably anti-aircraft gunnery. Whilst staff at the RAF and the NPL were engaged in developing ever more sophisticated aircraft, their counterparts employed by the Munitions Inventions Department and based at HMS Excellent at Whale Island, near Portsmouth, aided by the Karl Pearson and his 'computers' at University College London (UCL), were producing a stream of high-angle range tables for gunnery. Pearson had employed women in his laboratories before the War, but work generated by the conflict led many of these women into new fields as well as bringing them into his workforce.

The majority of the women mathematicians engaged in such work were educated at either Cambridge University or at one of a number of colleges in the London area that were affiliated to the University of London. There were two main reasons for this, one academic and the other geographic. Much of the mathematics required for application to aerodynamics is relatively complex and, at that time, those women able to meet the exacting demands of a Cambridge education in the subject were the most likely to be able to offer a meaningful academic contribution to the ongoing aeronautical research. But there were practical considerations too. Since the Admiralty, RAF, and NPL were all located in and around London, it made sense to source staff from suitable candidates educated and living in the surrounding area.

Pivotal Centres of Higher Education in Mathematics for Women 1870–1918

During the nineteenth century, Cambridge University was the fulcrum of British mathematics, and its Mathematical Tripos the most prestigious and demanding examination in Britain. It was punishingly hard, both physically and mentally—students often broke down whilst studying for it or during the examination itself—but the rewards were great. The order of merit was reported nationally and students who came high on the list of wranglers (those in the first class) had a passport to the career of their choice, be it the law, the Church, medicine, mathematics, or on whatever they set their sights (Neale 1907; Barrow-Green 2019). It is hard to over-estimate the kudos attached to being the senior wrangler, the cachet that went far beyond the bounds of the University.

From the middle of the second half of the century, women could study mathematics at Cambridge—the women's colleges Girton and Newnham were founded in 1869 and 1872, respectively—but they had to obtain permission to sit the Tripos examination, they could not do so by right, and they could not be awarded a degree with its associated privileges and voting rights. For over three-quarters of a century the two colleges were not even officially part of the University; they would have to wait until 1948 to enjoy this status. That is not to say that all Cambridge men were against women studying mathematics, and indeed some of them would travel to Girton or Newnham to hold classes or to give private tuition, or 'coaching' as it was called.[2]

Initially, the women studying mathematics caused little stir. But in 1880 when Charlotte Scott was judged equal to the eighth wrangler she created a sensation. The newspapers and periodicals were full of her success—she had done better than 93 of the 102 men taking the examination. Scott's achievement generated a growth in support for women students and from 1881 women were given the right to take the Tripos examination and to have their results published, albeit separately from the men. They were still prohibited from being awarded degrees, however. Ten years later, there was an even greater sensation when Philippa Fawcett was judged to be above the senior wrangler, reports being published as far afield as *The New York Times* ("Miss Fawcett's Honour" 1890).[3] She had scored thirteen per cent more marks than the highest ranked man, G.T. Bennett, achieving what many had believed impossible. Nevertheless, when the Tripos list was published, her name (and that of the other sixteen women) still appeared below that of all the men.[4]

Between 1890 and 1909, the last year in which there was an official order of merit, around twenty or so women sat the examination each year. After the change in regulations the number of both women and men fell, with the number of men falling even more during the war period, especially once conscription began to bite (see Table 26.1). Nevertheless, as the century progressed, the number of women completing the Mathematical Tripos course each year generated an increasing pool of mathematical talent, a talent that could be utilized for the war effort. Not only were these women highly trained

Table 26.1 The numbers of women and men sitting the Cambridge mathematical tripos examination, 1908 to 1919, with concomitant output of wranglers

	Women Wranglers	Women Total including wranglers	Men Wranglers	Men Total including wranglers
1908	3	18	28	83
1909	1	10	31	74
1910	1	11	23	46
1911	1	12	24	52
1912	2	16	27	54
1913	1	8	30	66
1914	2	10	25	58
1915	2	8	14	28
1916	2	12	11	20
1917	2	13	5	11
1918	0	7	3	4
1919	2	17	9	25

mathematically but by succeeding in the Mathematical Tripos they had proved themselves hard workers under pressure, an essential quality in a time of crisis.

The first woman's university college in the United Kingdom was Bedford College, which was founded as a higher education college for women in 1849 by Elizabeth Jesser Reid, and it became part of the University of London in 1900. During the War, the number of women in teaching roles at the College increased dramatically, notably in the chemistry department, which was requisitioned by the government to conduct research for the military on gas warfare. Chemistry was another area of academic and practical expertise in high demand for application to aeronautics, primarily in the specialization of materials science. The integrity and efficacy of aircraft construction hardware was a salient factor in the success or otherwise of new designs. Bedford thus educated a number of women who would make significant contributions to the war effort. They were not all thoroughbred mathematicians, but they certainly required, and would have doubtless received, a solid grounding in mathematics to underpin their respective work in physics, chemistry, and engineering.

Another academic institution that produced suitable female candidates for employment in the field of aeronautics was East London College, which was born during the latter part of the nineteenth century out of the People's Palace, a philanthropic organization designed to bring culture, recreation, and education to the East End of London. Funded by the Beaumont Trust and the Drapers' Company, one of the Livery Guilds of the City of London, The People's Palace established Technical Schools to train a new generation of tradespeople, but it soon became obvious that there was a need for higher education opportunities in this area of the capital. Teaching was expanded to include the sciences, arts, and humanities, and the Schools were formerly admitted into the University of London in 1915 under the new name East London College (ELC). The College itself became a centre for teaching and research in the subject of aeronautics[5] and, in the 1930s, became Queen Mary College.

Royal Holloway College would also provide many suitably qualified women to feed the growing demand for mathematics-based technical expertise. Holloway was founded as a women-only institution. Funded by entrepreneur Thomas Holloway, the College officially opened the doors to its spectacular Founder's Building in 1886, with the first students arriving the following year. Like Bedford, the College became a constituent part of the University of London in 1900.[6]

A better understanding of the nature of the work being undertaken by women mathematicians and scientists during the cataclysm of WWI can be gained by considering a selection of individual case studies of some of those employed at the Admiralty, NPL, RAF, and UCL. Whilst not exhaustive, these few examples cast light on the nature and variability of these women's backgrounds, education, and pathways into the realms of aeronautics and anti-aircraft gunnery. They also illustrate their diverse journeys and interests beyond the War.

The Admiralty Air Department

In 1917, as the air war raged in Europe and involved ever-increasing numbers of aircraft and aircrew, the need to ensure the structural integrity of existing and new aircraft designs became paramount. In London, Wing Commander Alec Ogilvie led the Technical Section of the Admiralty Air Department, which included a number of individuals dedicated to addressing aircraft structural issues; amongst them were a number of women mathematicians (Royle 2017, p. 348). A photograph held in the archive of Annie Trout, who was a member of the team, offers an intriguing insight into the number of women being employed in the Department at the time and reveals their identities (Fig. 26.1).[7] They were an eclectic group, but all shared the common goal of wanting to apply their academic groundings in mathematics and science to contribute to the structural integrity and safety of RFC aircraft.

At least three of the women shown in the photograph were educated at Royal Holloway: Olive Moger, Annie Trout, and Dorothy Chandler. Moger attended Bath High School as a teenager and was awarded a scholarship to study mathematics at Holloway, graduating from London University in 1903

Fig. 26.1 The Admiralty Air Department 1918. Image from the Annie Trout Archive held at the Hartley Library at the University of Southampton. Reproduced by kind permission of the University of Southampton Archives

with a second-class honours degree. She then moved into the world of topography, working in London on the Victorian Histories of the Counties of England project,[8] and in 1917 was at the Wheat Commission working as a statistical clerk. As applies to a number of the women mentioned in this narrative, the exact nature of how she was recruited to work at the Admiralty is unclear. In Moger's case it is also difficult to determine with any certainty her precise role within the department, although her academic qualifications and previous employment imply she would likely have been working in some mathematical context. Following the War, however, her achievements are better documented. She became a genealogist renowned for her passion for the transcription and cataloguing of wills.[9]

Annie Trout graduated in 1907 from the University of London with a first-class honours degree in mathematics. As was common for women holding such a qualification, she moved into education, cementing her credentials in 1908 with a certificate of efficiency in teaching from the University of Cambridge. After the War she took up a post at the University of Southampton as a lecturer in mathematics. As evidenced by the material held in her archive at the University, Trout was a mathematician who engaged completely with the task in hand at the Admiralty. Her personal copy of the core text used to inform the Department's modus operandi, the *Handbook of Strength Calculations* (Pippard and Pritchard 1918), is embellished with copious handwritten comments to elaborate and clarify various mathematical techniques. She also compiled her own set of detailed notes, complete with sketches and diagrams, covering a diverse range of technical material associated with aeronautics. There is also evidence of her excitement and pride in being part of the rather exclusive group at the Admiralty. Amongst her reminiscences of the period, she recalls a special show of captured German aircraft put on in the London Borough of Islington in 1918:

> The variety in structure was far in advance of our own, which had remained as far as wing structures were concerned very much the same in fundamentals as in 1914. One machine had markedly swept back wings with the depth diminishing towards the wing tip. This came to mind when I first saw a Delta wing. The show was very secret and confidential, but our own Section was allowed to see it! (Trout 1918)

Something else Trout reveals in her notes is that Ogilvie was happy to authorize flying experience excursions for anyone in the Stressing office who might be interested in taking to the air, although she gives no indication as to whether or not she took advantage of this rather tempting offer. Women were apparently viewed as being preferentially suited to some tasks over their male counterparts. Relating a test she attended of a new aircraft component that had been designed at the RAF, she notes: 'One test I watched was of a new construction [...] with many small rivets. I was told then that women were far better for this riveting than men' (Trout 1918). Hidden away in Trout's archive is a small postcard from someone who signs off, 'Salutations, H.P.H' (Trout 1926). The content comprises the resolution of a problem in the field of Cremona transformations, hinting that the author has to be Trout's co-worker at the Admiralty, Hilda Hudson. The card is dated 26 October 1926, which tells us that the two women had remained in touch after the War and were clearly drawing upon each other's mathematical expertise to resolve conundrums. There are two remarkable photographs in the archive. One is Trout's own immaculate copy of the departmental photograph from the Admiralty, and the other is an image from 1938 showing the attendees at a national conference held in Oxford for university lecturers in engineering and mathematics. Of the over one hundred delegates, the only woman present is Annie Trout, hinting at her rather unique status in academia at the time, and highlighting the lack of inter-war progress made in establishing women mathematicians in academic posts in higher education in Britain.

Dorothy Chandler was an undergraduate at Holloway from 1907 until 1910, graduating in mathematics with second-class honours having first studied for the Intermediate Arts exam which covered Latin, English, Pure and Applied Mathematics. Chandler worked on aircraft design, and is known to have flown as a passenger in 1918 aboard one of the largest aircraft of the time, the Handley Page V/1500 (Fig. 26.2), sitting in the tail gunner's

Fig. 26.2 Handley Page V/1500 aircraft: wikimedia commons. https://commons.wikimedia.org/wiki/File:Handley_Page_V-1500_in_1918.jpg. Accessed 6 January 2021

cockpit, which must have been quite an experience for the office-based worker.

Cambridge University provided two of the key members of the group who appear in the Admiralty photograph, Hilda Hudson and Letitia Chitty, and another important contributor who is not shown, Beatrice Cave-Browne-Cave. Hudson was *chef d'équipe* to the other women. She arrived directly from a teaching post at West Ham Technical College, whilst Chitty came directly from the university, interrupting her undergraduate studies. Beatrice Cave-Browne-Cave transferred directly from Karl Pearson's laboratory at UCL under slightly acrimonious circumstances. It is worth considering the backgrounds and achievements of these women in more detail, since their stories exemplify the extent of the mathematical expertise they brought to the Admiralty.

Hilda Hudson was born into mathematics. Her father William lectured in mathematics at Cambridge and was subsequently appointed Professor of Mathematics at King's College, London, shortly after Hilda's birth in 1881. Her mother read mathematics at Newnham College, Cambridge, and her elder brother excelled at St. John's, Cambridge, achieving the coveted accolade of senior wrangler in the Mathematical Tripos of 1898; her sister would, very creditably, be ranked alongside the equal-eighth wranglers on the list of 1900. What pressure then on Hilda to shine when she arrived at Newnham in that same year. Like her family before, however, she rose to the challenge and held bragging rights over her sister by achieving a mark equivalent to the seventh wrangler on the list of 1903 following Part I of the Mathematical Tripos examinations. A year later she took Part II of the Tripos—the only woman to do so—and was placed in the third division of the First Class. After leaving Cambridge in the summer of 1904 she spent the winter semester at the University of Berlin where she attended lectures given by Hermann Amandus Schwarz, Friedrich Schottky, and Edmund Landau.[10] Her return from Germany saw her back at Cambridge, first as a lecturer and later as an Associate Research Fellow working on 'Birational Transformations in Three Dimensions', where she was looking to improve and develop work done previously by Arthur Cayley and Corrado Segre (Royle 2020, pp. 152–162).

Perhaps a defining moment in Hudson's life came in 1912 when she became the first woman to deliver a communication at the International Congress of Mathematicians (Hudson 1912). A short spell in the United States at Bryn Mawr College working with Charlotte Scott preceded Hudson's appointment as a lecturer in mathematics at West Ham, which coincided with the advent of the Institute's Junior Engineering School for Boys. Situated in the East End of London, the facility opened in 1900 to provide courses in Science, Engineering, and Art for Boys. Later, secretarial and trade courses for girls were added and eventually, in 1913, came the Junior Engineering School for Boys. It was the seedling of what has now grown into the University of East London. Hudson resigned from her research fellowship in Cambridge at the end of March 1913, two months before its formal conclusion, to facilitate the move to London.

The exact reason Hudson decided to leave her teaching post in 1917 and join the civil service is unclear, but the government had been actively running recruitment drives to draw women into the vacuum created in the traditionally male-dominated professions by conscription, which had been introduced for men in 1916. Following a recommendation from the Royal Society, however, Hudson had been working with polymath Sir Ronald Ross on deriving mathematical models to describe the spread of malaria, so her potential for bringing weight to bear on the mathematics needed to assist in the war effort must have been clear.[11] She was immediately drafted into the Admiralty to mentor the group of women that would become an essential cog in the wheel of the Stressing Section of the Structures office. She was slightly older and more experienced than most of her female colleagues and had the presence and work ethic to set a fine example, soon earning herself the title of Sub-section Director. She also demonstrated her mathematical flexibility, temporarily casting aside her passion for, and expertise in, geometry to enter the applied world of moments, stresses, and strains. This transition should not be underplayed; these are disparate disciplines within mathematics. In addition to acting as the pivot between the key men in the department, Hudson would individually author two notable pieces of work that were published after the War. The first article, 'The Strength of Laterally Loaded Struts', appeared in *Aeronautical Engineering* in June of 1920 (Hudson 1920a). Her second piece, 'Incidence Wires', appeared in the *Aeronautical Journal* in the same year (Hudson 1920b).

Letitia Chitty, unlike Hudson, was not surrounded at home by a family of high-achieving mathematicians, although they were all academically minded, having a particular affiliation with Balliol College, Oxford. Having been schooled by private tutors at Winchester College,[12] she went up to Newnham, Cambridge, in 1916 to read mathematics. Her talent for the subject soon became apparent and, as the demand grew in London for competent mathematicians to assist with the war effort, it was agreed that she could be released from her undergraduate studies with the promise of a return once the conflict had ended. Rumours of Hudson's work had already filtered back to Newnham via the dons' network, so the Admiralty was where Chitty wanted to be. By a somewhat convoluted route, in August 1917, the now twenty-year-old Chitty presented herself for work in London. She was allocated a shared room with Chandler and Mary Hutchison (Royle 2020, pp. 162–168).[13]

Some insight into the mathematics being used and mathematical methods being employed by the Stressing Section at that time can be gleaned from Chitty's later recollections of her time at the Admiralty.[14] 'We relied upon our slide rules and arithmetic in the margins, supported by the theorem of 3 moments and Southwell's curves for struts', she states, indicating the manual nature of calculations, and a reliance on the adaption and application of known mathematics to the novel situation an aircraft in flight presented.[15] The curves she mentions relate to the contemporary preference for mathematical equations to be offered in graphical form for ease of use. It is clear that Chitty was inspired by Section leader Sutton Pippard during her time at the Admiralty, and there was certainly an element of mutual respect.[16] Such was Pippard's influence that, after the War, Chitty returned to Cambridge University and immediately transferred from mathematics to engineering, later being placed in the First Class in the Mechanical Sciences Tripos, the first woman ever to achieve this distinction. She would later team-up with Pippard to undertake stress analysis on all manner of objects, including arches, wheels, dams, and extensible cables. Pippard was placed in charge of the Civil Engineering Department at Imperial College in 1933 and was reunited with Chitty when she was appointed his Research Assistant in the following year. Then began a fruitful academic partnership, as witnessed by the many subsequent joint publications spanning the years 1936–1960 (Skempton 1970). Prior to this symbiotic collaboration she also found occasion to work with in the field of hydrodynamic stability, emphasizing her remarkable flexibility.

Beatrice Cave-Browne-Cave was one of the first acquaintances made by Chitty upon the latter's arrival in London. Cave-Browne-Cave was already working alongside the Holloway-educated mathematician, Eleanor Lang,[17] for Leonard Bairstow, who had himself recently moved to the Admiralty from the NPL. Bairstow was recognized as one of Britain's leading aerodynamicists and became the first Zaharoff Professor of Aerodynamics at Imperial College following the War. These three academics formed a close partnership and would continue their association after the conflict, publishing two papers in the field of fluid dynamics (see Bairstow et al. 1922, 1923). Cave-Browne-Cave had been educated at home and, like Hudson, was surrounded by siblings who shared her passion for mathematics. She would eventually go up to Girton in 1895 and come away in 1899 having been placed in the Third Class in Part II of the Mathematical Tripos, perhaps, with hindsight, a rather modest reflection of her mathematical potential. She immediately took one of the few options open to female mathematicians at the time and became a teacher, taking a position at Clapham High School, but it would be an opening at UCL just before the War began that would launch Cave-Browne-Cave's career in mathematics. Her sister Frances was employed as a lecturer of mathematics at Girton but had established a concurrent working relationship with statistician Karl Pearson at UCL, and so likely played some part in Beatrice's appointment. Beatrice's initial work was statistical in nature but, as the War intensified, and much to Pearson's chagrin, in 1916 she took an opportunity to earn more money by working at the Admiralty. In Pearson's view, she was 'not playing the game' and had acted unfairly towards the laboratory.[18] At the Admiralty, Beatrice worked on aircraft tail loading analysis and the study of aircraft oscillations, an endeavour that resulted in a sole-authored paper that was published as an Advisory Committee for Aeronautics (ACA) *Technical Report* (Cave-Browne-Cave 1918), and served to demonstrate her grasp of the relevant mathematics.

The National Physical Laboratory (NPL)

The NPL, located at Bushy House, Teddington, was established in 1900 to set standards in science and engineering. Various rumblings and statements of intent had been made at a number of the British Association for the Advancement of Science meetings during the 1890s as unease grew at Germany's advantage in this field, and so the arrival of the British equivalent of the *Physikalisch-Technische Reichsanstalt* [Physical and Technical Institute of the German Reich], established in the mid-1880s, was long overdue (Royle 2020, p. 9). Initially the predominant technical institutions and societies dictated the specializations at the Laboratory; mechanics and engineering, electricity, optics, chemistry, metrology, terrestrial magnetism, and thermometry were the chosen demarcations. The nineteenth century had witnessed the founding of a number of professional trades associations, the Institutions of Civil Engineers (1818), Mechanical Engineers (1847), and Electrical Engineers (1889) which were doubtless the most influential in determining the NPL's initial technical structure and focus. As time passed, however, its remit broadened and, in 1909, the Aeronautical Division appeared.

Two women who typified the throughput from Bedford College into the NPL were Dorothy Marshall and Marie Gayler. Marshall was born in London and educated at King Edward VI High School for Girls in Birmingham, which was founded in 1883, before enrolling at Bedford College in 1886. She later transferred to UCL, where she studied electrical technology, physics, and chemistry. She graduated in 1891 with a B.Sc. and then remained at UCL until 1894 as a postgraduate research student. Her time was spent studying the effect of heat on liquids, and she published three papers in connection with this work, one co-authored with renowned chemist William Ramsey (Marshall and Ramsay 1896). In 1896 she accepted a one-year Demonstratorship at Newnham College and then followed a career in education that took her to Clapham High School via Girton College, Avery Hill Training College, and Huddersfield Municipal High School. The demand generated by the War for skilled chemists drew Marshall to the NPL in 1916 where she was appointed as a Scientific Research Assistant. She remained at the Teddington facility for the balance of her working life, and published numerous papers on fluid motion in conjunction with Thomas Stanton, the Superintendent of the Engineering Department and Fellow of the Royal Society who was later knighted in recognition of his work during the War (Rayner-Canham 2008, p. 229). A number of these joint papers were penned during the War, although not published until peace had been brokered (Marshall and Stanton 1920; Marshall et al. 1921). Marshall and Stanton also teamed up with Oxford University physics graduate, Constance Jones, to do research into the properties of fluids in turbulent motion. Jones added essential expertise to the team's efforts to study the enigmatic properties of the boundary layer of such fluids with their bounding surface. Assisted by Jones, Marshall would eventually develop a new and extremely sensitive pitot device to improve the accuracy of data gathered in this pursuit, details of which appeared in an article in the *Proceedings of the Royal Society* in August 1920 (Stanton et al. 1920); a technical drawing of the device can be seen in Fig. 26.3. Part way through the War, Jones was moved out of the fluid dynamics laboratory into another department at the NPL concerned with testing and improving aerofoil design. It was during this period she collaborated with aerodynamicist L.W. Bryant and contributed to numerous technical reports relating to wing design.[19]

Marie Gayler was educated at St. Mary's College school in Gerrards Cross, prior to Bedford, and was awarded a B.Sc. in 1912 that combined chemistry and mathematics. She then taught Science at Colston's Girls' School in Bristol until 1915 when she moved into her post at the NPL. Once there, she worked with Walter Rosenhain in the field of metallurgy. She was one of the first two women to be employed in the Metallurgy Department,[20] and made such

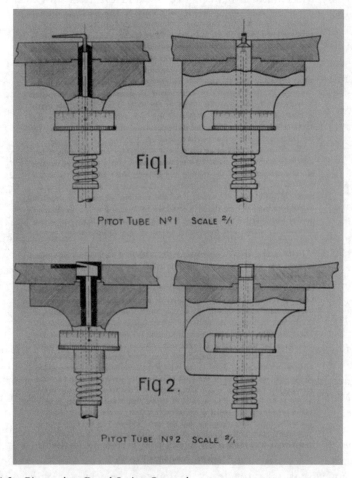

Fig. 26.3 Pitot tube, *Royal Society Journal*

an impression that she was kept on after the War to continue her work; she went on to have a full career at the NPL. Of great note was her work on the composition of amalgams used in dentistry (Baker 2019).[21]

An alumna who typified the output from ELC was Isabel Hadfield, who graduated in 1914, a year later augmenting her degree with a Diploma of Education that she used to become a Chemistry Mistress with the Birmingham Education Council. In 1917 Hadfield was asked to join the staff of the NPL where she looked at chemical problems relating to the new field of aeronautics. Her main work focused on the type and treatment of fabrics used to cover the wings and fuselage of the aircraft. These materials had to be light enough so as not to degrade aircraft performance in flight, but sufficiently resilient to withstand the aerodynamic forces and weather to which they might be subjected. Hadfield was an important member of the Fabrics Research Coordinating Committee of the Department of Scientific and Industrial Research. Her post-war endeavour focused on the field of micro-analytical chemistry (Rayner-Canham 2008, p. 55).[22]

The Royal Aircraft Factory

The Royal Aircraft Factory at Farnborough was the jewel in the crown of the ACA. The Factory went through something of an identity crisis during the immediate pre-war years; the Royal Balloon Factory, as it was known in 1909, became the Army Aircraft Factory in 1911 before completing its metamorphosis into the RAF in 1912 (Royle 2020, p. 12). Its main purpose in its latter guise was to be a source of innovation in design in aeronautics and a primary research establishment, working in parallel with the NPL. The RAF was where new aircraft for the RFC were designed and tested, and also a centre of innovation and exploration in the field of aeronautics. During WWI it witnessed the gathering together of some of the finest mathematicians, scientists, and engineers that Britain could muster, as the tactical advantage of controlling the air over a battlefield became apparent to the military hierarchy. The contributions of many of these academics were not confined to being made in the offices scattered across the RAF's real estate, some of them actually learned to fly so that they could conduct their own experiments in the air, whilst others volunteered as observers to amass in-flight data. Inevitably some of these brave individuals lost their lives in this pursuit of knowledge, but the impact they had on driving forward aeronautics in Britain during this period cannot be overstated.

It was into this cauldron of aeronautical progress and excellence that a number of women mathematicians and scientists were added as the War progressed. The RAF's central role in research and development at the time invoked rapid expansion and, coupled with the demand for men elsewhere, created a void that suitably qualified women were able to fill. Of the women who worked at the RAF during the War, one of the most mathematically talented was Lorna Swain, who had completed the Mathematical Tripos in

1913. Swain, the only woman wrangler of her year, had been coached by George Birtwhistle, a former senior wrangler, who himself had been prepared for Cambridge by hydrodynamicist Horace Lamb. Birtwhistle in turn instilled in Swain an interest in fluid motion and after the Tripos, prior to taking up a post at Newnham, she went on a research visit to Göttingen where Ludwig Prandtl had established a centre for the study of aerodynamics, and Felix Klein had been promoting German interest in applied mathematics (Kennedy 1934). Swain's decision to go to Göttingen may well have been influenced by Arthur Berry, one of the mathematical lecturers who tutored at Newnham, who had studied with Klein in the 1880s; or it may have been the indirect influence of Lamb, via Birtwhistle, since Lamb and Klein were good friends.

With the outbreak of war, Swain had to make a hasty departure from Germany, her visit having been cut short. To continue her research, she went to Manchester University to work with Lamb, who had been appointed to a chair there in 1885, and together they published an article on tidal motion. The following year she returned to Newnham, but by 1917 she was to be found at Farnborough. There she worked on problems of propeller vibration and was another of the few women to have her name attached to one of the Reports and Memoranda of the ACA. The article in question, which appeared in 1919, was written together with Hanor A. Webb, another mathematician employed at the RAF (Webb and Swain 1919).

After the War, Swain returned to Newnham and continued her research in fluid dynamics. This resulted in two papers, both published by the Royal Society. The first was on the motion of a viscous fluid, which she wrote jointly with Arthur Berry (Berry and Swain 1923). Her work in this area linked closely to the work of Bairstow, Cave-Brown-Cave, and Lang noted above. The second (Swain 1929) was the result of a return visit to Göttingen in 1928–1929 where she went to work with Prandtl at his Institute for Technical Physics, by then the world's leading establishment for the study of aerodynamics. In this paper she extended results of Prandtl's on the shape of the turbulent wake behind a body of revolution placed in a uniform stream, also including a discussion of the distribution of the velocity in the wake. It is a testament to the regard in which she was held by Prandtl that it was he who encouraged her to attack the problem and who communicated her results to the Royal Society. Highly respected as a teacher and with a reputation for encouraging women to study mathematics, Swain had the distinction of being appointed a University Lecturer at Cambridge in 1926, one of the first women in Britain to hold such a position, lecturing regularly on hydromechanics and dynamics (Kennedy 1934).

Not all the women doing work for the RAF were located at the Farnborough facility. Ethel Elderton, for example, was based at UCL with Karl Pearson, where she had been since before the War. Elderton had attended Bedford College but on the death of her father had left early, without a degree, and became a schoolteacher. She did, however, have mathematics in her background. Her father and a younger brother were Cambridge wranglers and an elder brother was an actuary and a friend of Pearson's, and it was her mathematical ability that in 1905 led to her employment at UCL where she worked in the Eugenics Laboratory of Francis Galton. According to the science writer Rosaleen Love, Elderton was the woman who did most to wear down 'Galton's prejudice that women were intuitive unintellectual creatures incapable of sound academic work' (Love 1979, p. 152). And she certainly proved herself to be an extremely capable statistician, publishing a *Primer of Statistics* together with her elder brother in 1909, as well as many papers as a sole author and with Pearson both before and after the War.

Elderton produced an important joint paper with Pearson and another mathematician, Andrew Young, which considered torsion in the propellers being used by aircraft (Young et al. 1918). The relevance of this research was that it related to the effect of the flexing force of the air on an aeroplane propeller rotating at high speed. Since the flexure of a beam of asymmetrical cross-section produces a significant torsion, there were serious implications regarding the efficiency and speed of a rotating propeller because of the changes to the angle of attack of the blade to the relative airflow that such torsion could produce. The team's analysis concluded with a formula that could be applied to calculate the approximate torsion per unit length on a propeller due to the unbalanced shear couple it experienced when rotating at high speed. The arguments offered to justify the formula involve some relatively complex mathematics, mostly based upon and developing original work by Saint-Venant,[23] but there is also a large element of numerical computation. It is not clear how the work involved in producing the paper was divided amongst the authors, but since she had a reputation as an energetic computer and much of her other work with Pearson concerned statistical analysis it is likely that she contributed substantially to the numerical work.

A woman mathematician actually based at Farnborough was Annie Betts. Graduating in 1906 with a B.Sc. from London, Betts, who came from a family with an engineering background in railway construction, was also involved in the development of aircraft propellers, co-authoring a paper with H.A. Mettam that derived formulae for predicting certain aspects of aircraft performance (Betts and Mettam 1918).[24] She took a keen interest in aeronautics beyond the War, combining this with her passion for apicology. Editing *Bee World* for many years, she would write no fewer than 170 articles for the publication.

HIGH-ANGLE RANGE TABLES AT UNIVERSITY COLLEGE LONDON (UCL)

As well as working on aeronautics and helping to keep aircraft in flight stable, controllable, and structurally sound, mathematicians were also engaged in what might be considered the converse: shooting aircraft down. World War One was the first conflict to be fought in the air in addition to land and sea, and at its outbreak the ballistics of anti-aircraft gunnery was a subject in its infancy. England had not been invaded since the Dutch fleet had sailed up the Thames in the seventeenth century but now there was the threat of invasion by air, and not only by German aircraft but, and terrifyingly, by Zeppelins. The situation is well-captured by a letter to Karl Pearson from Adelaide Davin, one of Pearson's statistical staff at UCL, written on 9 September 1915:

> The whole of London is in a state of subdued excitement today as a result of the raid last night. We are all congratulating ourselves that we have seen a Zeppelin at last, although the N. Londoners had a much better view than the S. From all accounts the damage appears to have been greatest just at the back of College. A bomb fell in the centre of Queen's Square, and nearly every window is smashed – numerous shops were destroyed in Theobalds Road, and several houses in Red Lion Square have been seriously damaged. I was coming home in a tram just before 11 o'c. when the driver called out that there was a Zep, and that it had been fired at twice – then the tram was stopped, and the lights turned out, whereupon several women began to shriek. I got out and walked home, to find all the neighbours in the street gazing heavenwards. Nobody obeyed the Instructions to seek shelter. We could see the flashes from the anti-aircraft guns, but they all went very wide of the mark.[25]

To combat the threat, the Ministry of Munitions set up the Anti-Aircraft Experimental Section (AAES). At Whale Island near Portsmouth, mathematicians were deployed carrying out ballistics experiments and gathering data, whilst at UCL, Pearson and his human computers were converting that data into anti-aircraft graphic range tables for use by soldiers at the front (Fig. 26.4). It was complex and detailed work, far more complex than that required for flat fire which essentially involves shooting at a stationary target rather than one moving fast in the air.[26] Pearson had employed women in his two statistical laboratories before the War, so it was natural that he should continue to do so after he had made himself and his laboratories available for the war effort.[27] Initially members of staff from the laboratories produced unemployment charts for the Board of Trade, calculated the torsional strain in the blades of aeroplane propellers for the RAF (as previously described), and calculated bomb trajectories for the Admiralty Air Department. But it was the work for the AAES during 1917 that occupied almost everyone in the laboratories and made the most substantial contribution in service of the State, and much of it was done by women. In January of that year, Pearson had two men on his staff and four women. By the end of the year the staff had increased in number to twenty, of whom nine were women, some doing draughtsmanship and others engaged in computing.

One of the most valued members of Pearson's staff during the War was Adelaide Davin, who had obtained a first-class degree in science at UCL in 1912 before beginning work with Pearson on statistical research connected with the evolution of mammals.[28] An able draughtswoman, a skill honed during her earlier biological work, she was responsible for the 'most beautifully prepared' ballistics charts. These charts were not only praised by the

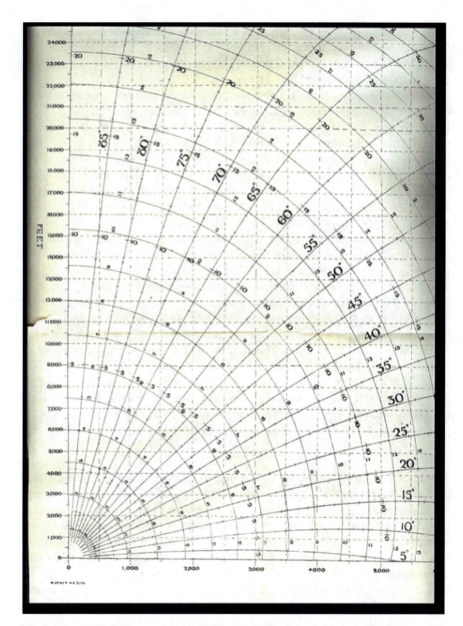

Fig. 26.4 H.W. Richmond (ed.) 1925. *Textbook of Anti-Aircraft Gunnery*, vol. 1, HMSO. London: H.M.S.O., facing p. 534 (Image owned by June Barrow-Green)

mathematicians working in Portsmouth, but they were highly regarded by the British Expeditionary Force in France (Barrow-Green 2014). In November 1917, the Comptroller of the Ministry of Munitions wrote to Pearson to say that he very much appreciated Davin's 'public spirited action' in remaining in his team 'in spite of the fact that she could probably obtain a higher salary elsewhere'.[29] When Pearson handed over the responsibility for the AAES work to the Ministry of Munitions in the spring of 1918, he wrote in his journal that, 'The only member [of staff] who will be a serious loss is Miss Davin, who has been invaluable for the whole of our war work'.[30] After the War, Davin made a career in botany, being awarded a Ph.D. in 1928, and working for several years in Ireland. She returned to England to work as part of the team at the Norfolk Flax Establishment at Sandringham, which made an important contribution to the war effort in World War Two.

Another woman employed on AAES work was Ethel Elderton who moved into it at the beginning of 1917 along with the rest of Pearson's team. Her enthusiasm and capacity for hard work remained undimmed, and, in the summer of 1917, Pearson singled her out, together with Davin, for additional pay for working extended hours and for giving up their customary twelve weeks of holiday. She got on very well with Pearson, being the only member of his team to remain with him after the transfer of the AAES work to the Ministry of Munitions. The two of them continued working alone in the laboratory 'finishing up one or two special gunnery problems' (Pearson 1938, p. 92). Elderton remained at UCL for the rest of her career, being promoted to a readership in 1931. Writing in 1930, Pearson considered her employment at UCL to be 'a most happy choice' (Pearson 1930, p. 258).

Conclusions

Mathematics, particularly engineering mathematics, was not a field populated by many women in the early 1900s. There was little societal precedence, encouragement, or support for it to be otherwise. The women discussed in this chapter, however, certainly stand out as clear examples of those who swam against the prevailing tide of prejudice and conventional expectation. Many of them benefited from familial stimuli and sound financial backing, traits that were almost obligatory prerequisites for any young woman aspiring to go up to a college such as Newnham or Girton. But infringing the male-dominated world of mathematics also demanded other intrinsic qualities. These women were eminently capable of original and independent mathematical thought and had the strength of character to overcome the tradition standing in their way. Their mathematical flexibility was also a testament to their broad understanding of the subject. Many of these women were also groundbreakers, fine examples being Hudson, the first woman to deliver a paper at an ICM, Chitty, the first woman to be placed in the First Class in the Mechanical Science Tripos, and Cave-Browne-Cave, one of the few women to sole author an ACA *Technical Report*.

Objectively, much of the mathematics and science being undertaken and addressed by these women whilst working for the Admiralty, RAF, NPL, or AAES during the War could not always be described as exceptional, but it was certainly necessary. With each new design of aircraft came the concomitant demand for the calculations that would determine its structural integrity and limitations. With each new range table came the need for greater accuracy. Without competent and dedicated mathematicians and scientists, the new breed of industrial aeronautical engineers would have been exposed to relying entirely on the rather blunt tools of judgement and experience. That said, whilst the bulk of the mathematics may have been well established, its specific application to aircraft or gunnery was not, and it was here where the challenges lay.

The pathway leading to a career in applied mathematics outside of education for female mathematicians at the start of the twentieth century was strewn with all manner of obstacles, but the women discussed here were clearly able to negotiate them admirably. Did the demands and urgency of the War assist their career progressions? The conflict certainly created employment opportunities as men were taken in ever-increasing numbers by conscription, but these women nevertheless seized their chance to infiltrate traditionally maledominated domains. Their calculations, expertise, and diligence will inevitably have prevented many unsafe aircraft designs moving from the drawing board into production and provided information to make gunnery a more exact science; these were their practical legacies. However, these women represented more than just a technical check and balance in the chain of aircraft design and production, or efficient computers of range tables. They were pioneers who demonstrated that it was possible for women to overcome the dogmatic, institutionalized prejudices of the time, and they earned the right to stand tall as credible applied mathematicians during the First World War and beyond.

NOTES

1. On 28 October 1915, *The Times*, in an article on 'The Demand for Qualified Women', reported that, 'Already posts have been found in aircraft works where a Mathematics Tripos was of value' (p. 5). Two years later, on 31 August 1917, the same newspaper, in an article on 'Women's War Work', noted the 'big demand' for women with degrees in mathematics (p. 3).
2. In order to win a high place in the order of merit, coaching was generally deemed essential. Amongst the most high-profile of the mathematicians who supported women students was the Sadleirian professor, Arthur Cayley, who taught at Girton and for several years was chair of the Council of Newnham.
3. Scott later made her career in the United States as the inaugural professor of mathematics at Bryn Mawr College.

4. After Fawcett's success, the clamour for women to be awarded degrees grew louder but still it was not loud enough. The University did not fully open its doors to women until May 1948. Those who wanted undergraduate degrees had to go to London or, from 1920, Oxford. Those who wanted higher degrees had to go abroad—the Ph.D. did not come to Britain until after the First World War.
5. The aeronautical engineering department at East London College was the first of its kind in Britain. It was set up in 1909 by Albert Thurston, a graduate of the College.
6. From 1945 the College admitted male postgraduates, with the first male undergraduates arriving in 1965. Royal Holloway and Bedford Colleges merged in 1985 and became the single entity, Royal Holloway and Bedford New College.
7. Photograph of The Admiralty Air Department, 1918, by kind permission of the Hartley Library, University of Southampton. MS112 Papers of Miss A.M. Trout LF 780 UNI 2/7/75.
8. The Project began in 1899 and employed approximately 20 women to write the parish histories for the various counties of Britain.
9. Moger compiled abstracts from about 6600 Devonshire wills, and listed some 5000 others, information that was bound into 22 volumes. The originals were destroyed in the Second World War, but copies survived and are held in the Devon Heritage Centre in Exeter. She also worked on similar projects in the counties of Somerset and Cornwall.
10. It is likely that Schwarz and his colleagues were major influences in developing Hudson's interest in conformal transformations, a topic initially introduced to her by mathematician Arthur Berry during her time at Cambridge, and one that would eventually dominate her mathematical research, which culminated in 1927 with the publication of her comprehensive and well-respected treatise, 'Cremona transformations in plane and space' (Hudson 1927).
11. Sir Ronald Ross (1857–1932) was a British doctor who won the Nobel Prize for the Physiology of Medicine in 1902 for his studies in the transmission of malaria. In 1916 Hudson was appointed by the Royal Society to help him with the mathematics in his work which resulted in two co-authored papers (Ross and Hudson 1917). In 1917 he was appointed to the British War Office as a Consultant in Malariology.
12. Despite Winchester College being a school only for boys at the time, Chitty's access to tuition there would likely have been a boon derived from her father's position as archivist there. Winchester College was one of the top private schools in the country and it is therefore likely that Chitty was tutored to a high standard in mathematics.
13. Mary Hutchison was a product of the ELC, graduating in 1913 with a B.Sc. Honours in Mathematics; she and Chandler both arrived at the Admiralty a week before Chitty.

14. Chitty contributed to the centenary edition of *The Royal Aeronautical Society's Journal* in 1966 (Chitty 1966), an article from which all the quotes used above are taken.
15. The three moments theorem here refers to Émile Clapeyron's work in the middle part of the nineteenth century, which addresses the relationship between the bending moments at three consecutive supports of a horizontal beam (Clapeyron 1858).
16. Pippard, who studied engineering at Bristol, was appointed as technical advisor to the Director of the Air Department of the Admiralty in 1915. After the War he became an influential member of staff at Imperial College, London.
17. Lang, a student at Royal Holloway, obtained a third-class honours degree in mathematics in 1905, and went on to be awarded an M.A. from UCL in 1911.
18. Pearson's *Journal of the Galton Laboratory 1915–1918*. Pearson Papers, UCL Pearson 4/17, p. 101.
19. Jones co-authored numerous Reports and Memoranda (R&M) at the NPL, including R&Ms 355, 366, 375, and 418. She and her colleague L.W. Bryant married during their time at the NPL, her name appearing as C.N. Bryant in subsequently published work.
20. The other was Isabel Hadfield.
21. She was awarded an M.Sc. in 1921, and a DSc in 1924, both from Bedford, along with MISI/MIM Hon MBDA as her career progressed.
22. A comprehensive account of the role of British women chemists during WWI, including Isabel Hadfield, can be found in Rayner-Canham and Rayner-Canham (2008).
23. Adhémar Jean Claude Barré de Saint–Venant (1797–1886) was a French mathematician and mechanician who worked in the fields of stress analysis and hydrodynamics.
24. Mettam was a Cambridge mathematician who was a member of the post-war Cambridge Aeronautical Club. Newnham and Girton Colleges were allowed to be members of the Club and, given her experience in aeronautics, Letitia Chitty was elected onto the Club's Committee.
25. A. Davin to K. Pearson, 9 September 1915, Pearson Papers, UCL, 674/9.
26. Each table supplied the ballistic and fuze data for a gun (with a given muzzle velocity) in graphical form. The fuze time scale is arbitrary (i.e. not in seconds). Usually, as in the range table shown in Fig. 26.4, the maximum setting was 22, which represented something between 40 and 60 s, depending on the type of fuze. The horizontal range and height are the x- and y-axes, and the curves plotted represent actual trajectories for different elevations of the gun.
27. For example, in 1908, five out of fourteen research workers in the laboratories were women, and Pearson wrote to Galton that, 'their work is equal at the very least to that of the men. They are women who in many

cases have taken higher academic honours than the men and are intellectually their peers. They were a little tried therefore when your name appeared on the Committee of the Anti-Suffrage Society!' (as quoted in Love 1979, p. 146). It should be noted, however, that at this time women were paid considerably less than men, which might well have been a factor.

28. After the War, Davin published papers with Pearson. We are very grateful to Adelaide Davin's great niece, Alison Pearson (no relation to Karl Pearson) for supplying us with information about her great aunt's education and later life.

29. A.E. Moore to Karl Pearson, 16 November 1917, Pearson Papers, UCL, Pearson 9/8/1.

30. Pearson's *Journal of the Galton Laboratory 1915–1918*, Pearson Papers, UCL Pearson 4/17, p. 108.

References

Bairstow, Leonard, Beatrice Cave, and Eleanor Lang. 1922. "The Two-Dimensional Slow Motion of Viscous Fluids." *Proceedings of the Royal Society of London A* 100, no. 705: 394–413.

Bairstow, Leonard, Beatrice Cave, and Eleanor Lang. 1923. "The Resistance of a Fluid Moving in a Viscous Fluid." *Philosophical Transactions of the Royal Society of London A* 223, no. 614: 383–432.

Baker, Nina. 2019. Magnificent Women in Engineering: Engineer of the Week Series, No. 25. https://www.magnificentwomen.co.uk. Accessed 6 January 2021.

Barrow-Green, June. 2014. "Cambridge Mathematicians' Responses to the First World War." In *The War of Guns and Mathematics*, ed. David Aubin and Catherine Goldstein, 59–124. Providence, Rhode Island: American Mathematical Society.

Barrow-Green, June. 2019. "'Stokes of Pembroke S.W. & a Very Good One': The Mathematical Education of George Gabriel Stokes." In *George Gabriel Stokes: Life, Science and Faith*, ed. Mark McCartney, Andrew Whitaker, and Alastair Wood, 47–62. Oxford: Oxford University Press.

Berry, Arthur, and Lorna M. Swain. 1923. "On the Steady Motion of a Cylinder through Infinite Viscous Fluid." *Proceedings of the Royal Society A* 102, no. 719: 766–78.

Betts, A., and H. Mettam. 1918. "Empirical Formulae for a Variable Pitch Airscrew, with Applications to the Prediction of Aeroplane Performance." *Technical Report of the Advisory Committee for Aeronautics 1918–1919*, II: 761–84.

Cave-Browne-Cave, Beatrice. 1918. "The Calculations of the Periods and Damping Factors of Aeroplane Oscillations and a Comparison with Observations." *Technical Report of the Advisory Committee for Aeronautics, 1918–1919*, II: 811–34.

Chitty, Letitia. 1966. "Contribution from Miss L. Chitty." *Journal of The Royal Aeronautical Society*, 70: 67–68.

Clapeyron, Paul Émile. 1858. "Mémoire sur le travail des forces élastiques dans un corps solide élastique déformé par l'action de forces extérieures." *Comptes Rendus* 46: 208–12.

Hudson, Hilda Phoebe. 1912. "On Binodes and Nodal Curves." *Proceedings of the Fifth International Congress of Mathematicians*, II: 118–21. Cambridge: Cambridge University Press.
Hudson, Hilda Phoebe. 1920a. "The Strength of Laterally Loaded Struts." *The Aeroplane* 18: 1178–180.
Hudson, Hilda Phoebe. 1920b. "Incidence Wires." *Aeronautical Journal* 24: 505–16.
Hudson, Hilda Phoebe. 1927. *Cremona Transformations in Plane and Space*. Cambridge: Cambridge University Press.
Kennedy, M.D. 1934. "Lorna Mary Swain." *Journal of the London Mathematical Society* 9: 155–57.
Love, Rosaleen. 1979. "'Alice in Eugenics-Land': Feminism and Eugenics in the Scientific Careers of Alice Lee and Ethel Elderton." *Annals of Science* 36: 145–58.
Marshall, Dorothy, and William Ramsay. 1896. "A Method of Comparing Directly the Heats of Evaporation of Different Liquids at Their Boiling Points." *The London, Edinburgh, and Dublin Philosophical Magazine and Journal of Science* 41: 38–52.
Marshall, Dorothy, and Thomas Stanton. 1920. "Preliminary Note on the Effect of Surface Roughness on the Heat Transmitted from Hot Bodies to Fluids Flowing over Them." *Technical Report of the Advisory Committee for Aeronautics for the Year 1916–1917* I: 31–35.
Marshall, Dorothy, Thomas Stanton, and Ernest Griffiths. 1921. "On the Dissipation of Heat from the Surface on an Air-cooled Engine when Running and at Rest." *Technical Report of the Advisory Committee for Aeronautics for the year 1917–1918* II: 734–47.
"Miss Fawcett's Honour: The Sort of Girl this Lady Wrangler Is." 1890. *The New York Times*, 8 June: 5.
Neale, Charles Montague. 1907. *The Senior Wranglers of the University of Cambridge, from 1748 to 1907*. Bury St Edmunds: F.T. Groom & Sons.
Pearson, Egon. 1938. *Karl Pearson. An Appreciation of Some Aspects of His Life*. Cambridge: Cambridge University.
Pearson, Karl. 1930. *The Life, Letters and Labours of Francis Galton*. Cambridge: Cambridge University Press.
Pippard, A., and J. Pritchard. 1918. *Handbook of Strength Calculations*. London: Ministry of Munitions Technical Department (Aircraft Production).
Rayner-Canham, Marelene, and Geoffrey Rayner-Canham. 2008. *Chemistry Was Their Life: Pioneer British Women Chemists, 1880–1949*. London: Imperial College Press.
Ross, Ronald, and Hilda P. Hudson. 1917. "An Application of the Theory of Probabilities to the Study of a Priori Pathometry. Parts II & III." *Proceedings of the Royal Society A* 93(650): 212–40.
Royle, Tony. 2017. "The Impact of the Women of the Technical Section of the Admiralty Air Department on the Structural Integrity of Aircraft during World War One." *Historia Mathematica* 44: 342–66.
Royle, Tony. 2020. *The Flying Mathematicians of World War I*. Montreal: McGill-Queen's University Press.
Skempton, Alec. 1970. "Alfred John Sutton Pippard." *Biographical Memoirs of the Royal Society* 16: 463–78.
Stanton, Thomas E., Dorothy Marshall, and Constance N. Bryant. 1920. "On the Conditions at the Boundary of a Fluid in Turbulent Motion." *Proceedings of the Royal Society A* 97: 413–34.

Swain, Lorna Mary. 1929. "On the Turbulent Wake Behind a Body of Revolution." *Proceedings of the Royal Society* 125: 647–59.

Trout, Annie Mary. 1918. Notes on mathematical subjects. MS112 Papers of Miss A.M. Trout. Hartley Library, University of Southampton.

Trout, Annie Mary. 1926. Postcard to Hilda Hudson. MS112 Papers of Miss A.M. Trout. Hartley Library, University of Southampton.

Webb, Hanor A., and Lorna Mary Swain. 1919. "Vibration Speeds of Airscrew Blades." *Technical Report of the Advisory Committee for Aeronautics for the Year 1917–1918* II: 753–60. London: Her Majesty's Stationary Office.

Young, Andrew, Ethel Elderton, and Karl Pearson. 1918. "On the Torsion Resulting from Flexure in Prisms with Cross-Sections of Uni-axial Symmetry Only." *Drapers' Company Research Memoirs Technical Series, VII*. Cambridge: Cambridge University Press.

CHAPTER 27

More Than Pioneers—How Women Became Professional Engineers Before the Mid-Twentieth Century

Nina Baker

Second-wave feminism is generally considered to be the upwelling of interest in women's rights which came about in the late 1960s and early 1970s. Many women entering fields of work which had been traditionally male-majority, were often seen (and publicised) in that period as pioneers: 'first woman to do a thing' (Stanley 2016, p. 70). Many of them were indeed 'firsts' but typically earlier women's stories had by this time become forgotten and only started to re-emerge with the introduction of Women's History as an academic topic, generally credited to Gerda Hedwig Lerner's two-volume study *Women & History* (Lerner 1986, 1993). Even with that renaissance, it would be some while before women engineers' histories were being considered methodically, as for example Canel et al. in 2000, which comments (p. xiii) that feminist publishers initially dismissed women engineers as too boring to be publishable. Even in the early twenty-first century, the reaction to expressing an interest in women's history in engineering was generally that there was no history worth knowing. Fortunately, such attitudes have pretty much died out and a wider interest in women's part in engineering history has grown in recent years.

Those wishing to encourage girls to choose Science, Technology, Engineering and Maths (STEM) subjects at school and college, will often cite pioneering women in those fields as early role models. There are many claims as to who was the first woman engineer, and the historic printed word is

N. Baker (✉)
Scotland, UK
e-mail: nina.baker1@btinternet.com

© The Author(s), under exclusive license to Springer Nature Switzerland AG 2022
C. G. Jones et al. (eds.), *The Palgrave Handbook of Women and Science since 1660*, https://doi.org/10.1007/978-3-030-78973-2_27

not always to be trusted in this, since it seems that fact-checking has not always been rigorous, especially in volunteer-edited feminist periodicals of the early twentieth century, such as *The Woman Engineer* (WES 1919–) or *The Vote* (WFL 1909–33). This chapter will therefore discount Hypatia, fourth century Graeco-Egyptian astronomer and mathematician, and Ada Lovelace, the nineteenth century English mathematician credited with the invention of algorithms or programs for computers, on the grounds that they were not true engineers as the term is understood today. This then leaves us with a variety of women who are credited with inventions over the centuries, although not with actual careers or qualifications. A reasonably well-known example of such a woman would be Sarah Guppy (1770–1852), wife of a Bristol shipping agent, who took out a patent[1] for a means of securing the cables for suspension bridges (Dresser 2016). Although she did take out other patents, she had no training in engineering and never progressed her ideas beyond allowing famous male engineers of the period, such as Thomas Telford, to use them for their own bridge designs. Feminist engineering historians like to claim her as an early British woman engineer but this seems to be stretching definitions to a large extent and so such early innovators will also not be examined in detail here. The question then arises as to who or what is or was an engineer? This varies with time, place and culture. The term itself was barely in use (in printed English) before the end of the eighteenth century (Google Books Ngram reader), but the coming of the Industrial Revolution saw its use massively increase. For centuries, the term was mainly in military use relating to the building of fortifications and 'machines of war', such as ballistae. The first engineer to make the distinction from military engineering was John Smeaton who called himself a 'civil engineer' and founded the Society of Civil Engineers in 1771. The spectra of skill levels, education, gender and specialisations of those calling themselves engineers have fluctuated over time and place. In Britain, the term engineer has meant, variously at different times, a skilled working-class man who made metal machinery (as compared to a 'wright' who made wooden items—as in millwright or wheelwright) or operated such machines; a professional who designed such items or supervised those making them; and—today—anyone at the professional level in a vast array of different branches from biomedical to aerospace. In some countries, such as the USA or Germany, no one may describe themselves as an engineer without having completed many years of post-university experience and qualifications. In some countries there are more limited understandings of engineering and the general societal view may, as for instance in Syria, be that engineering means mainly civil engineering. The Women's Engineering Society, the first such society in the world, founded in 1919, was open initially to any woman who had worked at any skill level in engineering. However, it soon restricted full membership to professionals and graduates of engineering, only later coming to welcome women who were skilled technicians or scientists in any of the multiplicity of engineering fields.

The heroic 'muscular' model of the male engineer is much celebrated in the UK, with figures such as James Watt, Isambard Brunel and George Stephenson being revered during and after their lifetimes, often to the point of inaccurate hagiography (see for example Smiles 1865, 1879). In this model of public persona for engineers and their work, they are portrayed as achieving amazing, almost miraculous, things in the face of significant physical adversity. In comparison, women engineers' stories have generally been omitted until recently and their technical contributions perceived as minor in comparison. The first systematic history of women engineers was published in 2000 (Canel et al. 2000), looking at the variety of routes by which women found their way into engineering in the late nineteenth and early twentieth centuries in various countries.

This chapter will control the otherwise impossibly broad range of histories by focussing on women who worked at a professional level, however they may have reached that level. In the early period covered by this chapter, the opportunities for university training were limited and many women found other routes into engineering. The majority of case histories will be from the United Kingdom but comparisons with other countries will be used to demonstrate just how significant cultural norms have been in circumscribing or enabling women's entry into and career progression in engineering.

Educating Girls and Women

To enter any professional occupation has generally been defined as requiring a high level of education and to enter engineering requires a basic grounding in mathematics and sciences. The ancient UK universities prepared men mainly for the law or the church and later for teaching and medicine. The earliest schools, often church-sponsored, prepared boys for work or university but very few young men attended the latter and even fewer actually graduated. Most middle- and upper-class children were educated at home by their parents or by tutors or governesses until the latter half of the nineteenth century and even boys often got little to no science education during their schooling.

Education for girls was intended to equip them to be better mothers or servants. Women of the lower classes had always had to work hard, often even after marriage or motherhood, to help the household income. However, as the eighteenth century moved into the nineteenth, it became increasingly clear that such women were ill-equipped for either new industrial work or traditional domestic tasks, such as cleaning, laundry, cooking or nursing. In addition, the demand for governesses for higher class homes exposed the fact that even middle-class women had insufficient education to pass onto the next generation.

Although we often think of the period after the terrible losses of the First World War as one when many women who might have expected to marry could not do so, there were earlier periods when this was also a concern for fathers of daughters. The census returns from, for instance, Scotland in

1871 showed that there were twice the percentage of women who had never married, compared with men who had never married, in the over-forty-five age group (HMSO 1881), amounting to over 80,000 Scotswomen who had to fend for themselves or depend upon relatives in middle and old age. The probable causes of this gender imbalance were the 'pull' of men to the army, navy, empire service or emigration for work. This was despite efforts to ship women to the colonies, as by the Female Middle-Class Emigration Society, founded in 1862 and which merged into the British Women's Emigration Association in 1908. Thus, the imbalance was such that many women of all classes could have no expectation of marrying and needed to be prepared to support themselves in adult life, especially after their fathers had died. In the census of 1911, about 500,000 women were not reliant on a male provider. The need for girls and women to be better educated was evident to all classes.

Middle-class unmarried women were particularly vulnerable, socially and financially, if there was no man to provide for them, as paid employment had to be socially acceptable, 'appropriate' work. This could limit them to posts as governesses, companions or (by lowering their sights) ladies' maids or dressmakers. These badly paid options revealed how under-educated many girls were, even for the most traditionally gendered occupations, so training for governesses was one of the reasons for the first efforts to provide better education for girls and women. London's Queen's College was established in 1845 with the patronage of Queen Victoria, initially as a hostel for unemployed governesses, but soon began offering lectures and by 1900 was a proper school. By that time, feminist educators such as Dorothea Beale and Frances Buss had opened seriously academic schools for girls, which encouraged more middle-class families to consider that their daughters might be as academically able as their sons.

The next stage was the campaign to allow British women access to universities. Although Oberlin College, USA, admitted women in 1833, the University of London took another forty-five years before it became the first UK university to do so and the Scottish universities a further ten years. In the meantime, if women could not get into universities, many progressively minded male academics set up schemes, such as the Edinburgh Ladies' Educational Association lectures, to take university level courses out to interested women. Many men, even some actively supporting some education outside the university degree system, such as Dr. William Milligan of Aberdeen, considered that the physical rigours of university study would be too severe for women, who might literally 'simply be killed off' (Milligan 1877, p. 6). Mathematics in particular, a subject we perhaps now consider to be entirely in the realm of the mind, was considered extremely demanding physically on its students.

Meanwhile boys had had access not only to centuries of good schools and universities, but also to those which taught higher mathematics and physical sciences. The 1851 census reported, for instance, that although similar numbers of girls and boys attended schools, only 0.1% of girls were taught any mathematics (as contrasted with basic arithmetic) compared with 2.8% of boys

(HMSO 1854). Even working-class males could get a part-time education as the network of Mechanics' Institutes spread, such as those originating from the School of Arts of Edinburgh (established in 1821 and later to become Heriot-Watt University). The vast majority of even the best girls' schools taught no science, unless a little botany. As late as the mid-twentieth century many girls' schools still did not offer the higher maths or physical sciences essential for engineering studies. An Englishwoman who gained her engineering degree during the Second World War had attended two girls' secondary schools where the only science taught was biology. When she arrived at Cambridge University, in 1941, she had only had very limited teaching of physics, given by the cookery mistress, and had not heard of Ohm's Law (Gardiner 1990). When the first women did get access to universities, the majority took degrees which would equip them for the areas of paid employment where they knew there were opportunities for women: medicine or teaching. Only tiny numbers took pure sciences or the applied sciences for engineering. Whilst a complete survey of the Scottish universities' early female engineering graduates (Baker 2009) has been undertaken, there has not been a similarly systematic survey of the rest of the UK.

Women Becoming Engineers

In the late nineteenth and early twentieth centuries it was commonplace for aspiring engineers to be pupils of an established engineer (to whom a fee was paid) and to become professionals without having been to college or university. It was also still possible for shop-floor apprentices with ambition and talent to work their way up through the supervisory chain into management of a works. Today the routes into professional engineering are tightly controlled in almost all nations, with professional bodies stipulating criteria for accredited university degrees, followed by specified levels of early career responsibilities, as the usual way to become a chartered engineer. In many countries there are legal controls on what qualifies a person to call themselves an engineer.

The first woman in the world to gain an engineering degree is believed to have been Elizabeth Bragg who graduated in civil engineering in 1876, from the University of California, Berkeley, but she never worked as an engineer (Weingardt 2014). The first woman in the British Isles to get an engineering degree is thought to have been Alice Jacqueline Perry (1885–1969), who graduated from Queen's College Galway in 1906, at a time when the whole of Ireland was still part of the United Kingdom (Irish Architectural Archive, n.d.). Perry had assisted her engineer father during her studies and replaced him (at the same salary) as county surveyor for Galway on his death just after she graduated. Nina Cameron Graham Walley (1891–1974) graduated in civil engineering in 1912 from the University of Liverpool (University of Liverpool 1912) but, like Bragg, married immediately and only practised engineering informally, assisting her husband.

Two other notable pioneers in the UK from approximately the same period, who did practice as engineers, took the non-academic route into the profession: Dorothée Pullinger Martin MBE (1896–1982) (whose career is discussed in more detail later) and Victoria Drummond MBE (1894–1978). They were both very forceful characters who were determined to enter engineering in the face of considerable opposition. Each made use of their families' connections to push their way into training opportunities and ultimately employment in professional engineering. Pullinger's father (Thomas C. Pullinger, a notable early car designer) was managing director of Arrol-Johnston Ltd., Scottish automobile makers, and her use of his connections and status to start an apprenticeship at the car works was a reasonable choice at the time (Clarsen 2003). Drummond, the UK's first female marine engineer, was almost exactly Dorothée's contemporary in Scotland and made use of her own aristocratic, but impoverished, family's connections to get the practical training in workshops and shipyards she needed in order to become the first woman to become a ship's engineering officer (Drummond 1994). Drummond faced enormous prejudice afloat and ashore, with the Board of Trade examiners infamously refusing to give her a pass on her chief engineer's examination thirty-one times. However, she sailed as a chief engineer with many non-UK shipping lines and won an MBE and Lloyds' medal for bravery following a wartime incident at sea in 1940.

Whilst, from the social viewpoint of today, these women might be seen as exploiting nepotism to achieve their aims, it should be recognised that such social advantages alone would not have been sufficient. Even after the social upheavals of the First World War life was only very slowly opening-up for middle-class women, and only those with ability and considerable determination would make it into the harshly male engineering environment. As Drummond was to discover the hard way, the key test of a professional engineer in British industry then was, perhaps still is, how well one could 'manage the men' working on the factory floor, or down the ship's engine room.

Everything about the world they wanted to enter would have been against them. The necessary education was difficult to get, the essential first job was difficult to get unless you had contacts, and indeed for many young women this remains a barrier even today in such areas as the skilled construction trades (Bagilhole 2014). Prior to the industrial upheavals of the First World War, the only industrial settings in which significant numbers of women worked were in textile mills and industrial laundries, in which they were semi-skilled machine operators rather than engineers (Malcolmson 1986). Women were unlikely to feel welcome in the wholly male environment of the manufacturing workshop, even if the owner was sufficiently sympathetic to offer a woman a start. The trades unions were highly protective of their trades and many larger works had their own freemasonry lodges and other social groups which were closed to women. Until the 1920s scarcely any of the local or national professional and scientific bodies admitted women to their lectures, let alone membership, so that the networking normal to men of all social classes would also have been prohibited to women seeking to become engineers. Men controlled all the entry points and mostly kept them locked against women in the first half of the twentieth century, hence the importance of Pullinger's father's high-status position in getting her the start she needed.

Even the practical matter of what clothing to wear would have been an issue, even though women munitions workers had worn coveralls of many types and even trousers. We do not know what Pullinger wore to work but her clothes would have needed to be both smart enough for her status in management and also practical and hard-wearing. Drummond had to wear boilersuits to negotiate the ladders and crawl-spaces of engine rooms, even though it was close to scandalous even in the 1920s for a woman to wear trousers (Smith and Greig 2003).

Early Graduate: Chemical Engineering—Elisa Leonida-Zamfirescu

Elisa Leonida (1887–1973) is an example of a truly pioneering woman who became a professional engineer by the formal route: getting an engineering degree and then working her way up the profession over a lifetime career, even after marriage. She is generally considered to be the first woman to achieve this.[2] Leonida, from an engineering family, was born in the Romanian town of Galați. She was refused admission to Bucharest Bridges and Highways School (a males-only military school) and she obtained her degree in engineering from the Royal Academy of Technology Berlin in 1912, although the authorities there faced objections too and described her as a 'special case'. Her professors tried to ignore her presence in class or else shouted at her to go back to the kitchen, but her patience and talent overcame many prejudices, and she was widely admired when she did graduate. Although Englishwoman Nina Graham obtained her engineering degree in the same year, Graham immediately married and did not have a further engineering career.

Hence Leonida is generally acknowledged as the world's first qualified and practising female engineer. Her first job was at the Geological Institute of Romania, to which she returned after the war during which she married Constantin Zamfirescu. Her continuing work was on identifying new resources of fossil fuels and metal ores and included field surveys, something which must have been challenging in a developing but largely traditional and rural country (Enciclopedia României 2009). She was the first female member of the Romanian society for engineers.

1910–20s: Mechanical Engineering—Verena Holmes

Verena Holmes B.Sc. (Eng.), AMIMechE, AIMarE. MILocEng (1889–1964) was born in Kent, where her father was a schools' inspector. She became the first president of the Women's Engineering Society who was not just a qualified engineer but also practiced engineering her whole working life (Stanley

2010). Holmes found her true vocation when the First World War opened the door for women to enter engineering. Her first job was building wooden propellers for Integral Propeller Co., Hendon, then at Ruston and Hornsby, an aero engine firm in Lincoln, as their Lady Superintendent responsible for the selection and welfare of 1500 female employees. As her real interest was engineering, she persuaded the company to let her start as an apprentice in the fitting shops. Holmes gained experience as a turner and completed an apprenticeship as a draughtsman, so that when the war ended, she was the only woman who was allowed to stay on with the firm.

Holmes gained her B.Sc. (Engineering) degree extramurally from London University in 1922, and from then produced many inventions, of which seventeen were patented, including the Holmes-Wingfield pneumo-thorax apparatus. This equipment was for inserting a long metal hypodermic needle through the chest wall in order to collapse the lung under a controlled pressure as a rather terrifying treatment for tuberculosis. She designed and made this apparatus for Dr. R. C. Wingfield's work with tuberculosis patients in Scotland. In the early 1920s she worked with various consultant engineers before setting up her own design consultancy in 1925. Her patents for locomotive valves got her a job at the North British Locomotive Works, in Glasgow (1928–31), and from 1932–39 she was a design engineer at Research Engineers Ltd. When the Second World War started, Verena instigated the Women's Engineering Society's training scheme for women, later adopted by the government. She also advised Ernest Bevin, the Minister of Labour, on the training of munition workers. In 1946 she and Sheila Leather formed Holmes and Leather, a company employing women to make Bantam metal shearing machines, and other Holmes inventions, in Kent.

Having been a founder member of the Women's Engineering Society, Verena Holmes was its President in 1931–32, and her legacy to the society was used to start a Verena Holmes Memorial Lecture series which toured schools for many years.

Progressing in a Career—Or Not

Whether with a degree qualification or informal practical training, entry into engineering employment was of course only the first step. Holmes and Drummond were rare examples of women whose determination and skills enable them to carve out lifetime careers as practical engineers. Getting a job in the first place was hard enough, but then progressing in the profession was another situation generally stacked against women. In many, perhaps most, cases of women who had successful engineering careers before the late twentieth century, they benefitted from individual male managers ('gatekeepers') who made it their business to encourage and support women whose skills they valued. The emergence of the military-industrial complex in the industrialised countries also supported new fields of engineering, such as electronics, aeronautical and chemical engineering. The pressure to provide talented staff and the lack of embedded historical prejudicial structures led to early women engineers becoming concentrated in those areas. It continues to be the case that women are clustered in the newer branches of engineering but that their progression is often frustrated in mid-career by hidden barriers or 'glass ceilings', characterised by hidden and implicit biases in institutions and individuals.

One of the most obvious barriers in the early twentieth century was the 'marriage bar' (Morgan 2015). Operating in most industrialised countries, this was variously either a statutory requirement or a customary expectation that women would leave their paid employment on marriage and very definitely on becoming pregnant. The UK Civil Service did regulate that its female staff should leave on marriage (until 1947) but examples at the Royal Aircraft Establishment (and probably other professional level situations) indicate that there must have been a 'blind eye' turned since a number of women aeronautical engineers, such as Beatrice Shilling (Mrs. Naylor) and Muriel Barker (Mrs. Glauert), continued to work during the 1920s and 1930s after their marriages. Mrs. Naylor was kept on a 'temporary staff' contract for ten years until the marriage bar was lifted for civil servants (Freudenberg 2004).

Once a woman started to have children that was generally the end of her career, and indeed it was common to read in firms' newsletters of Miss X 'retiring' from her job on marriage or Mrs. Y doing so because of starting a family. For well-educated, experienced women engineers who had put considerable effort to get into engineering at all, it must have been very frustrating to have to turn their backs on all they had achieved, unless they could afford paid servants to look after the household and children, as Pullinger did. The problem of the career-killing 'career break' was not significantly addressed until pioneering nuclear engineer and president (1984) of the Women's Engineering Society, Professor Daphne Jackson OBE (1936–91), set up a fund for fellowships to enable women to return to science or engineering after such breaks. Since her premature death, the Daphne Jackson Trust has continued to provide such fellowships, now open to male returners too.

1920–30s: Aeronautical Engineering—Hilda Margaret Lyon

Hilda Margaret Lyon M.A., M.Sc., AFRAeS (1896–1946) was a talented mathematician who became not 'merely' a trailblazing Yorkshire woman in engineering at a time when women were not expected to be engineers, but also a technical pioneer who contributed significant technological advances during the early days of aviation. From a family of shopkeepers, she had no relatives in anything technical as role models, somewhat unusually for early women in engineering. Her 1918 Maths Tripos from Cambridge led her immediately into structural analysis of early aeroplanes, initially for aircraft firms but later for the government's research station, Royal Aircraft Establishment Farnborough. She did Master's research on the effects of fan turbulence on wind tunnel models, at MIT in the USA, and then worked for the world's top aerodynamicist of the time, Professor Ludwig Prandtl at Göttingen in Germany during the Nazi rise to power. When her mother became ill, Lyon had to leave her work in Germany to nurse her, fulfilling a traditional daughterly role despite other family members living in the same house as her mother. Then, as now, such a 'career break' would generally kill a career but Lyon was sufficiently respected in her field that her former colleagues at the Royal Aircraft Establishment, and others, helped her to find work she could do from home, sometimes even being able to pay her for small pieces of work (Lyon 1944). Eventually the RAE found a post for her and she remained with them until her premature death.

Lyon helped design the R101 airship, improved the safety of submarines in the Second World War and did fundamental work on the longitudinal stability of aircraft in flight, much-cited and the basis for flight safety software today. In her own lifetime she was well known but now, some seventy years after her untimely death, her story is only just beginning to re-emerge into the public arena. As with many women who earned good careers in engineering, Lyon had started out in maths, but in later life would say that if she had to do it all again she would have followed her time at Cambridge with a spell learning practical engineering at a technical college.

1940s: The Second World War—Déjà vu All Over Again: Dorothée Pullinger Martin

The lack of understanding of both how to manage women workers *en masse* and how quickly they could be trained to do a 'skilled' man's job, was apparently a lesson difficult to learn (in WW1) and easily forgotten (by WW2). Women with industrial experience were thinner on the ground in the first war than the second but had to be brought in to advise both government and industry in both conflicts.

Dorothée Aurelie Marianne Pullinger MBE (1894–1986) was a very determined woman who pushed her way into the career she wanted in engineering and later as a successful serial entrepreneur at a time when neither was expected of women. Her father, Thomas, was a well-known early car designer who worked all over France when she was growing up and became managing director of the New Arrol-Johnston Car Company, in Paisley, not long before she left school. Against his very traditional views of women's roles, she entered the company to train as an engineer, initially at Paisley and then at a bespoke new factory near Dumfries. In 1916 she became the Lady Superintendent in charge of thousands of women munitions workers at Vickers' factory in Barrow in Furness. All major munitions contractors were required to have these Lady Superintendents, to overcome many early problems when factories found themselves unable to manage a female workforce. The male management and shop-floor foremen had no experience of managing women and the majority of the women had never worked in a factory either, creating a lot of unexpected labour unrest and welfare problems. Although Dorothée was only twenty-two when she went to Vickers, she made a triumphant success of her role there and must also have learned a lot about managing and training masses of women. She returned from her war work to find her father's views on women's abilities much changed, as he had built a second factory at Tongland, in the South of Scotland, explicitly to be 'A fine university for women engineers' (Clarsen 2003). After the war father and daughter, noting how many women had learned to drive in war service, set about designing and manufacturing a light car aimed at the female market: the Galloway cars. Having completed her practical training by running the foundry, she joined both the management and the board of directors of Galloway Motors Ltd (Clarsen 2004). Despite the company overcoming the near-ubiquitous application of the Restoration of Pre-War Practices Act (HMSO 1919) and retaining its women workers, Dorothée found herself constantly criticised for 'taking a man's job'. After she married in 1924, she first moved to the sales side and then, as the motor industry went into decline, set up her own business in London: an industrial-scale steam laundry which she ran successfully for about 20 years, whilst also raising two children.

When the Second World War started, the government and industry found themselves in the same position as in the First World War, with regard to introducing women into munitions factories. There were the same labour unrest and welfare issues as before and veteran women of the earlier conflict were brought in to sort things out. Mrs. Dorothée Pullinger Martin, with many other women of her generation from the Women's Engineering Society, were drafted in to advise the Ministry of Production and industry leaders, including Lord Nuffield who had failed to make good on unrealistic promises of Spitfire production, due to labour unrest and chaotic organisation. She somewhat disparagingly commented that 'Men, it seems both bosses and foremen, are afraid of the women' (*Power* Laundry 1942). After the war she sold up her business in London and built another industrial laundry, in Guernsey, to meet the needs of the burgeoning post-war tourism boom there.

Cultures and Contexts: How Do Women Engineers' Experiences Vary in Different Countries and Eras?

When advocating the desirability of a higher proportion of engineers being women in a country with a very low proportion compared to global figures, such as the United Kingdom, comments are sometimes made that 'women just don't want to do engineering and so don't apply for jobs/courses', it is worth considering the huge variability in the statistics. UNESCO's report (Huyer 2015) on women in the sciences assembled statistics for women in engineering globally which places the UK twelfth from the bottom out of eighty-three nations, with 11%. Malaysia tops the list but other nations in the >40% range are mainly former-Soviet nations. This demonstrates that women not only want to be engineers but are engineers in approaching equal numbers to men in many countries. The earliest comparator document is probably the conference report from the 1st International Conference of Women Engineers and Scientists (ICWES 1964), at which representatives from many countries supplied statistics of the proportion of engineers who were women. Only nineteen years after the turmoil of the Second World War, many nations were still in reconstruction but already it is possible to see the cultural differences embedding which will lead to the contrasts seen in the UNESCO report. Nation-building, post-colonial independence, the high status of engineering, all are already key features in the countries with higher take-up by women in the 1960s.

Another time-related context which is frequently raised relating to the history of women entering engineering (and other traditionally male-majority occupations), is wartime. Whilst it is true that many women had their first opportunities to enter engineering trades or professions during the two world wars, these should be regarded as periods of 'false entry', since they were almost entirely forced out of that work as soon as peace arrived. This was achieved either by legislation (HMSO 1919) or by media efforts to change the message from 'women doing heroic work in the munitions factories' to 'evil women stealing jobs from our brave boys back from the front'. In Europe and the USA women were first encouraged and soon forced to leave their engineering work and return to the domestic environment after both wars.

In many countries, organisations set up during the nineteenth-century professionalisation period were initially exclusively male, but in those nations where there were parallel organisations promoting women's rights—for example campaigns to get the vote—bridges were sometimes created between the two types of organisations and these ultimately helped women's access to professional engineering societies and other routes to professional recognition. This did not happen in France where the eighteenth-century origins of engineering professional identity meant it was heavily militarised. On the other hand, the militarised engineering systems in both France and Russia were partly responsible for the establishment of the unique women-only engineering schools in those countries. In contrast neither the UK nor the USA had such centralised state control and entry to engineering was initially strongly linked to class status but with some flexibility to admit men of talent and ambition even if of lower classes. Admission to professional engineering in the UK and USA, whether by university qualification or by pupillage with a senior man, thereby remained controlled by male protection, jurisdiction and privilege, even well into the twentieth century. In the USA, with its mass middle-class from the nineteenth century from which engineers were drawn, nevertheless fewer than half had any qualifications until the early twentieth century. The UK valued its multiplicity of small firms, craft skills and kinship networks, in a cultural framework that persists in excluding women and other 'Others' even now, especially in the less formal construction sector. In yet another system, both France and Sweden had highly centralised systems which would admit males, so as to groom them for a limited number of high-status government posts. In Sweden such engineers were literally seen as heroes of the nation's modernisation effort. Their training was densely timetabled such that the young men spent all their waking hours with their cohort of students (Canel et al. 2000) and were introduced to future employers as they progressed through the course.

Nazi Germany, despite its otherwise traditional view that women should spend their time on children, church and cooking, were in such need of technically skilled people to build up their military and its industrial supply chain, that the elite Kaiser Wilhelm Institutes were admitting women from 1911 onwards, with some 225 gaining Ph.D.s between 1911–45. Many overseas engineers visited or studied there too, including some British women, such as Hilda Lyon who was at Göttingen working for Professor Prandtl as the Nazis power rose in 1934. Many women who worked in the aeronautical and ballistics departments of the twenty Kaiser Wilhelm Institutes were recruited by the Allies at the war's end, most ending up in the UK or USA's weapons or aerospace programmes. One such woman was Dr. Johanna Weber (1910–2014), who was one of Prandtl's students at Göttingen when Lyon was there. During the war she worked on ballistics for weapons manufacturer Krupp and then returned to the *Aerodynamische Versuchsanstalt Göttingen* [Aerodynamics Research Institute in Göttingen]. At the end of the war, she was the only German woman scientist recruited (with many of her male colleagues) to work at the Royal Aircraft Establishment, where her expertise on wing shapes contributed to the design of the Victor bomber and the VC10 airliner (Green 2015).

In Germany, as in the UK, the defence industries were at the forefront of the development of aviation and later of electronics. These were both new fields of engineering which did not carry the traditions of the older fields (such as civil and mechanical engineering) and were more welcoming to women from the outset. Traditionalist structures still generally limited promotion prospects and women did not have the chance to rise to the tops of these industries until the late twentieth century, but nevertheless there were more women higher up in these fields than in other branches of engineering at comparable periods.

1950s: The Cold War—Peggy, Betty and Joan in the Defence Industries

Peggy Hodges OBE (1921–2008), Betty Laverick OBE (1925–2010) and Joan Lavender MBE (1928–2008) are examples of a generation of women whose talent and determination to push their way into engineering gave them successful careers in the emerging electronics and computers of the post-WW2 defence industries. Peggy and Joan both came from working-class backgrounds and were scholarship recipients at their schools. Whilst Peggy and Betty found their way to university degrees, Joan climbed the ladder from apprentice, via part time technical college (IMechE 1948) study to senior engineering roles.[3] This latter career path would be almost impossible today, but she rose to become the Computer Aided Design and Manufacture Controller, in British Aerospace's Air Weapons Division, helping design Blue Streak rockets and the Excalibur guided artillery shell.

Betty Laverick, on the other hand, came from a family of manufacturing chemists who encouraged her interest in science, but she too needed scholarships to get her to the best local grammar school, followed by a degree in physics and radio and a Ph.D. in electronics. Although she then married, Betty was lucky that the marriage bar had gone by the early 1950s and so she was able to work for Marconi Defence Systems Ltd. as a microwave engineer, working on guided weapons systems. She moved to Elliott Automation, rising to become the general manager of Elliott Automation Radar Systems (WES 1965, Vol. 9) where she was involved in the development of the airborne Early Warning System later known as the Nimrod.

Peggy Hodges' maths degree from Cambridge took her to work on airborne communications, the Instrument Landing System blind beacon landing equipment and missile projects such as the Red Dean and Sea Dart, which relied heavily on the systems assessments produced by Hodges and her team. She became an expert on simulation and systems, including assessments of random aberrations, types of dish stabilisation, target glint and sea reflection problems (WES 1971, Vol. 11). She progressed to become Systems Manager and then Project Manager of the Guided Weapons Project, was consulted by other laboratories and government departments and was sent on government missions to the USA. Among other projects, Hodges worked in the Underwater Weapons Division on trials planning and analysis for air-launched guided torpedoes, and later worked on simulation, identifying problems affecting guided weapons systems. Both Peggy and Betty were presidents of the Women's Engineering Society (in 1972 and 1968 respectively). Although this period was generally very difficult for women wishing to be engineers, they were able to open doors for others.

Time, as well as place, can shape cultures: *when* changes occur can be significant in affecting what those changes are. In the case of industrialisation this has varied widely from nation to nation. In those in the earliest waves of the industrial revolution, such as Britain, France, Germany and the USA, the professionalisation of engineering and its associated institutions, societies etc., are of older vintage than, for example, Turkey, Syria and Israel and other Middle and Far Eastern countries, which did not industrialise until the mid-twentieth century, as they emerged as new nations. Zengin (2010) observed that this later industrialisation, coupled with revolutionary and modernising attitudes under Kemal Attaturk (1881–1938), combined to place Turkish women in engineering in higher percentages than most other far more 'highly developed' nations. In Austria, whilst other European nations started grudgingly to admit women to engineering university courses, it took the complete collapse of their political system before women could enter such programmes (Canel et al. 2000).

1960–80s: Modern Structures—Marjem Chatterton

Cultural expectations of family and surrounding society have a strong influence on young people; these expectations often survive at least one generation when imported from former environments when people migrate to areas where there are different attitudes and gendered expectations.

Although Marjem (née Marynia Znamirowska) B.Sc., FIStructE (1916–2010) qualified before the Second World War, and hence seems to have been one of the pioneers of an earlier age, her career and professional field were solidly of the mid- to late-twentieth century. Other than aerospace, the modern skyscraper is probably one of the icons of the period and she made herself their expert in the countries in which she worked. Born in Warsaw, Poland, in September 1916, she grew up in a large Orthodox Jewish family, where girls' education was considered important (Flood 2010). She had planned to return to Poland to study chemical engineering at the University of Warsaw but, in 1934, it was clear that the situation for Poland's Jews was worsening, so she enrolled at the Hebrew Technical Institute in Haifa, in what was then called Palestine. Known as the 'Technion', one of her aunts—Rachel Shalon, the first female engineer in the country—was a faculty assistant there. Graduating in civil engineering in 1939, with the first distinction in engineering awarded by the Technion, Znamirowska took a job that had been offered to her by a faculty member, Josef Edelman, who ran the Technical Office of the Collective Settlements Association, building some of the country's largest kibbutzim.

After the war, Mrs. Chatterton—now married and with a young daughter—decided with her British husband to leave Palestine. She emigrated to Southern Rhodesia in 1947 and found a job as a reinforced concrete designer within two days of arriving in the country. Her experience with reinforced concrete structures in Palestine was particularly useful, as at that time it was nearly impossible to get hold of heavy steel sections locally. Specialising in multi-storey structures, her buildings still define the skyline of Zimbabwe's capital Harare, with her last major project—the Reserve Bank—being the tallest office building in the country. In 1969 she established her own consulting firm, M. Chatterton and Partners. As well as prestigious urban skyscrapers, she designed many industrial facilities for the cotton, fertiliser and sugar industries. In 1976, the worsening political situation in Zimbabwe led to her moving to Leeds University as a lecturer, but in 1984 she returned to consultancy in Zimbabwe, also teaching at the national university. Zimbabwe's independence in 1980 heralded a flurry of work as sanctions were lifted and investment poured in and she gained a lot of work before her retirement and return to the UK in 1999. She was the first woman fellow of the Institute of Structural Engineers.

Some Thoughts on Researching Women Engineers

Because so few women engineers' histories have been known, the principal approach to date has been biographical, making use of family history or local history types of sources. This approach typically examines its subject's family background and education and often focuses strongly on her early life and how she became an engineer against the social cultural norms of her time and place. Many Wikipedia entries are of this sort and have often been the result of Wikithons hosted by feminist organisations aiming to bring more women's stories into Wikipedia.

Another approach could be to look at an entire branch of engineering and consider women's involvement over time. Without falling into a morass of statistics, this approach can be very helpful in demonstrating the breadth of women's involvement, say when writing for or presenting to organisations associated with that branch of engineering. Since the more modern fields of engineering, such as aeronautics, computers or bioengineering, were often less bound up with convention when they were established, both employers and professional bodies were more open to women's involvement from the outset. However, in some cases this apparent equality did not last, and male domination could assert itself after some years, in which case the historic survey of women's involvement can help to highlight how such cultures change.

A further way of looking into women in engineering is to delve into their technical achievements. Criticism is sometimes levelled at feminist historical accounts of women in engineering on the grounds that women's technical achievements are not as significant as those of men. Whilst this may or may not be true, as measured against some empirical objective datum of changes to the world, there is often so much hidden from view about women's achievements that who can be sure? Women who were well known in their own lifetimes for their technical contributions sometimes then fell from view and the effort to bring such stories to light can be rewarding on many levels. Their home or work localities, today's workers in their fields or national institutions are often delighted to learn of and make use of the 'new' story about an interesting woman's work, resulting in commemorations, publicity etc. Other women's work may have been subsumed within the male colleague's or superior's public recognition, as infamously happened in the case of scientist Rosalind Franklin's DNA structure discoveries. As discussed above, the typical model of a single (male) project leader being credited with all the success, regardless of which of the project's workers contributed what, is the easiest way for the media to tell a story.

To bring technical achievements of named individual women into the foreground is certainly not easy and, in some cases, may no longer be possible. Even major business empires often leave few or no archive resources behind after their demise. As we have seen, this approach is often impeded by the secret nature of work in the defence industries in which so many mid-twentieth century women engineers congregated. Many engineers work on government contracts under conditions of legal secrecy, having signed the Official Secrets Act, and neither the individuals nor their employing organisations can reveal what they worked on until that security barrier is officially lifted. This is generally at least thirty years after the event, by which time much in the way of memories of living people, or hard copy archival material may have vanished anyway.

However, if it is thought to be desirable to have a fully diverse workforce at all levels in fields previously traditionally associated with one gender (or ethnicity etc.), then to attract new demographics to that occupation requires that the target audience sees that others like them do and have previously done such work. One of the ways to help girls and women to make choices that can lead to engineering careers is for them to see that women have done so in the past and present. Perhaps surprisingly, it is still the case that girls and young women—on announcing an interest in engineering careers—are still being told by those around them that this is an unusual choice for a girl. It is therefore also essential that the general public around them knows that women do and have done engineering. Bringing stories of women in engineering out into the mainstream is one of the tools that can overcome gendered assumptions. Historic examples have their place in that 'toolbox', alongside current young role models.

Notes

1. Patent no. 3405, 1811, for 'a new mode of constructing and erecting bridges and rail-roads without arches or sterlings, whereby the danger of their being washed away by floods is avoided'.
2. The biographical notes of individual women typically involve multiple sources to piece together their stories, so only the principal sources have been given here, for brevity.
3. Information from an unpublished obituary booklet for her funeral, by her friend John Hill. Joan Lavender's application form for membership of the Institution of Mechanical Engineers can be found in their archives.

References

Bagilhole, Barbara. 2014. "Equality and Opportunity in Construction." In *Building the Future: Women in Construction*, ed. Meg Munn. The Smith Institute. Chapter 2. https://www.women-into-construction.org/wp-content/uploads/2019/08/building-the-future-women-in-construction.pdf. Accessed 29 December 2020.

Baker, Nina. 2009. "Early Women Engineering Graduates from Scottish Universities." *Women's History Magazine* 60: 21–30.

Canel, Annie, Ruth Oldenzeil, and Karin Zachmann, eds. 2000. "Crossing Boundaries, Building Bridges: Comparing the History of Women Engineers, 1870s–1990s." *Studies in the History of Science, Technology and Medicine* 12. Australia: Harwood Academic Publishers.

Clarsen, Georgine. 2003. "'A Fine University for Women Engineers': A Scottish Munitions Factory in World War I." *Women's History Review* 12, no. 3: 333–56.

Clarsen, Georgine. 2004. *Pullinger [married name Martin], Dorothée Aurélie Marianne (1894–1986)*. Oxford Dictionary of National Biography. https://doi.org/10.1093/ref:odnb/71798. Accessed 29 December 2020.

Dresser, Madge. 2016. "Guppy [née Beach; other married name Coote], Sarah." *Oxford Dictionary of National Biography*. https://doi.org/10.1093/ref:odnb/109112. Accessed 29 December 2020.

Drummond, Cherry. 1994. *The Remarkable Life of Victoria Drummond—Marine Engineer*. London: Institute of Marine Engineers.

Enciclopedia României. 2009. *Elisa Leonida Zamfirescu*. http://enciclopediaromaniei.ro/wiki/Elisa_Leonida_Zamfirescu. Accessed 29 December 2020.

Flood, Z. 2010. "Obituary: Marynia Chatterton." *The Guardian*, 22 March.

Freudenberg, Matthew. 2004. *Negative Gravity, the Life of Beatrice Shilling*. Creech St. Michael: Charlton Publications.

Gardiner, Jane. 1990. *Memoirs of a Woman Engineer*. Kibworth: The Book Guild Ltd.

Google Books Ngram Viewer (analysis of use of certain words in books published between set dates, here 'engineer' between 1700–2008). https://books.google.com/ngrams/graph?content=engineer&case_insensitive=on&year_start=1700&year_end=2008&corpus=15&smoothing=7&share=&direct_url=t4%3B%2Cengineer%3B%2Cc0%3B%2Cs0%3B%3Bengineer%3B%2Cc0%3B%3BEngineer%3B%2Cc0%3B%3BENGINEER%3B%2Cc0. Accessed 29 December 2020.

Green, John. 2015. "Obituary: Dr Johanna Weber." *Royal Aeronautical Society*, 12 January 2015. https://www.aerosociety.com/news/obituary-dr-johanna-weber/. Accessed 29 December 2020.

HMSO. 1854. *Census of Great Britain, 1851. Education. England and Wales. Report and Tables*. HMSO, p. xxvii, Table 4.

HMSO. 1881. *Census of Scotland (1881) Report. Edinburgh*. HMSO. Appendix, p. xxxvii, Tables xxxix and xl.

HMSO. 1919. *Restoration of Pre-war Practices Act 1919*.

Huyer, Sophia. 2015. "Is the Gender Gap Narrowing in Science and Engineering?" *UNESCO Science Report Towards 2030. United Nations Educational, Scientific and Cultural Organization*. Second revised edition 2016, Section 03: 84–104. http://www.unesco.org/new/en/media-services/single-view/news/women_still_a_minority_in_engineering_and_computer_science/. Accessed 29 December 2020.

ICWES. 1964. *Proceedings of the First International Conference of Women Engineers and Scientists, Focus for the Future, Developing Engineering and Scientific Talent*. The Society of Women Engineers & National Science Foundation. June 15–21, 1964, section V-1.

Irish Architectural Archive. n.d. Perry, Alice Jacqueline. *Dictionary of Irish Architects 1720–1940*. https://www.dia.ie/architects/view/4332/PERRY-ALICEJACQUELINE. Accessed 29 December 2020.

Lerner, Gerda. 1986. *Women & History, vol. 1. The Creation of Patriarchy*. New York: Oxford University Press.

Lerner, Gerda. 1993. *Women & History, vol. 2. The Creation of Feminist Consciousness: From the Middle Ages to 1870*. New York: Oxford University Press.

Lyon, Hilda M. 1944. "Adventures in Aeronautical Design and Research." *The Woman Engineer* 5: 291–95.

Malcolmson, Patricia E. 1986. *English Laundresses: A Social History, 1850–1930*. Urbana: University of Illinois Press.

Milligan, William. 1877. "The Higher Education of Women." *Aberdeen Press and Journal*, Saturday, 3 November: 6.

Morgan, Bea. 2015. "A History of Women in the UK Civil Service. For Home Office Women." https://www.civilservant.org.uk/women-history.html. Accessed 11 March 2019.

Power Laundry. 1942. "She Will Advise Government on Women War Workers." *Power Laundry*, July: 67.

Smiles, Samuel. 1865. *Lives of Boulton and Watt*. London: John Murray.

Smiles, Samuel. 1879. *Lives of the Engineers, the Locomotive, George and Robert Stephenson*. London: John Murray.

Smith, Catherine, and Cynthia Greig. 2003. *Women in Pants: Manly Maidens, Cowgirls, and Other Renegades*. New York: Harry N. Abrams.

Stanley, Autumn. 2010. "Holmes, Verena Winifred." *Oxford Dictionary of National Biography*. https://doi.org/10.1093/ref:odnb/66362. Accessed 29 December 2020.

Stanley, Jo. 2016. *From Cabin 'Boys' to Captains. 250 Years of Women at Sea*. Stroud: The History Press.

University of Liverpool. 1912. University Calendars PUB/1/1/3/18, 1924–25.

Weingardt, R.G. 2014. "The First Lady of Structural Engineering." *Civil + Structural Engineer Magazine*, 19 February 2014. https://csengineermag.com/article/the-first-lady-of-structural-engineering/. Accessed 21 April 2019.

WES. 1919–. *The Woman Engineer Journal*. Women's Engineering Society. 1919–present [including downloadable names index]. https://www.theiet.org/publishing/library-archives/the-iet-archives/online-exhibitions/women-and-engineering/the-woman-engineer-journal/. Accessed 29 December 2020.

WES. 1965. Mrs Elizabeth Laverick. *The Woman Engineer Journal* 9, no. 17: 13–14. Women's Engineering Society.

WES. 1971. Miss Peggy Hodges. *The Woman Engineer Journal* 11, no. 3: 2–3. Women's Engineering Society.

WFL. 1909–33. *The Vote, the Organ of the Women's Freedom League*. London. https://www.britishnewspaperarchive.co.uk/titles/id/vote. Accessed 29 December 2020.

Zengin, Berna. 2010. *Women Engineers in Turkey. Gender, Technology, Education and Professional Life*. Saarbrücken: LAP Lambert Academic Publishing.

CHAPTER 28

Women and Surgery After the Great War

Claire Brock

While the Great War provided unprecedented opportunities for women in surgery both at home and near the front line, the years after the Armistice have always been seen by historians as a time of stasis, retrogression even, especially in women's surgical progress. This was even starker considering the developments in women's social and political position at this point; when women over twenty-one were enfranchised in 1928, the co-educational experiment of the war years in medical schools was entering its death throes, declared a spectacular failure by detractors. Those who examined the situation after the war, however, remarked differently upon surgical prospects, current and future. Many developments came in the unsurprising fields of gynaecology and obstetrics, but this should not detract from the fact that others gained recognition in, for example, orthopaedics or ophthalmology. Women adapted to the post-war world by taking their surgical skills in new directions, rather than giving up completely. Although they faced significant challenges in this period, medical women were an acknowledged fact. While door-after-door was slammed in their faces in the 1920s and 1930s, other portals opened for women surgeons to explore. This chapter will examine the surgical possibilities which were available to women in the 1920s and 1930s, through case studies of Louisa Martindale (1872–1966), Louise McIlroy (1874–1968), and Maud Forrester-Brown (1885–1970). By considering the avenues open to women

C. Brock (✉)
School of Arts, University of Leicester, Leicester, UK
e-mail: cb178@leicester.ac.uk

© The Author(s), under exclusive license to Springer Nature Switzerland AG 2022
C. G. Jones et al. (eds.), *The Palgrave Handbook of Women and Science since 1660*, https://doi.org/10.1007/978-3-030-78973-2_28

surgeons in the interwar decade, as well as the changing response to them from the lay press and the general public, a new way of thinking about the 1920s and 1930s will be provided which acknowledges the vital importance of a specifically female surgical tradition emerging in this period from which future generations could build. Rather than exploring a gulf, always seen as representative of these years, this chapter traces the emergence of a narrative which learned carefully from the past, while striding strongly into the future, taking opportunities as they came.

'Very little changed after the war: post-war employment prospects for women in medicine were essentially the same as in 1914', concluded Jennian Geddes in her assessment of the fortunes of the Women's Hospital Corps after the closure of the Endell Street Military Hospital in 1919. Marriage, retirement, or employment in a women's hospital was the fate of the majority: 'none went into surgery, the area in which many had the greatest expertise' (2006, pp. 115–16). Similarly, Mary Ann Elston has remarked that for those who experienced co-education from 1916 the move towards equality and the opportunities in public health and general practice meant that the separatism of the female-run hospitals was increasingly anachronistic (2001, pp. 96–97). This sense of a gap between women trained in the interwar years and their predecessors, however, was less apparent for those aspiring to a surgical career. In their global surgical journeying, for example, Louisa Aldrich-Blake and Louisa Martindale had been educating themselves unofficially alongside men since before the Great War, bringing their findings to bear upon their own practice. Far from exacerbating the differences between the first two generations of women surgeons and the next, the 1920s and 1930s witnessed the consolidation of a specifically female surgical tradition. This operated either within or without a hospital run by women and drew both upon the surgical skills of female practitioners well practised in the discipline itself. Ways of practising were passed on through observation, but also through public and professional stance. More and more women sought the ultimate accolade, Fellowship of the Royal College of Surgeons, England, to bolster their professional prospects. By 1923, of 1736 Fellows, ten were women; although only 0.58 per cent of the total, their number was growing slowly but steadily after first being admitted in 1911 (*Calendar of the Royal College of Surgeons* 1923, p. 117). Indeed, twenty-three women became Fellows between 1919 and 1930; another thirty joined their ranks in the 1930s (Ghilchik 2011). The female surgical tradition taught the means by which women could operate in the theatre, but also in an occasionally hostile society. As such, the interwar woman surgeons learned from the past, engaged actively with current opportunities, while always moving towards the future.

The Post-War World for Medical Women

It is important to acknowledge that the period saw testing times for the medical profession as a whole; for example, fears of overcrowding, largely due to the influx of students to the medical schools directly after the end of the conflict, dogged the twenties. Such changes for medical women were seen as even more rapid and profound since the war years, and encouraged far greater competition among themselves. Women were blamed, indeed, both for contributing to an increase in numbers and wasting opportunities through marriage and maternity if they qualified. While detractors criticised the excess, supporters bemoaned the lack. There were simultaneously too many and not enough medical women. The spatial metaphors throughout the 1920s and 1930s were dizzying, their dimensions shifted from miniscule to vast depending on the perspective of the writer or speaker. Arguments were essentially contradictory, therefore: women were both without practice and patients and robbed medical men, their wives, and their children, of their daily bread (Chadburn 1930, p. 108). In October 1923, the surgeon Maud Chadburn, in her Presidential Address to the London Association of the Medical Women's Federation, provided the succinct statement that: 'We shall always be criticised as doctors, as women, and as women doctors'. She continued: 'perhaps there is some good in this, but it is tiresome that people won't let us alone' (1923, p. 178). Half a century of fending off continual criticism had led, Chadburn concluded, to thickened skins, but also unwelcome distraction from quotidian practice.

Chadburn's address was a partial response to the ill-advised choice of Sir Humphrey Rolleston, President of the Royal College of Physicians, to open the academic year at the London (Royal Free Hospital) School of Medicine for Women (LSMW). In the light of the reaction to his speech, it was an evidently uncomfortable and hostile occasion for speaker and audience alike. Rolleston's controversial topic of 'The Problem of Success for Medical Women' elicited an enormous press response to the views of this famous medical man. While commending women as students, he derided their wasting of careers through marriage and subsequent retirement, and their lack of capacity as original researchers. Indeed, as the *Daily Express* proclaimed: 'Women As Surgeons. Not So Good As Man. Less Original' ("Women as Surgeons" 1923, n.p.). Another simply summed up the profession as 'A Man's Business', whereby the 'undoubted fact' of women's inferiority as surgeons was down to their impatience and man's greater capacity for taking pains in his practice ("A Man's Business" 1923, n.p.). The riskiness of this statement in the 1920s was made clear by the writer's willingness to emphasise in exclamatory fashion the fact that these were the views of a male medical friend and not, presumably, her own. Key to the reaction to Rolleston's speech, however, was that, even though he did not discuss women in surgery at all, this was not how the lay press perceived it.

The abilities or otherwise of women in the operating theatre had been a dominant theme since women had entered the medical profession in the mid-nineteenth century (see Brock 2017). That arguments were being formed around female surgical skills in the interwar years shows that surgery was still shorthand for complexity. What differed, however, was that women had proved themselves as surgeons during the Great War; such derogatory comments were now attacked more than they were produced. Hence, of course, the defensive timidity shown in the 'man's business' vignette. This is not to state that doubt about capabilities did not linger, but rather that it was more vigorously set upon as retrograde. Rolleston's comments came at a time when those medical schools which had admitted women as a financial necessity during wartime were finding excuses to close the doors. Given the occasional attack upon female competence, the lay press was surprisingly supportive of women's surgical abilities in the 1920s. Instead, trenchant criticism was directed at the medical schools themselves. The London withdrew its offer in 1922, citing initially the indelicacy of teaching 'unpleasant' subjects, including surgery, to mixed classes. This, as the *Observer* put it, 'singularly unfortunate' phrase led to a backlash against the hospital's dated and unscientific approach and a mocking of its weak and blushing male students ("Women Medicals" 1922, n.p.). Such was the response that the Chairman of the hospital, Lord Knutsford, was forced into a volte face involving a public defence of women's surgical abilities, if not a reversal of the decision. Before 'the storm had calmed', Knutsford proclaimed in the *Times* that he was 'strongly in favour of women being doctors, converted to this opinion by what [he] saw of Dr Flora Murray's and Miss Garrett Anderson's War Hospital in Endell-street': surgeons who 'made history by their pioneer work' ("Letter to the Editor" 1922, p. 8). In doing so, Knutsford repelled criticism through praising the very surgical skills his hospital no longer sought to teach to women students. A moral victory even if not an actual one.

If male medical students blanched at women's presence, they were also mocked for their chivalrous sensitivity at operations. The problem with female students, concluded the metropolitan medical schools, was that, ironically, they received the best vantage point when observing surgical procedures or dissections, allowing them unfair advantages over their accommodating male compatriots. Women had been given the front row since Elizabeth Garrett Anderson eagerly stepped forward to watch her first operation at the Middlesex Hospital in 1860, but this precedence was now seen as disadvantageous to those who stood behind (Letter from Elizabeth Garrett Anderson to Emily Davies 1860). 'Anxiety about competition' was at the root of this, noted the *Daily Mail*, citing the professional and lay 'ridicule' stemming from the London's behaviour ("Sex Rivalry in Medicine. Case Against Women Doctors" 1922, n.p.). The unfairness of a chivalric 'poor view' at operations was still being referred to five years later, when the doors were closed to women at all medical schools except University College Hospital, King's College Hospital and, of course, the LSMW (*Manchester Guardian* 1928, n.p.). Such 'unsporting' female dominance and sense of masculine delicacy received as short a shrift as the argument about women students contributing to sporting losses at co-educational schools. The latter was received as without a 'whit' of 'logic': an 'absurd contention'. As the *Daily Telegraph* reminded its readers: 'The fact that a surgeon excites our admiration by his skill on the football field is no guarantee whatever that he will show an equal skill in the operating theatre' ("Women Medical Students" 1928, n.p.). In the year of 'votes for flappers', the *Nation and Athenaeum* summed up the whole action of the London hospitals as 'anti-feminist prejudice, but of also anti-social conduct' ("Women, the Vote and the Hospitals" 1928, p. 958). 'Miss 1928', claimed Valentine Williams in the *Sunday Graphic*, was being treated appallingly by a 'positively pathetic' attitude: 'It is pathetic because it is old-fashioned, like the spectacle of a solitary hansom nowadays trying to pick up a fare among the whirling taxis of the Strand, and because it is foredoomed, like any other endeavour to stay the progress of humanity, to failure' ("Professions for Men Only?" 1928, n.p.). If medical women were unable to keep what had always been labelled as a 'temporary measure' of co-education going in most instances, they must have been heartened to see the press support, which lambasted the lumbering, archaic, and unsupportable in modern Britain.

'Picking up fares' depended, of course, on the market after qualification. While medical women were a fact, it did not necessarily mean everyone had to revel in this. The most virulent attacks often came from women themselves. In among discussions of the value of co-education and the teamwork seen to characterise post-war practice, some women turned on each other in public. An 'Ex-Surgical Sister' presented 'The Case Against Women Doctors', angered by the assumption that what was inappropriate for female doctors was fine for nurses. Bitter invective ran through this piece, which destroyed the work of women surgeons who, while offering respectable qualifications, were unreliable as far as their skills were concerned: 'They themselves are nervous, uneasy. They vex the patient, even when he or she is about to be placed under anaesthetics, with needless questions, so that when the unconscious patient is on the operating table the nerves cause involuntary movements of the limbs. And they do not seem to know how to delegate authority to the surgical sister, which is an essential power of a good surgeon. They do not seem capable of the calm, almost the callousness which a surgeon should possess' ("The Case Against Women Doctors" 1922, n.p.). So worrisome was the behaviour of female surgeons, argued the 'Ex-Surgical Sister', that the patient reacted nervously to the prospect of being operated upon by them, even when unconscious. Without confidence in themselves, incapable of transferring any authority to their team, at one level terrified, at another individualist, women were a liability in the operating theatre. No one wanted them, not even patients professing to be 'Suffragists', claimed the 'Ex-Surgical Sister'. They instead sought male doctors and especially the skills of a male surgeon rather than submit themselves to the terrors of a procedure by a member of their own sex.

Women patients added another level of scrutiny. While these can be viewed comically in the 'Overheard' at hospital section of the *LSMW Magazine*, features in the lay press confirmed that not every woman was happy to be seen by a surgeon of her own sex. As a 'Stout Old Lady' confided to Sister in a typical 'Overheard' feature of 1923: '"Well, of course, the Lady Doctors, they be very nice and all that, but Sister do you really think they are as *deficient* as the Gentleman Doctors?" (Reassurance from Sister)' ("Overheard" 1923, p. 134). An unidentified official from a London medical school remarked categorically in the same year that 'It is very seldom that women make really good surgeons': 'female patients won't employ them. If a woman is ill, unless it is something very trivial or very intimate, she prefers to go to a man doctor'. They simply have 'more confidence' in them, based upon centuries of practise ("Male Doctors Preferred" 1923, n.p.). Their 'deficiencies', as the Stout Old Lady might have had it, paled into insignificance when positioned next to male superiority and skill. Yet, criticism of women's real benefit to their own sex was regularly placed next to resounding confidence in their judgement. Even the anonymous official left the reader with the conclusion that there is no possibility of women abandoning the profession and that 'a few outstanding women' will 'cope with the men in surgery and research work'. The *Hospital*, in July 1920, published two contributions from women patients with experience of women surgeons. The first, a Scottish lady, deplored any attack on 'the wise and intimate counsel, and the cool confident skill' of her own 'most valuable' woman surgeon ("The Public Well-Being" 1920, p. 459). The other, from Somersetshire, took a different stance. Seriously ill, while away from her home town, she was forced to call in a woman, 'who lost her head and nearly lost me my life'; she was without 'temperament or nerve'. The *Hospital* concluded its editorial, however, by blaming the second patient, who, they suspected, was the one full of nerves and temperament. 'Doctors of the future' should be judged 'more and more on the grounds of skill, judgment and personality', irrespective of sex (1920, p. 459). Another Somerset source noted in 1924 that her friends 'having once trusted a woman surgeon, tell me they would not change for any consideration, and that the operations they have performed for them have proved most wonderfully effective and skilful' ("Our Special Lady Correspondent" 1924, n.p.). Nearly a decade later, in September 1932, when John England asked, 'Can You Trust a Woman Doctor?', the answer was a resounding 'yes' (1932, n.p.). Women surgeons no longer had to rush into print to defend themselves; very frequently their gratified patients did it for them in the interwar years.

'One of the Most Skilful Surgeons in the Country': Louise McIlroy

John England's article provided examples of remarkably successful medical women, naming '[o]ne of the most skilful surgeons in the country' as Professor Dame Louise McIlroy. When all the controversy about co-education was occurring in the decade after the Great War, Louise McIlroy's elevation, in 1921, to a University of London Chair in Obstetrics and Gynaecology was treated with fascination in the lay press. The majority of articles focussed upon three main things: firstly, her youth; secondly, the fact that the post was open to both sexes; and thirdly, but most prominently, her salary. Indeed, the latter featured in every headline proclaiming her achievement. The ways in which she was publicly discussed give an insight into the post-war understanding of the woman surgeon.

The University of London introduced the clinical unit as a new system of medical education in thes early 1920, indicating the increasing importance of specialisation. That the chair in gynaecology and obstetrics was awarded to a woman at the hospital which had supported the medical education of women for nearly fifty years was a triumph: 'a telling confirmation', as the *Manchester Guardian* put it, 'of the position which women have now for themselves' in the field ("Education of Women Doctors" 1921, n.p.). The press made much of McIlroy's skills, citing her education (she was a Glasgow graduate, not a LSMW one), as well as her Scottish Women's Hospitals' service as Chief Surgeon of the Girton and Newnham Unit, during the war and her stint in Constantinople in charge of the Royal Army Medical Corps' 82nd General Hospital. McIlroy was labelled a 'War Heroine', more than deserving of her post ("£2,000 a Year Woman Professor" 1921, n.p.). However, what really fascinated was, as a *Daily Express* headline succinctly put it, '£2,000 a Year for Woman Doctor. Men Beaten in Open Competition' (1921, n.p.). The *Evening News* addressed McIlroy as 'the £2,000 a year girl', full of 'youth and grace', and a 'brilliant surgeon in a becoming hat and veil' ("London's £2,000-A-Year Girl Professor" 1921, n.p.). Although on the surface patronising, the comments make McIlroy's achievements all the more startling. She was forty-six, so hardly a girl, but her smart and attractive appearance smoothed over the enormous salary and the sparkling reputation was bolstered even further by her apparent youth. Not only could women compete with men as students; they could also do so at the top of the profession, and the financial reward for this added another layer of importance. For McIlroy, in an interview with the *Manchester Guardian*, her appointment was key to the future development of a field in which women were underrepresented. Rather than viewing this post as a demotion from the heights of the war years, McIlroy saw this as a 'new opportunity' for 'others to specialise in these subjects' ("Professor McIlroy in Her New Post" 1921, n.p.). In addition to widening prospects in an area in which women had much to contribute, McIlroy believed there was a 'great deal to be done' for women's health in general.

Despite her extremely active war service with the Scottish Women's Hospitals, Louise McIlroy did not lament peacetime conditions nor perceive a gaping loss in her return to work in obstetrics and gynaecology (see Brock 2017). In her farewell address to the Medical Society at the Royal Free Hospital in 1934, McIlroy enthused her students: 'With luck you may become a H[ouse].P[hysician]. or H[ouse].S[urgeon]. in your own hospital. Even with greater luck you may become an obstetric H.S.!' (McIlroy 1934b, p. 102). Indeed, McIlroy saw the 'exhaustive research' and 'the laborious work' for which she was famed as key to improving the health and prospects of mothers and children (McIlroy 1911, p. 227). A prolific researcher, before the war McIlroy was Senior Assistant to the Muirhead Professor of Obstetrics and Gynaecology at the University of Glasgow and Assistant Gynaecological Surgeon at the Glasgow Royal Infirmary. She published widely, before and after the war, very often with an eye on the ways in which surgery had an impact on social concerns. She wrote papers, for example, on problems in the treatment of venereal disease (McIlroy 1914), toxaemia in pregnancy (McIlroy 1922), maternal mortality (McIlroy 1930), caesarean section (McIlroy 1932), and contraception.[1] Such problems plagued the interwar years particularly, when maternal mortality actually increased, and McIlroy's Unit at the Royal Free Hospital researched conditions closely which led to unnecessary deaths (see Loudon 1992). In the light of the recent Government Departmental Committee on Maternal Mortality, McIlroy proudly reported the timeliness of her then decade-long department: 'The chief aim of the teaching in the Unit is the study of normal conditions and the prevention and diagnosis of complications. The adaptation of institutional to domestic conditions is of the utmost importance to students in teaching them self-reliance. Successful midwifery means the expenditure of time and infinite patience, with accuracy of diagnosis. It is the teacher of the undergraduate who is responsible for the future success of a national maternity service' (McIlroy 1931, p. 680). For McIlroy, transmitting skills carefully and precisely to the next generation had the potential to save countless lives, and she experimented with novel teaching methods to do so, including the use of cinematic film. When she left the Unit for consulting practice in 1934, Louise McIlroy drew repeatedly on the importance of the team and their 'loyalty and reward' 'splendid loyalty' and 'loyal co-operation' (*LSMW Magazine* 1934a, p. 138). In this evidently tightly knit female-only Unit, reputation had been built on sympathetic understanding and sharing of expertise.

The Benefits of Global Experience: Louisa Martindale

Shared experience characterised the career of Louisa Martindale. Martindale felt keenly 'the great necessity of international friendship' and her global outlook fed directly and fruitfully into her own surgical experience from the very outset of her career (Martindale 1924, p. 70). She was a permanent 'post-graduate', always eager to see rather than read about new developments and changing techniques. With Louisa Aldrich-Blake, Martindale travelled to watch the best surgeons in the world, male or female, and incorporated knowledge into her own practice. In her simply, but emphatically, titled autobiography, *A Woman Surgeon*, published in 1951, Martindale reeled off her international influences, a veritable pantheon of early twentieth-century scientists and surgeons: from America, Sir William Osler, Howard Kelly, Harvey Cushing, the Mayo brothers, and Crile; in Europe, Schauta, Hitschmann, Wertheim, Doederlein, Krönig, Gauss, Wintz, Holfelder, Olshausen, Heyman, Forsell, Regaud, Beclère, Madame Curie, Stoeckel, and Veit; and in England, Lord Moynihan, Professor Grey Turner, and above all Dame Louisa Aldrich-Blake (p. 10). Vitally, Martindale learned from everyone, whatever their specialty or background; indeed, whatever their sex. That she then chose to apply her knowledge to the benefit of women patients pointed to a wide-ranging experience filtered for a specific goal. The *British Medical Journal* (*BMJ*) stated that as Secretary and then President of the Medical Women's Federation, and later the Medical Women's International Association, her 'social vision and professional ability' made a distinct impact on the position of women in medicine ("Obituary" 1966, p. 548). Although she operated in a series of women-run hospitals, there was no narrowness of outlook in Louisa Martindale's career and she transmitted this broad surgical understanding through her practice as well as her written works.

Martindale is best known for her pioneering therapeutic work, using x-rays in the treatment of cancer (Moscucci 2007, 2017). She also employed her apparatus, however, in non-malignant uterine conditions and after-treatment for those who had undergone surgery for cancer of the breast. Unsurprisingly, she observed applications of these methods while on her 'busman's holidays'. Inspired to buy her own equipment, Martindale purchased her first machine, fortuitously, on a trip to Germany in 1913. From that point she began to treat conditions such as fibromyomata of the uterus, climacteric haemorrhage, and carcinoma of the uterus and ovaries by intensive x-ray therapy. However, as she was keen to point out, 'this was only one method of treatment' (Martindale 1925, pp. 690–719). Martindale favoured a combined approach: operable cancer cases were treated with x-rays after they had been operated upon; while inoperable cases received deep therapy or radium, or both. In doing so, Louisa Martindale explored and experimented with the ways in which surgical procedures could be successfully enhanced through close association with contemporary scientific developments to tackle especially one of the most feared of twentieth-century diseases. Early and adequate 'post-operative radiotherapy' in breast cancer, for example, could improve the results of surgery (Martindale 1931, p. 229). She also developed a clear system of surgical note-taking, inspired by her American visits. Card indexes were utilised with each patient entered on four different cards, with name, disease, operation, and name of referring doctor. She also kept a large folder with full notes and correspondence in addition to files for pathological slides. Operative technique and the end results were recorded. All of these approaches provided an excellent model for any surgeon, and facilitated the unfailingly clear and logical presentation of Martindale's published papers (Martindale 1951, pp. 187–88).[2]

From the beginning of the establishment of the New Sussex Hospital for Women in Brighton, Martindale was determined to incorporate paying patients into her institution, fees ranging from 15 to 84 shillings a week ("The Newest Hospital for Women" 1919, p. 203). A large proportion of her cases were professional women; for example, headmistresses, doctors, or nurses, which belied the sense that the more educated would not choose their own sex for surgery. Not only did they do so, but they were willing to pay for consultation and treatment too. Although accounts in the lay press, as we have seen, joked that medical women were only visited by their own sex for the trivial or the embarrassing, Martindale expressed clear satisfaction when she gained the trust of her patients. Throughout her long career, as she recorded in her autobiography, few were 'unco-operative' and only occasionally were postponements of surgery required. No necessary operation, however, remained incomplete (1951, p. 237). Martindale relates the narrative of her companion, Ismay Fitzgerald's, surgery which took place at her own home with Louisa Aldrich-Blake operating. One of Fitzgerald's nurses 'cheered her up by telling her that nothing would induce her to have a woman surgeon, but she soon became a convert and, like other converts, a very ardent one'. Indeed, the same nurse later employed Martindale to operate upon her (p. 236). After her death, an obituary in the *BMJ* provided reasons for Martindale's excellent surgical outcomes ("Obituary" 1966, p. 548). As a surgeon, her technique was deft and rapid; she stood strain remarkably. In addition to this, her 'great success' was her understanding of human nature, which was translated into the special appeal of the New Sussex Hospital, with its comfortable private wards for professional women. Additionally, preparation and after-care of patients was key to an easier return to health and activity. As a surgeon, Louisa Martindale considered far more than simply the operation itself. From consultation to discharge, detail and order were vital to success.

Hospital architecture also fascinated Martindale, and this led to a special surgical ward being built at the New Sussex Hospital in 1931. This development was not without opposition, however, and Martindale described the difficulties she encountered from members of the hospital board, as well as nursing staff, when trying to move away from 'old accepted standards of ward planning' (1951, p. 132). To develop the new arrangement, Martindale had requested that the architects visit some of the newest hospitals both in Sweden and the United States. The surgical ward was singled out as a special feature of the hospital's new building:

> In the usual hospital ward the beds are placed at right angles to the walls so that the patient is facing the light from the opposite windows and has the radiator near the head instead of the foot of the bed. On the new principle the beds are placed in pairs parallel to the wall, with a window to each bed, separated by a series of partitions of metal and glass connected by rods which enable curtains to be drawn on the corridor side, thus obviating the need of portable screens.

Channels in the screens carry the wires for light and wireless and everything is given a flush finish to facilitate cleaning. In addition to the lights over the beds, there are points a few inches from the floor to enable the night nurses to see their patients without turning on the main lights. (p. 131)

Practical and modern, these changes to layout and structure made day-to-day life on the ward more palatable to the patient and easier to manage for the staff. By incorporating ideas from other countries and bringing them to bear upon her own establishment, Martindale sought out the best ways to make the surgical experience as pleasant as it could be for all concerned. Such measures fed very much into Martindale's desires to encourage confidence between surgeon and patient, steps which would culminate in her involvement with the Marie Curie Hospital, Hampstead, set up in 1929. With her expertise in x-ray and radium therapy, Martindale was a natural choice, and the hospital was a culmination of women surgeons' efforts to convince female patients of the advantages of early diagnosis (see Moscucci 2007, pp. 139–63). Despite being established at such a difficult economic time, the unique cause garnered enormous support in the 1930s. In 1935, when a questionnaire was sent to hospitals regarding the ways in which they treated cancer, the verdict was that the 'skill, care, and judgement' of the Marie Curie Hospital ensured they 'stood out head and shoulders above all the other centres in the country'. Forty per cent of the women treated were still living and the result could be better if earlier treatment was obtained (Mellanby 1939).[3] For Louisa Martindale and her colleagues, the successes were very much attributed to the 'Marie Curie woman surgeon': 'not only because she is consulted by the patient earlier in the case when the symptoms are only slight, but because she agrees to follow out a certain technique and is meticulous in its application' (1951, pp. 209–10). The woman surgeon of the 1930s embodied exceptional care, precision, and certainty, achieving acclaim because of her combination and replication of these exemplary qualities.

New Surgical Horizons: Maud Forrester-Brown

The youngest surgeon to be considered here, Maud Forrester-Brown, would have surprised those male students who believed that their women compatriots expressed no interest in the athletic side of the medical school. A champion swimmer, in 1910 Forrester-Brown won the Ladies' Race at the London University Athletic Union Swimming Gala, and followed this up with the more local LSMW Swimming Club Cup ("Swimming Club" 1910, p. 329). By 1915, in the 'diving for plates' competition, she set a fine example 'plunging to the bottom and swimming round there for an interminable length of time picking up innumerable plates; unfortunately none of the other competitors had the power—or breath—to follow her lead' ("Swimming Club" 1915, p. 41). Her athletic prowess was coupled with academic excellence; the former would be remembered in her later career, when she recommended sport for rehabilitation or disability. Indeed, she practised what she preached: horse-riding, for example, had been utilised successfully in recovery from her own injuries, as well as employed for a number of selected patients ("Horse-riding for the Disabled" 1965, p. 1382). Forrester-Brown was also a keen supporter of co-education, while simultaneously believing that her own alma mater, the LSMW, with its 'high standard of efficiency' would be able 'to compete fearlessly' with others ("Letter to the Editor" 1910, p. 326). In many ways, this was an apt description of Maud Forrester-Brown herself, whose career in orthopaedics, set her on a new path for the woman surgeon of the interwar years.

Forrester-Brown took great advantage of the chances thrown up by the Great War to gain surgical experience across the country, from her own Royal Free Hospital to Dundee and Sheffield (Kirkup 2008, p. 2). But it was her appointment in 1916 as Resident Surgeon to the Orthopaedic Wards at the Bangour Military Hospital, Edinburgh, which formed the basis of her future surgical trajectory ("Recent Appointments" 1916, p. 138). For Forrester-Brown, the war was not the end of her experience nor her career as a surgeon. Her earliest research was based on procedures at Bangour, where she worked with Sir Harold Stiles, who himself had been a house surgeon under Lister (Ghilchik 2011, p. 216). Stiles and Forrester-Brown later described their surgery in *Treatment of Injuries of the Peripheral Spinal Nerves* (1922), and Forrester-Brown would contribute her own chapter, 'A Study of the Results of Operations for Nerve Injury at Bangour', which explored the cases operated on between 1916 and 1919, to Sir Robert Jones' *Orthopaedic Surgery of Injuries* (1921, pp. 415–30). Her years at Bangour also gained Forrester-Brown her MD in 1920, with a thesis on 'The results of operations for peripheral nerve injuries' (Kirkup 2008, p. 3). From what was described as 'the exceptionally abundant material' observed during the war, Stiles and Forrester-Brown noted that their specialist work was intended for the general surgeon so that 'a successful result for himself and his patient' could be achieved (1922, p. 1). Although 'very few cases' returned 'to the absolute normal after a nerve injury', Forrester-Brown remarked upon the value of surgical procedures for rehabilitation and return to work (1921, pp. 419–30).[4] It was this value of transmitting surgical knowledge more widely from the specialist to the general for the benefit of the patient which characterised the rest of Forrester-Brown's career as a surgeon.

The *LSMW Magazine* followed Maud Forrester-Brown's next steps with interest, remarking upon her success in the Primary Fellowship of the Royal College of Surgeons in November 1921 (p. 104), appointment as Honorary Visiting Orthopaedic Surgeon at Fairview School in 1923 (p. 131), Surgeon to the Children's Orthopaedic Hospital, Bath, in 1925 (p. 114), and then Senior Surgeon at the Bath and Wessex Orthopaedic Hospital in 1936 (p. 82). By 1934, she had also been appointed to the Hospital for Adult Surgical Tuberculosis at Chard and Honorary Orthopaedic Surgeon to the Swanage Red Cross Hospital in Dorset. She was a member of the Council of the Bath and

Bristol branch of the British Medical Association, and also had been elected a member of the International Orthopaedic Association the year before 1934 (p. 25). As this last accolade suggests, and in similar fashion to Louisa Martindale, Forrester-Brown was a great surgical traveller, winning a William Gibson Research Scholarship from the Royal Society of Medicine, which allowed her to study orthopaedic surgery in cities such as Paris, Bologna, and Boston from 1924 to 1926 ("School and Hospital Notes" 1923, pp. 190–91). After settling in the West Country, Forrester-Brown took charge of a number of major and minor children's clinics across the region (Ghilchik 2011, p. 216). As she had noted in her war work, what mattered to Forrester-Brown was the sharing of knowledge. Her 'thoughtful' ideas about continuity of treatment and co-operation between general practitioners and specialists in an age when no one practitioner could 'be efficient and up-to-date in all branches' of medicine and surgery were praised by the *BMJ* in 1926 ("Research in General Practice" 1926, p. 844). Throughout most of her published research, Forrester-Brown reiterated the necessity for shared principles, and for early communication, resulting in action to avoid disability (see for example Forrester-Brown 1938, pp. 454–61 and 1929).

Forrester-Brown drew on her training as a former pupil of Sir Robert Jones and her experience as an orthopaedic consultant to transmit her legacy to the next generation through the *LSMW Magazine* in 1934. Forrester-Brown feared that sensible advice had not been passed on in a previous issue concerned with osteopathy and wanted to correct assumptions. Firstly, she noted, while the general practitioner might be happy with the restoration of structure after fracture or accident, the patient desired function. The correct manipulations were vital for this to occur. Lasting damage could mar any erroneous attempt, so the orthopaedic surgeon should be consulted over persistent disability of limbs or spine. If only, Forrester-Brown remarked, general practitioners did not dismiss the patient's wish for a specialist, even if they felt the injury was trivial. Indeed, it was the most 'minor ills of life' that caused the 'most misery'. Secondly, and in addition to midwifery, general practitioners were likely to be faced with the majority of cases in this area. Education in manipulation was therefore vital to success in the profession. The patient's restoration to function and the reputation of the practitioner went hand-in-hand. Meaningful dialogue between specialist surgeon, generalist, and patient was key to treatment which lasted and worked. Maud Forrester-Brown was ever keen to pass on her expertise in orthopaedics, gained working alongside male colleagues, to ensure that when the surgeon received the patient, everything that could be done for them had been carried out as effectively and as efficiently as possible.

'LET YOUR EXPERIENCE FILL THEM IN LATER': TOWARDS A FEMALE SURGICAL LEGACY

While the majority of this chapter has looked at the interwar generation, some of the earliest and pioneering female surgeons died in the 1920s and 1930s. Tributes poured in especially for Louisa Aldrich-Blake in 1925 and Mary Scharlieb in 1930. What emerged from a mass of heartfelt dedications was the transfer of particular skills from one generation to the other through observance and assistance, as well as the written word. As one of Aldrich-Blake's 'Post', or a junior member of her surgical team, K. Stuart Harris, remarked: 'she once said to us, "Never mind about detail now, while you are learning as students, get broad outlines, and let your experience fill them in later"' (Harris 1926, pp. 36–37). This haptic knowledge marked the beginnings of shared female surgical expertise.[5] In the 1920s and 1930s, experience was gained through observing surgery, whether that was by men or by women, of any sort, at home or indeed across the globe, and brought back to be assimilated into one's own specialist practice and communicated to others. Whereas doors had to be forced open in the past to see what went on behind them, with more confidence, the female surgeon of this period was willing and able to communicate her techniques and findings for the benefit of others. The interwar woman surgeon, despite an assumption by historians that her world narrowed in the 1920s and 1930s, in fact looked increasingly outward and globally. This is very much compounded by the contemporary female patient's assumption that, somehow, women operated differently. As 'A Grateful Patient' acknowledged after the death of Aldrich-Blake, 'I owe my life, not only to her remarkable surgery [...], but to her unerring judgment. Her surgical skill and extraordinary wisdom as to what risks she could not or could not take with a patient's life are known to all' (1926, n.p.). Although hard to define, impossible to put into words with precision, the woman surgeon simply knew how to perform.

While some questioned women's achievements in surgery during this period, an uncertainty mirrored by historians, the interwar decades saw a moment when British women surgeons began to develop their own legacy, and consolidate their own traditions, their own ways of operating which drew upon shared experience. That this took place in perceived restricted locations should not be viewed as unfortunate or narrowing, because this is not how the woman surgeon viewed herself at this point. Public support increased in the 1920s and 1930s, and derided the increasingly unsupportable and retrograde criticism of her skills. As Mabel Ramsay noted of her fellow surgical colleague Louisa Aldrich-Blake after her death: 'Wisdom, courage and calmness were always hers in difficult situations and I felt she was a firm rock on which to rest' (1926, pp. 58–59). 'A Grateful Patient' utilised the same metaphor: she 'was like a rock to which one could cling with absolute confidence and moreover with a certainty that she would never let one down' (1926). The woman surgeon retained her fascination, but it was a wonder supported by confidence in her abilities to perform effectively and efficiently in the space of the operating theatre.

NOTES

1. McIlroy and Mary Scharlieb testified against Marie Stopes in Stopes V Sutherland and Another in 1923. Both believed that women's lack of education in the use of certain types of contraception led to more harm than good without proper medical advice. See "Medico-Legal" (1923), McIlroy on p. 446.
2. Very sadly for the historian, Martindale 'devoted several weeks to tearing up old notes of thousands of patients', p. 245.
3. Sir Edward Mellanby was chief administrative officer of the Medical Research Council.
4. For the result of varied procedures, see Tables i–vii, 419–30.
5. For more on surgical skills, see the special issue of *Medical History*, eds. Nicholas Whitfield and Thomas Schlich, 59, no. 3 (July 2015).

REFERENCES

"A Grateful Patient." 1926. *Time and Tide*, 15 January. In *Medical Women: Album of Newspaper Cuttings*, vol. vi. Part ii, H72/SM/Y/02/007. London Metropolitan Archives.
"An Ex-Surgical Sister." March 1922. "The Case Against Women Doctors. Members of their Own Sex Hesitate to Trust Female Practitioner." In *Medical Women: Album of Newspaper Cuttings*, vol. vi. Part i. H72/SM/Y/02/006. London Metropolitan Archives.
"Ban on Women Medical Students. The Campaign to Get It Lifted." 1928. *Manchester Guardian*, 8 December. In *Medical Women: Album of Newspaper Cuttings*, vol. vii. Part i. H72/SM/Y/02/008. London Metropolitan Archives.
Brock, Claire. 2017. *British Women Surgeons and Their Patients, 1860–18*. Cambridge: Cambridge University Press.
Calendar of the Royal College of Surgeons, England, 1923. 1923. London: Taylor and Francis.
Chadburn, Maud. 1923. "Presidential Address." *London Association of the Medical Women's Federation*, 11 October. Reproduced in *London (Royal Free Hospital) School of Medicine for Women Magazine* xviii, no. 86: 177–88.
Chadburn, Maud. 1930. "Inaugural Address.'" 1 October, *London (Royal Free Hospital) School of Medicine for Women Magazine* xxv, no. 107: 107–18.
Editorial. 1926. "Research in General Practice." *British Medical Journal* 2, no. 3435 (6 November): 844.
"Education of Women Doctors. London's Important New Step." 1921. *Manchester Guardian*, 1 February. In *Medical Women: Album of Newspaper Cuttings*, vol. vi, Part i. H72/SM/Y/02/006. London Metropolitan Archives.

Elizabeth Garrett Anderson to Emily Davies, Bayswater, Wednesday 5 September 1860. HA436/1/1/1: Letters from Elizabeth Garrett Anderson to Emily Davies: June-December 1860. Ipswich Record Office: Suffolk.

Elston, Mary Ann. 2001. "'Run by Women (Mainly) for Women': Medical Women's Hospitals in Britain, 1866–1948." In *Women and Modern Medicine*, eds. Anne Hardy and Lawrence Conrad, 73–107. Amsterdam: Rodopi.

England, John. 1932. "Can You Trust a Woman Doctor? A Hard, Uphill Fight to Get Recognition—Cold-Shouldered by Men Doctors. Hospitals Run Entirely by Women." In *Medical Women: Album of Newspaper Cuttings*, vol. vii. Part ii. H72/SM/Y/02/009. London Metropolitan Archives.

Forrester-Brown, Maud. 1921. "A Study of the Results of Operations for Nerve Injury at Bangour." In *Orthopaedic Surgery of Injuries*, ed. Sir Robert Jones, vol. ii: 415–30. London: Henry Frowde / Hodder & Stoughton.

Forrester-Brown, Maud F. 1929. *Diagnosis and Treatment of Deformities in Infancy and Early Childhood*. London: Humphrey Milford/Oxford University Press.

Forrester-Brown, Maud F. 1965. "Horse-Riding for the Disabled." *British Medical Journal* 1, no. 5446 (22 May): 1382.

Forrester-Brown, Maud F. 1938. "The Lorenz Bifurcation Osteotomy for Irreducible Congenital Dislocation of Hip." *Proceedings of the Royal Society of Medicine* 31, no. 5: 454–61.

Geddes, Jennian. 2006. "The Women's Hospital Corps: Forgotten Surgeons of the First World War." *Journal of Medical Biography* 14: 109–17.

Ghilchik, Margaret. 2011. *The Fellowship of Women: Two Hundred Surgical Lives*. St Ives: Smith-Gordon.

Harris, K. Stuart. 1926. "On Louisa Aldrich-Blake." *London (Royal Free Hospital) School of Medicine for Women Magazine* xxi, no. 93: 36–37.

Kirkup, John. 2008. "Maud Frances Forrester-Brown (1885–1970). Britain's First Woman Orthopaedic Surgeon." *West of England Medical Journal: Bristol Medico-Historical Society Proceedings* 114, no. 1: 1–12.

Letter to the Editor. 1910. *London (Royal Free Hospital) School of Medicine for Women Magazine* v, no. 47: 326.

Letter to the Editor. 1922. "London Hospital's Decision. Lord Knutsford Explains." *Times*, 10 March, 8.

"London's £2,000 a Year Woman Professor." 1921. *Evening News*, 17 February. In *Medical Women: Album of Newspaper Cuttings*, vol. vi. Part i. H72/SM/Y/02/006. London Metropolitan Archives.

"London's £2,000 a Year for Woman Doctor." 1921. *Daily Express*, 18 February. In *Medical Women: Album of Newspaper Cuttings*, vol. vi. Part i. H72/SM/Y/02/006. London Metropolitan Archives.

"London's £2,000-A-Year Girl Professor." 1921. *Evening News*, 31 March. In *Medical Women: Album of Newspaper Cuttings*, vol. vi. Part i. H72/SM/Y/02/006. London Metropolitan Archives.

Loudon, Irvine. 1992. *Death in Childbirth: An International Study of Maternal Care and Maternal Mortality, 1800–1950*. Oxford: Clarendon Press.

"Male Doctors Preferred. Women Who Do Not Trust Skill of Their Own Sex." 1923. *Evening Standard*, 3 October. In *Medical Women: Album of Newspaper Cuttings*, vol. vi. Part. i. H72/SM/Y/02/006. London Metropolitan Archives.

Martindale, Louisa. 1924. "Three Weeks in America." *London (Royal Free Hospital) School of Medicine for Women Magazine* xix, no. 88: 63–70.

Martindale, Louisa. 1925. "Fibromyomata of the Uterus: A Series of 252 Cases Treated Either by Surgical Operation or by Intensive X-ray Therapy." *Journal of Obstetrics and Gynaecology of the British Empire* 32, no. 4: 690–719.
Martindale, Louisa. 1931. "Treatment of Cancer of the Breast. A Clinical Review of 150 Cases." *Lancet* 5605 (31 January): 229–35.
Martindale, Louisa. 1951. *A Woman Surgeon*. London: Victor Gollancz.
McIlroy. A. Louise 1911. "A Demonstration on the Origin of the Follicle Cells of the Ovary." *Proceedings of the Royal Society of Medicine* 4: 226–27.
McIlroy, A. Louise. 1914. "Some Problems in the Treatment of Venereal Diseases." *British Medical Journal* 1, no. 2776 (14 March): 579–82.
McIlroy, A. Louise. 1922. "Some Observations on the Investigation of Toxaemias in Pregnancy." *British Medical Journal* 1, no. 3912 (4 March): 335–38.
McIlroy, A. Louise. 1930. "An Address on Maternal Mortality." *British Medical Journal* 1, no. 3606 (15 February): 269–73.
McIlroy, A. Louise. 1931. "Education in Obstetrics and Gynaecology. System of Teaching in the Royal Free Hospital Unit." *British Medical Journal* 1, no. 3667: 679–80.
McIlroy, A. Louise. 1932. "Indications for Caesarean Section." *Postgraduate Medical Journal*, 8, no. 82: 310–17.
McIlroy, Louise. 1934a. "Letters from Louise McIlroy, 29 September, to past and present students of the RFH and to past and present assistants of the Unit." *London (Royal Free Hospital) School of Medicine for Women Magazine* xxix, no. 119: 138.
McIlroy, A. Louise. 1934b. "Domiciliary Midwifery: Being the Remarks of What Was Styled a Lecture to the Medical Society at the Royal Free Hospital." *London (Royal Free Hospital) School of Medicine for Women Magazine* xxix, no. 119: 101–9.
"Medico-Legal. Birth Control Libel Action. Stopes v. Sutherland and Another". 1923. *British Medical Journal*, 1, no. 3245 (10 March): 445–448.
Mellanby, Sir Edward. 1939. "Cancer Treatment for Women. Tribute to the Marie Curie Hospital." *Times*, 48256 (17 March): 18.
Moscucci, Ornella. 2007. "The 'Ineffable Freemasonry of Sex': Feminist Surgeons and the Establishment of Radiotherapy in Early Twentieth-Century Britain.' *Bulletin of the History of Medicine* 81, no. 1: 139–63.
Moscucci, Ornella. 2017. *Gender and Cancer in England, 1860–1948*. Basingstoke: Palgrave.
"Obituary: Louisa Martindale." 1966 *British Medical Journal* 1, no. 5486 (26 February): 547–49.
Our Special Lady Correspondent. 1924. "Whispers for Women." *Somerset County Gazette*. 27 September. In *Medical Women: Album of Newspaper Cuttings*, vol. vii. Part ii. H72/SM/Y/02/009. London Metropolitan Archives.
"Overheard." 1923. *London (Royal Free Hospital) School of Medicine for Women Magazine* xviii, no. 85: 134.
"Professor McIlroy in Her New Post. 'Team Work' in Hospitals." 1921. *Manchester Guardian*. In *Medical Women: Album of Newspaper Cuttings*, vol. vi. Part i. H72/SM/Y/02/006. London Metropolitan Archives.
Ramsay, Mabel L. 1926. "On Louisa Aldrich-Blake." *London (Royal Free Hospital) School of Medicine for Women Magazine* xxi, no. 93: 38–39.
"Recent Appointments: Miss Forrester-Brown." 1916. *London (Royal Free Hospital) School of Medicine for Women Magazine* xi, no. 65: 138.

"School and Hospital Notes". 1923. *London (Royal Free Hospital) School of Medicine for Women Magazine*, xviii, no. 86: 190–191.

"Sex Rivalry in Medicine. Case Against Women Doctors." 1922. *Daily Mail*. In *Medical Women: Album of Newspaper Cuttings*, vol. vi. Part I. H72/SM/Y/02/006. London Metropolitan Archives.

"Swimming Club." 1910. *London (Royal Free Hospital) School of Medicine for Women Magazine*, v, no. 47: 329.

"Swimming Club." 1915. *London (Royal Free Hospital) School of Medicine for Women Magazine*, x, no. 60: 41.

Stiles, Sir Harold J., and Maud Forrester-Brown. 1922. *Treatment of Injuries of the Peripheral Spinal Nerves*. London: Henry Frowde and Hodder & Stoughton.

"The Newest Hospital for Women." 1919. *Hospital* 66, no. 1721: 203.

"The Public Well-Being: Women Doctors. Two Points of View." 1920. *Hospital* 68, no. 1782: 459.

Williams, Valentine. 1928. "Professions for Men Only? Miss 1928 Knocks at the Door in Harley Street and Downing Street and Finds it Shut." *Sunday Graphic*, 25 March. In *Medical Women: Album of Newspaper Cuttings*, vol. vii. Part i. H72/SM/Y/02/008. London Metropolitan Archives.

"Women Medicals." 1922. *Observer*, 5 March. In *Medical Women: Album of Newspaper Cuttings*, vol. vi. Part i. H72/SM/Y/02/006. London Metropolitan Archives.

"Women as Surgeons." 1923. *Daily Express.*, 2 October, "A Man's Business." In *Medical Women: Album of Newspaper Cuttings*, vol. vi. Part i, H72/SM/Y/02/006. London Metropolitan Archives.

"Women Medical Students." 1928. *Daily Telegraph*, 21 March. In *Medical Women: Album of Newspaper Cuttings*, vol. vii. Part i. H72/SM/Y/02/008. London Metropolitan Archives.

"Women, the Vote and the Hospitals." 1928. *Nation and Athenaeum* 958 (31 March). In *Medical Women: Album of Newspaper Cuttings*, vol. vii. Part i: 958. H72/SM/Y/02/008. London Metropolitan Archives.

CHAPTER 29

Technology Users vs. Technology Inventors and Why We Should Care

Wendy M. DuBow

For those of us who live in information economies, technology undergirds every aspect of our contemporary lives—from media to healthcare, from toys to scientific research. Most of us not only work with computers no matter what our job, but we also drive computers in the form of automobiles, wear computers in the form of watches or digital activity monitors, and carry computers everywhere we go in the form of cellular phones. Many household items as basic as refrigerators, thermostats, and televisions are run by computers. The omnipresence of technology makes Computer and Information Sciences (CIS) one of the most ubiquitous STEM fields in the twenty-first century. Thus, it is important to understand who is creating our technology and who is not.

In the United States (US), and globally, women comprise a majority of the consumer base. In 2009, women controlled about $20 trillion in annual consumer spending globally (Silverstein and Sayre 2009), with numbers rising as women have begun earning more income. In the US, women controlled 75 per cent of consumer spending, meaning women determined how to spend $4.3 trillion of the $5.9 trillion being spent on consumer goods. Although recent reliable data is scarce, we know that women controlled more than half of their household's spending on electronics in 2008. Among women of

W. M. DuBow (✉)
National Center for Women & Information Technology (NCWIT), University of Colorado, Boulder, CO, USA
e-mail: wendy.dubow@colorado.edu

© The Author(s), under exclusive license to Springer Nature Switzerland AG 2022
C. G. Jones et al. (eds.), *The Palgrave Handbook of Women and Science since 1660*, https://doi.org/10.1007/978-3-030-78973-2_29

colour, the combined buying power in the US has been estimated at $1 trillion (Prudential 2015). With population growth in the US leaning heavily towards people of colour in coming decades (Humphreys 2018), women of colour's buying power is destined to rise. And yet these powerful consumers remain largely just that in the technology sector: consumers. Women, and people of colour, are proportionately under-represented as designers and inventors of technology. Indeed, for the last several decades, the computing workforce that has been creating these technologies embedded in all of our lives has been strikingly homogenous.

The Gender and Racial-ethnic Composition of the US Technology Workforce

In 2019, women comprised 57 per cent of the professional workforce in the US, but only 26 per cent of the computing workforce—a disproportion that has been consistent throughout the 2000s. The tech sector is unevenly distributed by ethnicity as well. While Black/African-American people made up about 13 per cent of the US population in 2019, they held only 9 per cent of the computing jobs. A significantly growing group in the US, Latinx individuals, comprised 19 per cent of the population in 2019, but held only 8 per cent of the computing jobs. In contrast, Asians who comprise only 6 per cent of the US population, held 23 per cent of positions in computing occupations (Bureau of Labor Statistics 2019a; US Census Bureau 2019). There are many reasons to explain this lack of proportional representation in the field of technology, which we will explore later in this chapter; in any case, numbers are only part of the story.

An equally important part of the story is the level of power and influence people have in a given industry. If women and people of colour are minorities in technology, are they also less influential in terms of innovation and decision making? Given the many high-profile cases in the US news about harassment and discrimination based on gender or ethnicity in the tech sector (see Wakabayashi 2017; Dunlevy 2019), we are left to wonder whether those individuals who identify as a woman, and/or identify as a person of colour, have equal influence and participation in technology innovation? Do they provide meaningful input into new technologies, or are they somehow excluded from the innovation brain trust? In sum, who is driving technology innovation?

Historically, most US-based technology companies have been loath to share their employees' demographic characteristics, but there have been some in the US who have demanded transparency. In October 2013, one person from within the technology industry started making waves. Tracy Chou, a software engineer, wrote a blog in Medium.com lamenting the obfuscation of statistics regarding women in tech (Chou 2013). In her blog, she shared her then-employer's tally of women in technical positions (Pinterest: eleven out of eighty-nine engineers). Shortly after that, she created a list in GitHub, a site for open-source software development, where hundreds of individuals have since put in their counts of women engineers in their companies. Her guerrilla approach to getting real diversity data was very controversial, and it seemed to instigate a shift.

Around the same time, Reverend Jesse Jackson Sr.'s Rainbow PUSH Coalition chose Equity in Technology as one of its reform foci, bringing public attention to the 'veil of secrecy' surrounding who was making design and usability decisions about the technology we all buy and use. On 2 June 2014, Reverend Jackson made a call to action in front of twenty-five technology companies, including Google, Microsoft, LinkedIn, Pandora, eBay, Apple, Salesforce, Amazon, Yelp, Yahoo!, Facebook, Verizon, Comcast, AT&T, VMWare, Groupon and others—by personally addressing some of these executives and publishing an open letter asking them 'to voluntarily and publicly disclose your EEO-1 report and the racial and gender make up of your workforce in an expeditious manner' (Jackson 2014). The EEO-1 report is a survey mandated by the US Federal government, for employers with 100 or more employees. The regulations require company employment data to be categorized by race/ethnicity, gender and job category. Typically, individual reports are kept confidential, with only aggregate data being shared with the public. Jackson commended companies such as Intel, HP and Cisco that had long been releasing their employee demographic compositions, and praised Google for recently doing so and for acknowledging that the proportion of women and people of colour in their workforce was insufficient. On their website, the Rainbow PUSH Coalition pronounced, 'Silicon Valley must evolve and expand to look like America' (https://rainbowpush.org/). As a result of Chou, Jackson and other's efforts, much media interest was generated, and social pressure on technology companies to 'release diversity numbers' became a rallying cry.

As companies began releasing their employee demographics, what many had suspected to be true was confirmed: technology companies' US-based employees are largely white or Asian-American and predominantly men, particularly when it comes to the technical positions. Of the companies that have publicly released their statistics, it appears that under 5 per cent of their professional workforce is comprised of individuals from historically underrepresented minority groups (Evans and Rangarajan 2017). In these reports, 'underrepresented minorities' includes African-American/Black, Latinx, Native Hawaiian or Pacific Islander, American Indian or Alaskan Native, and two or more races; it excludes those who are White or Asian. Many companies hedged when sharing their demographics by not providing gender breakdowns by occupation, just giving overall percentages in the workforce, thus including, for example, human resources and administration areas that tend to have largely women workers. So, even those companies who shared data did not necessarily share all the relevant data, presumably because the statistics did not render them favourably.

Thus, while attention has focused intermittently on the lack of diversity in tech, the fact remains true that as of this writing, the technology workforce in the United States is 90 per cent White or Asian, and 75 per cent men (Scott et al. 2018). New tech companies are not on track to change the status quo either. Technology startup entrepreneurs are overwhelmingly White (87 per cent) and comprised of men (83 per cent) (CB Insights 2010), while just 1 per cent of all entrepreneurs launching venture capital-backed companies are African-American/Black (Finney and Rencher 2016). Women of colour comprise less than 1 per cent of all founders receiving venture capital funding. In some regions and in some subfields, Blacks/African-Americans and Hispanics have made inroads into technology, but there is still a long way to go until we achieve proportional representation and parity (Muro et al. 2018).

Undergraduate Degree Diversity in Computing

The relative lack of diversity in the technology industry is not a surprise for several reasons; chief among them that post-secondary degree data broken down by gender and race has long been publicly available in the US. We already knew that women and racial/ethnic minorities were underrepresented among graduates with Computer and Information Sciences bachelor's degrees from US four-year (non-profit) universities. In 2019, women earned only 21 per cent of all computer and information sciences bachelor's degrees from non-profit post-secondary institutions in the US (the 2718 for-profit post-secondary institutions are not included in this statistic). But aligned with professional workforce proportions, in that same year, women earned more than half (58 per cent) of all bachelor's degrees (National Center for Education Statistics 2019). Some argue that women are simply 'not choosing' science, technology, engineering and mathematics (STEM) fields. We will return to the issue of choice later in this chapter, but for now let us say that this argument used to be made for STEM fields in general. And yet there are STEM fields such as biology, maths and statistics, and chemistry, where women earn about half of the bachelor's degrees in the US (National Center for Education Statistics 2019). Moreover, women have earned about 40 per cent of the physical sciences bachelor's degrees for the past two decades.

The statistics for students of colour in the US tell a slightly different story (Fig. 29.1). In 2019, men and women of colour earned about 40 per cent of all bachelor's CIS degrees conferred by non-profit institutions, close to their proportion in the US population. As is evident in the chart below, participation by gender within each ethnic group is not equivalent to women's proportion of these populations. Other than Asian, Americans of colour are not represented in the computing workforce even to the extent they have earned the requisite degrees, regardless of gender.

Fig. 29.1 Race and gender composition of CIS bachelor's degrees 2019 (Reproduced from Cierra Kelley, Lyn Swackhamer and Wendy DuBow. 2021. *National Center for Women & Information Technology* [Based on National Center for Educational Statistics: Integrated Post-secondary Education System. 2019. CIP 11-Computer and Information Sciences], by kind permission of the National Center for Women and Information Technology, University of Colorado)

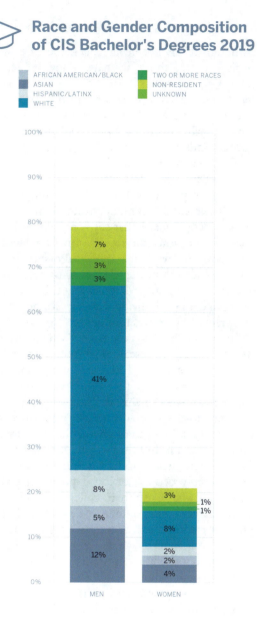

Because of rounding to the nearest whole, Native American students are not included in this chart, since their representation is only 0.5%. Rounding also means totals add to more than 100 per cent.

Making 'The Choice' to Go into Technology

Although it has not always been the case, most technology companies now profess to care about diversity. Many publicly announce funding set aside for increasing the diversity of their own employees or of feeder post-secondary institutions. Most post-secondary departments of computer science and engineering would also like to see more women and people of colour in their ranks, so what is stopping progress? Why are so few women of all racial/ethnic backgrounds and men of colour choosing to go into the field of computing?

To begin with, choice is not as simple as it may first appear. Constrained choice theory is a framework that has had great influence on our understanding of public health, and it has implications for this topic as well (Bird and Rieker 2008). In public health, constrained choice refers to the individual choices people make that turn out not to be simply freely decided individual choices, but rather the results of national level social policy, the physical environment, and workplace policy. Policies and physical environments actually limit and lead individuals' choices of what they eat, how much they exercise, and even how much responsibility they take for their own health. Thus, what seems like simple individual choices for good or poor health habits are in reality constrained by external influences.

In a discipline like computing, which has long been dominated by a single demographic, the research suggests that individuals may not 'choose' to be part of the tech world because of societal beliefs about who belongs in the precursor activities that lead to a computing major or job. Sociologist Shelley Correll (2001) explains that career choices actually happen quite early. If students do not take advanced mathematics in secondary school for instance, they are unlikely—and often unable—to pursue a major in STEM because they cannot catch up. In other words, according to multiple studies (Cheryan et al. 2017), their major, and consequently their career choice, has already been made, simply because of the opportunities afforded them earlier. Thus, their choices of what to study in college and what career to pursue are constrained by the external environment. So, whereas in public health, it is policy and the physical environment that constrains individual choices, in tech, it is societal expectations, such as gender and race roles that constrain individual choices.

These constraints can sometimes come from surprising places. There is research suggesting that adult influencers, such as teachers, parents, and school counsellors may actually be dissuading girls and all students of colour from these classes because of internalized beliefs about who 'belongs' in computing. Internalized beliefs influence which academic classes and extra-curricular activities students are encouraged to do by the adult influencers in their lives. This becomes very important in terms of who ends up in computing in later years. We know that early access and exposure to computing experiences can spark interest in computing for girls and boys alike. We also know that youth of different backgrounds have different degrees of access to and experience with computing. Research has shown patterns of inequity—in both access and experience—based on gender, race, socio-economic status, geographic location, primary language, and (dis)ability (Kanny et al. 2014; Margolis et al. 2008; Margolis and Fischer 2002; Goode 2010; Light 2001; Selwyn 2004; Valadez and Duran 2007; Warschauer et al. 2004).

In their book *Unlocking the Clubhouse*, which reshaped computing educators' and researchers' conceptions, researchers Margolis and Fisher offer a compelling description of social factors that heavily influence who ends up in computing. These include family expectations, peer groups and academic environments that converge to stimulate a technology identity. For boys, these factors often align to reinforce the dominant narrative that boys are naturally attracted to computing (Margolis and Fischer 2002). For girls, the story is somewhat different. There are girls who might tinker with technology, or partake in computing-related activities such as videogames or robotics; however, these early experiences do not typically create their identity, in part because these activities conflict with commonly held notions about what girls should like and do (Cheryan et al. 2015).

Early computing access and exposure builds what is called preparatory privilege. This privilege enables youth to seamlessly develop an identity as a person who 'does computing' and quietly sets them up for later success in technology fields (Margolis et al. 2008). Youth from less educated or less wealthy families, or from communities with low to no access to out-of-school opportunities, are challenged to obtain the skills and knowledge needed to succeed in computing classes. At the same time, they also must work through the cultural stereotypes about computer scientists and computing fields (Margolis et al. 2008). Additionally, they must contend with both implicit and explicit biases (Goode 2010), social or cultural isolation, and an environment that often does not recognize their competence (Carlone and Johnson 2007; Espinosa 2011).

For those not in the majority who do venture into computer classrooms or school-based labs, they are often met with an environment that feels unfamiliar, or even, exclusionary. Sexist posters, 'nerdy' gadgets or other 'geeky' decor in rooms or offices can suggest that there is an 'in group'; historically marginalized groups may feel they do not belong to it (Barker et al. 2006; Cheryan and Plaut 2010). Worse, teachers, counsellors or parents often guide girls, and boys and girls of colour, to non-technical classes. This guidance in itself communicates that technology is not appropriate for them, even when they have a budding interest or established aptitude. These adult influencers may be influenced by stereotypes they have about White and Asian boys' 'natural' talent for computing (Blickenstaff 2005; Voyles et al. 2007; Ong et al. 2011; Lent et al. 2000).

In secondary school, adolescents are constructing their identity. All the challenges and supports they face along the way through adolescence contribute or detract from their identity as someone who does computing (Goode 2010; Espinosa 2011; Carlone and Johnson 2007; Patton 1990). A sense of belonging, or fitting in, is critical for supporting students' interest and persistence at this stage. It is not surprising then that access to computing education opportunities and support from adult influencers in secondary school is correlated with computing interest and computing self-efficacy (DuBow et al. 2017; Cheryan et al. 2017). In fact, one study has shown that if high school girls declare interest in pursuing computing while they are in high school, the majority of them actually do elect a computing major when they get to college. These formative experiences in secondary school thus constrain individual 'choice' and this subtle sifting of students at the secondary school level results in a more homogeneous group who go on to pursue computing in post-secondary education. Subsequently, what they study in college sets them up for what they pursue as a career, thus replicating the gender and racial segregation from earlier educational stages.

Not Only Pipeline, but also Culture

While the so-called talent pipeline is a very important predictor of who ends up in the computing field, it is not the only reason for the underrepresentation of women and people of colour in computing. We also know from social science research that the culture of academic computing departments and technology companies has often been unwelcoming to those who do not fit the stereotypical mould. As already mentioned, these stereotypes can be reinforced by a department's décor in labs, classrooms or even hallways, as well as by the instructors' pedagogical practices. Inadvertently, instructors may create homework assignments in computer science classes that rely on stereotypes for what might interest students—video games, *Star Wars*, etc. Comments from fellow classmates who are part of the majority can be off-putting and exacerbate nascent feelings of exclusion. On the job market, the lack of welcome can start as early as the recruiting phase. Research conducted on recruiters and job fairs concluded that companies are undermining their efforts to diversify their candidate pool when they send narrow-minded, untrained recruiters to talk to potential candidates. Recruiters have been known to overlook students from schools not considered élite, to make jokes about pornography, or reference stereotypically geeky culture as an enticement (Wynn and Correll 2018).

In both work and academic contexts, individuals typically find themselves in group work or team situations. Poorly managed group work situations can enable the damage from stereotypes to flourish. Consider for example the use of pair programming in a computer science classroom. This is a practice recommended by experts to facilitate equal participation among students, deepen learning, and enhance computing confidence. Using pair programming in introductory computing courses actually increases the percentage of students declaring a computer science major, and those students are more likely to still be in the major one year later (Hanks et al. 2011; McDowell et al. 2006). However, it only works if it is highly structured.

To achieve the true pedagogical potential of pair programming, experts recommend that one student be designated as the 'driver' and the other the 'navigator'. Different types of learning occur in each role. The 'driver' performs the 'on computer' tasks needed to complete the programme, including controlling the mouse and keyboard to enter, compile, and run the code. The 'navigator' actively watches the driver's activities, makes suggestions, points out errors and problems, asks questions, and thinks about slightly longer-term strategies. But imagine if these roles are never switched, and the dominant student is always the driver? The learning opportunity for both is lost since the less experienced student, or quieter student, never has a chance to see what it is like to write code, and the dominant student never sees other ways of solving problems. Not surprisingly, men students often assume the dominant role in these and other group situations, unless the instructor mandates that roles are switched regularly.

Next, imagine if the instructor does not pair the students in some structured way? Without formal pairings, students will tend to choose those who seem most like them, resorting to a natural 'in group' preference (Godsil et al. 2016). This tends to leave out the students who do not fit into an obvious grouping—the only woman, or the only black person in the class, for example. This is just one example—within an otherwise solid pedagogical approach—of how stereotypes can be reinforced even with the best intentions of not doing so.

Not surprisingly, there are pertinent examples in the workplace as well. Without attention to culture and the ways in which it can reinforce bias and squelch diversity of thought and experiences, people from minority groups will likely have different experiences in their jobs than people from majority groups. For example, women in software development meetings may be interrupted more, their ideas restated by men without acknowledgement, and they may be given secretarial roles in a team's technical work despite having equivalent credentials. While sometimes these slights stem from overt bias, often they come from implicit, or unconscious bias.

Implicit, or unconscious bias, are automatic associations we all make based on stereotypes about particular groups. They arise without us even recognizing their presence and without us noticing that they have influenced our thoughts and behaviours. In fact, we may have conscious values that are in direct contradiction to our implicit biases. However, unless checked, our implicit biases rule the day because, according to social science research, they directly 'affect behavior and are far more predictive than self-reported racial attitudes' (Godsil et al. 2014, p. 10). These biases and stereotypes are so insidious that they actually change individuals' performances if they simply expect to be perceived as the negative stereotype about their identity. This phenomenon is called stereotype threat. Numerous studies (Beasley and Fischer 2012; Blascovich et al. 2001; Cheryan et al. 2009; Steele 1997) have shown that when a certain identity is activated for individuals, and that identity is associated with poorer performance on the task, then those groups actually perform poorly, compared to their performance without the negative stereotype invoked. For example, when mathematically-proficient Asian and white men are both given difficult mathematics tests, their performances are affected depending on which types of stereotypes are invoked (Aronson et al. 1999). When Asian men are primed to believe Asians are good at mathematics, they literally perform better than their non-Asian counterparts do. And white men perform worse when this stereotype about Asians being good at mathematics is invoked prior to test-taking.

These and other empirical studies have well established that our psychological frameworks have real-world impacts, and yet they often operate unfettered in academic and workplace settings. As this chapter has explained, the real-world impact is that we end up reinforcing differential participation in computing for women and other groups that have historically been underrepresented. These groups' lower participation means the people making our technology draw from limited experiences that do not necessarily reflect the varied consumers who use the technologies every day.

Innovation Mishaps Because of Homogeneity

While it is impossible to know what new innovations we miss out on when large portions of the population are absent, we can identify from recent history some technologies that have suffered from a lack of diversity at the development table.

When voicemail was invented, the voice recognition software was not initially tested with women, and therefore, it did not pick up on women's voices, which are often higher and softer than men's (Margolis and Fischer 2002). In fact, there is a long history of speech recognition technology performing better for men than women. When automobiles began using voice recognition systems, women drivers' instructions were unable to be processed because the systems did not recognize their voices. Even more egregious, in response to this shortcoming being made public, auto industry representatives allegedly recommended that women learn to speak differently, louder and lower (i.e. more like men) so that the technology would respond better (McMillan 2010).

Voice recognition also has been important in the medical toolkit. In a study of medical technology used for inputting patient data (American Roentgen Ray Society 2007), error rates were found to be higher for women physicians' dictation than for men's. Again, the proposed solution was for women to train more on the software. There are other studies showing that voice-based, medical tracking software was not created with the woman user in mind (Rodger and Pendharkar 2004). Likely, this is because there were no women technicians at the innovation table, or even in the testing environment, despite the fact that women are a majority of healthcare providers. The implications of voice recognition systems are not, of course, restricted to automobile use and medical documentation. Our phones, our smart speakers, even our Internet searches (Lawson 2017), are increasingly voice activated. Linguist

Rachael Tatman points out that women's voices should actually be more easily recognized than men's because of their characteristics, especially pitch, cadence and breathiness (2017).

As studies have shown, bias is actually built into the development schema through imbalances in the training data used during product development (Bock and Shamir 2015; Torralba and Efros 2011). Indeed, as Tatman points out, including a 'socially stratified sampling of speakers has historically not been the priority during corpus construction for computational applications' (2017, p. 57). In all of these cases, the result is that the woman end user is at a disadvantage; she must either use more of her time to train the software, not be able to use the software, or have her work be more error-ridden than her men counterparts.

Built-in bias is not, of course, restricted to voice recognition. A video that went viral in summer 2017 shows a hand soap dispenser that literally did not work for darker-skinned hands (Futureism 2017). In the month it was posted, the tweet about it was shared more than 93,000 times, and the video had more than 1.86 million views (Lazzaro 2017). The optical recognition at the heart of this technology, an infrared sensor, was apparently not tested on people of colour; therefore, it did not recognize darker-skinned hands and did not perform as designed for a large portion of the population who would be using it. Comments in social media indicated that people wondered whether this was a function of 'racism' or a 'technical glitch'. To consider these mishaps an innocent technical glitch ignores the realities of systemic racism and sexism in the US, and thus, also in tech development settings.

Personal consumer items have been affected by a lack of diversity as well. Early releases of Apple's iPhone Health App allowed users to track dozens of self-entered health measures but ignored one of the most commonly tracked data for women: menstruation. This oversight—another faux pas that completely neglected to include a large portion of users—was quickly fixed. In another example, wearable heart rate trackers have been shown to be inaccurate for people with darker skin (Hailu 2019). While the medical field uses infrared light to obtain people's heart rate and other readings, the consumer products use green light because it is cheaper. The light in these gadgets functions as the optical sensor that tracks the volume of blood at the wrist at any given moment. But as the medical literature shows, melanin interferes with green light, thus people with darker skin do not get as accurate readings as do people with lighter skin (Kollias and Baqer 1985). This deficiency has massive implications. As of 2018, more than forty million people wear some sort of heart-tracking device, out of health consciousness. While this device may be part of a sports regimen for some, for others with heart issues it is actually a life and death situation. Many people are counting on their heart rate monitoring to signal that they need to take medicine or use a defibrillator. If those monitor readings are wrong for them, and only them, this is not only a potential individual tragedy, but also surfaces as a grave public health inequity—and all because the developers did not understand how the technology would work with all users, but focused only on light-skinned users.

Many in the field of computer science understand that societal biases are baked directly into technology. Joanna Bryson, a computer scientist who studies artificial intelligence (AI), reflected on the evidence of racial and gender bias in machine learning algorithms by noting, 'A lot of people are saying this is showing that AI is prejudiced. No. This is showing we're prejudiced and that AI is learning it' (The Guardian View 2018). There are some who believe that seeing biases in algorithms may actually be a positive phenomenon because it provides humans a chance to fix the biases. As computer scientist and ethicist Sandra Wachter points out, 'At least with algorithms, we can potentially know when the algorithm is biased. Humans, for example, could lie about the reasons they did not hire someone. In contrast, we do not expect algorithms to lie or deceive us' (Devlin 2017).

In fact, the obviousness of bias in some computing applications has made it possible to both call attention to societal bias in general and to fix it in the technology. Take for example the case of Amazon using artificial intelligence to automatically review résumés. Amazon was implementing new technology in order to expedite the screening process for incoming job applications. In 2018, Amazon announced it would stop using computer programmes to sift through résumés because the programme was consistently choosing men candidates over women candidates. The automated process had been based on training the computer programme to identify potentially successful and unsuccessful candidates by using résumés of those who had been hired in the past. The training, therefore, included feeding the machine mostly men's résumés, since they came from the existing IT workforce, which is dominated by men. Setting the characteristics of the men's résumés as the ideal simply begot more men, as only their résumés were considered the ideal, or standard.

Amazon's public course shift is a welcome decision that illuminates two important facts about machine learning, the most widely used technique of artificial intelligence. Because of how these programmes learn, they can only learn from whatever data is fed to them. If the data reflects discriminatory patterns, the intelligent programmes will simply continue those patterns. Because these biases are literally built into the technology, they can be diminished considerably, or perhaps even eradicated, by more diversity on the development team, as well as more attention given to the diversity of the end users. The basic technologies that contribute to our mundane lives in the US would be much improved with the simple presence of women and people of colour on the design team and in the testing environments (The Guardian View 2018). The influence of different people's characteristics and life experiences would likely help to spark innovation in other ways that have yet to be discovered.

Why Diversity in Tech Is Important: Job Stability and Economic Potential for Women and People of Colour

Innovation is not the only thing affected by the tech sector's current imbalance. The future economic stability of women and people of colour also could be affected. Indeed, economic status can be greatly improved through becoming part of the technology workforce. With a brief exception in the technology bust of the US recession around 2000, over the last several decades, employment in the technology sector has provided people with wealth and job security in a growing industry. The US Department of Labor predicts that by 2029, there will be nearly 4 million job openings in computing in the US (Bureau of Labor Statistics 2019b). For the last three projections made by the US Department of Labor, computing occupations have been included in the highest growth fields. In addition, with a median annual income of $88,340 USD per year, computing occupations pay higher than any other occupational group except management (Bureau of Labor Statistics 2019c). Thus, including more women and people of colour in technology fields is seen by some in the US as a path for achieving greater economic and social equity.

We can all think of problematic situations where due to input from people who did not think like us, we came to a better solution than if we had been solving the problem on our own. This can be called 'diversity of thought'. But in reality, most of us surround ourselves by people who look like us and have a similar racial/ethnic and even socio-economic background. In academic and workplace environments, the concept of 'diversity of thought', while it sounds laudable, actually can be a distraction that allows leaders and organizations to escape the hard work of true diversity and inclusion (Bastian 2019). More important than diversity of thought among otherwise like people is diversity of gender, race and ethnic background. Environments inclusive of all races, ethnicities and genders is the appropriate outcome for a moral and just society. Creating inclusive workplaces also corrects historic wrongs against women and people of colour. And as discussed earlier in this chapter, it means our technology workforce will better reflect our technology consumers. Companies with diverse leaders, who have thriving teams that integrate individuals with varied backgrounds, will have the edge in a competitive global industry.

According to a plethora of social science research, these companies will also enjoy greater creativity and faster problem-solving. For example, one international study surveyed 1,400 team members from 100 teams and found that gender-balanced teams were the most likely to experiment, be creative, share knowledge, and fulfil tasks (Gratton et al. 2007). The data was taken from twenty-one companies in seventeen different countries. The study also found that the most confident teams had a slight majority of women (60 per cent). There has also been research conducted on group problem-solving and what has come to be known as 'collective intelligence' (Wooley et al. 2015). Collective intelligence refers to the capacity of a group to perform well on tasks requiring intelligence, such as brainstorming, moral reasoning, and mathematical reasoning. Results of studies have shown that when the average social perceptiveness of group members is high, the group performs better than groups with lower averages. Due to gender socialization, women tend to score higher than men do on these tests, therefore having women on teams is a strong predictor of effective group processing. Generally, the research has found that groups whose members are too similar to each other lack the variety of perspectives and skills needed to perform well on a broad range of tasks; at the same time, groups who are too different from one another tend to have difficulties communicating. As with the pair programming example mentioned earlier, these drawbacks can be mitigated through structuring teamwork processes, dividing projects into modular tasks, and explicitly valuing group collaboration (Page 2008). All of this suggests—indeed demonstrates—that diversity is beneficial to academic and industry environments, and, in turn, to innovation.

Why Diversity Is Important: Improves the Bottom Line

Social science research also demonstrates that companies' profits improve with increased diversity in management and leadership. In addition, mixed gender teams, when managed correctly, demonstrate superior team dynamics and productivity by staying on schedule and under budget, with improved employee performance, thus contributing to an improved financial outcome. One study done in the US with 500 companies reported that those with more diverse teams in terms of race and gender had higher sales revenue, more customers, greater market share and greater profits than did those companies with less diverse teams (Herring 2009).

The gender and racial-ethnic composition of leadership matters as well. A large study of global companies found that those with women on their executive boards did better on return on equity, debt/equity ratios and average growth than did companies with all male executive boards (Curtis et al. 2012). This was true even after the economic crash of 2008. A different study of 100 organizations—private, public and non-profit—determined that those with three or more women on their boards did better than those with fewer or no women on all measures, including leadership, accountability, coordination and work environment (Desvaux et al. 2007). Research has also shown that companies with gender diversity at the top levels had higher than average financial performances compared to others in their sectors, with better return on equity, earnings and stock price increases (Desvaux et al. 2007; Herring 2009). In analysing the gender composition of top management at a sample of Fortune 1000 companies, researchers found that those with greater representation of women had a greater return on assets (Krishnan and Park 2005). Given these benefits to innovation and financial return, many organizations are devoting resources and time to making changes. But is all this energy being focused in the best direction?

What Will Not Work to Increase Diversity

In the United States, there are many books, groups and websites devoted to what some call a 'fix the woman' mindset. This refers to the idea that women (and other minorities) can take personal steps, on their own, to change inequitable treatment in academic settings or in the workplace. Common advice includes learning how to act more confident, how to make yourself heard in meetings, how to negotiate for salary, find mentors or sponsors to support you, and so on. While these strategies are important for anyone's academic trajectory or industry career, these efforts are typically aimed at women so they can better grapple with environments that are exclusive or biased towards traditionally 'male' ways of communicating and acting. The trouble with these approaches is that they put the onus on the individual to fix an inherently flawed system. Consider that even if a large number of individuals fixed themselves, that would not change anything about the system for incoming generations who would then work or study within the same old constraints. These 'fix the woman' approaches also fail to take into account the very real power differentials between women, which varies by what background they come from.

In the wake of Facebook's former chief operating officer Sheryl Sandberg's 2013 book, *Lean In: Women, Work and the Will to Lead*, hundreds, perhaps thousands, of Lean In groups popped up around the nation, and a shot of steroids was given to the self-help industry. Since then, many have pointed out that women are not a monolith, and therefore, no single solution will work for all women. In fact, as many have pointed out, leaning in can have dire consequences for women of colour, who are exorcised even more from most work and academic environments in the US than are white women (see for example Gibson 2018). Ultimately, for the participation of women to change in computing, the locus of change cannot be women, or other individuals from racial/ethnic groups who have been underrepresented in computing. All individuals need tools to cope in work and academic environments, but people also need environments that welcome their diverse experiences and perspectives and appreciate the innovative edge they can bring to the technology industry.

What Organizations Can Do to Increase Diversity

During the second decade of the twenty-first century, awareness of the importance of diversity to innovation and profits has been increasing, with more media attention on the research showing its benefits. What remains to be seen is whether this awareness will lead to corporate changes in recruiting, hiring and company culture. Some research has shown that diversity training is not enough and can even backfire (Chang et al. 2019). Top leadership commitment to diversity, along with diversity training, needs to be coupled with clear explanations about how everyone reaps benefits from any new policies and from including a wider variety of people. Some researchers contend that initiatives are most likely to succeed in changing culture not only by instituting policies, but also by attending to concerns and fears among the majority (Chang et al. 2019). This education must be done by the organization and not left to minority individuals to bear the burden.

Having inclusive policies in the workplace is not enough; visible leaders need to take advantage of them as a model for the company's employees. For example, a policy that allows for flexible work hours to accommodate personal hobbies or families is an important mainstay for companies that wish to be inclusive. But everyone—including top leaders—should take advantage of this flexibility and time off in order for the policy to infiltrate company culture. Research done on men who were leaders in corporate organizations suggests that when top leaders in a company take family time, it not only benefits the leader himself but also sends a strong message to the organization. The message is dual-pronged, that personal lives are important, and that flextime is not something that benefits only women, who are often the primary caretakers of children even in dual-income households (DuBow and Ashcraft 2016).

Companies can also change whom they hire. They can train recruiters and hiring committees in how to recruit and hire for diversity. Learning to go to new places to recruit (not just the ivy league schools for example), making sure recruiters are not in themselves turning off potential candidates, and that committees are not screening out applicants because they fail to fit a preconceived mould, would go a long way towards diversifying the computing workforce. Policies that ensure there is always a certain number of women and people of colour in the interview pool can help raise the odds that new positions will be filled by individuals from groups historically underrepresented in computing.

There are also many research-based strategies that computing departments in secondary and post-secondary schools can use to recruit more girls and women, and more people of colour, into computing classes and programmes. These include inviting all kinds of students into introductory level computing classes by overcoming the stereotypes about who does computing. Instructors then must level the playing field and make sure that previous experience is not needed to succeed in the class, or in the major. At some American colleges and universities, there are now different tracks for the introductory computing sequence so that those without the preparatory privilege of experience with rigorous computing can catch up and still declare a college major in computer science if they so desire (Margolis et al. 2008). Still, bringing individuals into an unwelcoming environment—whether overtly hostile or simply exclusionary—will not shift the status quo. That ultimately can be shifted only by a commitment to change the conscious and subconscious practices that got us into this situation in the first place.

When language and environment are not deliberately inclusive of all people, then inevitably some people feel excluded, even when that is not the intent. Feeling excluded leads to isolation and undermines a sense of belonging, which is essential for individuals to succeed and persist in any professional field. There are many research-based ways secondary schools, post-secondary schools, and industry workplaces can make their environments more welcoming and inclusive to all kinds of people. It takes effort, and a commitment to change, at least by leadership if not by the rank and file. Importantly, everyone stands to benefit, not just those who have historically been excluded. When more minds, with different experience bases and perspectives, are engaged in coming up with technology-based solutions to the world's issues, we all win.

REFERENCES

American Roentgen Ray Society. 2007. "Voice Recognition Systems Seem to Make More Errors with Women's Dictation." *ScienceDaily*. www.sciencedaily.com/releases/2007/05/070504133050.htm. Accessed 17 October 2019.

Aronson, Joshua, Michael J. Lustina, Catherine Good, Kelli Keough, Claude M. Steele and Joseph Brown. 1999. "When White Men Can't Do Math: Necessary and Sufficient Factors in Stereotype Threat." *Journal of Experimental Social Psychology* 35: 29–46.

Barker, Lecia J., Eric Snow, Kathy Garvin-Doxas, and Tim Weston. 2006. "Recruiting Middle School Girls into IT: Data on Girls' Perceptions and Experiences from a Mixed-demographic Group." In *Women and Information Technology: Research on Underrepresentation*, ed. Joanne Cohoon and William Aspray, 115–36. Cambridge, MA: The MIT Press.

Bastian, Rebekah. 2019. "Why We Need to Stop Talking about Diversity of Thought." *Forbes*. https://www.forbes.com/sites/rebekahbastian/2019/05/13/why-we-need-to-stop-talking-about-diversity-of-thought/#276b4e2567c3. Accessed 10 October 2019.

Beasley, Maya A., and Mary J. Fischer. 2012. "Why They Leave: The Impact of Stereotype Threat on the Attrition of Women and Minorities from Science, Math and Engineering Majors." *Social Psychology of Education* 15, no. 4: 427–48.

Bird, Chloe E., and Patricia P. Rieker. 2008. *Gender and Health: The Effects of Constrained Choices and Social Policies*. Cambridge: Cambridge University Press.

Blascovich, Jim, Steven J. Spencer, Diane Quinn, and Claude Steele. 2001. "African Americans and High Blood Pressure: The Role of Stereotype Threat." *Psychological Science* 12, no. 3: 225–29.

Blickenstaff, Jacob Clark. 2005. "Women and Science Careers: Leaky Pipeline or Gender Filter?" *Gender and Education* 17, no. 4: 369–86.

Bock, Benjamin, and Lior Shamir. 2015. "Assessing the Efficacy of Benchmarks for Automatic Speech Accent Recognition." In *Proceedings of the 8th International*

Conference on Mobile Multimedia Communications, 133–36. Institute for Computer Sciences, Social-Informatics and Telecommunications Engineering.

Bureau of Labor Statistics. 2019a. *Current Population Study: Characteristics of the Employed, Household Data Annual Averages*. Table 1: Employed persons by detailed occupation, sex, race, and Hispanic or Latino ethnicity.

Bureau of Labor Statistics. 2019b. *Employment Projections*. Table 1.2: Employment by Detailed Occupation, 2019 and projected 2029.

Bureau of Labor Statistics. 2019c. *Employment Projections*. Table 1.7: Occupational Projections, 2019–29, and Worker Characteristics, 2019.

Carlone, Heidi B., and Angela Johnson. 2007. "Understanding the Science Experiences of Successful Women of Color: Science Identity as an Analytic Lens." *Journal of Research in Science Teaching* 44, no. 8: 1187–1218.

CB Insights. 2010. *Venture Capital Human Capital Report*. https://app.cbinsights.com/research/venture-capital-human-capital-report/ Accessed 10 October 2019.

Chang, Edward H., Katherine L. Milkman, Dena M. Gromet, Robert W. Rebele, Cade Massey, Angela L. Duckworth, and Adam M. Grant. 2019. "The Mixed Effects of Online Diversity Training." *Proceedings of the National Academy of Sciences* 116, no. 16: 7778–83.

Cheryan, Sapna, and Victoria C. Plaut. 2010. "Explaining Underrepresentation: A Theory of Precluded Interest." *Sex Roles* 63, no. 7–8: 475–88.

Cheryan, Sapna, Allison Master, and Andrew N. Meltzoff. 2015. "Cultural Stereotypes as Gatekeepers: Increasing Girls' Interest in Computer Science and Engineering by Diversifying Stereotypes." *Frontiers in Psychology* 6, no. 49: 1–8.

Cheryan, Sapna, Victoria C. Plaut. Paul G. Davies, and Claude M. Steele. 2009. "Ambient Belonging: How Stereotypical Cues Impact Gender Participation in Computer Science." *Journal of Personality and Social Psychology* 97, no. 6: 1045–60.

Cheryan, Sapna, Sianna A. Ziegler, Amanda K. Montoya, and Lily Jiang. 2017. "Why Are Some STEM Fields More Gender Balanced than Others?" *Psychological Bulletin* 143, no. 1: 1–35.

Chou, Tracy. 2013. "Where Are the Numbers?" Medium.com. https://medium.com/@triketora/where-are-the-numbers-cb997a57252. Accessed 10 October 2019.

Correll, Shelley J. 2001. "Gender and the Career Choice Process: The Role of Biased Self-assessments." *American Journal of Sociology* 106, no. 6: 1691–1730.

Curtis, Mary, Christine Schmid, and Marion Struber. 2012. *Gender Diversity and Corporate Performance*. Credit Suisse Research Institute.

Desvaux, George, Sandrine Devillard-Hoellinger, and Pascal Baumgarten. 2007. *Women Matter: Gender Diversity, a Corporate Performance Driver*. New York: McKinsey.

Devlin, Hannah. 2017. "AI Programs Exhibit Racial and Gender Biases, Research Reveals." *The Guardian*. https://www.theguardian.com/technology/2017/apr/13/ai-programs-exhibit-racist-and-sexist-biases-research-reveals. Accessed 10 October 2019.

DuBow, Wendy M., and Catherine Ashcraft. 2016. "Male Allies: Motivations and Barriers for Participating in Diversity Initiatives in the Technology Workplace." *International Journal of Gender, Science and Technology* 8, no. 2: 160–80.

DuBow, Wendy M., Alexis Kaminsky, and Joanna Weidler-Lewis. 2017. "Multiple Factors Converge to Influence Women's Persistence in Computing: A Qualitative Analysis." *Computing in Science & Engineering* 19, no. 3: 30–39.

Dunlevy, Leah. 2019. "More than a Third of Tech Industry Employees Have Experienced or Witnessed Sexism, a New Survey Finds." *Pacific Standard Magazine*. https://psmag.com/news/sexism-in-the-tech-industry. Accessed 17 October 2019.

Espinosa, Lorelle. 2011. "Pipelines and Pathways: Women of Color in Undergraduate STEM Majors and the College Experiences that Contribute to Persistence." *Harvard Educational Review* 81, no. 2: 209–41.

Evans, Will, and Sinduja Rangarajan. 2017. *Hidden Figures: How Silicon Valley Keeps Diversity Data Secret*. Reveal. https://www.revealnews.org/article/hidden-figures-how-silicon-valley-keeps-diversity-data-secret/. Accessed 10 October 2019.

Finney, Kathryn, and Marlo Rencher. 2016. *The Real Unicorns of Tech: Black Women Founders. The Project Diane Report*. https://www.digitalundivided.com/projectdiane-report-main/the-projectdiane-report-2016-the-real-unicorns-of-tech-black-women. Accessed 10 October 2019.

Futureism. 2017. "This 'Racist Soap Dispenser' at Facebook Office Does Not Work for Black People" [Video file]. https://www.youtube.com/watch?v=YJjv_OeiHmo. Accessed 10 October 2019.

Gibson, Caitlin. 2018. "The End of Leaning in: How Sheryl Sandberg's Message of Empowerment Fully Unraveled." *Washington Post*. https://www.washingtonpost.com/lifestyle/style/the-end-of-lean-in-how-sheryl-sandbergs-message-of-empowerment-fully-unraveled/2018/12/19/9561eb06-fe2e-11e8-862a-b6a6f3ce8199_story.html. Accessed 10 October 2019.

Godsil, Rachel D., Linda R. Tropp, Phillip Atiba Goff, and John A. Powell. 2014. *Addressing Implicit Bias, Racial Anxiety, and Stereotype Threat in Education and Healthcare. The Science of Equality, Volume 1*. https://equity.ucla.edu/wp-content/uploads/2016/11/Science-of-Equality-Vol.-1-Perception-Institute-2014.pdf. Accessed 10 October 2018.

Godsil, Rachel D., Linda R. Tropp, Phillip Atiba Goff, John A. Powell, and Jessica MacFarlane. 2016. *The Effects of Gender Roles, Implicit Bias, and Stereotype Threat on the Lives of Women and Girls. The Science of Equality, Volume 2*. https://equity.ucla.edu/wp-content/uploads/2016/11/Science-of-Equality-Volume-2.pdf. Accessed 10 October 2018.

Goode, Joanna. 2010. "The Digital Identity Divide: How Technology Knowledge Impacts College Students." *New Media & Society* 12, no. 3: 497–513.

Gratton, Lynda, Elisabeth Kelan, Andreas Voigt, Lamia Walker, and Hans-Joachim Wolfram. 2007. "Innovative Potential: Men and Women in Teams." *The Lehman Brothers Centre for Women in Business*. London: London Business School.

Hailu, Ruth. 2019. "Fitbits and Other Wearables May Not Accurately Track Heart Rates in People of Color." *STAT News*. https://www.statnews.com/2019/07/24/fitbit-accuracy-dark-skin/. Accessed 10 October 2019.

Hanks, Brian, Sue Fitzgerald, Renée McCauley, Laurie Murphy, and Carol Zander. 2011. "Pair Programming in Education: A Literature Review." *Computer Science Education* 21, no. 2: 135–173. https://doi.org/10.1080/08993408.2011.579808.

Herring, C. 2009. "Does Diversity Pay?: Race, Gender, and the Business Case for Diversity." *American Sociological Review* 74, no. 2: 208–224. https://doi.org/10.1177/000312240907400203.

Humphreys, Jeffrey M. 2018. *Selig Center for Economic Growth Report: The Multicultural Economy*. https://estore.uga.edu/C27063_ustores/web/classic/product_detail.jsp?PRODUCTID=6376. Accessed 10 October 2019.

Jackson, Jesse L. 2014. "Open Letter to Silicon Valley: 'Be Transparent, Release your EEO1 Data'" [Blog post]. http://www.rainbowpushsv.org/2014/06/open-letter-to-silicon-valley-be-transparent-release-your-eeo1-data/. Accessed 10 October 2019.

Kanny, Mary Allison, Linda J. Sax, and Tiffani A. Riggers-Piehl. 2014. "Investigating Forty Years of STEM Research: How Explanations for the Gender Gap Have Evolved over Time." *Journal of Women and Minorities in Science and Engineering* 20, no. 2: 127–48.

Kollias, Nikiforos, and Ali Baqer. 1985. "Spectroscopic Characteristics of Human Melanin in Vivo." *Journal of Investigative Dermatology* 85, no. 1: 38–42.

Krishnan, Hema A., and Daewoo Park. 2005. "A Few Good Women—On Top Management Teams." *Journal of Business Research* 58, no. 12: 1712–20.

Lawson, Matt. 2017. *4 Guidelines for the Future of Marketing.* https://www.thinkwithgoogle.com/marketing-resources/micro-moments/future-of-marketing-mobile-micro-moments/. Accessed 17 October 2019.

Lazzaro, Sage. 2017. "Is This Soap Dispenser RACIST? Controversy as Facebook Employee Shares Video of Machine That only Responds to White Skin." *Daily Mail UK.* https://www.dailymail.co.uk/sciencetech/article-4800234/Is-soap-dispenser-RACIST.html. Accessed 10 October 2019.

Lent, Robert W., Steven D. Brown, and G. Hackett. 2000. "Contextual Supports and Barriers to Career Choice: A Social Cognitive Analysis." *Journal of Counseling Psychology* 47, no. 1: 36.

Light, Jennifer. 2001. "Rethinking the Digital Divide." *Harvard Educational Review* 71, no. 4: 709–34.

Margolis, Jane, Rachel Estrella, Joanna Goode, Jennifer Jellison Holme and Kimberly Nao. 2017 [2008]. *Stuck in the Shallow End: Education, Race, and Computing.* Cambridge, MA: MIT Press.

Margolis, Jane, and Allan Fisher. 2002. *Unlocking the Clubhouse: Women in Computing.* Cambridge, Massachusetts: MIT Press.

McDowell, Charlie, Linda Werner, Heather E. Bullock, and Julian Fernald. 2006. "Pair Programming Improves Student Retention, Confidence, and Program Quality." *Communications of the ACM* 49, no. 8: 90–95.

McMillan, Graeme. 2010. "It's Not You, It's It: Voice Recognition Doesn't Recognize Women. Online Article." *Time Magazine.* http://techland.time.com/2011/06/01/its-not-you-its-it-voice-recognition-doesnt-recognize-women/. Accessed 10 October 2019.

Muro, Mark, Alan Berube, and Jacob Whiton. 2018. *Black and Hispanic Underrepresentation in Tech: It's Time to Change the Equation.* Washington, DC: The Brookings Institution.

National Center for Educational Statistics: Integrated Post-Secondary Education System. 2019. CIP 11-Computer & Information Services.

Ong, Maria, Carol Wright, Lorelle Espinosa, and Gary Orfield. 2011. "Inside the Double Bind: A Synthesis of Empirical Research on Undergraduate and Graduate Women of Color in Science, Technology, Engineering, and Mathematics." *Harvard Educational Review* 81, no. 2: 172–209.

Page, Scott E. 2008. *The Difference: How the Power of Diversity Creates Better Groups, Firms, Schools, and Societies.* New Edition. Princeton: Princeton University Press.

Patton, Michael Quinn. 1990. *Qualitative Evaluation and Research Methods.* New York: SAGE Publications.

Prudential. 2015. "A Total Market Approach: Winning with Women and Multicultural Consumers, 16." http://www.prudential.com/media/managed/totalmarket/TotalMarketStrategy.pdf. Accessed 17 October 2019.

Rodger, James A., and Parag C. Pendharkar. 2004. "A Field Study of the Impact of Gender and User's Technical Experience on the Performance of Voice-activated Medical Tracking Application." *International Journal of Human-Computer Studies* 60, no. 5–6: 529–44.

Scott, Allison, Freada Kapor Klein, Frieda McAlear, Alexis Martin, and Sonia Koshy. 2018. *The Leaky Pipeline: A Comprehensive Framework for Understanding and Addressing the Lack of Diversity across the Tech Ecosystem*: 1–32. https://www.kaporcenter.org/wp-content/uploads/2018/02/KC18001_report_v6.pdf. Accessed 10 October 2019.

Selwyn, Neil. 2004. "The Information Aged: A Qualitative Study of Older Adults' Use of Information and Communications Technology." *Journal of Aging Studies* 18, no. 4: 369–84.

Silverstein, Michael J., and Kate Sayre. 2009. "The Female Economy." *Harvard Business Review* 87, no. 9: 46–53.

Steele, Claude M. 1997. "A Threat in the Air: How Stereotypes Shape Intellectual Identity and Performance." *American Psychologist* 52, no. 6: 613.

Tatman, Rachael. 2017. "Gender and Dialect Bias in YouTube's Automatic Captions." In *Proceedings of the First ACL Workshop on Ethics in Natural Language Processing*, 53–59.

"The Guardian View on Artificial Intelligence: Human Learning." 2018. *The Guardian*, 14 October. https://www.theguardian.com/commentisfree/2018/oct/14/the-guardian-view-on-artificial-intelligence-human-learning. Accessed 10 October 2019.

Torralba, Antonio, and Alexei A. Efros. 2011. *Unbiased Look at Dataset Bias. Computer Vision and Pattern Recognition* 1, no. 2: 1521–28.

US Census Bureau. 2019. *National Population by Characteristics 2010–2019*. Washington, DC: US.

Valadez, James R., and Richard Duran. 2007. "Redefining the Digital Divide: Beyond Access to Computers and the Internet." *The High School Journal* 90, no. 3: 31–44.

Voyles, M.M., Susan M. Haller, and Timothy Fossum. 2007. "Teacher Responses to Student Gender Differences." *SIGCSE* 39, no. 3: 226–30.

Wakabayashi, Daisuke. 2017. "Google Fires Engineer Who Wrote Memo Questioning Women in Tech." *New York Times*. https://www.nytimes.com/2017/08/07/business/google-women-engineer-fired-memo.html?_r=0. Accessed 17 October 2019.

Warschauer, Mark, Michele Knobel, and Leeann Stone. 2004. "Technology and Equity in Schooling: Deconstructing the Digital Divide." *Educational Policy* 18, no. 4: 562–88.

Woolley, Anita Williams, Ishani Aggarwal, and Thomas W. Malone. 2015. "Collective Intelligence and Group Performance." *Current Directions in Psychological Science* 24, no. 6: 420–24.

Wynn, Alison T., and Shelley J. Correll. 2018. "Puncturing the Pipeline: Do Technology Companies Alienate Women in Recruiting Sessions?" *Social Studies of Science* 48, no. 1: 149–64.

Index

A
Academic Assistance Council (ACC), 135
Admiralty, 9, 86–90, 92, 93, 96, 137, 549, 550, 552, 554–558, 567–569
  Air Department, 20, 137, 553, 564, 568, 569
  Ship Welding Committee, 139
adolescents, 617
Adolf Frederik of Holstein-Gottrop, 403
Advice to Young Mothers (Mason), 71, 72
Aeronautical Engineering, 557
Aeronautical Society, 137
aeronautics, 19, 20, 549, 550, 552, 554, 561, 563, 569, 588
Afanasyevna, Tatyana T. *See* Ehrenfest-Afanasyevna, Tatyana T.
Agassiz, Charles, 424
Agents of the S.H.I.E.L.D., 488
agriculture, 6, 54, 117, 204, 206, 208, 214, 215, 469, 473
Airy family, 250
Airy, Professor George, 92–94, 98
Aldrich-Blake, Louisa, 594, 600–602, 606
Algarotti, Francesco, 55, 56, 388
algorithms, 277, 313, 314, 574, 622
Althoff, Friedrich, 227, 228
Amazon, 149, 344, 471, 613, 622
American Mathematical Society (AMS), 229, 230, 235

analytical chemical work, 15, 356, 357, 373
analytical engine, 5, 313, 314
anatomy, 9, 30, 47, 51, 61, 70, 119, 177, 290, 540
anatomy dissections, 530, 533, 539
Anatomy of Human Bones (Monro), 58
Anderson, Mary, 529
Animal Ecology (Elton), 466
animated photography, 180
Anita Borg Institute, 325
Annihilation, 484, 488, 494–496
anonymity, 56, 58
anonymous publications, 48
Ansichten der Natur (Humboldt), 392
anthropology, 296, 298, 334
Anti-Aircraft Experimental Section (AAES), 564, 566, 567
anti-aircraft gunnery, 550, 552, 563
Antiquities Society, 102
Antonelli, Kathleen McNaulty Mauchly, 321
apogamy, 181
Apollo missions, 489, 490
Apple, 613, 621
aptitude tests, 322
Arab countries, 309, 310, 327
Arago, François, 270, 506
Arber, Agnes, 102, 110, 119
Arber, Edward, 118, 119
archaeology, 177, 178, 186
Archbold, Helen, 139, 140

architecture, 4, 173, 177, 212, 347, 602
archives
 Imperial College, London; day nursery, 142; development, 131, 140; family connections, 140; identifying individuals, 140; male mentors, 136, 140; matriculated students, 134; student support, 142
Archives for Women in Medicine, 130
aristocratic women, 17, 399, 408
Armstrong, Henry Edward, 134
Army Aircraft Factory, 561
Arnold, Matthew, 254
Arrival, 495
Arrol-Johnston Car Company, 582
Arrow, 488
art, botanical, 177
art degrees, 325, 342
artificial intelligence (AI), 311, 314, 622
artists, 17, 177–179, 322, 416, 426, 428, 431, 437, 440, 442, 472, 478, 509, 518, 525. See also illustrators; painters
Ascent of Man (Blind), 301
Asian men, 620
Aspects of Nature (Humboldt), 392
astronomical drawings, 278
astronomical observatory, 270, 273, 277, 278, 401, 402, 413
astronomy
 Caroline Lucretia Herschel, 279
 Elisabetha Koopman Hevelius, 270
 Maddalena and Teresa Manfredi, 275
 Maria Clara Eimmart, 277
 Maria Margarethe Winkelmann-Kirch, 273
 Marie-Jeanne Amélie Harlay, 278
athletics, 540, 603
atomic weight determinations, 357, 358, 363, 364
atomism, 29, 32, 33
atoms, 29, 31–33, 35, 36, 355, 357, 358, 362, 363, 373
Attaturk, Kemal, 586
Audubon Society, 470, 473
Austin, Elen Elaine, 138
Austin, Sarah, 390
Australia, 17, 95, 130, 427, 430

Australian Women's Register, 130
Austria, 196, 310, 371, 372, 586
Ayrton, Hertha, 134, 136, 145, 171, 175, 185
Ayrton, William Edward, 134, 136

B

Baba, Hitoe, 206, 207
Babbage, Charles, 5, 313–315, 506
'babe scientist', 491
Bailey, Anita, 140
Bairstow, Leonard, 558, 562
Bakerian Lecture, 139
Baker, Inezita Hilda, 154
Baker, John, 139
Baker, Lily, 538
Bangor University, 139
Bangour Military Hospital, Edinburgh, 604
Barker, Muriel, 581, 617
Barrington, Amy, 182
Bartik, Betty Jean Jennings, 321
Bassi, Laura, 7
Bate, Dorothea, 182
Bates, Henry, 424
Bateson, William, 181, 182
Beale, Dorothea, 174, 191, 576
Beaufort, Captain Francis, 88, 90, 93, 96, 97
Becquerel, Henri, 5, 362
Bedford College, 20, 102, 103, 110, 112, 114, 132, 139, 157, 182, 552, 559, 562, 568
Beebe, William, 469
Behn, Aphra, 384, 389, 390
Belgium, 87, 310
Bell, Alexander, 316
Bell, Gertrude, 155
Bell telephone company, 317
Benson, Margaret Jane, 183
Bergman, Torbern, 406, 407
Bernert, Traude, 371, 372
Bernoulli numbers, 314
Berry, Arthur, 562, 568
Berzelius, J. Jac., 357
Besterman, Theodore, 52, 55, 59
Betts, Annie, 563
Bevin, Ernest, 580

bias, 20, 60, 285, 293, 311, 345, 374, 484, 511, 581, 617, 619, 621, 622, 625
Big Bang Theory, 483, 492, 495
Bigelow, Henry, 469
Big Science, 368, 373
Biological Bulletin, 453–455, 460
biological determinism, 290, 297, 301, 303
  feminist challenges, 291, 303
biological sciences, 4, 186, 291, 449, 471
biopics, 491
Bird, Isabella, 151, 152, 167, 616
Bird, The (Michelet), 518
Birtwhistle, George, 562
Black Panther, 484, 488, 496
Blackwell, Dr. Elizabeth, 530, 536, 538, 539
black women scientists, 485, 488. *See also* women of colour
Blazing World (Cavendish), 35, 39, 40
Blind, Mathilde, 300–302, 304
Blomefield, Leonard, 514
blood transfusion, 141
Blunt, Wilfrid, 425
Boivin, Marie, 70
Bolza, Oskar, 226, 227
Bones, 493
Book of Sun-dials, The (Gatty), 517, 526
Borg, Anita, 15, 325
Bortkiewicz, Helene von, 231
Bosse, Robert, 229
Bosworth, Anne Lucy, 231
botanical art, 177
botanical illustrators, 177, 178, 430, 433. *See also* North, Marianne
botanists, 17, 53, 54, 119, 178, 180, 181, 304, 387, 407, 411, 423–426, 428–432, 440
botany, 4, 51, 54, 62, 112, 113, 118, 119, 176, 183, 208, 259, 357, 387, 411, 414, 428, 506, 566, 577
Bourdieu, Pierre, 384
Bovell, Emily, 529
Bowdich, Sarah, 19, 383, 384, 506, 512–516, 522
Boyle, Robert, 27, 37, 39, 40
Bragg, Elizabeth, 577

Bragg, William, 371
breastfeeding ("suckling"), 76
Brenchley, Winifred, 183
Breton, Adela, 178
Brett, John, 431
Briggs, Shirley, 467, 472, 478
British Association for the Advancement of Science (BAAS), 19, 168, 181, 224, 509, 516, 559
British Astronomical Association, 10, 173, 184
British Federation of University Women (BFUW), 101, 109–114, 123, 136
British Medical Journal (*BMJ*), 532, 534, 601, 602, 605
British Medical Register, 133
British Women's Emigration Association, 576
Brooks, Harriet, 324, 363, 364, 367, 374
Brooks, Paul, 466–477, 479
Browne, Thomas, 32
Brück, Mary, 87, 89
Bryan, Margaret, 16
Bryant, L.W., 560, 569
Bryson, Joanna, 622
Buchan, William, 73, 74
Buckland, William, 512, 516
Bulgaria, 310
Burdett-Coutts, Angela, 180
Busk, Ellen Martha, 179
Busk, George, 179
Buss, Frances, 576
Butler, Josephine, 299
buying power, 612
Byron, Lord George Gordon, 5, 6, 312

C
Calder, Charles, 251
Call, The (Zangwill), 137, 171
Calopteryx virgo, 411
Cambridge University
  admission of women, 101, 117, 550, 556, 558
  Caroline Herschel, 264
  colleges for women, 102
  degrees, 165, 577
  doctoral degrees, 225

Faculty of Engineering, 139
John Herschel, 248, 260
Letitia Chitty, 137, 556, 569
Lorna Swain, 561
Mathematical Tripos, 19, 550, 556
mathematics, 20, 137
Sedgwick Club, 101, 116
societies, Sedgwick Club, 114
Cameron, Julia Margaret, 259, 261
Canadian Archive of Women in STEM, 131
cancer treatment, 20
Cannizzaro, Stanislao, 358
career breaks, 581, 582
career choices, 335, 616
career progression, 575
Carnegie, Andrew, 316
carnivorous plants, 437
Carothers, Julia Sarah, 202
Carpenter, Harold, 139
Carson, Rachel, 18, 465, 478
  education and early career, 466
  scientific credibility, 18, 470, 475, 476
  Silent Spring and its aftermath, 473
  writing career, 468
Carter, Elizabeth, 388, 389
Cartmell, Martha J., 202
Cassini, Jean-Dominique, Comte de, 279
Caton-Thompson, Gertrude, 11, 152, 156, 165–167
Cats (Michelet), 521
Cave-Browne-Cave, Beatrice, 556, 558, 566
Cavendish, Margaret, Duchess of Newcastle
  Blazing World, 35, 39, 40
  early particle theory, 31
  education for women, 40
  later science, 34
  Observations Upon Experimental Philosophy, 28, 35, 36, 38, 40, 42, 43
  scientific revolution, 27
Cavendish, William, Marquis of Newcastle, 31
Cayley, Arthur, 224, 225, 229, 556
CBS Access, 493
CBS Reports, 475
Chadburn, Maud, 595

Chandler, Dorothy, 553, 555, 557, 568
Chaplin, Matilda, 529
Chapman, Olive Murray, 43, 162
Chardin, Jean-Baptiste-Siméon, 410
Charleton, Walter, 27–30, 40, 42
Chatterton, Marjem, 587
Cheesman, Evelyn, 156, 165
chemical companies, 473–475
chemical engineering, 143, 365, 579, 580, 587
Chemical Lectures (Shaw), 54, 57
Chemical Society, 102, 358
chemistry
  Bedford College, 552, 559
  female leadership, 367
  Marie-Geneviève-Charlotte D'Arconville, 47
  opportunities and limitations through marriage, 364
  radioactivity research, 361, 363
  see also periodic system
Chernobyl, 489–491
Chers Disparus (Pujade-Renaud), 520
Chester Society of Natural Science (CSNS), 101, 115, 116
childbirth, 57, 67, 68, 73, 75, 76, 278. *See also* midwifery
children
  beauties of nature, 472
  breastfeeding ("suckling"), 76
  day nursery, 142
  health, 67, 68, 76–78, 531, 600
  and women's careers, 449
children's books, 68, 70, 79
Chinese Academy of Science, 10, 22
Chisholm, Grace E., 225, 228, 229, 231, 237
Chitty, Letitia, 137, 138, 556–558, 566, 568, 569
choice. *See* career choices
Chou, Tracy, 612, 613
Christianity, 29, 49, 202, 250, 251
Church, Frederic, 430
Church of Scotland, 251
City and Guilds College (C&G), 132, 134, 136
Civil List pension, 85, 93, 94
civil service, 96, 257, 467, 557, 581

class, 4, 9, 76–78, 85, 93–95, 116, 134, 151, 154, 156, 162, 172, 184–186, 206, 248, 250, 257, 258, 299, 326, 334, 338, 341–343, 404, 414, 427, 437, 486, 495, 496, 508, 511, 512, 516, 530, 534, 537–539, 550, 551, 575, 576, 578, 579, 584, 596, 616–619, 626
classical atomism, 31
Clay, Beatrice, 115, 116
Clayton, Robert, 370
clergyman science, 516
Cleve, Astrid, 357, 360, 367, 375
Cleve, Per, 357, 360
climate change, 471, 477, 478
clouds, 181, 259, 471
Clough, Anne Jemima, 254
Clouston, Thomas S., 297
Coignou, Caroline, 102, 105, 109, 110, 112–114
Cold War, 473
Collaboration (Michelet), 520
collections, 16, 17, 62, 69, 80, 86, 113, 116, 119, 130, 131, 143, 174, 265, 266, 334, 355, 360, 387, 392, 399, 403, 404, 406–408, 410–418, 424, 425, 430, 431, 437, 454, 461, 513, 515–518, 520. *See also* Swedish Royal Collection
collective intelligence, 623
comets, 247, 274, 277–279, 281–285
communication. *See* phatic communication, and ethos
companionate wives, 251, 253–255, 265
compass adjusting, 86, 92, 93
complementarianism, 49
Complete practice of midwifery (Stone), 75
Computations Inc., 323
Computer and Information Science (CIS), 611, 614
  bachelor's degree, 614
computers, 5, 13–15, 22, 307–309, 311, 312, 320–325, 342, 343, 345–347, 487, 489, 550, 563, 564, 567, 574, 585, 588, 611, 616–619, 622, 626
computing
  Ada Lovelace, 14, 307, 308
  career choice, 616
  culture as deterrent, 544
  diversity in undergraduate degrees, 309, 614
  international comparison, 309, 311
  intersectionality, 326
  social forces, 321
  see also information and communication technologies
computing classes, 617, 626
computing workforce, 21, 612, 614, 626
Condorcet, Marquis de, 49, 50
Conrady, Alexander Eugen, 141
Conrady, Doris, 141
Conrady, Hilda, 141
Conrady, Irene, 141
Conservation in Action (Carson), 467
constrained choice theory, 616
consumer spending, 611
conversaziones, 172
Conway, Anne, 28
Cooke, M.C., 436, 437
Copernicus, Nicolas, 386
Correll, Shelley, 616, 618
Corrigan, Dominic, 534
Corson, Dale, 371
cosmochemistry, 370
Costello, Donald Paul
  relationship with Catherine Henley, 452, 453, 457; phatic communication, 452, 457; phatic communication and ethos, 457
Cotter, Charles H., 90–93
County Council Act 1888, 102
Courant, Richard, 234
Courtauld, Mollie, 154
Cowley, Elizabeth Buchanan, 235
Cowper-Temple, William Francis, 537
Cox, Emily M., 388, 394
Cremona, Luigi, 224, 568
Cremona transformations, 555
Cressy-Marcks, Violet, 156
Crosby Cantrill, Thomas, 107
Crosfield, Margaret, 105, 107
Cudworth, Ralph, 32, 33
cultural history, 300
cultural norms, 47, 341, 575, 588
Curie, Irène, 490
Curie Laboratory, 369
Curie, Pierre, 5, 361, 362

Cuvier, George, 512
  Memoirs of Baron Cuvier, French translation, 512
cycles of credit, 17

D

Daily Express, 595, 599
Daily Mail, 596
Daily News, 185, 431
Daily Telegraph, 597
Dale, Elizabeth, 110
Dalle Donne, Maria, 71
Dalton, John, 357
Darboux, Gaston, 224
D'Arconville, Louis-Lazare Thiroux, 51
D'Arconville, Marie-Geneviève-Charlotte Darlus Thiroux
  education, 47
  male mentor, 8
  meat conservation and decomposition, 60
  personal networks, 53
  translations, 54, 56–58
Darwin, Charles
  biological determinism, 290, 291, 301, 303; feminist challenges, 298
  female inferiority, 289, 292, 294, 301
  Marianne North, 424
  translations, 393
Darwin, Erasmus, 434
Darwin family, 250
data processing, 324
Davenport Adams, W.H., 518, 519
Davies, Emily, 254, 324, 530, 596
Davin, Adelaide, 564, 566, 569, 570
Davy, Humphrey, 132, 512, 525
Davy, Lady Jane, 132, 512
Dawson, Dame Sandra, 132
day nursery, 142
Debierne, André-Louis, 361, 368, 369
de Commerçon, Philibert, 414
De la Beche, Henry T., 511, 512
De la Croze, Maturin Veyssière, 401
Delille, Jacques, 387
De Reuck, Marjorie, 132, 143
Descartes, R., 27, 29, 30, 33, 36, 43

Descent of Man (Darwin), 289, 434, 438, 439
  feminist challenges, 298
Desolate Year, The (Monsanto), 475
'Devonian Controversy', 512
Difference Engine, 313, 314
Digby, Kenelm, 28–30, 32, 42
Digby, Lettice, 136, 181
discrimination, 9, 103, 108, 109, 167, 308, 324, 325, 458, 496, 545, 612
dissection, 19, 57, 534, 596
diversity
  computing degrees, 308, 325, 348
  importance of, 625
  innovation mishaps due to lack of, 620
  US technology workforce, 614
'diversity of thought', 619, 623
divorce, 312, 367, 512
domestic abuse, 521
Domestic Cookery (Rundell), 75, 80
domestic medicine, 75
Domestic midwife (Stephen), 73, 75
Domvile, William, 389
dowry, 280, 337, 343, 345
Drew, Helen, 107
Drummond, Victoria, 578–580
Dublin Medical Press, 542, 543
Dublin Women's Suffrage Association, 535
Du Châtelet-Lomont, Gabrielle Émilie Le Tonnelier de Breteuil
  anonymous publications, 48
  education, 47, 51–53
  male mentor(s), 8, 48, 61
  personal networks, 56
  Royal Academy of Sciences, 52
  translation, 8, 50, 56, 57, 386, 387
Dunraven, Earl of, 177
Durham, Florence Margaret, 182
dynamical meteorology, 138

E

Early Years Education Service, Imperial College, 142
earthquakes, 178, 319, 511
East India Company, 88, 89, 255, 257, 410

East London College (ELC), 20, 552, 561, 568
ecology, 18, 33, 36, 41, 468–472, 474, 476–478
economic potential, 21, 623
economic stability, 623
Edelman, Josef, 587
Eden, Horatia R.F., 517
Edge of the Sea, The (Carson), 470
Edgerton, Winifred Harington, 225
Edgeworth, Maria, 251, 254–257, 259, 266
Edinburgh Ladies' Educational Association, 576
Edinburgh School of Medicine for Women, 532, 544
Edinburgh Seven, 529, 537
Edison, Thomas, 316
education
 aristocratic women, 399, 408
 Elizabeth Garrett Anderson, 299, 530
 Gabrielle Émilie Du Châtelet-Lomont, 51–53
 George Romanes, 294
 Henry Maudsley, 296
 Herschel children, 256
 India, 15, 255, 333, 336, 344
 Lovisa Ulrika of Sweden, 416
 Margaret Cavendish, 7
 Margaret Mason, 8
 Marie-Geneviève-Charlotte D'Arconville, 47
 mathematics, 201, 203, 205, 231, 232, 251, 262, 312, 550, 554, 567
 medical, 70, 141, 529–533, 535, 539, 599
 public, 467, 468
 see also women-only universities, Japan
Education Act 1870, 102
educational hypergamy, 343
educational reform, 12, 232, 233, 240
Ehrenfest-Afanasyevna, Tatyana T., 237
Eibe, Thyra, 225
eidophone, 184
Eimmart, Georg Christoph, 277
Eimmart, Maria Clara, 277
Einstein, Albert, 234
eka-manganeses, 364–366

Elderton, Ethel, 562, 563, 566
electrical engineering, 347
element separation and positioning, 358
Elles, Gertrude, 102, 106, 107, 117, 119, 123
Elliott Automation, 586
emanation, 362, 363
embryology, 29, 452, 453, 457
Emett, Delta, 184
Enabling Act 1876, 537
Encyklopädie der mathematischen Wissenschaften, 237
Endell Street Military Hospital, 594
engineering, 4, 10, 12, 18, 20, 109, 131, 137, 159, 197, 210, 212–216, 257, 309, 312, 315, 321, 322, 325, 334–336, 339, 340, 342, 343, 345–348, 350, 351, 483, 484, 492, 494, 549, 552, 555, 558, 559, 563, 566, 573–575, 577–589, 616
 India, 334
engineering degrees, 20, 339, 577, 579
engineers
 career progression, 567
 Cold War, 586
 cultures and contexts, 583
 Dorothée Pullinger Martin, 578
 education, 196, 248, 257, 315
 Elisa Leonida-Zamfirescu, 579
 Hilda Margaret Lyon, 581
 Marjem Chatterton, 587
 research approach, 558
 training, 257, 578
 use and meaning of the term, 574
 Verena Holmes, 579, 580
English Short Title Catalogue (ESTC), 384
ENIAC, 308, 321, 322, 326
Enlightenment, 5, 7, 8, 47–50, 53, 55, 60, 73, 290, 382, 384, 387, 388, 402, 406, 408, 516
 cultural norms, 47, 48
environmentalism, 33, 476, 478
Ephemeris calculators, 275
Epicurean atomism, 29, 31, 34
epigenesis, 29
Ernst August of Hannover, 401
Erxleben, Dorothea, 70
ethnic minorities, 21, 489, 614

US technology workforce, 612
ethnic-minority women, 487
eugenics, 182, 303, 424
Eugenics Laboratory, 563
evangelicalism, 251
Evans, Helen, 529
Evening News, 599
'Evolutionary Erotics' (Naden), 302
evolutionary theory, 5, 14, 290–292, 295, 300

F
family connections, 140, 181
Faraday, Michael, 16
Farmer, J.B., 181, 319
Fawcett, Millicent, 541
Fawcett, Nina, 150
Fawcett, Percy, 149
Fawcett, Philippa, 551
Fayum dig dispute, 157
female body, medical view of, 296
female inferiority, 289, 290, 292, 294, 298, 301. *See also* biological determinism
female leadership, 367
female mentors, 209, 450
Female Middle-Class Emigration Society, 576
femininity, 4, 7, 15, 16, 95, 163, 467, 484, 487, 488, 539
feminisation of translation, 386
feminism, 290, 291, 303
feminist critiques, 7
  biological determinism, 290, 297, 301
Fenster, Della D., 225, 226
Fermi, Enrico, 366
Ficino, Marsilio, 32
film
  annihilating/stereotyping women scientists, 487
  genre and women scientists, 484, 491
  hidden figures, 484, 486, 489, 491
  media production and representation, 494
  normalisation of women scientists, 484, 486
filmmaking, 180
financial independence, 344

First World War
  women doctors, 531, 542
  women mathematicians; Admiralty Air Department, 553; higher education centres, 555; National Physical Laboratory (NPL), 20; Royal Aircraft Factory (RAF), 20; University College London (UCL), 175, 550
Fitzgerald, Caroline, Lady Kingsbourough, 68
Fitzgerald, Ismay, 602
Flamsteed, John, 248, 284
flexible work hours, 626
Flicker, Eva, 484, 488
Fontenelle, Bernard de, 52, 389, 390
Forbes, Rosita, 155, 160, 357
foreign languages, 51
Forester, Lady Maria, 133
Forkel, Margarete, 388
form, 16, 27, 29–31, 33, 35, 36, 39, 42, 69, 107, 109, 112, 131, 132, 135, 164, 179, 199, 249, 250, 257, 258, 263, 289, 304, 319, 365, 388, 389, 392, 402, 407, 417, 423, 430, 437, 440, 443, 447, 452, 455, 472, 485, 487, 507, 509, 517, 519, 521, 536, 557, 569, 589, 611
Forrester-Brown, Maud, 20, 593, 603–605
Forster, Georg, 388
Foundations of Physics (Du Châtelet), 52, 58
Fountaine, Dr. E.C., 154
Fourcroy, Antoine François de, 53, 54
France, 31, 47, 50, 58, 61, 62, 70, 95, 98, 210, 224, 225, 272, 279, 309, 414, 506, 508, 512, 513, 516, 519, 525, 566, 582, 584, 586
Frances, Betty Jean Jennings, 321
Franklin, Rosalind, 448, 588
Frederick II of Prussia, 402
Frederick IV of Denmark, 403
Fredrik I of Sweden, 404
Freelance Programmers Ltd, 323
Freeman's Journal, 535, 536
Freeman, Dorothy, 473, 474, 478
French Academy of Science, 9

French Enlightenment, 389, 403, 409, 416
French Mathematical Society, 225, 235
Fresh-Water Fishes of Great Britain (Bowdich), 515
Freytag, Thekla, 233
Fridericianum, 407, 408
Frisch, Otto, 366
Fry, Mariabella, 181
Fukui, Shinji, 209
Furtwängler, Philipp, 236, 238

G
Gadow, Clara Maud, 183
Gadow, Hans, 183
Gahan, Mary Anne, 134
galleries, 407, 408, 410
Galloway Motors Ltd, 583
Galton, Francis, 173, 182, 293, 424, 432, 438, 439, 563
Gardner, E.W., 156–159, 163, 165, 166
Garrett Anderson, Elizabeth, 133, 299, 304, 530, 596
Garrett Anderson, Louisa, 542
gatekeepers, 151, 156, 580. *See also* male mentors
Gatty, Alfred, 516
Gatty, Charles Henry, 518
Gatty, Margaret, 19, 506, 516, 518
Gatty Marine Laboratory, 518
Gayler, Marie, 559, 560
Geddes, Patrick, 296
Geiringer, Hilda, 236, 240
gender, 4, 7, 9, 11, 13, 15, 21, 40, 42, 47, 49, 58–60, 78, 85, 93–95, 129, 139, 142, 143, 150, 163, 164, 167, 195, 196, 199, 211, 227, 248, 257, 290–293, 300, 303, 304, 308, 309, 324, 326, 327, 333–336, 338, 339, 341, 343, 345–347, 350, 351, 382, 383, 414, 427, 437, 438, 478, 484, 486–489, 491–493, 495, 507, 508, 517, 522, 523, 525, 545, 574, 576, 589, 612–614, 616–618, 622–624
 CIS Bachelor's Degree, 614
gender diversity, 624
gendered space, 371

gender equality, 11, 195, 197, 213, 215, 218, 300–302, 487
gender gap, 4, 15, 196, 333–336, 338, 339, 349, 350
 and patrifocality, 338
gender identity, 138, 496
gender norms, 8, 59, 60
gender specialization, 291
General Council of Medical Education and Registration, 533
genre, 35, 39, 384, 390, 484, 486, 491, 492, 521
Gentry, Ruth, 227, 230
Geographical Journal (GJ), 152, 153, 156, 159, 162, 164
geological societies, 101–103, 105, 107, 108, 110, 112, 123
 international, 103
 national, 104
Geological Society of London (GSL), 101, 103–105, 116, 511, 512
 Lyell medal, 104
 Transactions of the Geological Society of London, 511
geological time, 518
geologists
 local societies, 103, 114, 120
 networking societies, 108
Geologists' Association (GA), 104
geology, 9, 10, 102–104, 110, 112–119, 133, 157, 176, 181, 214, 251, 387, 477, 505, 511
George III, 87, 281, 283, 387
German Mathematical Society (DMV), 235, 236
Germany
 computing, 327
 educational reform, 233
 engineers, 196, 574, 586
 women mathematicians; *Encyklopädie der mathematischen Wissenschaften*, 238; male mentors, 12, 136, 217; *Mathematische Annalen*, 237
 see also Nazi Germany
Gernet, Maria, 224
Girton College, Cambridge, 102, 110, 181–183, 185, 229, 254, 304, 569
Glanvill, John, 389

Glanvill, Joseph, 27, 28, 30
Glasgow Royal Infirmary, 600
'glass ceilings', 513, 581
'glass cliff', 513, 525
Gleditsch, Ellen, 357, 362–364
Goethe, Johann Wolfgang von, 388, 508
Goldstine, Herman, 322
Gordon-Cumming, Constance Frederica, 178
Gordon, Maria Ogilvie, 102–104
Göttingen University, 12, 223, 225–228, 230–235, 240
   Habilitation, 234
   male mentors, 12
government contracts, 589
government service, 467
Grace Hopper conferences, 325
Graduate Women International. *See* International Federation of University Women (IFUW)
Graham, Maria, 506, 511, 521
Graham, Nina Cameron, 577
gravity, 386
Gravity, 495
Greece, 310
Greenly, Annie, 119
Greenough, George Bellas, 511, 512, 521
Gregory, Professor F.G., 139
Grew, Nehemiah, 30
Grey's Anatomy, 492
group work, 618
gunnery, 550, 566, 567
Gunze Girls' School, 200
Guppy, Sarah, 574
Gurney, Russell, 530, 537
Gustav IV Adolf of Sweden, 399, 414
Gwatkin, Henry Melvill, 263
gynaecology, 20, 296, 297, 532, 593, 599

H
Hadfield, Isabel, 561, 569
haematology, 141
Hahn, Otto, 53, 361, 362, 366
Haller, Albrecht von, 387
hand soap dispensers, 621
Hansmann, Frieda, 230

Hardcastle, Henry, 260
Harlay, Marie-Jeanne Amélie, 278
Harris, K. Stuart, 606
Harrison, Robert, 174, 179, 191
Harvard University Medical School, 130
Haslam, Thomas, 535
Hassanein Ahmed Bey, 155
Hasselquist, Fredrik, 413, 414
Hatoyama, Haruko, 204
Haynes, Roslyn, 485, 487
Healey, Elizabeth, 134
Health App, 621
heart rate trackers, 621
Heath, A. Grace, 134
Hebrew Technical Institute, Haifa, 587
heliocentric theory, 386
Hemsley, William Botting, 429, 430, 432
Henley, Catherine
   interpersonal relationships, 449, 460
   phatic communication and ethos, 17, 449, 451–454, 456, 460
   working with precision, 454
heredity, 181, 182
Hermite, Charles, 237
Herschel, Amelia, 259
Herschel, Caroline (b. 1750), 247
Herschel, Caroline (b. 1830), 255
Herschel, Constance, 262, 267
Herschel family, 14, 249, 250, 255, 265
   education, 14, 249
Herschel, Francisca, 259, 261, 262, 264
Herschel, Isaac, 280
Herschel, Isabella, 255, 258, 261, 264
Herschel, John, 248, 250, 251, 253, 258, 260, 279, 284, 315
Herschel, Julia, 261
Herschel, Louisa, 266
Herschel, Margaret, 258, 261
Herschel, Maria Sophia, 259, 266
Herschel, Matilda Rose, 259, 261, 262, 267
Herschel, William (b. 1738), 280
Herschel, William (b.1833), 247, 248, 256, 280–283
Hevelius, Elisabetha Koopman, 272, 273
Hevelius, Johannes, 270–273
Hicks, M.A., 179
Hicks, William M., 179

hidden figures, 486, 489, 490
Hidden Figures, 326, 484, 487–491, 497
Hilbert, David, 223, 230–232, 235–238
Hill, Edward John, 141
Hill, Violet (Peggy), 141
Hinks, Arthur, 155–161, 165–167
Hirano, Chiyoko, 207
Histoire Céleste Francaise, 279
Histoire naturelle des poissons (Cuvier), 514
HMS Challenger, 261
Hodges, Peggy, 585, 586
Hoffman, Darleane, 373
Hofmann, August von, 140
Hokkaidō University, 198, 205, 208, 212, 218
Holberton, Frances Snyder, 321
Holmes, Verena, 579, 580
homogeneity, 249
Hönigschmid, Otto, 357, 363
Honma, Yasu, 208
Hooker, Joseph, 425–428, 439
Hooke, Robert, 27, 37–40
Hooker, William, 424, 428
Horder, Lord Thomas, 540
Horner, Mary, 516
horology, 517
Horovitz, Stefanie, 357, 363, 364
Howe, Kay, 467
Hudson, Hilda, 363, 555–558, 566, 568
Hueper, Dr. Wilhelm C., 475
Hugonin, Charlotte, 512
Humboldt, Alexander von, 390–393
Hurwitz, Adolf, 235, 238
husbands, 13, 34, 55, 62, 86, 87, 93, 105, 118, 119, 132–134, 136, 150, 153, 154, 179, 180, 183, 185, 186, 224, 227, 232, 236, 237, 251–254, 257, 265, 270–274, 278, 280, 336, 337, 361, 365, 367, 392, 402, 404, 415, 478, 519, 520, 525, 577, 587. *See also* male collaborators
Huxley, T.H., 105, 134, 254, 290, 291
Huygens, Constantijn, 27–29

I
IBM Programmer Aptitude Test, 322

Ichibangase, Yasuko, 212
illustrators, 58, 176–178, 382, 467, 505, 518. *See also* North, Marianne
Imitation Game, The, 489
Imperial College, London
  archives; day nursery, 142; development, 131, 140; family connections, 140; identifying individuals, 140; male mentors, 136, 140; matriculated students, 134; student support, 142
Imperial Women's Medical and Pharmaceutical College (IWMPC), Japan, 202
Index to Flansteed's Observations (Herschel), 284
India
  computing, 311
  female research scientists, 349
  Institutions of National Importance (INI), 339
  patrifocality and educational decisions, 336
  STEM, 15, 333–336, 345, 346, 348; benefits, 343, 345; cost, 339, 343; social dangers and marriageability risks, 340
  STEM gender gap, and patrifocality, 15, 333, 335
  women doctors, 346, 535, 536
Indian Institutes of Technology (IIT), 335
infant care, 76
infinitesimals, 30, 33
information and communication technologies, 307, 308, 309, 311, 315, 316, 320, 326. *See also* computing
innovation mishaps, 620
Insect, The (Michelet), 518
insecticides, 473, 474
Institute for Women and Technology, 325
Institution of Civil Engineers (ICE), 137, 138
Institution of Electrical Engineers (IEE), 136, 145
Institution of Mechanical Engineers, 589
Institutions de Physique (De Châtelet), 52, 56, 58, 416

Institutions of National Importance (INI), India, 339
intellectual abilities, 182, 249, 293, 294
intellectual companionship, 253
intelligence, 30, 34, 35, 39, 41, 43, 88, 293, 294, 296, 300, 343, 404, 409, 452, 457, 488, 624
International Conference of Women Engineers and Scientists (ICWES), 584
International Congress of Mathematics, 228, 235, 236, 556
International Federation of University Women (IFUW), 109, 114, 362, 371
international geological societies, 103
International Network of Women Engineers and Scientists (INWES) Research and Education Institute, Canada, 131
intersectionality, 326, 486
Interstellar, 495
interwar period, 20, 197, 203, 211
Ionn, Frederick Peter, 86, 95
Ionn, Mathew Seymour, 87
Ionn, Peter, 87, 96, 97
Iowa State University, 130
iPhone, 621
Iran, 311
Ireland, 19, 68, 76, 310, 389, 494, 531, 535–537, 540, 541, 543–545, 566, 577
Irish Association of Women Graduates and Candidate Graduates, 535
Irish Examiner, 540
Irish Times, 532, 539
I, Robot, 492
isotopes, 363, 368, 370, 371
Israel, 586

J
Jablonski, Johann, 274
Jackson, Jesse, 613
Jacquier, Francois, 56
Jann, Rosemary, 292
Japan
  women-only universities; historical development, 197; present day support, 213; research environment, 197; STEM pioneers, 197, 203, 209
Japan Women's University (JWU), 197, 201–203, 205, 208, 211–213, 216
Jardin du Roi, 51, 53–55
Jebb, Louisa, 117
Jenyns, Leonard, 514–517, 523, 526
Jex-Blake, Sophia, 299, 529–531, 533, 535, 537, 538, 545
job stability, 623
Johns Hopkins University, 204, 225, 450, 452, 466
Johnson, Katherine, 490
Joliot-Curie, Irène, 15, 210, 356, 368
Joliot-Curie, Jean Frédéric, 210, 369
Jones, Constance, 559
Jones, Sir Robert, 604, 605
Joshigakuin, 202
Jurassic Park, 495
Jussieu, Bernard de, 48
justice, 110, 117, 295, 299, 300. *See also* social justice

K
Kahn, Margarete, 232
Kaiser Wilhelm Institutes, 585
Kalm, Pehr, 413, 414
Kant, Immanuel, 49
Karlik, Berta, 362, 371, 372
Katō, Sechi, 207
Kawashima, Keiko, 52, 59, 63
Kellner, Charlotte, 135
Kennedy, John F., 474
Keogh, Sir Alfred, 135
Kew Gardens, 181, 427, 439
  Marianne North Gallery, 425
King and Queen's College of Physicians of Ireland (KQCPI), 537, 545
Kingsborough, Lady (Caroline Fitzgerald), 69
King's College Hospital, 596
Kingslake, Rudolf, 141
Kingsley, Charles, 115, 116, 424, 434
Kingsley, Mary, 162
King, William, 313
Kirch, Christine, 283
Kirch, Gottfried, 273

Kirch, Margaretha, 275
Klein, Felix, 12, 223–240, 562
    Encyklopädie der mathematischen Wissenschaften, 237
    Mathematische Annalen, 237
Klingenstierna, Samuel, 406
knowledge, social nature of, 381
Knutsford, Lord, 596
Königsberger, Leo, 224
Kornfeld, Dr. Gertrude, 136
Kosmos (Humboldt), 159, 392, 393
Kovalevskaya, Sofia, 223–225, 227
Kronecker, Leopold, 227
Kummer, Ernst Eduard, 227
Kuroda, Chika, 203, 204, 211
Kyōritsu Girls' Vocational School, 200, 204
Kyushu University, 206, 207, 212

L
Laboratory Life (Latour and Woolgar), 447
Ladd-Franklin, Christine, 227
Ladies' Society for Physical Sciences, Netherlands, 416
ladies' soirées, Royal Society
    context and description, 172
    women exhibitors, 176, 188
Lagerborg, Ebba Louise Nanny, 225
Lalande, Joseph Jérôme, 278, 279
Lalande, Marie-Jeanne de, 279
Lamarr, Hedy, 490
Lamb, Horace, 562
Lamplugh, George, 107
Lancet, 296, 541, 544
Lang, Eleanor, 558, 562, 569
Langham Place Circle, 290, 530
languages, 6, 16, 18, 56, 67, 72–75, 157, 202, 231, 251, 255, 257, 262, 271, 296, 297, 311–313, 317, 322, 324, 325, 344, 381–384, 387, 393, 394, 401, 406, 411, 460, 469, 476, 617, 627
Laplace, Pierre Simon, 390, 511
Latour, Bruno, 17, 447–449, 456
Lavender, Joan, 585, 589
Laverick, Betty, 585, 586
Lawes, Caroline, 179

Lawes, Sir John Bennet, 179
lead, 11, 49, 54, 76, 78, 143, 153, 206, 308, 310, 322, 346, 363, 364, 374, 404, 407, 447, 455, 473, 485, 584, 589, 603, 616, 625
leadership, female, 367
Lean In groups, 625
Lebedeva, Vera, 232, 237
Lee, Alice, 182
Lee, Mrs. *See* Bowdich, Sarah
Leeds University, 587
Leibniz, Gottfried Wilhelm, 55, 56, 273, 274, 401, 402, 407, 409, 417
Leonida-Zamfirescu, Elisa, 579
Lermontova, Julia, 358–360
Lerner, Gerda Hedwig, 573
Lexis, Wilhelm, 231
'Life's Gifts' (Schreiner), 303
Lindley, John, 428
linguistic turn, 381
Linnaeus, Carl, 400, 406, 407, 410–416, 418
Linnean Society, 102, 181, 388, 417
Litvinova, Elizaveta Fedorovna, 224, 225
Lloyd, Eleanor, 517
Löbenstein, Klara, 232
local societies, 103, 114, 120
London Hospital Medical College, 543
London Mathematical Society, 179, 224, 235
London School of Medicine for Women (LSMW), 133, 531, 532, 536, 538, 540, 541, 544, 595, 596, 599, 603, 604
Lost City of Z, The, 149, 150, 162, 165
Love and Mr Lewisham (Wells), 134
Lovelace, Ada, 5, 14, 307, 308, 311, 383, 574
Lovelace, Roberta, 454
Lovén, Sven, 404, 412, 414, 415, 417
Lovisa Ulrika of Sweden
    collection's modern day location, 399
    collection's significance to science, 412
    collector, 404, 415
    current scientific references to collections, 415
    early life and education, 400
    philosophe, 404, 409, 415
Lowe, Richard, 516

LSMW Magazine, 598, 600, 604, 605
Lubbock, Neville, 264
Lucretius, 28, 31–33, 52
Lyceum Ladies' Club, 175
Lyell, Charles, 105, 434
Lyell, Mary (Horner), 516
Lyell medal, 104
Lyon, Hilda Margaret, 581, 582, 585
Lytle, Mark Hamilton, 465, 467–471, 473–477, 479

M

Machinae Coelestis (Hevelius), 270–272
machine learning, 622
Machke, Heinrich, 226, 227
MacKenzie, Kenneth, 371
MacKinnon, Annie Louise, 225, 229
Maclear, Jack, 261, 262
MacMillan, Edwin, 373
Macnamara, Rawdon, 534
Ma Collaboration (Michelet), 519, 520
Macquer, Pierre-Joseph, 48, 52–54, 57, 60–62
MacRoberts, Rachel, 103, 105
Maddison, Ada Isabel, 229, 230
Magazine of Art, 431, 433
magazines, 86, 88, 209, 300, 428, 466, 468, 471, 475, 540, 541
Makin, Bathsua, 41
Makita, Raku, 203
Malaysia, 311, 327, 583
male collaborators, 13, 186, 460
male mentors
  chemistry, 8
  Imperial College, London, 137
  mathematics, 12
  see also gatekeepers
Maltby, Margaret Eliza, 96, 98, 228
Manchester Guardian, 596, 599
Manfredi, Agnese, 276
Manfredi, Eustachio, 275, 276
Manfredi, Maddalena and Teresa, 275, 276
Manhattan Project, 372, 373
Mansfield, Professor Averil Olive, 142
Manual of British Vertebrate Animals (Jenyns), 514
Mapother, E.D., 540

Marcet, Jane, 16
Marianne North Gallery, Kew Gardens, 425
Marie Curie Hospital, 603
Mariette, Auguste, 521
Mariner's Calculator, 89–91
marriage, 11, 51, 69, 70, 79, 86, 87, 95, 96, 118, 119, 136, 141, 184, 204, 249, 251–254, 257, 258, 260–264, 270, 273, 278, 283, 292, 318, 336, 337, 342–346, 350, 351, 364, 366, 367, 403, 404, 516, 517, 575, 579, 581, 594, 595
marriage bar, 20, 581, 586
Married Women's Property Acts 1870 and 1882, 102
Marr, John, 117–119
Marshall, Dorothy, 559
Martindale, Louisa, 20, 593, 594, 600–603, 605
Martin, Emilie Norton, 230
Mary, A Fiction (Wollstonecraft), 69
masculine tropes, 539
masculinity, 312, 316, 319, 484, 488, 540
Maskelyne, Nevil, 248, 283, 284
Mason College, Birmingham, 102
Mason, Margaret, Lady Mount Cashell
  Advice to Young Mothers, 8, 71, 72
  children's books, 68, 70
  medical education, 70
  medical practice, 68, 73, 346
  midwifery, 8, 71, 73, 74
  upbringing and personal life, 68–72
Masters, Maxwell T., 388
maternal healthcare, 72
maternal mortality, 600
mathematical instruments, 9, 85, 87, 89, 97, 403
Mathematical Practitioners of Hanoverian England 1714-1840 (Taylor), 86
Mathematical Principles of Natural Philosophy (Newton), 386
mathematicians, First World War
  Admiralty Air Department, 20, 137, 553
  higher education centres, 555

National Physical Laboratory (NPL), 20, 549
Royal Aircraft Factory (RAF), 20
University College London (UCL), 175
mathematics
  Cambridge University, 20, 137, 550
  Germany; educational reform, 233; *Encyklopädie der mathematischen Wissenschaften*, 238; German Mathematical Society (DMV), 235–237; male mentors, 12; *Mathematische Annalen*, 237
  Tokyo Women's Christian University (TWCU), 202, 213, 218
Mathematische Annalen, 237
Matilda effect, 373, 374, 448
Matsuo, Yukari, 216
matter as "self-knowing", 33
Matthew/Matilda effect, 13. See also Matilda effect
Maudsley, Henry, 296–299, 304
Maunder, Annie, 173, 183, 186
Maunder, Edward, 183
Maupertuis, Pierre Louis Moreau de, 55
Mayeda, Toshiko Kuki, 369–371
McIlroy, Louise, 20, 593, 598–600, 607
McKenny Hughes, Prof T., 105, 106, 115, 118, 119
McPhee, Margaret, 118
meat conservation and decomposition, 60
mechanical engineering, 346–348, 585
media, 17, 18, 440, 443, 458, 468, 471, 475, 484–487, 489, 493–496, 584, 588, 611, 613, 621, 625. See also film; magazines; newspapers; television
media production, 494
Medical Act 1858, 529
medical degrees, 70, 346, 537, 538
medical dramas, 485, 492, 493
medical education, 141, 529–533, 535, 539, 599
medical empowerment, 67. See also *Advice to Young Mothers* (Mason)
medical knowledge, 67, 68, 71, 73–75, 78, 79

medical profession, 19, 132, 142, 296, 529–538, 544, 545, 595, 596
  admission of women, 535; arguments in favour and against, 531; experiences and careers, 538
medical professionalization, 72
medical schools, 19, 132, 133, 141, 202, 529–532, 537–540, 543–545, 593, 595, 596, 598, 603
medical students, 70, 71, 530, 531, 534, 538–543, 545, 596
  drop-out rates, 543
medical technology, 620
Medical Times and Gazette, 532
Medical Women's Federation, 20, 595, 601
Medical Women's International Association, 601
'medical-women' question, 532
Medici, Anna Maria Louisa de', 408
Medici-Lorraine Family Pact, 408
medicine
  Archives for Women in Medicine, 130
  female body, 291, 295
  post-war world, 595
  see also London School of Medicine for Women (LSMW); Mason, Margaret, Lady Mount Cashell; surgery
Meier, Ernst von, 227, 228
Meitner, Lise, 15, 22, 356, 361, 362, 366, 367
Meltzer, Marlyn Wescoff, 321
Mémoires d'une enfant [*Memoirs of a Girl Child*] (Michelet), 519
Menabrea, Luigi, 314
men, collaborations with, 48, 285. See also male mentors
Mendeleev, Dmitri Ivanovich, 356, 358–360, 364, 365
Mendelson, Sara, 40, 42
menstruation, 621
mercantile marine, 92
Merian, Maria Sibylla, 416, 426, 441
Merrill, Helen Abbot, 235
Merritt, Anna Lee, 177, 178
metallurgy, 139, 365, 560
Metcalf, Ida Martha, 225
Meteorological Office, 138

Mettam, H.A., 563, 569
Meyer, Lothar, 358
Michelet, Athénaïs, 19, 506, 518, 520, 526
Michelet, Jules, 518, 519, 521
microscopy, 37, 450
midwifery, 8, 71–75, 80, 600, 605
midwives book, The (Sharp), 74
Milbanke, Lady Annabella, 5, 312
military hospitals, 542
Millar, Hugh, 516
Milligan, Dr William, 576
Millingen, John Gideon, 296
Mill, John Stuart, 292
Minerva, 406, 408
missionaries, 202, 207
Mitchell, Margaret O., 180, 360
Miyazaki, Yukio, 209
modesty topos, 507, 513
Moger, Olive, 553, 554, 568
Moldenhauer, Helen, 140
Molyneux, William, 389
Monograph on British Graptolites (Elles and Wood), 107
Montolieu, Maria Henrietta, 387
Moore, Stephen, 2nd Earl of Mount Cashell, 69
Morandi Manzolini, Anna, 71
More, Henry, 28, 32
More Madden, Dr. Thomas, 535, 536
Moritzen, Anna Ilse, 280
Morland, Mary, 512
Morning Toilet (Chardin), 410
Morse, Samuel, 316, 319
Moscucci, Ornella, 73, 296, 601, 603
Moseley, Henry, 364
Motte, Andrew, 386
Mountain, The (Michelet), 518–520
Mozens, H.J., 6, 275
Mueller, Sir Ferdinand, 178
Müller, Johann Heinrich, 278
Munitions Inventions Department, 550
Murchison, Charlotte, 506, 511, 523
Murchison, Roderick, 511, 512
Murray, George, 180
Myers, Frederic William Henry, 179
Myller, Alexandru, 232

N

Naden, Constance, 298, 302
Nakamura, Masanao, 198
Nara Women's University, 197, 211–214
Naruse, Jinzō, 201, 203
National Blood Transfusion Service, 141
national geological societies, 104. *See also* Geological Society of London (GSL)
National Physical Laboratory (NPL), 549, 550, 552, 558–561, 567, 569
Nation and Athenaeum, 597
natural history, 10, 16, 52, 116, 133, 176, 183, 184, 186, 255–257, 355, 387, 399, 405–408, 410–413, 415, 416, 427, 429, 431, 434, 469, 514, 519–521
Natural History Museum, 104, 114, 141, 182, 184, 191, 414
Natural History of Fishes (Cuvier), 514
nature, 4, 13, 18, 27–29, 32, 34–39, 41, 48, 49, 56, 59, 75, 76, 113, 149, 151, 160, 173, 184, 186, 233, 257, 259, 289, 296, 297, 299, 300, 302, 303, 309, 323, 356, 362, 370, 371, 374, 383, 385, 389–392, 394, 401, 406, 425, 427, 433–435, 453, 460, 465–475, 478, 479, 485, 518–521, 525, 534, 535, 544, 552, 554, 557, 558, 588, 602
Nature, 363, 374, 432, 519–521
Nature: The Poetry of Earth and Sea (Michelet), 519
Natuurkundig Genootschap der Dames, 416
Nautical Academy, 85, 86, 89
nautical instruments, 9, 86, 88–90, 97, 172
nautical tables, 313
navigational warehouse, 86
navigation, publications in, 278
navigation technology, 9
Navy timetables, 279
Nazi Germany, 585
Nernst, Walther, 229
Ness, Wilhelmina Elizabeth, 155, 161, 165–167
Netflix, 493, 494

networking, 8, 9, 48, 53, 101, 104, 105, 107–109, 113, 114, 118, 120, 123, 162, 173, 186, 311, 466, 477, 578
networking societies, 109
Neumann, Elsa, 230
New Hospital for Women, 531
Newnham College, Cambridge, 102, 105, 107, 110, 112, 113, 119, 138, 154, 180, 556
newspapers, 78, 88, 203, 227, 347, 467, 471, 475, 541, 551, 567
New Sussex Hospital for Women, 602
Newtonian Science, 57, 388
Newton, Isaac, 38, 50, 55–58, 386–388, 407
New Yorker, 474
New York Times, 311, 312, 551
Nitobe, Inazō, 202, 207
Nobel Prize winners, 13, 22, 225, 362, 367, 368, 370, 373
Noddack, Ida (née Tacke), 364–367
Noddack, Walter, 364–367, 378
Noether, Emmy, 224, 234, 236–239
Noether, Max, 237
North, Frederick, 424
North, Marianne
   autobiography, 423, 424
   Charles Darwin, 423, 424
   life and context, 427
   reception of paintings, 431
   travels, 17
Norton, Arthur, 533
Notes on Irish Architecture (Dunraven), 177
Nuffield, Lord, 583

O
Oberlin College, 576
Observations Upon Experimental Philosophy (Cavendish), 28, 35, 38, 43
Observer, 596
obstetrics, 20, 68, 71, 73, 74, 346, 532, 535, 536, 593, 599
ocean ecosystem, 469
Oceans 8, 488
Ochanomizu University, 199, 211, 212, 214, 216

Official Secrets Act, 589
Ogilvie, Alec, 553
Oman, 311
Omnibus, 471
On the Connexion of the Physical Sciences. By Mary Somerville (review essay by Whewell), 507
On the Nature of Things (Lucretius), 31
On the Origin of Species (Darwin), 291
On the Power of Movement in Plants (Darwin), 436
Oommen, Mariam, 135
optical recognition technology, 621
optics, 16, 141, 506, 559
Original Stories from Real Life (Wollstonecraft), 69, 79
Ormerod, Eleanor, 6, 22
Orphan Black, 493–495
Ortner, Gustav, 371
Osaka University, 197
Otsuma Women's University, 214
Otté, Elise C., 392, 393
Owens College, Manchester, 102
Oxford University, 203, 559
oxygen thermometer, 370

P
Pagett, Rose, 183
Pagett, Sir George E., 183
painters, 31, 177, 178, 277, 407, 417, 423, 425, 426, 428, 430, 440. *See also* illustrators
painting(s), 177, 178, 179, 181, 256, 257, 293, 404, 406, 407, 410, 417, 418, 423–426, 429–433, 437–441. *See also* botanical art
pair programming, 618, 624
palaeobotany, 109, 119, 176, 183
palaeontologists, 107, 115, 119, 182
paratext, 388
Paris Academy of Sciences, 58
Parshall, Karen H., 225, 226
patents, 89, 90, 136, 317, 318, 344, 359, 448, 574, 580
pathology, 141, 143, 296, 347
patrifocality, 336, 337, 345, 348, 350, 525
   and educational decisions, 336, 350

and STEM, 347
patrilineality, 336
Pearson, Karl, 182, 550, 556, 558, 562, 564, 569, 570
Pechey, Edith, 529
people of colour, 21, 311, 326, 489, 612, 613, 616, 618, 621–623, 626
  job stability and economic potential, 623
People's Palace, 552
Perey, Marguerite, 368, 369, 371
periodic system
  analytical chemical work, 15, 357
  element separation and positioning, 358
  undetected elements, 364
peripheral nerve injuries, 604
Perry, Alice Jacqueline, 577
personality tests, 322, 324
Personal Narrative (Humboldt), 391
pesticides, 18, 473–475, 478
Peterson, Jeanne, 183, 253, 254
pharmacy, 51, 54, 57, 347, 407
phatic communication, and ethos, 457
Philosophiae Naturalis Principia Mathematica (Newton), 57
Philosophical and Physical Opinions (Cavendish), 30, 41
Philosophical Fancies (Cavendish), 30, 32, 34
photography, 176–180, 186, 277
physical sciences
  On the Connexion of the Physical Sciences. By Mary Somerville (review essay by Whewell), 506
  women scientists, 483
physics, 4, 5, 12, 18, 30, 32, 47, 51–53, 55, 57, 112, 133, 135, 141, 143, 197, 205–207, 210, 211, 214, 216, 225, 229–231, 234, 239, 278, 309, 339, 348, 350, 355, 362, 416, 476, 477, 484, 492, 507, 508, 552, 559, 577, 586
physiology, 31, 182, 290, 296, 299, 423, 432, 540
'pipeline' model, 506, 510, 521, 523
Pippard, A.J. Sutton, 137, 138, 554, 558, 569
Pisati, Laura, 236

Pitot Tube, 559
platinum, 358–360, 364, 365
Plumptre, Anne, 390
Poems and Fancies (Cavendish), 29–31, 33, 36, 40
poetry, 8, 14, 62, 69, 293, 300, 387, 466, 471, 520
Polak, Professor Julia, 143
pollution scandals, 476
Porter, Helen. *See* Archbold, Helen
Portugal, 310
Pouchet, Georges, 521
Poullain, François, 249
Prandtl, Ludwig, 562, 582, 585
Prefectural Medical Schools, 202
preformation, 29
Presbyterian Mission Female Seminary, 202
Pre-Raphaelite Movement, 431
Prichard, Augustin, 392, 393
Priestley, Joseph, 516
Principia Mathematica (Newton), 416
Pringle, Sir John, 54, 61, 63
Private School Order, 199
professional communication, 451. *See also* phatic communication, ethos
professionalization, 19, 68, 70, 73, 78, 308, 325, 505, 522
Professional School Order, 199
property, 102, 227, 290, 312, 327, 336, 361, 362, 369, 381, 387, 403, 466
psychiatry, 299
puberty, 77, 336
publication, anonymous, 48
public education, 467
public health, 543, 544, 594, 616, 621
Pujade-Renaud, Claude, 520
Pullinger Martin, Dorothée, 578, 583
Pullinger, Thomas C., 578
Pulteney, Richard, 411, 413
Punch, 428
Pupil of Nature (Mears), 75, 80
Putrefaction, 54, 60

Q
Qatar, 311
Queen's College Belfast, 538
Queen's College Cork, 537

Queen's College Galway, 537, 577
Queen's College, London, 576
queer TV characters, 486, 493

R
race, 7, 292, 297, 299, 300, 326, 340, 394, 484, 486, 493, 495, 496, 613, 616, 617, 623, 624
  CIS Bachelor's Degree, 614
  US technology workforce, 613
radioactivity, 13, 210, 361, 362, 364, 368, 371, 373
radioactivity research, 361–364
radio technology, 308, 316, 319
radiotherapy, 601
Ragsdale, Virginia, 230
Rainbow PUSH Coalition, 613
Raisin, Catherine, 103, 105, 107, 110, 112, 114
Ramsey, Dr. Mabel, 544
Ramsey, William, 559
rare earth elements, 358–361
recruiters, 618, 626
Reeks, Margaret, 146
Reeks, Maria Ellen, 140
Reeks, Trenham, 140
Reeks, Trenham Howard, 140
Reform Act 1832, 102, 290
Reid, Elizabeth Jesser, 552
Reineck, Annie, 224
Relation historique (Humboldt), 391
religious belief, 477. *See also* Christianity
renaissance, 32, 383, 386, 402, 573
research associates, 449–451, 453–455, 457, 459, 460
Reserve Bank, Zimbabwe, 587
Reynolds, Doris, 110
rhetorical work, 450
Rhodes, Miss, 179, 191
Richards, Evelleen, 290–292, 439, 440
Richards, Robert, 360
Richards, Theodore William, 357, 363
RIKEN, 197, 204, 205, 207–209, 217
Rippon, Gina, 5
Rix, Herbert, 173, 190, 191
Robertson, Agnes, 110, 118
Robinson, Ron, 90
Robinson, William, 544

Rodell, Marie, 473, 475, 478
Rolleston, Dr., 535
Rolleston, Sir Humphrey, 595
Romanes, George, 289, 294
Romanes, James, 118
Romania, 232, 310, 579
Róna, Elisabeth, 373
Rosenhain, Walter, 560
Rosen, Nils, von Rosenstein, 406
Rossetti, Frances, 78
Rossiter, Margaret
  adaptive behavior, 447
  Felix Klein, 226
  Mathew/Matilda effect, 13, 373, 374, 448
  research associates, 449, 451, 457
  working with precision, 454
  zoology, 451
Ross, Sir Ronald, 557, 568
Ross, Thomasina, 391
Rouelle, François-Guillaume, 52
Rousseau, Jean-Jacques, 49, 249
Routledge, Katherine, 152, 163
Routledge, William Scoresby, 153, 164
Rowan, Marian Ellis, 178
Royal Academy of Sciences, 47, 52, 56, 58, 229, 402, 406, 407, 418
Royal Academy of Technology Berlin, 579
Royal Aircraft Establishment (RAE), 581, 582, 585
Royal Aircraft Factory (RAF), 137, 549, 550, 552, 554, 561, 562, 564, 567
Royal Army Medical Corps, 599
Royal Astronomical Society (RAS), 10, 186, 264, 283, 284
Royal Balloon Factory, 561
Royal Berlin Academy of Sciences, 274
Royal College of Chemistry (RCC), 132, 140, 145
Royal College of Physicians, 141, 142, 537, 595
Royal College of Science (RCS), 132, 134, 135, 145, 181
  women associates, 141
Royal College of Surgeons, Dublin, 539, 540
Royal College of Surgeons, England, 142, 594

Royal Flying Corps (RFC), 549
Royal Geographical Society (RGS), 10, 11, 145, 149–153, 155–165, 167
　in *The Lost City of Z*, 149, 150, 162
　women's lectures; audience, 151, 160, 161, 164; contributors, 151; lecture locations, 162
Royal Holloway College, 114, 552
Royal Microscopical Society, 10
Royal Navy, 92, 516
Royal School of Mines (RSM), 132, 133, 139–141, 145
Royal Society of London
　Bakerian Lecture, 139
　The Blazing World (Cavendish), 35, 39, 40
　Caroline Herschel, 247
　founding year, 269
　Helen Archbold, 139
　Hertha Ayrton, 136, 171, 175
　ladies' soirées, 11, 171–174, 178, 179, 182, 185, 186; context and description, 172; women exhibitors, 11, 176
　women's lectures, 152; forms of lectures, 162; social element, 162
Royal Society of Medicine, 605
Royal Swedish Academy of Sciences, 360, 406, 413, 414, 417
Royer, Clémence, 384, 393
rugby, 540, 545
Rundell, Maria Eliza, 75, 80
Runge, Carl, 234
Runge, Iris, 234
Rutherford, Ernest, 362
Ruysch, Rachel, 431

S
Sabine, Elizabeth, 383, 384, 392
salaries, 110, 345
samarskite, 360
Sandberg, Sheryl, 625
Sandford, Kenneth, 158, 166
Sargant, Ethel, 119, 180, 181
Saruhashi, Katsuko, 212
Satō, Shōsuke, 207
Saudi Arabia, 311
Saunders, Edith Rebecca, 181

Sawayanagi, Masatarō, 203, 218
scanning electron microscopy (SEM), 139
Scharlieb, Dr. Mary, 536, 538, 606, 607
Schiebinger, Londa, 7, 9, 22, 43, 48, 49, 51, 58, 249, 273, 274
Schönflies, Arthur, 229
Schreiner, Olive, 303
science
　coinage of term, 525
　feminist critiques, 7
　hard and soft, 13
　modern understandings, 3
　nineteenth century conception, 3, 79, 85, 248, 250, 265, 285
　women in, 5, 22, 367, 458, 460, 491, 494, 505, 506, 516, 522, 523
science advisors, 494, 495
Science Council of Japan (SCJ), 211, 212, 216
science fiction literature, 494
Science Teacher in Training Scholarships, 133
science writing, 3, 186, 523
scientific artists, 178
scientific drawings, 277. *See also* botanical art
scientific education, 7, 14, 114, 133, 136, 181, 234
scientific instruments, 9, 36, 91, 136, 408, 413, 517
Scientific Revolution, 27
scientific time, 517
scientific translation. *See* translation
scientific travel writing, 390
scientific writing, 382, 471, 477
scientists, coinage of term, 19, 525
Scotland, 102, 103, 518, 530, 575, 578, 580, 583
Scott, Charlotte Angas, 225, 230, 235
Scott, Henderina, 180, 192
Scottish Oceans Institute, 518
Scottish Women's Hospitals, 542, 599
Scott, Margaret. *See* Gatty, Margaret
Sea, The (Michelet), 518, 520
Sea Around Us, The (Carson), 469, 471
Seaborg, Glenn, 373
sea creatures, 262, 468
seaweeds, 180, 440, 516–518

Seba, Albert, 404
second-wave feminism, 382, 573
Second World War, 152, 568, 577, 580, 582–584, 587
Secord, James, 381
Sedgwick Club, Cambridge University, 101, 114, 116, 120, 121
Segrè, Emilio, 371, 372
Sense8, 493
separate spheres, 289, 290, 299
Serbia, 310
Sex Disqualification (Removal) Act 1919, 102, 105
sextants, 90, 91, 270–272, 276
sexuality, 336, 340, 341, 493–496
sexual selection, 291–294, 302–304, 439, 440
Shakespear, Gilbert, 118
Shalon, Rachel, 587
Sharpey, Dr., 533
Shaw, William Napier, 138
Shelley, Mary, 72, 77
Shelley, Percy, 71
Shilling, Beatrice, 581
shipping, 9, 85, 91, 574, 578
Shirley, Stephanie, 323
Shove, Edith, 538
Shteir, Ann, 4, 176, 385, 428, 443
Shutt, Elsie, 323
Silent Spring (Carson), 18, 465, 473–479
Simcox, Edith, 300
Singer, Sandra L., 226, 235
Sinnett, Jane Percy, 390
Skeat, Ethel, 102, 107, 110, 113, 115–118
skin colour, 621, 622
Skinker, Mary, 466, 478
Skłodowska-Curie, Marie
  biopics, 491
  collaborative work, 5, 6, 9, 13, 15
  laboratory, 362, 364, 368, 369
  ladies' soirées, 171, 177
  Timeless, 490, 491
skyscrapers, 587
Slater, Ida, 107, 119
Smeaton, John, 574
Smedley, Ida, 109
Smellie, William, 73, 75

social class. *See* class
social justice, 311
societal bias, 622
Society of Apothecaries, 133, 530
Society of Civil Engineers, 574
sociological turn, 384
Soddy, Frederick, 362, 363
soft science, 13
software crisis, 14, 308, 321, 323, 324
software engineering, 308, 321, 323, 324, 326
Somerville, Mary
  and Ada Lovelace, 383
  Civil List pension, 93, 94
  income, 315
  and Margaret Herschel, 251
  and Matilda Rose Herschel, 259
  Royal Astronomical Society (RAS), 10, 264, 284
  translation, 16, 384, 390, 513
  writing, 16, 94
  see also On the Connexion of the Physical Sciences. By Mary Somerville (review essay by Whewell)
'Somerville effect', 19, 521
Someya, Akiko, 208
Sommerfeld, Christoph Arnold, 273
'Song of the Willi' (Blind), 300
Sophie Charlotte von Hannover, 401
Southwell, Richard V., 137, 557
Soviet scientists, 490
space, gendered, 371
Spain, 310
speech recognition technology, 620
Spence, Frances Bilas, 321
Spencer, Herbert, 295–299, 438
sports, 341, 540, 621
Sri Lanka, 311
Stanton, Thomas, 559, 560
star catalogue, 272
Stark, Freya, 160, 163
Star Trek, 18, 485, 487
Star Trek: Discovery, 488, 493
Stebnitzky, Alexandrine von, 231
Steinke, Jocelyn, 485, 486, 495
STEM
  Canadian Archive of Women in STEM, 131

in film and television, 18; annihilating/stereotyping women scientists, 487; genre and women scientists, 18, 209, 483, 495; hidden figures, 486; media production and representation, 494; normalisation of women scientists, 484
India; benefits, 343, 345; cost, 343; gender gap, 15, 196, 333, 336; patrifocality, 347; social dangers and marriageability risks, 340
Japan; pioneers, 197; present day support, 213; research environment, 197, 210
STEM education, 11, 12, 195, 196, 340
STEM gender gap, 15, 333, 338
and patrifocality, 338, 348
Stephen, Margaret, 73, 75
stereoscopic photography, 179
stereotypes, 195, 302, 322, 324, 325, 351, 478, 486, 488, 489, 492, 495, 506, 523, 617–620, 626
stereotype threat, 619
stereotyping, 320, 485
Stewart, Alexander, 250, 251
Stewart, Margaret Brodie. *See* Herschel, Margaret
St. George's Hospital Medical College, 543
Stiles, Sir Harold, 604
St. Mary's Dispensary for Women and Children, 530
St Mary's Hospital Medical School, 132, 540
St Nicholas Magazine, 466
Stockholm International Geological Congress, 103
Stokes, Dr., 534
Stokes, Margaret McNair, 177
Stopes, Marie, 109, 176, 183, 607
Strassmann, Fritz, 366
streaming services, 493
students of colour, 614, 616
student support, 142
Sue, Jean Joseph, 58
suffrage, 102, 136, 137, 175, 292, 297, 298, 532, 541
suffragists, 133, 174, 541, 597

Suge, Dr. Yoshio, 209
Sunday Graphic, 597
Sundström, Anna, 357
surgery
female legacy, 20
Louisa Martindale, 20, 601
Louise McIlroy, 593
Maud Forrester-Brown, 20
post-war world, 593
surgical sciences, 20
Swab, Anton von, 406
Swain, Lorna, 561, 562
Swallow, Ellen, 360, 367
Sweden, 16, 224, 310, 357, 360, 399, 400, 403, 404, 406, 407, 409–411, 413, 416, 417, 584, 585, 602.
See also Royal Swedish Academy of Sciences
Swedish Royal Academy of Letters, 406, 408
Swedish Royal Collection
modern day location, 404
significance to science, 404, 412–416
swimming, 341, 603
Switzerland, 224, 310, 414
Sydney Sussex College, Cambridge, 132
Sylvester, James Joseph, 224, 225, 239
sympathy, 32, 33, 42, 431
Syria, 574, 586
Systers mailing list, 325

T
Tacke, Ida. *See* Noddack, Ida (née Tacke)
Tait, Lawson, 297
talent pipeline, 618
Tanahashi, Ayako, 204
Tange, Ume, 203, 204
Tärnström, Christopher, 413, 415
Taussky, Olga, 236, 237
Tavarez, Paola, 486
Taylor, Geoffrey Ingram, 139, 145
Taylor Jane, George, 86, 87
Taylor, Janet
biographical sources, 86
Civil List pension, 85, 93, 94
compass adjusting business, 86
early life, 87, 95

mathematical and nautical instruments, 9, 90, 91
obituary, 94, 95
publications, 86
Taylor, John Ellor, 435
T.C.D, 540
teacher training colleges, 199, 202, 208
technicians, 4, 17, 184, 264, 368–371, 448, 449, 453, 574, 620
technology
 career choices, 4
 culture as deterrent, 544
 importance of diversity, 625
 innovation mishaps, 620
 see also computing
technology workforce, 614, 623
 gender and racial-ethnic composition, 612, 624
Teitelbaum, Ruth Lichterman, 321
telegraph technology, 314, 316, 317
telephone operators, 317, 318, 320
telephone technology, 316, 318, 319
telephonic communication, 174
telephony, 316–318, 326
television
 annihilating/stereotyping women scientists, 487
 genre and women scientists, 484
 hidden figures, 484, 486, 490
 media production and representation, 494
 normalisation of women scientists, 484
Telford, Thomas, 137, 574
Tessin, Carl Gustav, 404
Textbook of Anti-Aircraft Gunnery, 565
Thompson, Arthur, 296
Thomson, J.J., 183, 364
Thor, 489
Thorne, Anne, 142
Thorne, Isabel, 529
Tighe, George, 70
Tighe, Laurette, 70, 72, 80
Timeless, 489–491
time measurement, 517
Times
 ladies' soirées, Royal Society, 172
 Marianne North, 177
 Marianne North Gallery, Kew Gardens, 425

women doctors, 531, 542
women mathematicians, 550
Times of India, 345
Tinné, Alexandra, 151
Tipper, Constance Elam Fligg, 138, 139, 145
'Tipper Test', 139
Todd, Margaret, 363
Tōhoku Imperial University, 203–205, 207, 218
Tōhō University, 202, 211
Tokyo Academy of Physics, 203
Tokyo Institute of Technology, 198, 215, 216
Tokyo Liberal Arts and Science University, 210
Tokyo Woman's Christian University (TWCU), 202
Tokyo Women's Medical School (TWMS), 202
Tokyo Women's Medical University (TWMU), 197, 202, 213
Tokyo Women's Normal School (TWNS), 199
Toshima, Michiko, 206
Tōyō Eiwa Jogakuin, 202
Transactions of the Geological Society of London, 511
translation
 Ada Lovelace, 383
 Gabrielle Émilie Du Châtelet-Lomont, 48, 50, 56, 57, 387
 Marie-Geneviève-Charlotte D'Arconville, 54, 57
 Memoirs of Baron Cuvier, 512
 Nature: The Poetry of Earth and Sea (Michelet), 518
 and popularisation of science, 390
 visibility and paratext, 388
 women's contribution, 385
Trinity College, Dublin, 112, 113, 116, 540
Trinity House, 88
Trout, Annie, 553–555
Tsuda College, 211
Tsuda, Umeko, 202, 203, 213
Tsuda University, 202, 214
Tsujimura, Michiyo, 204, 208
Tuchman, Gaye, 487

Turkey, 413, 586
Turner-Warwick, Professor Dame Margaret Elizabeth Harvey, 142

U
Uffizi gallery, 407
Under the Sea Wind (Carson), 468
'Undersea' (Carson), 468
United Arab Emirates, 311, 327
United States (US)
  computing, 309, 321, 327
  consumer spending, 611
  engineers, 231
  female STEM representation, 347
  information and communication technologies, 308, 311, 316, 320
  Oberlin College, 576
  technology workforce, 614
UNIVAC computer, 322
University College Hospital, 596
University College London (UCL), 182, 183, 550, 552, 556, 558, 559, 562–564, 566, 569, 570
University of Berlin, 227, 230, 236, 401, 402, 556
University of Bern, 224, 230, 530, 537
University of Birmingham, 532
University of Bristol, 538
University of Cambridge. *See* Cambridge University
University of Durham, 97, 538
University of Edinburgh, 6, 529, 537
University of Glasgow, 538, 600
University of Heidelberg, 224
University of London, 20, 183, 225, 428, 538, 550, 552, 554, 576, 599
University of Melbourne, 130
University of North Carolina (UNC), 449–452, 455, 459, 460
University of Rochester, 141
University of Southampton, 554, 568
University of St Andrews, 518
University of Tokyo, 197, 198, 203, 212
University of Zurich, 530
university ranking, 213
Unlocking the Clubhouse (Margolis and Fisher), 617

Uppsala University, 16, 399, 406, 407, 410–415
Uranus, 281, 282
Urey, Harold C., 370
US Environmental Protection Agency (EPA), 476
US federal service, 467
US Fish and Wildlife Service (FWS), 465, 467, 471, 476
US Liberty ships, 139
US technology workforce, 20

V
Vaccà Berlinghieri, Andrea, 70
Vaerting, Mathilde, 235
Vaughan, Janet Maria, 141
Verena Holmes Memorial Lectures, 580
Vickers' factory, 582
Victoria, Queen, 289, 576
Views of Nature (Humboldt), 392
voice figures, 184
voicemail, 620
'Voices of Science', 143
void, 31, 32, 561
Voltaire
  attitude towards women, 542
  and Frederick II of Prussia, 402
  and Gabrielle Émilie Du Châtelet-Lomont, 48, 50, 52, 55, 59, 62
  and Lovisa Ulrika of Sweden, 409
  Principia Mathematica (Newton), 416
von Gernet, Nadeschda Nikolaevna, 232, 235
von Neumann, John, 322
Vote, The, 574
Votes for Women, 541

W
Wachter, Sandra, 622
Wade, Thomas-Francis, 261
Wager, Phyllis, 154, 155
Walker Dunbar, Eliza Louisa, 537
Wallace, Alfred Russell, 301, 304, 424, 434
Wallerius, Johan Gottschalk, 406, 407

Walley, Nina Cameron. *See* Graham, Nina Cameron
Wallington, Emma, 298
Waloff, Nadia, 136
Waseda University, 215
Watson, Janet Vida, 130, 140, 144
Watts-Hughes, Megan (Margaret), 184
Webb, Hanor A., 562
Weber, Dr Johanna, 585
Wedderburn, Dorothy, 132
Wedell, Charlotte, 235
Weierstrass, Karl, 227
Western Union, 316
Westminster Medical College, 543
wet-chemical analyses, 15, 356
Whaley, Leigh, 5, 8, 48, 49
Wheatstone, Charles, 314
Whewell, William, 19, 251, 505–512, 514–517, 521–523, 525
  On the Connexion of the Physical Sciences. By Mary Somerville (review essay), 506
Whiteford, Marion, 135
White, Gilbert, 516
Whiteley, Martha Annie, 135, 139, 145
Whiteside, Ida, 236
Whitting, Frances G., 180
Wikipedia, 525, 588
Willembourg, Walther, 319, 320
Willey, Dr. Florence, 541
Williams, Clara, 539
Williams, Helena Maria, 384, 391
Winchester College, 557, 568
Wingfield, Dr R.C., 580
Winkelmann-Kirch, Maria Margarethe, 273
Winston, Mary F., 225, 227–231, 237
Winter, Alison, 91
Wirtinger, Wilhelm, 236, 238
Wollstonecraft, Mary, 69, 70, 74, 77, 79, 249, 292
Woman Engineer, The, 574
womanhandling of text, 383
'Woman Question', 290, 295, 303
Woman Surgeon, A (Martindale), 600
women
  attitudes towards, 27, 50, 431. *See also* biological determinism
  enlightenment cultural norms, 48
  in science, scepticism towards, 6, 10, 14, 15, 21, 49, 58, 129–131, 248, 249, 253, 265, 367, 448, 458, 460, 491, 494, 505, 506, 512, 516, 521–523
Women in Science and Engineering (WISE) archive, 130
women of colour, 311, 485, 486, 612, 614, 625
  buying power, 612
  see also black women scientists
women-only universities, Japan
  historical development, 197
  STEM pioneers, 197; present day support, 213; research environment, 197
Women's Engineering Society, 574, 579–581, 583, 586
women's history, 150, 505, 506, 573
Women's Home Companion, 472
Women's Institute, 538
Women's Institute for English Studies, Japan, 202
Women's Land Army, 117
women's movement, 290, 291, 296
  arguments against, 291
Wood, Dr. Andrew, 533
Wood, Ethel, 102, 107, 117–119
Woods, Henry, 118
Woodward, Joan, 132
Woolgar, Steve, 17, 447–449
Woolhope Society, 116
workforce diversity, 589
Workman, Rachel, 103, 104, 123
World's Fair 1893, 227
World War I. *See* First World War
World War II. *See* Second World War
Worthington, Stella, 153
Wrangel, Margarete von, 235
Wrinch, Dorothy, 236
writers, 16, 28, 31, 32, 74, 163, 264, 292, 299, 382, 390, 408, 451, 466, 471, 472, 493, 494, 515. *See also* Carson, Rachel
writing, scientific, 53, 382, 471, 477. *See also* scientific travel writing

X
X-Files, 495
x-ray therapy, 601, 603

Y
Yasui, Tetsu, 203
Yasukawa, Etsuko, 212
Yokota, Yurie, 208
York, Miss, 180
Yoshimura, Fuji, 208
Yoshioka, Yayoi, 202
Young, Andrew, 563
ytterbium, 360, 361

Yuasa, Toshiko, 210, 211

Z
Zahm, Rev. John Augustus, 6
Zapolskaya, Ljubov Nikolaevna, 231, 232
Zehui He, 9, 22
Zeuthen, Hieronymus G., 224, 225
Zimbabwe, 587
Znamirowska, Marynia. *See* Chatterton, Marjem
zoology, 184, 415, 449, 451, 462

Printed in the United States
by Baker & Taylor Publisher Services